MATLAB
FOR MECHANICAL ENGINEERS

MATLAB FOR MECHANICAL ENGINEERS

RAO V DUKKIPATI

PhD, PE

Fellow of ASME and CSME
Ex-Professor and Chairman
Department of Mechanical Engineering
Ex-Director, Graduate Programme
Fairfield University
Fairfield, Connecticut
USA

NEW AGE REFERENCE

An Imprint of

NEW AGE INTERNATIONAL (P) LIMITED, PUBLISHERS

LONDON • NEW DELHI • NAIROBI

Bangalore • Chennai • Cochin • Guwahati • Hyderabad • Kolkata • Lucknow • Mumbai

Visit us at **www.newagepublishers.com**

Published by New Age International (P) Ltd., Publishers
First Edition: 2008
Reprint: 2018

GLOBAL OFFICES

- **New Delhi** **NEW AGE INTERNATIONAL (P) LIMITED, PUBLISHERS**
  7/30 A, Daryaganj, New Delhi-110002, (INDIA)
  Tel.: (011) 23253771, 23253472, **Telefax:** 23267437, 43551305
  E-mail: contactus@newagepublishers.com • Visit us at www.newagepublishers.com
- **London** **NEW AGE INTERNATIONAL (UK) LTD.**
  27 Old Gloucester Street, London, WC1N 3AX, UK
  E-mail: info@newacademicscience.co.uk • Visit us at www.newacademicscience.co.uk
- **Nairobi** **NEW AGE GOLDEN (EAST AFRICA) LTD.**
  Ground Floor, Westlands Arcade, Chiromo Road (Next to Naivas Supermarket), Westlands, Nairobi, KENYA
  Tel.: 00-254-713848772, 00-254-725700286
  E-mail: kenya@newagepublishers.com

BRANCHES

- **Bangalore** 37/10, 8th Cross (Near Hanuman Temple), Azad Nagar, Chamarajpet, Bangalore- 560 018
  Tel.: (080) 26756823, **Telefax:** 26756820, **E-mail: bangalore@newagepublishers.com**
- **Chennai** 26, Damodaran Street, T. Nagar, Chennai-600 017, **Tel.:** (044) 24353401, **Telefax:** 24351463
  E-mail: chennai@newagepublishers.com
- **Cochin** CC-39/1016, Carrier Station Road, Ernakulam South, Cochin-682 016, **Tel.:** (0484) 2377303, **Telefax:** 4051304
  E-mail: cochin@newagepublishers.com
- **Guwahati** Hemsen Complex, Mohd. Shah Road, Paltan Bazar, Near Starline Hotel, Guwahati-781 008,
  Tel.: (0361) 2513881, **Telefax:** 2543669, **E-mail: guwahati@newagepublishers.com**
- **Hyderabad** 105, 1st Floor, Madhiray Kaveri Tower, 3-2-19, Azam Jahi Road, Near Kumar Theater, Nimboliadda Kachiguda, Hyderabad-500 027, **Tel.:** (040) 24652456, **Telefax:** 24652457
  E-mail: hyderabad@newagepublishers.com
- **Kolkata** RDB Chambers (Formerly Lotus Cinema) 106A, 1st Floor, S N Banerjee Road, Kolkata-700 014
  Tel.: (033) 22273773, **Telefax:** 22275247, **E-mail: kolkata@newagepublishers.com**
- **Lucknow** 16-A, Jopling Road, Lucknow-226 001, **Tel.:** (0522) 2209578, 4045297, **Telefax:** 2204098
  E-mail: lucknow@newagepublishers.com
- **Mumbai** 142C, Victor House, Ground Floor, N.M. Joshi Marg, Lower Parel, Mumbai-400 013
  Tel.: (022) 24927869, **Telefax:** 24915415, **E-mail: mumbai@newagepublishers.com**
- **New Delhi** 22, Golden House, Daryaganj, New Delhi-110 002, **Tel.:** (011) 23262368, 23262370, **Telefax:** 43551305
  E-mail: sales@newagepublishers.com

ISBN: 978-81-224-2270-2

C-17-07-10614

Printed in India at Glorious Printers, Delhi.
Typeset at Goswami Associates, Delhi.

NEW AGE INTERNATIONAL (P) LIMITED, PUBLISHERS
7/30 A, Daryaganj, New Delhi-110002
Visit us at **www.newagepublishers.com**
(CIN: U74899DL1966PTC004618)

I dedicate this book

to

Lord Sri Venkateswara

Preface

The main objective of this book is to provide the students with the opportunity to improve their programming skills using the MATLAB environment to implement algorithms and to teach the use of MATLAB as a tool in solving problems in engineering.

This book includes the coverage of basics of MATLAB and application of MATLAB software to solve problems in engineering: control systems, engineering mechanics: statics and dynamics, mechanical vibrations, electrical circuits, and numerical methods. With this foundation of basic MATLAB applications in engineering problem solving, the book provides opportunities to explore advanced topics in application of MATLAB as a tool.

Chapter 1 presents a brief introduction to control systems, engineering mechanics: statics and dynamics, mechanics of materials, mechanical vibrations, electrical circuits, and numerical methods. An introduction to MATLAB basics is presented in Chapter 2. Chapter 2 also presents MATLAB commands. MATLAB is considered as the software of choice. MATLAB can be used interactively and has an inventory of routines, called as functions, which minimize the task of programming even more. Further information on MATLAB can be obtained from: The MathWorks, Inc., 3 Apple Hill Drive, Natick, MA 01760. In the computational aspects, MATLAB has emerged as a very powerful tool for numerical computations involved in engineering problems. The idea of computer-aided design and analysis using MATLAB with the Symbolic Math Tool box, and the Control System Tool box has been incorporated. Chapter 3 consists of many solved problems that demonstrate the application of MATLAB to the analysis of control systems, engineering mechanics: statics and dynamics, mechanics of materials, mechanical vibrations, electrical circuits, and numerical methods. In chapter 4, we have briefly reviewed the direct numerical integration methods for the solution of a single or system of differential equations. Many numerical methods are available for the solutions of the response of dynamic systems. We have discussed several widely used step-by-step numerical integration methods for linear dynamic response analysis. A brief description of these integration methods is presented and their application is illustrated. The integration schemes considered were three explicit and four implicit methods. They are the explicit schemes (the central difference method, two-cycle interaction with trapezoidal rule and fourth order Runge-Kutta method) and the implicit schemes (Houbolt method, Wilson Theta method, Newmark Beta method and the Park Stiffly stable method). Application of these direct numerical integration methods is illustrated with a case study of a linear dynamic system.

Presentations are limited to very basic topics to serve as an introduction to advanced topics in those areas of discipline. Chapters 2, 3 and 4 include a great number of worked examples and unsolved exercise problems to guide the student to understand the basic principles, concepts, and use of MATLAB in solving a variety of engineering problems.

I sincerely hope that the final outcome of this book helps the students in developing an appreciation for the topics of control systems and signal processing.

An extensive bibliography to guide the student to further sources of information on control systems and signal processing engineering is provided at the end of the book. All end-of chapter problems are fully solved in the Solution Manual available only to Instructors.

—Author

Acknowledgements

I am grateful to all those who have had a direct impact on this work. Many people working in the general areas of application of MATLAB software to solve engineering problems have influenced the format of this book. I greatly owe my indebtedness to all the authors of the articles listed in the bibliography of this book. Finally, I would very much like to acknowledge the encouragement, patience, and support provided by my family members: my wife, Sudha, my family members, Ravi, Madhavi, Anand, Ashwin, Raghav, and Vishwa who have also shared in all the pain, frustration, and fun of producing a manuscript.

I would appreciate being informed of errors, or receiving other comments about the book. Please write to the authors' Fairfield University address or send e-mail to **Rdukkipati@mail.fairfield.edu** or **Professordukkipati@yahoo.com.**

Rao V. Dukkipati

Contents

Chapter 1

Introduction

In this chapter, we introduce control systems, engineering mechanics, mechanical vibrations, electrical circuits, numerical methods, optimization and symmetric matrix eigen value problems.

1.1 CONTROL SYSTEMS

1.1.1 Introduction

This section presents a brief introduction and overview of control systems. Some of the terms commonly used to describe the operation, analysis, and design of control systems are presented.

1.1.2 Control Systems

Control systems is an interdisciplinary field covering many areas of engineering and sciences. Control systems exist in many systems of engineering, sciences, and in human body. *Control* means to regulate, direct, command, or govern. A *system* is a collection, set, or arrangement of elements (subsystems). A *control system* is an interconnection of components forming a system configuration that will provide a desired system response. Hence, a control system is an arrangement of physical components connected or related in such a manner as to command, regulate, direct, or govern itself or another system.

In order to identify, delineate, or define a control system, we introduce two terms: *input* and *output* here. The *input* is the stimulus, excitation, or command applied to a control system, and the *output* is the actual response resulting from a control system. The *output* may or may not be equal to the specified response implied by the input. Inputs could be physical variables or abstract ones such as *reference, set point* or *desired* values for the output of the control system. Control systems can have more than one input or output. The input and the output represent the desired response and the actual response respectively. A control system provides an output or response for a given input or stimulus, as shown in Fig. 1.1.

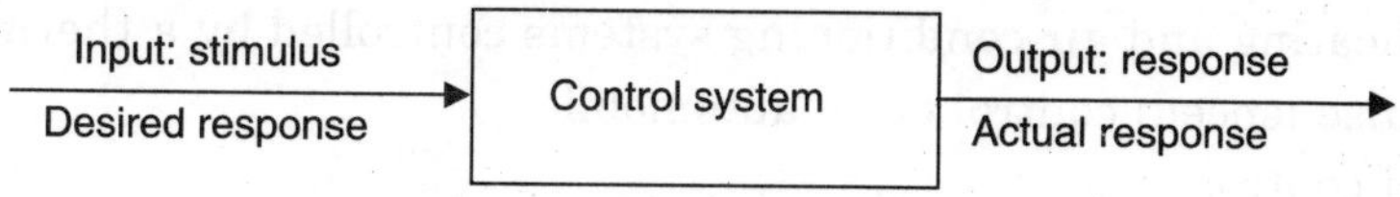

Fig. 1.1 Description of a control system.

The output may not be equal to the specified response implied by the input. If the output and input are given, it is possible to identify or define the nature of the system's components. Broadly speaking, there are three basic types of control systems:

(*a*) Man-made control systems

(*b*) Natural, including biological control systems

(*c*) Control systems whose components are both man-made and natural.

An electric switch is a man-made control system controlling the electricity-flow. The simple act of pointing at an object with a finger requires a biological control system consisting chiefly of eyes, the arm, hand and finger and the brain of a person, where the input is precise-direction of the object with respect to some reference and the output is the actual pointed direction with respect to the same reference. The control system consisting of a person driving an automobile has components, which are clearly both man-made and biological. The driver wants to keep the automobile in the appropriate lane of the roadway. The driver accomplishes this by constantly watching the direction of the automobile with respect to the direction of road. Figure 1.2 is an alternate way of showing the basic entities in a general control system.

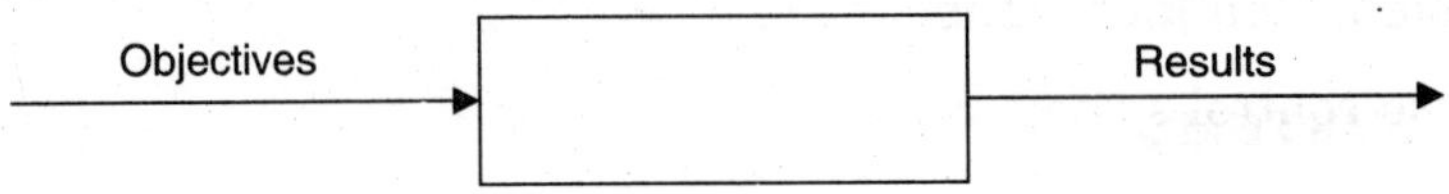

Fig. 1.2 Components of a control system.

In the steering control of an automobile for example, the direction of two front wheels can be regarded as the result or controlled output variable and the direction of the steering wheel as the actuating signal or objective. The control-system in this case is composed of the steering mechanism and the dynamics of the entire automobile. As another example, consider the idle-speed control of an automobile engine, where it is necessary to maintain the engine idle speed at a relatively low-value (for fuel economy) regardless of the applied engine loads (like air-conditioning, power steering, etc.). Without the idle-speed control, any sudden engine-load application would cause a drop in engine speed that might cause the engine to stall. In this case, throttle angle and load-torque are the inputs (objectives) and the engine-speed is the output. The engine is the controlled process of the system. A few more applications of control-systems can be found in the print wheel control of an electronic typewriter, the thermostatically controlled heater or furnace which automatically regulates the temperature of a room or enclosure, and the sun tracking control of solar collector dish.

Control system applications are found in robotics, space-vehicle systems, aircraft autopilots and controls, ship and marine control systems, intercontinental missile guidance systems, automatic control systems for hydrofoils, surface-effect ships, and high-speed rail systems including the magnetic levitation systems.

1.1.2.1 Examples of control systems

Control systems find numerous and widespread applications from everyday to extraordinary in science, industry, and home. Here are a few examples:

(*a*) Home heating and air-conditioning systems controlled by a thermostat

(*b*) The cruise (speed) control of an automobile

(*c*) Manual control:

- (*i*) Opening or closing of a window for regulating air temperature or air quality
- (*ii*) Activation of a light switch to regulate the illumination in a room
- (*iii*) Human controlling the speed of an automobile by regulating the gas supply to the engine

(*d*) Automatic traffic control (signal) system at roadway intersections

(e) Control system which automatically turns on a room lamp at dusk, and turns it off in daylight
(f) Automatic hot water heater
(g) Environmental test-chamber temperature control system
(h) An automatic positioning system for a missile launcher
(i) An automatic speed control for a field-controlled dc motor
(j) The attitude control system of a typical space vehicle
(k) Automatic position-control system of a high speed automated train system
(l) Human heart using a pacemaker
(m) An elevator-position control system used in high-rise multilevel buildings.

1.1.3 Control System Configurations

There are two control system configurations: open-loop control system and closed-loop control system.

(a) Block

A block is a set of elements that can be grouped together, with overall characteristics described by an input/output relationship as shown in Fig. 1.3. A block diagram is a simplified pictorial representation of the cause-and-effect relationship between the input(s) and output(s) of a physical system.

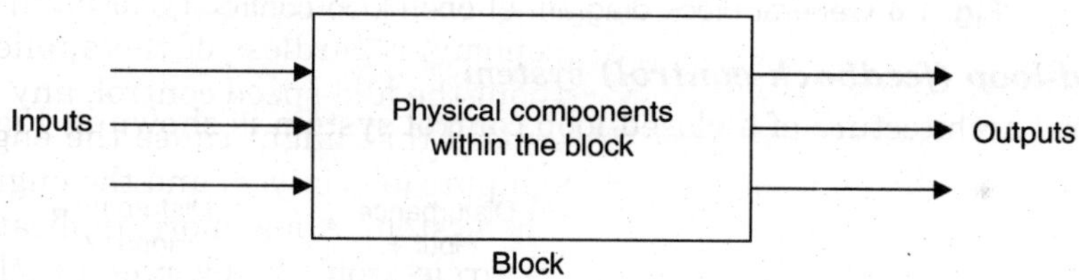

Fig. 1.3 Block diagram.

The simplest form of the block diagram is the single block as shown in Fig. 1.3. The input and output characteristics of entire groups of elements within the block can be described by an appropriate mathematical expressions as shown in Fig. 1.4.

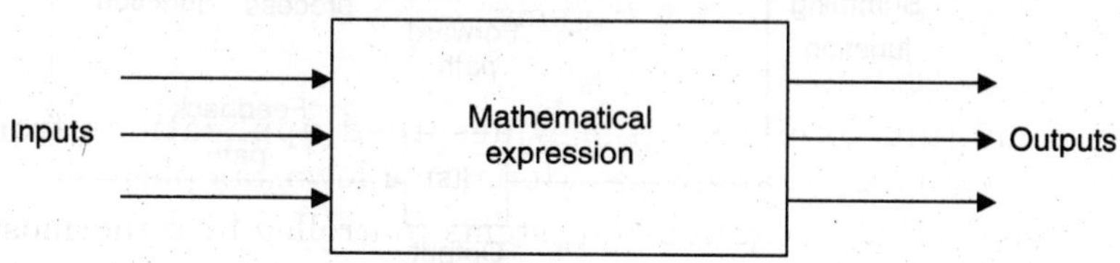

Fig. 1.4 Block representation.

(b) Transfer function

The transfer function of a system (or a block) is defined as the ratio of output to input as shown in Fig. 1.5.

$$\text{Transfer function} = \frac{Output}{Input}$$

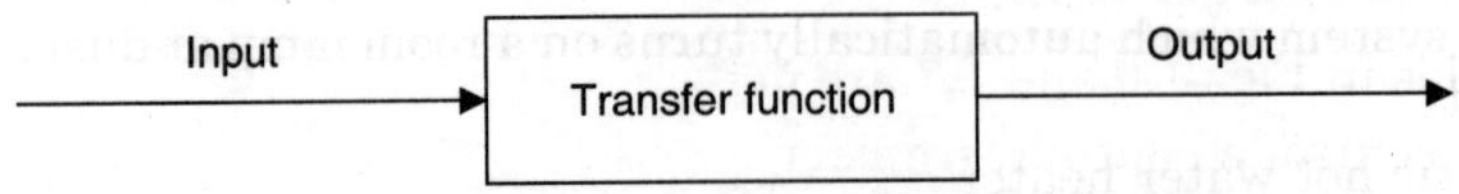

Fig. 1.5 Transfer function.

Transfer functions are generally used to represent a mathematical model of each block in the block diagram representation. All the signals are transfer functions on the block diagrams. For instance, the time function reference input is $r(t)$, and its transfer function is $R(s)$ where t is time and s is the Laplace transform variable or complex frequency.

(c) Open-loop control system

A general block diagram of open-loop system is shown in Fig. 1.6.

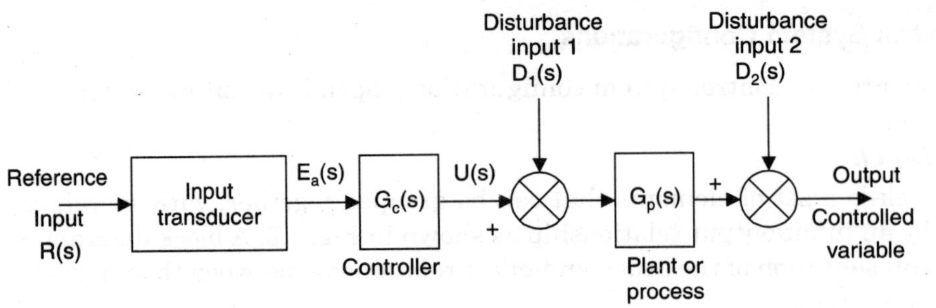

Fig. 1.6 General block diagram of open-loop control system.

(d) Closed-loop (feedback control) system

The general architecture of a closed-loop control system is shown in Fig. 1.7.

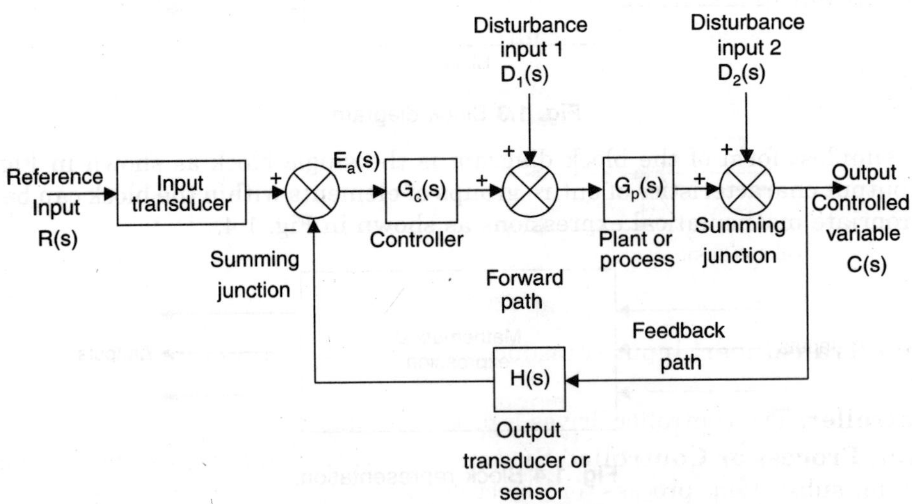

Fig. 1.7 General block diagram of closed-loop control system.

1.1.4 Control System Terminology

The variables in Figs.1.6 and 1.7 are defined as follows:

$C(s)$	controlled output, transfer function of $c(t)$
$D(s)$	disturbance input, transfer function of $d(t)$
$E_a(s)$	actuating error, transfer function of $e_a(t)$
$G_a(s)$	transfer function of the actuator
$G_c(s)$	transfer function of the controller
$G_p(s)$	transfer function of the plant or process
$H(s)$	transfer function of the sensor or output transducer = $G_s(s)$
$R(s)$	reference input, transfer function of $r(t)$.

Summing Point: As shown in Fig. 1.8 the block is a small circle called a summing point with the appropriate plus or minus sign associated with the arrows entering the circle. The output is the algebraic sum of the inputs. There is no limit on the number of inputs entering a summing point.

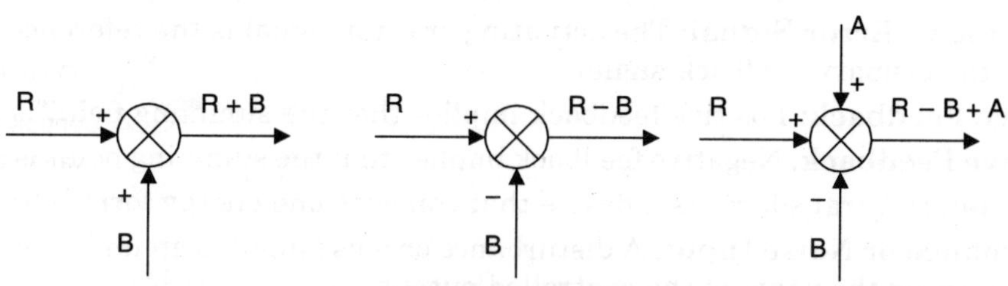

Fig. 1.8 Summing point.

Takeoff Point: A takeoff point allows the same signal or variable as input to more than one block or summing point, thus permitting the signal to proceed unaltered along several different paths to several destinations as shown in Fig. 1.9.

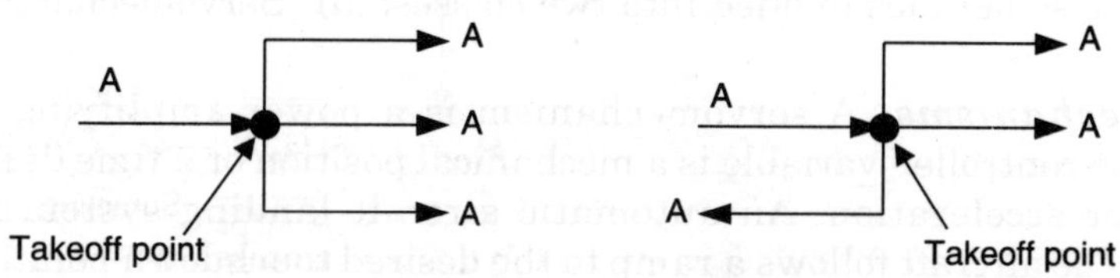

Fig. 1.9 Takeoff point.

Input Transducer: Input transducer converts the form of input to that used by the controller.

Controller: The controller drives a process or plant.

Plant, Process or Controlled System $G_p(s)$: The plant, process, or controlled system is the system, subsystem, process, or object controlled by the feedback control system. For example, the plant can be a furnace system where the output variable is temperature.

Controlled Output C(s): The controlled output $C(s)$ is the output variable of the plant under the control of the control system.

Forward Path: The forward path is the transmission path from the summing point to the controlled output.

Feedback Path: The feedback path is the transmission path from the controlled output back to the summing point.

Feed Forward (Control) Elements: The feed forward (control) elements are the components of the forward path that generate the control signal applied to the plant or process. The feed forward (control) elements include controller(*s*), compensator(*s*), or equalization elements, and amplifiers.

Feedback Elements: The feedback elements establish the fundamental relationship between the controlled output $C(s)$ and the primary feedback signal $B(s)$. They include sensors of the controlled output, compensators, and controller elements.

Reference Input R(s): The reference input is an external signal applied to the control system generally at the first summing input, so as to command a specified action of the process or plant. It typically represents ideal or desired process or plant output response.

Primary Feedback Signal: The primary feedback signal is a function of the controlled output summed algebraically with the reference input to establish the actuating or error signal. An open-loop system has no primary feedback signal.

Actuating or Error Signal: The actuating or error signal is the reference input signal plus or minus the primary feedback signal.

Positive Feedback: Positive feedback implies that the summing point is an adder.

Negative Feedback: Negative feedback implies that the summing point is a subtractor.

Transducer: A transducer is a device that converts one energy form into another.

Disturbance or Noise Input: A disturbance or noise input is an undesired stimulus or input signal affecting the value of the controlled output.

Time Response: The time response of a system, subsystem, or element is the output as a function of time, generally following the application of a prescribed input under specified operating conditions.

1.1.5 Control System Classes

Control systems are sometimes divided into two classes: (*a*) Servomechanisms, and (*b*) Regulators.

(a) Servomechanisms. A servomechanism is a power-amplifying feedback control system in which the controlled variable is a mechanical position or a time derivative of position such as velocity or acceleration. An automatic aircraft landing system is an example of servomechanism. The aircraft follows a ramp to the desired touchdown point. Another example is the control system of an industrial robot in which the robot arm is forced to follow some desired path in space.

(b) Regulators. A regulator or regulating system is a feedback control system in which the reference input or command is constant for long periods of time, generally for the entire time interval during which the system is operational. Such an input is known as *set point*. An example of a regulator control system is the human biological system that maintains the body temperature at approximately 98.6°F in an environment that usually has a different temperature.

1.1.5.1 Supplementary terminology

(*a*) *Linear System*: A linear system is a system where input/output relationships may be represented by a linear differential equation. The plant is linear if it can be accurately described using a set of linear differential equations. This attribute indicates that system parameters do not vary as a function of signal level.

Similarly, the plant is a lumped-parameter (rather than distributed parameter) system if it can be described using ordinary (rather than partial) differential equations. This condition is generally accomplished if the physical size of the system is very small in comparison to the wavelength of the highest frequency of interest.

(*b*) *Time-Variant System:* A time-variant is a system if the parameters vary as a function of time. Thus, a time-variant system is a system described by a differential equation with variable coefficients. A linear time variant system is described by linear differential equations with variable coefficients. A rocket-burning fuel system is an example of time variant system since the rocket mass varies during the flight as the fuel is burned.

(*c*) *Time-Invariant System:* A time-invariant system is a system described by a differential equation with constant coefficients. Thus, the plant is time invariant if the parameters do not change as a function of time. A linear time invariant system is described by linear differential equations with constant coefficients. A single degree of freedom spring mass viscous damper system is an example of a time-invariant system provided the characteristics of all the three components do not vary with time.

(*d*) *Multivariable Feedback System:* The block diagram representing a multivariable feedback system where the interrelationships of many controlled variables are considered is shown in Fig. 1.10.

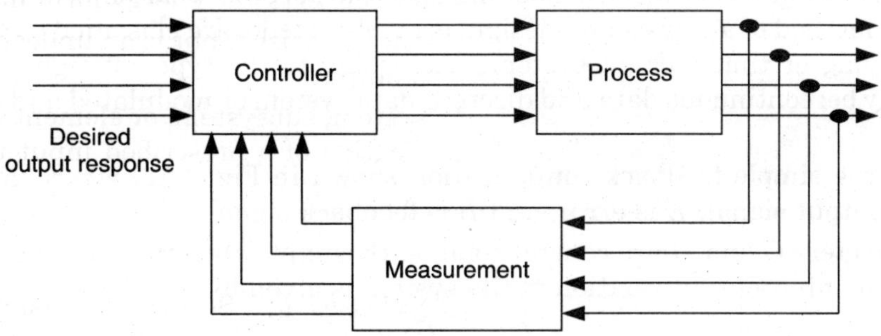

Fig. 1.10 Multivariable control system.

1.1.6 Feedback Systems

Feedback is the property of a closed-loop system, which allows the output to be compared with the input to the system such that the appropriate control action may be formed as some function of the input and output.

For more accurate and more adaptive control, a link or feedback must be provided from output to the input of an open-loop control system. So the controlled signal should be fed back and compared with the reference input, and an actuating signal proportional to the difference of input and output must be sent through the system to correct the error. In general, feedback is said to exist in a system when a closed sequence of cause-and-effect relations exists between system variables. A closed-loop idle-speed control system is shown in Fig. 1.11. The reference input N_r sets the desired idle-speed. The engine idle speed N should agree with the reference value N_r and any difference such as the load-torque T is sensed by the speed-transducer and the error detector. The controller will operate on the difference and provide a signal to adjust the throttle angle to correct the error.

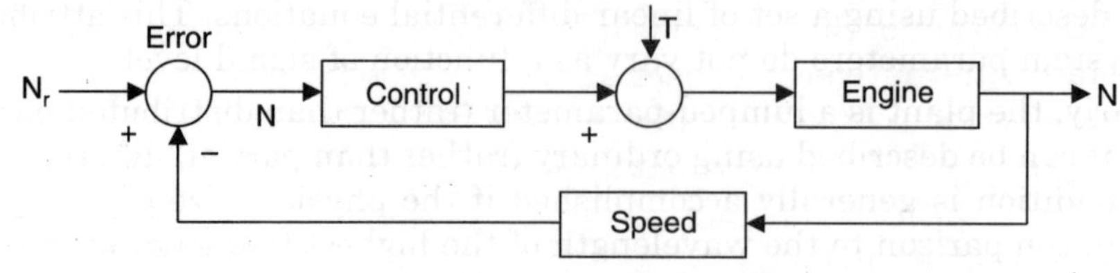

Fig. 1.11 Closed-loop idle-speed control system.

1.1.7 Analysis of Feedback

The most important features, the presence of feedback impacts to a system are the following:

(*a*) Increased accuracy: its ability to reproduce the input accurately

(*b*) Reduced sensitivity of the ratio of output to input for variations in system characteristics and other parameters

(*c*) Reduced effects of nonlinearities and distortion

(*d*) Increased bandwidth (bandwidth of a system that ranges frequencies (input) over which the system will respond satisfactorily)

(*e*) Tendency towards oscillation or instability

(*f*) Reduced effects of external disturbances or noise.

A system is said to be *unstable*, if its output is out of control. Feedback control systems may be classified in a number of ways, depending upon the purpose of classification. For instance, according to the method of analysis and design, control-systems are classified as linear or non-linear, time-varying or time-variant systems. According to the types of signals used in the system, they may be: continuous data and discrete-data system or modulated and unmodulated systems.

Consider the simple feedback configuration shown in Fig. 1.12, where R is the input signal, C is the output signal, E is error, and B is feedback signal.

The parameters G and H are constant-gains. By simple algebraic manipulations, it can be shown that the input-output relation of the system is given by

$$M = \frac{C}{R} = \frac{G}{1 + GH}$$

The general effect of feedback is that it may increase or decrease the gain G. In practical control-systems, G and H are functions of frequency, so the magnitude of $(1 + GH)$ is greater than 1 in one frequency range, but less than 1 in another. Thus feedback affects the gain G of a nonfeedback system by a factor $(1 + GH)$.

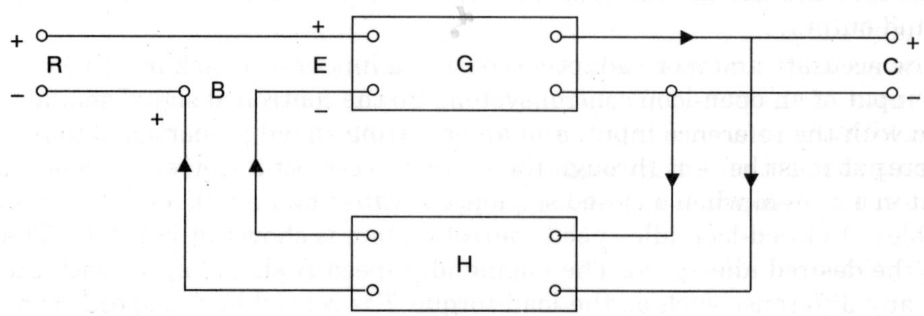

Fig. 1.12 Feedback system.

If $GH = -1$, the output of the system is infinite for any finite input, such a state is called unstable system-state. Alternatively feedback stabilizes an unstable system and the sensitivity of a gain of the overall system M to the variation in G is defined as:

$$S_G^M = \frac{\partial M/M}{\partial G/G} = \frac{\text{Percentage change in } M}{\text{Percentage change in } G}$$

where ∂M denotes incremental change in M due to incremental change in $G(\partial G)$. One can write sensitivity-function as:

$$S_G^M = \frac{\partial M/M}{\partial G/G} = \frac{1}{1+GH}$$

By increasing GH, the magnitude of the sensitivity-function is made arbitrarily small.

1.1.8 Control System Analysis and Design Objectives

Control systems engineering consists of *analysis* and *design* of control systems configurations. Control systems are *dynamic*, in that they respond to an input by first undergoing a transient response before attaining a steady-state response which corresponds to the input. There are three main objectives of control systems analysis and design. They are:

1. Producing the response to a transient disturbance which is acceptable
2. Minimizing the steady-state errors: Here, the concern is about the accuracy of the steady-state response
3. Achieving stability: Control systems must be designed to be stable. Their natural response should decay to a zero value as time approaches infinity, or oscillate.

Analysis is investigation of the properties and performance of an existing control system. Design is the selection and arrangement of the control system components to perform a prescribed task. The design of control systems is accomplished in two ways: *design by analysis* in which the characteristics of an existing or standard system configuration are modified, and *design by synthesis*, in which the form of the control system is obtained directly from its specifications.

1.2 ENGINEERING MECHANICS

1.2.1 Introduction

Engineering mechanics is the science that considers the motion of bodies under the action of forces and the effects of forces on that motion. Mechanics includes *statics*, which deals with the special case of a body at rest or a body that moves with constant velocity. A body at rest or moving with constant velocity is said to be in *equilibrium*.

The mechanics of rigid bodies may be divided into *statics* and *dynamics*. Each point in a rigid body is always at a constant distance from any other point in the body. Dynamics is further divided into *kinematics* and *kinetics*. *Kinematics* defines the relationships among displacement, velocity, and acceleration of a moving body. *Kinetics* defines the relationship between the forces that act on a body and the motion of the body. A particle is a body whose size does not affect its response to the forces that act on it.

1.2.2 Newtonian Mechanics

Collectively, the study of statics and dynamics is called *classical mechanics*. Classical mechanics treats the motion of bodies of ordinary size that move at speeds that are small compared to the speed of light. Newton (1642-1727) formulated the law of universal gravitation and the mathematics of calculus. Newton introduced the concepts of force and mass. Newton formulated the three laws of motion that are the basis of engineering applications of mechanics. The classical mechanics is often called *Newtonian-mechanics.*

1.2.3 Newton's Laws of Motion

Newton's laws apply directly to a particle, a body that may be treated as a point mass. Newton's laws may be stated in terms of a particle as follows:

First Law: In the absence of applied forces, a particle originally at rest or moving with constant speed in a straight line will remain at rest or continue to move with constant speed in a straight line.

Second Law: If a particle is subjected to a force, the particle will accelerate. The acceleration of the particle will be in the direction of the force, and the magnitude of the acceleration will be proportional to the force and inversely proportional to the mass of the particle.

Newton's second law can be expressed in the form

$$a = \frac{F}{m}$$

where F = force

m = mass

a = acceleration

Newton's second law is the basis for the study of kinetics of a particle.

Third Law: For every action, there is an equal and opposite reaction, or the mutual forces exerted by two particles on each other are always equal and oppositely directed.

1.2.4 Resultants of Coplanar Force Systems

The resultant R for a concurrent system is given by

$$R = \sqrt{(\Sigma F_x)^2 + (\Sigma F_y)^2}$$

and

$$\theta_x = \tan^{-1}\left[\frac{\Sigma F_y}{\Sigma F_x}\right]$$

where ΣF_x, ΣF_y = algebraic sums of the x and y components of the forces of the system respectively

θ_x = the angle that the resultant R makes with the x-axis.

For a parallel system, the resultant R is given by

$$R = \Sigma F$$

and

$$R\bar{a} = \Sigma M_o$$

where ΣF = algebraic sum of the forces of the system

O = any moment center in the plane

$\bar{a}$ = perpendicular distance from the moment center O to the resultant R

$R\bar{a}$ = moment of R with respect to O.

ΣM_o = algebraic sum of the moments of the forces of the system with respect to O

The resultant R for a nonconcurrent and nonparallel system is given by

$$R = \sqrt{(\Sigma F_x)^2 + (\Sigma F_y)^2 + (\Sigma F_z)^2}$$

and
$$\theta_x = \tan^{-1}\left[\frac{\Sigma F_y}{\Sigma F_x}\right]$$

where ΣF_x, ΣF_y = algebraic sums of the x and y components of the forces of the system respectively, θ_x = the angle that the resultant R makes with the x-axis.

The action of the line of the resultant force is given by

$$R\bar{a} = \Sigma M_o$$

where O = any moment center in the plane

$\bar{a}$ = perpendicular distance from the moment center O to the resultant R

$R\bar{a}$ = moment of R with respect to O

ΣM_o = algebraic sum of the moments of the forces of the system with respect to O.

1.2.5 Resultants of Noncoplanar Force Systems

In noncoplanar force systems, the forces are not all concurrent and not in parallel. The resultant R for a concurrent system is given by

$$R = \sqrt{(\Sigma F_x)^2 + (\Sigma F_y)^2 + (\Sigma F_z)^2}$$

with the direction cosines

$$\cos\theta_x = \frac{\Sigma F_x}{R}, \; \cos\theta_y = \frac{\Sigma F_y}{R} \quad \text{and} \quad \cos\theta_z = \frac{\Sigma F_z}{R}$$

where ΣF_x, ΣF_y, ΣF_z = algebraic sums of the x, y and z components of the forces of the system respectively.

For a parallel system, the resultant R is given by

$$R = \Sigma F$$

$$R\bar{x} = \Sigma M_z$$

$$R\bar{z} = \Sigma M_x$$

where ΣF = algebraic sum of the forces of the system

$\bar{x}$ = perpendicular distance from the yz plane to the resultant

$\bar{z}$ = perpendicular distance from the xy plane to the resultant

ΣM_x, ΣM_z = algebraic sum of the moments of the forces of the system about the x and z axes respectively.

If $\Sigma F = 0$, the resultant couple C when exists is given by

$$C = \sqrt{(\Sigma M_x)^2 + (\Sigma M_z)^2}$$

and
$$\phi = \tan^{-1}\left[\frac{\Sigma M_z}{\Sigma M_x}\right]$$

where ϕ is the angle that the vector representing the resultant couple makes with the x-axis.

The magnitude of the resultant R of the nonconcurrent system at the origin (x,y and z axes is placed with their origin at the base point) is given by

$$R = \sqrt{(\Sigma F_x)^2 + (\Sigma F_y)^2 + (\Sigma F_z)^2}$$

with the direction cosines

$$\cos \theta_x = \frac{\Sigma F_x}{R}, \cos \theta_y = \frac{\Sigma F_y}{R} \text{ and } \cos \theta_z = \frac{\Sigma F_z}{R}$$

The magnitude of the resulting couple C is given by

$$C = \sqrt{(\Sigma M_x)^2 + (\Sigma M_y)^2 + (\Sigma M_z)^2}$$

with the direction cosines

$$\cos \phi_x = \frac{\Sigma M_x}{C}, \cos \phi_y = \frac{\Sigma M_y}{C} \text{ and } \cos \phi_z = \frac{\Sigma M_z}{C}$$

where $\Sigma M_x, \Sigma M_y, \Sigma M_z$ = algebraic sums of the moments of the forces of the system about x, y and z axes respectively

ϕ_x, ϕ_y, ϕ_z = angles which the vector representing the couple C makes with the x, y and z axes respectively.

1.2.6 Equilibrium of Coplanar Force Systems

The necessary and sufficient conditions for the equilibrium of a coplanar force system are

$$R = \Sigma F = 0$$

and

$$C = \Sigma M = 0$$

where ΣF = vector sum of all forces of the system

ΣM = vector sum of the moments of all the forces of the system.

For concurrent system, any of the following sets of equations ensures equilibrium (the resultant is zero). The concurrency is assumed at the origin.

Set 1: $\Sigma F_x = 0$

$\Sigma F_y = 0$

Set 2: $\Sigma F_x = 0$

$\Sigma M_A = 0$

(A may be chosen any place in the plane except on the y-axis)

Set 3: $\Sigma M_A = 0$

$\Sigma M_B = 0$

(A and B may be chosen any place in the plane except A, B and the origin do not lie on the same straight line)

where $\Sigma F_x, \Sigma F_y$ = algebraic sum of the x and y components of the forces of the system respectively

$\Sigma M_A, \Sigma M_B$ = algebraic sum of the moments of the forces of the system about A and B respectively.

For parallel system, any of the following sets of equations ensures equilibrium (the resultant is neither a force nor a couple).

Set 1: $\Sigma F = 0$

$\Sigma M_A = 0$

Set 2: $\Sigma M_A = 0$

$\Sigma M_B = 0$

(A and B may chosen any place in the plane provided in the line joining A and B is not parallel to the forces of the system)

where ΣF = algebraic sum of the forces of the system parallel to the action lines of the forces

ΣM_A, ΣM_B = algebraic sum of the moments of the forces of the system about A and B respectively.

For nonconcurrent and nonparallel system, any of the following sets of equations ensures equilibrium (the resultant is neither a force nor a couple).

Set 1:
$$\Sigma F_x = 0$$
$$\Sigma F_y = 0$$
$$\Sigma M_A = 0$$

Set 2:
$$\Sigma F_x = 0$$
$$\Sigma M_A = 0$$
$$\Sigma M_B = 0$$

(Provided that the line joining A and B is not perpendicular to the x-axis)

Set 3:
$$\Sigma M_A = 0$$
$$\Sigma M_B = 0$$
$$\Sigma M_C = 0$$

(Provided that A, B and C do not lie on the same straight line)

where ΣF_x, ΣF_y, ΣF_z = algebraic sum of the x, y and z components of the forces respectively

ΣM_A, ΣM_B, ΣM_C = algebraic sum of the moments of the forces of the system about any three points A, B and C in the plane respectively.

1.2.7 Equilibrium of Noncoplanar Force System

The necessary and sufficient conditions that R and C be zero vectors are given by

$$R = \Sigma F = 0$$

and
$$C = \Sigma M = 0$$

where ΣF = vector sum of all the forces of the system

ΣM = vector sum of the moments of all the forces of the system relative to any point.

For a concurrent and noncoplanar system, the following set of equations must be satisfied:

$$\Sigma F_x = 0$$
$$\Sigma F_y = 0$$
$$\Sigma F_z = 0$$

where ΣF_x, ΣF_y, ΣF_z = algebraic sums of the x, y and z components of the forces of the system respectively.

For a parallel noncoplanar system, the set of equations to be satisfied for equilibrium are

$$\Sigma F_y = 0$$
$$\Sigma M_x = 0$$
$$\Sigma M_z = 0$$

where ΣF_y = algebraic sum of the forces of the system along the y-axis which is selected parallel to the system.

ΣM_x, ΣM_z = algebraic sums of the moments of the forces of the system about the x and z axes respectively.

The necessary and sufficient conditions required for equilibrium for a nonconcurrent noncoplanar system are given by

$$\Sigma F_x = 0$$
$$\Sigma F_y = 0$$
$$\Sigma F_z = 0$$
$$\Sigma M_x = 0$$
$$\Sigma M_y = 0$$
$$\Sigma M_z = 0$$

where ΣF_x, ΣF_y, ΣF_z = algebraic sums of the *x*, *y* and *z* components of the forces of the system respectively.

ΣM_x, ΣM_y, ΣM_z = algebraic sums of the components of the forces of the system about the *x*, *y* and *z* axes respectively.

1.2.8 Trusses

A *truss* is a system of slender members that are pinned together. The members are free to rotate at the pinned joints and carry forces only. All the external forces act at the joints.

Trusses are examples of coplanar force systems in equilibrium. Trusses are assumed to be rigid members all located in one plane. The weights of the truss members are neglected. Forces are transmitted from one member to another through pin joints. These members are called two-force members. A two-force member is in equilibrium under the effect of two resultant forces one at each end. The two-force members will be either in tension or compression. There are two methods available for the analysis of trusses.

For a stable planar truss, the following condition applies:

$$2n = m + p$$

where n = number of joints

m = number of members

p = number of unknown external forces

In a stable three-dimensional truss, the condition that holds is

$$3n = m + p$$

Method of Joints: In this method, a free-body diagram of any pin in the truss is drawn. Maximum of two unknown forces act on that pin. Proceed from one pin to another until all unknowns have been obtained.

Method of Sections: A free-body diagram of a section of the truss is drawn. The forces in the members cut act as external forces. The system is a nonconcurrent and nonparallel one. In any one section no more than three unknown forces are to be found.

1.2.9 Analysis of Beams

A *beam* is a long member that is subjected to external lateral forces, *F* and external lateral moments, *T*. These cause internal lateral forces *V* called *shear forces* and internal lateral moments *M* known as *bending moments*. The shear forces and the bending moments induce lateral deformations, also called *bending*.

Shear and Moment Diagrams: Shear and moment diagrams show the variation of *V* and *M* across a beam. Summation of the forces and moments of forces are used to plot *V* and *M* directly for the beams.

The slope of the shear diagram at any section along the beam is the negative of the load per unit length at that point. The change in shear between two sections of a beam carrying a distributed load equals the negative area of the load diagram between the two sections. The slope of the moment diagram at any section along the beam is the value of the shear at that section. The change in the moment between two sections of a beam equals the area of the shear diagram between the two sections.

1.2.10 Friction

The tangential force that opposes the sliding of one body relative to the other is the *static friction* between the two bodies. *Kinetic friction* is the tangential force between two bodies after motion begins.

Referring to Fig. 2.1, the following terms are defined:

Limiting friction F' is the maximum value of static friction that occurs when motion is impending. *Coefficient of static friction* is the ratio of the limiting friction F' to the normal force N, or $\mu = \left(\frac{F'}{N}\right)$. *Coefficient of kinetic friction* is the ratio of the kinetic friction to the normal force. *Angle of response*, α is the angle to which an inclined plane may be raised before an object resting on it will move under the action of the force of gravity and the reaction of the plane.

In Fig. 1.13, R is the resultant of F' and N and $\alpha = \phi$.

Belt Friction: When a belt passes over a pulley, the tensions in the belt on the two sides of the pulley will differ. When slip is about to occur, the tensions T_1 and T_2 are given by

$$T_1 = T_2 e^{\mu\beta}$$

where T_1 = larger tension

T_2 = smaller tension

μ = coefficient of friction

β = angle of wrap (radians)

e = 2.718 (base of natural logarithms)

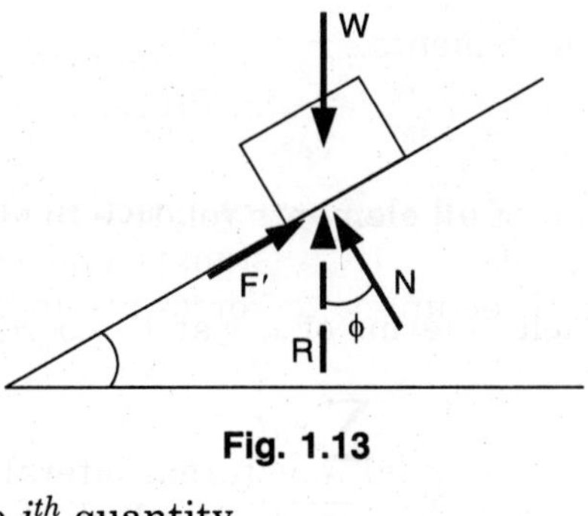

Fig. 1.13

where d_i = magnitude of the ith quantity

$\bar{x}, \bar{y}, \bar{z}$ = coordinates of centroid of the assemblage

x_i, y_i, z_i = coordinates of P_i at which d_i is concentrated.

The centroid of a continuous quantity of mass m is given by

$$\bar{\mathbf{r}} = \frac{\int \mathbf{r}\, dm}{\int dm}$$

or

$$\bar{x} = \frac{\int x\, dm}{\int dm} = \frac{Q_{yz}}{m}$$

$$\bar{y} = \frac{\int y\, dm}{\int dm} = \frac{Q_{xz}}{m}$$

$$\bar{z} = \frac{\int z\, dm}{\int dm} = \frac{Q_{xy}}{m}$$

where Q_{xy}, Q_{yz}, Q_{xz} = first moments with respect to xy, yz and xz planes respectively.

1.2.11 First Moments and Centroids

The centroid of n similar quantities, $d_1, d_2, d_3, ..., d_n$ located at points $P_1, P_2, P_3, ..., P_n$ for which the position vectors relative to a selected point O are $\mathbf{r}_1, \mathbf{r}_2, \mathbf{r}_3, ..., \mathbf{r}_n$ has a position. Vector $\bar{\mathbf{r}}$ is defined as

$$\bar{\mathbf{r}} = \frac{\sum_{i=1}^{n} \mathbf{r}_i d_i}{\sum_{i=1}^{n} d_i}$$

where $d_i = i^{th}$ quantity

$\mathbf{r}_i$ = position vector of the ith element

$\sum_{i=1}^{n} d_i$ = sum of all n elements

$\sum_{i=1}^{n} \mathbf{r}_i d_i$ = first moment of all elements relative to selected point O

The centroid can be written in terms of x, y and z coordinates as

$$\bar{x} = \frac{\sum_{i=1}^{n} x_i d_i}{\sum_{i=1}^{n} d_i}$$

$$\bar{y} = \frac{\sum_{i=1}^{n} y_i d_i}{\sum_{i=1}^{n} d_i}$$

$$\bar{z} = \frac{\sum_{i=1}^{n} z_i d_i}{\sum_{i=1}^{n} d_i}$$

1.2.12 Virtual Work

A *virtual displacement* δs of a particle is defined as any infinitesimal change in the position of the particle consistent with the constraints imposed on the particle. *Virtual work* δU done by a force, F_t is defined as $F_t\,\delta s$, where F_t is the magnitude of the component of the force along the virtual displacement δs. Similarly, virtual work δU done by a couple of moment M is defined as $M\delta\theta$, where $\delta\theta$ is the virtual angular displacement.

Equilibrium: The necessary and sufficient condition for the *equilibrium of a particle* is zero virtual work done by all the forces acting on the particle during any virtual displacement δs. The necessary and sufficient condition for the *equilibrium of a rigid body* is zero virtual work done by all the external forces acting on the body during any virtual displacement consistent with the constraints imposed on the body. The *equilibrium* of a system of rigid bodies exists if the potential energy V has a stationary value. *Stable equilibrium* occurs if the potential energy V is a minimum. *Unstable equilibrium* occurs if the potential energy V is a maximum. *Neutral equilibrium* exists if a system remains in any position in which it is placed.

If x is the variable and V the potential energy of the system, then we have

$$\frac{d^2V}{dx^2} > 0 \quad \text{stable equilibrium}$$

$$\frac{d^2V}{dx^2} < 0 \quad \text{unstable equilibrium}$$

$$\frac{d^2V}{dx^2} = 0 \quad \text{neutral equilibrium}$$

1.2.13 Kinematics of a Particle

Kinematics is the study of motion without regard to the forces or other factors that influence the motion.

Rectilinear motion: For motion of a point along a straight line, the following formulas are valid for constant acceleration $a = k$:

$$v = v_0 + kt$$
$$v^2 = v_0^2 + 2ks$$
$$s = v_0 t + \frac{1}{2}kt^2$$

$$s = \frac{1}{2}(v + v_0)t$$

where v_0 = initial velocity
v = final velocity
k = constant acceleration
t = time
s = displacement

Curvilinear motion: In a plane, curvilinear motion is motion along a plane curve (path). The velocity and acceleration of a point on such a curve can be expressed as

(*a*) rectangular components
(*b*) tangential and normal components
(*c*) radial and transverse components

(*a*) Rectangular components

The position vector $\mathbf{r}(t)$, the velocity vector $\mathbf{v}(t)$, and the acceleration vector $\mathbf{a}(t)$ are given by

$$\mathbf{r}(t) = x(t)i + y(t)j + z(t)k$$

$$\mathbf{v}(t) = v_x(t)i + v_y(t)j + z_z(t)k = \dot{x}(t)i + \dot{y}(t)j + \dot{z}(t)k$$

$$\mathbf{a}(t) = a_x(t)i + a_y(t)j + a_z(t)k = \ddot{x}(t)i + \ddot{y}(t)j + \ddot{z}(t)k$$

where the over-dot represents time differentiation.

The rectangular components of velocity and acceleration are given by

$$\mathbf{v}(t) = \dot{x}(t),\ \mathbf{v}_y(t) = \dot{y}(t),\ \mathbf{v}_z(t) = \dot{z}(t)$$

$$\mathbf{a}(t) = \ddot{x}(t),\ \mathbf{a}_y(t) = \ddot{z}(t),\ \mathbf{a}_z(t) = \ddot{y}(t)$$

The components of position are given by

$$\mathbf{x}(t) = x(t_1) + \int_{t_1}^{t} v_x(s)ds$$

$$\mathbf{y}(t) = y(t_1) + \int_{t_1}^{t} v_y(s)ds$$

$$\mathbf{z}(t) = z(t_1) + \int_{t_1}^{t} v_z(s)ds$$

The components of velocity are given by

$$\mathbf{v}_x(t) = v_x(t_1) + \int_{t_1}^{t} a_x(s)ds$$

$$\mathbf{v}_y(t) = v_y(t_1) + \int_{t_1}^{t} a_y(s)ds$$

$$\mathbf{v}_z(t) = v_z(t_1) + \int_{t_1}^{t} a_z(s)ds$$

Also $$v_x^2(x) = v_x^2(x_1) + 2\int_{x_1}^{x} a_x(x)dx$$

(*b*) Tangential and normal components

Referring to Fig. 1.14, the velocity vector **v** can be written as

$$\mathbf{v} = \mathbf{u}\mathbf{n}_t$$

where u = magnitude of the velocity vector

$\mathbf{n}_t$ = tangential unit vector directed along the velocity vector.

Also $$\dot{n} = \dot{\theta}\mathbf{n}_n$$

and $$\dot{\mathbf{n}}_n = -\dot{\theta}\mathbf{n}_t$$

where $\mathbf{n}_n$ = the normal unit vector defined to be perpendicular to the tangential unit vector.

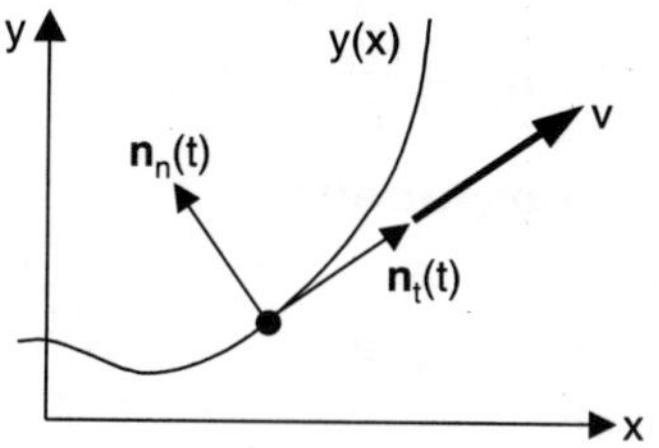

Fig. 1.14

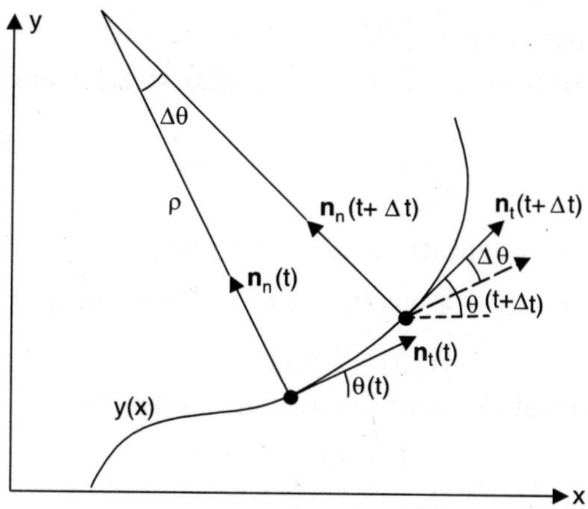

Fig. 1.15

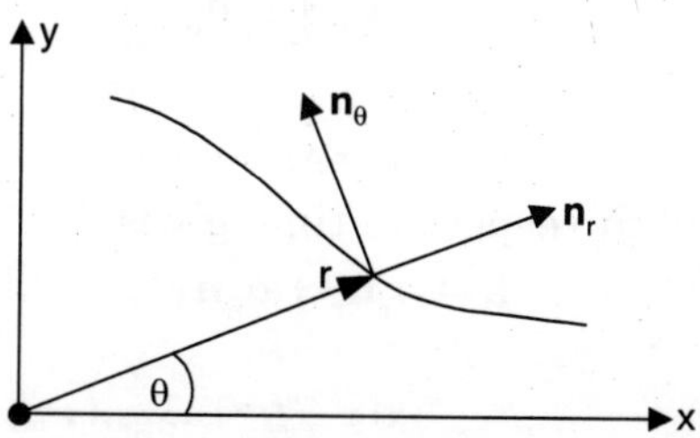

Fig. 1.16

In Fig. 1.15, the radius of curvature ρ of the path at time t is shown which is obtained by the intersection of the lines extending from $\mathbf{n}_t(t)$ and $\mathbf{n}_t(t + \Delta t)$ where Δt is time increment. The angle θ changes an incremental amount Δθ and the point moves an incremental amount Δs.

Now $$\rho\Delta\theta = \Delta s$$

or $$\rho\dot{\theta} = v$$

where $$\dot{\theta} = \frac{d\theta}{dt}$$

$$v = \frac{ds}{dt}$$

Differentiating (2.31) with respect to time gives

$$\mathbf{a} = a_t\mathbf{n}_t + a_n\mathbf{n}_n$$

where $$a_t = \dot{v}$$

$$a_n = \frac{v^2}{\rho}$$

(c) Radial and transverse components

The polar form of a position vector is

$$\mathbf{r} = r\mathbf{n}_r \tag{1.1}$$

where $r = |\ \mathbf{r}\ |$ is the magnitude of r

$\mathbf{n}_r$ = unit vector in the direction of $\mathbf{r}$

$\mathbf{n}_r$ = the radial unit vector

$\mathbf{n}_\theta$ = the circumferential unit vector

The circumferential unit vector is perpendicular to $\mathbf{n}_r$. Hence $\mathbf{n}_r$ and $\mathbf{n}_\theta$ are related to i and j in Fig. 1.16 as

$$\mathbf{n}_r(t) = \cos\theta(t)i + \sin\theta(t)j$$

$$\mathbf{n}_\theta(t) = -\sin\theta(t)i + \cos\theta(t)j \tag{1.2}$$

Differentiating the above equations with respect to time gives

$$\dot{\mathbf{n}}_r = -\dot{\theta}\sin\theta i + \dot{\theta}\cos\theta j = \dot{\theta}(-\sin\theta i + \cos\theta j) = \dot{\theta}\,\mathbf{n}_\theta$$

and $$\dot{\mathbf{n}}_\theta(t) = -\dot{\theta}\mathbf{n}_r \tag{1.3}$$

Hence, the derivatives of the unit vectors are given by

$$\dot{\mathbf{n}}_r = \dot{\theta}\mathbf{n}_\theta$$

$$\dot{\mathbf{n}}_\theta = -\dot{\theta}\mathbf{n}_r \tag{1.4}$$

Similarly, from Eqs. (1.1) and (1.4), we get

$$\mathbf{v} = v_r\mathbf{n}_r + v_\theta\mathbf{n}_\theta \tag{1.5}$$

where $v_r = \dot{r}$

$$v_\theta = r\dot{\theta}$$

Differentiating Eq. (1.5) with respect to time gives

$$\mathbf{a} = a_r\mathbf{n}_r + a_\theta\mathbf{n}_\theta \tag{1.6}$$

where $a_r = \ddot{r} - r\dot{\theta}^2$

$$a_\theta = r\ddot{\theta} + 2\dot{r}\dot{\theta} \tag{1.7}$$

1.2.14 D'Alembert's Principle

D'Alembert's principle states that

$$\Sigma F - ma = 0$$

where ΣF = vector sum of all the forces acting on the particle

m = mass of the particle

a = acceleration of the particle

Hence, an imaginary force (also known as inertia force) which is collinear with ΣF but opposite in sense and of magnitude ma would cause it to be in equilibrium if applied to the particle. All the equations of equilibrium are then applicable.

1.2.15 Kinematics of a Rigid Body in Plane Motion

In plane motion of a rigid body, every point in the body remains at a constant distance from a fixed plane. In Fig. 1.17, B is an arbitrary point in the body and xyz is a non-rotating reference frame.

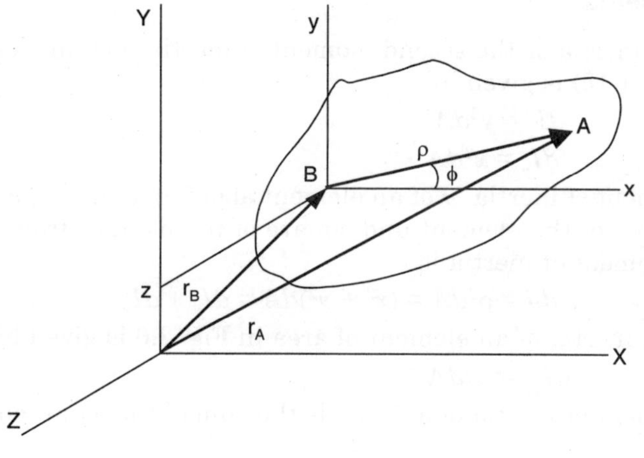

Fig. 1.17

The position vector $\mathbf{r}_A$ of any point A (fixed or moving) in the lamina is given by

$$\mathbf{r}_A = \mathbf{r}_B + \rho \tag{1.8}$$

where $\mathbf{r}_B$ = position vector of B

ρ = vector BA

Differentiating Eq. (1.8) gives

$$\mathbf{v}_A = \dot{r}_A = \dot{\mathbf{r}}_B + \dot{\rho} = \dot{r}_B + \rho\omega\mathbf{e}_\phi \tag{1.9}$$

where $\bar{\mathbf{r}}_B$ = linear velocity of B relative to the fixed axes X, Y and Z

ω = angular velocity (magnitude) of ρ about any line parallel to Z-axis

$\mathbf{e}_\phi$ = unit vector perpendicular to ρ (in the direction of increasing ϕ)

The acceleration $\mathbf{a}_A$ is given by

$$a_A = \dot{\mathbf{v}}_A = \ddot{\mathbf{r}}_A = \ddot{\mathbf{r}}_B - \rho\omega^2 e_r + \rho\alpha\, e_f \tag{1.10}$$

where $\ddot{\mathbf{r}}_B$= acceleration of B relative to fixed axes X, Y and Z

$\mathbf{e}_r$ = unit vector along ρ directed from B toward A

$\mathbf{e}_\phi$ = unit vector perpendicular to ρ (in the direction of increasing ϕ)

α = angular acceleration (magnitude) of ρ about any line parallel to the Z-axis.

Equations (1.9) and (1.10) can also be written as

$$\mathbf{v_A} = \mathbf{v_B} + \omega \mathbf{X} \rho$$

or

$$\mathbf{v_A} = \mathbf{v_B} + \mathbf{v_{A/B}} \tag{1.11}$$

and

$$\mathbf{a_A} = \mathbf{a_B} + \omega \mathbf{X}(\omega \mathbf{X} \rho) + \alpha \mathbf{X} \rho$$

$$\mathbf{a_A} = \mathbf{a_B} + \mathbf{a_{A/B}} \tag{1.12}$$

where $\omega = \dot{\phi} k = \omega \mathbf{k}$

$$\alpha = \ddot{\phi}\mathbf{k} = \dot{\omega}\mathbf{k} = \alpha\mathbf{k}$$

$$\rho\omega\mathbf{e}_\phi = \omega\mathbf{X}\rho$$

$$\rho\alpha\mathbf{e}_\phi = \alpha\mathbf{X}\rho$$

$$-\rho\omega^2\mathbf{e} = \omega\mathbf{X}(\omega\mathbf{X}\rho)$$

in which $\mathbf{v}_{A/B}$ = relative velocity of A as it rotates around B

$\mathbf{a}_{A/B}$ = relative acceleration of A as it rotates around B

1.2.16 Moments of Inertia

The axial moment of inertia or the second moment of inertia, I of an element of area about an axis in its plane (Fig. 1.18) is given by

$$dI_x = y^2 dA$$

$$dI_y = x^2 dA$$

The polar moment of inertia, J of an element about an axis perpendicular to its plane is the product of the area of the element and square of its distance from the axis. Referring to Fig. 2.6, the polar moment of inertia is

$$dJ = \rho^2 dA = (x^2 + y^2)dA = dI_y + dI_x$$

The product of inertia of an element of area in Fig. 2.6 is given by

$$dI_{xy} = xydA$$

The axial moment of inertia of an area is the sum of the axial moments of inertia of its elements

$$I_x = \int y^2 \, dA$$

$$I_y = \int x^2 \, dA$$

The radius of gyration of an area with respect to an axis is given by

$$k = \sqrt{I/A}$$

The polar moment of inertia of an area is the sum of the polar moments of inertia of its elements

$$J = \int \rho^2 \, dA$$

The product of inertia of an area is the sum of the products of inertia of its elements

$$I_{xy} = \int xy \, dA$$

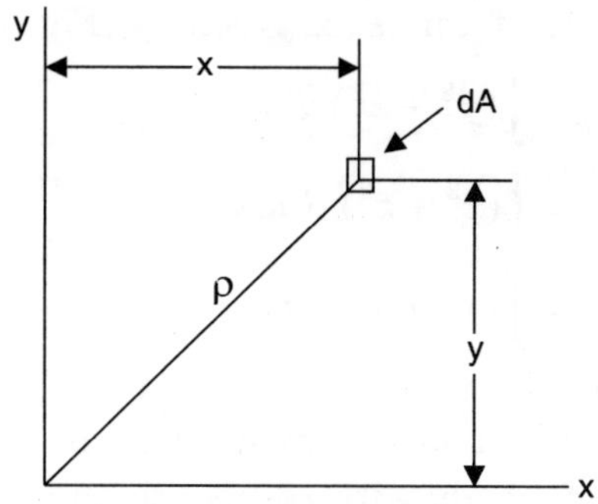

Fig. 1.18

The parallel-axis theorem states that the axial or polar moment of inertia of an area about any axis equals the axial or polar moment of inertia of the area about a parallel axis through the centroid of the area plus the product of the area and the square of the distance between the two parallel axes. Referring to Fig. 1.19, we have

$$I_x = I_{x'} + Am^2$$
$$I_y = I_{y'} + An^2$$
$$I_z = J + Ar^2$$
$$I_{xy} = I_{xy} + Amn$$

where A = area

I_{xy} = product of inertia of the area with respect to x and y axes

$I_{x'y'}$ = product of inertia about two parallel centroid axes x' and y'

m, n = coordinates of G relative to the (x, y) axes through O or the coordinates of O relative to the (x', y') axes through G.

x, y = any axes through O

$x'y'$ = coplanar parallel axes through the centroid G

The axial or polar moment of inertia, or product of inertia of a composite area is the sum of the axial or polar moments of inertia, or products of inertia, of the component areas of the whole area.

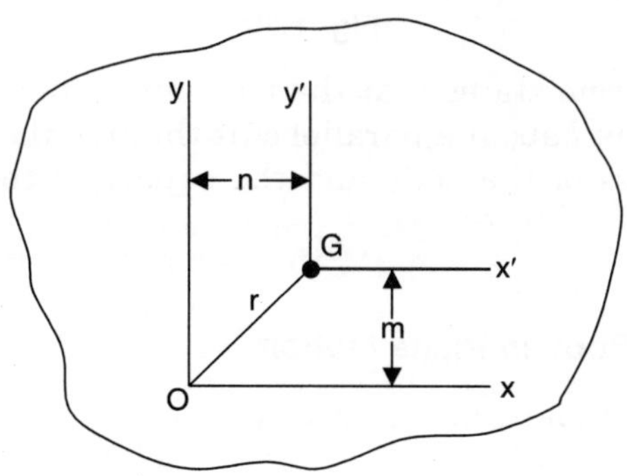

Fig. 1.19

The axial moment of inertia of a mass is given by (Fig. 1.20)

$$I_x = \int (y^2 + z^2)\, dm$$

$$I_y = \int (x^2 + z^2)\, dm$$

$$I_z = \int (x^2 + y^2)\, dm$$

where dm = mass of element

I_x, I_y, I_z = axial moments of inertia with respect to the x, y and z axes respectively.

The radius of gyration k of a body with respect to an axis is given by

$$k = \sqrt{I/m}$$

The product of inertia of a mass is given by

$$I_{xy} = \int xy\, dm$$

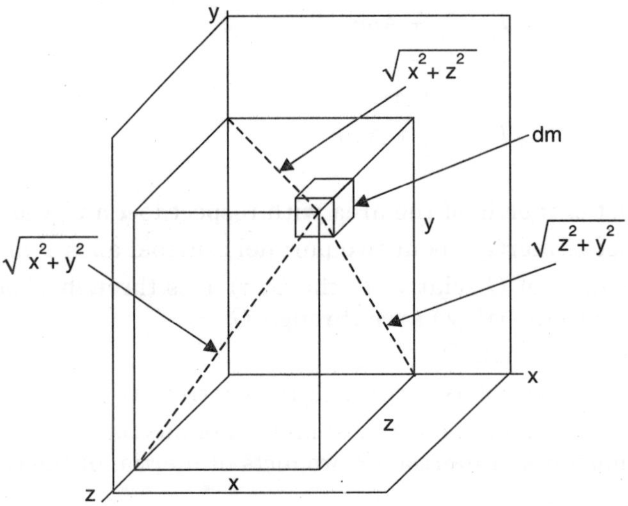

Fig. 1.20

The parallel axis theorem states that the moment of inertia of a body about an axis is equal to the moment of inertia I about a parallel axis through the center of gravity of the body plus the product of the mass of the body and the square of the distance between the two parallel axes.

1.2.17 Dynamics of a Rigid Body in Plane Motion

The vector equations of plane motion are given by

$$\Sigma \mathbf{F} = m\,\bar{\mathbf{a}}$$

$$\Sigma \mathbf{M}_O = I_O \alpha \mathbf{k} + m\mathbf{r}_{G/O} \times \mathbf{a}_O = (I_O\,\alpha + m\,\bar{x}\,a_{Oy} - m\,\bar{y}\,a_{Ox})\mathbf{k}$$

where $\Sigma \mathbf{F}$ = resultant of the external forces acting on the body

$\Sigma \mathbf{M}_O$ = resultant of the external moments acting on the body

m = mass of the body

$\bar{\mathbf{a}}$ = acceleration of the mass center of the body

$\mathbf{a}_O$ = acceleration of reference point O

α = angular acceleration of the body

I_O = moment of inertia of the body relative to the reference point O

$\bar{x}, \bar{y}$ = coordinates of the mass center relative to the reference point O

$\mathbf{r}_{G/O}$ = position vector of the mass center relative to the reference point O

a_{Ox}, a_{Oy} = magnitude of the components of the acceleration of the reference point O along the x and y axes

The scalar equations of the plane motion are given by

$$\Sigma F_x = m\bar{a}_x, \quad \Sigma F_y = m\bar{a}_y, \quad \Sigma \overline{M} = \bar{I}\alpha$$

where $\Sigma F_x, \Sigma F_y$ = algebraic sums of the magnitudes of the components of the external forces along the x and y axes respectively

m = mass of the body

$\bar{a}_x, \bar{a}_y$ = components of the linear acceleration of the mass center in x and y directions respectively

$\Sigma \overline{M}$ = algebraic sum of the moments of the external forces about the mass center

$\bar{I}$ = moment of inertia of the body about the mass center

α = magnitude of the angular acceleration of the body.

The scalar equations for translation of a rigid body are given by

$$\Sigma F_x = ma_x$$

$$\Sigma F_y = ma_y$$

and

$$\Sigma M = 0$$

where $\Sigma F_x, \Sigma F_y$ = algebraic sums of the components of the external forces in the x and y directions respectively

m = mass of the body

a_x, a_y = acceleration components in the x and y directions respectively

ΣM = sum of the moments of the external forces about the mass center of the body

The scalar equations of motion of a rigid body under the action of an unbalanced force system for a body with a plane of symmetry and rotates about a fixed-axis perpendicular to the plane are given by

$$\Sigma F_n = m\bar{r}\omega^2$$

$$\Sigma F_t = m\bar{r}\alpha$$

$$\Sigma M_o = I_O\alpha$$

where ΣF_n = algebraic sum of the components of all external forces (which are the applied forces F_1, F_2, F_3, etc., the gravitational force on the body, and the reaction R of the axis on the body) along the n axis, which is the line drawn between the center of rotation O and the mass center G; note that the positive sense is from G toward O because $\bar{a}_n = \bar{r}\omega^2$ has that sense

ΣF_t = algebraic sum of the components of the external forces along the t axis, which is perpendicular to the n axis at O; note that the positive sense along this axis agrees with that of $\bar{a}_t = \bar{r}\alpha$

ΣM_O = algebraic sum of the moments of the external forces about the axis of rotation through O; note that positive sense agrees with the assumed sense of the angular acceleration α

m = mass of the body

G = center of mass of the body

$\bar{r}$ = distance from the center of rotation O to the mass center G

$\bar{I}_O$ = moment of inertia of the body about the axis of rotation

ω = angular speed of the body

α = magnitude of the angular acceleration of the body

This type of rotation is called *non-centroidal rotation.*

When G and O coincide (rotation about a fixed axis through G) and $F = 0$, the equations of motion are given by

$$\Sigma F_x = 0 \quad \Sigma F_y = 0 \quad \Sigma \overline{M} = \bar{I}\alpha$$

where ΣF_x = algebraic sum of the components of the external forces along any axis chosen as the x-axis

ΣF_y = algebraic sum of the components of the external forces along the y-axis

$\Sigma \overline{M}$ = algebraic sum of the moments of the external forces about the axis of rotation through the mass center G (axis of symmetry)

$\bar{I}$ = moment of inertia of the body about the axis of rotation through the mass center G

α = magnitude of the angular acceleration of the body

The type of rotation is called *centroidal rotation.*

1.2.18 Work and Energy

The work done to move particle from point 1 to point 2 by the resultant force F acting on the particle is given by

$$U_{1-2} = \int_1^2 \mathbf{F} \bullet dr$$

The kinetic energy T of a particle with mass m and moving with speed v is defined as $\frac{1}{2}mv^2$.

The work done on a particle by the resultant force as it moves from point 1 to point 2 is equal to the change in kinetic energy.

$$U_{1-2} = T_2 - T_1$$

where T_1, T_2 = initial and final kinetic energy respectively at points 1 and 2.

The kinetic energy T of a rigid body in translation is

$$T = \frac{1}{2} mv^2.$$

The kinetic energy T of a rigid body in rotation is

$$T = \frac{1}{2} I_o \omega^2$$

where I_o = mass moment of inertia of the body about the axis of rotation

ω = angular speed

The kinetic energy T of a body in plane motion is given by

$$T = \frac{1}{2} \bar{I} \omega^2.$$

where ω = angular speed

$\bar{I}$ = moment of inertia about an axis through the mass center parallel to the z-axis.

The potential energy may be defined as the negative of the work done by the conservative force acting on the body in bringing it from the datum position to the position in question. The selection of the datum is arbitrary. The principle of work and energy states that the work done by the external forces acting on a rigid body during a displacement is equal to the change in kinetic energy of the body during the same displacement. The sum of the work done by the non-conservative external forces and the work done by the internal forces acting on a system of particles is equal to the change in the total (kinetic and potential) energy of the system of the particles over the time interval of the action

or $$E = T + V$$

where E = total energy (kinetic and potential)

T = kinetic energy

V = potential energy

The law of conservation of energy states that if a particle (or body) is acted upon by a conservation force system, the sum of the kinetic energy and potential energy is a constant.

1.2.19 Impulse and Momentum

The *linear momentum* of the ith particle is defined as

$$\mathbf{L}_i = m_i \mathbf{v}_i$$

where m_i = mass of the ith particle

$\mathbf{v}_i$ = i^{th} particle's velocity

The linear momentum of the system of particles is the sum of the linear momenta of the particles

or $$\mathbf{L} = \sum_{i=1}^{n} \mathbf{L}_i$$

The linear impulse acting on the system imparted over a time interval is given by

$$\mathbf{G}_{1-2} = \mathbf{L}_2 - \mathbf{L}_1 \qquad (1.13)$$

where $\mathbf{L}$ = linear momentum of the system

$$\mathbf{L} = \sum_{i=1}^{n} m_i \mathbf{v}_i$$

$$\mathbf{G}_{1-2} = \int_{t_1}^{t_2} \mathbf{F}\, dt$$

Equation (1.13) states that the linear impulse acting on the system is equal to the change in the system's linear momentum over the time interval.

The momentum of the linear momentum is given by

$$\mathbf{H}_O = \mathbf{r} \times \mathbf{L}$$

$\mathbf{H}_O$ is called the angular momentum about the point O.

The angular momentum of a system of n particles about a point A is given by

$$\mathbf{H}_A = \sum_{i=1}^{n} \mathbf{H}_{Ai} = \sum_{i=1}^{n} \mathbf{r}_{i/A} \times \mathbf{L}_i = \sum_{i=1}^{n} \mathbf{r}_{i/A} \times m_i \mathbf{v}_i$$

where A is any point.

The resultant external moment about the point A is given by

$$\mathbf{M}_A = \sum_{i=1}^{n} M_{Ai} = \sum_{i=1}^{n} \mathbf{r}_{i/A} \times \mathbf{F}_i$$

Differentiating Eq. (2.76) with respect to time yields

$$\dot{\mathbf{H}}_A = -\mathbf{v}_A \times \mathbf{L} + \mathbf{M}_A$$

If the point A is fixed then $\mathbf{v}_A = 0$ and if the point A is the mass center C then $\mathbf{v}_A \times \mathbf{L} = \mathbf{v}_C \times m\mathbf{v}_C = 0$. Whether the point A represents a fixed point O or the mass center C, then

$$\dot{\mathbf{H}}_A = \mathbf{M}_A \tag{1.14}$$

Equation (1.14) states that the time derivative of the angular momentum about a fixed point O (or the mass center O) is equal to the resultant external moment about the fixed point O (or the mass center C). Integrating Eq. (1.14) yields

$$\begin{aligned}\mathbf{N}_{A1-2} &= \int_{t_1}^{t_2} \mathbf{M}_A\, dt = \int_{t_1}^{t_2} \dot{\mathbf{H}}_A\, dt = \mathbf{H}_A(t_2) - \mathbf{H}_A(t_1) \\ &= \mathbf{H}_{A2} - \mathbf{H}_{A1} \end{aligned} \tag{1.15}$$

where $\mathbf{N}_{A1-2} = \int_{t_1}^{t_2} \mathbf{M}_A dt$ = angular impulse imparted over the time interval

Hence $\mathbf{N}_{A1-2} = \mathbf{H}_{A2} - \mathbf{H}_{A1}$ (1.16)

Equation (1.16) states that the angular impulse acting on the system about the fixed point O (or the mass center C) is equal to the change in the system's angular momentum about the point O (or the mass center C) over the time interval.

Conservation of linear momentum in a given direction occurs if the sum of the external forces in that direction is zero. Conservation of angular momentum about an axis occurs if the sum of the moments of the external forces about that axis is zero.

1.3 MECHANICAL VIBRATIONS

Vibration is the motion of a particle or a body or system of connected bodies displaced from a position of equilibrium. Most vibrations are undesirable in machines and structures because they produce increased stresses, energy losses, cause added wear, increase bearing loads, induce fatigue, create passenger discomfort in vehicles, and absorb energy from the system. Rotating machine parts need careful balancing in order to prevent damage from vibrations.

Vibration occurs when a system is displaced from position of stable equilibrium. The system tends to return to this equilibrium position under the action of restoring forces (such as the elastic forces, as for a mass attached to a spring, or gravitational forces, as for a simple pendulum). The system keeps moving back and forth across its position of equilibrium. A *system* is a combination of elements intended to act together to accomplish an objective. For example, an automobile is a system whose elements are the wheels, suspension, car body, and so forth. A *static* element is one whose output at any given time depends only on the input at that time while a *dynamic* element is one whose present output depends on past *inputs.* In the same way we also speak of *static* and *dynamic systems.* A *static system* contains all elements while a *dynamic system* contains at least one dynamic element.

A physical system undergoing a time-varying interchange or dissipation of energy among or within its elementary storage or dissipative devices is said to be in a *dynamic state.* All of the elements in general are called *passive, i.e.,* they are incapable of generating net energy. A dynamic system composed of a finite number of storage elements is said to be *lumped* or *discrete,* while a system containing elements, which are dense in physical space, is called *continuous.* The analytical description of the dynamics of the discrete case is a set of ordinary differential equations, while for the continuous case it is a set of partial differential equations. The analytical formation of a dynamic system depends upon the kinematic or geometric constraints and the physical laws governing the behavior of the system.

1.3.1 Classification of Vibrations

Vibrations can be classified into three categories: *free, forced,* and *self-excited. Free vibration* of a system is vibration that occurs in the absence of external force. An external force that acts on the system causes forced vibrations. In this case, the exciting force continuously supplies energy to the system. Forced vibrations may be either deterministic or random (see Fig. 1.21). *Self-excited vibrations* are periodic and deterministic oscillations. Under certain conditions, the equilibrium state in such a vibration system becomes unstable, and any disturbance causes the perturbations to grow until some effect limits any further growth. In contrast to forced vibrations, the exciting force is independent of the vibrations and can still persist even when the system is prevented from vibrating.

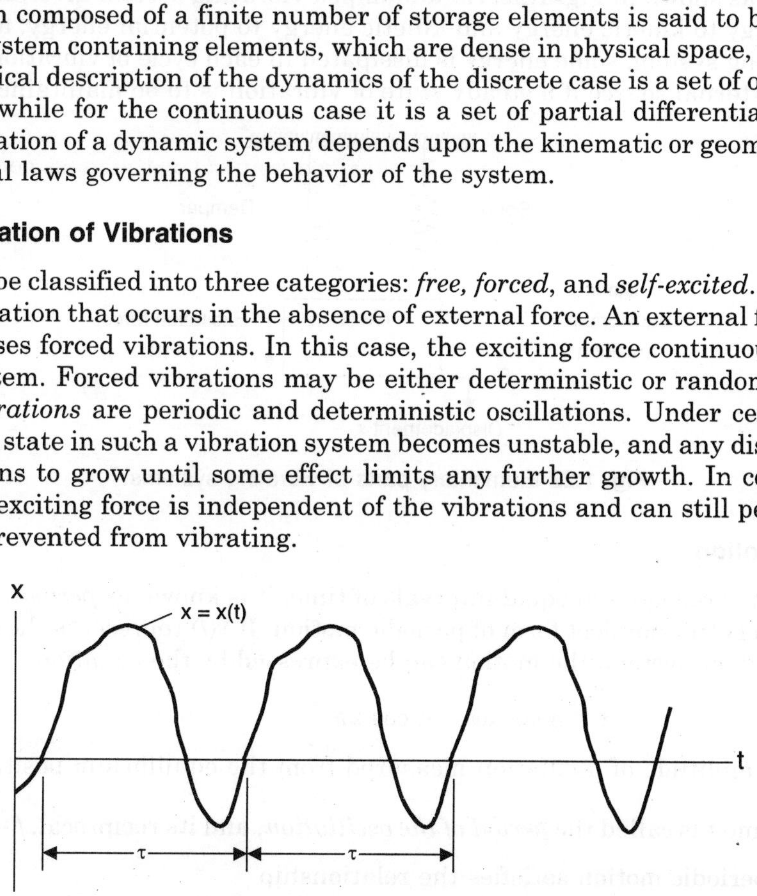

Fig. 1.21(a) A deterministic (periodic) excitation.

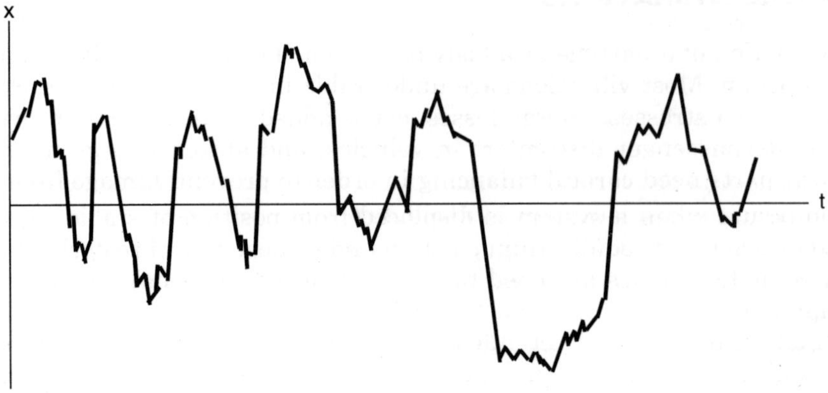

Fig. 1.21(b) Random excitation.

1.3.2 Elementary Parts of Vibrating Systems

In general, a vibrating system consists of a spring (a means for storing potential energy), a mass or inertia (a means for storing kinetic energy), and a damper (a means by which energy is gradually lost) as shown in Fig. 1.22. An undamped vibrating system involves the transfer of its potential energy to kinetic energy and kinetic energy to potential energy, alternatively. In a damped vibrating system, some energy is dissipated in each cycle of vibration and should be replaced by an external source if a steady state of vibration is to be maintained.

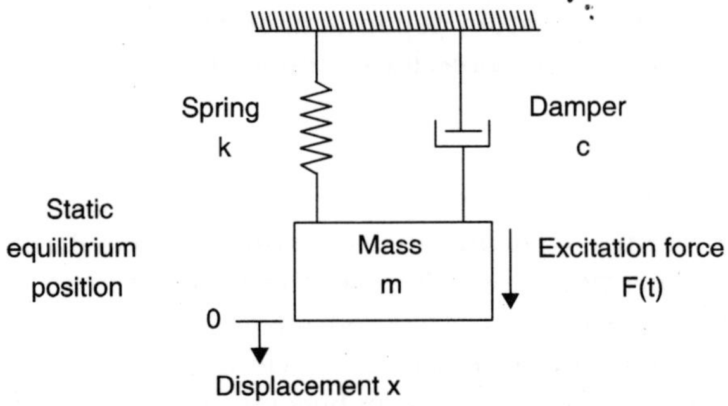

Fig. 1.22 Elementary parts of vibrating systems.

1.3.3 Periodic Motion

When the motion is repeated in equal intervals of time, it is known as *periodic motion*. Simple harmonic motion is the simplest form of periodic motion. If $x(t)$ represents the displacement of a mass in a vibratory system, the motion can be expressed by the equation

$$x = A \cos \omega t = A \cos 2\pi \frac{t}{\tau}$$

where A is the amplitude of oscillation measured from the equilibrium position of the mass. The repetition time τ is called the *period of the oscillation*, and its reciprocal, $f = \frac{1}{\tau}$, is called the *frequency*. Any periodic motion satisfies the relationship

$$x(t) = x(t + \tau)$$

i.e., Period, $\tau = \frac{2\pi}{\omega}$ s/cycle

Frequency, $f = \frac{1}{\tau} = \frac{\omega}{2\pi}$ cycles/s, or Hz

ω is called the *circular frequency* measured in rad/sec.

The velocity and acceleration of a harmonic displacement are also harmonic of the same frequency, but lead the displacement by $\pi/2$ and π radians, respectively. When the acceleration $\ddot{x}$ of a particle with rectilinear motion is always proportional to its displacement from a fixed point on the path and is directed towards the fixed point, the particle is said to have *simple harmonic motion*.

The motion of many vibrating systems in general is not harmonic. In many cases the vibrations are periodic as in the impact force generated by a forging hammer. If $x(t)$ is a periodic function with period τ, its Fourier series representation is given by

$$x(t) = \frac{a_0}{2} + \sum_{n=1}^{\infty} (a_n \cos n\omega t + b_n \sin n\omega t)$$

where $\omega = 2\pi/\tau$ is the fundamental frequency and $a_0, a_1, a_2, \ldots, b_1, b_2, \ldots$ are constant coefficients, which are given by:

$$a_0 = \frac{2}{\tau}\int_0^{\tau} x(t)\, dt$$

$$a_n = \frac{2}{\tau}\int_0^{\tau} x(t) \cos n\omega t\, dt$$

$$b_n = \frac{2}{\tau}\int_0^{\tau} x(t) \sin n\omega t\, dt$$

The exponential form of $x(t)$ is given by:

$$x(t) = \sum_{n=-\infty}^{\infty} c_n e^{in\omega t}$$

The Fourier coefficients c_n can be determined, using

$$c_n = \frac{1}{\tau}\int_0^{\tau} (x)t\, e^{-in\omega t}\, dt$$

The harmonic functions $a_n \cos n\omega t$ or $b_n \sin n\omega t$ are known as the *harmonics of order n* of the periodic function $x(t)$. The harmonic of order n has a period τ/n. These harmonics can be plotted as vertical lines in a diagram of amplitude (a_n and b_n) versus frequency ($n\omega$) and is called *frequency spectrum*.

1.3.4 Discrete and Continuous Systems

Most of the mechanical and structural systems can be described using a finite number of degrees of freedom. However, there are some systems, especially those include continuous elastic members, have an infinite number of degree of freedom. Most mechanical and structural systems have elastic (deformable) elements or components as members and hence have an infinite number of degrees of freedom. Systems which have a finite number of degrees of freedom are known as *discrete* or *lumped parameter systems*, and those systems with an infinite number of degrees of freedom are called *continuous* or *distributed systems*.

1.3.5 Vibration Analysis

The outputs of a vibrating system, in general, depend upon the initial conditions, and external excitations. The vibration analysis of a physical system may be summarised by the four steps:

1. Mathematical Modelling of a Physical System
2. Formulation of Governing Equations
3. Mathematical Solution of the Governing Equations
4. Physical Interpretation of the Results

1. Mathematical modelling of a physical system

The purpose of the mathematical modelling is to determine the existence and nature of the system, its features and aspects, and the physical elements or components involved in the physical system. Necessary assumptions are made to simplify the modelling. Implicit assumptions are used that include:

(*a*) A physical system can be treated as a continuous piece of matter

(*b*) Newton's laws of motion can be applied by assuming that the earth is an internal frame

(*c*) Ignore or neglect the relativistic effects

All components or elements of the physical system are linear. The resulting mathematical model may be linear or non-linear, depending on the given physical system. Generally speaking, all physical systems exhibit non-linear behaviour. Accurate mathematical modelling of any physical system will lead to non-linear differential equations governing the behaviour of the system. Often, these non-linear differential equations have either no solution or difficult to find a solution. Assumptions are made to linearise a system, which permits quick solutions for practical purposes. The advantages of linear models are the following:

(1) their response is proportional to input

(2) superposition is applicable

(3) they closely approximate the behaviour of many dynamic systems

(4) their response characteristics can be obtained from the form of system equations without a detailed solution

(5) a closed-form solution is often possible

(6) numerical analysis techniques are well developed, and

(7) they serve as a basis for understanding more complex non-linear system behaviours.

It should, however, be noted that in most non-linear problems it is not possible to obtain closed-form analytic solutions for the equations of motion. Therefore, a computer simulation is often used for the response analysis.

When analysing the results obtained from the mathematical model, one should realise that the mathematical model is only an approximation to the true or real physical system and therefore the actual behaviour of the system may be different.

2. Formulation of governing equations

Once the mathematical model is developed, we can apply the basic laws of nature and the principles of dynamics and obtain the differential equations that govern the behaviour of the system. A basic law of nature is a physical law that is applicable to all physical systems irrespective of the material from which the system is constructed. Different materials behave differently under different operating conditions. Constitutive equations provide information about the materials of which a system is made. Application of geometric constraints such as the kinematic relationship between displacement, velocity, and acceleration is often necessary to complete the mathematical modelling of the physical system. The application of geometric constraints is necessary in order to formulate the required boundary and/or initial conditions.

The resulting mathematical model may be linear or non-linear, depending upon the behaviour of the elements or components of the dynamic system.

3. Mathematical solution of the governing equations

The mathematical modelling of a physical vibrating system results in the formulation of the governing equations of motion. Mathematical modelling of typical systems leads to a system of differential equations of motion. The governing equations of motion of a system are solved to find the response of the system. There are many techniques available for finding the solution, namely, the standard methods for the solution of ordinary differential equations, Laplace transformation methods, matrix methods, and numerical methods. In general, exact analytical solutions are available for many linear dynamic systems, but for only a few non-linear systems. Of course, exact analytical solutions are always preferable to numerical or approximate solutions.

4. Physical interpretation of the results

The solution of the governing equations of motion for the physical system generally gives the performance. To verify the validity of the model, the predicted performance is compared with the experimental results. The model may have to be refined or a new model is developed and a new prediction compared with the experimental results. Physical interpretation of the results is an important and final step in the analysis procedure. In some situations, this may involve (*a*) drawing general inferences from the mathematical solution, (*b*) development of design curves, (*c*) arrive at a simple arithmetic to arrive at a conclusion (for a typical or specific problem), and (*d*) recommendations regarding the significance of the results and any changes (if any) required or desirable in the system involved.

1.3.5.1 Components of vibrating systems

(*a*) **Stiffness elements**

Sometimes it requires finding out the equivalent spring stiffness values when a continuous system is attached to a discrete system or when there are a number of spring elements in the system. Stiffness of continuous elastic elements such as rods, beams, and shafts, which produce restoring elastic forces, is obtained from deflection considerations.

The stiffness coefficient of the rod (Fig. 1.23) is given by $k = \dfrac{EA}{l}$

The cantilever beam (Fig. 1.24) stiffness is $k = \dfrac{3EI}{l^3}$

The torsional stiffness of the shaft (Fig. 1.25) is $K = \dfrac{GJ}{l}$

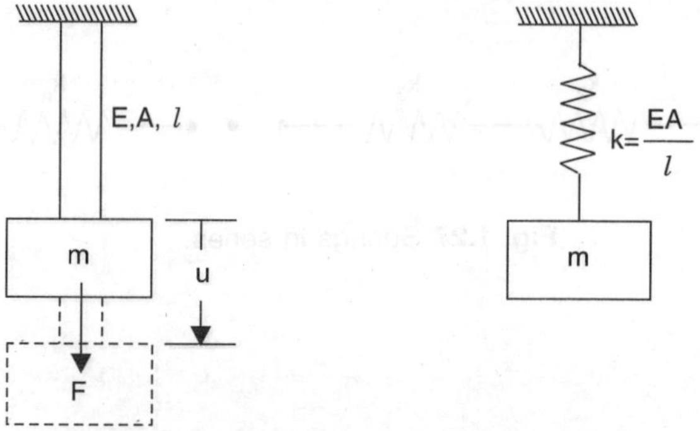

Fig.1.23 Longitudinal vibration of rods.

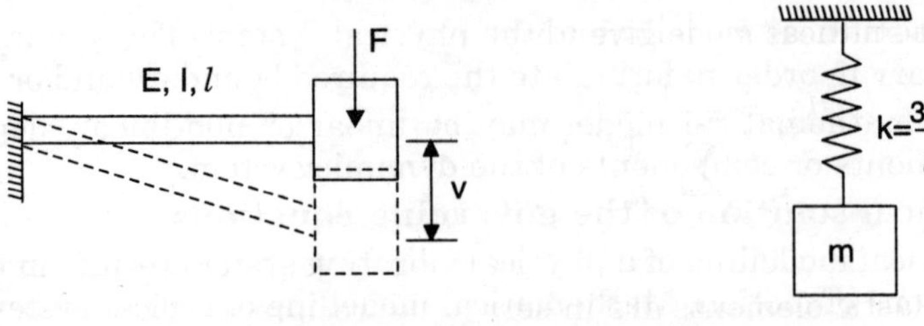

Fig. 1.24 Transverse vibration of cantilever beams.

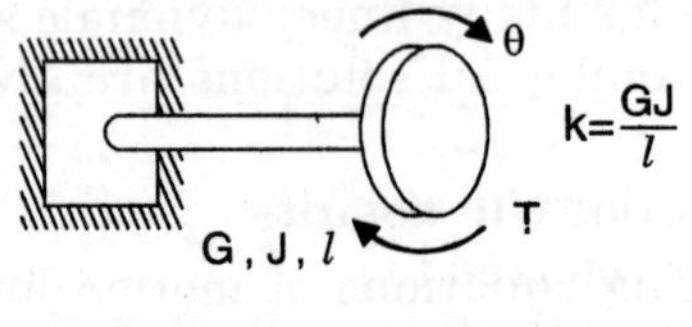

Fig. 1.25 Torsional system.

When there are several springs arranged in parallel as shown in Fig. 1.26, the equivalent spring constant is given by algebraic sum of the stiffness of individual springs. Mathematically,

$$k_{eq} = \sum_{i=1}^{n} k_i$$

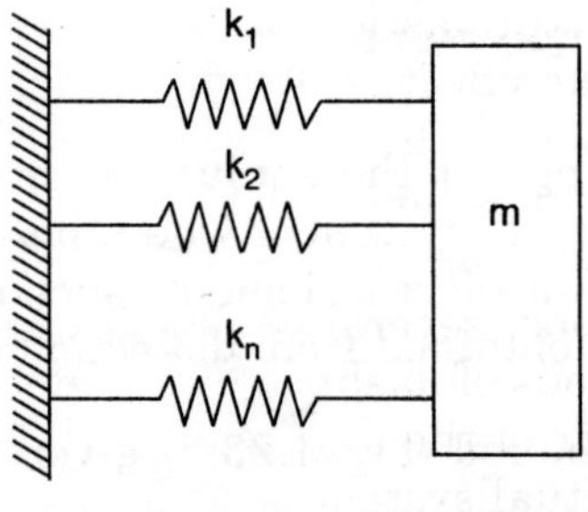

Fig. 1.26 Springs in parallel.

When the springs are arranged in series as shown in Fig. 1.27, the same force is developed in each spring and is equal to the force acting on the mass.

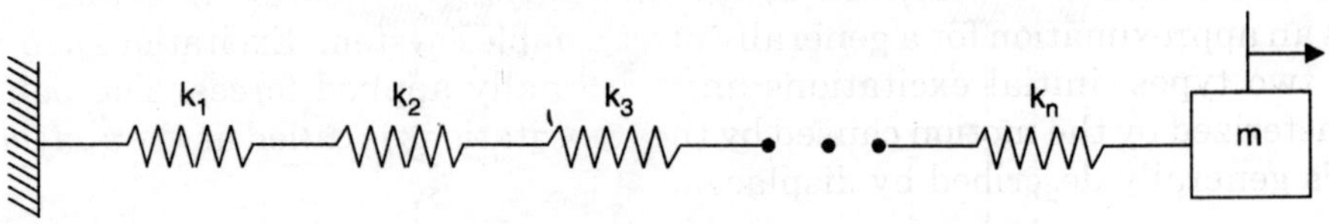

Fig. 1.27 Springs in series.

The equivalent stiffness k_{eq} is given by:

$$1/k_{eq} = \frac{1}{\sum_{i=1}^{n} \frac{1}{k_i}}$$

Hence, when elastic elements are in series, the reciprocal of the equivalent elastic constant is equal to the reciprocals of the elastic constants of the elements in the original system.

(*b*) **Mass or inertia elements**

The mass or inertia element is assumed to be a rigid body. Once the mathematical model of the physical vibrating system is developed, the mass or inertia elements of the system can be easily identified.

(*c*) **Damping elements**

In real mechanical systems, there is always energy dissipation in one form or another. The process of energy dissipation is referred to in the study of vibration as *damping*. A damper is considered to have neither mass nor elasticity. The three main forms of damping are *viscous damping*, *Coulomb* or *dry-friction damping*, and *hysteresis damping*. The most common type of energy-dissipating element used in vibrations study is the *viscous damper*, which is also referred to as a *dashpot*. In viscous damping, the damping force is proportional to the velocity of the body. Coulomb or dry-friction damping occurs when sliding contact that exists between surfaces in contact are dry or have insufficient lubrication. In this case, the damping force is constant in magnitude but opposite in direction to that of the motion. In dry-friction damping energy is dissipated as heat.

Solid materials are not perfectly elastic and when they are deformed, energy is absorbed and dissipated by the material. The effect is due to the internal friction due to the relative motion between the internal planes of the material during the deformation process. Such materials are known as visco-elastic solids and the type of damping which they exhibit is called as *structural* or *hysteretic damping*, or *material* or *solid damping*.

In many practical applications, several dashpots are used in combination. It is quite possible to replace these combinations of dashpots by a single dashpot of an equivalent damping coefficient so that the behaviour of the system with the equivalent dashpot is considered identical to the behaviour of the actual system.

1.3.6 Free Vibration of Single Degree of Freedom Systems

The most basic mechanical system is the *single-degree-of-freedom system*, which is characterized by the fact that its motion is described by a single variable or coordinates. Such a model is often used as an approximation for a generally more complex system. Excitations can be broadly divided into two types, initial excitations and externally applied forces. The behavior of a system characterized by the motion caused by these excitations is called as the *system response*. The motion is generally described by displacements.

1.3.6.1 Free vibration of an undamped translational system

The simplest model of a vibrating mechanical system consists of a single mass element which is connected to a rigid support through a linearly elastic massless spring as shown in Fig. 1.28. The mass is constrained to move only in the vertical direction. The motion of the system is described by a single coordinate *x*(*t*) and hence it has one degree of freedom (DOF).

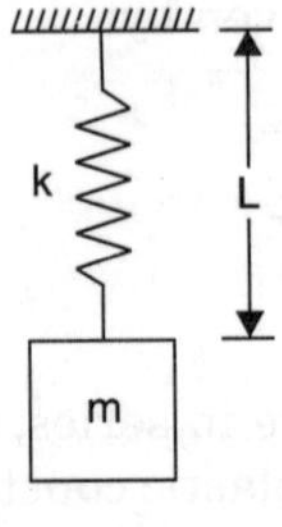

Fig. 1.28 Spring mass system.

The equation of motion for the free vibration of an undamped single degree of freedom system can be rewritten as

$$m\ddot{x}(t) + kx(t) = 0$$

Dividing through by m, the equation can be written in the form

$$\ddot{x}(t) + \omega_n^2 x(t) = 0$$

in which $\omega_n = \sqrt{k/m}$ is a real constant. The solution of this equation is obtained from the initial conditions

$$x(0) = x_0, \; \dot{x}(0) = v_0$$

where x_0 and v_0 are the initial displacement and initial velocity, respectively.

The general solution can be written as

$$x(t) = A_1 e^{i\omega_n t} + A_2 e^{-i\omega_n t}$$

in which A_1 and A_2 are constants of integration, both complex quantities. It can be finally simplified as:

$$x(t) = \frac{X}{2}\left[e^{i(\omega_n t - \phi)} + e^{-i(\omega_n t - \phi)}\right] = X\cos(\omega_n t - \phi)$$

so that now the constants of integration are X and ϕ.

This equation represents harmonic oscillation, for which reason such a system is called a *harmonic oscillator*.

There are three quantities defining the response, the *amplitude* X, the *phase angle* ϕ and the *frequency* ω_n, the first two depending on external factors, namely, the initial excitations, and the third depending on internal factors, namely, the system parameters. On the other hand, for a given system, the frequency of the response is a characteristic of the system that stays always the same, independently of the initial excitations. For this reason, ω_n is called the *natural frequency* of the harmonic oscillator.

The constants X and ϕ are obtained from the initial conditions of the system as follows:

$$X = \sqrt{x_0^2 + \left(\frac{v_0}{\omega_n}\right)^2}$$

and
$$\phi = \tan^{-1}\left[\frac{v_0}{x_0\omega_n}\right]$$

The *time period* τ, is defined as the time necessary for the system to complete one vibration cycle, or as the time between two consecutive peaks. It is related to the natural frequency by

$$\tau = \frac{2\pi}{\omega_n} = 2\pi\sqrt{\frac{m}{k}}$$

Note that the natural frequency can also be defined as the reciprocal of the period, or

$$f_n = \frac{1}{\tau} = \frac{1}{2\pi}\sqrt{\frac{k}{m}}$$

in which case it has units of cycles per second (cps), where one cycle per second is known as one Hertz (Hz).

1.3.6.2 Free vibration of an undamped torsional system

A mass attached to the end of the shaft is a simple torsional system (Fig. 1.29). The mass of the shaft is considered to be small in comparison to the mass of the disk and is therefore neglected.

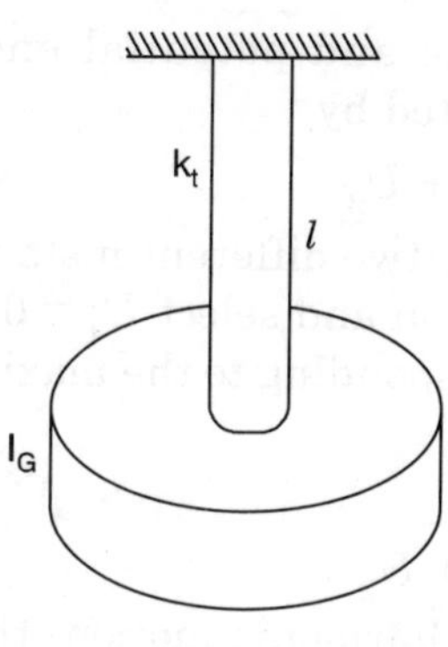

Fig. 1.29 Torsional system.

The torque that produces the twist M_t is given by

$$M_t = \frac{GJ}{l}$$

where J = the polar mass moment of inertia of the shaft $\left(J = \frac{\pi d^4}{32}\right.$ for a circular shaft of diameter $\left.d\right)$

G = shear modulus of the material of the shaft.

l = length of the shaft.

The torsional spring constant k_t is defined as

$$k_t = \frac{T}{\theta} = \frac{GJ}{l}$$

The equation of motion of the system can be written as:

$$I_G\ddot{\theta} + k_t\theta = 0$$

The natural circular frequency of such a torsional system is $\omega_n = \left(\frac{k_t}{I_G}\right)^{1/2}$

The general solution of equation of motion is given by

$$\theta(t) = \theta_0 \cos \omega_n t + \frac{\dot{\theta}_0}{\omega_n} \sin \omega_n t$$

1.3.6.3 Energy method

Free vibration of systems involves the cyclic interchange of kinetic and potential energy. In undamped free vibrating systems, no energy is dissipated or removed from the system. The kinetic energy T is stored in the mass by virtue of its velocity and the potential energy U is stored in the form of strain energy in elastic deformation. Since the total energy in the system is constant, the principle of conservation of mechanical energy applies. Since the mechanical energy is conserved, the sum of the kinetic energy and potential energy is constant and its rate of change is zero. This principle can be expressed as

$$T + U = \text{constant}$$

or
$$\frac{d}{dt}(T + U) = 0$$

where T and U denote the kinetic and potential energy, respectively. The principle of conservation of energy can be restated by

$$T_1 + U_1 = T_2 + U_2$$

where the subscripts 1 and 2 denote two different instances of time when the mass is passing through its static equilibrium position and select $U_1 = 0$ as reference for the potential energy. Subscript 2 indicates the time corresponding to the maximum displacement of the mass at this position, we have then

$$T_2 = 0$$

and
$$T_1 + 0 = 0 + U_2$$

If the system is undergoing harmonic motion, then T_1 and U_2 denote the maximum values of T and U, respectively and therefore last equation becomes

$$T_{max} = U_{max}$$

It is quite useful in calculating the natural frequency directly.

1.3.6.4 Stability of undamped linear systems

The mass/inertia and stiffness parameters have an effect on the stability of an undamped single degree of freedom vibratory system. The mass and stiffness coefficients enter into the characteristic equation which defines the response of the system. Hence, any changes in these coefficient will lead to changes in the system behavior or response. In this section, the effects of the system inertia and stiffness parameters on the stability of the motion of an undamped single degree of freedom system are examined. It can be shown that by a proper selection of the inertia and stiffness coefficients, the instability of the motion of the system can be avoided. A stable system is one which executes bounded oscillations about the equilibrium position.

1.3.6.5 Free vibration with viscous damping

Viscous damping force is proportional to the velocity $\dot{x}$ of the mass and acting in the direction opposite to the velocity of the mass and can be expressed as

$$F = c\dot{x}$$

where c is the damping constant or coefficient of viscous damping. The differential equation of motion for free vibration of a damped spring-mass system (Fig. 1.30) is written as:

$$\ddot{x} + \frac{c}{m}\dot{x} + \frac{k}{m}x = 0$$

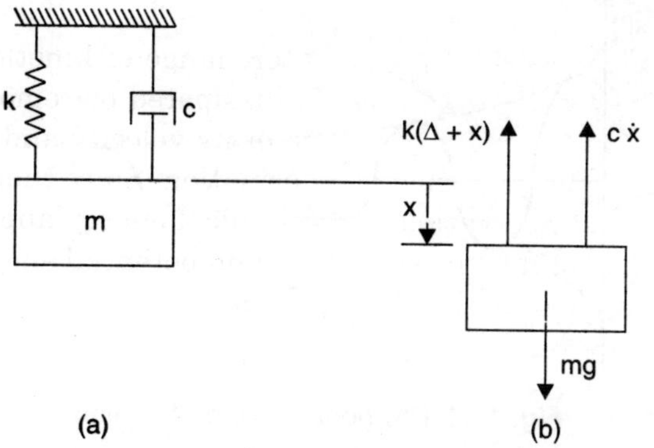

Fig. 1.30 Damped spring-mass system.

By assuming $x(t) = Ce^{st}$ as the solution, the auxiliary equation obtained is

$$s^2 + \frac{c}{m}s + \frac{k}{m} = 0$$

which has the roots

$$s_{1,2} = -\frac{c}{2m} \pm \sqrt{\left(\frac{c}{2m}\right)^2 - \frac{k}{m}}$$

The solution takes one of three forms, depending on whether the quantity $(c/2m)^2 - k/m$ is zero, positive, or negative. If this quantity is zero,

$$c = 2m\omega_n$$

This results in repeated roots $s_1 = s_2 = -c/2m$, and the solution is

$$x(t) = (A + Bt)e^{-(c/2m)t}$$

As the case in which repeated roots occur has special significance, we shall refer to the corresponding value of the damping constant as the *critical damping constant*, denoted by $C_c = 2m\omega_n$. The roots can be written as:

$$s_{1,2} = \left(-\zeta \pm \sqrt{\zeta^2 - 1}\right)\omega_n$$

where $\omega_n = (k/m)^{1/2}$ is the circular frequency of the corresponding undamped system, and

$$\zeta = \frac{c}{C_c} = \frac{c}{2m\omega_n}$$

is known as the *damping factor*.

If $\zeta < 1$, the roots are both imaginary and the solution for the motion is

$$x(t) = Xe^{-\zeta\omega_n t}\sin(\omega_d t + \phi)$$

where $\omega_d = \sqrt{1-\zeta^2}\,\omega_n$ is called the damped circular frequency which is always less than ω, and φ is the phase angle of the damped oscillations. The general form of the motion is shown in Fig. 1.31. For motion of this type, the system is said to be *underdamped.*

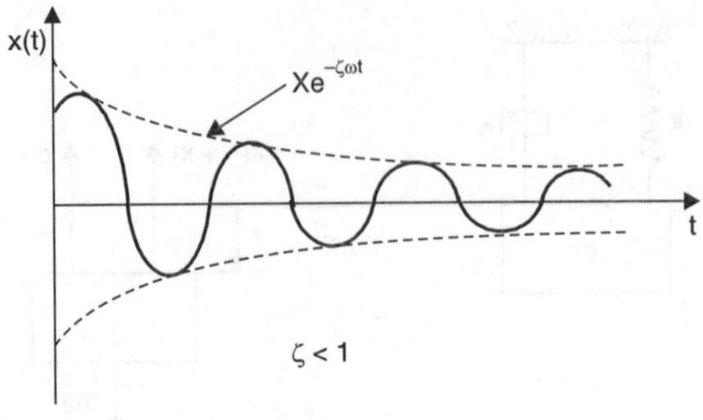

Fig. 1.31 The general form of motion.

If $\zeta = 1$, the damping constant is equal to the critical damping constant, and the system is said to be *critically damped*. The displacement is given by

$$x(t) = (A + Bt)e^{-\omega_n t}$$

The solution is the product of a linear function of time and a decaying exponential. Depending on the values of A and B, many forms of motion are possible, but each form is characterized by amplitude which decays without oscillations, such as is shown in Fig. 1.32.

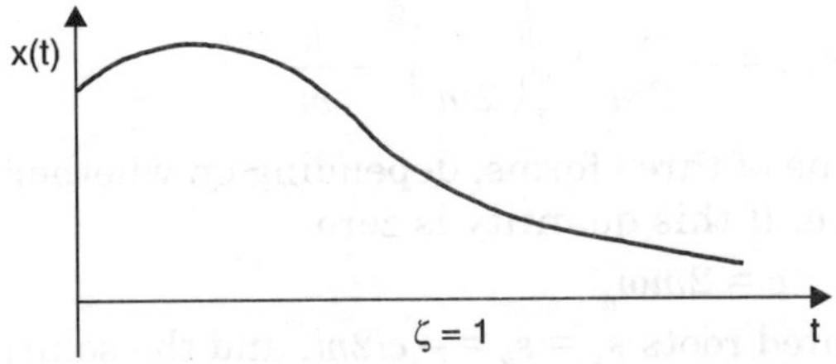

Fig. 1.32 Amplitude decaying without oscillations.

In this case $\zeta > 1$, and the system is said to be *overdamped*. The solution is given by:

$$x(t) = C_1 e^{(-\zeta + \sqrt{\zeta^2 - 1})\,\omega_n t} + C_2 e^{(-\zeta - \sqrt{\zeta^2 - 1})\,\omega_n t}$$

The motion will be non-oscillatory and will be similar to that shown in Fig. 1.33.

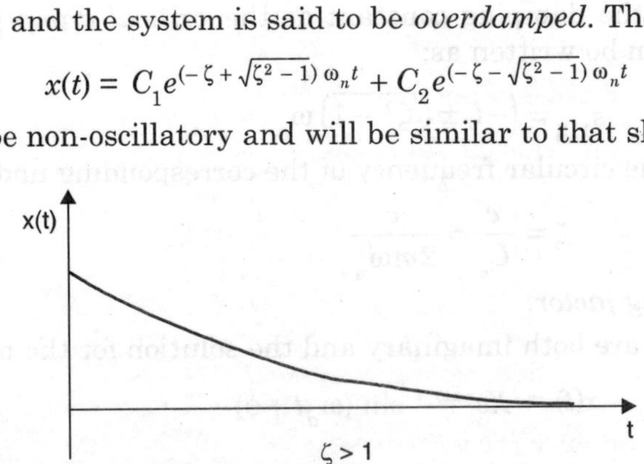

Fig. 1.33 Non-oscillatory motion.

1.3.6.6 Logarithmic decrement

The logarithmic decrement represents the rate at which the amplitude of a free damped vibration decreases. It is defined as the natural logarithm of the ratio of any two successive amplitudes.

The ratio of successive amplitudes is

$$\frac{x_i}{x_{i+1}} = \frac{Xe^{-\zeta\omega_n t_i}}{Xe^{-\zeta\omega_n (t_i + \tau_d)}} = e^{\zeta\omega_n \tau_d} = \text{constant}$$

The logarithmic decrement

$$\delta = \ln \frac{x_i}{x_{i+1}} = \ln e^{\zeta\omega_n \tau_d} = \zeta\omega_n \tau_d$$

Substituting $\tau_d = 2\pi/\omega_d = 2\pi/\omega_n \sqrt{1-\zeta^2}$ gives

$$\delta = \frac{2\pi\zeta}{\sqrt{1-\zeta^2}}$$

1.3.6.7 Torsional system with viscous damping

The equation of motion for such a system can be written as

$$I\ddot{\theta} + c_t\dot{\theta} + k_t\theta = 0$$

where I is the mass moment of inertia of the disc, k_t is the torsional spring constant (restoring torque for unit angular displacement), and θ is the angular displacement of the disc.

1.3.6.8 Free vibration with coulomb damping

Coulomb or dry-friction damping results when sliding contact exists between two dry surfaces. The damping force is equal to the product of the normal force and the coefficient of dry friction. The damping force is quite independent of the velocity of the motion. Consider a spring-mass system in which the mass slides on a horizontal surface having coefficient of friction f, as in Fig. 1.34.

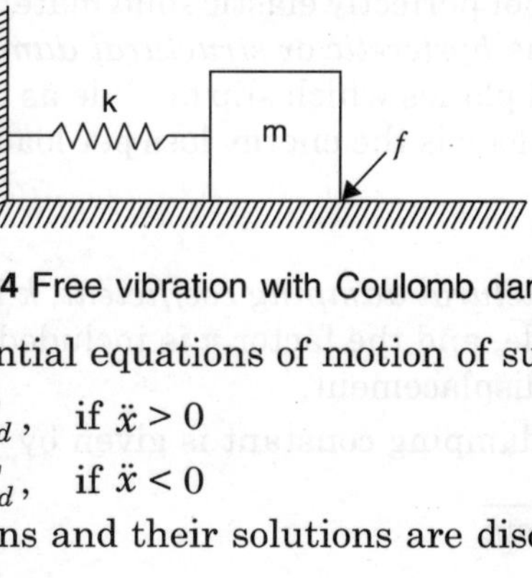

Fig. 1.34 Free vibration with Coulomb damping.

The corresponding differential equations of motion of such system are

$$m\ddot{x} = -kx - F_d, \quad \text{if } \ddot{x} > 0$$

$$m\ddot{x} = -kx + F_d, \quad \text{if } \ddot{x} < 0$$

These differential equations and their solutions are discontinuous at the end points of their motion.

The general solution is then

$$x = A\sin\omega t + B\cos\omega t + \frac{F_d}{k} \qquad (\dot{x} < 0)$$

for motion toward the left. For the initial conditions of $x = x_0$ and $\dot{x} = 0$ at $t = 0$ for the extreme position at the right, the solution becomes

$$x = \left(x_0 - \frac{F_d}{k}\right)\cos\omega t + \frac{F_d}{k} \qquad (\dot{x} < 0)$$

This holds for motion toward the left, or until $\dot{x}$ again becomes zero.

Hence the displacement is negative, or to the left of the neutral position, and has a magnitude $2F_d/k$ less than the initial displacement x_0.

A constant amplitude loss of $4F_d/k$ occurs for each cycle of motion as shown in Fig. 1.35. The motion is a linearly decaying harmonic function of time, consisting of one-half sine wave parts which are offset successively up or down by F_d/k depending on whether the motion is to the left or to the right.

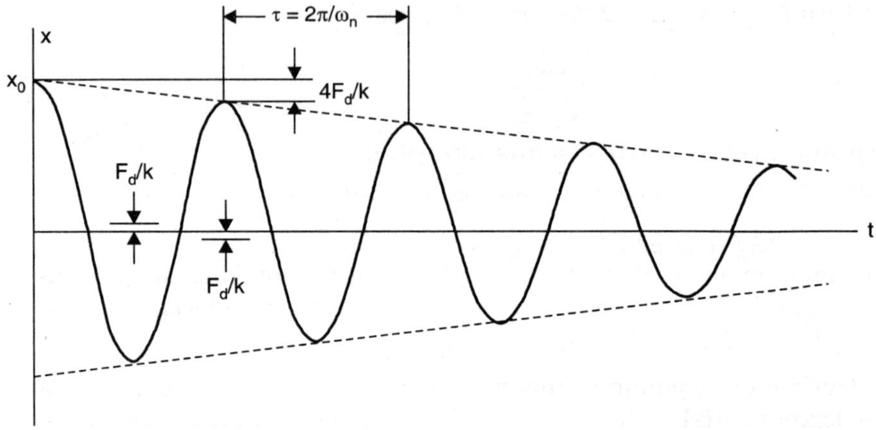

Fig. 1.35 Response of system subjected to Coulomb damping.

1.3.6.9 Free vibration with hysteretic damping

In general, solid materials are not perfectly elastic solid materials, in particular, metals, exhibit what is commonly referred to as *hysteretic* or *structural damping*. The hysteresis effect is due to the friction between internal planes which slip or slide as the deformations takes place. The enclosed area in the hysteresis loop is the energy loss per loading cycle. The energy loss ΔU can then be written as

$$\Delta U = \pi\beta\, kX^2$$

where β is a dimensionless *structural damping coefficient*, k is the equivalent spring constant, X is the displacement amplitude, and the factor π is included for convenience. The energy loss is a nonlinear function of the displacement.

The equivalent viscous damping constant is given by

$$c_e = \frac{\beta k}{\omega} = \beta\sqrt{mk}$$

1.3.7 Forced Vibration of Single-degree-of-freedom Systems

A mechanical or structural system is often subjected to external forces or external excitations. The external forces may be harmonic, non-harmonic but periodic, non-periodic but having a defined form or random. The response of the system to such excitations or forces is called *forced response*. The response of a system to a harmonic excitation is called *harmonic response*. The non-periodic excitations may have a long or short duration. The response of a system to suddenly applied non-periodic excitations is called *transient response*. The sources of harmonic excitations are unbalance in rotating machines, forces generated by reciprocating machines, and the motion of the machine itself in certain cases.

1.3.7.1 Forced vibration of damped system

Consider a viscously damped single degree of freedom spring mass system shown in Fig. 1.36, subjected to a harmonic function $F(t) = F_0 \sin \omega t$, where F_0 is the force amplitude and ω is the circular frequency of the forcing function.

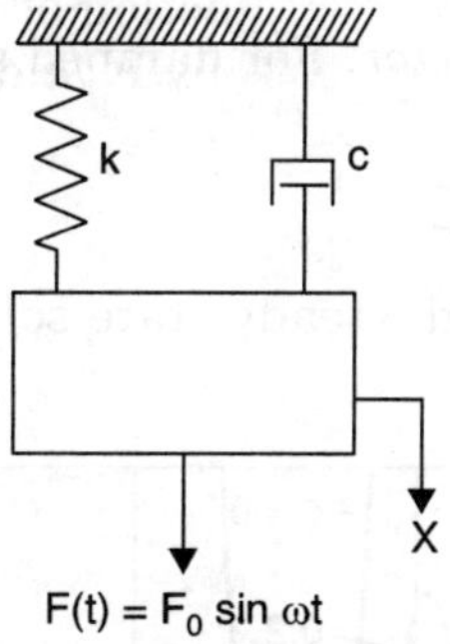

Fig. 1.36 Forced vibration of single degree of freedom system.

The equations of motion of the system is $\ddot{x} + \frac{c}{m}\dot{x} + \frac{k}{m}x = \left(\frac{F_0}{m}\right)\sin \omega t$

The solution of the equation contains two components, complimentary function x_h and particular solution x_p. That is

$$x = x_h + x_p$$

The particular solution represents the response of the system to the forcing function. The complementary function x_h is called the *transient response* since in the presence of damping, the solution dies out. The particular integral x_p is known as the steady state solution. The steady state vibration exists long after the transient vibration disappears.

The particular solution or the steady state solution x_p can be assumed in the form

$$x_p = A_1 \sin \omega t + A_2 \cos \omega t$$

By defining $r = \frac{\omega}{\omega_n}, \zeta = \frac{c}{C_c} = \frac{c}{2m\omega}$, and $X_0 = F_0/k$ the amplitudes A_1 and A_2 are obtained as follows:

$$A_1 = \frac{(1 - r^2)\, X_0}{(1 - r^2)^2 + (2r\zeta)^2}$$

and

$$A_2 = \frac{-(2r\zeta)\, X_0}{(1 - r^2)^2 + (2r\zeta)^2}$$

The steady state solution x_p can be written as

$$x_p = \frac{X_0}{(1 - r^2)^2\ (2r\zeta)^2}\left[(1 - r^2)\sin \omega t - (2r\zeta)\cos \omega t\right]$$

which can also be written as

$$x_p = \frac{X_0}{\sqrt{(1 - r^2)^2\ (2r\zeta)^2}}\sin(\omega t - \phi)$$

where X_0 is the forced amplitude and ϕ is the phase angle defined by

$$\phi = \tan^{-1}\left(\frac{2r\zeta}{1-r^2}\right)$$

It can be written in a more compact form as

$$x_p = X_0\beta \sin(\omega t - \phi)$$

where β is known as *magnification factor*. For damped systems β is defined as

$$\beta = \frac{1}{(1-r^2)^2 + (2r\zeta)^2}$$

This forced response is called steady state solution, which is shown in Figures 1.37 and 1.38.

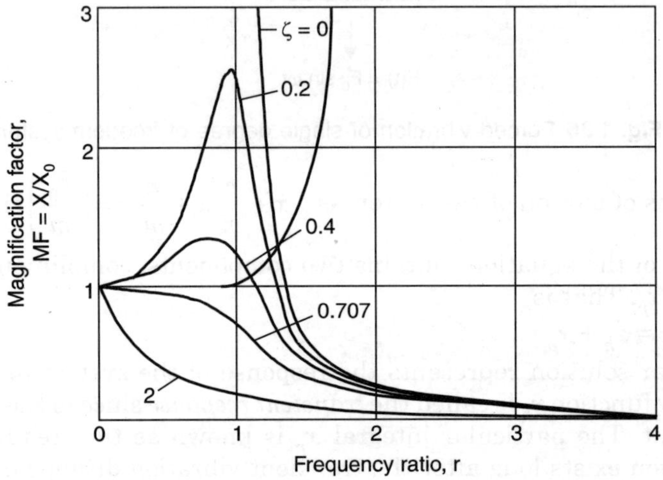

Fig. 1.37 Non-dimensional amplitude versus frequency-ratio.

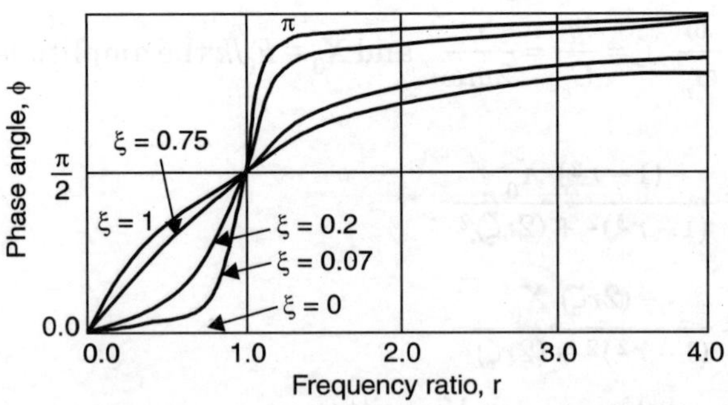

Fig. 1.38 Phase angle versus frequency-ratio.

The magnification factor β is found to be maximum when

$$r = \sqrt{1-2\zeta^2}$$

The maximum magnification factor is given by:

$$\beta_{max} = \frac{1}{2\zeta\sqrt{1-\zeta^2}}$$

In the undamped systems, the particular solution reduces to

$$x_p(t) = \frac{\dfrac{F_0}{k}}{\left[1 - \left(\dfrac{\omega}{\omega_n}\right)^2\right]} \sin \omega t$$

The maximum amplitude can also be expressed as

$$\frac{X}{\delta_{st}} = \frac{1}{1 - \left(\dfrac{\omega}{\omega_n}\right)^2}$$

where $\delta_{st} = F_0/k$ denotes the static deflection of the mass under a force F_0 and is sometimes known as *static deflection* since F_0 is a constant static force. The quantity X/δ_{st} represents the ratio of the dynamic to the static amplitude of motion and is called the *magnification factor, amplification factor*, or *amplitude ratio.*

1.3.7.2 Resonance

The case $r = \dfrac{\omega}{\omega_n} = 1$, that is, when the circular frequency of the forcing function is equal to the circular frequency of the spring-mass system is referred to as *resonance*. In this case, the displacement $x(t)$ goes to infinity for any value of time t.

The amplitude of the forced response grows with time as in Fig. 1.39 and will eventually become infinite at which point the spring in the mass-spring system fails in an undesirable manner.

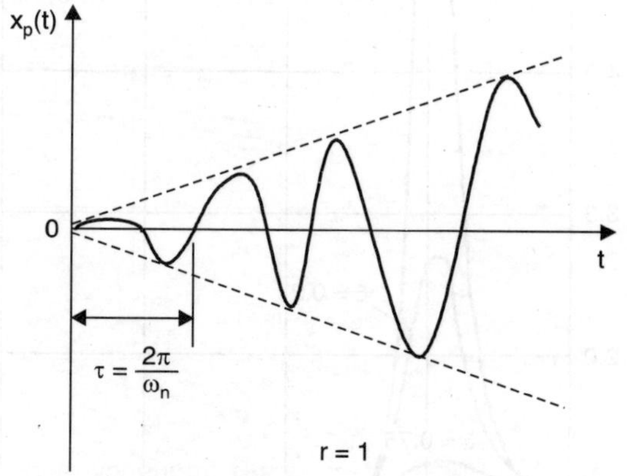

Fig. 1.39 Resonance response.

1.3.7.3 Beats

The phenomenon of *beating* occurs for an undamped forced single degree of freedom spring-mass system when the forcing frequency ω is close, but not equal, to the system circular frequency ω_n. In this case, the amplitude builds up and then diminishes in a regular pattern. The phenomenon of beating can be noticed in cases of audio or sound vibration and in electric power generation when a generator is started.

1.3.7.4 Transmissibility

The forces associated with the vibrations of a machine or a structure will be transmitted to its support structure. These transmitted forces in most instances produce undesirable effects such as noise. Machines and structures are generally mounted on designed flexible supports known as *vibration isolators* or *isolators*.

In general, the amplitude of vibration reduces with the increasing values of the spring stiffness k and the damping coefficient c. In order to reduce the force transmitted to the support structure, a proper selection of the stiffness and damping coefficients must be made.

From regular spring-mass-damper model, force transmitted to the support can be written as

$$F_T = k\,x_p + c\dot{x}_p = X_0\beta\sqrt{k^2 + (c\omega)^2}\,\sin(\omega t - \bar{\phi})$$

where $\bar{\phi} = \phi - \phi_t$

and ϕ_t is the phase angle defined as

$$\phi_t = \tan^{-1}\left(\frac{c\omega}{k}\right) = \tan^{-1}(2r\zeta)$$

Transmitted force can also be written as:

$$F_T = F_0\beta_t \sin(\omega t - \bar{\phi})$$

where $$\beta_t = \frac{\sqrt{1 + (2r\zeta)^2}}{\sqrt{(1 - r^2)^2 + (2r\zeta)^2}}$$

The *transmissibility* β_t is defined as the ratio of the maximum transmitted force to the amplitude of the applied force. Figure 1.40 shows a plot of β_t versus the frequency ratio r for different values of the damping factor ζ.

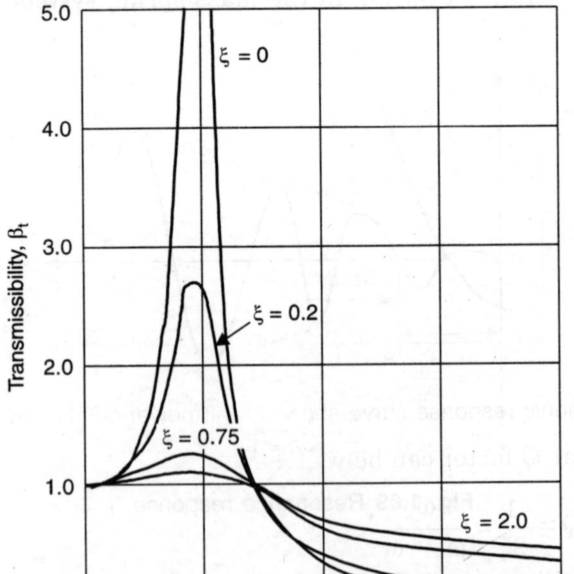

Fig. 1.40 Non-dimensional force transmitted vs. frequency ratio.

It can be observed from Fig. 1.43, that $\beta > 1$ for $r < \sqrt{2}$ which means that in this region the amplitude of the transmitted force is greater than the amplitude of the applied force. Also, the $r < \sqrt{2}$, the transmitted force to the support can be reduced by increasing the damping factor ζ. For $r = \sqrt{2}$, every curve passes through the point $\beta_t = 1$ and becomes asymptotic to zero as the frequency ratio is increased. Similarly, for $r > \sqrt{2}$, $\beta_t < 1$, hence, in this region the amplitude of the transmitted force is less than the amplitude of the applied force. Therefore, the amplitude of the transmitted force increases by increasing the damping factor ζ. Thus, vibration isolation is best accomplished by an isolator composed only of spring-elements for which $r > \sqrt{2}$ with no damping element used in the system.

1.3.7.5 Quality factor and bandwidth

The value of the amplitude ratio at resonance is also known as the *Q factor* or *Quality factor* of the system in analogy with the term used in electrical engineering applications. That is,

$$Q = \frac{1}{2\zeta}$$

The points R_1 and R_2, whereby the amplification factor falls to $Q/\sqrt{2}$, are known as *half power points*, since the power absorbed by the damper responding harmonically at a given forcing frequency is given by

$$\Delta W = \pi c \omega X^2$$

The *bandwidth* of the system is defined as the difference between the frequencies associated with the half power points R_1 and R_2 as depicted in Fig. 1.41.

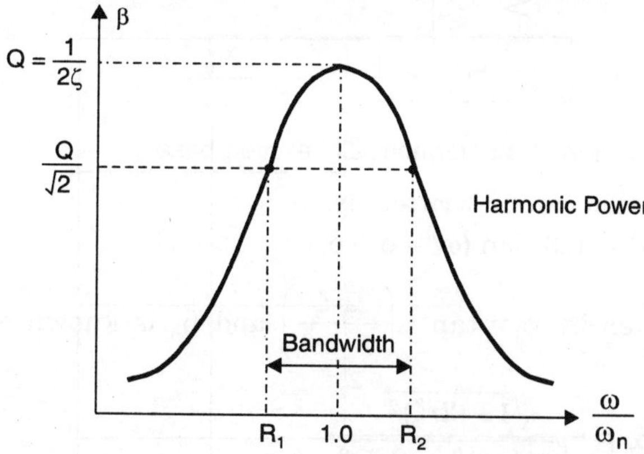

Fig. 1.41 Harmonic response curve showing half power points and bandwidth.

It can be shown that Q-factor can be written as:

$$Q = \frac{1}{2\zeta} = \frac{\omega_n}{\omega_2 - \omega_1}$$

The quality factor Q can be used for estimating the equivalent viscous damping in a vibrating system.

1.3.7.6 Rotating unbalance

Unbalance in many rotating mechanical systems is a common source of vibration excitation which may often lead to unbalance forces. If M is the total mass of the machine including an eccentric mass m rotating with an angular velocity ω at an eccentricity e, it can be shown that the particular solution takes the form:

$$x_p(t) = \left(\frac{me}{M}\right)\beta_r \sin(\omega t - \phi)$$

where β_r is the magnification factor which is given by

$$\beta_r = \frac{r^2}{\sqrt{(1-r^2)^2 + (2r\zeta)^2}}$$

The steady state vibration due to unbalance in rotating component is proportional to the amount of unbalance m and its distance e from the center of the rotation and increases as the square of the rotating speed. The maximum displacement of the system lags the maximum value of the forcing function by the phase angle ϕ.

1.3.7.7 Base excitation

In many mechanical systems such as vehicles mounted on a moving support or base, the forced vibration of the system is due to the moving support or base. The motion of the support or base causes the forces being transmitted to the mounted equipment. Figure 1.42 shows a damped single degree of freedom mass-spring system with a moving support or base.

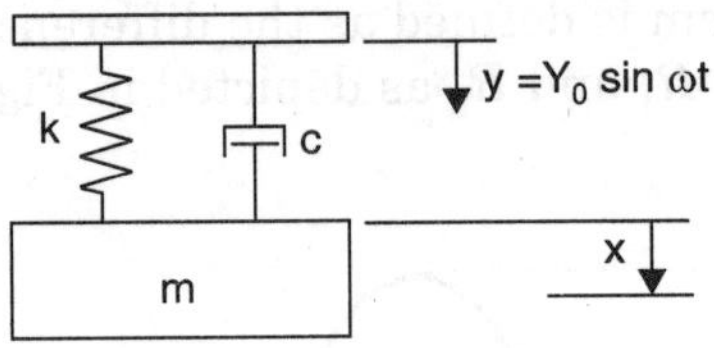

Fig. 1.42 Harmonically excited base.

The steady state solution can be written as:

$$x_p(t) = Y_0\beta_b \sin(\omega t - \phi + \phi_b),$$

where phase angle ϕ is given by $\phi = \tan^{-1}\left(\dfrac{2r\zeta}{1-r^2}\right)$ and β_b is known as the displacement transmissibility given by: $\beta_b = \dfrac{\sqrt{1+(2r\zeta)^2}}{\sqrt{(1-r^2)^2+(2r\zeta)^2}}$

The motion of the mass relative to the support denoted by z can be written as

$$z = x - y$$

$$= \frac{Y_0 r^2}{\sqrt{(1-r^2)^2 + (2r\zeta)^2}} \sin(\omega t - \phi)$$

1.3.7.8 Response under Coulomb damping

When a single-degree-of-freedom with Coulomb damping is subjected to a harmonic forcing conditions, the amplitude relationship is written as:

$$X = \frac{X_0}{\sqrt{(1 - r^2)^2 + (4F/\pi Xk)^2}}$$

which gives $X = X_0 \dfrac{\sqrt{1 - (4F/\pi F_0)^2}}{1 - r^2}$

This expression for X has a real value, provided that

$$4F < \pi F_0 \quad \text{or} \quad F < \frac{\pi}{4} F_0$$

1.3.7.9 Response under hysteresis damping

The steady-state motion of a single degree of freedom forced harmonically with hysteresis damping is also harmonic. The steady-state amplitude can then be determined by defining an equivalent viscous damping constant based on equating the energies.

The amplitude is given in terms of hysteresis damping coefficient β as follows

$$X = \frac{X_0}{\sqrt{(1 - r^2)^2 + \beta^2}}$$

1.3.7.10 General forcing conditions and response

A general forcing function may be periodic or nonperiodic. The ground vibrations of a building structure during an earthquake, the vehicle motion when it hits a pothole, are some examples of general forcing functions. Nonperiodic excitations are referred to as *transient*. The term transient is used in the sense that nonperiodic excitations are not steady state.

1.3.7.11 Fourier series and harmonic analysis

The Fourier series expression of a given periodic function $F(t)$ with period T can be expressed in terms of harmonic functions as

$$F(t) = \frac{a_0}{2} + \sum_{n=1}^{\infty} a_n \cos n\omega t + \sum_{n=1}^{\infty} b_n \sin n\omega t$$

where $\omega = \dfrac{2\pi}{T}$ and a_0, a_n and b_n are constants.

$F(t)$ can also be written as follows:

$$F(t) = F_0 + \sum_{n=1}^{\infty} F_n \sin(\omega_n t + \phi_n)$$

where $F_0 = a_0/2$, $F_n = \sqrt{a_n^2 + b_n^2}$, with $\omega_n = n\omega$ and $\phi_n = \tan^{-1}\left(\dfrac{a_n}{b_n}\right)$

1.3.8 Harmonic Functions

Harmonic functions are periodic functions in which all the Fourier coefficients are zeros except one coefficient.

1.3.8.1 Even functions

A periodic function $F(t)$ is said to be even if $F(t) = F(-t)$. A cosine function is an even function since $\cos\theta = \cos(-\theta)$. If the function $F(t)$ is an even function, then the coefficients b_m are all zeros.

1.3.8.2 Odd functions

A periodic function $F(t)$ is said to be odd if $F(t) = -F(-t)$. The sine function is an odd function since $\sin\theta = -\sin(-\theta)$. For an odd function, the Fourier coefficients a_0 and a_n are identically zero.

1.3.8.3 Response under a periodic force of irregular form

Usually, the values of periodic functions at discrete points in time are available in graphical form or tabulated form. In such cases, no analytical expression can be found or the direct integration of the periodic functions in a closed analytical form may not be practical. In such cases, one can find the Fourier coefficients by using a numerical integration procedure. If one divides the period of the function T into N equal intervals, then length of each such interval is $\Delta t = T/N$.

The coefficients are given by

$$a_0 = \frac{2}{N}\sum_{i=1}^{N} F(t_i)$$

$$a_n = \frac{2}{N}\sum_{i=1}^{N} F(t_i)\cos n\omega t_i$$

$$b_n = \frac{2}{N}\sum_{i=1}^{N} F(t_i)\sin n\omega t_i$$

1.3.8.4 Response under a general periodic force

To find the response of a system under general periodic force, consider a single degree of freedom system shown in Fig. 1.43.

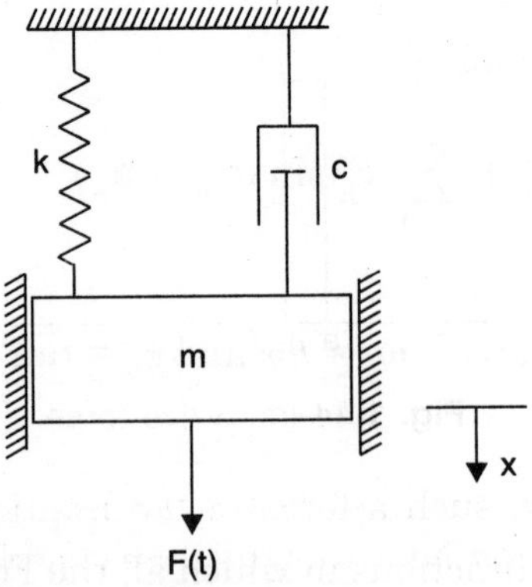

Fig. 1.43 Single degree of freedom system.

Let the periodic force $F(t)$ can be expressed in terms of harmonic functions by the use of Fourier series as follows:

$$F(t) = \frac{a_0}{2} + \sum_{n=1}^{\infty} (a_n \cos n\omega t + b_n \sin n\omega t)$$

Then steady-state solution can be written as

$$x_p(t) = \frac{a_0}{2k} + \sum_{n=1}^{\infty} \frac{a_n / k}{\sqrt{(1 - r_n^2)^2 + (2\xi r_n)^2}} \cos (n\omega t - \psi_n)$$

$$+ \sum_{n=1}^{\infty} \frac{b_n / k}{\sqrt{(1 - r_n^2)^2 + (2\xi r_n)^2}} \sin (n\omega t - \psi_n)$$

In most cases, the first two or three terms of this series are sufficient to describe the response of the system. If one of the harmonic frequencies $n\omega$ is close to or equal to ω, then $r \approx 1$, and the corresponding amplitude ratio can become large and resonance can occur.

1.3.8.5 Transient vibration

When a mechanical or structural system is excited by a suddenly applied nonperiodic excitation $F(t)$, the response to such excitation is called *transient response*, as the steady-state oscillations are generally not produced.

1.3.8.6 Unit impulse

Impulse is time integral of the force which is finite and is written as

$$\hat{F} = \int F(t)\, dt$$

where $\hat{F}$ is the linear impulse (in pound seconds or Newton seconds) of the force.

Figure 1.44 shows an impulsive force of magnitude $F = \hat{F}/\epsilon$ acting at $t = a$ over the time interval ϵ . As ϵ approaches zero, the magnitude of the force becomes infinite but the linear impulse $\hat{F}$ is well defined.

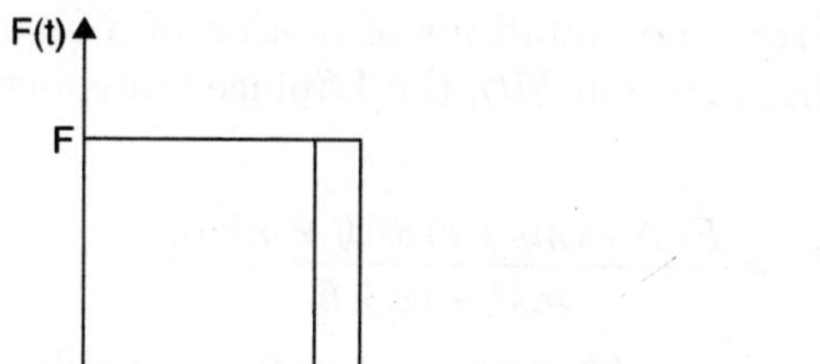

Fig. 1.44 Impulsive force.

When $\hat{F}$ is equal to unity, such a force in the limiting case ($\epsilon \to 0$) is called the *unit impulse*, or the *Direc delta function* $\delta(t - a)$, which has the following properties:

$$\delta(t - a) = 0, \quad \text{for } t \neq a$$

$$\int_0^{\infty} \delta(t - a)\, dt = 1$$

$$\int_0^\infty \delta(t-a)\, F(t)\, dt = F(a)$$

where $0 < a < \infty$. By using these properties, an impulsive force $F(t)$ acting at $t = a$ to produce a linear impulsive $\hat{F}$ of arbitrary magnitude can be expressed as

$$F(t) = \hat{F}\,\delta(t-a)$$

1.3.8.7 Impulsive response of a system

The response of a damped spring-mass system to an impulsive force is given by

$$x(t) = \hat{F}\,H(t)$$

where $H(t)$ is called the *impulse response function* and can be written as

$$H(t) = \frac{1}{m\omega_d} e^{-\zeta\omega_n t} \sin \omega_d t,$$ where ω_d is damped natural frequency.

If the force applied at a time $t = \tau$, this can be written as:

$$H(t-\tau) = \frac{1}{m\omega_d} e^{-\zeta\omega_n (t-\tau)} \sin \omega_d (t-\tau)$$

1.3.8.8 Response to an arbitrary input

The total response is obtained by finding the integration

$$x(t) = \int_0^t F(\tau)\, H(t-\tau)\, d\tau$$

This is called the convolution integral or Duhamel's integral and is sometimes referred as the superposition integral.

1.3.8.9 Laplace transformation method

The Laplace transformation method can be used for calculating the response of a system to a variety of force excitations, including periodic and nonperiodic. The Laplace transformation method can treat discontinuous functions with no difficulty and it automatically takes into account the initial conditions. The usefulness of the method lies in the availability of tabulated Laplace transform pairs. From the equations of motion of a single degree of freedom system subjected to a general forcing function $F(t)$, the Laplace transform of the solution $x(t)$ is given by

$$\bar{x}(s) = \frac{\bar{F}(s) + (ms + c)\, x(0) + m\dot{x}(0)}{ms^2 + cs + k}$$

The method of determining $x(t)$ given $\bar{x}(s)$ can be considered as an inverse transformation which can be expressed as

$$x(t) = L^{-1}\{\bar{x}(s)\}$$

1.3.9 Two Degree of Freedom Systems

Systems that require two independent coordinates to describe their motion are called *two degree of freedom systems.* Some examples of two degree of freedom models of vibrating systems are shown in Figs. 1.45 (*a*) and (*b*).

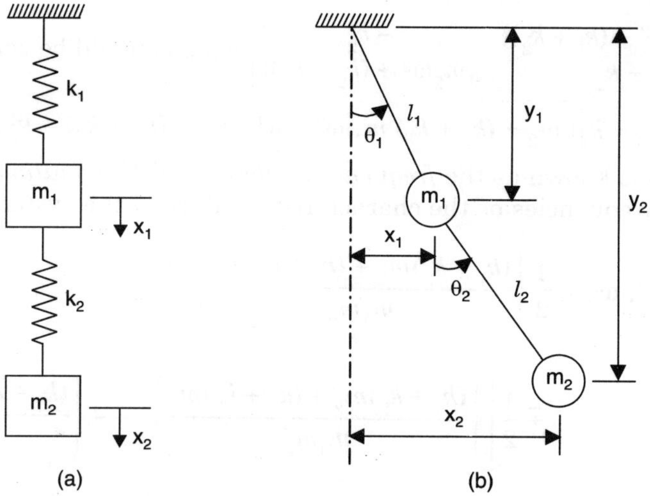

Fig. 1.45 Two degree of freedom systems.

1.3.9.1 Equations of motion

Consider the viscously damped two-degree of freedom spring mass system shown in Fig. 1.46.

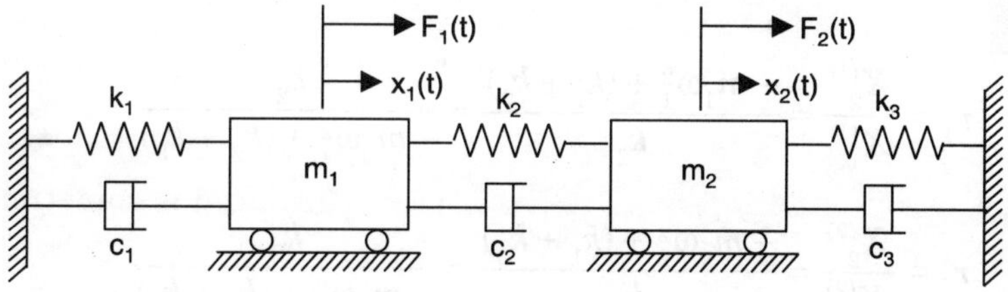

Fig. 1.46 Two-degree of freedom damped spring-mass damper.

The system is completely described by the two coordinates $x_1(t)$ and $x_2(t)$, which define the positions of the two masses m_1 and m_2, respectively, for any arbitrary time t, from the respective equilibrium positions. The external forces acting on the masses m_1 and m_2 of the system are $F_1(t)$ and $F_2(t)$ respectively.

Applying Newton's second law of motion to each of the masses m_1 and m_2 we can write the two equations of motion as:

$$m_1\ddot{x}_1(t) + (c_1 + c_2)\dot{x}_1(t) - c_2\dot{x}_2(t) + (k_1 + k_2)\, x_1(t) - k_2x_2(t) = F_1(t)$$

$$m_2\ddot{x}_2(t) - c_2\dot{x}_1(t) + (c_2 + c_3)\dot{x}_2(t) - k_2x_1(t) + (k_2 + k_3)\, x_2(t) = F_2(t)$$

These equations reveal that the motion of m_1 will influence the motion of mass m_2, and vice versa.

1.3.9.2 Free vibration analysis

Let the free vibration solution of the equations of motion be

$$x_1(t) = X_1 \cos(\omega t + \phi)$$

$$x_2(t) = X_2 \cos(\omega t + \phi)$$

where X_1 and X_2 are constants, which denote the maximum amplitudes of $x_1(t)$ and $x_2(t)$, and ϕ is the phase angle. Substituting these expressions in equations of motion leads to a characteristic determinant

$$\det \begin{bmatrix} \{-m_1\omega^2 + (k_1 + k_2)\} & -k_2 \\ -k_2 & \{m_2\omega^2 + (k_2 + k_3)\} \end{bmatrix}$$ which should be zero for consistency.

or $$(m_1 m_2)\omega^4 - \{(k_1 + k_2)\, m_2 + (k_2 + k_3)\, m_1\}\omega^2 + \{(k_1 + k_2)\,(k_2 + k_3) - k_2^2\} = 0$$

This equation is known as the *frequency or characteristic equation.* The solution of this equation yields the frequencies or the characteristic values of the system.

$$\omega_1^2, \omega_2^2 = \frac{1}{2}\left\{\frac{(k_1 + k_2)m_2 + (k_2 + k_3)m_1}{m_1 m_2}\right\}$$

$$\mp \frac{1}{2}\left[\left\{\frac{(k_1 + k_2)m_2 + (k_2 + k_3)m_1}{m_1 m_2}\right\}^2 - 4\left\{\frac{(k_1 + k_2)\,(k_2 + k_3) - k_2^2}{m_1 m_2}\right\}\right]^{1/2}$$

ω_1 and ω_2 are called the *natural frequencies* of the system.

The values of X_1 and X_2 depend on the natural frequencies ω_1 and ω_2. By denoting the values of X_1 and X_2 corresponding to ω_1 as $X_1^{(1)}$ and $X_2^{(1)}$ and those corresponding to ω_2 as $X_1^{(2)}$ and $X_2^{(2)}$:

$$r_1 = \frac{X_2^{(1)}}{X_1^{(1)}} = \frac{-m_1\omega_1^2 + (k_1 + k_2)}{k_2} = \frac{k_2}{-m_2\omega_1^2 + (k_2 + k_3)}$$

$$r_2 = \frac{X_2^{(2)}}{X_1^{(2)}} = \frac{-m_1\omega_2^2 + (k_1 + k_2)}{k_2} = \frac{k_2}{-m_2\omega_2^2 + (k_2 + k_3)}$$

The normal modes of vibration corresponding to ω_1^2 and ω_2^2 can be expressed, respectively, as

$$\{X^{(1)}\} = \begin{Bmatrix} X_1^{(1)} \\ X_2^{(1)} \end{Bmatrix} = \begin{Bmatrix} X_1^{(1)} \\ r_1 X_1^{(1)} \end{Bmatrix}$$

and $$\{X^{(2)}\} = \begin{Bmatrix} X_1^{(2)} \\ X_2^{(2)} \end{Bmatrix} = \begin{Bmatrix} X_1^{(2)} \\ r_2 X_1^{(2)} \end{Bmatrix}$$

The vectors $\{X^{(1)}\}$ and $\{X^{(2)}\}$, which denote the normal modes of vibration, are known as the *modal vectors* of the system.

1.3.9.3 Torsional system

Consider the torsional system shown in Fig. 1.47, consisting of two disks on a shaft supported in frictionless bearings at the ends.

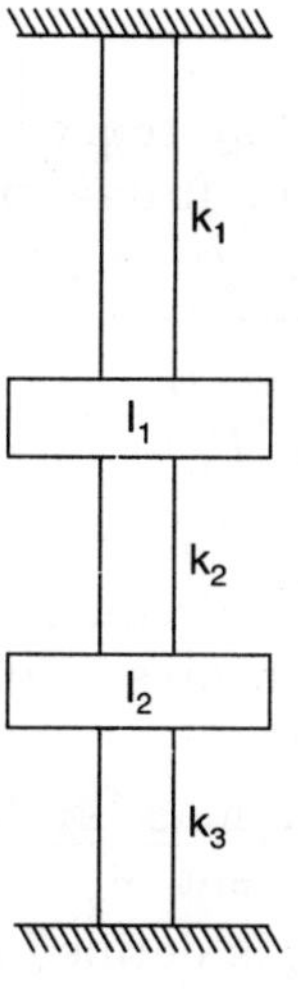

Fig. 1.47 Torsional system.

The differential equations of motion as

$$I_1\ddot{\theta}_1 + (k_1 + k_2)\theta_1 - k_2\theta_2 = 0$$

$$I_2\ddot{\theta}_2 + (k_2 + k_3)\,\theta_2 - k_2\theta_1 = 0$$

where k_i is the torsional stiffness of shaft i, $i = 1, 2, 3$, defined as

$$k_i = \frac{G_i J_i}{l_i}$$

where G_i is the modulus of rigidity, J_i is the polar moment of inertia, and l_i is the length of the shaft. By using the matrix notation, the differential equations of motion can be written in matrix form as

$$\begin{bmatrix} I_1 & 0 \\ 0 & I_2 \end{bmatrix}\begin{bmatrix} \ddot{\theta}_1 \\ \ddot{\theta}_2 \end{bmatrix} + \begin{bmatrix} k_1 + k_2 & -k_2 \\ -k_2 & k_2 + k_3 \end{bmatrix}\begin{bmatrix} \theta_1 \\ \theta_2 \end{bmatrix} = \begin{bmatrix} 0 \\ 0 \end{bmatrix}$$

1.3.9.4 Coordinate coupling and principal coordinates

The term *coupling* is used in vibration analysis to indicate a connection between equations of motion. In general an n degree of freedom vibration system requires n independent coordinates to describe completely its configuration. Often, it is quite possible to find some other set of n coordinates to describe the same configuration of the system completely. Each of these sets of n coordinates is called the *generalized coordinates.*

In the dynamic equations of motion, if the mass matrix $[M]$ is non-diagonal, then mass or *dynamic coupling* exists and if the stiffness matrix $[K]$ is non-diagonal then *stiffness* or *static coupling* exists. In general, it is possible to find a coordinate system that has neither *mass* or *dynamic coupling* nor *stiffness* or *static coupling.* Then the equations are decoupled into two independent equations and can be solved independently of the other. Such coordinates are called *principal coordinates* or *normal coordinates.*

1.3.9.5 Forced vibrations

When a two degree of freedom undamped system is subjected to the harmonic forces, $F_1(t) = F_1 \sin \omega t$ and $F_2(t) = F_2 \sin \omega t$, then the amplitudes of displacement of masses is given by

$$X_1 = \frac{a_{22}F_1 - a_{12}F_2}{a_{11}a_{22} - a_{12}a_{21}}$$

and
$$X_2 = \frac{a_{11}F_2 - a_{21}F_1}{a_{11}a_{22} - a_{12}a_{21}}$$

The denominator defines the natural frequencies of the system ω_1 and ω_2. The motions of the system are coupled and hence each mass will exhibit resonance even if the resonant force acts on only one mass of the system.

For a damped two-degree of spring-mass system under external forces the solution is obtained from mechanical impedance concept.

The mechanical impedance $Z_{rs}(i\omega)$ is defined as

$$Z_{rs}(i\omega) = -\omega^2 m_{rs} + i\omega c_{rs} + k_{rs}, \quad (r, s = 1, 2)$$

1.3.9.6 Orthogonality principle

If ω_1 and ω_2 are two eigenvalues (natural frequencies) and $X^{(1)}$ and $X^{(2)}$ are the corresponding eigenvectors (natural modes) they must satisfy

$$\omega_1^2[M]\,X^{(1)} = [K]\,X^{(1)}$$
$$\omega_2^2[M]\,X^{(2)} = [K]\,X^{(2)}$$

Then it can be shown that

For $\omega_1 \neq \omega_2$, $[X^{(2)}]^T\,[M]X^{(1)} = 0$

This property is very useful, as for example to check the accuracy of computation of normal modes by its application.

1.3.10 Multi-degree-of-freedom System

A *multi-degree-of-freedom system* is defined as a system whose motion is described by more than one generalized coordinate. In general, n coordinates are needed in order to describe the motion of an n-degree-of-freedom system. Figure 1.48 shows some examples of multi degrees of freedom systems.

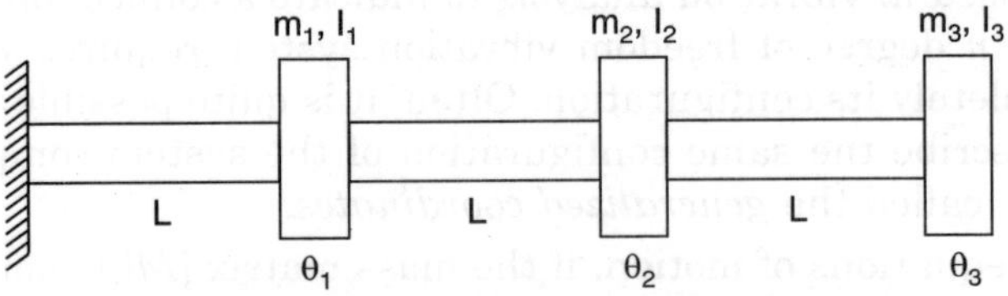

(a) Three-degree-of-freedom torsional system.

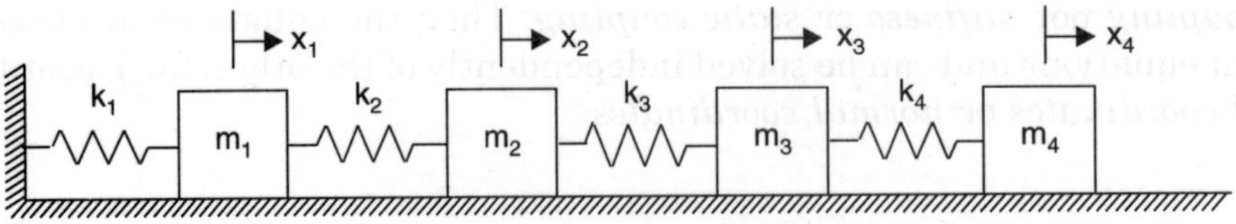

(b) Four-degree-of-freedom spring mass system.

Fig.1.48 Multi-degree of freedom systems.

An n degree-of-freedom system is governed by n coupled differential equations and has n natural frequencies. The solution of coupled differential equations can be written as the sum of a homogeneous solution and a particular solution. The free-vibration properties of the system are represented by the homogeneous solution while the particular solution represents the forced response.

1.3.10.1 Equations of motion

Consider the motion of an n-degree of freedom system whose motion is described by the generalized coordinates, $x_1, x_2, ..., x_n$ as shown in Fig. 1.49.

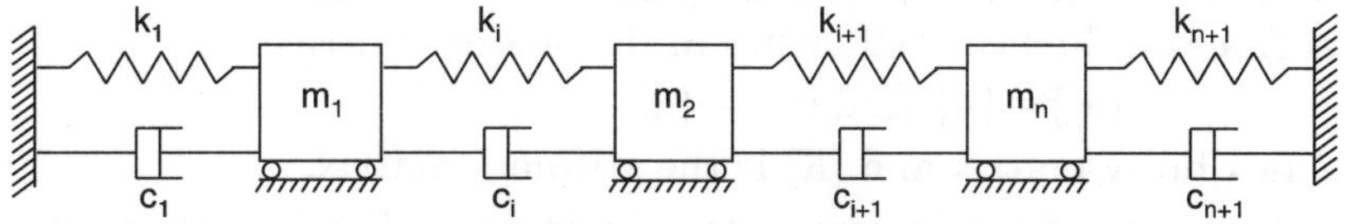

Fig. 1.49 Multi-degree of freedom system.

Applying the Newton's second law to mass m_i ($i = 1, 2, ..., n$), one can write the differential equation of motion as:

$$m_i\ddot{x}_i(t) - c_{i+1}\dot{x}_{i+1} + (c_i + c_{i+1})\dot{x}_i - c_i\dot{x}_{i-1} - k_{i+1}x_{i+1} + (k_i + k_{i+1})x_i - k_ix_{i-1} = 0$$

For general use, it is convenient to write this equation as in the following matrix form

$$[M]\ddot{x}(t) + [C]\dot{x}(t) + [K]x(t) = 0$$

with $[M]$, $[C]$ and $[K]$ being square matrices containing the coefficients m_{ij}, c_{ij} and k_{ij} respectively.

In this particular case, the mass-matrix is diagonal. For a different set of coordinates, $[M]$ is not necessarily diagonal.

1.3.10.2 Stiffness influence coefficients

For a linear system, inertial, damping and stiffness properties enter explicitly in the differential equations through the mass-coefficients m_{ij}, damping-coefficients c_{ij} and stiffness coefficients k_{ij} ($i, j = 1,2,..., n$) respectively. Of the three, stiffness coefficients are the elastic properties causing a dynamic system to vibrate, *e.g.*, restoring-forces. *Stiffness coefficients* are also known as *stiffness influence coefficients.* Stiffness influence coefficients k_{ij} is defined as the force required at $x = x_i$ to produce a unit displacement $u_j = 1$ at point $x = x_j$ and also the displacements at all other points for which $x \neq x_j$ are zero. In other words, they define a relation between the displacement at a point and the forces acting at various other points of system. Invoking the superposition principle, the force at $x = x_i$ producing displacements u_j at $x = x_j$ ($j = 1,2,..., n$) is

$$F_i = \sum_{j=1}^{n} k_{ij}u_j$$

1.3.10.3 Flexibility influence coefficients

Let the system be acted upon by a single-force F_j at $x = x_j$ and consider the displacement of any arbitrary point $x = x_i$ ($i = 1, 2, ..., n$) due to force F_j. Flexibility influence coefficient is defined as the *displacement of the point* $x = x_i$ due to unit force $F_j = 1$ applied at the point $x = x_j$. Invoking the principle of superposition and obtaining displacement u_i at $x = x_i$ resulting from all forces F_j ($j = 1, 2, ..., n$) by simply summing up the individual contributions

$$U_i = \sum_{j=1}^{n} a_{ij} F_j$$

Note that the units of a_{ij} are m/N.

For a single-degree of freedom system with only one spring, the stiffness influence coefficient is merely the spring-constant, whereas the flexibility influence coefficient is its reciprocal.

1.3.10.4 Matrix formulation

For multi-degree of freedom systems, a more general formulation is employed. Arranging the flexibility and stiffness influence coefficients in the square matrices as

$$[a_{ij}] = [A], \text{ and } [k_{ij}] = [K]$$

where $[A]$ is the flexibility matrix and $[K]$ is the stiffness matrix.

The flexibility and stiffness matrices are the inverse of one another. Often the stiffness coefficients are easier to evaluate than the flexibility coefficients. When the stiffness matrix is singular, the flexibility matrix does not exist. This implies that the system admits rigid-body motions, in which the system undergoes no elastic deformations. This can happen when supports do not fully restrain the system from moving. Thus in the absence of adequate supports, the definition of flexibility coefficients cannot be applied, so that the coefficients are not defined.

1.3.10.5 Inertia influence coefficients

The mass-matrix is associated with the kinetic energy. For a multi-degree of freedom system with $\dot{x}_i$ as the velocity of mass m_i (i = 1, 2, ..., n.), the kinetic energy is given by

$$T = \frac{1}{2} \dot{x}^T [M] \dot{x}$$

where $[M]$ is the mass-matrix or inertia matrix.

The elements of the mass-matrix m_{ij} are known as the inertia influence coefficients. The coefficients m_{ij} can be obtained using the impulse-momentum relations. The inertia influence coefficients m_{1j}, m_{2j}, ..., m_{nj} are defined as the set of impulses applied at points 1, 2, ..., n respectively, to produce a unit velocity at points 1, 2, ..., n respectively, to produce a unit velocity at point j and zero velocity at every other point. Thus, for a multi degree of freedom system, the total impulse at point i, can be found by summing up the impulses causing the velocities $\dot{x}_j$ (j = 1, 2, ..., n) as

$$\tilde{F} = [M]\dot{X}$$

where $[M]$ is the mass matrix, $\dot{X}$ and $\tilde{F}$ are the velocity and impulse vectors of size $n \times 1$ respectively.

1.3.10.6 Normal mode solution

The general formulation of the differential equations governing the free-vibrations of a linear-undamped n-degree-of-freedom system can be written as

$$[M]\ddot{x} + [K]x = 0$$

where $[M]$ and $[K]$ are symmetric $n \times n$ mass and stiffness matrices respectively and x is the n-dimensional column-vector of generalized coordinates.

Free vibrations of a multi-degree-of-freedom system are initiated by the presence of an initial potential or kinetic energy.

The normal-mode solution in the form of

$$x(t) = Xe^{i\omega t}$$

where ω the frequency of vibration and X is an n-dimensional vector called a *mode shape*. Each natural frequency has at least one corresponding mode shape. The general solution is a linear superposition over all possible modes.

The frequency or eigenvalue equation is defined as

$$-\omega^2 [M]X + [K]X = 0$$

The trivial solution (X = 0) is obtained unless

$$\det[[M]^{-1} [K] - \omega^2 I] = 0$$

Thus ω^2 must be an eigenvalue of $[M]^{-1} [K]$. This form is called characteristic equation. The square of a real positive eigenvalue has two possible values, one positive and one negative. While both are used to develop the general solution, the positive square root is identified as a natural frequency. The mode shape is the corresponding eigenvector.

1.3.10.7 Natural frequencies and mode shapes

Generally in vibration problems, the characteristic equation has only real-roots since the matrices under consideration are symmetric. Assuming that all the eigenvalues of $[M]^{-1} [K]$ corresponding to the symmetric mass and stiffness matrices are non-negative. Then there exist n-real natural frequencies that can be arranged by $\omega_1 \le \omega_2 \le \ldots \le \omega_n$. Each distinct eigenvalue ω_i^2, $i = 1, 2, \ldots, n$, has a corresponding non-trivial eigenvector X_i, which satisfies

$$[M]^{-1}[K]X_i = \omega_i^2 X_i$$

This mode shape X_i is an n-dimensional column vector of the form

$$X_i = \begin{bmatrix} X_{i1} \\ X_{i2} \\ \vdots \\ X_{in} \end{bmatrix}$$

This mode shape is not unique. The eigenvector is unique only to arbitrary multiplicative constant. Normalization schemes exist such that the constant is chosen so the eigenvector satisfies an externally imposed condition. The algebraic complexity of the solution grows exponentially with the number of degrees of freedom. Hence, numerical methods, which do not require the evaluation of the characteristic equation, are used for systems with a large number of degrees of freedom.

1.3.10.8 Mode shape orthogonality

In the solution of problems involving multi-degree-of-freedom vibration, one useful fundamental relation exists between the principal modes. Consider any two principal modes of oscillation of a system of several degrees of freedom. Let these be r^{th} and s^{th} modes and the corresponding eigenvalues be ω_r^2 and ω_s^2, then it can shown that

$$\{X\}_r^T [M]\{X\}_s = 0, \quad r \ne s$$

$$\{X\}_r^T [K]\{X\}_s = 0 \quad r \ne s$$

These define the matrix form of the orthogonal relationships between principal modes of vibration. Since $[M]$ is often a diagonal matrix and $[K]$ is not, it is usually simpler to write the orthogonality matrix with respect to $[M]$. The orthogonality relation with respect to $[M]$ is written in expanded form as

$$\sum_{i=1}^{n}\sum_{j=1}^{n} m_{ij}\, x_i^r\, x_j^s = 0, \quad r \neq s$$

Thus the orthogonality relation for the principal modes of vibration is essentially a relation between the amplitudes of two principal modes. These are not necessarily successive modes but any two modes. It is convenient to normalize mode shapes by requiring that the kinetic energy scalar product of a mode shape with itself is equal to one.

1.3.10.9 Response of a system to initial conditions

Response of multi-degree-of-freedom system subjected to initial excitations $x(0)$ and $\dot{x}(0)$ in the general form can be written as

$$x(t) = \sum_{r=1}^{n}\left[U_r^T\, Mx(0)\cos\omega_r t + \frac{1}{\omega_r}U_r^T\, M\dot{x}(0)\sin\omega_r t\right]U_r$$

Here each of the natural modes can be excited independently of the other.

1.3.11 Free Vibration of Damped Systems

In the equations of free motion including viscous damping, we can assume a harmonic form for the response. Due to the presence of damping, the characteristic equation will be a polynomial that has complex conjugate roots. For a given complex conjugate eigenvalues there are conjugate eigenvectors. The normal mode method or modal analysis applies only to undamped systems or systems where the damping can be made mathematically equivalent to the mass or stiffness matrix. Sometimes damping can be ignored in the forced response of a vibrating system.

1.3.12 Proportional Damping

For some special systems, where the damping matrix is linearly related to the mass and stiffness matrices, the simultaneous diagonalization of the stiffness and mass matrices can be accomplished along with that of the damping matrix. Such systems are called *proportional damping systems.*

Here $[C] = \alpha[K] + \beta[M]$

where α and β are constants.

Differential equations governing the free vibrations of a linear system with proportional damping can be written as

$$[M]\ddot{X} + (\alpha[K] + \beta[M])\dot{X} + [K]X = 0$$

If $\omega_1 \leq \omega_2 \leq \ldots \leq \omega_n$ are the natural frequencies of an undamped system whose mass-matrix is $[M]$ and stiffness matrix $[K]$ and $U_1, U_2, \ldots, U_n$ are the corresponding normalized mode shapes. The expansion-theorem implies that X can be written as a linear combination of the mode shape vector.

$$X = \Sigma p_i\, U_i$$

The matrix triple products possessing orthogonality properties, is written as

$$[U]^T[M][U]\{\ddot{P}\} + [U]^T(\alpha[K] + \beta[M])\,[U]\{\dot{P}\} + [U]^T[K][U]\{P\} = 0$$

The orthogonality of modes with respect to mass and stiffness permits the following substitutions:

$$[U]^T [M] [U] = [I]$$

and

$$[U]^T [K] [U] = [\text{diag } \omega^2] = [\Omega]$$

The equations can now be decoupled into governing equations for each degree of freedom. Mathematically,

$$\ddot{p}_i + (\alpha\omega_i^2 + \beta)\, \dot{p}_i + \omega_i^2\, p_i = 0$$

In this connection, modal damping ratio is defined as $\xi_i = \dfrac{1}{2}\left(\alpha\omega_i + \dfrac{\beta}{\omega_i}\right)$

The general solution for free vibration problem under $\xi_i < 1$ is given by

$$p_i(t) = A_i e^{-\xi_i \omega_i t} \sin (\omega_i \sqrt{1 - \xi_i^2}\; t - \phi_i)$$

where A_i and ϕ_i are constants determined from the initial conditions. Finally the solution is obtained in terms of generalized coordinates.

1.3.13 General Viscous Damping

The differential equations governing the free-vibrations of a multi-degree-of-freedom system with viscous damping are given by

$$[M]\ddot{X} + [C]\dot{X} + [K]\, X = 0$$

If the damping is arbitrary, then the principal coordinates of the undamped system do not uncouple the above equation. The equation can be reformulated as $2n$ first-order differential equations by writing

$$[\tilde{M}]\, \dot{y} + [\tilde{K}]\, y = 0$$

where

$$[\tilde{M}] = \begin{bmatrix} O & [M] \\ [M] & [C] \end{bmatrix},\ [\tilde{K}] = \begin{bmatrix} -[M] & O \\ O & [K] \end{bmatrix},\ y = \begin{bmatrix} \dot{X} \\ X \end{bmatrix}$$

If the values of γ are complex-conjugate eigenvalues of $[\tilde{M}]^{-1} [\tilde{K}]$ and ϕ is a corresponding eigenvector, then the solution takes the form as

$$y = \phi\, e^{-\gamma t}$$

1.3.14 Harmonic Excitations

Differential equations governing the motion of an n-degree of freedom undamped system subject to a single-frequency excitation with all excitation terms at the same phase can be written as:

$$[M]\ddot{x} + [K]x = F \sin \omega t$$

where F is an n-dimensional vector of constant forces. A particular solution of the form is assumed as follows:

$$x(t) = U \sin \omega t$$

where U is an n-dimensional vector of undetermined coefficients.

It results in by usual method as a solution

$$U = (-\omega^2 [M] + i\omega[C] + [K])^{-1}\, F$$

Alternative to this method of undetermined coefficients, Laplace transform method can also be employed.

1.3.15 Modal Analysis for Undamped Systems

The differential equations governing the forced vibration motion of an undamped linear n-degree-of-freedom system are

$$M\ddot{X} + KX = F$$

The method of modal analysis uses the principal coordinates of the system to uncouple this equation as follows:

$$\sum_{i=1}^{n} \ddot{p}_i (X_j\ MX_i) + \sum_{i=1}^{n} p_i (X_j KX_i) = X_j F$$

Application of mode shape orthogonality leads to only one non-zero term in each summation, *i.e.*, the term corresponding to $i = j$. Since the mode shapes are normalized, the following set of equations are obtained

$$\ddot{p}_j + \omega_j^2\, p_j = g_j(t)$$

where $g_j(t) = X_j\, F$

If the initial conditions for p_i are both zero, then the convolution integral solution is given by

$$p_i(t) = \frac{1}{\omega_i} \int_0^t g_i(\tau) \sin\,[\omega_i (t - \tau)]\, d\tau$$

Once the solution for each p_i is obtained, the original generalized coordinates can be determined.

The same methodology can be applied to systems having proportional damping.

Here it leads to the differential equations for the principal coordinates as

$$\ddot{p}_i + 2\xi_i \omega_i \dot{p}_i + \omega_i^2\, p_i = g_i(t)$$

where ξ_i is modal damping-ratio.

In this case, the convolution-integral solution is given by

$$p_i(t) = \frac{1}{\omega_{d_i}} \int_0^t g_i(\tau) e^{-\xi_i \omega_i (t-\tau)} \sin \omega_{d_i}\, (t - \tau)\, d\tau$$

where $\omega_{d_i} = \omega_i \sqrt{1 - \xi_i^2}$

1.3.16 Lagrange's Equation

There are two general approaches to classical dynamics: *vectorial dynamics* and *analytical dynamics*. *Vectorial dynamics* is based directly on the application of Newton's second law of motion, concentrating on forces and motions. *Analytical dynamics* treats the system as a whole dealing with scalar quantities such as the kinetic and potential energies of the system. Lagrange proposed an approach, which provides a powerful and versatile method for the formulation of the equations of motion for any dynamical system. *Lagrange's equation* obtains the equation of motion in generalized coordinates approaching the system from the analytical dynamics point of view. Lagrange's equations are differential equations in which one considers the energies of the system and the work done instantaneously in time.

1.3.16.1 Generalized coordinates

The coordinates used to describe the motion in each degree of freedom of a system are termed as *generalized coordinates*. They may be Cartesian, polar, cylindrical or spherical coordinates, provided any one of them can be used to describe the configuration of the system where the motion along any one coordinate direction is independent of others. But, sometimes they may not have such simple physical or geometrical meaning. For example, the deflections of a string, stretched between two points, can be expressed in the form of trigonometric Fourier series, and the coefficients of all the terms in the series can be considered as a generalized coordinate set. This is because each trigonometric function in the series may be considered as a unique degree of freedom and the coefficients describe the extent of deflection in each degree of freedom.

It is possible to transform the coordinates from any one system to the generalized coordinate system or vice versa, through coordinate transformation. Consider a mechanical system consisting of N particles whose positions are (x_i, y_i, z_i), $i = 1,2,\ldots, N$, in a Cartesian coordinate system. The motion of the mechanical system is completely defined if the variation with time of these positions *i.e.* $x_i = x_i(t)$, $y_i = y_i(t)$, $z_i = z_i(t)$, are known. These $3N$ coordinates completely define a representative space. If it is possible to find another set of generalized coordinates, q_i, $i = 1,2,\ldots.n$, where $n = 3N$, then these two coordinate systems are related by the following:

$$x_i(t) = x_i(q_1, q_2, \ldots, q_n, t)$$
$$y_i(t) = y_i(q_1, q_2, \ldots, q_n, t)$$
$$z_i(t) = z_i(q_1, q_2, \ldots, q_n, t)$$

1.3.17 Principle of Virtual Work

The principle of virtual work is essentially a statement of the static or dynamic equilibrium of a mechanical system. A *virtual displacement*, denoted by δr, is an imaginary displacement and it occurs without the passage of time. The virtual displacement being infinitesimal obeys the rules of differential calculus.

Consider a mechanical system with N particles in a three-dimensional space whose Cartesian coordinates are $(x_1, y_1, z_1,\ldots,z_n)$. Suppose the system is subject to k constraints $\phi_j (x_1, y_1, z_1,\ldots,z_n, t) = 0$, $j = 1,2,\ldots,k$. The virtual displacements $\delta x_1, \delta y_1, \delta z_1$ etc. are said to be *consistent* with the system constraints if the constraint equations are still satisfied.

The virtual work performed by the resultant force vector $\overline{F}_i$ over the virtual displacement vector δr_i of particle i is

$$\delta W = \sum_{i=1}^{N} \overline{F}_i \cdot \delta \overline{r}_i$$

When the system is in equilibrium, the resultant force acting on each particle is zero. The resultant force is the sum of the applied force and the reaction force or the constraint force. The virtual work done by all the forces in moving through an arbitrary virtual displacement consistent with the constraints is zero.

1.3.18 D'Alembert's Principle

The principle of virtual work is extended to dynamics, in which form it is known as D' Alembert's principle. The principle of virtual work is extended to the dynamic case by considering the inertia forces and considering the systems to be in dynamic equilibrium.

The generalized principle of D'Alembert states that *the virtual work performed by the effective forces through infinitesimal virtual displacements compatible with the system constraints is zero.*

1.3.19 Lagrange's Equations of Motion

If Q_i is called the *generalized force* in the direction of the ith generalized coordinate, T is the kinetic energy and V is potential energy, then Lagrange's equation is given by

$$\frac{d}{dt}\left(\frac{\partial T}{\partial \dot{q}_i}\right) - \frac{\partial T}{\partial q_i} + \frac{\partial V}{\partial q_i} = Q_i$$

Expressing $T - V = L$, called the *Lagrangian*, the equation can be written as

$$\frac{d}{dt}\left(\frac{\partial L}{\partial \dot{q}_i}\right) - \frac{\partial L}{\partial q_i} = Q_i$$

1.3.20 Variational Principles

An alternative approach to the study of motion is the use of variational principle, which views the motion as a whole from the beginning to the end. This involves a search for the path in the configuration space, which yields a stationary value for a certain integral. Unlike as in the case of differential equations, the initial and final points in the configuration space are fixed in this approach. The most celebrated variational principle in dynamics is the *Hamilton's principle.*

1.3.21 Hamilton's Principle

Hamilton's principle is the most important and powerful variational principle in dynamics. It is derived from the generalized D'Alembert's principle. The generalized version of Hamilton's principle can be written as

$$\int_{t_0}^{t_1} (\delta T + \delta W)\, dt = 0 \quad \text{or} \quad \int_{t_0}^{t_1} \delta\,(T - V)\, dt = 0, \ \delta \int_{t_0}^{t_1} L\, dt = 0$$

where $L = T - V$.

The usual form of Hamilton's principle applies to a more restricted class of systems, which are called *conservative systems*. In these systems all the applied forces are derivable from a potential function $V(q, t)$.

The usual form of Hamilton's principle states that: The actual path in the configuration space followed by a holonomic system from t_0 and t_1 is such that the integral

$$I = \int_{t_0}^{t_1} L\, dt$$

is stationary with respect to any path variations, which vanish at the end points.

1.4 ELECTRICAL CIRCUITS

1.4.1 Electrical Elements

Here, we briefly review the three types of basic passive electrical elements: resistor, inductor and capacitor.

Resistance elements: *Ohm's Law:* The voltage drop V_R across a linear resistor is proportional to the current i_R flowing through the resistor, where the constant of proportionality is the resistance R as shown in Fig. 1.50.

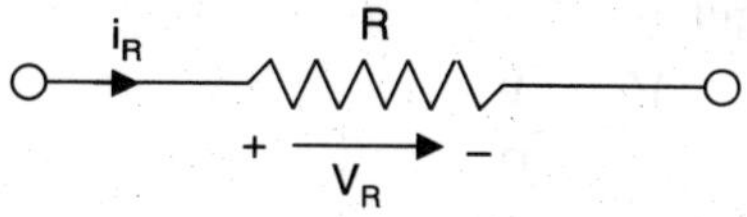

Fig. 1.50

$$V_R = R\, i_R$$

Resistors do not store electrical energy in any form but dissipate it as heat. The rate of energy dissipated (power consumed) by a resistor is given by

$$P = i_R^{\,2}\, R = \frac{V_R^{\,2}}{R} \quad \text{(W or J/s)}$$

Resistors in series: The current i passes through each element as shown in Fig. 1.51(*a*). The total voltage drop is given by the sum of the voltage drops across each element, or

$$V = V_1 + V_2$$

Applying Ohm's law, we obtain

$$iR_{eq} = iR_1 + iR_2$$

or

$$R_{eq} = R_1 + R_2$$

The voltage drop across each resistor is then given by

$$V_1 = \left(\frac{R_1}{R_1 + R_2}\right) V$$

and

$$V_2 = \left(\frac{R_2}{R_1 + R_2}\right) V$$

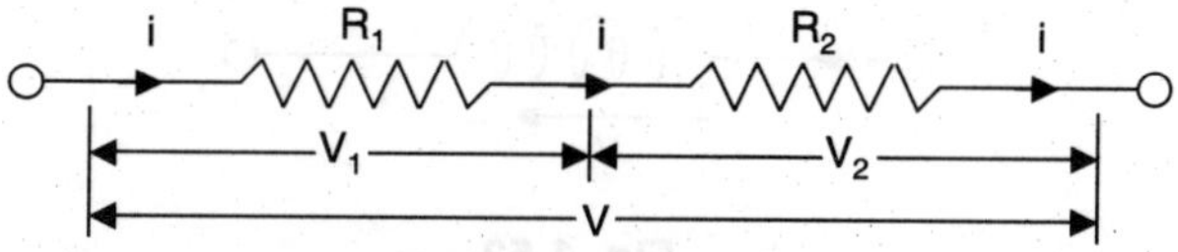

(a) Resistors in series

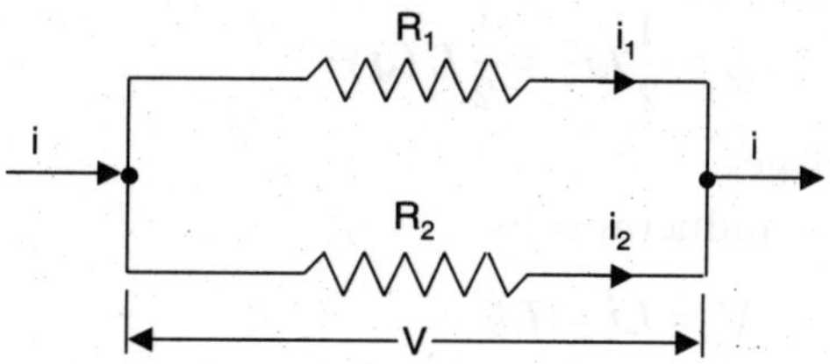

(b) Resistors in parallel

Fig. 1.51

Resistors in parallel: All the elements in the case have the same voltage applied across them as shown in Fig. 1.51 (*b*).

$$i = i_1 + i_2$$

Applying ohm's law, we get

$$\frac{V}{R_{eq}} = \frac{V}{R_1} + \frac{V}{R_2}$$

or

$$R_{eq} = \frac{R_1 R_2}{R_1 + R_2} \quad (1.17)$$

If there are *n* resistors, we can write

$$\frac{1}{R_{eq}} = \frac{1}{R_1} + \frac{1}{R_2} + \ldots + \frac{1}{R_n} \quad (1.18)$$

Solving Eq. (1.17) for the currents i_1 and i_2, we obtain

$$i_1 = \left(\frac{R_2}{R_1 + R_2}\right) i \quad (1.19)$$

$$i_2 = \left(\frac{R_1}{R_1 + R_2}\right) i \quad (1.20)$$

Inductance elements: The voltage V_L across the inductor *L* is given by (see Fig. 1.52)

$$V_L = L\frac{di_L}{dt}$$

where i_L is the current through the inductor.

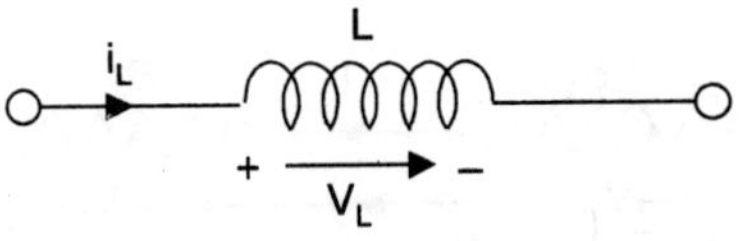

Fig. 1.52

The energy stored in an inductance is

$$E = \frac{1}{2}Li^2 = \frac{1}{2}L(\dot{q})^2$$

where *q* is the electrical charge.

The voltage drop in an inductance is

$$V_i = L\dot{i} = L\ddot{q}$$

Inductances in series: Since the voltage drop through an inductor is proportional to the inductance *L*, we have (Fig. 1.53 (*a*))

$$L_{eq} = L_1 + L_2$$

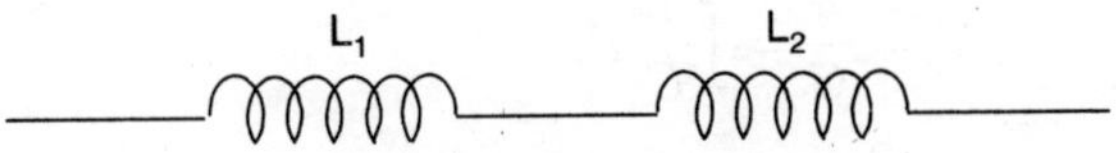

(a) Inductance in series

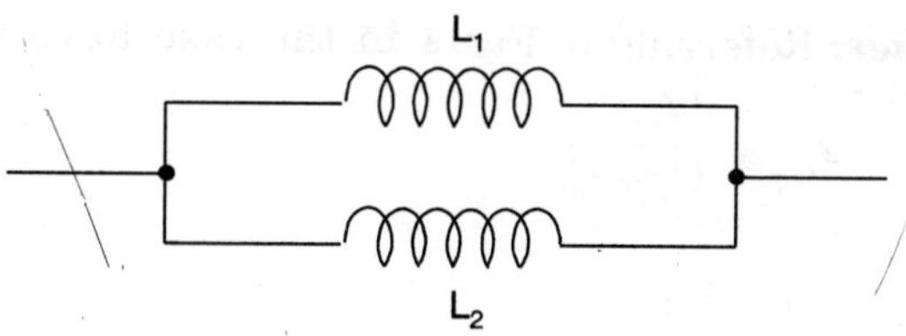

(b) Inductances in parallel

Fig. 1.53 Inductances.

Inductances in parallel: Referring to Fig.1.53 (*b*), we have

$$L_{eq} = \frac{L_1 L_2}{L_1 + L_2}$$

Similarly, for n inductors

$$\frac{1}{L_{eq}} = \frac{1}{L_1} + \frac{1}{L_2} + \ldots + \frac{1}{L_n}$$

Capacitance elements: Capacitance C is a measure of the quantity of charge that can be stored for a given voltage across the plates. The capacitance C of a capacitor is given by

$$C = \frac{q}{V_c}$$

where q is the quantity of charge stored and V_c is the voltage across the capacitor.

Since $i = \frac{dq}{dt}$ and $V_c = q/C$, we have

$$i = C\frac{dV_c}{dt}$$

or $$dV_c = \frac{1}{C} i\, dt$$

Hence $$V_c = \frac{1}{C}\int i_c\, dt$$

This is shown in Fig. 1.54.

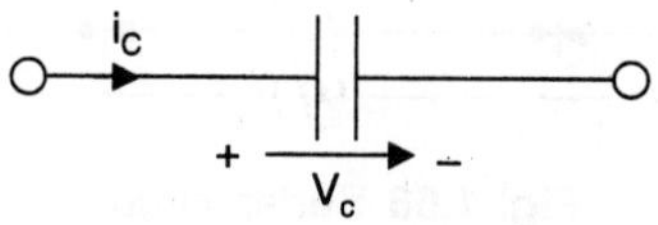

Fig. 1.54 Capacitor.

The energy stored in a capacitor is given by

$$E = \frac{1}{2} CV^2$$

The voltage drop across capacitor is given by

$$V_c = \frac{q}{C} = \int \frac{i}{C}\, dt$$

Capacitors in series: Referring to Fig. 1.55 (*a*) we have

$$C_{eq} = \frac{C_1 C_2}{C_1 + C_2}$$

For n capacitors

$$\frac{1}{C_{eq}} = \frac{1}{C_1} + \frac{1}{C_2} + \ldots + \frac{1}{C_n}$$

Capacitors in parallel: For capacitors in parallel (see Fig.1.55 (*b*))

$$C_{eq} = C_1 + C_2$$

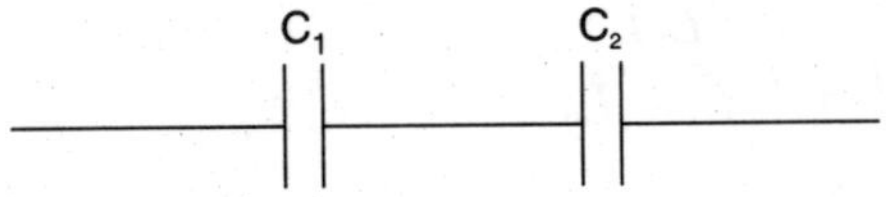

(a) Capacitors in series

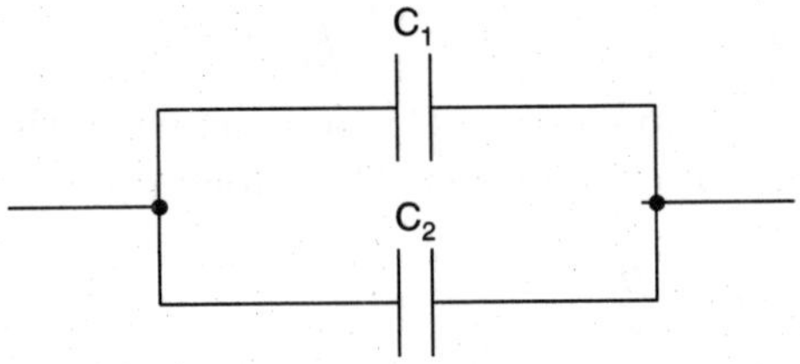

(b) Capacitors in parallel

Fig. 1.55 Capacitor.

1.4.2 Electrical Circuits

In this section, we apply ohm's law to series and parallel circuits to determine the combined resistance of the given circuit.

Series circuits: The combined resistance of series-connected resistors of a simple series circuit is given by the sum of the individual resistances. The voltage between points A and B of the simple series circuit shown in Fig. 1.56 is given by

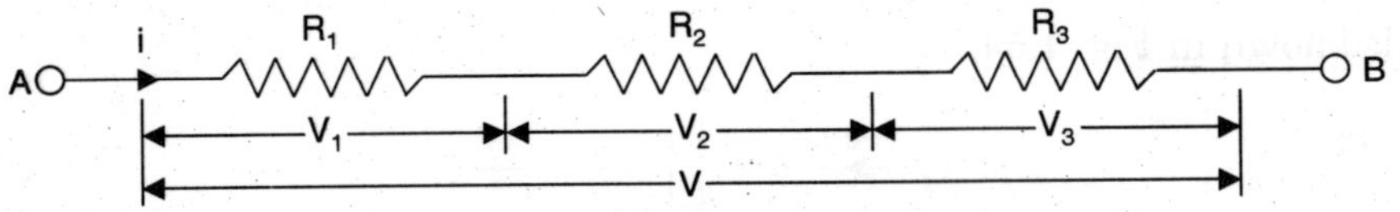

Fig. 1.56 Series circuit.

$$V = V_1 + V_2 + V_3$$

where $V_1 = iR_1$

$V_2 = iR_2$

and $V_3 = iR_3$

Hence $$\frac{V}{i} = R_1 + R_2 + R_3$$

Therefore, the combined resistance R of the series circuit is given by

$$R = R_1 + R_2 + R_3$$

Parallel circuits: For the parallel electrical circuit shown in Fig. 1.57, we can write

$$i_1 = \frac{V}{R_1}$$

$$i_2 = \frac{V}{R_2}$$

and

$$i_3 = \frac{V}{R_3}$$

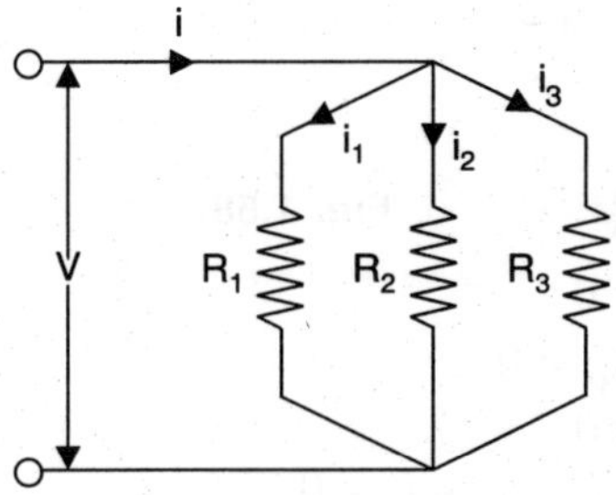

Fig. 1.57 Parallel circuit.

Now $\quad i = i_1 + i_2 + i_3$

Therefore

$$i = \frac{V}{R_1} + \frac{V}{R_2} + \frac{V}{R_3} = \frac{V}{R}$$

where R is the combined resistance. Hence

$$\frac{1}{R} = \frac{1}{R_1} + \frac{1}{R_2} + \frac{1}{R_3}$$

Therefore

$$R = \frac{1}{\frac{1}{R_1} + \frac{1}{R_2} + \frac{1}{R_3}} = \frac{R_1 R_2 R_3}{R_1 R_2 + R_2 R_3 + R_3 R_1}$$

1.4.3 Kirchhoff's Laws

Kirchhoff's laws are the two most useful physical laws for modeling electrical systems. It is necessary to apply Kirchhoff's laws in solving electric circuit problems as they involve many electromotive forces such as resistance, capacitance and inductance.

The Kirchhoff's laws are stated as follows:

1. *Kirchhoff's current law (node law):* The algebraic sum of all the currents flowing into a junction (or node) is zero (node analysis).

In other words, the sum of currents entering a node is equal to the sum of the currents leaving the same node. A *node* is an electrical circuit is a point where three or more wires are joined together. Currents going toward a node are considered positive while currents leaving a node are treated as negative.

The algebraic sum of all currents (*in*) a circuit node is zero. That is,

$$\sum_n (i_n)_{in} = 0$$

Referring to Fig. 1.58, Kirchhoff's current law states that

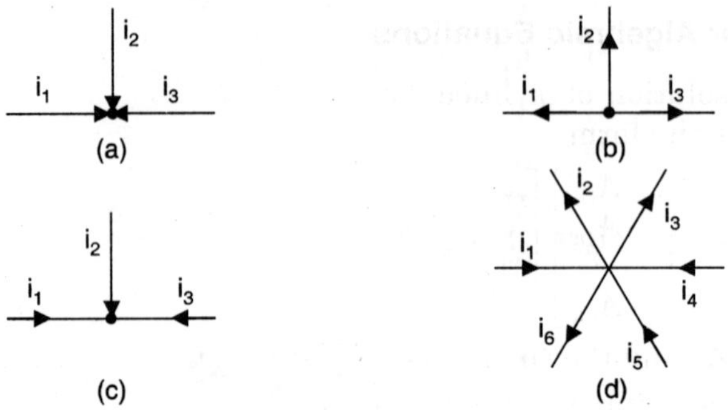

Fig. 1.58

Fig. 1.58 (*a*) $i_1 + i_2 + i_3 = 0$
Fig. 1.58 (*b*) $-(i_1 + i_2 + i_3) = 0$
Fig. 1.58 (*c*) $i_1 + i_2 - i_3 = 0$
Fig. 1.58 (*d*) $i_1 - i_2 - i_3 + i_4 + i_5 - i_6 = 0$

2. *Kirchhoff's voltage law (loop law):* The algebraic sum of all the potential drops around a closed loop (or closed circuit) is zero (loop analysis).

In other words, the sum of the voltage drops is equal to the sum of the voltage rises around a loop. That is, the sum of all voltage drops around a circuit loop is zero. Hence

$$\Sigma V_{drop} = 0$$

or

$$\Sigma V_{gain} = 0$$

The voltage drops or voltage gains should be appropriately indicated for loop analysis. Figure 1.59 shows examples with useful sign convention.

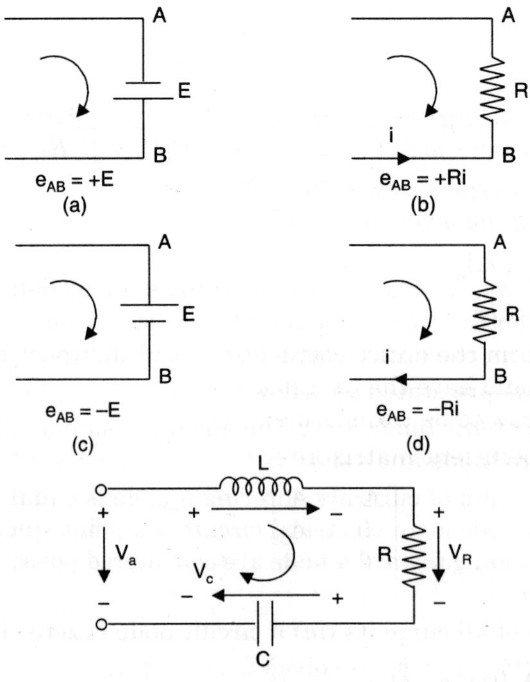

Fig. 1.59

1.5 NUMERICAL METHODS

1.5.1 System of Linear Algebraic Equations

Here we consider the solution of n linear, algebraic equations in n unknowns. A system of algebraic equations has the form

$$\begin{bmatrix} A_{11} & A_{12} & \dots & A_{1n} \\ A_{21} & A_{22} & \dots & A_{2n} \\ \vdots & \vdots & \vdots & \vdots \\ A_{n1} & A_{n2} & \dots & A_{nn} \end{bmatrix} \begin{bmatrix} x_1 \\ x_2 \\ \vdots \\ x_n \end{bmatrix} = \begin{bmatrix} b_1 \\ b_2 \\ \vdots \\ b_n \end{bmatrix} \tag{1.21}$$

where the coefficients A_{ij} and the constants b_j are known, and x_i represent the unknowns.

Equation (1.21) is simply written as

$$Ax = b \tag{1.22}$$

1.5.2 Gauss Elimination Method

Consider the equations at some instant during the elimination phase.

$$\begin{bmatrix} A_{11} & A_{12} & A_{13} & \dots & A_{1k} & \dots & A_{1j} & \dots & A_{1n} \\ 0 & A_{22} & A_{23} & \dots & A_{2k} & \dots & A_{2j} & \dots & A_{2n} \\ 0 & 0 & A_{33} & \dots & A_{3k} & \dots & A_{3j} & \dots & A_{3n} \\ \vdots & \vdots & \vdots & \vdots & \vdots & \vdots & \vdots & \vdots & \vdots \\ 0 & 0 & 0 & \dots & A_{kk} & \dots & A_{kj} & \dots & A_{kn} \\ \hline \vdots & \vdots & \vdots & \vdots & \vdots & \vdots & \vdots & \vdots & \vdots \\ 0 & 0 & 0 & \dots & A_{ik} & \dots & A_{ij} & \dots & A_{in} \\ \vdots & \vdots & \vdots & \vdots & \vdots & \vdots & \vdots & \vdots & \vdots \\ 0 & 0 & 0 & \dots & A_{nk} & \dots & A_{nj} & \dots & A_{nn} \end{bmatrix} \begin{bmatrix} b_1 \\ b_2 \\ b_3 \\ \dots \\ b_k \\ \dots \\ b_i \\ \dots \\ b_n \end{bmatrix} \begin{matrix} \\ \\ \\ \leftarrow \text{pivot row} \\ \\ \leftarrow \text{row being} \\ \text{transformed} \\ \\ \end{matrix} \tag{1.23}$$

In the above Eq. (1.23), the first k rows of A have already been transformed to upper triangular form. Hence, the current pivot equation is the kth equation and all the equations below it are still to be transformed.

Let the ith row be a typical row below the pivot equation that is to be transformed. We obtain this by multiplying the pivot row by $\lambda = A_{ik}/A_{kk}$ and subtracting it from the ith row. Then

$$A_{ij} \leftarrow A_{ij} - \lambda A_{kj} \qquad j = k, k+1, \dots, n \tag{1.24}$$

$$b_i \leftarrow b_i - \lambda b_k \tag{1.25}$$

In order to transform the entire coefficient matrix to upper triangular form, k and I in Eqs. (1.24) and (1.25) should have the ranges $k = 1, 2, \dots, n - 1$ (choose the pivot row), $i = k + 1$, $K + 2, \dots, n$ (selects the row to be transformed).

The augmented coefficient matrix after Gauss elimination has the form

$$[A/b] = \begin{bmatrix} A_{11} & A_{12} & A_{13} & \dots & A_{1n} \\ 0 & A_{22} & A_{23} & \dots & A_{2n} \\ 0 & 0 & A_{33} & \dots & A_{3n} \\ \vdots & \vdots & \vdots & \vdots & \vdots \\ 0 & 0 & 0 & \dots & A_{nn} \end{bmatrix} \begin{bmatrix} b_1 \\ b_2 \\ b_3 \\ \vdots \\ b_n \end{bmatrix} \tag{1.26}$$

The last equation, $A_{nn}\, x_n = b_n$, is solved first, giving

$$x_n = b_n/A_{nn} \tag{1.27}$$

Now conducting the back substitution, we have the solution as

$$x_k = \left(b_k - \sum_{j=k+1}^{n} A_{kj}\, x_j \right) \frac{1}{A_{kk}} \qquad k = n-1, n-2, \ldots, 1 \tag{1.28}$$

1.5.3 LU Decomposition Methods

Any square matrix A can be written as a product of a lower triangular matrix L and an upper triangular matrix U.

$$A = LU \tag{1.29}$$

The process of computing L and U for a given A is known as *LU decomposition* or *LU factorization*.

The given equations can be rewritten as $LUx = b$ and using the notation $Ux = y$, then

$$Ly = b \tag{1.30}$$

which can be solved for y by forward substitution.

Hence $Ux = y$ which gives x by the back substitution process.

1.5.4 Choleski's Decomposition

Choleski's decomposition $A = LL^T$ requires that A should be symmetric. The decomposition process involves taking square roots of certain combinations of the elements of A.

A typical element in the lower triangular portion of LL^T is of the form,

$$(LL^T)_{ij} = L_{i1}\, L_{j1} + L_{i2}\, L_{j2} + \ldots + L_{ij}\, L_{ij} = \sum_{k=1}^{j} L_{ik} L_{jk} \qquad i \geq j \tag{1.31}$$

Equating this term to the corresponding element of A gives

$$A_{ij} = \sum_{k=1}^{j} L_{ik}\, L_{jk} \qquad i = j, j+1, \ldots, n; \quad j = 1, 2, \ldots, n \tag{1.32}$$

Taking the term containing L_{ij} outside the summation in Eq. (1.32), we obtain

$$A_{ij} = \sum_{k=1}^{j-1} L_{ik}\, L_{jk} + L_{ij} L_{jj} \tag{1.33}$$

If $i = j$, then the solution is

$$L_{ij} = \sqrt{A_{ij} - \sum_{k=1}^{j-1} L^2{}_{jk}} \qquad j = 2, 3, \ldots, n \tag{1.34}$$

For a non-diagonal term, we get

$$L_{ij} = \frac{\left(A_{ij} - \sum_{k=1}^{j-1} L_{ik}\, L_{jk} \right)}{L_{jj}} \qquad j = 2, 3, \ldots, n-1;\ i = j+1, j+2, \ldots, n \tag{1.35}$$

1.5.5 Gauss-Seidel Method

The equations $Ax = b$ can be written in scalar form as

$$\sum_{j=1}^{n} A_{ij} X_j = b_i \qquad i = 1, 2, \ldots, n \tag{1.36}$$

Extracting the term containing x_i from the summation sign gives

$$A_{ii} x_i + \sum_{\substack{j=1 \\ j \neq i}}^{n} A_{ij} x_j = b_i \qquad i = 1, 2, \ldots, n \tag{1.37}$$

Solving for x_i, we obtain

$$x_i = \frac{1}{A_{ii}} \left(b_i - \sum_{\substack{j=1 \\ j \neq i}}^{n} A_{ij} x_j \right) \qquad i = 1, 2, \ldots, n \tag{1.38}$$

Hence, the iterative scheme is

$$x_i \leftarrow \frac{1}{A_{ii}} \left(b_i - \sum_{\substack{j=1 \\ j \neq i}}^{n} A_{ij} x_j \right) \qquad i = 1, 2, \ldots, n \tag{1.39}$$

We start by choosing the starting vector x. The procedure for Gauss-Seidel algorithm is summarized here with relaxation:

(*a*) conducting k iterations with $\omega = 1$ (or $k = 10$). After the kth iteration record $\Delta x^{(k)}$.

(*b*) carryout additional p iterations ($p \geq 1$) and record $\Delta x^{(k+p)}$ after the last iteration.

(*c*) perform all subsequent iterations with $\omega = \omega_{opt}$, where

$$\omega_{opt} = \frac{2}{1 + \sqrt{1 - (\Delta x^{(k+p)} / \Delta x^{(k)})^{1/p}}}$$

1.5.6 Gauss-Jordan Method

Let us consider a system of linear algebraic equations, in the matrix form

$$[A]\{x\} = \{b\} \tag{1.40}$$

where, for simplification, $[A]$ is of order 3×3. The augmented matrix is

$$\begin{bmatrix} a_{11} & a_{12} & a_{13} & b_1 \\ a_{21} & a_{22} & a_{23} & b_2 \\ a_{31} & a_{32} & a_{33} & b_3 \end{bmatrix} \tag{1.41}$$

The solution of equation (1.40) is

$$[I]\{x\} = [A]^{-1}\{b\} = \{\alpha\} \tag{1.42}$$

where $[I]$ is the identity matrix. The augmented matrix is

$$\begin{bmatrix} 1 & 0 & 0 & \alpha_1 \\ 0 & 1 & 0 & \alpha_2 \\ 0 & 0 & 1 & \alpha_3 \end{bmatrix} \tag{1.43}$$

In the Gauss-Jordan method the augmented matrix (1.41) is converted to the augmented matrix (1.43) by a series of operations similar to the Gaussian elimination method. In the Gaussian elimination method an upper triangular matrix is derived while in the Gauss-Jordan method an identity matrix is derived.

1.5.7 Jacobi Method

This is an iterative technique solving with an assumed solution vector and successive refinement by iteration. The system of equations for consideration is

$$\begin{array}{l} a_{11}x_1 + a_{12}x_2 + a_{13}x_3 + \ldots + a_{1n}x_n = b_1 \\ a_{21}x_1 + a_{22}x_2 + a_{23}x_3 + \ldots + a_{2n}x_n = b_2 \\ \ldots \quad \ldots \quad \ldots \quad \ldots \\ \ldots \quad \ldots \quad \ldots \quad \ldots \\ a_{i1}x_1 + a_{i2}x_2 + a_{i3}x_3 + \ldots + a_{in}x_n = b_i \\ \ldots \quad \ldots \quad \ldots \quad \ldots \\ a_{n1}x_1 + a_{n2}x_2 + a_{n3}x_3 + \ldots + a_{nn}x_n = b_n \end{array} \tag{1.44}$$

Rewriting the above equations

$$\begin{array}{l} x_1 = (b_1 - a_{12}x_2 - a_{13}x_3 - \ldots - a_{1n}x_n)/a_{11} \\ x_2 = (b_2 - a_{21}x_1 - a_{23}x_3 - \ldots - a_{2n}x_n)/a_{22} \\ \ldots \quad \ldots \quad \ldots \quad \ldots \\ x_i = (b_i - a_{i1}x_i - a_{i2}x_2 - \ldots - a_{ii-1}n_{i-1} - a_{ii+1}x_{i+1} - \ldots - a_{in}x_n)/a_{ii} \\ \ldots \quad \ldots \quad \ldots \quad \ldots \quad \ldots \\ x_n = (b_n - a_{n1}x_1 - a_{n2}x_2 - \ldots - a_{nn-1}x_{n-1})/a_{nn} \end{array} \tag{1.45}$$

This procedure is valid only if all the diagonal elements are non zero. The equations are to be rearranged suitably to avoid the non zero elements in the main diagonal.

Substituting the values of x_i^r at any stage in the iterative process on the right hand side of equations (1.45) gives the values to the next stage, *i.e.* x_i^{r+1}. In other words, the scheme is given by the system of equations (1.46) with a superscript r on the right side and a superscript $r + 1$ on the left hand side. Rewriting the equations,

$$\begin{array}{l} x_1^{r+1} = (b_1 - a_{12}x_2^r - a_{13}x_3^r - \ldots - a_{1n}\, x_n^r)/a_{11} \\ x_2^{r+1} = (b_2 - a_{21}x_1^r - a_{23}x_3^r - \ldots - a_{2n}\, x_n^r)/a_{22} \\ \ldots \quad \ldots \quad \ldots \quad \ldots \\ x_i^{r+1} = (b_i - a_{i1}x_i^r - a_{i2}x_2^r - \ldots - a_{ii-1}x_{i-1}^r - a_{ii+1} - \ldots - a_{in}x_n^r)/a_{ii} \\ \ldots \quad \ldots \quad \ldots \quad \ldots \\ x_n^{r+1} = (b_n - a_{n1}x_1^r - a_{n2}x_2^r - \ldots - a_{n,n-1}x_{n-1}^r)/a_{nn} \end{array} \tag{1.46}$$

The sequence $x^0, x^1, x^2, \ldots$ generated by the equations of (1.46) gives a sequence which converges to the solutions vector x which satisfies the set of equations given in Eq.(1.44), *i.e.* $[A]\{x\} = \{b\}$.

Equation (1.46) in the matrix form is as follows

$$x^{r+1} = \{V\} + [B]\{x\}^r \tag{1.47}$$

where $\{V\} = \begin{Bmatrix} \dfrac{b_1}{a_{11}} \\ \dfrac{b_2}{a_{22}} \\ \cdots \\ \dfrac{b_i}{a_{ii}} \\ \cdots \\ \dfrac{b_n}{a_{nn}} \end{Bmatrix}$

and $[B] = \begin{bmatrix} 0 & a_{12}/1_{11} & 1_{13}/a_{11} & \cdots & \cdots & \cdots & a_{1n}/a_{11} \\ a_{21}/a_{22} & 0 & a_{23}/a_{22} & \cdots & \cdots & \cdots & a_{2n}/a_{22} \\ \cdots & \cdots & \cdots & \cdots & \cdots & \cdots & \cdots \\ a_{i1}/a_{ii} & a_{i2}/a_{ii} & \cdots & 0 & a_{ii+1}/a_{ii} & \cdots & a_{ii}/a_{nn} \\ \cdots & \cdots & \cdots & \cdots & \cdots & \cdots & \cdots \\ a_{n1}/a_{nn} & \cdots & \cdots & \cdots & a_{n,n-1}/a_{nn} & \cdots & 0 \end{bmatrix}$

The system of equations given in Eq. (1.44) can be written in the form

$$\{x\} = \{v\} + [B]\{x\} \tag{1.48}$$

Now we can construct expressions for $x^{(1)} \ldots x^{(2)} \ldots$ in terms of $x^{(0)}$

$$\{x\}^1 = \{v\} + [B]x^0$$

$$\{x\}^2 = \{v\} + [B]x^1$$
$$= \{v\} + [B](\{v\} + [B]x^{(0)})$$
$$= \{v\} + [B](\{v\} + [B]^2x^0$$

$$\{x\}^3 = \{v\} + [B]\{x\}^{(2)}$$
$$= \{v\} + [B](\{v\} + [B]\{v\} + [B]^2x^{(0)})$$
$$= \{v\} + [B]\{v\} + [B]^2\{v\} + [B]^3x^0$$

Generalizing

$$x^{(r)} = \{v\} + [B]\{v\} + [B]^2\{v\} + \ldots [B]^{r-1}\{v\} + [B]^r x^{(0)}$$

Finally

$$x^{(r)} = ([I] + [B] + [B]^2 + [B]^3 \ldots [B]^{r-1})\{v\} + [B]^{(r)}x^{(0)} \tag{1.49}$$

Here we notice that $([I] + [B] + [B]^2 + [B]^3 \ldots [B]^{r-1})$ is a matrix geometric progression. It is possible to obtain the sum of r terms of the above expressions. Let us denote

$$s_r = [I] + [B] + [B]^2 + [B]^3 \ldots [B]^{r-1}$$
$$= ([I] - [B]^r)([I] - [B])^{-1}$$

provided $([I] - [B])$ is non-singular. Here we see that as $r \to \infty$ the limit of the sum exists if $[B]^r \to 0$, in this case $s_r \to s = ([I] - [B])^{-1}$.

Hence in Eq.(1.49) x^r will converge to the vector $\{x\}$ provided $[B]^r \to 0$ as $r \to \infty$. Then the vector $\{x\}$ can be written as

$$\{x\} = ([I] - [B]^{-1}\{v\}$$

This means that $([I] - [B])\ \{x\} = (v)$ and hence $\{x\} = [B]\{x\} + \{v\}$. This is just the equation (1.48).

1.5.8 The Householder Factorization

This method transforms the matrix into an upper triangular form by making use of reflection matrices; we have

$$[P_H]^T[A] = [S_H] \tag{1.50}$$

where $[P_H]^T = [P_{n-1}]^T \ldots [P_2]^T[P_1]^T$ (1.51)

The $[S_H]$ is an upper triangular matrix. The matrices $[P_i]^T$, i = 1...n – 1 are reflection matrices computed in such a way that $[P_i]^T$ reduce the subdiagonal elements in column i of the coefficient matrix.

Then we have

$$[P_i]T = \begin{bmatrix} [I_{i-1} & 0 \\ 0 & [P_i] \end{bmatrix} \tag{1.52}$$

here $[I_{i-1}]$ is the identity matrix of size i – 1 and

$[P_i] = 1 - \theta\{w\}\{w\}^T$;

where $\theta = \dfrac{2}{\{W_i\}^T\{W_i\}}$

Here $[P_i]$ is a symmetric matrix and $\{w_i\}$ is a vector of size $n - i + 1$. The vector $\{w_i\}$ will be chosen as explained later.

Because $[P_i]$ is symmetric, $[P_i]^T = [P_i]$.

Now solution of the equation $[A]\{x\} = \{b\}$ can be obtained by the following equation

$$[A] = [P_H][S_H]$$

where $\{v\} = [S_H]\{x\}$.

The vector $\{v\}$ is obtained as $\{v\} = [P_H]^T\{b\}$.

To explain the transformation, as a first step let us compute the $[P_1]$,

i.e., $$[\tilde{A}] = [P_1][A] \tag{1.53}$$

and $$[A] = [a_1\ ;\ A_1] \tag{1.54}$$

where a_1 is the first column of $[A]$. Then we have

$$[P_1] = I - \theta\{w_1\}\{w_1\}^T$$

Here we have the first element of the vector $[P_1][a_1]$ to be non zero since the sub-diagonal elements of column 1 of matrix $[\tilde{A}]$ is required to be zero.

Now choose the vector $\{w_1\}$ such that it must fulfill the condition

$$(I - \theta\{w_1\}\{w_1\}^T)[a_1] = \pm \|a_1\|_2\, e_1 \tag{1.55}$$

where e_1 is a nondimensional unit vector from eq. (1.55)

$$[a_1] - \theta\{w_1\} = \pm \|a_1\|_2\, e_1 \tag{1.56}$$

where θ is a constant and equal to $e\{w_1\}^T\{w_1\}$.

Let us set θ = 1.0, then we obtain

$$\{w_1\} = [a_1] + \text{sign}(a_{11})\, \|a_1\|_2\, e_1 \tag{1.57}$$

From equation (1.53) we can get

$$\{v_1\}^T = \{w_1\}^T[A] \tag{1.58}$$

hence, $$[\tilde{A}] = [A] - \theta(\{w_1\}\{v_1\}^T)$$

The factorization is explained in the example.

1.6 OPTIMIZATION

Optimization is minimizing or maximizing a function. The function $F(x)$ is called the *merit function* or *objective function*. The components of x are known as the design variables.

The minimum point must be bracketed before a minimization algorithm can be used. The bracketing procedure consists of starting with an initial value of x_0 and moving downhill computing the functions at $x_1, x_2, x_3, \ldots$ until the point x_n is reached where $f(x)$ increases for the first time. The minimum point is now bracketed in the interval (x_{n-2}, x_n). The increasing in step size follows as $h_{i+1} = c\, h_i$ where $c > 1$.

Suppose the minimum of $f(x)$ has been bracketed in the interval (a, b) of length h. To telescope the interval, the function at $x_1 = b - Rh$ and $x_2 = a + Rh$ is evaluated. If $f_1 > f_2$, then the minimum lies in (x_1, b); otherwise it is located in (a, x_2).

Next, we evaluate the function at $x_2 = a + Rh'$ and repeat the process. We noted that $x_2 - x_1 = 2\,Rh - h$ and $x_1 - a = h' - Rh'$ or $2Rh - h = h' - Rh'$ and substituting $h' = Rh$, we obtain $R = 0.618033989$. The number of telescopings required to reduce h from $|b - a|$ to an error tolerance $\in$ is given by

$$n = \frac{\ln(\in/|b-a|)}{\ln R} = -\,2.078087 \ln \frac{\in}{|b-a|} \tag{1.59}$$

1.6.1 Conjugate Gradient Methods

The objective here is to minimize $F(x)$, where the components of x are the n independent design variables. Consider the quadratic function

$$F(x) = c - \sum_i b_i x_i + \frac{1}{2}\sum_i \sum_j A_{ij} x_i x_j$$

$$= c - b^T x + \frac{1}{2} x^T A x \tag{1.60}$$

Differentiating Eq.(1.60) *w.r.t.* x_i gives

$$\frac{\partial F}{\partial x_i} = -\,b_i + \sum_j A_{ij} x_j$$

or in vector notation

$$\Delta F = -\,b + Ax \tag{1.61}$$

where ΔF is the *gradient* of F.

The gradient along u when the motion takes place along the line $x = x_0 + su$, where s is the distance moved, is given by

$$\nabla F|_{x_0 + su} = -\,b + A(x_0 + su) = \nabla F|_{x_0} + sAu$$

If the change in the gradient sAu is perpendicular to a vector V, then

$$\nabla\,^T Au = 0 \tag{1.62}$$

The directions of u and V are said to be mutually *conjugate*.

1.6.2 Newton's Method

Newton's method is a gradient method and can be conveniently used to optimize functions with several parameters. Let the function to be optimized be $U(\bar{x})$, where $\bar{x}$ is the vector of parameters $x_1, x_2, \ldots, x_n$. The function $U(\bar{x})$ can be expanded in the Taylor's series about a point $\bar{x}^*$ as

$$U(\bar{x}) = U(\bar{x}^*) + \sum_{i=1}^{n} \frac{\partial U(\bar{x}^*)}{\partial X_i}(x_i - x_i^*) \tag{1.63}$$

Newton's method uses only two terms in the series. Expressing in concise form, the above series can be written as

$$U(\bar{x}) = U(\bar{x}^*) + g^T(\bar{x}^*)(\bar{x} - \bar{x}^*) \tag{1.64}$$

where $\bar{g}(\bar{x}^*)$ is the vector of first derivatives given by

$$\bar{g}(\bar{x}^*) = \left[\frac{\partial U(\bar{x}^*)}{\partial x_1}, \frac{\partial U(\bar{x}^*)}{\partial x_2}, \ldots\right]^T \tag{1.65}$$

The minimum of $U(\bar{x})$is obtained by setting

$$\frac{\partial U}{\partial x_i} = 0 \tag{1.66}$$

in Eq. (1.64) which yields the set of equations

$$\bar{g}(\bar{x}^*) + \bar{J}(\bar{x}^*)(\bar{x} - \bar{x}^*) = 0 \tag{1.67}$$

where $\bar{J}(\bar{x}^*)$ is the Jacobian matrix of second derivatives given by

$$\bar{J}(\bar{x}^*) = \begin{vmatrix} \dfrac{\partial^2 U(\bar{x}^*)}{\partial x_1^2} & \dfrac{\partial^2 U(\bar{x}^*)}{\partial x_1 \partial x_2} & \cdots & \dfrac{\partial^2 U(\bar{x}^*)}{\partial x_1 \partial x_n} \\ \dfrac{\partial^2 U(\bar{x}^*)}{\partial x_n \partial x_1} & & \cdots & \dfrac{\partial^2 U(\bar{x}^*)}{\partial x_n^2} \end{vmatrix} \tag{1.68}$$

A set of linear algebraic equations in the unknown x_i's are given by Eq. (1.67). If $\bar{x}^*$ is taken as $\bar{x}_k$, the kth point in the step by step search for the minimum, then solution of Eq. (1.67) is given (k + 1)st approximation. Writing Eq.(1.67) in the form

$$\bar{g}(\bar{x}_k) = (\bar{J}(\bar{x}_k)(\bar{x}_{k+1} - \bar{x}_k) = 0 \tag{1.69}$$

it is possible to obtain an iterative form for the solution as

$$\bar{x}_{k+1} = \bar{x}_k - [\bar{J}(\bar{x}_k)]^{-1}\bar{g}(\bar{x}_k) \tag{1.70}$$

It is not actually necessary to obtain the inverse of the matrix $\bar{J}[\bar{x}_k]$ to solve for the new approximation $\bar{x}_k$. It is possible to write Eq. (1.69) in the form

$$\bar{J}(\bar{x}_k)\bar{\delta}_k = -\bar{g}(\bar{x}_k) \tag{1.71}$$

where $\bar{\delta}_k = \bar{x}_{k+1} - \bar{x}_k$

1.6.3 The Concept of Quadratic Convergence

Conjugate Directions for a Quadratic Function

The gradient method is expressed as

$$\bar{x}_{k+1} = \bar{x}_k - \lambda_k \bar{g}(\bar{x}_k) \tag{1.72}$$

or in the form

$$\overline{\delta}_k = -\lambda_k \overline{g}(\overline{x}_k) \tag{1.73}$$

which is similar in form to Eq. (1.62).

During each iteration λ_k is selected to minimize $U(\overline{x}_{k+1})$ in the gradient direction. A gradient search tends to zig-zig quite badly particularly for quadratic functions. If it is possible to establish the best direction to take for a quadratic function, it would likely also be better for most other non-linear functions. This is called quadratic convergence and the approach is rather surprisingly successful. Equation (1.72) has the general form, in this case, given by

$$\overline{x}_{k+1} = \overline{x}_k + \lambda_k \overline{C}(\overline{x}_k) \tag{1.74}$$

where $\overline{C}(\overline{x}_k)$ defines the conjugate direction vector at each step. The general quadratic form is

$$U(\overline{x}) = \frac{1}{2}\overline{x}^T \overline{A}\overline{x} + \overline{B}^T \overline{x} + d \tag{1.75}$$

where $\overline{A}$ is a matrix and $\overline{B}$ is a vector.

It is required to determine a means of establishing conjugate directions for an optimization function of this form, and show that they give convergence to the minimum. For this we require $\overline{A}$ to be a positive-define matrix, which means it must be such that all quadratic terms of Eq. (1.75) are positive. This is equivalent to requiring Eq. (1.75) to be convex so that it has a minimum. $\overline{A}$ will also be symmetric since

$$\frac{\partial^2 U}{\partial x_i \partial x_j} = A_{ij} = \frac{\partial^2 U}{\partial x_j \partial x_i} = A_{ji} \tag{1.76}$$

and consequently $\overline{A}^{-1}$ is also symmetric. For convenience, let

$$\overline{g}(\overline{x}_k) = \overline{g}_k \; ; \; \overline{g}(\overline{x}) = \overline{g} \; ; \overline{C}(\overline{x}_k) = \overline{C}_k$$

Note that the gradient vector of the quadratic (1.75) is

$$\overline{g} = \overline{A}\overline{x} + \overline{B} \tag{1.77}$$

Equation (1.74) is iterated for successive steps from i to $(n-1)$, to obtain

$$\overline{x}_n = \overline{x}_i + \sum_{k=1}^{n-1} \lambda_k \overline{C}_k \tag{1.78}$$

Premultiplying both sides by $\overline{A}$ and adding $\overline{B}$, we get

$$\overline{A}\overline{x}_n = \overline{B} = \overline{A}\overline{x}_i + B + \sum_{k=i}^{n-1} \lambda_k \overline{A}\overline{C}_k \tag{1.79}$$

Using Eq. (1.77), this becomes

$$\overline{g}_n = \overline{g}_i + \sum_{k=i}^{n-1} \lambda_k \overline{A}\overline{C}_k \tag{1.80}$$

Finally premultiplying by $\overline{C}_{i-1}^T$,

$$\overline{C}_{i-1}^T \overline{g}_n = \overline{C}_{i-1}^T \overline{g}_1 + \sum_{k=1}^{n-1} \lambda_k \overline{C}_{i-1}^T \overline{A}\overline{C}_k \tag{1.81}$$

It can be demonstrated intuitively that $\overline{C}_{i-1}^T \overline{g}_i$ is zero by the following reasoning. In the $i-1$ step, $\overline{C}_{i-1}$ is followed to a minimum for U, with takes us to $\overline{x}_i$ and therefore the gradient $\overline{g}_i$ is normal to $\overline{C}_{i-1}$. Consequently,

$$\overline{C}_{i-1}^T \overline{g}_i = 0 \tag{1.82}$$

This can be demonstrated more rigorously as follows. Consider λ_{i-1} as a variable that is being adjusted to minimize U and bring the search to $\overline{x}_i$. Consequently,

$$\frac{dU}{d\lambda_{j-1}} = \frac{dU}{dx_1}\frac{dx_1}{d\lambda_{i-1}} + \frac{\partial U}{\partial x_2}\frac{dx_2}{d\lambda_{i-1}} + \ldots + \frac{\partial U}{\partial x_n}\frac{dx_n}{d\lambda_{i-1}} \tag{1.83}$$

Referring to Fig. 1.60, the search vector $\overline{C}_{i-1}$ at the origin has components $\overline{C}_{1,i-1}$, ..., $C_{n,i-1}$, and the search moves along vector $\lambda_{i-1}\overline{C}_{i-1}$ from x_{i-1} to x_1. In general, then, for any value of λ_{i-1},

$$\overline{x} = \overline{x}_{i-1} + \lambda_{i-1}\overline{C}_{i-1} \tag{1.84}$$

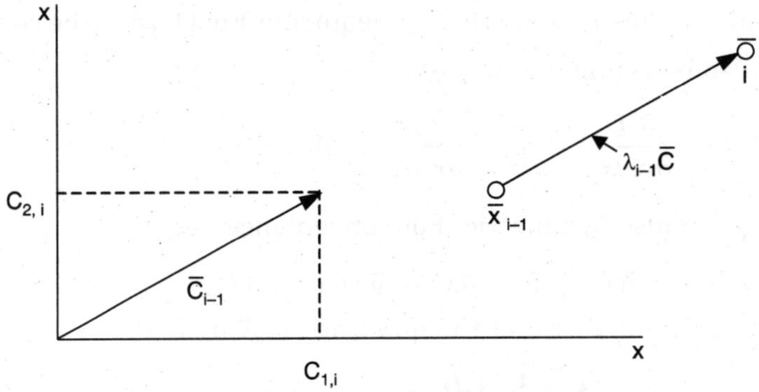

Fig. 1.60 The (*i* – 1)th search step.

Taking the derivative with respect to λ_{i-1} results in

$$\frac{d\overline{x}}{d\lambda_{i-1}} = \frac{d\overline{x}_{i-1}}{d\lambda_{i-1}} + \overline{C}_{i-1} \tag{1.85}$$

Since $\overline{x}_{i-1}$ is constant at this state in the search, we have

$$\overline{C}_{i-1} = \frac{d\overline{x}}{d\lambda_{i-1}} \tag{1.86}$$

Thus Eq. (1.83) can be written for any value of λ_{i-1}, as

$$\frac{dU}{d\lambda_{i-1}} = \overline{C}_{i-1}^T \overline{g} \tag{1.87}$$

Now at $\bar{x}_1$, $dU/d\lambda_{i-1}$ must be zero for a minimum U. Thus, we have

$$\bar{C}_{i-1}^T \bar{g} = 0 \tag{1.88}$$

Consequently, Eq.(1.81) reduces to

$$\bar{C}_{i-1}^T \bar{g}_n = \sum_{k=i}^{n-1} \lambda_k \bar{C}_{i-1}^T \bar{A} \bar{C}_k \tag{1.89}$$

The conjugate vectors are defined as those satisfying

$$\bar{C}_i^T \bar{A} \bar{C}_j = 0 \tag{1.90}$$

For $i \neq j$. Since $\bar{A}$ must be a positive-define matrix as defined above, the summation term of Eq. (1.89) is zero so that

$$\bar{C}_{i-1}^T \bar{g}_n = 0 \tag{1.91}$$

The theory of n-dimensional vectors states that if we construct a set of n-vectors all orthogonal or conjugate to each other, then any other vector can be written as a linear combination of these vectors. Therefore, no other vector can be orthogonal to all of the original n vectors other than the zero vector. Since Eq. (1.91) is an expression of orthogonality of the nth gradient vector with all n conjugate vectors, then $\bar{g}_n$ must be zero, which is the condition for the minimum of the quadratic. Thus the minimum of the quadratic can be found in the n steps if the search directions are conjugate.

The conjugate directions in two dimensions are shown in Fig. 1.61. The first search direction is the gradient vector, which is chosen arbitrarily. The second direction is conjugate to the first one. Several methods have been proposed for generating the first search direction and these methods are applied to nonquadratic functions also. It should be noted that Newton's method would go to the minimum in one step if the function to be minimized is quadratic.

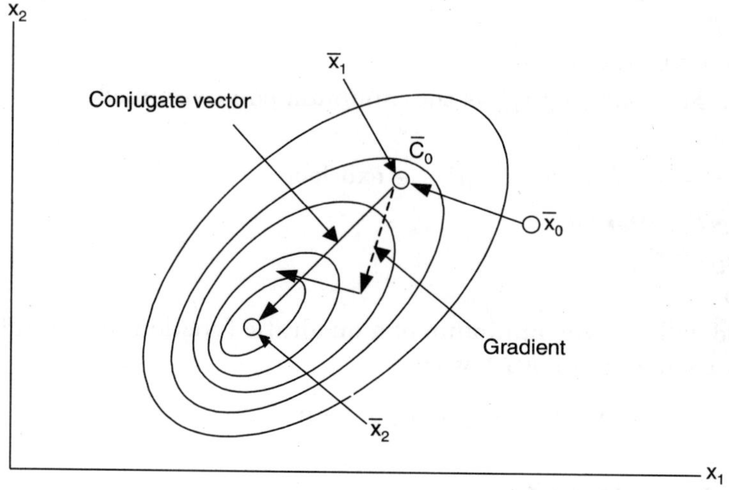

Fig. 1.61 Conjugate directions.

1.6.4 Powell's Method

For an optimization problem involving n design variables, the basic algorithm is described below:

(*a*) select a point x_0 in the design space

(*b*) select the starting vectors V_i, i = 1, 2, ..., n

(*c*) do with $i = 1, 2, \ldots, n$

minimize $F(x)$ along the line thro x_{i-1} in the direction of V_i.

assume the minimum point as x_i

end do

(*d*) $V_{n+1} \leftarrow x_0 - x_n$

minimize $F(x)$ along the line thro x_0 in the direction of V_{n+1}

assume the minimum point as x_{n+1}

if $|x_{n+1} - x_0| < \in$ exit loop

do with $i = 1, 2, \ldots, n$

$V_i \leftarrow V_{i+1}$

end do

end cycle

The minimum point of a quadratic surface is reached in n cycles.

1.6.5 Fletcher-Reeves Method

The algorithm is described below:

(*a*) select a starting point x_0

(*b*) $g_0 \leftarrow -\nabla F(x_0)$

(*c*) $V_0 \leftarrow g_0$

(*d*) loop with $i = 0, 1, 2, \ldots$

minimize $F(x)$ along V_i; x_{i+1} is the minimum point assumed

$g_{i+1} \leftarrow -\nabla F(x_{i+1})$

if $|g_{i+1}| < \in$ or $|F(x_{i+1}) - F(x_i)| < \in$ exit loop

$r \leftarrow (g_{i+1} \bullet g_{i+1})/(g_i \bullet g_i)$

$V_{i+1} \leftarrow g_{i+1} + rV_i$

end loop

The method will find the minimum of a quadratic function in n iterations. Also, V_i and V_{i+1} are mutually conjugate. In other words,

$$V_i^T A V_{i+1} = 0 \text{ and } g_i g_{i+1} = 0.$$

1.6.6 Hooke and Jeeves Method

Hooke and Jeeves method is a sequential technique each step of which consists of two kinds of moves, the exploratory move and the pattern move.

General procedure:

1. Start with an arbitrarily chosen point $X_1 = \begin{Bmatrix} x_1 \\ x_2 \\ \vdots \\ x_n \end{Bmatrix}$, called the starting base point and prescribed step lengths Δx_i in each of the coordinate directions u_i, $i = 1, 2, \ldots, n$. Set $k = 1$.
2. Compute $f_k = f(\mathbf{X}_k)$. Set $i = 1$, $\mathbf{Y}_{k0} = \mathbf{X}_k$, and start the exploratory move as stated in step 3.
3. The variable x_i is perturbed about the current temporary base point $\mathbf{Y}_{k,i-1}$ to obtain the new temporary base point as

$$\mathbf{Y}_{k,i} = \begin{cases} \mathbf{Y}_{k,i-1} + \Delta x_i \mathbf{u}_i & \text{if} \quad f^+ = f(\mathbf{Y}_{k,i-1} + \Delta x_i \mathbf{u}_i) \\ & \qquad < f = f(\mathbf{Y}_{k,i-1}) \\ \mathbf{Y}_{k,i-1} - \Delta x_i \mathbf{u}_i & \text{if} \quad f^- = f(\mathbf{Y}_{k,i-1} - \Delta x_i \mathbf{u}_i) \\ & \qquad < f = f(\mathbf{Y}_{k,i-1}) \\ & \text{if} \quad < f^+ = f(\mathbf{Y}_{k,i-1} + \Delta x_i \mathbf{u}_i) \\ \mathbf{Y}_{k,i-1} & \text{if} \quad f = f(\mathbf{Y}_{k,i-1}) < \min(f^+, f^-) \end{cases} \tag{1.92}$$

   This process is continued for $i = 1, 2, \ldots$, until x_n is perturbed to find $\mathbf{Y}_{k,n}$.
4. If the point $\mathbf{Y}_{k,n}$ remains same as $\mathbf{X}_k$, reduce the step lengths Δx_i (say, by a factor of 2), set $i = 1$ and go to step 3. If $\mathbf{Y}_{k,n}$ is different from $\mathbf{X}_k$, obtain the new base point as

$$\mathbf{X}_{k+1} = \mathbf{Y}_{k,n}$$

   and go to step 5.
5. With the help of the base points $\mathbf{X}_k$ and $\mathbf{X}_{k+1}$, establish a pattern direction $\mathbf{S}$ as

$$\mathbf{S} = \mathbf{X}_{k+1} - \mathbf{X}_k \tag{1.93}$$

   and find a point $\mathbf{Y}_{k+1,0}$ as

$$\mathbf{Y}_{k+1,0} = \mathbf{X}_{k+1} + \lambda \mathbf{S} \tag{1.94}$$

   where λ is the step length, which can be taken as 1 for simplicity. Alternatively, we can solve a one-dimensional minimization problem in the direction $\mathbf{S}$ and use the optimum step length λ^* in place of λ in Eq. (1.94).
6. Set $k = k + 1$, $f_k = f(\mathbf{Y}_{k0})$, $i = 1$, and repeat step 3. If at the end of step 3, $f(\mathbf{Y}_{k,n}) < f(\mathbf{X}_k)$, we take the new base point as $\mathbf{X}_{k+1} = \mathbf{Y}_{k,n}$ and go to step 5. On the other hand, if $f(\mathbf{Y}_{k,n}) \geq f(\mathbf{X}_k)$, set $\mathbf{X}_{k+1} \equiv \mathbf{Y}_k$, reduce the step lengths Δx_i, set $k = k + 1$, and go to step 2.
7. The process is assumed to have converged whenever the step lengths fall below a small quantity ε. Thus the process is terminated if

$$\max(\Delta x_i) < \varepsilon \tag{1.95}$$

1.6.7 Interior Penalty Function Method

In the interior penalty function methods, a new function (ϕ function) is constructed by augmenting a penalty term to the objective function. The penalty term is chosen such that its value will be small at points away from the constraint boundaries and will tend to infinity as the constraint boundaries are approached. Thus once the unconstrained minimization of $\phi(\mathbf{X}, r_k)$ is started from any feasible point $\mathbf{X}_1$, the subsequent points generated will always lie within the feasible domain since the constraint boundaries act as barriers during the minimization process. The ϕ function is defined as

$$\phi(\mathbf{X},r_k) = f(\mathbf{X}) - r_k \sum_{j=1}^{m} \frac{1}{g_j(\mathbf{X})} \tag{1.96}$$

Since the above equation does not allow any constraint to be violated, it requires a feasible starting point for the search toward the optimum point. The iteration procedure of this method can be summarized as follows:

Iterative process:

1. Start with an initial feasible point $\mathbf{X}_1$ satisfying all the constraints with strict inequality sign, that is, $g_j(\mathbf{X}_1) < 0$ for $j = 1, 2, \ldots, m$ and an initial value for $r_1 > 0$. Set $k = 1$.
2. Minimize $\phi(\mathbf{X}, r_k)$ by using any of the unconstrained minimization methods and obtain the solution $\mathbf{X}^*_k$.
3. Test whether $\mathbf{X}^*_k$ is the optimum solution of the original problem. If $\mathbf{X}^*_k$ is found to be optimum, terminate the process or else, go to the next step.
4. Find the value of the next penalty parameter, r_{k+1}, as
$$r_{k+1} = cr_k$$
where $c < 1$.
5. Set the new value of $k = k + 1$, take the new starting point as $\mathbf{X}_1 = \mathbf{X}^*_k$, and go to step 2.

All these aspects are discussed in the following paragraphs.

Starting feasible point X_1:

1. Select an arbitrary point $\mathbf{X}_1$ and evaluate the constraints $g_j(\mathbf{X})$ at the point $\mathbf{X}_1$. Since the point $\mathbf{X}_1$ is arbitrary, it may not satisfy all the constraints with strict inequality sign. If r out of a total of m constraints are violated, renumber the constraints such that the last r constraints will become the violated ones, that is,
$$g_j(\mathbf{X}_1) < 0, \quad j = 1, 2, \ldots, m - r$$
$$g_j(\mathbf{X}_1) \geq 0, \quad j = m - r + 1, m - r + 2, \ldots, m \tag{1.97}$$
2. Identify the constraint that is violated most at the point $\mathbf{X}_1$, that is obtain the integer k such that
$$g_k(\mathbf{X}_1) = \max[g_j(\mathbf{X}_1)] \quad \text{for } j = m - r + 1, m - r + 2, \ldots, m \tag{1.98}$$
3. Formulate a new optimization problem as
Find $\mathbf{X}$ which minimizes $g_k(\mathbf{X})$
subject to
$$g_j(\mathbf{X}) \leq 0, \qquad j = 1, 2, \ldots, m - r$$
$$g_j(\mathbf{X}) - gk\,(\mathbf{X}_1) \leq 0, \quad j = m - r + 1, m - r + 2, \ldots, k - 1, k + 1, \ldots, m \tag{1.99}$$
4. Solve the optimization problem formulated in step 3 by taking the point $\mathbf{X}_1$ as a feasible starting point using the interior penalty function method. Note that this optimization method can be terminated whenever the value of the objective function $g_k(\mathbf{X})$ drops below zero. The solution obtained $\mathbf{X}_M$ will satisfy at least one more constraint than did the original point $\mathbf{X}_1$.
5. If all the constraints are not satisfied at the point $\mathbf{X}_M$, set the new starting point as $\mathbf{X}_1 = \mathbf{X}_M$, and renumber the constraints such that the last r constraints will be the unsatisfied ones (this value of r will be different from the previous value), and go to step 2.

This procedure is repeated until all the constraints are satisfied and a point $\mathbf{X}_1 = \mathbf{X}_M$ is obtained for which $g_j(\mathbf{X}_1) < 0$, $j = 1, 2, \ldots, m$.

Initial value of the penalty parameter (r_1): Since the unconstrained minimization of $\phi(\mathbf{X}, r_k)$ is to be carried out for a decreasing sequence of r_k, it might appear that by choosing a very small value of r_1, we can avoid an excessive number of minimizations of the function ϕ. But from a computational point of view, it will be easier to minimize the unconstrained function $\phi(\mathbf{X}, r_k)$ if r_k is large. A moderate value has to be chosen for the initial penalty parameter (r_1). In practice, a value of r_1 that gives the value of $\phi(\mathbf{X}_1, r_1)$ approximately equal to 1.1 to 2.0 times the value of $f(\mathbf{X}_1)$ has been found to be quite satisfactory in achieving quick convergence of the process. Hence, for any initial feasible starting point $\mathbf{X}_1$, the value of r_1 can be taken as

$$r_1 < 0.1 \text{ to } 1.0 \frac{f(\mathbf{X}_1)}{-\sum_{j=1}^{m} 1/g_j(\mathbf{X}_1)} \tag{1.100}$$

Subsequent values of the penalty parameter: Once the initial value of r_k is chosen, the subsequent values of r_k is selected so that

$$r_{k+1} < r_k \tag{1.101}$$

The values of r_k are chosen such that

$$r_{k+1} < c r_k \tag{1.102}$$

where $c < 1$. The value of c can be taken as 0.1, 0.2, or 0.5.

Convergence criteria: The process will be terminated whenever the following conditions are satisfied.

The relative difference between the values of the objective function obtained at the end of any two consecutive unconstrained minimizations falls below a small number ε_1, that is,

$$\left| \frac{f(\mathbf{X}_k^*) - f(\mathbf{X}_{k-1}^*)}{f(\mathbf{X}_k^*)} \right| \le \varepsilon_1 \tag{1.103}$$

The difference between the optimum points $\mathbf{X}_k^*$ and $\mathbf{X}_{k-1}^*$ becomes very small. This can be judged in several ways. Some of them are given below:

$$|(\Delta\mathbf{X})_i| \le \varepsilon_2 \tag{1.104}$$

where $\Delta\mathbf{X} = \mathbf{X}_k^* - \mathbf{X}_{k-1}^*$, and $(\Delta\mathbf{X})_i$ is the ith component of the vector $\Delta\mathbf{X}$.

$$\max|(\Delta\mathbf{X})_i| \le \varepsilon_3 \tag{1.105}$$

$$|\Delta\mathbf{X}| = [(\Delta\mathbf{X})_1^2 + (\Delta\mathbf{X})_2^2 + \ldots + (\Delta\mathbf{X})_n^2]^{1/2} \le \varepsilon_4 \tag{1.106}$$

It is advisable to normalize the constraints so that they vary between –1 and 0 as far as possible.

If the constraints are not normalized, the problem can still be solved effectively by defining different penalty parameters for different constraints as

$$\phi(\mathbf{X}, r_k) = f(\mathbf{X}) - r_k \sum_{j=1}^{m} \frac{R_j}{g_j(\mathbf{X})} \tag{1.107}$$

where $R_1, R_2, \ldots, R_m$ are selected such that the contributions of different $g_j(\mathbf{X})$ to the ϕ function will be approximately the same at the initial point $\mathbf{X}_1$.

1.7 SYMMETRIC MATRIX EIGENVALUE PROBLEMS

The standard matrix eigenvalue problem is

$$Ax = \lambda x \tag{1.108}$$

where A is a given $n \times n$ matrix. The objective here is to find the scalar λ and the vector x.

Equation (1.108) can be rewritten as

$$(A - \lambda I)\, x = 0 \tag{1.109}$$

A nontrivial solution exists only if

$$|A - \lambda I| = 0 \tag{1.110}$$

Expansion of Eq. (1.46) gives the polynomial equation called the *characteristic equation*.

$$a_1\lambda^n + a_2\lambda^{n-1} + \ldots + a_n\lambda + a_{n+1} = 0$$

which has the roots λ_i, $i = 1, 2, \ldots, n$, called the *eigenvalues* of the matrix A. The solutions of x_i of $(A - \lambda_i I)x = 0$ are known as *eigenvectors*.

1.7.1 Jacobi Method

Applying the transformation $x = Px^*$ in Eq.(1.44) where P is a nonsingular matrix, we can write

$$P^{-1}APx^* = \lambda P^{-1}Px^*$$

or

$$A^*x^* = \lambda x^* \tag{1.111}$$

where $A^* = P^{-1}AP$.

Matrices that have the same eigenvalues are deemed to be *similar* and the transformation between them is called a *similarity transformation*. Diagonalizing A^*, Eq.(1.111) can be written as

$$\begin{bmatrix} A_{11}^* - \lambda & 0 & \ldots & 0 \\ 0 & A_{22}^* - \lambda & \ldots & 0 \\ \vdots & \vdots & \vdots & \vdots \\ 0 & 0 & \ldots & A_{nn}^* - \lambda \end{bmatrix} \begin{bmatrix} x_1^* \\ x_2^* \\ \vdots \\ x_n^* \end{bmatrix} \begin{bmatrix} 0 \\ 0 \\ \vdots \\ 0 \end{bmatrix} \tag{1.112}$$

Solving Eq.(1.112), we obtain

$$\lambda_1 = A_{11}^*, \lambda_2 = A_{22}^*, \ldots, \lambda_n = A_{nn}^* \tag{1.113}$$

$$x_1^* = \begin{bmatrix} 1 \\ 0 \\ \vdots \\ 0 \end{bmatrix}, \quad x_2^* = \begin{bmatrix} 0 \\ 1 \\ \vdots \\ 0 \end{bmatrix}, \ldots, \quad x_n^* = \begin{bmatrix} 0 \\ 0 \\ \vdots \\ 1 \end{bmatrix}$$

or

$$x^* = [x_1^*\; x_2^* \ldots x_n^*] = I$$

Therefore, the eigenvector matrix of A is

$$X = PX^* = PI = P$$

The transformation matrix P is the eigenvector matrix of A and the eigenvalues of A are the diagonal terms of A^*.

Jacobi rotation matrix:

Consider the special transformation in the plane rotation

$$x = Rx^*$$

where

$$R = \begin{bmatrix} & & k & & & l & & \\ 1 & 0 & 0 & 0 & 0 & 0 & 0 & 0 \\ 0 & 1 & 0 & 0 & 0 & 0 & 0 & 0 \\ 0 & 0 & c & 0 & 0 & s & 0 & 0 \\ 0 & 0 & 0 & 1 & 0 & 0 & 0 & 0 \\ 0 & 0 & 0 & 0 & 1 & 0 & 0 & 0 \\ 0 & 0 & -s & 0 & 0 & c & 0 & 0 \\ 0 & 0 & 0 & 0 & 0 & 0 & 1 & 0 \\ 0 & 0 & 0 & 0 & 0 & 0 & 0 & 1 \end{bmatrix} \begin{matrix} \\ k \\ \\ \\ \\ \\ l \\ \\ \\ \end{matrix} \tag{1.114}$$

R is called the *Jacobi rotation matrix* and $R^{-1} = R^T$.

Also $$A^* = R^{-1}AR = R^TAR \tag{1.115}$$

The matrix A^* has the same eigenvalues as the original matrix A and it is also symmetric.

Jacobi diagonalization:

The Jacobi diagonalization procedure uses only the upper half of the matrix and is summarized below:

(*a*) Obtain the largest absolute value off-diagonal element A_{kl} in the upper half of A.

(*b*) Compute ϕ, t, c and s

$$\phi = \frac{A_{kk} - A_{ll}}{2A_{kl}}$$

$$t = \frac{1}{2\phi}$$

$$c = \frac{1}{\sqrt{1+t^2}}$$

$$s = tc$$

(*c*) Compute τ from $\tau = \dfrac{s}{1+c}$

(*d*) Modify the elements in the upper half of A:

$$A^*_{kk} = A_{kk} - t\,A_{kl}$$

$$A^*_{ll} = A_{ll} - t\,A_{kl}$$

$$A^*_{kl} = A^*_{lk} = 0$$

$$A^*_{ki} = A^*_{ik} = A_{ki} - s(A_{li} + \tau A_{ki}), \;\; i \neq k, i \neq l$$

$$A^*_{li} = A^*_{il} = A_{li} + s(A_{ki} - \tau A_{ki}),\ i \neq k, i \neq l$$

where $\tau = \dfrac{s}{1+c}$

(*e*) Update the matrix P

$$P^*_{ik} = P_{ik} - s(P_{il} + \tau P_{ik})$$

$$P^*_{il} = P_{il} + s(P_{ik} - \tau P_{il})$$

Repeat the procedure until the $A_{kl} < \in$, where $\in$ is the error tolerance.

1.7.2 Householder Reduction to Tridiagonal Form

The computational procedure is carried out with $i = 1, 2, \ldots, n - 2$ as described below:

(*a*) Define A′ as the $(n - i) \times (n - i)$ lower-right hand portion of A

(*b*) Let $x = \begin{bmatrix} A_{i+1,i} & A_{i+2,i} & \cdots & A_{n,i} \end{bmatrix}^T$

(*c*) Compute $|x|$. Let $k = |x|$ if $x_1 > 0$ and $k = -|x|$ if $x_1 < 0$

(*d*) Let $u = [k + x_1 \quad x_2 \quad x_3 \quad \ldots \quad x_{n-1}]^T$

(*e*) Compute $H = |u|^2 / 2$

(*f*) Compute $V = A'u/H$

(*g*) Compute $g = u^T v/2H$

(*h*) Compute $w = V - gu$

(*i*) Compute the transformation $A \leftarrow A' - w^T u - u^T w$

(*j*) Set $A_{i,\,i+1} = A_{i+1,i} = -k$

1.7.3 Sturn Sequence

Consider a symmetric tridiagonal matrix. Its characteristic polynomial can be computed with $3(n - 1)$ multiplications as described below:

$$P_n(\lambda) = |A - \lambda I| = \begin{bmatrix} d_1 - \lambda & c_1 & 0 & 0 & \cdots & 0 \\ c_1 & d_2 - \lambda & c_2 & 0 & \cdots & 0 \\ 0 & c_2 & d_3 - \lambda & c_3 & \cdots & 0 \\ 0 & 0 & c_3 & d_4 - \lambda & \cdots & 0 \\ \vdots & \vdots & \vdots & \vdots & \vdots & \vdots \\ 0 & 0 & \cdots & 0 & c_{n-1} & d_n - \lambda \end{bmatrix}$$

$P_0(\lambda) = 1$

$P_1(\lambda) = d_1 - \lambda$

$P_i(\lambda) = (d_i - \lambda)P_{i-1}(\lambda) - c_{i-1}^2 P_{i-2}(\lambda),\ i = 2, 3, \ldots, n$

The polynomials $P_0(\lambda), P_1(\lambda), \ldots, P_n(\lambda)$ form a *sturn sequence*. The number of sign changes in the sequence $P_0(a), P_1(a), \ldots, P_n(a)$ is equal to the number of roots of $P_n(\lambda)$ that are smaller than a. If a member $P_i(a)$ of the sequence is zero, its sign is to be taken opposite to that of $P_{i-1}(a)$.

1.7.4 QR Method

In the preceding section we have described the Jacob's method which is a reliable method but it requires a large computation time and is valid only for real symmetric matrices. The *QR* method on the other hand is numerically extremely stable and is applicable to a general matrix. This method also provides a basis for developing a general purpose procedure for determining the eigenvalues and eigenvectors.

The method utilizes the fundamental property that a real matrix can be written as

$$[A] = [Q][U]$$

where $[Q]$ is orthogonal and $[U]$ is upper triangular. This is contrary to the decomposition in the form $[L][U]$ where $[L]$ is the unit lower matrix and $[U]$ is the upper triangular matrix. The property can be proved as follows.

As in Jacobi's method we introduce here the rotations matrix and denote it by $[R.(p, q, \theta)]$ to indicate the rows that contain the non-zero, off-diagonal elements. Note that $[R]$ is orthogonal.

$$[R((p,q,\theta)] = \begin{bmatrix} 1 & & & & & & & \\ & 1 & & & & & & \\ & & \cos\theta & & & \sin\theta & & \\ & & & 1 & & & & \\ & & & & 1 & & & \\ & & -\sin\theta & & & \cos\theta & & \\ & & & & & & 1 & \\ & & p & & & q & & 1 \end{bmatrix} \begin{matrix} \\ \\ p \\ \\ \\ q \\ \\ \\ \end{matrix}$$

If any vector $\{x\}$ is multiplied by $[R.(p, q, \theta)]$ such that $y = [R]\{x\}$ then

$$y_p = r_{pp}x_p + r_{pq}x_q = \cos\theta x_p + \sin\theta x_q$$

and

$$\mathbf{Y}_q = r_{qp}x_p + r_{qq}x_q = -\sin\theta x_p + \cos\theta x_q$$

i.e., $y_i = x_i$ if $i \neq p$ or q. Therefore it is necessary to select θ such that y_p or y_q is equal to zero. It should also be noted that if $x_p = x_q = 0$ then $y_p = y_q = 0$. If we write

$$[A] = (x^{(1)}x^{(2)}\ldots x^{(n)})$$

then

$$[R\ (p,\ q,\ \theta)][A] = (y^{(1)}y^{(2)}\ldots y^{(n)})$$

An element in either p^{th} row or q^{th} row is reduced by a proper choice of θ. Thus all the elements in the first column except the first element can be made zero by suitable values of θ and they remain zero even after multiplying with $[R]$. Similarly for all other columns elements are made zero for the elements below the diagonal elements by proper choice of θ values. Thus we obtain an upper triangular matrix by forming the product

$$[u] = R\ (n,n-1,\ \theta_{n,n-1})R(n,n-2,\ \theta_{n,n-2})\ \ldots R(3,1,\ \theta_{3,1})R(2,1,\theta_{2,1})[A]$$

$$= \left(\prod_{j=1}^{n-1}\prod_{i=j+1}^{n} R\,(i,\ j,\ \theta_{ij})\right)[A]$$

Since $[R]$ is an orthogonal matrix, the products of $[R]$ is also orthogonal. Hence we have $[U] = [S][A]$ where $[S]$ is orthogonal an $d[U]$ is upper triangular. Since $[S]$ is orthogonal we have $[A] = [S]^{-1}[U] = [S]^T[U]$. If $[Q] = [S]^T$ then $[A] = [Q][U]$.

We can construct a sequence of matrices $[A_0]$, $[A_1][A_2]\ldots$ where

$$[A_0] = [A] = [Q_0][A_0],\ [Q_k][A_k] = [Q_{k-1}][A_{k-1}]$$

for all values of $k \geq 1$. That is, we start with $[A_0] = [A]$ and put it in the form $[Q_0][U_0]$ to obtain $[Q_0]$ and $[U_0]$ and then obtain $[A_1]$ which is equal to the product of $[U_0][Q_0]$ in reverse order. Repeating this procedure we can obtain any number of sequence.

$[A] = [A_0] = [Q_0][U_0]$	determines $[Q_0]$ and $[U_0]$
$[A_1] = [U_0][Q_0]$	determines $[A_1]$
$= [Q_1][U_1]$	determines $[Q_1]$ and $[U_1]$
$[A_2] = [U_1][Q_1]$	determines $[A_2]$
$= [Q_2][U_2]$	determines $[Q_2]$ and $[U_2]$
$[A_k] = [U_{k-1}][Q_{k-1}]$	determines $[A_k]$
$= [Q_k][U_k]$	determines $[Q_k]$ and $[U_k]$

From the above sequence we can show that the product $Q_0Q_1Q_2...Q_k$ converges as $k \to \infty$. Then $[A_k]$ converges to an upper triangular matrix with the eigenvalues of $[A]$ as its diagonal elements. This can be proved as follows. We have

$$[A_k] = [Q_k][U_k] = [U_{k-1}][Q_{k-1}] \tag{1.116}$$

$$[A_{k-1}] = [Q_{k-1}][U_{k-1}] \tag{1.117}$$

From which

$$[U_{k-1}] = [Q_{k-1}]^{-1}[A_{k-1}] \tag{1.118}$$

Substituting Eq. (1.54) in Eq. (1.52) we get

$$[A_k] = [Q_{k-1}]^{-1}\,[A_{k-1}][Q_{k-1}]$$

Hence $[A_k]$ is similar to $[A_{k-1}]$. This implies that $[A_k]$ is similar to $[A_0] = [A]$. Therefore it has the same eigenvalues as $[A]$. Also

$$[A_{k-1}] = [Q_{k-2}]^{-1}\,[A_{k-2}][Q_{k-2}]$$

Thus $[A_{k+1}] = [Q_k]^{-1}\,[Q_{k-1}]^{-1}\,[Q_0]^{-1}\,[A][Q_0][Q_1]...[Q_k] = [P_k]^{-1}\,[A][P_k]$

where $[P_k] = [Q_0][Q_1]...[Q_k]$

If $k \to \infty [P_k]$ exists and we denote it by $[P]$,

then
$$\lim_{k\to\infty} Q_k = \lim_{k\to\infty}\left([P_{k-1}]^{-1}[P_k]\right)$$

$$= \left(\lim_{k\to\infty}[P_{k-1}]^{-1}\right)\left(\lim_{k\to\infty}[P_k]\right) = [P]^{-1}\,[P] = [I]$$

Here we have two limiting conditions

1. $$[A_k] = [P_{k-1}]^{-1}\,[A][P_{k-1}]$$

which means

$$\lim_{k\to\infty}[A_k] = \left(\lim_{k\to\infty}[P_{k-1}]^{-1}\right)[A]\left(\lim_{k\to\infty}[P_{k-1}]\right) = [P]^{-1}[A][P]$$

Therefore in the limit $[A_k]$ is similar to $[A]$ and hence has the same eigenvalues as $[A]$.

2. $$[A_k] = [Q_k][U_k]$$

that is
$$\lim_{k\to\infty}[A_k] = \left(\lim_{k\to\infty}[Q_k]\right)\left(\lim_{k\to\infty}[U_k]\right)$$

Here we have

$$\lim_{k\to\infty}\;[Q_k] = [I]$$

which results in

$$\lim_{k \to \infty} [A_k] = \lim_{k \to \infty} [U_k]$$

Since every $[U_k]$ is an upper triangular matrix, the limit of $[A_k]$ is also an upper triangular matrix.

The accuracy of this method mainly depends on the effectiveness of the algorithm used for decompositions of $[A_k]$ into $[Q_k][U_k]$. The limit $k \to \infty$ can exist for large size problems.

REFERENCES

Bathe, K.J., and Wilson, E.L., *Numerical Methods in Finite Element Analysis,* Prentice Hall, Englewood Cliffs, NJ, 1976.

Beards, C.F., *Vibrations and Control System,* Ellis Horwood, 1988.

Bedford, A., and Fowler, W., *Engineering Mechanics: Statics & Dynamics,* 4th ed, Prentice Hall, New Jersey, 2005.

Beer, F.B., Russell Johnston, E., and DeWolf, J.T., *Mechanics of Materials,* 3rd ed, McGraw Hill, New York, 2001.

Beer, F.B., Russell Johnston, E., and Eisenberg, E.R., *Vector Mechanics for Engineers-Statics and Dynamics,* 8th edn, McGraw Hill, New York, 2007.

Bhat, R.B., and Dukkipati, R.V., *Advanced Dynamics,* Narosa Publishing House, New Delhi, India, 2001.

Bhat, R.B., and Chakraverty, S., *Numerical Analysis in Engineering,* Narosa Publishing House, New Delhi, India, 2004.

Boresi, A., and Schmidt, R.J., Engineering *Mechanics: Statics,* 2nd edn, Brooks/Cole Publishing Company, New York, 2001.

Cernica, J.N., *Strength of Materials,* Holt, Rinehart and Winston, Inc, New York, 1966.

Chesmond, C.J., *Basic Control System Technology,* Edward Arnold, 1990.

Cogdell, J.R., *Foundations of Electrical Circuits,* Prentice Hall, Englewood Cliffs, NJ, 1999.

Dorf, R.C., Svoboda, J.A., *Introduction to Electric Circuits,* Wiley, New York, NY, 2006.

Dorf, R.C., and Bishop, R.H., *Modern Control Systems,* 9th edn, Prentice Hall, Upper Saddle River, NJ, 2001.

Dorsey, John., *Continuous and Discrete Control Systems,* McGraw Hill, New York, NY, 2002.

Dukkipati, R.V., Ananda Rao, M., and Bhat, R.B., *Computer Aided Analysis and Design of Machine Elements,* New Age International (P) Ltd, New Delhi, India, 2nd edition, 2006.

Dukkipati, R.V., and Amyot, J.R., *Computer Aided Simulation in Railway Dynamics,* Marcel Dekker Inc, New York, 1988.

Dukkipati, R.V., and Ray, P.K., *Product and Process Design for Quality, Economy and Reliability,* Narosa Publishing House, New Delhi, India, 2006.

Dukkipati, R.V., and Ray, P.K., *Quality Control, Improvement and Management,* Narosa Publishing House, New Delhi, India, 2007.

Dukkipati, R.V., and Srinivas, J., *An Introduction to Robotics*, McGraw Hill, New Delhi, India, 2007.

Dukkipati, R.V., and Srinivas, J., *Vibrations: Problem Solving Companion*, Narosa Publishing House, New Delhi, India, ISBN# 81-7319-526-9, India, 2006.

Dukkipati, R.V., Pang, J., Qatu, M., Sheng, G., and Zuo, S., *Problems and Solutions in Road Vehicle Dynamics,* SAE, Warrandale, PA, 2007.

Dukkipati, R.V., Pang, J., Qatu, M., Sheng, G., and Zuo, S., *Road Vehicle Dynamics,* SAE, Warrandale, PA, 2006.

Dukkipati, R.V., and Srinivas, J., *Applied Numerical Methods*, Narosa Publishing House, New Delhi, India, 2008.

Dukkipati, R.V., *Advanced Engineering Analysis*, Narosa Publishing House, New Delhi, India, 2006.

Dukkipati, R.V., Advanced *Mechanical Vibrations*, Narosa Publishing House, New Delhi, India, 2006.

Dukkipati, R.V., *Analysis and Design of Control Systems using MATLAB*, New Age International (P) Ltd, New Delhi, India, 2006.

Dukkipati, R.V., *Control Systems*, Narosa Publishing House, New Delhi, India, 2005.

Dukkipati, R.V., *Engineering System Dynamics*, Narosa Publishing House, New Delhi, India, 2004.

Dukkipati, R.V., *MATLAB for Electrical Engineers*, New Age International (P) Ltd, New Delhi, India, 2008.

Dukkipati, R.V., *Solving Engineering Mechanics Problems with MATLAB*, New Age International (P) Ltd, New Delhi, India, 2007.

Dukkipati, R.V., *Solving Engineering System Dynamics Problems with MATLAB*, New Age International (P) Ltd, New Delhi, India, 2007.

Dukkipati, R.V., *Solving Vibration Analysis Problems using MATLAB*, New Age International (P) Ltd, New Delhi, India, 2007.

Dukkipati, R.V., *Textbook of Mechanical Vibrations,* Prentice Hall of India, New Delhi, India, 2004.

Dukkipati, R.V., *Vehicle Dynamics,* Narosa Publishing House, New Delhi, India, 2000.

Dukkipati, R.V., *Vibration Analysis*, Narosa Publishing House, New Delhi, India, 2004.

Fogiel, M., Ogden, J.R., *The Electric Circuits Problem Solver: A Complete Solution Guide to Any Text Book,* Research & Education Association, 1998.

Gerald, C.F., and Wheatley, P.O., *Applied Numerical Analysis*, 5th ed, Addison-Wesley, Reading, MA, 1994.

Gladwell, J., and Wait, R., *A Survey of Numerical Methods of Partial Differential Equations,* Oxford University Press, New York, 1979.

Hayt, W.H., and Kemmerly, J.E., *Engineering Circuit Analysis*, 5th edn, McGraw Hill, New York, 1993.

Hibbler, R.C., *Engineering Mechanics: Statics & Dynamics,* Prentice Hall, Englewood Cliffs, NJ, 1995.

Johnson, D.E., Johnson, J.R., Hilburn, J.L., Scott, P.D., *Electric Circuit Analysis*, 3rd edn, Wiley, New York, 1997.

Johnson, D.E., *Basic Electric Circuit Analysis*, 5th edn, Wiley, New York, 2006.

Johnson, D.E, Hilburn, J.L, Johnson, J.R., and Scott, P.D., *Electric Circuit Analysis*, 3rd edn, Prentice Hall, Englewood Cliffs, NJ, 1990.

Meirovitch, L., *Elements of Vibration Analysis*, 2nd edn, McGraw Hill, New York, 1986.

Meriam, J.L, and Kraige, L.G., *Engineering Mechanics: Dynamics*, 5th edn, Wiley, New York, 2002.

Milne, W.E., *Numerical Solution of Differential Equations*, Wiley, New York, 1953.

Moin, P., *Fundamentals of Engineering Numerical Analysis*, Cambridge University Press, New York, 2001.

Nahvi, M., *Schaums Outline of Electrical Circuits,* 4th edn, McGraw Hill, New York, 2002.

Nash, W., *Schaum's Outlines: Statics and Mechanics of Materials,* McGraw Hill, New York, 1992.

Nash, W., *Schaum's Outlines: Strength of Materials,* 4th ed, McGraw Hill, New York, 1998.

Nelson, E.W., Best, C.L., and McLean, W.G., *Schaum's Outlines: Engineering mechanics- Statics and Dynamics,* 5th edn, McGraw Hill, New York, 1998.

Nilsson, J.W., and Riedel, S.A., *Electric Circuits,* 6th edn, Prentice Hall, Englewood Cliffs, NJ, 1999.

Nilsson, J.W., *Electrical Circuits,* 7th edn, Prentice Hall, Englewood Cliffs, NJ, 2004.

Paul, C.R., *Fundamentals of Electric Circuit Analysis*, Wiley, New York, 2000.

Rao, J.S., and Dukkipati, R.V., *Mechanism and Machine Theory*, Wiley Eastern Limited, New Delhi, India, Second Edition (with solutions manual – 1992), 1992.

Riley, W., Sturges, L., and Morris, D., *Mechanics of Materials*, Wiley, New York, 2007.

Seto, W.W., *Theory and Problems of Mechanical Vibrations*, Schaum series, McGraw Hill, New York, 1964.

Sharan, A., and Dukkipati, R.V., *Efficient Methods in Planar Linkage Kinematic Analysis*, Memorial University of Newfoundland Press, St. Johns, Canada, 1991.

Singh Guru, B., Warrier, R., *Electric Circuits-Analysis and Design*, Oxford University Press, Oxford, 2005.

Smith, R.J., and Dorf, R.C., Circuits, Systems, and Devices, 5th edn, Wiley, New York, 1992.

Starr, A.T., *Electrical Circuits and Wave Filters,* Pitman Publishing, New York, 1938.

Steidel, R.F., *An Introduction to Mechanical Vibrations,* 3rd ed, Wiley, New York, 1981.

Thomson, W.T., and Dahleh, M.D., *Theory of Vibrations with Applications,* 5th edn, Prentice Hall, Englewood Cliffs, NJ, 199.

Tse, F.S., Morse, I.E., and Hinkle, R.T., *Mechanical Vibrations,* Allyn and Bacon, Boston, MA, 1963.

Vierck, R.K., *Vibration Analysis,* 2nd edn, Harper & Row, New York, 1979.

CHAPTER 2

MATLAB BASICS

2.1 INTRODUCTION

This chapter is a brief introduction to **MATLAB** (an abbreviation of **MAT**rix **LAB**oratory) basics, registered trademark of computer software, version 4.0 or later developed by the Math Works Inc. The software is widely used in many of science and engineering fields. MATLAB is an interactive program for numerical computation and data visualization. MATLAB is supported on Unix, Macintosh, and Windows environments. For more information on MATLAB, contact **The MathWorks.Com**. A Windows version of MATLAB is assumed here. The syntax is very similar for the DOS version.

MATLAB integrates mathematical computing, visualization, and a powerful language to provide a flexible environment for technical computing. The open architecture makes it easy to use MATLAB and its companion products to explore data, create algorithms, and create custom tools that provide early insights and competitive advantages.

Known for its highly optimized matrix and vector calculations, MATLAB offers an intuitive language for expressing problems and their solutions both mathematically and visually. Typical uses include:

- Numeric computation and algorithm development
- Symbolic computation (with the built-in Symbolic Math functions)
- Modeling, simulation, and prototyping
- Data analysis and signal processing
- Engineering graphics and scientific visualization

In this chapter, we will introduce the MATLAB environment. We will learn how to create, edit, save, run, and debug m-files (ASCII files with series of MATLAB statements). We will see how to create arrays (matrices and vectors), and explore the built-in MATLAB linear algebra functions for matrix and vector multiplication, dot and cross products, transpose, determinants, and inverses, and for the solution of linear equations. MATLAB is based on the language C, but is generally much easier to use. We will also see how to program logic constructs and loops in MATLAB, how to use subprograms and functions, how to use comments (%) for explaining the programs and tabs for easy readability, and how to print and plot graphics both two and three dimensional. MATLAB's functions for symbolic mathematics are presented. Use of these functions to perform symbolic operations, to develop closed form expressions for solutions to algebraic equations, ordinary differential equations, and system of equations was presented. Symbolic mathematics can also be used to determine analytical expressions for the derivative and integral of an expression.

2.1.1 Starting and Quitting MATLAB

To start MATLAB click on the MATLAB icon or type in MATLAB, followed by pressing the *enter* or *return* key at the system prompt. The screen will produce the MATLAB **prompt** >> (or EDU >>), which indicates that MATLAB is waiting for a command to be entered.

In order to quit MATLAB, type **quit** or **exit** after the prompt, followed by pressing the *enter* or *return* key.

2.1.2 Display Windows

MATLAB has three display windows. They are

1. A *Command Window* which is used to enter commands and data to display plots and graphs.
2. A *Graphics Window* which is used to display plots and graphs
3. An *Edit Window* which is used to create and modify M-files. M-files are files that contain a program or script of MATLAB commands.

2.1.3 Entering Commands

Every command has to be followed by a carriage return <cr> (enter key) in order that the command can be executed. MATLAB commands are case sensitive and *lower case* letters are used throughout.

To execute an *M-file* (such as Project_1.m), simply enter the name of the file without its extension (as in Project_1).

2.1.4 MATLAB Expo

In order to see some of the MATLAB capabilities, enter the *demo* command. This will initiate the *MATLAB EXPO. MATLAB Expo* is a graphical demonstration environment that shows some of the different types of operations which can be conducted with MATLAB.

2.1.5 Abort

In order to *abort* a command in MATLAB, hold down the control key and press c to generate a local abort with MATLAB.

2.1.6 The Semicolon (;)

If a semicolon (;) is typed at the end of a command, the output of the command is not displayed.

2.1.7 Typing %

When percent symbol (%) is typed in the beginning of a line, the line is designated as a comment. When the *enter* key is pressed, the line is not executed.

2.1.8 The clc Command

Typing *clc* command and pressing *enter* cleans the command window. Once the *clc* command is executed a clear window is displayed.

2.1.9 Help

MATLAB has a host of built-in functions. For a complete list, refer to MATLAB user's guide or refer to the *on-line Help*. To obtain help on a particular topic in the list, *e.g.*, inverse, type *help inv*.

2.1.10 Statements and Variables

Statements have the form

```
>> variable = expression
```

The equals ("=") sign implies the assignment of the expression to the variable. For instance, to enter a 2 × 2 matrix with a variable name *A*, we write

```
>> A == [1 2 ; 3 4] <ret>
```

The statement is executed after the carriage return (or enter) key is pressed to display

```
A =
        1    2
        3    4
```

2.2 ARITHMETIC OPERATIONS

The symbols for arithmetic operations with scalars are summarized below in Table 2.1.

Table 2.1

Arithmetic operation	*Symbol*	*Example*
Addition	+	6 + 3 = 9
Subtraction	–	6 – 3 = 3
Multiplication	*	6 * 3 = 18
Right division	/	6/3 = 2
Left division	\	6\3 = 3/6 = 1/2
Exponentiation	^	6 ^ 3 (6^3 = 216)

2.3 DISPLAY FORMATS

MATLAB has several different screen output formats for displaying numbers. These formats can be found by typing the help command: help format in the Command Window. A few of these formats are shown in Table 2.2 for 2π.

Table 2.2 Display formats

Command	*Description*	*Example*
format short	Fixed-point with 4 decimal digits	>> 351/7 ans = 50.1429
format long	Fixed-point with 14 decimal digits	>> 351/7 ans = 50.14285714285715
format short e	Scientific notation with 4 decimal digits	>> 351/7 ans = 5.0143e+001

Command	*Description*	*Example*
format long e	Scientific notation with 15 decimal digits	>> 351/7 ans = 5.014285714285715e001
format short g	Best of 5 digit fixed or floating point	>> 351/7 ans = 50.143
format long g	Best of 15 digit fixed or floating point	>> 351/7 ans = 50.1428571428571
format bank	Two decimal digits	>> 351/7 ans = 50.14
format compact	Eliminates empty lines to allow more lines with information displayed on the screen	
format loose	Adds empty lines (opposite of compact)	

2.4 ELEMENTARY MATH BUILT-IN FUNCTIONS

MATLAB contains a number of functions for performing computations which require the use of logarithms, elementary math functions, and trigonometric math functions. List of these commonly used elementary MATLAB mathematical built-in functions are given in Tables 2.3 to 2.8.

Table 2.3 Common Math Functions

Function	*Description*
abs(x)	Computes the absolute value of **x**.
sqrt(x)	Computes the square root of **x**.
round(x)	Rounds **x** to the nearest integer.
fix(x)	Rounds (or truncates) **x** to the nearest integer toward 0.
floor(x)	Rounds **x** to the nearest integer toward $-\infty$.
ceil(x)	Rounds **x** to the nearest integer toward ∞.
sign(x)	Returns a value of -1 if **x** is less than 0, a value of 0 if **x** equals 0, and a value of 1 otherwise.
rem (x, y)	Returns the remainder of *x/y*. For example, **rem(25, 4)** is 1, and **rem(100, 21)** is 16. This function is also called a **modulus** function.
exp(x)	Computes e^x, where *e* is the base for natural logarithms, or approximately 2.718282.
log (x)	Computes ln **x**, the natural logarithm of **x** to the base *e*.
log 10(x)	Computes $\log_{10}$ **x**, the common logarithm of **x** to the base 10.

Table 2.4 Exponential functions

Function	*Description*
exp(x)	Exponential (e^x)
log(x)	Natural logarithm
log10(x)	Base 10 logarithm
sqrt(x)	Square root

Table 2.5 Trigonometric and hyperbolic functions

Function	*Description*
sin(x)	Computes the sine of **x**, where **x** is in radians.
cos(x)	Computes the cosine of **x**, where **x** is in radians.
tan(x)	Computes the tangent of **x**, where **x** is in radians.
asin(x)	Computes the arcsine or inverse sine of **x**, where **x** must be between – 1 and 1. The function returns an angle in radians between – π/2 and π/2.
acos(x)	Computes the arccosine or inverse cosine of **x**, where **x** must be between – 1 and 1. The function returns an angle in radians between 0 and π.
atan(x)	Computes the arctangent or inverse tangent of **x**. Thefunction returns an angle in radians between – π/2 and π/2.
atan2(y,x)	Computes the arctangent or inverse tangent of the value *y/x*. The function returns an angle in radians that will be between – π and π, depending on the signs of **x** and **y**.
sinh(x)	Computes the hyperbolic sine of **x**, which is equal to $\frac{e^{x} - e^{-x}}{2}$.
cosh(x)	Computes the hyperbolic cosine of **x**, which is equal to $\frac{e^{x} + e^{-x}}{2}$.
tanh(x)	Computes the hyperbolic tangent of **x**, which is equal to $\frac{\sinh x}{\cosh x}$.
asinh(x)	Computes the inverse hyperbolic sine of **x**, which is equal to $\ln\left(x + \sqrt{x^2 + 1}\right)$.
acosh(x)	Computes the inverse hyperbolic cosine of **x**, which is equal to $\ln\left(x + \sqrt{x^2 - 1}\right)$.
atanh(x)	Computes the inverse hyperbolic tangent of **x**, which is equal to $\ln \sqrt{\frac{1+x}{1-x}}$ for $\lvert x \rvert \le 1$.

Table 2.6 Round-off functions

Function	*Description*	*Example*
round(x)	Round to the nearest integer	>> round(20/6) ans = 3
fix(x)	Round towards zero	>> fix(13/6) ans = 2
ceil(x)	Round towards infinity	>> ceil(13/5) ans = 3
floor(x)	Round towards minus infinity	>> floor(– 10/4) ans = –3
rem(x,y)	Returns the remainder after **x** is divided by **y**	>> rem(14,3) ans = 2
sign(x,y)	Signum function. Returns 1 if **x** > 0, – 1 if **x** < 0, and 0 if **x** = 0.	>> sign(7) ans = 1

Table 2.7 Complex number functions

Function	*Description*
conj(x)	Computes the complex **conjugate** of the complex number **x**. Thus, if **x** is equal to $a + i\, b$, then **conj(x)** will be equal to $a - i\, b$.
real(x)	Computes the real portion of the complex number **x**
imag(x)	Computes the imaginary portion of the complex number **x**.
abs(x)	Computes the absolute value of **magnitude** of the complex number **x**.
angle(x)	Computes the angle using the value of **atan2(imag(x), real(x))**; thus, the angle value is between $-\pi$ and π.

Table 2.8 Arithmetic operations with complex numbers

Operation	*Result*
$c_1 + c_2$	$(a_1 + a_2) + i(b_1 + b_2)$
$c_1 + c_2$	$(a_1 - a_2) + i(b_1 - b_2)$
$c_1 \bullet c_2$	$(a_1a_2 - b_1b_2) + i(a_1b_2 - a_2b_1)$
$\dfrac{c_1}{c_2}$	$\left(\dfrac{a_1a_2 + b_1b_2}{a_2^2 + b_2^2}\right) + i\left(\dfrac{a_2b_1 + b_2a_1}{a_2^2 + b_2^2}\right)$
$\lvert c_1 \rvert$	$\sqrt{a_1^2 + b_1^2}$ (magnitude or absolute value of c_1)
$c_1{}^*$	$a_1 - ib_1$ (conjugate of c_1)
(Assume that $c_1 = a_1 + ib_1$ and $c_2 = a_2 + ib_2$.)	

2.5 VARIABLE NAMES

A variable is a name made of a letter or a combination of several letters and digits. Variable names can be up to 63 (in MATLAB 7) characters long (31 characters on MATLAB 6.0). MATLAB is case sensitive. For instance, *XX*, *Xx*, *xX*, and *xx* are the names of four different variables. It should be noted here that we should not use the names of a built-in function for a variable. For instance, avoid using: sin, cos, exp, sqrt, ..., etc. Once a function name is used to define a variable, the function cannot be used.

2.6 PREDEFINED VARIABLES

MATLAB includes a number of predefined variables. Some of the predefined variables that are available for use in MATLAB programs are summarized in Table 2.9.

Table 2.9 Predefined variables

Predefined variable in MATLAB	*Description*
ans	Represents a value computed by an expression but not stored in variable name.
pi	Represents the number π.
eps	Represents the floating-point precision for the computer being used. This is the smallest difference between two numbers.
inf	Represents infinity which for instance occurs as a result of a division by zero. A warning message will be displayed or the value will be printed as ∞.
i	Defined as $\sqrt{-1}$ which is: $0 + 1.0000i$.
j	Same as *i*.
NaN	Stands for Not a Number. Typically occurs as a result of an expression being undefined, as in the case of division of zero by zero.
clock	Represents the current time in a six-element row vector containing year, month, day, hour, minute, and seconds.
date	Represents the current date in a character string format.

2.7 COMMANDS FOR MANAGING VARIABLES

Table 2.10 lists commands that can be used to eliminate variables or to obtain information about variables that have been created. The procedure is to enter the command in the Command Window and the *Enter* key is to be pressed.

Table 2.10 Commands for managing variables

Command	*Description*
clear	Removes all variables from the memory.
clear x, y, z	Clears/removes only variables **x, y,** and **z** from the memory.
who	Lists the variables currently in the workspace.
whos	Displays a list of the variables currently in the memory and their size together with information about their bytes and class.

2.8 GENERAL COMMANDS

In Tables 2.11 to 2.15 the useful general commands on on-line help, workspace information, directory information, and general information are given.

Table 2.11 On-line help

Function	*Description*
help	Lists topics on which help is available.
helpwin	Opens the interactive help window.
helpdesk	Opens the web browser based help facility.
help *topic*	Provides help on *topic*.
lookfor *string*	Lists help topics containing *string*.
demo	Runs the demo program.

Table 2.12 Workspace information

Function	*Description*
who	Lists variables currently in the workspace.
whos	Lists variables currently in the workspace with their size.
what	Lists m-, mat-, and mex-files on the disk.
clear	Clears the workspace, all variables are removed.
clear *x y z*	Clears only variables *x*, *y*, and *z*.
clear all	Clears all variables and functions from workspace.
mlock *fun*	Locks function *fun* so that **clear** cannot remove it.
munlock fun	Unlocks function *fun* so that **clear** can remove it.
clc	Clears command window, command history is lost.
home	Same as **clc.**
clf	Clears figure window.

Table 2.13 Directory information

Function	*Description*
pwd	Shows the current working directory.
cd	Changes the current working directory.
dir	Lists contents of the current directory.
ls	Lists contents of the current directory, same as **dir.**
path	Gets or sets MATLAB search path.
editpath	Modifies MATLAB search path.
copyfile	Copies a file.
mkdir	Creates a directory.

Table 2.14 General information

Function	*Description*
computer	Tells you the computer type your are using.
clock	Gives you wall clock time and date as a vector.
date	Tells you the date as a string.
more	Controls the paged output according to the screen size.
ver	Gives the license and the version information about MATLAB installed on your computer.
bench	Benchmarks your computer on running MATLAB compared to other computers.

Table 2.15 Termination

Function	*Description*
c (Control –c)	Local abort, kills the current command execution.
quit	Quits MATLAB.
exit	Same as **quit.**

2.9 ARRAYS

An array is a list of numbers arranged in rows and/or columns. A one-dimensional array is a row or a column of numbers and a two-dimensional array has a set of numbers arranged in rows and columns. An array operation is performed *element-by-element.*

2.9.1 Row Vector

A vector is a row or column of elements.

In a row vector the elements are entered with a space or a comma between the elements inside the square brackets. For example,

$$x = [7\ \ -1\ \ 2\ \ -5\ \ 8]$$

2.9.2 Column Vector

In a column vector the elements are entered with a semicolon between the elements inside the square brackets. For example,

$$x = [7;\ -1;\ 2;\ -5;\ 8]$$

2.9.3 Matrix

A matrix is a two-dimensional array which has numbers in rows and columns. A matrix is entered row-wise with consecutive elements of a row separated by a space or a comma, and the rows separated by semicolons or carriage returns. The entire matrix is enclosed within square brackets. The elements of the matrix may be real numbers or complex numbers. For example to enter the matrix,

$$A = \begin{bmatrix} 1 & 3 & -4 \\ 0 & -2 & 8 \end{bmatrix}$$

The MATLAB input command is

$$A = [1\ \ 3\ \ -4\ ;0\ \ -2\ \ 8]$$

Similarly for complex number elements of a matrix B

$$B = \begin{bmatrix} -5x & \ln 2x + 7\sin 3y \\ 3i & 5 - 13i \end{bmatrix}$$

The MATLAB input command is

$$B = [-5*x \quad \log(2*x) + 7*\sin(3*y);\ 3i \quad 5 - 13i]$$

2.9.4 Addressing Arrays

A colon can be used in MATLAB to address a range of elements in a vector or a matrix.

2.9.4.1 Colon for a vector

Va(:) – refers to all the elements of the vector **Va** (either a row or a column vector).

Va(*m*:*n*) – refers to elements *m* through *n* of the vector **Va**.

For instance

```
>>   V = [2  5  -1  11  8  4  7  -3  11]
>>   u = V (2:8)
     u =
         5  -1  11  8  4  7  -3  11
```

2.9.4.2 Colon for a matrix

Table 2.16 gives the use of a colon in addressing arrays in a matrix.

Table 2.16 Colon use for a matrix

Command	*Description*
A(:, *n*)	Refers to the elements in all the rows of a column *n* of the matrix *A*.
A(*n*, :)	Refers to the elements in all the columns of row *n* of the matrix *A*.
A(:, *m*:*n*)	Refers to the elements in all the rows between columns *m* and *n* of the matrix *A*.
A(*m*:*n*, :)	Refers to the elements in all the columns between rows *m* and *n* of the matrix *A*.
A(*m*:*n*, *p*:*q*)	Refers to the elements in rows *m* through *n* and columns *p* through *q* of the matrix *A*.

2.9.5 Adding Elements to a Vector or a Matrix

A variable that exists as a vector, or a matrix, can be changed by adding elements to it. Addition of elements is done by assigning values of the additional elements, or by appending existing variables. Rows and/or columns can be added to an existing matrix by assigning values to the new rows or columns.

2.9.6 Deleting Elements

An element, or a range of elements, of an existing variable can be deleted by reassigning blanks to these elements. This is done simply by the use of square brackets with nothing typed in between them.

2.9.7 Built-in Functions

Some of the built-in functions available in MATLAB for managing and handling arrays as listed in Table 2.17.

Table 2.17 Built-in functions for handling arrays

Function	*Description*	*Example*
length(A)	Returns the number of elements in the vector *A*.	>> *A* = [5 9 2 4]; >> length(*A*) ans = 4
size(A)	Returns a row vector [*m*, *n*], where *m* and *n* are the size $m \times n$ of the array *A*.	>> *A* = [2 3 0 8 11 ; 6 17 5 7 1] *A* = 2 3 0 8 11 6 17 5 7 1 >> size(*A*) ans = 2 5

(Contd.)

Function	*Description*	*Example*
reshape(A, m, n)	Rearrange a matrix A that has r rows and s columns to have m rows and n columns. r times s must be equal to m times n.	>> A = [3 1 4 ; 9 0 7] A = 3 1 4 9 0 7 >> B = reshape(A, 3, 2) B = 3 0 9 4 1 7
diag(v)	When v is a vector, creates a square matrix with the elements of v in the diagonal	>> v = [3 2 1]; >> A = diag(v) A = 3 0 0 0 2 0 0 0 1
diag(A)	When A is a matrix, creates a vector from the diagonal elements of A.	>> A = [1 8 3 ; 4 2 6 ; 7 8 3] A = 1 8 3 4 2 6 7 8 3 >> vec = diag(A) vec = 1 2 3

2.10 OPERATIONS WITH ARRAYS

We consider here matrices that have more than one row and more than one column.

2.10.1 Addition and Subtraction of Matrices

The addition (the sum) or the subtraction (the difference) of the two arrays is obtained by adding or subtracting their corresponding elements. These operations are performed with arrays of identical size (same number of rows and columns).

For example if A and B are two arrays (2 × 3 matrices).

$$A = \begin{bmatrix} a_{11} & a_{12} & a_{13} \\ a_{21} & a_{22} & a_{23} \end{bmatrix} \text{ and } B = \begin{bmatrix} b_{11} & b_{12} & b_{13} \\ b_{21} & b_{22} & b_{23} \end{bmatrix}$$

Then, the matrix addition $(A + B)$ is obtained by adding A and B is

$$\begin{bmatrix} a_{11} + b_{11} & a_{12} + b_{12} & a_{13} + b_{13} \\ a_{21} + b_{21} & a_{22} + b_{22} & a_{23} + b_{23} \end{bmatrix}$$

2.10.2 Dot Product

The dot product is a scalar computed from two vectors of the same size. The scalar is the sum of the products of the values in corresponding positions in the vectors.

For n elements in the vectors A and B:

$$\text{dot product} = \mathrm{A} \bullet \mathrm{B} = \sum_{i=1}^{n} a_i b_i$$

dot(A, B) computes the dot product of A and B. If A and B are matrices, the dot product is a row vector containing the dot products for the corresponding columns of A and B.

2.10.3 Array Multiplication

The value in position $c_{i,j}$ of the product C of two matrices, A and B, is the dot product of row i of the first matrix and column j of the second matrix:

$$c_{i,j} = \sum_{k=1}^{n} a_{i,k}\, b_{k,j}$$

2.10.4 Array Division

The division operation can be explained by means of the identity matrix and the inverse matrix operation.

2.10.5 Identity Matrix

An identity matrix is a square matrix in which all the diagonal elements are 1's, and the remaining elements are 0's. If a matrix A is square, then it can be multiplied by the identity matrix, I, from the left or from the right:

$$AI = IA = A$$

2.10.6 Inverse of a Matrix

The matrix B is the inverse of the matrix A if when the two matrices are multiplied the product is the identity matrix. Both matrices A and B must be square and the order of multiplication can be AB or BA.

$$AB = BA = I$$

2.10.7 Transpose

The transpose of a matrix is a new matrix in which the rows of the original matrix are the columns of the new matrix. The transpose of a given matrix A is denoted by A^T. In MATLAB, the transpose of the matrix A is denoted by A'.

2.10.8 Determinant

A determinant is a scalar computed from the entries in a square matrix. For a 2 × 2 matrix A, the determinant is

$$|A| = a_{11}\, a_{22} - a_{21}\, a_{12}$$

MATLAB will compute the determinant of a matrix using the **det** function:

det(A) Computes the determinant of a square matrix A.

2.10.9 Array Division

MATLAB has two types of array divisions, which are the left division and the right division.

2.10.10 Left Division

The left division is used to solve the matrix equation $Ax = B$ where x and B are column vectors. Multiplying both sides of this equation by the inverse of A, A^{-1}, we have

$$A^{-1}Ax = A^{-1}B$$

or

$$Ix = x = A^{-1}B$$

Hence $x = A^{-1}B$

In MATLAB, the above equation is written by using the left division character:

$$x = A\backslash B$$

2.10.11 Right Division

The right division is used to solve the matrix equation $xA = B$ where x and B are row vectors. Multiplying both sides of this equation by the inverse of A, A^{-1}, we have

$$x \bullet A\,A^{-1} = B \bullet A^{-1}$$

or

$$x = B \bullet A^{-1}$$

In MATLAB, this equation is written by using the right division character:

$$x = B/A$$

2.10.12 Eigenvalues and Eigenvectors

Consider the following equation,

$$AX = \lambda X \tag{2.1}$$

where A is an $n \times n$ square matrix, X is a column vector with n rows and λ is a scalar.

The values of λ for which X are nonzero are called the *eigenvalues* of the matrix A, and the corresponding values of X are called the *eigenvectors* of the matrix A.

Equation (2.1) can also be used to find the following equation

$$(A - \lambda I)X = 0 \tag{2.2}$$

where I is an $n \times n$ identity matrix. Equation (2.2) corresponding to a set of homogeneous equations and has nontrivial solutions only if the determinant is equal to zero, or

$$|A - \lambda I| = 0 \tag{2.3}$$

Equation (2.3) is known as the *characteristic equation* of the matrix A. The solution to Eq. (2.3) gives the eigenvalues of the matrix A.

MATLAB determines both the eigenvalues and eigenvectors for a matrix A.

eig(A) Computes a column vector containing the eigenvalues of A.

[Q, d] = eig(A) Computes a square matrix Q containing the eigenvectors of A as columns and a square matrix d containing the eigenvlaues (λ) of A on the diagonal. The values of Q and d are such that Q^*Q is the identity matrix and A^*X equals λ times X.

Triangular factorization or lower-upper factorization: Triangular or lower-upper factorization expresses a square matrix as the product of two triangular matrices – a lower triangular matrix and an upper triangular matrix. The **lu** function in MATLAB computes the *LU* factorization:

[L, U] = lu(A) Computes a permuted lower triangular factor in L and an upper triangular factor in U such that the product of L and U is equal to A.

QR factorization: The *QR* factorization method factors a matrix A into the product of an orthonormal matrix and an upper-triangular matrix. The **qr** function is used to perform the *QR* factorization in MATLAB:

[Q, R] = qr(A) Computes the values of Q and R such that $A = QR.Q$ will be an orthonormal matrix, and R will be an upper triangular matrix.

For a matrix A of size $m \times n$, the size of Q is $m \times m$, and the size of R is $m \times n$.

Singular Value Decomposition (SVD): Singular value decomposition decomposes a matrix A (size $m \times n$) into a product of three matrix factors.

$$A = USV$$

where U and V are orthogonal matrices and S is a diagonal matrix. The size of U is $m \times m$, the size of V is $n \times n$, and the size of S is $m \times n$. The values on the diagonal matrix S are called singular values. The number of non-zero singular values is equal to the rank of the matrix.

The *SVD* factorization can be obtained using the **svd** function:

[U, S, V] = svd(A) Computes the factorization of A into the product of three matrices, USV, where U and V are orthogonal matrices and S is a diagonal matrix.

svd(A) Returns the diagonal elements of S, which are the singular values of A.

2.11 ELEMENT-BY-ELEMENT OPERATIONS

Element-by-element operations can only be done with arrays of the same size. Element-by-element multiplication, division, and exponentiation of two vectors or matrices is entered in MATLAB by typing a period in front of the arithmetic operator. Table 2.18 lists these operations.

Table 2.18 Element-by-element operations

Arithmetic operators	
Matrix operators	*Array operators*
+ Addition	+ Addition
– Subtraction	– Subtraction
* Multiplication	•* Array multiplication
^ Exponentiation	•^ Array exponentiation
/ Left division	•/ Array left division
\ Right division	•\ Array right division

2.11.1 Built-in Functions for Arrays

Table 2.19 lists some of the many built-in functions available in MATLAB for analyzing arrays.

Table 2.19 MATLAB built-in array functions

Function	*Description*	*Example*
mean(A)	If *A* is a vector, returns the mean value of the elements	>> *A* = [3 7 2 16]; >> mean(*A*) ans = 14
C = max(A)	If *A* is a vector, *C* is the largest element in *A*. If *A* is a matrix, *C* is a row vector containing the largest element of each column of *A*.	>> *A* = [3 7 2 16 9 5 18 13 0 4]; >> *C* = max(*A*) *C* = 18
[d, n] = max(A)	If *A* is a vector, *d* is the largest element in *A*, *n* is the position of the element (the first if several have the max value).	>> [*d*, *n*] = max(*A*) *d* = 18 *n* = 7
min(A)	The same as **max(A),** but for the smallest element.	>> *A* = [3 7 2 16]; >> min(*A*) ans = 2
[d, n] = min(A)	The same as **[d, n] = max(A)**, but for the smallest element.	
sum(A)	If *A* is a vector, returns the sum of the elements of the vector.	>> *A* = [3 7 2 16]; >> sum(*A*) ans = 28
sort(A)	If *A* is a vector, arranges the elements of the vector in ascending order.	>> *A* = [3 7 2 16]; >> sort(*A*) ans = 2 3 7 16
median(A)	If *A* is a vector, returns the median value of the elements of the vector.	>> *A* = [3 7 2 16]; >> median(*A*) ans = 5
std(A)	If *A* is a vector, returns the standard deviation of the elements of the vector.	>> *A* = [3 7 2 16]; >> std(*A*) ans = 6.3770

Function	Description	Example
det(A)	Returns the determinant of a square matrix *A*.	>> *A* = [1 2 ; 3 4]; >> det(*A*) ans = − 2
dot(a, b)	Calculates the scalar (dot) product of two vectors *a* and *b*. The vector can each be row or column vectors.	>> *a* = [5 6 7]; >> *b* = [4 3 2]; >> dot(*a*, *b*) ans = 52
cross(a, b)	Calculates the cross product of two vectors *a* and *b*, (*a* × *b*). The two vectors must have 3 elements	>> *a* = [5 6 7]; >> *b* = [4 3 2]; >> cross(*a*, *b*) ans = − 9 18 − 9
inv(A)	Returns the inverse of a square matrix *A*.	>> *a* = [1 2 3; 4 6 8; − 1 2 3]; >> inv(*A*) ans = − 0.5000 0.0000 − 0.5000 − 5.0000 1.5000 1.0000 3.5000 − 1.0000 − 0.5000

2.12 RANDOM NUMBERS GENERATION

There are many physical processes and engineering applications that require the use of *random numbers* in the development of a solution.

MATLAB has two commands *rand* and *rand n* that can be used to assign random numbers to variables.

The *rand* command: The *rand* command generates uniformly distributed over the interval [0, 1]. A *seed value* is used to initiate a random sequence of values. The seed value is initially set to zero. However, it can be changed with the *seed* function.

The command can be used to assign these numbers to a scalar, a vector, or a matrix, as shown in Table 2.20.

Table 2.20 The *rand* command

Command	Description	Example
rand	Generates a single random number between 0 and 1.	>> rand ans = 0.9501
rand(1, n)	Generates an *n* elements row vector of random numbers between 0 and 1.	>> *a* = rand(1, 3) *a* = 0.4565 0.0185 0.8214

Command	*Description*	*Example*
rand(n)	Generates an $n \times n$ matrix with random numbers between 0 and 1.	>> b = rand(3) b = 0.7382 0.9355 0.8936 0.1763 0.9165 0.0579 0.4057 0.4103 0.3529
rand(m, n)	Generates an $m \times n$ matrix with random numbers between 0 and 1.	>> c = rand(2, 3) c = 0.2028 0.6038 0.1988 0.1987 0.2722 0.0153
randperm(n)	Generates a row vector with n elements that are random permutation of integers 1 through n.	>> randperm(7) ans = 5 2 4 7 1 6 3

2.12.1 The Random Command

MATLAB will generate Gaussian values with a mean of zero and a variance of 1.0 if a normal distribution is specified. The MATLAB functions for generating Gaussian values are as follows:

randn(n) Generates an $n \times n$ matrix containing Gaussian (or normal) random numbers with a mean of 0 and a variance of 1.

Randn(m, n) Generates an $m \times n$ matrix containing Gaussian (or normal) random numbers with a mean of 0 and a variance of 1.

2.13 POLYNOMIALS

A *polynomial* is a function of a single variable that can be expressed in the following form:

$$f(x) = a_0x^n + a_1x^{n-1} + a_2x^{n-2} + \ldots + a_{n-1}x^1 + a_n$$

where the variable is x and the coefficients of the polynomial are represented by the values a_0, a_1, ... and so on. The *degree* of a polynomial is equal to the largest value used as an exponent.

A vector represents a polynomial in MATLAB. When entering the data in MATLAB, simply enter each coefficient of the polynomial into the vector in descending order. For example, consider the polynomial

$$5s^5 + 7s^4 + 2s^2 - 6s + 10$$

To enter this into MATLAB , we enter this as a vector as

```
>>x = [5  7  0  2  -6  10]
x =
        5  7  0  2  -6  10
```

It is necessary to enter the coefficients of all the terms.

MATLAB contains functions that perform polynomial multiplication and division, which are listed below:

conv(a, b) Computes a coefficient vector that contains the coefficients of the product of polynomials represented by the coefficients in **a** and **b**. The vectors **a** and **b** do not have to be the same size.

[q, r] = deconv(n, d) Returns two vectors. The first vector contains the coefficients of the quotient and the second vector contains the coefficients of the remainder polynomial.

The MATLAB function for determining the roots of a polynomial is the roots function:

root(a) Determines the roots of the polynomial represented by the coefficient vector **a**.

The roots function returns a column vector containing the roots of the polynomial; the number of roots is equal to the degree of the polynomial. When the roots of a polynomial are known, the coefficients of the polynomial are determined when all the linear terms are multiplied, we can use the **poly** function:

poly(r) Determines the coefficients of the polynomial whose roots are contained in the vector **r**.

The output of the function is a row vector containing the polynomial coefficients.

The value of a polynomial can be computed using the *polyval* function, **polyval (a, x).** It evaluates a polynomial with coefficients **a** for the values in **x**. The result is a matrix the same size ad **x**. For instance, to find the value of the above polynomial at $s = 2$,

```
>>x = polyval([5  7  0   2  -6 10], 2)
x =
      278
```

To find the roots of the above polynomial, we enter the command **roots (a)** which determines the roots of the polynomial represented by the coefficient vector **a**.

```
>>roots([5  7  0  2  -6  10])
ans =
      -1.8652
      -0.4641 + 1.0832i
      -0.4641 - 1.0832i
      0.6967 + 0.5355i
      0.6967 - 0.5355i
% or
>> x = [5 7 0 2 -6 10]
x =
      5     7     0     2    -6    10
>> r = roots(x)
r =
      -1.8652
      -0.4641 + 1.0832i
      -0.4641 - 1.0832i
       0.6967 + 0.5355i
       0.6967 - 0.5355i
```

To multiply two polynomials together, we enter the command *conv*.

The polynomials are: $x = 2x + 5$ and $y = x^2 + 3x + 7$

```
>>x = [2   5];
>>y = [1   3   7];
>>z = conv(x, y)
z =
      2  11  29  35
```

To divide two polynomials, we use the command *deconv*.

```
z = [2 11 29 35]; x = [2 5]
        >> [g, t] = deconv (z, x)
        g = 1  3  7
        t =  0   0   0   0
```

2.14 SYSTEM OF LINEAR EQUATIONS

A system of equations is nonsingular if the matrix **A** containing the coefficients of the equations is nonsingular. A system of nonsingular simultaneous linear equations **(AX = B)** can be solved using two methods:

(*a*) Matrix Division Method.

(*b*) Matrix Inversion Method.

2.14.1 Matrix Division

The solution to the matrix equation **AX = B** is obtained using matrix division, or **X = A/B**. The vector **X** then contains the values of **x**.

2.14.2 Matrix Inverse

For the solution of the matrix equation **AX = B**, we premultiply both sides of the equation by $\mathbf{A}^{-1}$.

$$\mathbf{A}^{-1}\mathbf{AX} = \mathbf{A}^{-1}\mathbf{B}$$

or

$$\mathbf{IX} = \mathbf{A}^{-1}\mathbf{B}$$

where **I** is the identity matrix.

Hence

$$\mathbf{X} = \mathbf{A}^{-1}\mathbf{B}$$

In MATLAB, we use the command **x** = inv **(A)*B**. Similarly, for **XA = B**, we use the command **x = B***inv **(A)**.

The basic computational unit in MATLAB is the matrix. A matrix expression is enclosed in square brackets, [ ]. Blanks or commas separate the column elements, and semicolons or carriage returns separate the rows.

```
>>A = [1  2  3  4 ;  5  6  7  8 ;  9  10  11  12]
A =
        1    2    3    4
        5    6    7    8
        9    10   11   12
```

The transpose of a simple matrix or a complex matrix is obtained by using the *apostrophe* key

```
>>B = A'
B =
        1    5    9
        2    6    10
        3    7    11
        4    8    12
```

Matrix multiplication is accomplished as follows:

```
>>C = A * B
C =
      30      70      110
      70      174     278
      110     278     446
>>C = B * A
C =
      107     122     137     152
      122     140     158     176
      137     158     179     200
      152     176     200     224
```

The inverse of a matrix D is obtained as

```
>>D = [1  2 ; 3  4]
D =
      1   2
      3   4
>>E = inv (D)
E =
      -2.0000   1.0000
      1.5000    -0.5000
```

Similarly, its eigenvalue is

```
>>eig (D)
ans =
            -0.3723
             5.3723
```

Matrix operations require that the matrix dimensions be compatible. If A is an $n \times m$ and B is a $p \times r$ then $A \pm B$ is allowed only if $n = p$ and $m = r$. Similarly, matrix product $A * B$ is allowed only if $m = p$.

Example E2.1. *Consider the two matrices:*

$$\mathbf{A} = \begin{bmatrix} 1 & 0 & 1 \\ 2 & 3 & 4 \\ -1 & 6 & 7 \end{bmatrix} \text{ and } \mathbf{B} = \begin{bmatrix} 7 & 4 & 2 \\ 3 & 5 & 6 \\ -1 & 2 & 1 \end{bmatrix}$$

Using MATLAB, determine the following:

(*a*) $\mathbf{A + B}$

(*b*) $\mathbf{AB}$

(*c*) $\mathbf{A}^2$

(*d*) $\mathbf{A}^T$

(*e*) $\mathbf{B}^{-1}$

(*f*) $\mathbf{B}^T\mathbf{A}^T$

(*g*) $\mathbf{A}^2 + \mathbf{B}^2 - \mathbf{AB}$

(*h*) *determinant of* **A**, *determinant of* **B** *and determinant of* **AB**.

Solution:

```
>> A= [1 0 1; 2 3 4; -1 6 7]
     A =
      1      0      1
      2      3      4
     -1      6      7
>> B= [7 4 2; 3 5 6; -1 2 1]
     B =
            7     4     2
            3     5     6
           -1     2     1
```

(a)
```
>> C=A+B
     C =
            8      4      3
            5      8     10
           -2      8      8
```

(b)
```
>>D=A*B
     D =
            6      6      3
           19     31     26
            4     40     41
```

(c)
```
>> E=A^2
     E =
           0      6      8
           4     33     42
           4     60     72
```

(d)
```
>> % Let F= transpose of A
     >> F=A'
     F =
           1      2     -1
           0      3      6
           1      4      7
```

(e)
```
>> H = inv (B)
     H =
           0.1111     0.0000    -0.2222
           0.1429    -0.1429     0.5714
          -0.1746     0.2857    -0.3651
```

(f)
```
>> J=B'*A'
     J =
           6     19      4
           6     31     40
           3     26     41
```

(g)
```
>> K= A^2 + B^2 -A * B
     K =
          53     52     45
          15     51     58
          -2     28     42
```

(*h*)
```
det (A) =12
det (B) =-63
det (A*B) =-756
```

Example E2.2. *Determine the eigenvalues and eigenvectors of A and B using MATLAB*

$$\mathbf{A} = \begin{bmatrix} 4 & 2 & -3 \\ -1 & 1 & 3 \\ 2 & 5 & 7 \end{bmatrix} \qquad \mathbf{B} = \begin{bmatrix} 1 & 2 & 3 \\ 8 & 7 & 6 \\ 5 & 3 & 1 \end{bmatrix}$$

Solution:

% Determine the eigenvalues and eigenvectors

```
A=[4 2 -3 ; -1 1 3 ; 2 5 7]
    A =
            4        2        -3
           -1        1         3
            2        5         7
eig(A)
    ans =
           0.5949
           3.0000
           8.4051
lamda=eig(A)
    lamda =
           0.5949
           3.0000
           8.4051
[V,D]=eig(A)
    V =
           -0.6713    0.9163     -0.3905
           0.6713     -0.3984    0.3905
           -0.3144    0.0398     0.8337
    D =
           0.5949     0          0
           0          3.0000     0
           0          0          8.4051
```

Example E2.3. *Determine the values of x, y, and z for the following set of linear algebraic equations:*

$$x_2 - 3x_3 = -5$$
$$2x_1 + 3x_2 - x_3 = 7$$
$$4x_1 + 5x_2 - 2x_3 = 10$$

Solution:

Here

$$\mathbf{A} = \begin{bmatrix} 0 & 1 & -3 \\ 2 & 3 & -1 \\ 4 & 5 & -2 \end{bmatrix} \quad \mathbf{B} = \begin{bmatrix} 5 \\ 7 \\ 10 \end{bmatrix} \text{ and } \mathbf{X} = \begin{bmatrix} x_1 \\ x_2 \\ x_3 \end{bmatrix}$$

$$\mathbf{AX = B}$$
$$\mathbf{A^{-1}AX = A^{-1}B}$$
$$\mathbf{IX = A^{-1}B}$$

or

$$\mathbf{X = A^{-1}B}$$

```
>> A = [0 1 -3; 2 3 -1; 4 5 -2];
>> B = [-5; 7; 10]
>> x = inv (A) * B
x =
    -1.0000
    4.0000
    3.0000
>> check = A * x
check =
        -5
         7
        10
%   Alternative method
>> x = A\B
x =
    -1
     4
     3
```

2.15 SCRIPT FILES

A *script* is a sequence of ordinary statements and functions used at the command prompt level. A script is invoked at the command prompt level by typing the file-name or by using the pull down menu. Scripts can also invoke other scripts.

The commands in the Command Window cannot be saved and executed again. Also, the Command Window is not interactive. To overcome these difficulties, the procedure is first to create a file with a list of commands, save it, and then run the file. In this way the commands contained are executed in the order they are listed when the file is run. In addition, as the need arises, one can change or modify the commands in the file, the file can be saved and run again. The files that are used in this fashion are known as *script files*. Thus, a script file is a text file that contains a sequence of MATLAB commands. Script file can be edited (corrected and/or changed) and executed many times.

2.15.1 Creating and Saving a Script File

Any text editor can be used to create script files. In MATLAB script files are created and edited in the Editor/Debugger Window. This window can be opened from the Command Window. From the Command Window, select *File, New*, and then *M-file*. Once the window is open, the commands of the script file are typed line by line. The commands can also be typed in any text editor or word processor program and then copied and pasted in the Editor/Debugger Window.

The second type of M-files is the *function file*. Function file enables the user to extend the basic library functions by adding one's own computational procedures. Function M-files are expected to return one or more results. Script files and function files may include reference to other MATLAB toolbox routines.

MATLAB function file begins with a header statement of the form:

```
function (name of result or results) = name (argument list)
```

Before a script file can be executed it must be saved. All script files must be saved with the extension "*.m*". MATLAB refers to them as m-files. When using MATLAB M-files editor, the files will automatically be saved with a "*.m*" extension. If any other text editor is used, the file must be saved with the "*.m*" extension, or MATLAB will not be able to find and run the script file. This is done by choosing *Save As...* from the *File* menu, selecting a location, and entering a name for the file. The names of user defined variables, predefined variables, MATLAB commands or functions should not be used to name script files.

2.15.2 Running a Script File

A script file can be executed either by typing its name in the Command Window and then pressing the *Enter* key, directly from the Editor Window by clicking on the *Run* icon. The file is assumed to be in the current directory, or in the search path.

2.15.3 Input to a Script File

There are three ways of assigning a value to a variable in a script file.

1. The variable is defined and assigned value in the script file.

2. The variable is defined and assigned value in the Command Window.

3. The variable is defined in the script file, but a specified value is entered in the Command Window when the script file is executed.

2.15.4 Output Commands

There are two commands that are commonly used to generate output. They are the *disp* and *fprintf* commands.

1. The *disp* command

The *disp* command displays the elements of a variable without displaying the name of the variable, and displays text.

```
disp(name of a variable) or disp('text as string')
>>    A = [1 2 3 ; 4 5 6 ];
>> disp(A)
        1  2  3
        4  5  6
>> disp('Solution to the problem.')
        Solution to the problem.
```

2. The *fprintf* command

The *fprintf* command displays output (text and data) on the screen or saves it to a file. The output can be formatted using this command.

Example E2.4. *Write a function file Veccrossprod to compute the cross product of two vectors* **a**, *and* **b**, *where* **a** $= (a_1, a_2, a_3)$, $b = (b_1, b_2, b_3)$, *and* $a \times b = (a_2b_3 - a_3b_2, a_3b_1 - a_1b_3, a_1b_2 - a_2b_1)$. *Verify the function by taking the cross products of pairs of unit vectors: (i, j), (j, k), etc.*

Solution:

```
function c = Veccrossprod (a, b);
% Veccrossprod : function to compute c = a x b where a and b are 3D vectors
% call syntax:
% c = Veccrossprod (a, b);
c = [a(2) * b(3)-a(3) * b(2); a(3) * b(1)-a(1) * b(3); a(1) * b(2)-a(2) * b(1)];
```

2.16 PROGRAMMING IN MATLAB

One most significant feature of MATLAB is its extendibility through user-written programs such as the M-files. M-files are ordinary ASCII text files written in MATLAB language. A function file is a subprogram.

2.16.1 Relational and Logical Operators

A relational operator compares two numbers by finding whether a comparison statement is true or false. A logical operator examines true/false statements and produces a result which is true or false according to the specific operator. Relational and logical operators are used in mathematical expressions and also in combination with other commands, to make decisions that control the flow of a computer program.

MATLAB has six relational operators as shown in Table 2.21.

Table 2.21. Relational operators

Relational operator	*Interpretation*
<	Less than
<=	Less than or equal
>	Greater than
>=	Greater than or equal
= =	Equal
~ =	Not equal

The logical operators in MATLAB are shown in Table 2.22.

Table 2.22 Logical operators

Logical operator	*Name*	*Description*
& Example: *A&B*	AND	Operates on two operands (*A* and *B*). If both are true, the result is true (1), otherwise the result is false (0).
| Example: *A* | *B*	OR	Operates on two operands (*A* and *B*). If either one, or both are true, the result is true (1), otherwise (both are false) the result is false (0).
~ Example: ~ *A*	NOT	Operates on one operand (*A*). Gives the opposite of the operand. True (1) if the operand is false, and false (0) if the operand is true.

2.16.2 Order of Precedence

The following Table 2.23 shows the order of precedence used by MATLAB.

Table 2.23

Precedence	*Operation*
1 (highest)	Parentheses (If nested parentheses exist, inner have precedence).
2	Exponentiation.
3	Logical NOT (~).
4	Multiplication, Division.
5	Addition, Subtraction.
6	Relational operators (>, <, >=, <=, = =, ~=).
7	Logical AND (&).
8 (lowest)	Logical OR (|).

2.16.3 Built-in Logical Functions

The MATLAB built-in functions which are equivalent to the logical operators are:

and(A, B) Equivalent to *A* & *B*
or(A, B) Equivalent to *A* | *B*
not(A) Equivalent to ~*A*

List of the MATLAB logical built-in functions are described in Table 2.24.

Table 2.24 Additional logical built-in functions

Function	*Description*	*Example*
xor(a, b)	Exclusive or. Returns true (1) if one operand is true and the other is false	>>xor(8, – 1) ans = 0 >>xor(8, 0) ans = 1
all(A)	Returns 1 (true) if all elements in a vector *A* are true (nonzero). Returns 0 (false) if one or more elements are false (zero). If *A* is a matrix, treats columns of *A* as vectors, returns a vector with 1's and 0's.	>>*A* = [5 3 11 7 8 15] >>all(*A*) ans = 1 >>*B* = [3 6 11 4 0 13] >>all(*B*) ans = 0
any(A)	Returns 1 (true) if any element in a vector *A* is true (nonzero). Returns 0 (false) if all elements are false (zero). If *A* is a matrix, treats columns of *A* as vectors, returns a vector with 1's and 0's.	>>*A* = [5 0 14 0 0 13] >>any(*A*) ans = 1 >>*B* = [0 0 0 0 0 0] >>any(*B*) ans = 0

Function	*Description*	*Example*
find(A) **find(A>d)**	If *A* is a vector, returns the indices of the nonzero elements. If *A* is a vector, returns the address of the elements that are larger than *d* (any relational operator can be used).	>>*A* = [0 7 4 2 8 0 0 3 9] >>find(*A*) ans = 2 3 4 5 8 9 >>find(*A* > 4) ans = 4 5 6

The truth table for the operation of the four logical operators, AND, OR, *XOR*, and NOT are summarized in Table 2.25.

Table 2.25 Truth table

INPUT		*OUTPUT*				
A	*B*	*AND* *A&B*	*OR* *A|B*	*XOR* *(A,B)*	*NOT* *~A*	*NOT* *~B*
false	false	false	false	false	true	true
false	true	false	true	true	true	false
true	false	false	true	true	false	true
true	true	true	true	false	false	false

2.16.4 Conditional Statements

A conditional statement is a command that allows MATLAB to make a decision of whether to execute a group of commands that follow the conditional statement or to skip these commands.

if conditional expression consists of relational and/or logical operators

```
if   a < 30
     count = count + 1
     disp a
     end
```

The general form of a simple **if** statement is as follows:

```
if   logical expression
          statements
end
```

If the logical expression is true, the statements between the **if** statement and the **end** statement are executed. If the logical expression is false, then it goes to the statements following the **end** statement.

2.16.5 nested if Statements

Following is an example of *nested if* statements:

```
if   a < 30
     count = count + 1;
     disp(a);
     if   b > a
          b = 0;
     end
end
```

2.16.6 else and elseif Clauses

The else clause allows to execute one set of statements if a logical expression is true and a different set if the logical expression is false.

```
%    variable name inc
     if   inc < 1
          x_inc = inc/10;
          else
          x_inc = 0.05;
          end
```

When several levels of **if-else** statements are nested, it may be difficult to find which logical expressions must be true (or false) to execute each set of statements. In such cases, the **elseif** clause is used to clarify the program logic.

2.16.7 MATLAB while Structures

There is a structure in MATLAB that combines the for loop with the features of the if block. This is called the *while loop* and has the form:

while *logical expression*

This set of statements is executed repeatedly as long as the logical expressions remain true (equals +1) or if the expression is a matrix rather than a simple scalar variable, as long as *all* the elements of the matrix remain nonzero.

end

In addition to the normal termination of a loop by means of the **end** statement, there are additional MATLAB commands available to interrupt the calculations. These commands are listed in Table 2.26 below:

Table 2.26

Command	*Description*
break	Terminates the execution of MATLAB **for** and **while** loops. In nested loops, **break** will terminate only the innermost loop in which it is placed.
return	Primarily used in MATLAB functions, **return** will cause a normal return from a function from the point at which the return statement is executed.
error (*'text'*)	Terminates execution and displays the message contained in *text* on the screen. Note, the text must be enclosed in single quotes.

The MATLAB functions used are summarized in Table 2.27 below:

Table 2.27

Function	*Description*
Relational operators	A MATLAB *logical relation* is a comparison between two variables **x** and **y** of the same size effected by one of the six operators, <, <=, >, >=, = =, ~=. The comparison involves corresponding elements of **x** and **y**, and yields a matrix or scalar of the same size with values of "true" or "false" for each of its elements. In MATLAB, the value of "false" is zero, and "true" has a value of one. Any nonzero quantity is interpreted as "true".
Combinatorial operators	The operators **&** (AND) and **|** (OR) may be used to combine two logical expressions.

(Contd.)

Function	*Description*
all, any	If **x** is a vector, **all(x)** returns a value of one if *all* of the elements of **x** are nonzero, and a value of zero otherwise. When **X** is a matrix, all(**X**) returns a row vector of ones or zeros obtained by applying all to each of the columns of **X**. The function **any** operates similarly if any of the elements of **x are** nonzero.
find	If **x** is a vector, **i** = **find(x)** returns the indices of those elements of **x** that are nonzero (*i.e.*, true). Thus, replacing all the negative elements of **x** by zero could be accomplished by **i = find(x < 0);** **x(i) = zeros(size(i));** If **X** is a matrix, **[i,j]** = **find(X)** operates similarly and returns the row-column indices of nonzero elements.
if, else, elseif	The several forms of MATLAB **if** blocks are as follows: **if** *variable* **if** *variable* 1 **if** *variable* 1 block of statements block of statements block of statements executed if *variable* executed if *variable* 1 executed if *variable* 1 is "true", *i.e.*, nonzero is "true", *i.e.*, nonzero is "true", **end** **else** **elseif** *variable 2* block of statements block of statements executed if *variable l* executed if *variable* 2 is "false", i.e., zero is "true", **end** **else** **end** block of statements executed if neither *variable* is "true"
break	Terminates the execution of a **for** or **while** loop. Only the innermost loop in which **break** is encountered will be terminated.
return	Causes the function to return at that point to the calling routine. MATLAB *M-file* functions will return normally without this statement.
error (*'text'*)	Within a loop or function, if the statement **error**(*'text'*) is encountered, the loop or function is terminated, and the *text* is displayed.
while	The form of the MATLAB **while** loop is **while** *variable* block of statements executed as long as the value of *variable* is "true"; *i.e.*, nonzero **end**
	Useful when a function *F* itself calls a second "dummy" function "*f*". For example, the function *F* might find the root of an arbitrary function identified as a generic *f*(*x*). Then, the name of the actual *M-file* function, say **fname**, is passed as a *character string* to the function *F* either through its argument list or as a global variable, and the function is evaluated within *F* by means of **feval.** The use of **feval(name, x1, x2, ..., xn),** where **fname** is a variable containing the name of the function as a character string; *i.e.*, enclosed in single quotes, and **x1, x2, ..., xn** are the variables needed in the argument list of function **fname.**

2.17 GRAPHICS

MATLAB has many commands that can be used to create basic 2-D plots, overlay plots, specialized 2-D plots, 3-D plots, mesh, and surface plots.

2.17.1 Basic 2-D Plots

The basic command for producing a simple 2-D plot is

plot(x values, y values, 'style option')

where x values and y values are vectors containing the x- and y-coordinates of points on the graph.

style option is an optional argument that specifies the color, line-style, and the point-marker style.

The style option in the plot command is a character string that consists of 1, 2, or 3 characters that specify the color and/or the line style. The different color, line-style and marker-style options are summarized in Table 2.28.

Table 2.28 Color, line-style, and marker-style options

Color style-option		*Line style-option*		*Marker style-option*	
y	yellow	–	solid	+	plus sign
m	magenta	– –	dashes	O	circle
c	cyan	:	dotted	*	asterisk
r	red	–.	dash-dot	*x*	x-mark
g	green			.	point
b	blue			^	up triangle
w	white			*s*	square
k	black			*d*	diamond, etc.

2.17.2 Specialized 2-D Plots

There are several specialized graphics functions available in MATLAB for 2-D plots. The list of functions commonly used in MATLAB for plotting x-y data are given in Table 2.29.

Table 2.29 List of functions for plotting x-y data

Function	*Description*
area	Creates a filled area plot.
bar	Creates a bar graph.
barh	Creates a horizontal bar graph.
comet	Makes an animated 2-*D* plot.
compass	Creates arrow graph for complex numbers.
contour	Makes contour plots.
contourf	Makes filled contour plots.
errorbar	Plots a graph and puts error bars.
feather	Makes a feather plot.
fill	Draws filled polygons of specified color.
fplot	Plots a function of a single variable.
hist	Makes histograms.
loglog	Creates plot with log scale on both x and y axes.
pareto	Makes pareto plots.
pcolor	Makes pseudo color plot of matrix.

(Contd.)

Command	*Description*
pie	Creates a pie chart.
plotyy	Makes a double *y*-axis plot.
plotmatrix	Makes a scatter plot of a matrix.
polar	Plots curves in polar coordinates.
quiver	Plots vector fields.
rose	Makes angled histograms.
scatter	Creates a scatter plot.
semilogx	Makes semilog plot with log scale on the *x*-axis.
semilogy	Makes semilog plot with log scale on the *y*-axis.
stairs	Plots a stair graph.
stem	Plots a stem graph.

2.17.2.1 Overlay plots

There are three ways of generating overlay plots in MATLAB, they are:

(*a*) Plot command

(*b*) Hold command

(*c*) Line command

(*a*) **Plot command.** Example E2.7(*a*) shows the use of plot command used with matrix argument, each column of the second argument matrix plotted against the corresponding column of the first argument matrix.

(*b*) **Hold command.** Invoking hold on at any point during a session freezes the current plot in the graphics window. All the next plots generated by the plot command are added to the exiting plot. See Example E2.7(*a*).

(*c*) **Line command.** The line command takes a pair of vectors (or a triplet in 3-D) followed by a parameter name/parameter value pairs as argument. For instance, the command: *line* (*x data, y data, parameter name, parameter value*) adds lines to the existing axes. See Example E2.7(*a*).

2.17.3 3-D Plots

MATLAB provides various options for displaying three-dimensional data. They include line and wire, surface, mesh plots, among many others. More information can be found in the Help Window under Plotting and Data visualization. Table 2.30 lists commonly used functions.

Table 2.30 Functions used for 3-D graphics

Command	*Description*
plot3	Plots three-dimensional graph of the trajectory of a set of three parametric equations $x(t)$, $y(t)$, and $z(t)$ can be obtained using **plot3(x,y,z).**
meshgrid	If **x** and **y** are two vectors containing a range of points for the evaluation of a function, **[X,Y] = meshgrid(x, y)** returns two rectangular matrices containing the x and y values at each point of a two-dimensional grid.
mesh(X,Y,z)	If **X** and **Y** are rectangular arrays containing the values of the x and y coordinates at each point of a rectangular grid , and if **z** is the value of a function evaluated at each of these points, **mesh(X,Y,z)** will produce a three-dimensional perspective graph of the points. The same results can be obtained with **mesh(x,y,z)** which can also be used.

Command	*Description*
meshc, meshz	If the *xy* grid is rectangular, these two functions are merely variations of the basic plotting program **mesh,** and they operate in an identical fashion. **meshc** will produce a corresponding contour plot drawn on the *xy* plane below the three-dimensional figure, and **meshz** will add a vertical wall to the outside features of the figures drawn by **mesh.**
surf	Produces a three-dimensional perspective drawing. Its use is usually to draw surfaces, as opposed to plotting functions, although the actual tasks are quite similar. The output of **surf** will be a *shaded* figure. If row vectors of length *n* are defined by $x = r \cos \theta$ and $y = r \sin \theta$, with $0 \le \theta \le 2\pi$, they correspond to a circle of radius *r*. If $\vec{r}$ is a *column* vector equal to **r = [0 1 2]'**; then **z = r*ones(size(x))** will be a rectangular, 3 × *n*, arrays of 0's and 2's, and **surf(x, y, z)** will produce a shaded surface bounded by three circles; *i.e.*, a cone.
surfc	This function is related to **surf** in the same way that **meshc** is related to mesh.
colormap	Used to change the default coloring of a figure. See the MATLAB reference manual or the help file.
shading	Controls the type of color shading used in drawing figures. See the MATLAB reference manual or the help file.
view	**view(az,el)** controls the perspective view of a three-dimensional plot. The view of the figure is from angle "**el**" above the *xy* plane with the coordinate axes (and the figure) rotated by an angle **"az"** in a clockwise direction about the *z* axis. Both angles are in degrees. The default values are **az** = 37½° and **el** = 30°.
axis	Determines or changes the scaling of a plot. If the coordinate axis limits of a two-dimensional or three-dimensional graph are contained in the row vector $r = [x_{min}, x_{max}, y_{min}, y_{max}, z_{min}, z_{max}]$, **axis** will return the values in this vector, and **axis(r)** can be used to alter them. The coordinate axes can be turned *on* and *off* with **axis('on')** and **axis('off')**. A few other string constant inputs to **axis** and their effects are given below: **axis('equal')** — *x* and *y* scaling are forced to be the same. **axis('square')** — The box formed by the axes is square. **axis('auto')** — Restores the scaling to default settings. **axis('normal')** — Restoring the scaling to full size, removing any effects of **square** or **equal** settings. **axis ('image')** — Alters the aspect ratio and the scaling so the screen pixels are square shaped rather than rectangular.
contour	The use is **contour(x,y,z)**. A default value of *N* = 10 contour lines will be drawn. An optional fourth argument can be used to control the number of contour lines that are drawn. **contour(x,y,z,N),** if **N** is a positive integer, will draw **N** contour lines, and **contour(x,y,z,V),** if **V** is a vector containing values in the range of *z* values, will draw contour lines at each value of *z* = *V*.
plot3	Plots lines or curves in three dimensions. If **x, y,** and **z** are vectors of equal length, **plot3(x,y,z)** will draw, on a three-dimensional coordinate axis system, the lines connecting the points. A fourth argument, representing the color and symbols to be used at each point, can be added in exactly the same manner as with **plot.**
grid	**grid on** adds grid lines to a two-dimensional or three-dimensional graph; **grid off** removes them.
slice	Draws "slices" of a volume at a particular location within the volume.

Example E2.5. *(a) Generate an overlay plot for plotting three lines*

$$y_1 = \sin t$$
$$y_2 = t$$
$$y_3 = t - \frac{t^3}{3!} + \frac{t^5}{5!} + \frac{t^7}{7!}$$

Use (i) the plot command
(ii) the hold command
(iii) the line command

(b) Use the functions for plotting x-y data for plotting the following functions.
(i) $f(t) = t \cos t$
$0 \le t \le 10\pi$
(ii) $x = e^t$
$y = 100 + e^{3t}$
$0 \le t \le 2\pi.$

Solution:

(*a*) overlay plot

(*i*)

```
% using the plot command
t = linspace(0, 2*pi, 100);
y1 = sin(t); y2 = t;
y3 = t - (t.^3)/6 + (t.^5)/120 - (t.^7)/5040;
plot(t, y1, t, y2, '-', t, y3, 'o')
axis([0 5 -1 5])
xlabel('t')
ylabel('sin(t) approximation')
title('sin(t) function')
text(3.5,0, 'sin(t)')
gtext('Linear approximation')
gtext('4-term approximation')
```

Output is shown in Fig. E2.5 (a)

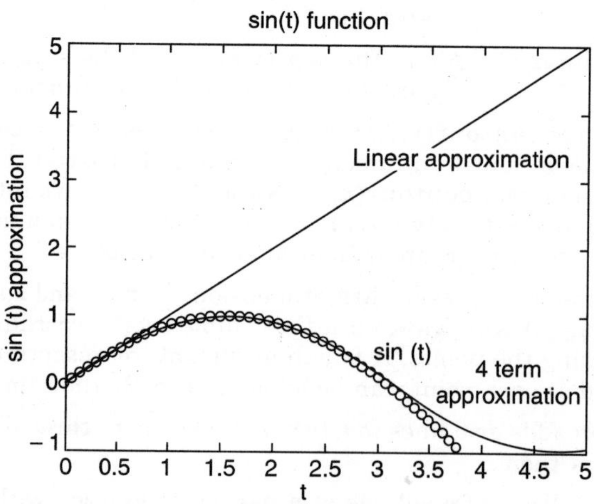

Fig. E2.5 (a)

(*ii*)

```
% using the hold command
x = linspace(0, 2*pi, 100); y1 = sin(x);
plot(x, y1)
hold on
y2 = x; plot(x, y2, '-' )
y3 = x - (x.^3)/6 + (x.^5)/120 - (t.^7)/5040;
plot(x, y3, 'o')
axis([0 5 -1 5])
hold off
```

Output is shown in Fig. E2.5(b)

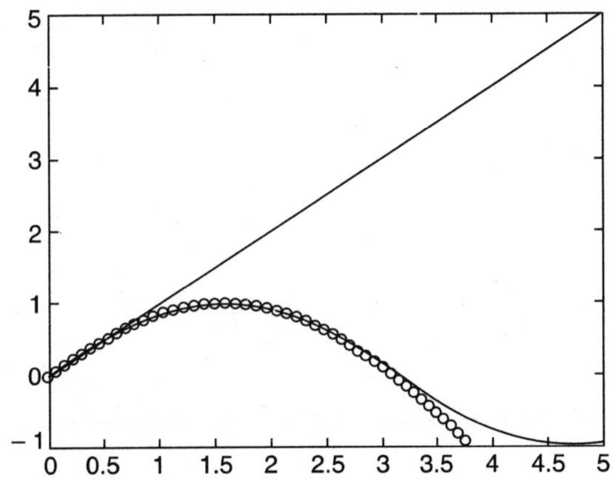

Fig. E2.5 (b)

(*iii*) % using the line command

```
t = linspace(0, 2*pi, 100);
y1 = sin(t);
y2 = t;
y3 = t - (t.^3)/6 + (t.^5)/120 - (t.^7)/5040;
plot(t, y1)
line(t, y2, 'linestyle', '-')
line(t, y3, 'marker', 'o')
axis([0 5 -1 5])
xlabel('t')
ylabel('sin(t) approximation')
title('sin(t) function')
legend('sin(t)', 'linear approx', '7th order approx')
```

Output is shown in Fig. E2.5 (c)

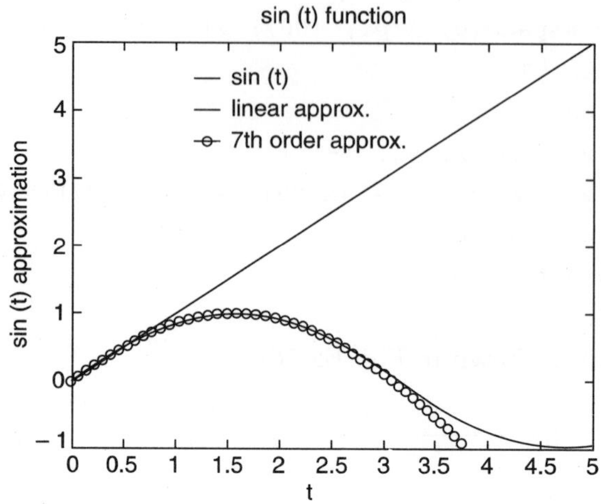

Fig. E2.5(c)

(*b*) Using Table 2.29 functions

(*i*) `fplot('x.*cos(x)', [0 10*pi])`

This will give the following figure (Fig. E2.5 (*d*))

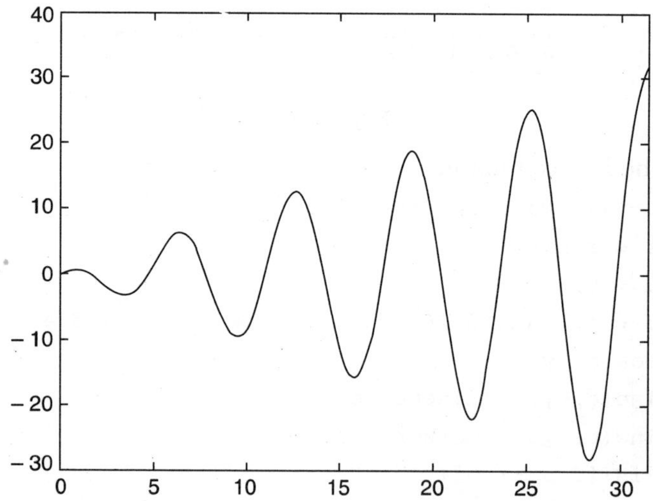

Fig. E2.5 (d)

(*ii*)

```
t = linspace(0, 2*pi, 200);
x = exp(t);
y = 100 + exp(3*t);
loglog(x, y), grid
```

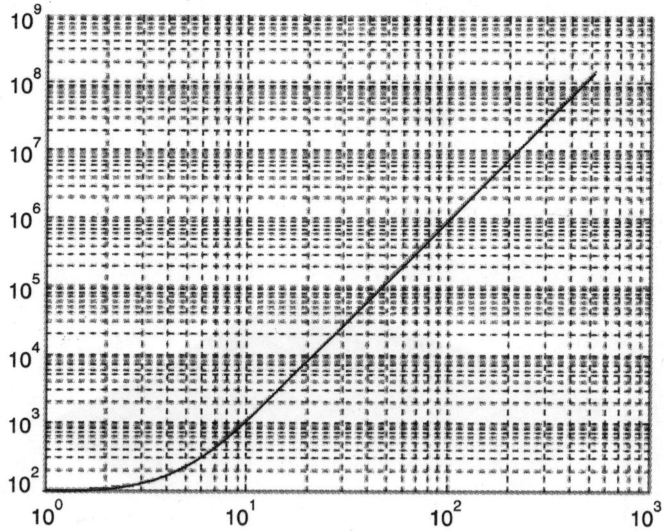

Fig. E2.5 (e)

Example E2.6. (*a*) *Plot the parametric space curve of*

$$x(t) = t$$
$$y(t) = t^2$$
$$z(t) = t^3 \qquad 0 \le t \le 2.0$$

(*b*) $z = -7/(1 + x^2 + y^2) \qquad |x| \le 5, |y| \le 5$

Solution:

(*a*)
```
>> t=linspace(0, 2,100);
>> x=t; y=t. ^2; z=t. ^3;
>> plot3(x, y, z), grid
```

The plot is shown in Figure E2.6 (*a*).

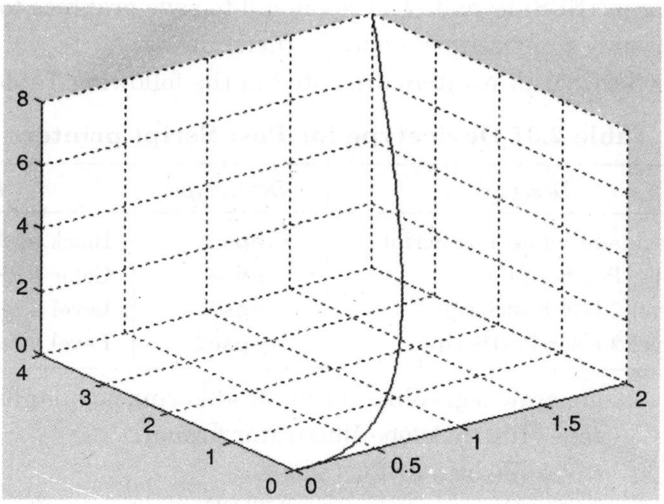

Fig. E2.6 (a)

```
(b) >> t=linspace(0, 2,100);
    >> x=t; y=t. ^2; z=t. ^3;
    >> plot3(x, y, z), grid
    >>  t=linspace(-5,5,50);y=x;
    >>  z=-7./(1+x.^2+y.^2);
    >> mesh(z)
```

The plot is shown in Figure E2.6(*b*).

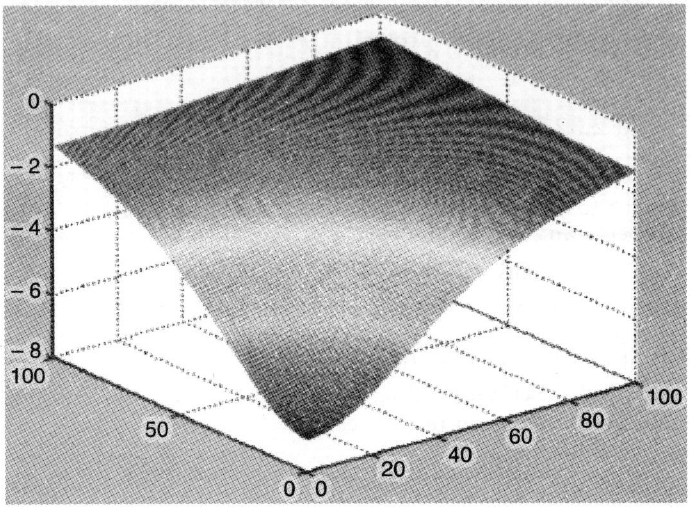

Fig. E2.6(b)

2.17.4 Saving and Printing Graphs

To obtain a hardcopy of a graph, type *print* in the Command Window after the graph appears in the Figure Window. The figure can also be saved into a specified file in the PostScripter or Encapsulated PostScript (EPS) format. The command to save graphics to a file is

print – *d* devicetype – options filename

where device type for PostScript printers are listed in the following Table 2.31.

Table 2.31 Devicetype for Post Script printers

Devicetype	*Description*	*Devicetype*	*Description*
ps	Black and white PostScript	**eps**	Black and white EPSF
psc	Color PostScript	**epsc**	Color EPSF
ps2	Level 2 BW PostScript	**eps2**	Level 2 black and white EPSF
psc2	Level 2 color PostScript	**epsc2**	Level 2 color EPSF

MATLAB can also generate a graphics file in the following popular formats among others.

–dill saves file in Adobe Illustrator format.
–djpeg saves file as a JPEG image.
–dtiff saves file as a compressed TIFF image.
–dmfile saves file as an M-file with graphics handles.

2.18 INPUT/OUTPUT IN MATLAB

In this section, we present some of the many available commands in MATLAB for reading data from an external file into a MATLAB matrix, or writing the numbers computed in MATLAB into such an external file.

2.18.1 The fopen Statement

To have the MATLAB read or write a separate data file of numerical values, we need to *connect* the file to the executing MATLAB program. The MATLAB functions used are summarized in Table 2.32.

Table 2.32 MATLAB functions used for input/output

Function	*Description*
fopen	Connects an existing file to MATLAB or to create a new file from MATLAB. **fid = fopen(**'*Filename*', *permission code*); where, if **fopen** is successful, **fid** will be returned as a positive integer greater than 2. When unsuccessful, a value of –1 is returned. Both the file name and the permission code are string constants enclosed in single quotes. The permission code can be a variety of flags that specify whether or not the file can be written to, read from, appended to, or a combination of these. Some common codes are: *Code* — *Meaning* **'r'** — read only **'w'** — write only **'r+'** — read and write **'a+'** — read and append The **fopen** statement positions the file at the beginning.
fclose	Disconnects a file from the operating MATLAB program. The use is **fclose(fid),** where **fid** is the *file identification number* of the file returned by **fopen.fclose('all')** will close all files.
fscanf	Reads opened files. The use is **A = fscanf(fid,** *FORMAT, SIZE*) where *FORMAT* specifies the types of numbers (integers, reals with or without exponent, character strings) and their arrangement in the data file, and optional *SIZE* determines how many quantities are to be read and how they are to be arranged into the matrix **A.** If *SIZE* is omitted, the entire file is read. The *FORMAT* field is a string (enclosed in single quotes) specifying the form of the numbers in the file. The *type* of each number is characterized by a percent sign (%), followed by a letter (**i** or **d** for integers, **e** or **f** for floating-point numbers with or without exponents). Between the percent sign and the type code, one can insert an integer specifying the maximum width of the field.
fprintf	Writes files previously opened. **fprintf(fid,** *FORMAT,* **A)** where **fid** and *FORMAT* have the same meaning as for **fscanf**, with the exception that for output formats the string must end with **\n**, designating the end of a line of output.

2.19 SYMBOLIC MATHEMATICS

In Secs. 2.1 to 2.18, the capability of MATLAB for numerical computations have been described. In this section some of MATLAB's capabilities for symbolic manipulations will be presented. Specifically, the symbolic expressions, symbolic algebra, simplification of mathematical expressions, operations on symbolic expressions, solution of a single equation or a set of linear algebraic equations, solutions to differential equations, differentiation and integration of functions using MATLAB are presented.

2.19.1 Symbolic Expressions

A symbolic expression is stored in MATLAB as a *character string*. A single quote marks are used to define the symbolic expression. For instance:

'sin(*y*/*x*)'; '*x*^4 + 5**x*^3 + 7**x*^2 – 7'

The independent variable in many functions is specified as an additional function argument. If an independent variable is not specified, then MATLAB will pick one. When several variables exist, MATLAB will pick the one that is a single lower case letter (except *i* and *j*), which is closest to *x* alphabetically.

The independent variable is returned by the function *symvar*,

symvar(s) Returns the independent variable for the symbolic expression *s*.

For example:

Expression s	symvar(s)
'5 * *c* * *d* + 34'	*d*
'sin(*y*/*x*)'	*x*

In MATLAB, a number of functions are available to simplify mathematical expressions by expanding the terms, factoring expressions, collecting coefficients, or simplifying the expression. For instance:

expand(s) Performs an expansion of s.

A summary of these expressions is given in Table 2.33. A summary of basic operations is given in Table 2.34. The standard arithmetic operation (Table 2.35) is applied to symbolic expressions using symbolic functions. These symbolic expressions are summarized in Table 2.36.

Table 2.33

	Simplification
collect	Collect common terms
expand	Expand polynomials and elementary functions
factor	Factorization
horner	Nested polynomial representation
numden	Numerator and denominator
simple	Search for shortest form
simplify	Simplification
subexpr	Rewrite in terms of subexpressions

Table 2.34

Basic Operations	
ccode	C code representation of a symbolic expression
conj	Complex conjugate
findsym	Determine symbolic variables
fortran	Fortran representation of a symbolic expression
imag	Imaginary part of a complex number
latex	LaTeX representation of a symbolic expression
pretty	Pretty prints a symbolic expression
real	Real part of an imaginary number
sym	Create symbolic object
syms	Shortcut for creating multiple symbolic objects

Table 2.35

Arithmetic Operations	
+	Addition
–	Subtraction
*	Multiplication
.*	Array multiplication
/	Right division
./	Array right division
\	Left division
.\	Array left division
^	Matrix or scalar raised to a power
.^	Array raised to a power
‘	Complex conjugate transpose
.‘	Real transpose

Table 2.36

Symbolic expressions	
horner(S)	Transposes **S** into its Horner, or nested, representation.
numden(S)	Returns two symbolic expressions that represent, respectively, the numerator expression and the denominator expression for the rational representation of **S**.
numeric(S)	Converts **S** to a numeric form (**S** must not contain any symbolic variables).
poly2sym(c)	Converts a polynomial coefficient vector **c** to a symbolic polynomial.
pretty(S)	Prints **S** in an output form that resembles typeset mathematics.
sym2poly(S)	Converts **S** to a polynomial coefficient vector.*
symadd(A,B)	Performs a symbolic addition, **A + B.**

(Contd.)

Symbolic expressions	
symdiv(A,B)	Performs a symbolic division, **A/B.**
symmul(A,B)	Performs a symbolic multiplication, **A * B.**
sympow(S,p)	Performs a symbolic power, **S^p.**
symsub(A,B)	Performs a symbolic subtraction, **A – B.**

2.19.2 Solution to Differential Equations

Symbolic math functions can be used to solve a single equation, a system of equations, and differential equations. For example:

solve(f) Solves a symbolic equation **f** for its symbolic variable. If **f** is a symbolic expression, this function solves the equation **f** = 0 for its symbolic variable.

solve(f1, ... fn) Solves the system of equations represented by **f1, ..., fn.**

The symbolic function for solving ordinary differential equation is **dsolve** as shown below:

dsolve('equation', 'condition') Symbolically solves the ordinary differential equation specified by '**equation**'. The optional argument **'condition'** specifies a boundary or initial condition.

The symbolic equation uses the letter **D** to denote differentiation with respect to the independent variable. A **D** followed by a digit denotes repeated differentiation. Thus, **Dy** represents *dy*/*dx*, and **D2y** represents d^2y/dx^2. For example, given the ordinary second order differential equation;

$$\frac{d^2x}{dt^2} + 5\frac{dx}{dt} + 3x = 7$$

with the initial conditions $x(0) = 0$ and $\dot{x}(0) = 1$.

The MATLAB statement that determine the symbolic solution for the above differential equation is the following:

```
x = dsolve('D2x = -5*Dx-3*x+7', 'x(0)=0', 'Dx(0)=1')
```

The symbolic functions are summarized in Table 2.37.

Table 2.37

Solution of Equations	
compose	Functional composition
dsolve	Solution of differential equations
finverse	Functional inverse
solve	Solution of algebraic equations

2.19.3 Calculus

There are four forms by which the symbolic derivative of a symbolic expression is obtained in MATLAB. They are:

diff(f) Returns the derivative of the expression **f** with respect to the default independent variable.

diff(f, 't') Returns the derivative of the expression **f** with respect to the variable t.

diff(f,n) Returns the nth derivative of the expression **f** with respect to the default independent variable.

diff(f, 't',n) Returns the nth derivative of the expression **f** with respect to the variable t.

The various forms that are used in MATLAB to find the integral of a symbolic expression **f** are given below and summarized in Table 2.38.

int(f) Returns the integral of the expression **f** with respect to the default independent variable.

int(f, 't') Returns the integral of the expression **f** with respect to the variable t.

int(f,a,b) Returns the integral of the expression **f** with respect to the default independent variable evaluated over the interval **[a,b]**, where **a** and **b** are numeric expressions.

int(f, 't',a,b) Returns the integral of the expression **f** with respect to the variable **t** evaluated over the interval **[a,b]**, where **a** and **b** are numeric expressions.

int(f, 'm', 'n') Returns the integral of the expression **f** with respect to the default independent variable evaluated over the interval **[m,n]**, where m and n are numeric expressions.

The other symbolic functions for pedagogical and graphical applications, conversions, integral transforms, and linear algebra are summarized in Tables 2.38 to 2.42.

Table 2.38

Calculus	
diff	Differentiate
int	Integrate
jacobian	Jacobian matrix
limit	Limit of an expression
symsum	Summation of series
taylor	Taylor series expansion

Table 2.39

Pedagogical and Graphical Applications	
ezcontour	Contour plotter
ezcontourf	Filled contour plotter
ezmesh	Mesh plotter
ezmeshc	Combined mesh and contour plotter
ezplot	Function plotter
ezplot	Easy-to-use function plotter

(Contd.)

Pedagogical and Graphical Applications	
ezplot3	Three-dimensional curve plotter
ezpolar	Polar coordinate plotter
ezsurf	Surface plotter
ezsurfc	Combined surface and contour plotter
funtool	Function calculator
rsums	Riemann sums
taylortool	Taylor series calculator

Table 2.40

Conversions	
char	Convert sym object to string
double	Convert symbolic matrix to double
poly2sym	Function calculator
sym2poly	Symbolic polynomial to coefficient vector

Table 2.41

Integral Transforms	
fourier	Fourier transform
ifourier	Inverse Fourier transform
ilaplace	Inverse Laplace transform
iztrans	Inverse Z-transform
laplace	Laplace transform
ztrans	Z-transform

Table 2.42

Linear Algebra	
colspace	Basis for column space
det	Determinant
diag	Create or extract diagonals
eig	Eigenvalues and eigenvectors
expm	Matrix exponential
inv	Matrix inverse
jordan	Jordan canonical form
null	Basis for null space

(Contd.)

Linear Algebra	
poly	Characteristic polynomial
rank	Matrix rank
rref	Reduced row echelon form
svd	Singular value decomposition
tril	Lower triangle
triu	Upper triangle

2.20 THE LAPLACE TRANSFORMS

The Laplace transformation method is an operational method that can be used to find the transforms of time functions, the inverse Laplace transformation using the partial-fraction expansion of $B(s)/A(s)$, where $A(s)$ and $B(s)$ are polynomials in s. In this Chapter, we present the computational methods with MATLAB to obtain the partial-fraction expansion of $B(s)/A(s)$ and the zeros and poles of $B(s)/A(s)$.

MATLAB can be used to obtain the partial-fraction expansion of the ratio of two polynomials, $B(s)/A(s)$ as follows:

$$\frac{B(s)}{A(s)} = \frac{\text{num}}{\text{den}} = \frac{b(1)s^n + b(2)s^{n-1} + \ldots + b(n)}{a(1)s^n + a(2)s^{n-1} + \ldots + a(n)}$$

where $a(1) \neq 0$ and num and den are row vectors. The coefficients of the numerator and denominator of $B(s)/A(s)$ are specified by the num and den vectors.

Hence num = $[b(1) \quad b(2) \quad \ldots \quad b(n)]$
den = $[a(1) \quad a(2) \quad \ldots \quad a(n)]$

The MATLAB command

r, p, k = residue(num, den)

is used to determine the residues, poles, and direct terms of a partial-fraction expansion of the ratio of two polynomials $B(s)$ and $A(s)$ is then given by

$$\frac{B(s)}{A(s)} = k(s) + \frac{r(1)}{s - p(1)} + \frac{r(2)}{s - p(1)} + \ldots + \frac{r(n)}{s - p(n)}$$

The MATLAB command **[num, den] = residue(r, p, k)** where r, p, k are the output from MATLAB which converts the partial fraction expansion back to the polynomial ratio $B(s)/A(s)$.

The command **printsys (num,den 's')** prints the num/den in terms of the ratio of polynomials in s.

The command **ilaplace** will find the inverse Laplace transform of a Laplace function.

2.20.1 Finding Zeros and Poles of B(s)/A(s)

The MATLAB command **[z,p,k] = tf2zp(num,den)** is used to find the zeros, poles, and gain K of $B(s)/A(s)$.

If the zeros, poles, and gain K are given, the following MATLAB command can be used to find the original num/den:

[num,den] = zp2tf (z,p,k)

2.21 CONTROL SYSTEMS

MATLAB has an extensive set of functions for the analysis and design of control systems. They involve matrix operations, root determination, model conversions, and plotting of complex functions. These functions are found in MATLAB's control systems toolbox. The analytical techniques used by MATLAB for the analysis and design of control systems assume the processes that are linear and time invariant. MATLAB uses models in the form of *transfer-functions or state-space equations.*

2.21.1 Transfer Functions

The transfer function of a linear time invariant system is expressed as a ratio of two polynomials. The transfer function for a single input and a single output (SISO) system is written as

$$H(s) = \frac{b_0 s^n + b_1 s^{n-1} + \ldots + b_{n-1}s + b_n}{a_0 s^m + a_1 s^{m-1} + \ldots + a_{m-1}s + a_m}$$

when the numerator and denominator of a transfer function are factored into the *zero-pole-gain form*, it is given by

$$H(s) = k\frac{(s - z_1)\,(s - z_2)\ldots(s - z_n)}{(s - p_1)\,(s - p_2)\ldots(s - p_m)}$$

The *state-space model* representation of a linear control system *s* is written as

$$\dot{x} = Ax + Bu$$
$$y = Cx + Du$$

2.21.2 Model Conversion

There are a number of functions in MATLAB that can be used to convert from one model to another. These conversion functions and their applications are summarized in Table 2.43.

Table 2.43 Model conversion functions

Function	*Purpose*
c2d	Continuous state-space to discrete state-space
residue	Partial-fraction expansion
ss3tf	State-space to transfer function
ss2zp	State-space to zero-pole-gain
tf2ss	Transfer function to state-space
tf2zp	Transfer function to zero-pole-gain
zp2ss	Zero-pole-gain to state-space
zp2tf	Zero-pole-gain to transfer function

Residue Function: The **residue** function converts the polynomial transfer function

$$H(s) = \frac{b_0 s^n + b_1 s^{n-1} + \ldots + b_{n-1}s + b_n}{a_0 s^m + a_1 s^{m-1} + \ldots + a_{m-1}s + a_m}$$

to the partial fraction transfer function

$$H(s) = \frac{r_1}{s - p_1} + \frac{r_2}{s - p_2} + \ldots + \frac{r_n}{s - p_n} + k(s)$$

[r,p,k] = residue(B, A) Determine the vectors **r, p,** and **k,** which contain the residue values, the poles, and the direct terms from the partial-fraction expansion. The inputs are the polynomial coefficients **B** and **A** from the numerator and denominator of the transfer function, respectively.

ss2tf Function: The **ss2tf** function converts the continuous-time, state-space equations

$$x' = Ax + Bu$$
$$y = Cx + Du$$

to the polynomial transfer function

$$H(s) = \frac{b_0 s^n + b_1 s^{n-1} + \ldots + b_{n-1} s + b_n}{a_0 s^m + a_1 s^{m-1} + \ldots + a_{m-1} s + a_m}$$

The function has two output matrices:

[num,den] = ss2tf(A,B,C,D,iu) Computes vectors **num** and **den** containing the coefficients, in descending powers of **s**, of the numerator and denominator of the polynomial transfer function for the **iu**^th^ input. The input arguments **A,B,C,** and **D** are the matrices of the state-space equations corresponding to the **iu**^th^ input, where **iu** is the number of the input for a multi-input system. In the case of a single-input system, **iu** is 1.

ss2zp Function: The **ss2zp** function converts the continuous-time, state-space equations

$$x' = Ax + Bu$$
$$y = Cx + Du$$

to the zero-pole-gain transfer function

$$H(s) = k\frac{(s - z_1)(s - z_2)\ldots(s - z_n)}{(s - p_1)(s - p_2)\ldots(s - p_m)}$$

The function has three output matrices:

[z,p,k] = ss2zp(A,B,C,D,iu) Determines the zeros **(z)** and poles **(p)** of the zero-pole-gain transfer function for the **iu**^th^ input, along with the associated gain **(k).** The input matrices **A, B, C,** and **D** of the state-space equations correspond to the **iu**^th^ input, where **iu** is the number of the input for a multi-input system. In the case of a single-input system **iu** is 1.

tf2ss Function: The **ts2ss** function converts the polynomial transfer function

$$H(s) = \frac{b_0 s^n + b_1 s^{n-1} + \ldots + b_{n-1} s + b_n}{a_0 s^m + a_1 s^{m-1} + \ldots + a_{m-1} s + a_m}$$

to the controller-canonical form state-space equations

$$x' = Ax + Bu$$
$$y = Cx + Du$$

The function has four output matrices:

[A,B,C,D] = tf2ss(num,den) Determines the matrices **A, B, C, and D** of the controller-canonical form state-space equations. The input arguments num and den contain the coefficients, in descending powers of **s**, of the numerator and denominator polynomials of the transfer function that is to be converted.

tf2zp Function: The **tf2zp** function converts the polynomial transfer function

$$H(s) = \frac{b_0 s^n + b_1 s^{n-1} + \ldots + b_{n-1}s + b_n}{a_0 s^m + a_1 s^{m-1} + \ldots + a_{m-1}s + a_m}$$

to the zero-pole-gain transfer function

$$H(s) = k\frac{(s - z_1)(s - z_2)\ldots(s - z_n)}{(s - p_1)(s - p_2)\ldots(s - p_m)}$$

The function has three output matrices:

[z,p,k] = tf2zp(num,den) Determines the zeros **(z)**, poles **(p)** and associated gain **(k)** of the zero-pole-gain transfer function using the coefficients, in descending powers of **s**, of the numerator and denominator of the polynomial transfer function that is to be converted.

zp2tf Function: The **zp2tf** function converts the zero-pole-gain transfer function

$$H(s) = k\frac{(s - z_1)(s - z_2)\ldots(s - z_n)}{(s - p_1)(s - p_2)\ldots(s - p_m)}$$

to the polynomial transfer function

$$H(s) = \frac{b_0 s^n + b_1 s^{n-1} + \ldots + b_{n-1}s + b_n}{a_0 s^m + a_1 s^{m-1} + \ldots + a_{m-1}s + a_m}$$

The function has two output matrices:

[num,den] = zp2tf(z,p,k) Determines the vectors num and den containing the coefficients, in descending powers of **s**, of the numerator and denominator of the polynomial transfer function. **p** is a column vector of the pole locations of the zero-pole-gain transfer function, **z** is a matrix of the corresponding zero locations, having one column for each output of a multi-output system, **k** is the gain of the zero-pole-gain transfer function. In the case of a single-output system, **z** is a column vector of the zero locations corresponding to the pole locations of vector **p**.

zp2ss Function: The **zp2ss** function converts the zero-pole-gain transfer function

$$H(s) = k\frac{(s - z_1)(s - z_2)\ldots(s - z_n)}{(s - p_1)(s - p_2)\ldots(s - p_m)}$$

to the controller-canonical form state-space equations

$$x' = Ax + Bu$$
$$y = Cx + Du$$

The function has four output matrices:

[A,B,C,D] = zp2ss(z,p,k) Determines the matrices **A**, **B**, **C**, and **D** of the control-canonical form state-space equations. **p** is a column vector of the pole locations of the zero-pole-gain transfer function, **z** is a matrix of the corresponding zero locations, having one column for each output of a multi-output system, **k** is the gain of the zero-pole-gain transfer function. In the case of a single-output system, **z** is a column vector of the zero locations corresponding to the pole locations of vector **p**.

2.22 THE LAPLACE TRANSFORMS

MATLAB can be used to obtain the partial-fraction expansion of the ratio of two polynomials, $B(s)/A(s)$ as follows:

$$\frac{B(s)}{A(s)} = \frac{num}{den} = \frac{b(1)s^n + b(2)s^{n-1} + \ldots + b(n)}{a(1)s^n + a(2)s^{n-1} + \ldots + a(n)}$$

where $a(1) \neq 0$ and num and den are row vectors. The coefficients of the numerator and denominator of $B(s)/A(s)$ are specified by the num and den vectors.

Hence num = $[b(1) \quad b(2) \quad \ldots \quad b(n)]$

den = $[a(1) \quad a(2) \quad \ldots \quad a(n)]$

The MATLAB command

r, p, k = residue(num, den)

is used to determine the residues, poles, and direct terms of a partial-fraction expansion of the ratio of two polynomials $B(s)$ and $A(s)$ which is then given by

$$\frac{B(s)}{A(s)} = k(s) + \frac{r(1)}{s - p(1)} + \frac{r(2)}{s - p(2)} + \ldots + \frac{r(n)}{s - p(n)}$$

The MATLAB command **[num, den] = residue(r, p, k)** where r, p, k are the output from MATLAB converts the partial fraction expansion back to the polynomial ratio $B(s)/A(s)$.

The command **printsys (num,den,'s')** prints the num/den in terms of the ratio of polynomials in s.

The command **ilaplace** will find the inverse Laplace transform of a Laplace function.

2.22.1 Finding Zeros and Poles of B(s)/A(s)

The MATLAB command **[z,p,k] = tf2zp(num,den)** is used to find the zeros, poles, and gain K of $B(s)/A(s)$.

If the zeros, poles, and gain K are given, the following MATLAB command can be used to find the original num/den:

[num,den] = zp2tf (z,p,k)

EXAMPLE PROBLEMS AND SOLUTIONS

Example E2.7. *Consider the function*

$$H(s) = \frac{n(s)}{d(s)}$$

where $n(s) = s^4 + 6s^3 + 5s^2 + 4s + 3$

$d(s) = s^5 + 7s^4 + 6s^3 + 5s^2 + 4s + 7$

(a) Find n(– 10), n(– 5), n(– 3) and n(– 1)

(b) Find d(– 10), d(– 5), d(– 3) and d(– 1)

(c) Find H(– 10), H(– 5), H(– 3) and H(–1)

Solution:

(*a*)

```
>> n=[1 6 5 4 3];      % n=s^4+6s^3+5s^2+4s+3
>> d=[1 7 6 5 4 7];     % d=s^5+7s^4+6s^3+5s^2+4s+7
>> n2=polyval(n,[-10])
n2 = 4463
>> nn10=polyval(n,[-10])
nn10 = 4463
>> nn5=polyval(n,[-5])
nn5 = -17
>> nn3=polyval(n,[-3])
nn3 = -45
>> nn1=polyval(n,[-1])
nn1 = -1
```

(*b*)

```
>> dn10=polyval(d,[-10])
dn10 = -35533
>> dn5=polyval(d,[-5])
dn5 = 612
>> dn3=polyval(d,[-3])
dn3 = 202
>> dn1=polyval(d,[-1])
dn1 = 8
```

(*c*)

```
>> Hn10=nn10/dn10
Hn10 = -0.1256
>> Hn5=nn5/dn5
Hn5 = -0.0278
>> Hn3=nn3/dn3
Hn3 = -0.2228
>> Hn1=nn1/dn1
Hn1 = -0.1250
```

Example E2.8. *Generate a plot of*

$$y(x) = e^{-0.7x} \sin \omega x$$

where $\omega = 15$ *rad/s, and* $0 \le x \le 15$. *Use the colon notation to generate the x vector in increments of 0.1.*

Solution:

```
>> x = [0 : 0.1 : 15];
>> w = 15;
>> y = exp(-0.7*x).*sin(w*x);
>> plot(x,y)
>> title('y(x)  = e^-^0^.^7^xsin \omegax')
>> xlabel('x')
>> ylabel('y')
```

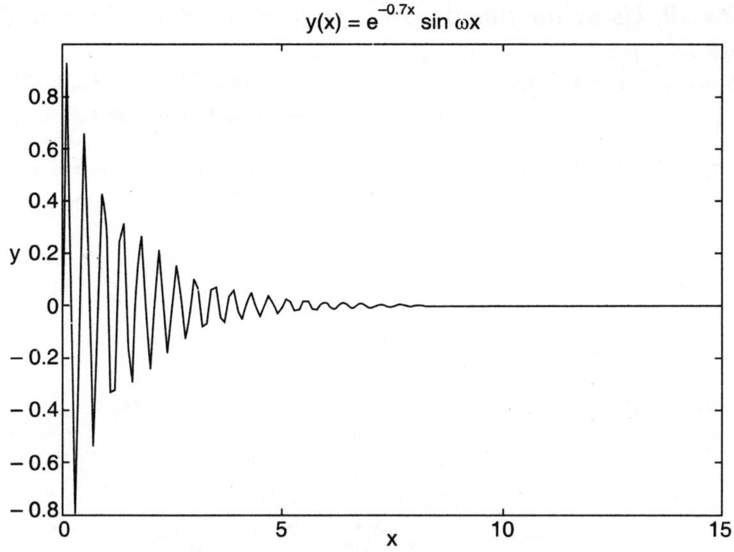

Fig. E2.8

Example E2.9. *Generate a plot of*

$$y(x) = e^{-0.6x} \cos \omega x$$

where $\omega = 10$ *rad/s, and* $0 \le x \le 15$*. Use the colon notation to generate the x vector in increments of 0.05.*

Solution:

```
>> x = [0 : 0.1 : 15];
>> w = 10;
>> y = exp(-0.6*x).*cos(w*x);
>> plot(x,y)
>> title('y(x) = e^-^0^.^6^xcos \omegax')
>> xlabel('x')
>> ylabel('y')
```

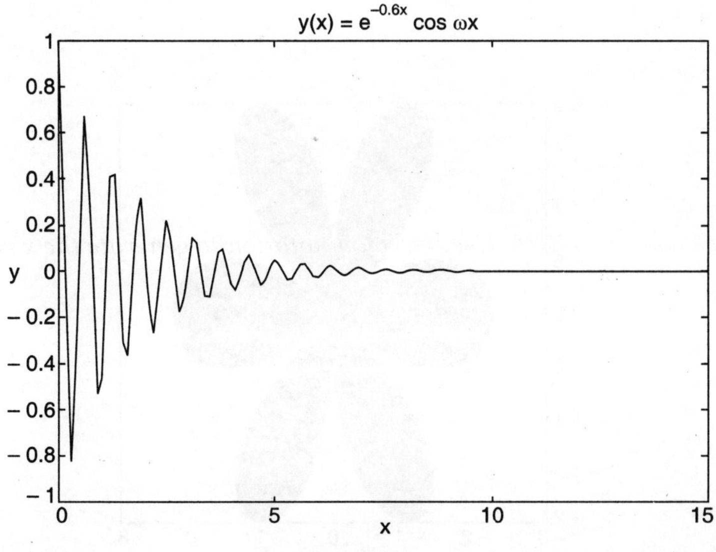

Fig. E2.9

Example E2.10. *Using the functions for plotting x-y data given in Table 2.29 plot the following functions.*

(*a*) $r^2 = 5 \cos 3t \quad 0 \le t \le 2\pi$

(*b*) $r^2 = 5 \cos 3t \quad 0 \le t \le 2\pi$

$x = r \cos t,\ y = r \sin t$

(*c*) $y_1 = e^{-2x} \cos x \quad 0 \le t \le 20$

$y_2 = e^{2x}$

(*d*) $y = \dfrac{\cos(x)}{x} \quad -5 \le x \le 5\pi$

(*e*) $f = e^{-3t/5} \cos t \quad 0 \le t \le 2\pi$

(*f*) $z = -\dfrac{1}{3}x^2 + 2xy + y^2 \quad |x| \le 7,\ |y| \le 7$

Solution:

(*a*)

```
t = linspace(0, 2*pi, 200);
r =sqrt(abs(5*cos(3*t)));
polar(t,r)
```

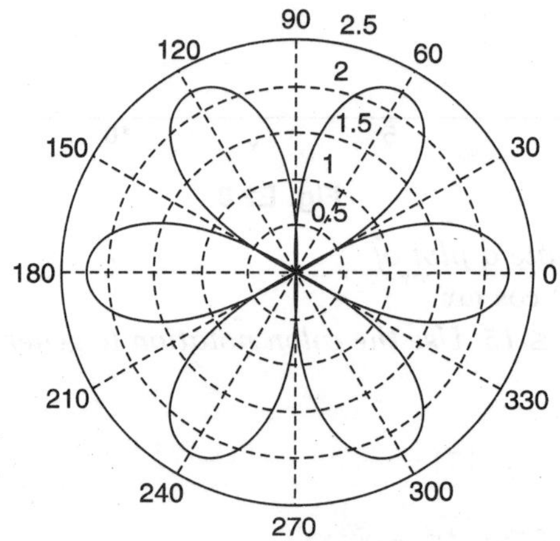

Fig. E2.10(a)

(*b*)

```
t = linspace(0, 2*pi, 200);
r =sqrt(abs(5*cos(3*t)));
x=r.*cos(t);
y=r.*sin(t);
fill(x,y,'k'),
axis('square')
```

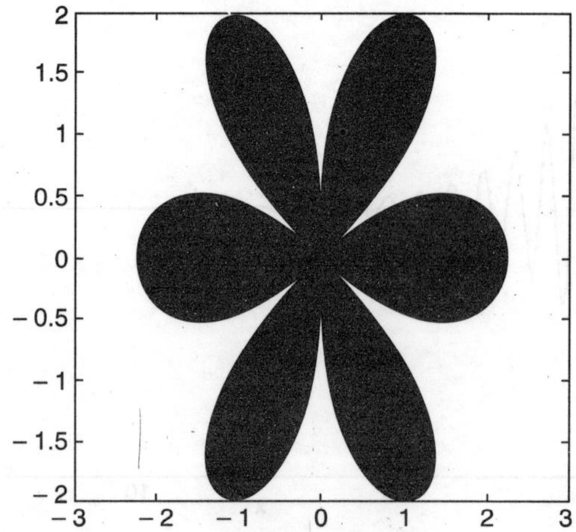

Fig. E2.10(b)

(*c*)
```
x=1:0.1:20;
y1=exp(-2*x).*cos(x);
y2=exp(2*x);
Ax=plotyy(x,y1,x,y2);
hy1=get(Ax(1),'ylabel');
hy2=get(Ax(2),'ylabel');
set(hy1,'string','exp(-2x).cos(x)')
set(hy2,'string','exp(-2x)');
```

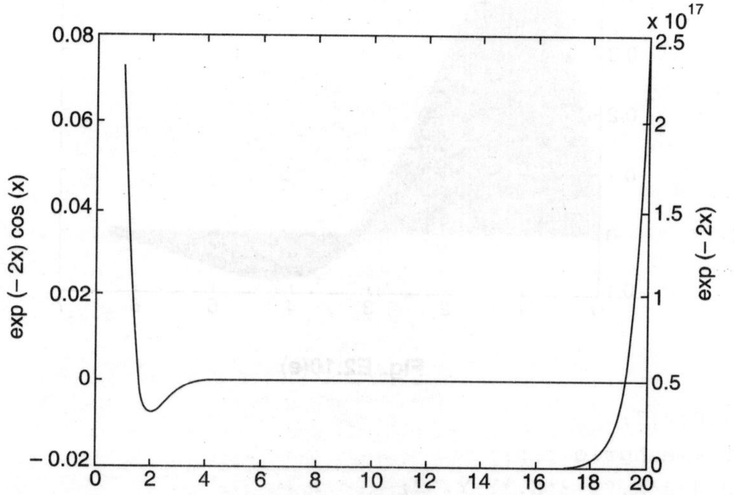

Fig. E2.10(c)

(*d*)
```
x=linspace(-5*pi,5*pi,100);
y=cos(x)./x;
area(x,y);
xlabel('x  (rad)'),ylabel('cos(x)/x')
hold on
```

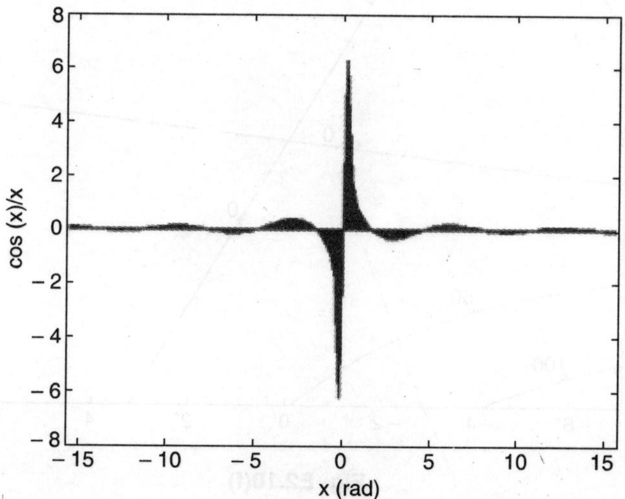

Fig. E2.10(d)

(*e*)
```
t=linspace(0,2*pi,200);
f=exp(-0.6*t).*sin(t);
stem(t,f)
```

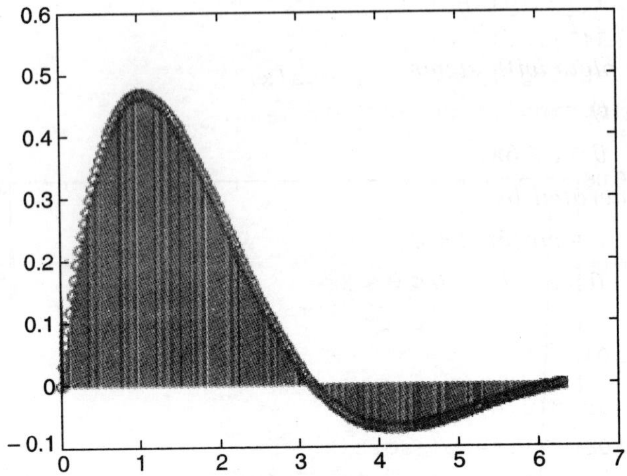

Fig. E2.10(e)

(*f*)
```
r=-7:0.2:7;
[X,Y]=meshgrid(r,r);
Z=-0.333*X.^2+2*X.*Y+Y.^2;
cs=contour(X,Y,Z);
label(cs)
```

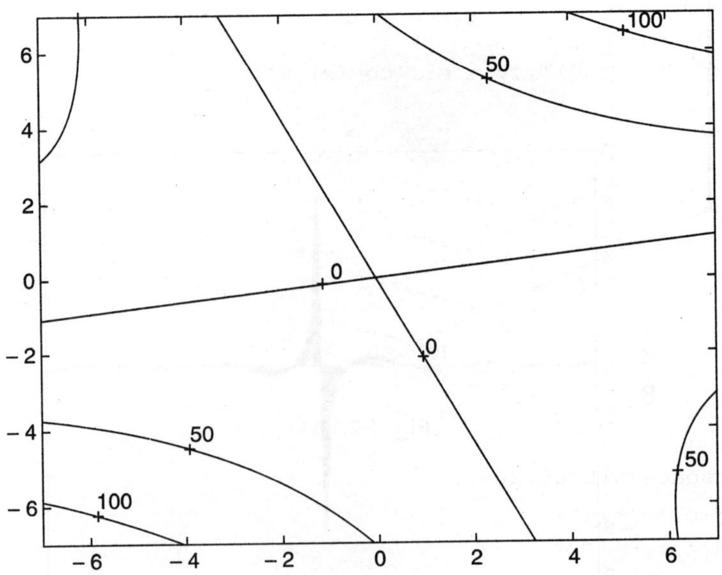

Fig. E2.10(f)

Example E2.11. *Use the functions listed in Table 2.30 for plotting 3-D data for the following.*

(*a*) $z = \cos x \cos y \; e^{-\frac{\sqrt{x^2 + y^2}}{5}}$

$|x| \leq = 7, \; |y| \leq 7$

(*b*) *Discrete data plots with stems*

$x = t, \; y = t \cos(t)$

$z = e^{t/5} - 2 \qquad 0 \leq t \leq 5\pi$

(*c*) *A cylinder generated by*

$r = \sin(5\pi z) + 3$

$0 \leq z \leq 1 \qquad 0 \leq \theta \leq 2\pi$

Solution:

(*a*)

```
u=-7:0.2:7;
[X,Y]=meshgrid(u,u);
Z=cos(X).*cos(Y).*exp(-sqrt(X.^2+Y.^2)/5);
surf(X,Y,Z)
```

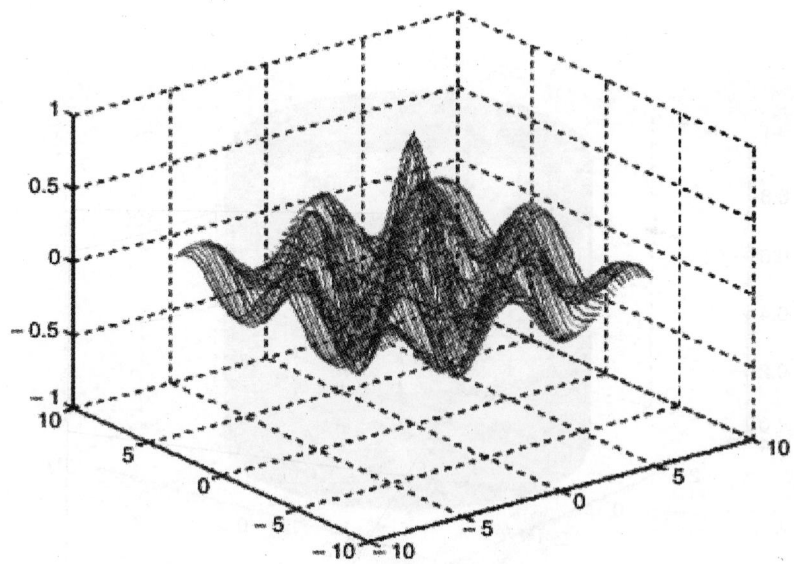

Fig. E2.11(a)

(*b*)

```
t=linspace(0,5*pi,200);
x=t;y=t.*cos(t);
z=exp(t/5)-2;
stem3(x,y,z,'filled');
xlabel('t'),ylabel('tcos(t)'),zlabel('e^t/5-1')
```

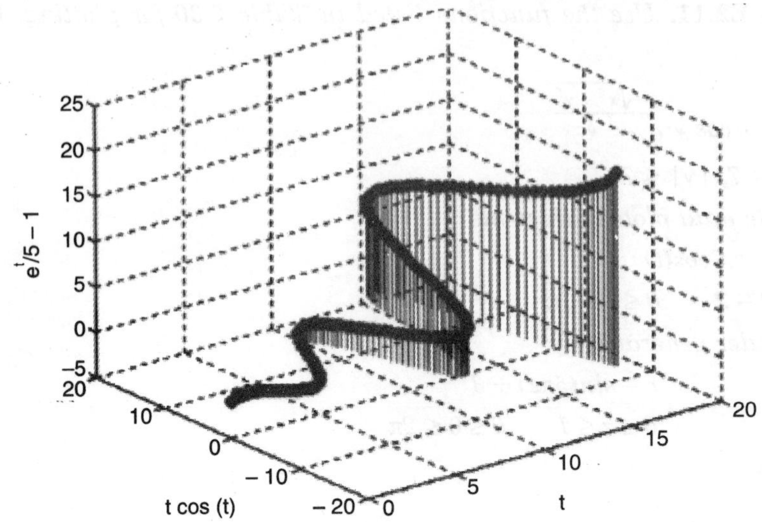

Fig. E2.11(b)

(*c*)
```
z=[0:0.2:1]';
r=sin(5*pi*z)+3;
cylinder(r)
```

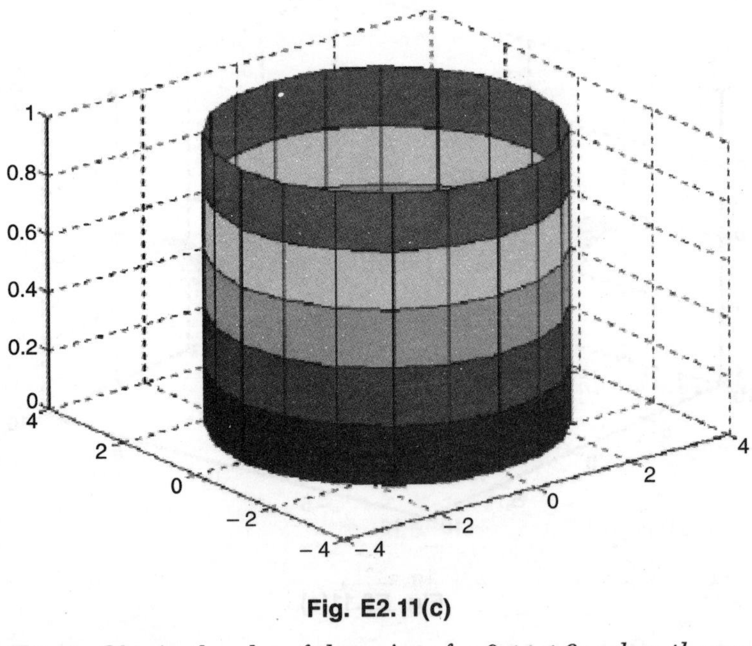

Fig. E2.11(c)

Example E2.12. *Obtain the plot of the points for* $0 \le t \le 6\pi$ *when the coordinates x,y,z are given as a function of the parameter t as follows:*

$$x = \sqrt{t}\ \sin(3t)$$

$$y = \sqrt{t}\ \cos(3t)$$

$$z = 0.8t$$

Solution:

% Line plots

```
>> t=[0:0.1:6*pi];
>> x=sqrt(t).*sin(3*t);
>> y=sqrt(t).*cos(3*t);
>> z=0.8*t;
>> plot3(x,y,z,'k','linewidth',1)
>> grid on
>> xlabel('x');ylabel('y');zlabel('z')
```

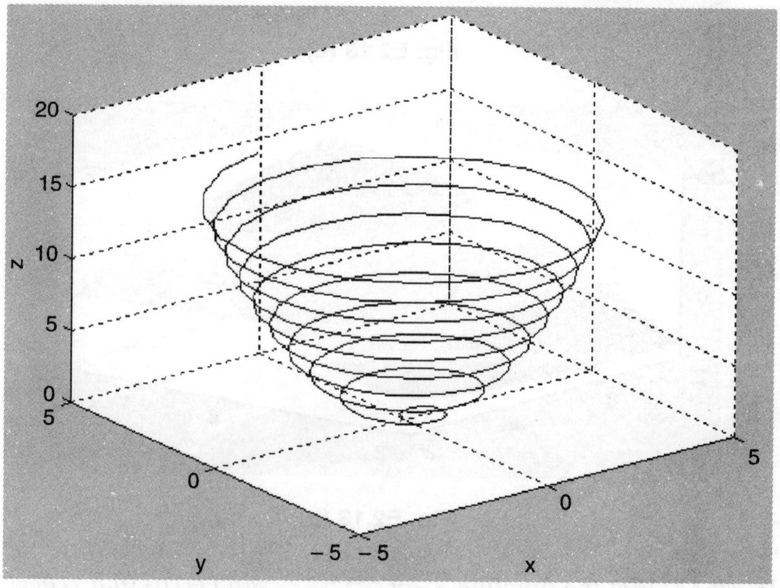

Fig. E2.12

Example E2.13. *Obtain the mesh and surface plots for the function* $z = \frac{2xy^2}{x^2 + y^2}$ *over the domain* $-2 \le x \le 6$ *and* $2 \le y \le 8$.

Solution:

% Mesh and surface plots

```
x=-2:0.1:6;
>> y=2:0.1:8;
>> [x,y]=meshgrid(x,y);
>> z=2*x.*y.^2./(x.^2+y.^2);
>> mesh(x,y,z)
>> xlabel('x');ylabel('y');zlabel('z')
>> surf(x,y,z)
>> xlabel('x');ylabel('y');zlabel('z')
```

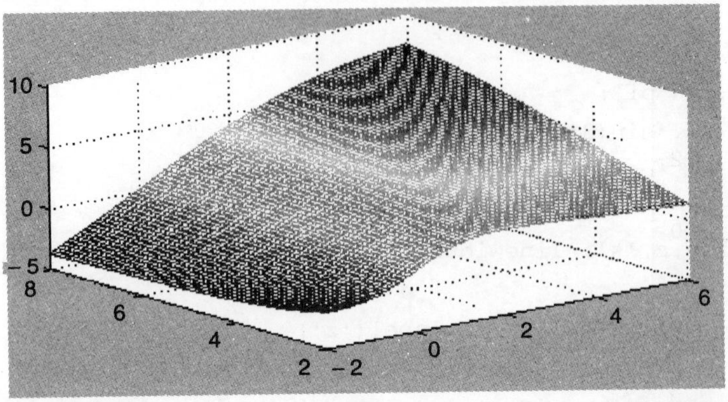

Fig. E2.13 (a)

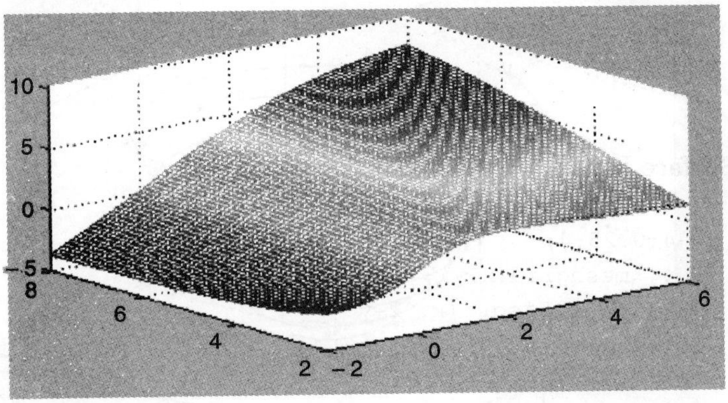

Fig. E2.13 (b)

Example E2.14. *Plot the function* $z = 2^{-1.5\sqrt{x^2 + y^2}}$ *sin (x) cos (0.5 y) over the domain* $-4 \le x \le 4$ *and* $-4 \le y \le 4$ *using Table 2.30.*

(*a*) *Mesh plot*

(*b*) *Surface plot*

(*c*) *Mesh curtain plot*

(*d*) *Mesh and contour plot*

(*e*) *Surface and contour plot*

Solution:

(*a*) % Mesh Plot

```
>> x=-4:0.25:4;
>> y=-4:0.25:4;
>> [x,y]=meshgrid(x,y);
>> z=2.^(-1.5*sqrt(x.^2 + y.^2)).*cos(0.5*y).*sin(x);
>> mesh(x,y,z)
>> xlabel('x');ylabel('y')
>> zlabel('z')
```

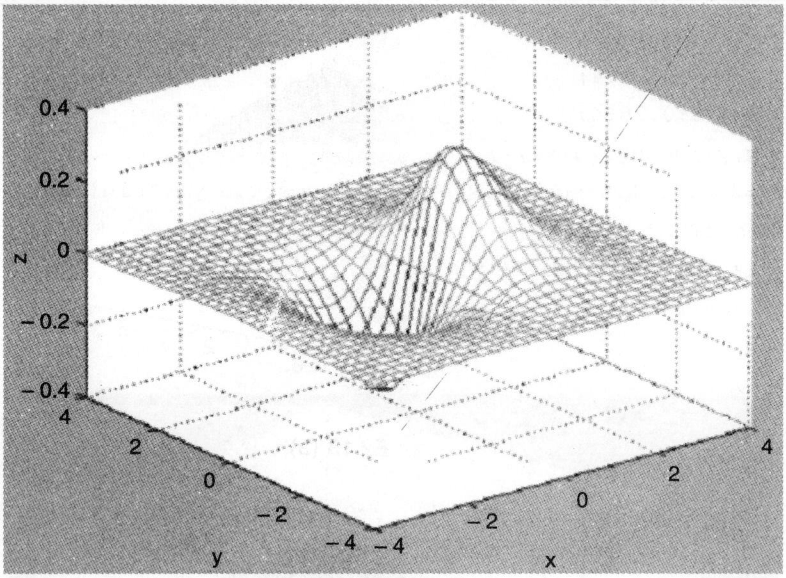

Fig. E2.14 (a)

(*b*) % Surface Plot

```
>> x=-4:0.25:4;
>> y=-4:0.25:4;
>> [x,y]=meshgrid(x,y);
>> z=2.0.^(-1.5*sqrt(x.^2+y.^2)).*cos(0.5*y).*sin(x);
>> surf(x,y,z)
>> xlabel('x');ylabel('y')
>> zlabel('z')
```

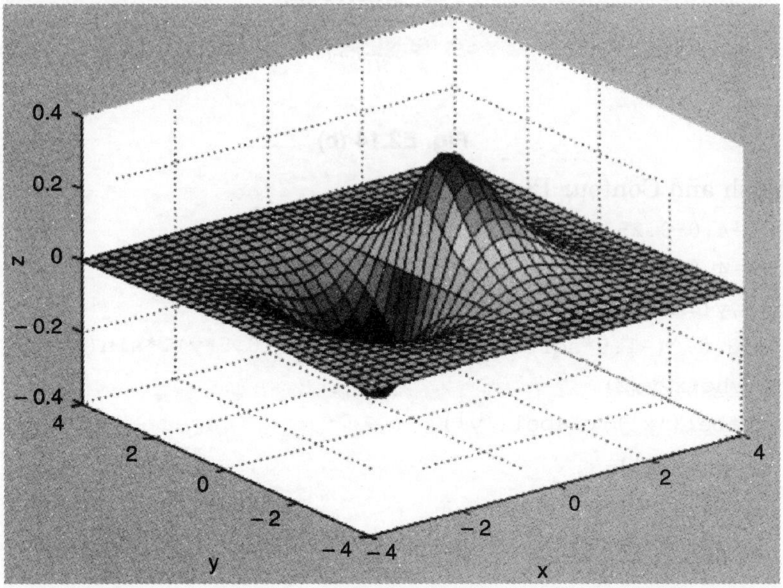

Fig. E2.14 (b)

(*c*) % Mesh Curtain Plot

```
>> x=-4.0:0.25:4;
>> y=-4.0:0.25:4;
>> [x,y]=meshgrid(x,y);
>> z=2.0.^(-1.5*sqrt(x.^2+y.^2)).*cos(0.5*y).*sin(x);
>> meshz(x,y,z)
>> xlabel('x');ylabel('y')
>> zlabel('z')
```

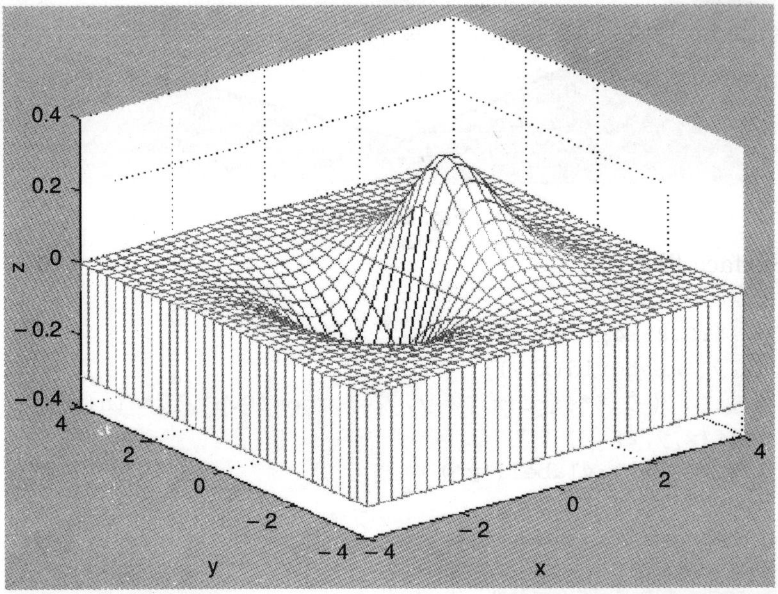

Fig. E2.14 (c)

(*d*) % Mesh and Contour Plot

```
>> x=-4.0:0.25:4;
>> y=-4.0:0.25:4;
>> [x,y]=meshgrid(x,y);
>> z=2.0.^(-1.5*sqrt(x.^2+y.^2)).*cos(0.5*y).*sin(x);
>> meshc(x,y,z)
>> xlabel('x');ylabel('y')
>> zlabel('z')
```

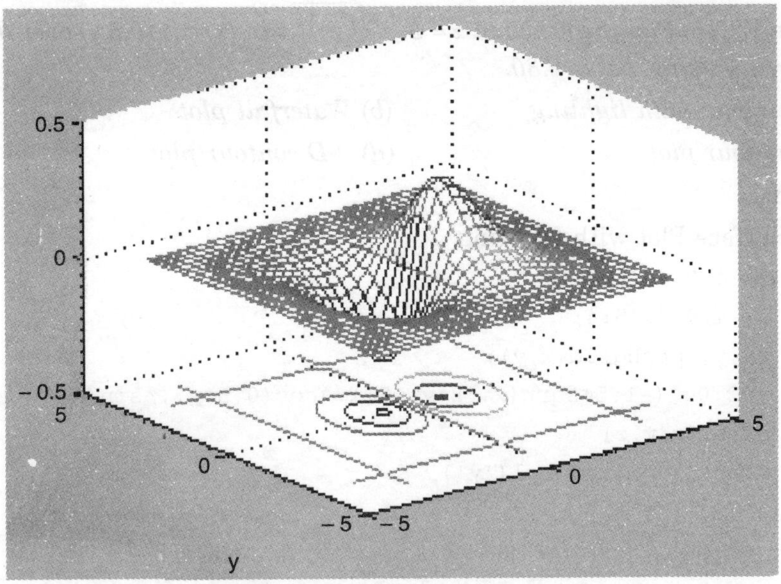

Fig. E2.14 (d)

(*e*) % Surface and Contour Plot

```
>> x=-4.0:0.25:4;
>> y=-4.0:0.25:4;
>> [x, y] =meshgrid(x, y);
>> z=2.0.^(-1.5*sqrt(x.^2+y.^2)).*cos|(0.5*y).*sin(x);
>> surfc(x,y,z)
>> xlabel('x');ylabel('y')
>> zlabel('z')
```

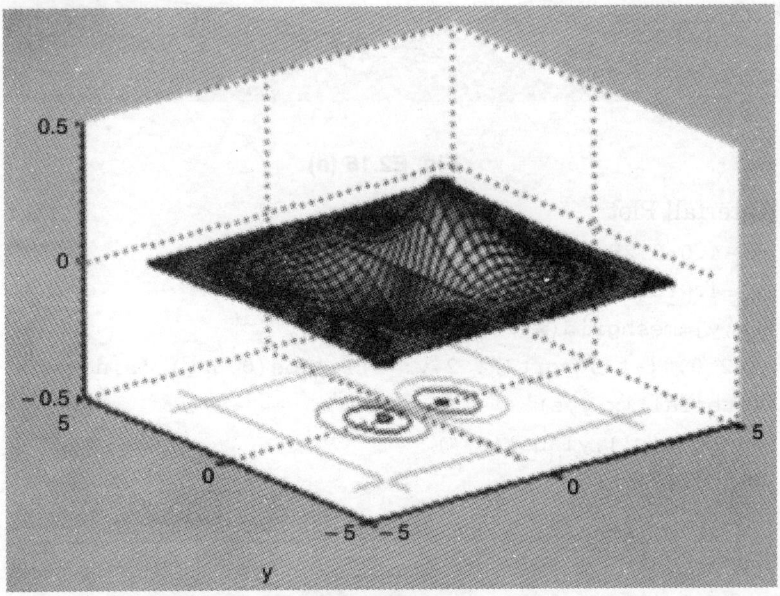

Fig. E2.14 (e)

Example E2.15. *Plot the function* $z = 2^{-1.5\sqrt{x^2 + y^2}}$ *sin (x) cos (0.5 y) over the domain* $-4 \le x \le 4$ *and* $-4 \le y \le 4$ *using Table 2.30.*

(*a*) *Surface plot with lighting* (*b*) *Waterfall plot*

(*c*) *3-D contour plot* (*d*) *2-D contour plot*

Solution:

(*a*) % Surface Plot with Lighting

```
>> x=-4.0:0.25:4;
>> y=-4.0:0.25:4;
>> [x,y]=meshgrid(x,y);
>> z=2.0.^(-1.5*sqrt(x.^2+y.^2)).*cos(0.5*y).*sin(x);
>> surfl(x,y,z)
>> xlabel('x');ylabel('y')
>> zlabel('z')
```

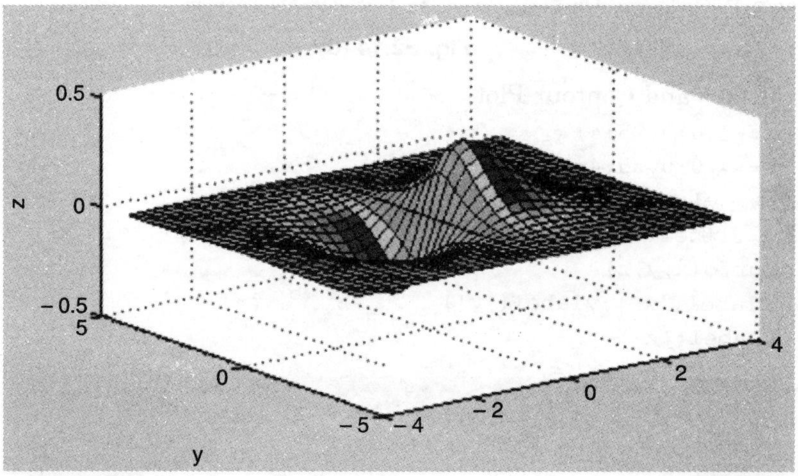

Fig. E2.15 (a)

(*b*) % Waterfall Plot

```
>> x=-4.0:0.25:4;
>> y=-4.0:0.25:4;
>> [x,y]=meshgrid(x,y);
>> z=2.0.^(-1.5*sqrt(x.^2+y.^2)).*cos(0.5*y).*sin(x);
>> waterfall(x,y,z)
>> xlabel('x');ylabel('y')
>> zlabel('z')
```

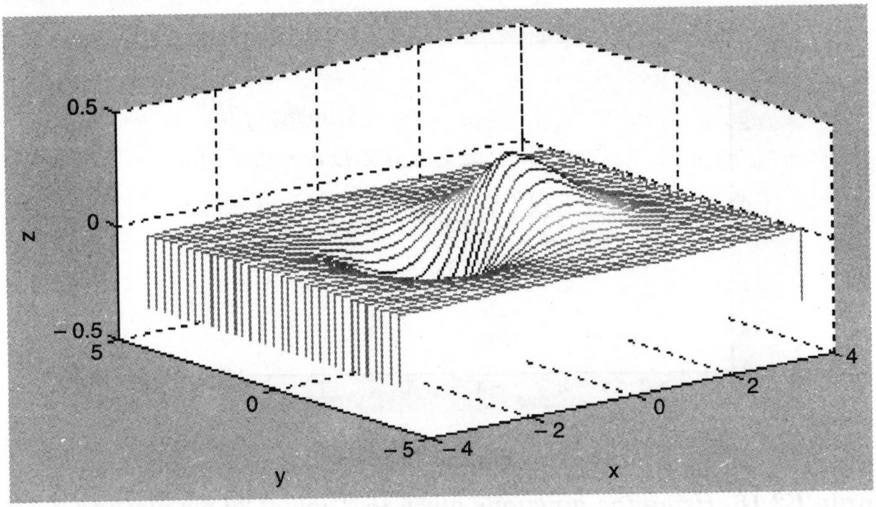

Fig. E2.15 (b)

(*c*) % 3-D Contour Plot

```
>> x=-4.0:0.25:4;
>> y=-4.0:0.25:4;
>> [x,y]=meshgrid(x,y);
>> z=2.0.^(-1.5*sqrt(x.^2+y.^2)).*cos(0.5*y).*sin(x);
>> contour3(x,y,z,15)
>> xlabel('x');ylabel('y')
>> zlabel('z')
```

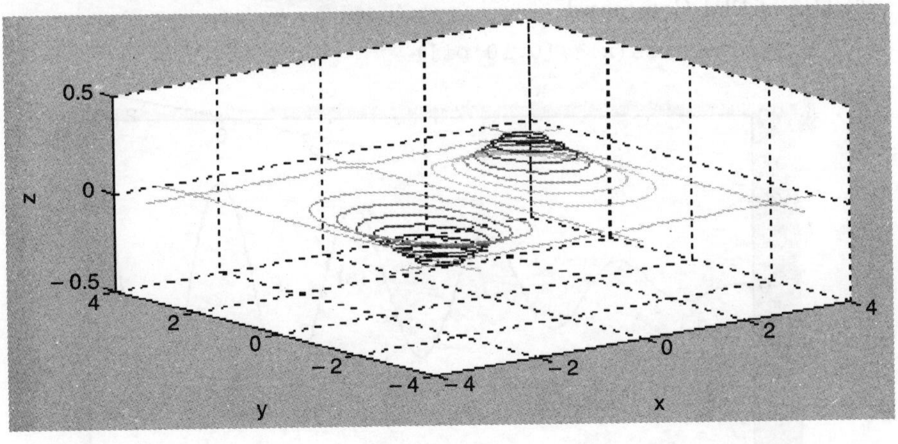

Fig. E2.15 (c)

(*d*) % 2-D Contour Plot

```
>> x=-4.0:0.25:4;
>> y=-4.0:0.25:4;
>> [x,y]=meshgrid(x,y);
>> z=2.0.^(-1.5*sqrt(x.^2+y.^2)).*cos(0.5*y).*sin(x);
>> contour(x,y,z,15)
>> xlabel('x');ylabel('y')
>> zlabel('z')
```

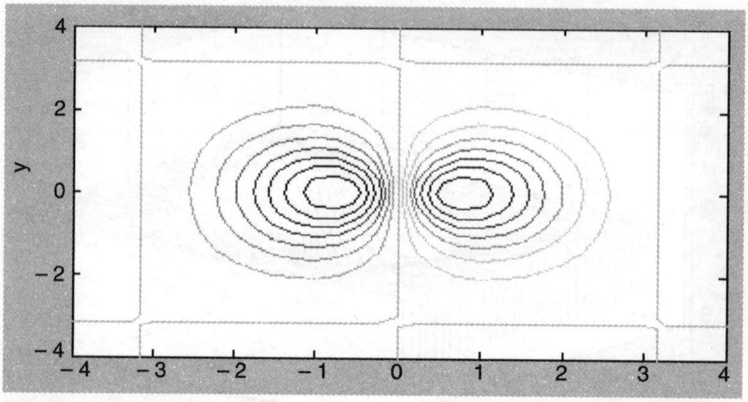

Fig. E2.15 (d)

Example E2.16. *Using the functions given in Table 2.29 for plotting x-y data, plot the following functions:*

(*a*) $f(t) = t \cos t$	$0 \le t \le 10\pi$	(*b*) $x = e^{-2t}, y = t$	$0 \le t \le 2\pi$
(*c*) $x = t, y = e^{2t}$	$0 \le t \le 2\pi$	(*d*) $x = e^{t}, y = 50 + e^{t}$	$0 \le t \le 2\pi$
(*e*) $r^2 = 3 \sin 7t$, $y = r \sin t$	$0 \le t \le 2\pi$	(*f*) $r^2 = 3 \sin 4t$, $y = r \sin t$	$0 \le t \le 2\pi$
(*g*) $y = t \sin t$	$0 \le t \le 5\pi$		

Solution:

(*a*) % Use of Plot Command

```
>> fplot('x.*cos(x)',[0,10*pi])
```

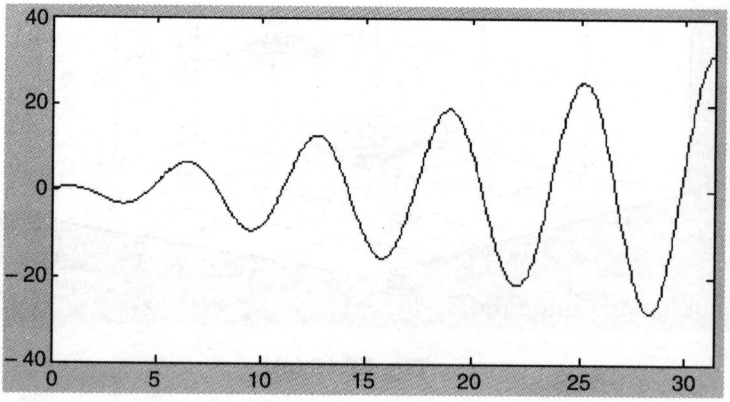

Fig. E2.16 (a)

(*b*) % Semilog x Command

```
>> t=linspace(0,2*pi,200);
>> x=exp(-2*t);y=t;
>> semilogx(x,y),grid
```

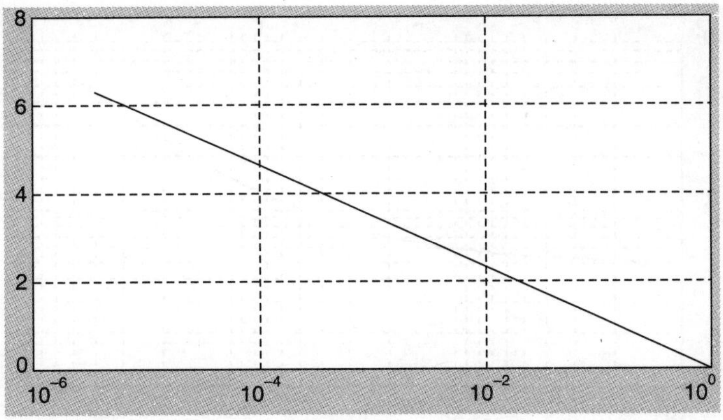

Fig. E2.16 (b)

(*c*) % Semilog *y* Command

```
t=linspace(0,2*pi,200);
>> semilogy(t,exp(-2*t)),grid
```

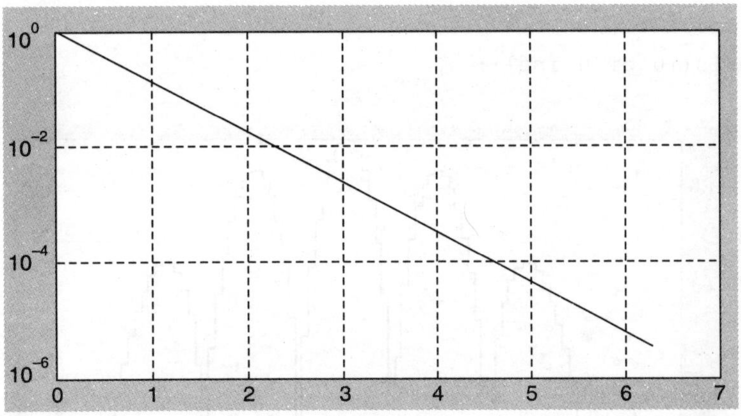

Fig. E2.16 (c)

(*d*) % Use of loglog Command

```
>> t=linspace(0,2*pi,200);
>> x=exp(t);
>> y=50+exp(t);
>> loglog(x,y),grid
```

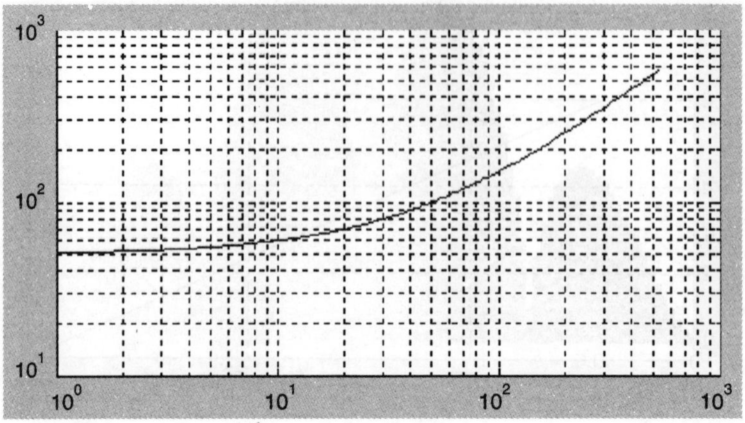

Fig. E2.16 (d)

(*e*) %Use of Stairs Command

```
>> t=linspace(0,2*pi,200);
>> r=sqrt(abs(3*sin(7*t)));
>> y=r.*sin(t);
>> stairs(t,y)
>> axis([0 pi 0 inf]);
```

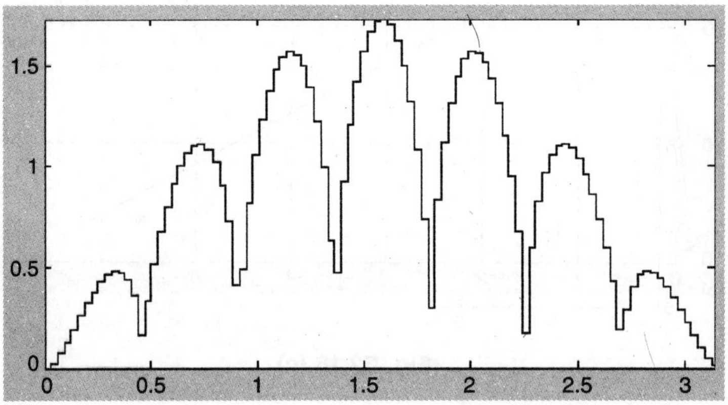

Fig. E2.16 (e)

(*f*) % Use of Bar Command

```
>> t=linspace(0,2*pi,200);
>> r=sqrt(abs(3*sin(4*t)));
>> y=r.*sin(t);
>> bar(t,y)
>> axis([0 pi 0 inf]);
```

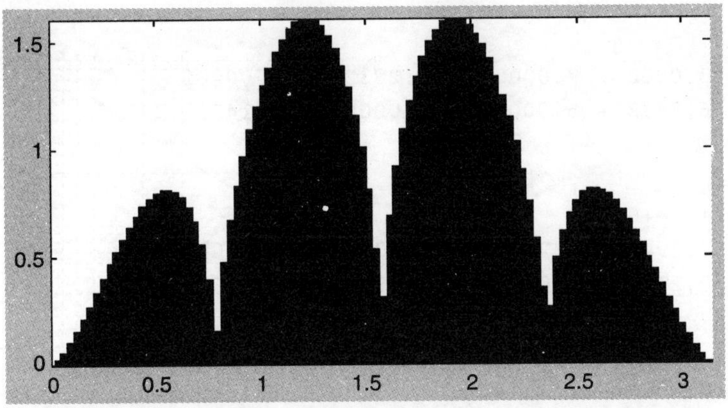

Fig. E2.16 (f)

(*g*) %use of comet Command

```
>> q=linspace(0,5*pi,200);
>> y=q.*sin(q);
>> comet(q,y)
```

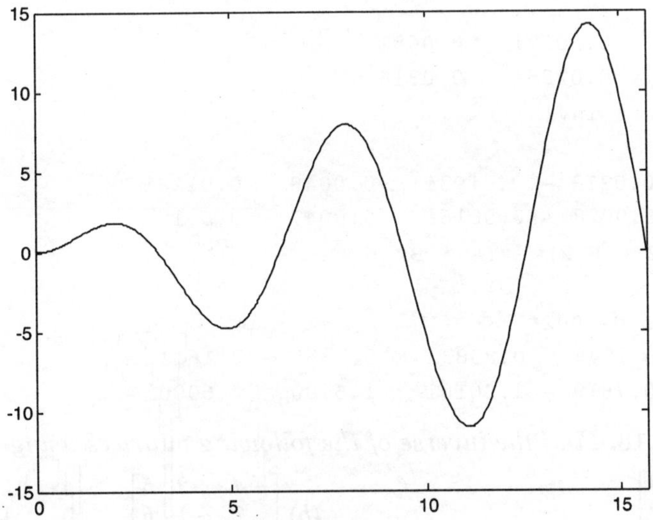

Fig. E2.16 (g)

Example E2.17. *Consider the two matrices*

$$A = \begin{bmatrix} 3 & 2\pi \\ 5j & 10+\sqrt{2}j \end{bmatrix}, \quad B = \begin{bmatrix} 7j & -15j \\ 2\pi & 18 \end{bmatrix}$$

Using MATLAB, determine the following:

(*a*) $A + B$ (*b*) AB (*c*) A^2

(*d*) A^T (*e*) B^{-1} (*f*) B^TA^T

(*g*) $A^2 + B^2 - AB$

Solution:

```
>> A = [3 2*pi;5j 10+sqrt(2)*j];
>> B = [7j -15j;2*pi 18];
```

(*a*) A + B

```
    ans =
          3.0000 + 7.0000i    6.2832 - 15.0000i
          6.2832 + 5.0000i   28.0000 + 1.4142i
```

(*b*) >> A * B

```
    ans =
          1.0e+002 *
          0.3948 + 0.2100i    1.1310 - 0.4500i
          0.2783 + 0.0889i    2.5500 + 0.2546i
```

(*c*) >> A^2

```
    ans =
          9.0000  + 3 1.4159i   81.6814 + 8.8858i
          -7.0711 + 65.0000i   98.0000 + 59.7002i
```

(*d*) >> inv(A)

```
    ans =
           0.1597 + 0.1917i   -0.1150 - 0.1042i
           0.0829 - 0.0916i    0.0549 + 0.0498i
```

(*e*) >> B^-1

```
    ans =
          0 - 0.0817i    0.0681
          0 + 0.0285i    0.0318
```

(*f*) >> inv(B) * inv(A)

```
    ans =
           0.0213 - 0.0193i   -0.0048 + 0.0128i
          -0.0028 + 0.0016i    0.0047 - 0.0017i
```

(*g*) >> (A^2 + B^2) - (A * B)

```
    ans =
           1.0e+002*
          -0.7948 - 0.8383i    0.7358 - 2.1611i
           0.7819 + 1.0010i    1.6700 - 0.6000i
```

Example E2.18. *Find the inverse of the following matrices using MATLAB:*

(*a*) $\begin{bmatrix} 3 & 2 & 0 \\ 2 & -1 & 7 \\ 5 & 4 & 9 \end{bmatrix}$ (*b*) $\begin{bmatrix} -4 & 2 & 5 \\ 7 & -1 & 6 \\ 2 & 3 & 7 \end{bmatrix}$

(*c*) $\begin{bmatrix} -1 & 2 & -5 \\ 4 & 3 & 7 \\ 7 & -6 & 1 \end{bmatrix}$

Solution:

```
>> clear % Clears the workspace
>> A = [3 2 0; 2 -1 7; 5 4 9]; % Spaces separate matrix columns - semicolons separate matrix rows
>> B = [-4 2 5; 7 -1 6; 2 3 7]; % Spaces separate matrix columns - semicolons separate matrix rows
>> C = [-1 2 -5; 4 3 7; 7 -6 1]; % Spaces separate matrix columns - semicolons separate matrix rows
```

```
>> inv(A); % Finds the inverse of the selected matrix
>> inv(B); % Finds the inverse of the selected matrix
>> inv(C) % Finds the inverse of the selected matrix
% Inverse of A
ans =
    0.4805    0.2338   -0.1818
   -0.2208   -0.3506    0.2727
   -0.1688    0.0260    0.0909
% Inverse of B
ans =
   -0.1773    0.0071    0.1206
   -0.2624   -0.2695    0.4184
    0.1631    0.1135   -0.0709
% Inverse of  C
ans =
    0.1667    0.1037    0.1074
    0.1667    0.1259   -0.0481
   -0.1667    0.0296   -0.0407
```

Example E2.19. *Determine the eigenvalues and eigenvectors of matrix* **A** *using MATLAB.*

(*a*) $A = \begin{bmatrix} 4 & -1 & 5 \\ 2 & 1 & 3 \\ 6 & -7 & 9 \end{bmatrix}$

(*b*) $A = \begin{bmatrix} 3 & 5 & 7 \\ 2 & 4 & 8 \\ 5 & 6 & 10 \end{bmatrix}$

Solution:

(*a*)
```
A = [4  - 1  5 ;  2  1  3 ;  6  - 7  9]
A =
     4    -1     5
     2     1     3
     6    -7     9
%The eigenvalues of A
format short e
eig(A)
ans =
  1.0000e+001
  5.8579e-001
  3.4142e+000
%The eigenvectors of A
[Q,d]=eig(A)
Q =
 -5.5709e-001 -8.2886e-001 -7.3925e-001
 -3.7139e-001 -3.9659e-002 -6.7174e-001
 -7.4278e-001  5.5805e-001 -4.7739e-002
d =
```

```
   1.0000e+001   0           0
       0   5.8579e-001       0
       0       0     3.4142e+000
```

(*b*)
```
A =
     3     5     7
     2     4     8
     5     6    10
%The eigenvalues of A
format short e
eig(A)
ans =
   1.7686e+001
  -3.4295e-001  +1.0066e+000i
  -3.4295e-001  -1.0066e+000i
%The eigenvectors of A
[Q,d]=eig(A)
Q =
Column 1
   5.0537e-001
   4.8932e-001
   7.1075e-001
Column 2
  -2.0715e-001  -5.2772e-001i
   7.1769e-001
  -3.3783e-001  +2.2223e-001i
Column 3
  -2.0715e-001  +5.2772e-001i
   7.1769e-001
  -3.3783e-001  -2.2223e-001i
d =
Column 1
1.7686e+001
        0
        0
Column 2
     0
-3.4295e-001  +1.0066e+000i
     0

Column 3
     0
     0
-3.4295e-001  -1.0066e+000i
```

Example E2.20. *Determine the eigenvalues and eigenvectors of* **AB** *using MATLAB.*

$$\mathbf{A} = \begin{bmatrix} 3 & 0 & 2 & 1 \\ 1 & 2 & 5 & 4 \\ 7 & -1 & 2 & 6 \\ 1 & -2 & 3 & 4 \end{bmatrix}$$

$$\mathbf{B} = \begin{bmatrix} 1 & 3 & 5 & 7 \\ 2 & -1 & -2 & 4 \\ 3 & 2 & 1 & 1 \\ 4 & 1 & 0 & 6 \end{bmatrix}$$

Solution:

% *MATLAB Program*

% The matrix "*a*" = *A***B*

```
>> A = [3 0 2 1; 1 2 5 4; 7 - 1 2 6; 1 - 2 3 4];
>> B = [1 3 5 7; 2 - 1 - 2 4; 3 2 1 1; 4 1 0 6];
>> a = A*B
a =
     13     14     17     29
     36     15     6      44
     35     32     39     83
     22     15     12     26
>> eig (a)
Ans. =
98.5461
2.2964
-1.3095
-6.5329
The eigenvectors are :
>> [Q, d] = eig (a)
Q =
     -0.3263    -0.2845     0.3908     0.3413
     -0.3619     0.7387    -0.7816    -0.9215
     -0.8168    -0.6026     0.4769     0.0962
     -0.3089     0.1016    -0.0950     0.1586
d =
     98.5461       0           0          0
           0       2.2964      0          0
           0       0          -1.3095     0
           0       0           0         -6.5329
```

Example E2.21. *Solve the following set of equations using MATLAB.*

(*a*) $x_1 + 2x_2 + 3x_3 + 5x_4 = 21$

$-2x_1 + 5x_2 + 7x_3 - 9x_4 = 18$

$5x_1 + 7x_2 + 2x_3 - 5x_4 = 25$

$-x_1 + 3x_2 - 7x_3 + 7x_4 = 30$

(*b*)
$$x_1 + 2x_2 + 3x_3 + 4x_4 = 8$$
$$2x_1 - 2x_2 - x_3 - x_4 = -3$$
$$x_1 - 3x_2 + 4x_3 - 4x_4 = 8$$
$$2x_1 + 2x_2 - 3x_3 + 4x_4 = -2$$

Solution:

(*a*)

```
>> A= [1 2 3 5;-2 5 7 -9; 5 7 2 -5;-1 -3 -7 7];
>> B = [21; 18; 25; 30];
>> S = A\B
S =
- 8.9896
 14.1285
- 5.4438
  3.6128
```

% Therefore $x_1 = -8.9896$, $x_2 = 14.12.85$, $x_3 = -5.4438$, $x_4 = 3.6128$.

(*b*)

```
>> A= [1 2 3 4; 2 -2 -1 1; 1 -3 4 -4; 2 2 -3 4];
>> B = [8;-3; 8;-2];
>> S=A\B
S =
  2.0000
  2.0000
  2.0000
- 1.0000
```

%Therefore $x_1 = 2.0000$, $x_2 = 2.0000$, $x_3 = 2.0000$, $x_4 = -1.0000$.

Example E2.22. *Use* diff *command for symbolic differentiation of the following functions:*

(*a*) $S_1 = e^{x^8}$

(*b*) $S_2 = 3x^3 e^{x^5}$

(*c*) $S_3 = 5x^3 - 7x^2 + 3x + 6$

Solution:

(*a*)

```
>> syms x
>> S1=exp(x^8);
>> diff (S1)
ans =
8*x^7*exp(x^8)
```

(*b*)

```
>> S2=3*x^3*exp(x^5);
>> diff (S2)
ans =
9*x^2*exp(x^5)+15*x^7*exp(x^5)
```

(*c*)

```
>> S3=5*x^3-7*x^2+3*x+6;
>> diff (S3)
ans =
15*x^2-14*x+3
```

Example E2.23. *Use MATLAB's symbolic commands to find the values of the following integrals.*

(*a*) $\int_{0.2}^{0.7} |x| \, dx$ (*b*) $\int_0^{\pi} (\cos y + 7y^2) \, dy$

(*c*) $\sqrt{x}$ (*d*) $7x^5 - 6x^4 + 11x^3 + 4x^2 + 8x + 9$

(*e*) $\cos a$

Solution:

(*a*)

```
>>syms x, y, a, b
>> S1=abs(x)
>> int (S1, 0.2, 0.7)
ans =
9/40
```

(*b*)

```
>> S2=cos (y) +7*y^2
>> int (S2, 0, pi)
ans =
7/3*pi^3
```

(*c*)

```
>> S3=sqrt (x)
>> int (S3)
ans =
2/3*x^ (3/2)
>> int (S3,'a','b')
ans =
2/3*b^ (3/2)-2/3*a^ (3/2)
>> int (S3, 0.4, 0.7)
ans =
7/150*70^ (1/2)-4/75*10^ (1/2)
```

(*d*)

```
>> S4=7*x^5-6*x^4+11*x^3+4*x^2+8*x-9
>> int (S4)
ans =
7/6*x^6-6/5*x^5+11/4*x^4+4/3*x^3+4*x^2-9*x
```

(*e*)

```
>> S5=cos (a)
>> int (S5)
ans =
sin (a)
```

Example E2.24. *Obtain the general solution of the following first order differential equations:*

(*a*) $\dfrac{dy}{dt} = 5t - 6y$ (*b*) $\dfrac{d^2y}{dt^2} + 3\dfrac{dy}{dt} + y = 0$

(*c*) $\dfrac{ds}{dt} = Ax^3$ (*d*) $\dfrac{ds}{dA} = Ax^3$

Solution:

(*a*)

```
>> solve('Dy=5*t-6*y')
ans =
5/6*t-5/36+exp(-6*t)*C1
```

(*b*)

```
>> dsolve('D2y+3*Dy+y=0')
ans =
C1*exp(1/2*(5^(1/2)-3)*t)+C2*exp(-1/2*(5^(1/2)+3)*t)
```

(*c*)

```
>> dsolve('Ds=A*x^3','x')
ans =
1/4*A*x^4+C1
```

(*d*)

```
>> dsolve('Ds=A*x^3','A')
ans =
1/2*A^2*x^3+C1
```

Example E2.25. *Determine the solution of the following differential equations that satisfies the given initial conditions.*

(*a*) $\dfrac{dy}{dx} = -7x^2 \quad y(1) = 0.7$ (*b*) $\dfrac{dy}{dx} = 5x\cos^2 y \quad y(0) = \pi/4$

(*c*) $\dfrac{dy}{dx} = -y + e^{3x} \quad y(0) = 2$ (*d*) $\dfrac{dy}{dt} + 5y = 35 \quad y(0) = 4$

Solution:

(*a*)

```
>> dsolve('Dy=-7*x^2','y(1)=0.7')
ans =
-7*x^2*t+7*x^2+7/10
```

(*b*)

```
>> dsolve('Dy=5*x*cos (y)^2','y(0)=pi/4')
ans =
atan (5*t*x+1)
```

(*c*)

```
>> dsolve('Dy=-y+ exp(3*x)','y(0)=2')
ans =
exp (3*x) +exp (-t)*(-exp (3*x) +2)
```

(*d*)

```
>> dsolve('Dy+5*y=35','y(0)=4')
ans =
7-3*exp(-5*t)
```

Example E2.26. *Given the differential equation*

$$\frac{d^2x}{dt^2} + 7\frac{dx}{dt} + 5x = 8u(t) \qquad t \geq 0$$

Using MATLAB program, find

(*a*) *x(t) when all the initial conditions are zero*

(*b*) *x(t) when x(0) = 1 and* $\dot{x}(0) = 2$.

Solution:

(*a*) $x(t)$ when all the initial conditions are zero

```
>> x = dsolve('D2x = -7*Dx - 5*x +8', 'x(0)= 0')
x =
8/5+ (-8/5-C2)*exp(1/2*(-7+29^(1/2))*t)+C2*exp(-1/2*(7+29^(1/2))*t)
```

(*b*) $x(t)$ when $x(0) = 1$ and $\dot{x}(0) = 2$

```
>> x = dsolve('D2x = -7*Dx - 5*x +8', 'x(0) = 1', 'Dx(0) = 2')
x =
8/5+ (-3/10-1/290*29^(1/2))*exp(1/2*(-7+29^(1/2))*t)-1/290*(-1+3*29^
(1/2))*29^(1/2)*exp(-1/2*(7+29^(1/2))*t)
```

Example E2.27. *Given the differential equation*

$$\frac{d^2x}{dt^2} + 12\frac{dx}{dt} + 15x = 35 \qquad t \geq 0$$

Using MATLAB program, find

(*a*) *x(t) when all the initial conditions are zero*

(*b*) *x(t) when x(0) = 0 and* $\dot{x}(0) = 1$.

Solution:

(*a*) $x(t)$ when all the initial conditions are zero

```
>> x = dsolve('D2x = -12*Dx - 15*x +35', 'x(0) = 0')
x =
7/3+ (-7/3-C2)*exp((-6+21^(1/2))*t)+C2*exp(-(6+21^(1/2))*t)
```

(*b*) $x(t)$ when $x(0) = 0$ and $\dot{x}(0) = 1$

```
>> x = dsolve('D2x = -12*Dx - 15*x + 35', 'x(0) = 0', 'Dx(0) = 1')
x =
7/3+(-7/6-13/42*21^(1/2))*exp((-6+21^ (1/2))*t)-1/126*
(-39+7*21^(1/2))*21^(1/2)*exp(-(6+21^ (1/2))*t)
```

Example E2.28. *Find the inverse of the following matrix using MATLAB.*

$$A = \begin{bmatrix} s & 2 & 0 \\ 2 & s & -3 \\ 3 & 0 & 1 \end{bmatrix}$$

Solution:

```
>> A = [s 2 0; 2 s -3; 3 0 1];
>> inv(A)
ans =
      [s/(s^2-22),  -2/(s^2-22),  -6/(s^2-22)]
      [-11/(s^2-22),  s/(s^2-22),  3*s/(s^2-22)]
      [-3*s/(s^2-22), 6/(s^2-22),  (s^2-4)/(s^2-22)]
```

Example E2.29. *Expand the following function F(s) into partial fractions using MATLAB. Determine the inverse Laplace transform of F(s).*

$$F(s) = \frac{1}{s^4 + 5s^3 + 7s^2}$$

The MATLAB program for determining the partial-fraction expansion is given below:

Solution:

```
>> b = [0 0 0 0 1];
>> a = [1 5 7 0 0];
>> [r, p, k] = residue (b, a)
r =
       0.0510 - 0.0648i
       0.0510 + 0.0648i
      -0.1020
       0.1429
p =
      -2.5000 + 0.8660i
      -2.5000 - 0.8660i
         0
         0
k = [ ]
```

% From the above MATLAB output, we have the following expression:

$$F(s) = \frac{r_1}{s-p_1} + \frac{r_2}{s-p_2} + \frac{r_3}{s-p_3} + \frac{r_4}{s-p_4} =$$

$$F(s) = \frac{0.0510 - 0.0648i}{s-(-2.5000+0.8660i)} + \frac{(0.0510+0.0648i}{s-(-2.5000-0.8660i)} + \frac{-0.1020}{s-0} + \frac{0.1429}{s-0}$$

% Note that the row vector *k* is zero implies that there is no constant term in this example problem.

% The MATLAB program for determining the inverse Laplace transform of *F*(*s*) is given below:

```
>> syms s
>> f = 1/(s^4 + 5*s^3 + 7*s^2);
>> ilaplace (f)
ans =
1/7*t-5/49+5/49*exp(-)*cos(1/2*3^(1/2)*t)+11/147*exp(-5/2*t)*3^
(1/2)*sin(1/2*3^(1/2)*t)
```

Example E2.30. *Expand the following function F(s) into partial fractions using MATLAB. Determine the inverse Laplace transform of F(s).*

$$F(s) = \frac{5s^2 + 3s + 6}{s^4 + 3s^3 + 7s^2 + 9s + 12}$$

Solution:

The MATLAB program for determining the partial-fraction expansion is given below:

```
>> b = [0 0 5 3 6];
>> a = [1 3 7 9 12];
>> [r,p,k] = residue(b,a)
r =
       -0.5357 - 1.0394i
       -0.5357 + 1.0394i
        0.5357 - 0.1856i
        0.5357 + 0.1856i
p =
       -1.5000 + 1.3229i
       -1.5000 - 1.3229i
       -0.0000 + 1.7321i
       -0.0000 - 1.7321i
k = [ ]
```

% From the above MATLAB output, we have the following expression:

$$F(s) = \frac{r_1}{s - p_1} + \frac{r_2}{s - p_2} + \frac{r_3}{s - p_3} + \frac{r_4}{s - p_4} =$$

$$F(s) = \frac{-0.5357 - 0.0394i}{s - (-1.500 + 1.3229i)} + \frac{(-0.5357 + 1.0394i)}{s - (-1.5000 + 1.3229i)}$$

$$+ \frac{0.5357 - 0.1856i}{s - (-0 + 1.7321i)} + \frac{0.5357 - 0.1856i}{s - (-0 - 1.7321i)}$$

% Note that the row vector k is zero implies that there is no constant term in this example problem.

% The MATLAB program for determining the inverse Laplace transform of *F*(*s*) is given below:

```
>> syms s
>> f = (5*s^2 + 3*s +6)/(s^4 + 3*s^3 + 7*s^2 + 9*s +12);
>> ilaplace(f)
ans =
11/14*exp(-3/2*t)*7^(1/2)*sin(1/2*7^(1/2)*t)-15/14*exp(-3/2*t)*cos
(1/2*7^(1/2)*t)+3/14*3^(1/2)*sin(3^(1/2)*t)+15/14*cos(3^(1/2)*t)
```

Example E2.31. *For the following function F(s):*

$$F(s) = \frac{s^4 + 3s^3 + 5s^2 + 7s + 25}{s^4 + 5s^3 + 20s^2 + 40s + 45}$$

Using MATLAB, find the partial-fraction expansion of F(s). Also, find the inverse Laplace transformation of F(s).

Solution:

$$F(s) = \frac{s^4 + 3s^3 + 5s^2 + 7s + 25}{s^4 + 5s^3 + 20s^2 + 40s + 45}$$

The partial-fraction expansion of $F(s)$ using MATLAB program is given as follows:

```
num = [ 1  3  5  7  25];
den = [1  5  20  40  45];
[r,p,k]  = residue(num,den)
r =
 -1.3849 + 1.2313i
 -1.3849 - 1.2313i
  0.3849 - 0.4702i
  0.3849 + 0.4702i
p =
 -0.8554 + 3.0054i
 -0.8554 - 3.0054i
 -1.6446 + 1.3799i
 -1.6446 - 1.3799i
k =
    1
```

From the MATLAB output, the partial-fraction expansion of $F(s)$ can be written as follows:

$$F(s) = \frac{r_1}{(s - p_1)} + \frac{r_2}{(s - p_2)} + \frac{r_3}{(s - p_3)} + \frac{r_4}{(s - p_4)} + k$$

$$F(s) = \frac{(-1.3849 + j1.2313)}{(s + 0.8554 - j3.005)} + \frac{(-1.3849 - j1.2313)}{(s + 0.8554 + j3.005)}$$

$$+ \frac{(0.3849 - j0.4702)}{(s + 1.6446 - j1.3799)} + \frac{(0.3849 + j0.4702)}{(s + 1.6446 + j1.3779)} + 1$$

Example E2.32. *Obtain the partial-fraction expansion of the following function using MATLAB:*

$$F(s) = \frac{8(s+1)(s+3)}{(s+2)(s+4)(s+6)^2}$$

Solution:

$$F(s) = \frac{8(s+1)(s+3)}{(s+2)(s+4)(s+6)^2} = \frac{(8s+8)(s+3)}{(s^2 + 6s + 8)(s^2 + 12s + 36)}$$

The partial fraction expansion of $F(s)$ using MATLAB program is given as follows:

```
EDU>> num=conv([8 8],[1 3]);
EDU>> den=conv([1 6 8],[1 12 36]);
EDU>>  [r,p,k]=residue(num,den)
r =
    3.2500
   15.0000
   -3.0000
   -0.2500
```

```
p =
    -6.0000
    -6.0000
    -4.0000
    -2.0000
k = [ ]
```

From the above MATLAB result, we have the following expansion:

$$F(s) = \frac{r_1}{(s - p_1)} + \frac{r_2}{(s - p_2)} + \frac{r_3}{(s - p_3)} + \frac{r_4}{(s - p_4)} + k$$

$$F(s) = \frac{3.25}{(s + 6)} + \frac{15}{(s - 15)} + \frac{-3}{(s + 3)} + \frac{-0.25}{(s + 0.25)} + 0$$

It should be noted here that the row vector k is zero, because the degree of the numerator is lower than that of the denominator.

$$F(s) = 3.25e^{-6t} + 15e^{15t} - 3e^{-3t} - 0.25e^{-0.25t}$$

Example E2.33. *Find the Laplace transform of the following function using MATLAB.*

(*a*) $f(t) = 7t^3 \cos(5t + 60°)$

(*b*) $f(t) = -7t\, e^{-5t}$

(*c*) $f(t) = -3 \cos 5t$

(*d*) $f(t) = t \sin 7t$

(*e*) $f(t) = 5\, e^{-2t} \cos 5t$

(*f*) $f(t) = 3 \sin(5t + 45°)$

(*g*) $f(t) = 5\, e^{-3t} \cos(t - 45°)$

Solution:

(*a*)
```
% MATLAB Program
>> syms t    %tell MATLAB that "t" is a symbol.
>> f = 7 * t^3*cos(5*t + (pi/3)); % define the function.
>> laplace(f)
ans =
-84/(s^2+25)^3*s^2+21/(s^2+25)^2+336*(1/2*s-5/2*3^(1/2))/(s^2+25)
^4*s^3-168*(1/2*s-5/2*3^(1/2))/(s^2+25)^3*s
>> pretty(laplace(f)) % the pretty function prints symbolic output
% in a format that resembles typeset mathematics.
```

```
                2                   1/2   3
               s        21    (1/2 s - 5/2 3) s
  - 84 ---------- + ---------- + 336 --------------------
         2    3       2    2           2    4
       (s + 25)     (s + 25)         (s + 25)
                            1/2
                 (1/2 s - 5/2 3) s
       - 168 --------------------
                 2    3
               (s + 25)
```

(*b*)
```
>>syms t x
>>f = -7*t*exp(-5*t);
>> laplace(f,x)
ans =
-7/(x+5)^2
```
(*c*)
```
>>syms t x
>>f = -3*cos(5*t);
>> laplace(f,x)
ans =
-3*x/(x^2+25)
```
(*d*)
```
>>syms t x
>>f = t*sin(7*t);
>> laplace(f,x)
ans =
1/(x^2+49)*sin(2*atan(7/x))
```
(*e*)
```
>>syms t x
>>f = 5*exp(-2*t)*cos(5*t);
>> laplace(f,x)
ans =
5*(x+2)/((x+2)^2+25)
```
(*f*)
```
>>syms t x
>>f = 3*sin(5*t + (pi/4));
>> laplace(f,x)
ans =
3*(1/2*x*2^(1/2)+5/2*2^(1/2))/(x^2+25)
```
(*g*)
```
>>syms t x
>>f = 5*exp(-3*t)*cos(t-(pi/4));
>> laplace(f,x)
ans =
      5*(1/2*(x+3)*2^(1/2)+1/2*2^(1/2))/((x+3)^2+1)
```

Example E2.34. *Generate partial-fraction expansion of the following function.*

$$F(s) = \frac{10^5\,(s+7)(s+13)}{s(s+25)(s+55)(s^2+7s+75)(s^2+7s+45)}$$

Solution:

Generate the partial fraction expansion of the following function:

```
numg=poly[-7 -13];
numg=poly([-7 -13]);
deng=poly([0 -25 -55 roots([1 7 75])' roots([1 7 45])']);
[numg,deng]=zp2tf(numg',deng',1e5);
Gtf=(numg,deng);
Gtf=tf(numg,deng);
G=zpk(Gtf);
[r,p,k]=residue(numg,deng)
```

```
r =
   1.0e-017 *
     0.0000
    -0.0014
     0.0254
    -0.1871
     0.1621
    -0.0001
     0.0000
     0.0011
p =
   1.0e+006 *
     4.6406
     1.4250
     0.3029
     0.0336
     0.0027
     0.0001
     0.0000
          0
k = [ ]
```

Example E2.35. *Determine the inverse Laplace transform of the following functions using MATLAB.*

(*a*) $F(s) = \dfrac{s}{s(s+2)(s+6)}$ (*b*) $F(s) = \dfrac{1}{s^2(s+5)}$

(*c*) $F(s) = \dfrac{3s+1}{(s^2+2s+9)}$ (*d*) $F(s) = \dfrac{s-25}{s(s^2+3s+20)}$

Solution:

(*a*)
```
>> syms s
>> f = s/(s*((s + 2)*(s + 6)));
>> ilaplace(f)
ans =
1/2*exp(-4*t)*sinh(2*t)
```

(*b*)
```
>> syms s
>> f = 1/((s^2)*(s + 5));
>> ilaplace(f)
ans =
1/3*t-2/9*exp(-3/2*t)*sinh(3/2*t)
```

(*c*)
```
>>syms s
>> f=(3*s+1)/(s^2+2*s+9);
>> ilaplace(f)
ans =
3*exp(-t)*cos(2*2^(1/2)*t)-1/2*2^(1/2)*exp(-t)*sin(2*2^(1/2)*t)
```

(*d*)
```
>>syms s
>> f = (s -25)/(s*(s^2 + 3*s +25));
>> ilaplace(f)
ans =
5/4*exp(-3/2*t)*cos(1/2*71^(1/2)*t)+23/284*71^(1/2)*exp(-3/2*t)
*sin(1/2*71^(1/2)*t)-5/4
```

Example E2.36. *Find the inverse Laplace transform of the following function using MATLAB.*

$$G(s) = \frac{(s^2 + 9s + 7)(s + 7)}{(s + 2)(s + 3)(s^2 + 12s + 150)}.$$

Solution:

% *MATLAB Program*

```
>> syms s    % tell MATLAB that "s" is a symbol.
>>G = (s^2 + 9*s +7)*(s + 7)/[(s + 2)*(s + 3)*(s^2 + 12*s + 150)]; % define
the function.
>>pretty(G) % the pretty function prints symbolic output
% in a format that resembles typeset mathematics.
            (s  + 9 s + 7) (s + 7)
        --------------------------------
        (s + 2) (s + 3) (s  + 12 s + 150)
>> g = ilaplace(G); % inverse Laplace transform
>>pretty(g)
                        44                      2915                 1/2
   - 7/26 exp(-2 t) + --- exp(-3 t) + ------ exp(-6 t) cos(114 t)
                       123                      3198
         889                   1/2      1/2
     + ------- exp(-6 t) 114  sin(114  t)
        20254
```

Example E2.37. *Generate the transfer function using MATLAB.*

$$G(s) = \frac{3(s + 9)(s + 21)(s + 57)}{s(s + 30)(s^2 + 5s + 35)(s^2 + 28s + 42)}$$

using

(*a*) *the ratio of factors* (*b*) *the ratio of polynomials*

Solution:

% *MATLAB Program:*

```
'a. The ratio of factors'
>>Gzpk = zpk([-9 -21 -57] , [0 -30 roots([1 5 35])'roots([1 28 42])'],3)
% zpk is used to create zero-pole-gain models or to convert TF or
% SS models to zero-pole-gain form.
'b. The ratio of polynomials'
>> Gp = tf(Gzpk)  % generate the transfer function
% Computer response:
```

```
ans =
```

(*a*) The ratio of factors

```
Zero/pole/gain:
                 3 (s+9) (s+21) (s+57)
  ------------------------------------------
  s (s+30) (s+26.41) (s+1.59) (s^2 + 5s + 35)
ans =
```

(*b*) The ratio of polynomials

Transfer function:

```
                  3 s^3 + 261 s^2 + 5697 s + 32319
  -----------------------------------------------------
  s^6 + 63 s^5 + 1207 s^4 + 7700 s^3 + 37170 s^2 + 44100 s
```

Example E2.38. *Generate the transfer function using MATLAB.*

$$G(s) = \frac{s^4 + 20s^3 + 27s^2 + 17s + 35}{s^5 + 8s^4 + 9s^3 + 20s^2 + 29s + 32}$$

using

(*a*) *the ratio of factors* (*b*) *the ratio of polynomials*

Solution:

% *MATLAB Program:*

```
% a. the ratio of factors
>>Gtf = tf([1 20 27 17 35], [1 8 9 20 29 32]) % generate the
% transfer function
% Computer response:
Transfer function:
       s^4 + 20 s^3 + 27 s^2 + 17 s + 35
       ---------------------------------
        s^4 + 8 s^3 + 9 s^2 + 20 s + 29
% b. the ratio of polynomials
>> Gzpk = zpk(Gtf)  % zpk is used to create zero-pole-gain models
% or to convert TF or SS models to zero-pole-gain form.
% Computer response:
Zero/pole/gain:
  (s+18.59) (s+1.623) (s^2 - 0.214s + 1.16)
 -------------------------------------------
  (s+7.042) (s+1.417) (s^2 - 0.4593s + 2.906)
```

2.23 SUMMARY

In this chapter the MATLAB environment which is an interactive environment for numeric computation, data analysis, and graphics was presented. Arithmetic operations, display formats, elementary built-in functions, arrays, scalars, vectors or matrices, operations with arrays including dot product, array multiplication, array division, inverse and transpose of a matrix, determinants, element by element operations, eigenvalues and eigenvectors, random number generating functions, polynomials, system of linear equation, script files, programming in

MATLAB, the commands used for printing information and generating 2-D and 3-D plots, input/output in MATLAB was presented with illustrative examples. MATLAB's functions for symbolic mathematics were introduced. These functions are useful in performing symbolic operations and developing closed-form expressions for solutions to linear algebraic equations, ordinary differential equations and systems of equations. Symbolic mathematics for determining analytical expressions for the derivative and integral of an expression was also presented.

REFERENCES

Chapman, S.J., *MATLAB Programming for Engineers,* 2nd edn., Brooks/Cole, Thomson Learning, Pacific Grove, CA, 2002.

Dukkipati, R.V., *Analysis and Design of Control Systems using MATLAB*, New Age International Publishers, New Delhi, India, 2006.

Dukkipati, R.V., and Shivakumar, M.R., *MATLAB for Electrical Engineers*, New Age International Publishers, New Delhi, India, 2007.

Dukkipati, R.V., and Srinivas, J., *Solving Engineering Mechanics Problems with MATLAB*, New Age International Publishers, New Delhi, India, 2007.

Dukkipati, R.V., *MATLAB for Engineers*, New Age International Publishers, New Delhi, India, 2006.

Dukkipati, R.V., *Solving Engineering System Dynamics Problems with MATLAB*, New Age International Publishers, New Delhi, India, 2006.

Dukkipati, R.V., *Solving Vibration Analysis Problems with MATLAB*, New Age International Publishers, New Delhi, India, 2006.

Etter, D.M., *Engineering Problem Solving with MATLAB,* Prentice-Hall, Englewood Cliffs, NJ, 1993.

Gilat, Amos., *MATLAB-An Introduction with Applications*, 2nd edn., Wiley, New York, 2005.

Hanselman, D., and Littlefield, B.R., *Mastering MATLAB 6,* Prentice Hall, Upper Saddle River, New Jersey, NJ, 2001.

Herniter, M.E., *Programming in MATLAB*, Brooks/Cole, Pacific Grove, CA, 2001.

Magrab, E.B., *An Engineers Guide to MATLAB*, Prentice Hall, Upper Saddle River, New Jersey, NJ, 2001.

Marchand, P., and Holland, O.T., Graphics and GUIs with MATLAB, 3rd ed, CRC Press, Boca Raton, FL, 2003.

Moler, C., *The Student Edition of MATLAB for MS-DOS Personal Computers with 3-1/2" Disks,* MATLAB Curriculum Series, The MathWorks, Inc., 2002.

Palm, W.J. III., *Introduction to MATLAB 7 for Engineers*, McGraw Hill, New York, NY, 2005.

Pratap, Rudra, *Getting Started with MATLAB- A Quick Introduction for Scientists and Engineers,* Oxford University Press, New York, NY, 2002.

Sigman, K., and Davis, T.A., *MATLAB Primer*, 6th edn, Chapman & Hall/CRC Press, Boca Raton, FL, 2002.

The MathWorks, Inc., *MATLAB*: *Application Program Interface Reference Version 6*, The MathWorks, Inc., Natick, 2000.

The MathWorks, Inc., *MATLAB*: *Creating Graphical User Interfaces, Version* 1, The MathWorks, Inc., Natick, 2000.

The MathWorks, Inc., *MATLAB: Function Reference,* The MathWorks, Inc., Natick, 2000.

The MathWorks, Inc., *MATLAB: Release Notes for Release 12,* The MathWorks, Inc., Natick, 2000.

The MathWorks, Inc., *MATLAB: Symbolic Math Toolbox User's Guide, Version 2,* The MathWorks, Inc., Natick, 1993-1997.

The MathWorks, Inc., MATLAB: *Using MATLAB Graphics, Version 6,* The MathWorks, Inc., Natick, 2000.

PROBLEMS

P2.1 Compute the following quantity using MATLAB in the Command Window:

$$\frac{17\,[\sqrt{5}-1]}{[15^2-13^2]}+\frac{5^7\log_{10}(e^3)}{\pi\sqrt{121}}+\ln(e^4)+\sqrt{11}$$

P2.2 Compute the following quantity using MATLAB in the Command Window:

$$B=\frac{\tan x+\sin 2x}{\cos x}+\log\,|\,x^5-x^2\,|+\cosh x-2\tanh x$$

for $x=\dfrac{5\pi}{6}$.

P2.3 Compute the following quantity using MATLAB in the Command Window:

$$x=a+\frac{ab}{c}\frac{(a+b)}{\sqrt{|ab|}}+c^a+\frac{\sqrt{14b}}{e^{3c}}+\ln(2)+\frac{\log_{10}c}{\log_{10}(a+b+c)}+2\sinh a-3\tanh b$$

for $a = 1$, $b = 2$ and $c = 1.8$.

P2.4 Use MATLAB to create

(*a*) a row and column vectors that has the elements: 11, – 3, $e^{7.8}$, ln (59), tan (π/3), 5 $\log_{10}$(26).

(*b*) a row vector with 20 equally spaced elements in which the first element is 5.

(*c*) a column vector with 15 equally spaced elements in which the first element is – 2.

P2.5 Enter the following matrix *A* in MATLAB and create:

$$A=\begin{bmatrix}1&2&3&4&5&6&7&8\\9&10&11&12&13&14&15&16\\17&18&19&20&21&22&23&24\\25&26&27&28&29&30&31&32\\33&34&35&36&37&38&39&40\end{bmatrix}$$

(*a*) a 4 × 5 matrix *B* from the 1st, 3rd, and the 5th rows, and the 1st, 2nd, 4th, and 8th columns of the matrix *A*.

(*b*) a 16 element-row vector **C** from the elements of the 5th row, and the 4th and 6th columns of the matrix *A*.

P2.6 Given the function $y=\left(x^{\sqrt{2}+0.02}+e^x\right)^{1.8}\ln x$. Determine the value of *y* for the following values of *x*: 2, 3, 8, 10, – 1, – 3, – 5, – 6.2. Solve the problem using MATLAB by first creating a vector **x**, and creating a vector **y**, using element-by-element calculations.

P2.7 Define a and b as scalars, $a = 0.75$, and $b = 11.3$, and x, y and z as the vectors, $x = 2, 5, 1, 9$, $y = 0.2, 1.1, 1.8, 2$ and $z = -3, 2, 5, 4$. Use these variables to calculate A given below using element-by-element computations for the vectors with MATLAB.

$$A = \frac{x^{1.1}y^{-2}z^5}{(a+b)^{b/3}} + a\frac{\left(\frac{z}{x}+\frac{y}{2}\right)}{z^a}$$

P2.8 Enter the following three matrices in MATLAB and show that

$$A = \begin{bmatrix} 1 & 2 & 3 \\ -8 & 5 & 7 \\ -8 & 4 & 6 \end{bmatrix} B = \begin{bmatrix} 12 & -5 & 4 \\ 7 & 11 & 6 \\ 1 & 8 & 13 \end{bmatrix} C = \begin{bmatrix} 7 & 13 & 4 \\ -2 & 8 & -5 \\ 9 & -6 & 11 \end{bmatrix}$$

(a) $A + B = B + A$

(b) $A + (B + C) = (A + B) + C$

(c) $7(A + C) = 7(A) + 7(C)$

(d) $A * (B + C) = A * B + A * C$

P2.9 Consider the polynomials

$p_1(s) = s^3 + 5s^2 + 3s + 10$

$p_2(s) = s^4 + 7s^3 + 5s^2 + 8s + 15$

$p_3(s) = s^5 + 15s^4 + 10s^3 + 6s^2 + 3s + 9$

Determine $p_1(2)$, $p_2(2)$, and $p_3(3)$

P2.10 The following polynomials are given:

$p_1(x) = x^5 + 2x^4 - 3x^3 + 7x^2 - 8x + 7$

$p_2(x) = x^4 + 3x^3 - 5x^2 + 9x + 11$

$p_3(x) = x^3 - 2x^2 - 3x + 9$

$p_4(x) = x^2 - 5x + 13$

$p_5(x) = x + 5$

Use MATLAB functions with polynomial coefficient vectors to evaluate the expressions at $x = 2$.

P2.11 Determine the roots of the following polynomials:

(a) $p_1(x) = x^7 + 8x^6 + 5x^5 + 4x^4 + 3x^3 + 2x^2 + x + 1$

(b) $p_2(x) = x^6 - 7x^6 + 7x^5 + 15x^4 - 10x^3 - 8x^2 + 7x + 15$

(c) $p_3(x) = x^5 - 13x^4 + 10x^3 + 12x^2 + 8x - 15$

(d) $p_4(x) = x^4 + 7x^3 + 12x^2 - 25x + 8$

(e) $p_5(x) = x^3 + 15x^2 - 23x + 105$

(f) $p_6(x) = x^2 - 18x + 23$

(g) $p_7(x) = x + 7$

P2.12 An aluminum thin-walled sphere is used as a marker buoy. The sphere has a radius of 65 cm and a wall thickness of 10 mm. The density of aluminum is 2700 kg/m^3. The buoy is placed in the ocean where the density of the water is 1050 kg/m^3. Determine the height H between the top of the buoy and the surface of the water.

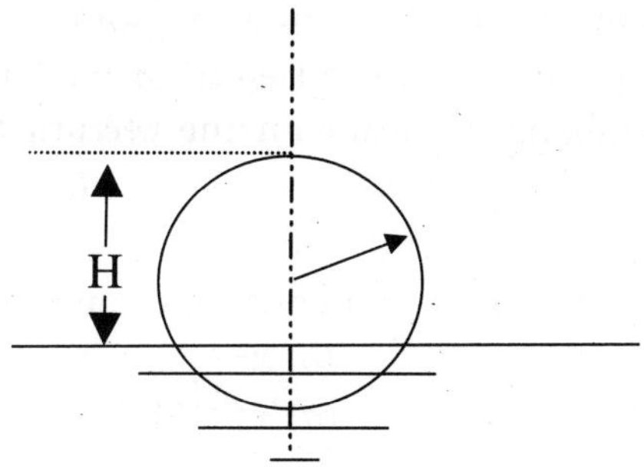

Fig. P2.12

P2.13 Determine the values of *x, y,* and *z* for the following set of linear algebraic equations:

$$x_2 - 3x_3 = -7$$
$$2x_1 + 3x_2 - x_3 = 9$$
$$4x_1 + 5x_2 - 2x_3 = 15$$

P2.14 Write a simple script file to find (*a*) dot product (*b*) cross-product of 2 vectors: $a = \hat{j} - \hat{k}$ and $b = 3\hat{i} - 2\hat{j}$.

P2.15 Write a function to find gradient of $f(x,y) = x^2 + y^2 - 2xy + 4$ at (*a*) (1,1) (*b*) (1, – 2) and (*c*) (0, – 3). Use the function name from command prompt.

P2.16 Write MATLAB functions $f = x^2 - 3x + 1$ and $g = e^x - 4x + 6$ and find the result $f(127)/g(5)$ from a script file.

P2.17 Plot the function $y = |x| \cos(x)$ for $-200 \le x \le 200$.

P2.18 Plot the following functions on the same plot for $0 \le x \le 2\pi$ using the plot function:

(*a*) $\sin^2(x)$ (*b*) $\cos^2 x$

(*c*) $\cos(x)$

P2.19 Plot a graph of the function $y = 45 \sin(0.4t)$ for $t \in [0, 3]$

P2.20 Consider the function $z = 0.56 \cos(xy)$. Draw a surface plot showing variation of *z* with *x* and *y*. Given $x \in [0, 10]$ and $y \in [0,100]$

P2.21 Figure P2.21 shows two boats: boat *A* travels south at a speed of 10 mph, and boat *B* travels east at a speed of 19 mph. The ships are positioned at 8 am are also shown in figure. Write a MATLAB program to plot the distance between the ships as a function of time for the next 5 hours.

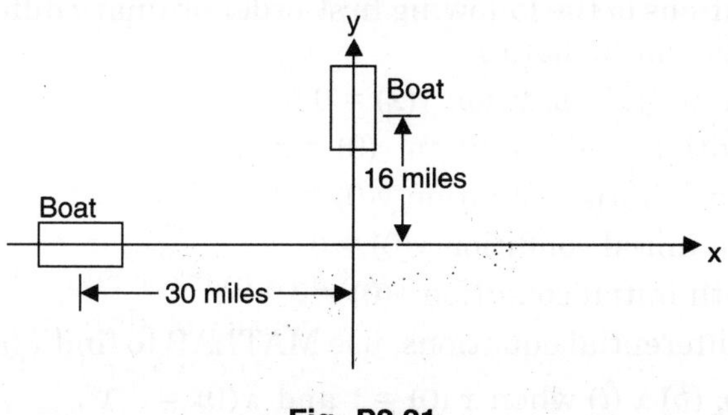

Fig. P2.21

P2.22 Consider the given symbolic expressions defined below:

S1 = '2/(x – 5)'; S2 = 'x ^ 5 + 9 * x – 15'; S3 = '(x ^ 3 + 2 * x +9) * (x * x – 5)';

Perform the following symbolic operations using MATLAB.

(*a*) S1S2/S3 (*b*) S1/S2S3 (*c*)S1/(S2)^2

(*d*) S1S3/S2 (*e*) (S2)^2/(S1S3)

P2.23 Solve the following equations using symbolic mathematics.

(*a*) $x^2 + 9 = 0$ (*b*) $x^2 + 5x - 8 = 0$

(*c*) $x^3 + 11x^2 - 7x + 8 = 0$ (*d*) $x^4 + 11x^3 + 7x^2 - 19x + 28 = 0$

(*e*) $x^7 - 8x^5 + 7x^4 + 5x^3 - 8x + 9 = 0$

P2.24 Determine the values of *x, y,* and *z* for the following set of linear algebraic equations:

$$2x + y - 3z = 11$$
$$4x - 2y + 3z = 8$$
$$-2x + 2y - z = -6$$

P2.25 Figure P2.25 shows a scale with two springs.

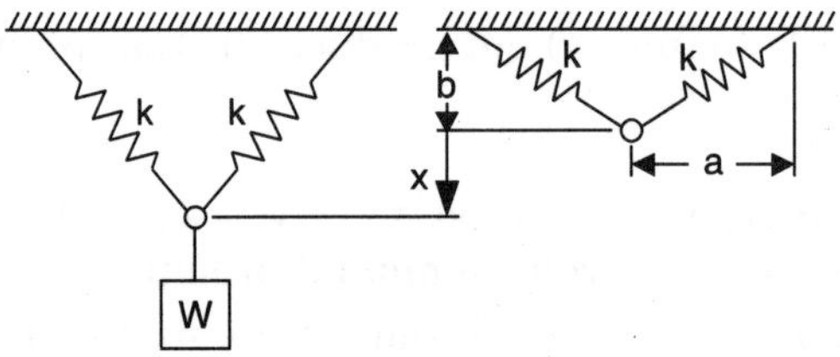

Fig. P2.25

The two springs are unstretched initially and will stretch when a mass is attached to the ring and the ring will displace downwards a distance of *x*. The weight *W* of the object is given by

$$W = \frac{2k}{l}(l - l_0)(b + x)$$

where l_0 = initial length of a spring = $\sqrt{a^2 + b^2}$ and l = the stretched length of the spring = $\sqrt{a^2 + (b + x)^2}$. If k = spring constant, write a MATLAB program to determine the distance x when W = 350 N. Given: a = 0.16 m, b = 0.045 m, and the spring constant k = 3000 N/m.

P2.26 Determine the solutions of the following first-order ordinary differential equations using MATLAB's symbolic mathematics.

(*a*) $y' = 8x^2 + 5$ with initial condition $y(2) = 0.5$.

(*b*) $y' = 5x \sin^2(y)$ with initial condition $y(0) = \pi/5$.

(*c*) $y' = 7x \cos^2(y)$ with initial condition $y(0) = 2$.

(*d*) $y' = -5x + y$ with initial condition $y(0) = 3$.

(*e*) $y' = 3y + e^{-5x}$ with initial condition $y(0) = 2$.

P2.27 For the following differential equations, use MATLAB to find $x(t)$ when (*a*) all the initial conditions are zero, (*b*) $x(t)$ when $x(0) = 1$ and $\dot{x}(0) = -1$.

(a) $\dfrac{d^2x}{dt^2} + 10\dfrac{dx}{dt} + 5x = 11$ (b) $\dfrac{d^2x}{dt^2} + 7\dfrac{dx}{dt} + 3x = 5$

(c) $\dfrac{d^2x}{dt^2} + 3\dfrac{dx}{dt} + 7x = -15$ (d) $\dfrac{d^2x}{dt^2} + \dfrac{dx}{dt} + 7x = 26$

P2.28 Figure P2.28 shows a water tank (shaped as an inverted frustum cone with a circular hole at the bottom on the side).

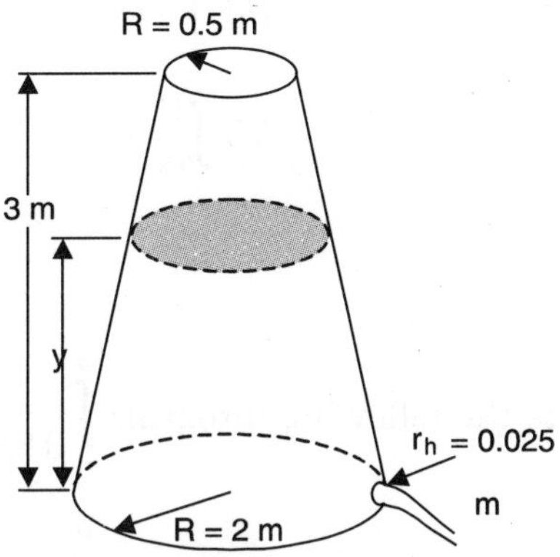

Fig. P2.28 Water-tank.

The velocity of water discharged through the hole is given by $v = \sqrt{2gy}$ where h = height of the water and g = acceleration due to gravity (9.81 m/s^2). The rate of discharge of water in the tank as the water drains out through the hole is given by:

$\dfrac{dy}{dt} = -\dfrac{\sqrt{2gyr_h^2}}{(2-0.5y)^2}$ where y = height of water and r_h=radius of the hole. Write a MATLAB program to solve and plot the differential equation. Assume, that the initial height of the water is 2.5 m.

P2.29 An airplane uses a parachute (See Fig. P2.29) and other means of braking as it slows down on the runway after landing. The acceleration of the airplane is given by $a = -0.005v^2 - 4$ m/s^2

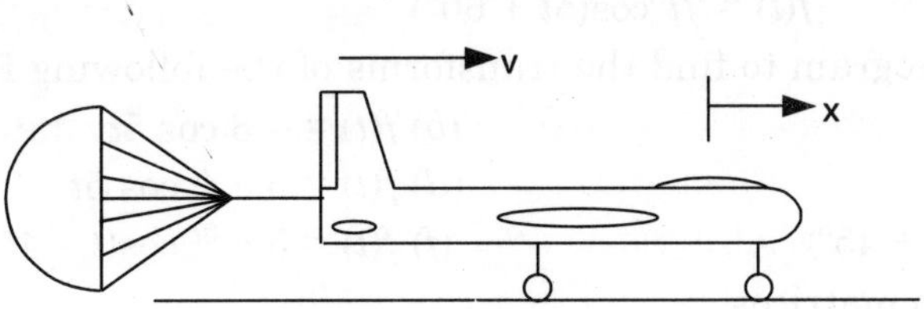

Fig. P2.29

Consider the airplane with a velocity of 500 km/h opens its parachute and starts decelerating at $t = 0$ seconds. Write a MATLAB program to solve the differential equation and plot the velocity from $t = 0$ seconds until the airplane stops.

P2.30 Obtain the first and second derivatives of the following functions using MATLAB's symbolic mathematics.

(*a*) $F(x) = x^5 - 8x^4 + 5x^3 - 7x^2 + 11x - 9$

(*b*) $F(x) = (x^3 + 3x - 8)(x^2 + 21)$

(*c*) $F(x) = (3x^3 - 8x^2 + 5x + 9)/(x + 2)$

(*d*) $F(x) = (x^5 - 3x^4 + 5x^3 + 8x^2 - 13)^2$

(*e*) $F(x) = (x^2 + 8x - 11)/(x^7 - 7x^6 + 5x^3 + 9x - 17)$

P2.31 Determine the values of the following integrals using MATLAB's symbolic functions.

(*a*) $\int (5x^7 - x^5 + 3x^3 - 8x^2 + 7)dx$ (*b*) $\int \sqrt{x \cos x}$

(*c*) $\int x^{2/3} \sin^2 2x$ (*d*) $\int_{0.2}^{1.8} x^2 \sin x \, dx$

(*e*) $\int_{-1}^{-0.2} |x| dx$

P2.32 Use MATLAB to calculate the following integral: $\int_0^5 \frac{1}{0.8x^2 + 0.5x + 2} dx$

P2.33 Use MATLAB to calculate the following integral: $\int_0^{10} \cos^2 (0.5x) \sin^4 (0.5x) \, dx$

P2.34 The variation of gravitational acceleration g with altitude y is given by:

$g = \frac{R^2}{(R + y)^2} g_0$, where $R = 6371$ km is radius of the earth and g_o=9.81 m/s^2 is gravitational acceleration at sea level. The change in the gravitational potential energy ΔU of an object that is raised up from the earth is given by: $\Delta U = \int_0^y mg \, dy$. Determine the change in the potential energy of a satellite with a mass of 500 kg that is raised from the surface of the earth to a height of 800 km.

P2.35 Find the Laplace transform of the following function using MATLAB:

$$f(t) = 7t^3\cos(5t + 60°)$$

P2.36 Use MATLAB program to find the transforms of the following functions.

(*a*) $f(t) = -7t\, e^{-5t}$ (*b*) $f(t) = -3 \cos 5t$

(*c*) $f(t) = t \sin 7t$ (*d*) $f(t) = 5\, e^{-2t} \cos 5t$

(*e*) $f(t) = 3 \sin(5t + 45°)$ (*f*) $f(t) = 5\, e^{-3t} \cos(t - 45°)$

P2.37 Consider the two matrices

$$\mathbf{A} = \begin{bmatrix} 1 & 0 & 2 \\ 2 & 5 & 4 \\ -1 & 8 & 7 \end{bmatrix} \text{ and } \mathbf{B} = \begin{bmatrix} 7 & 8 & 2 \\ 3 & 5 & 9 \\ -1 & 3 & 1 \end{bmatrix}$$

Using MATLAB, determine the following:

(*a*) $\mathbf{A + B}$ (*b*) $\mathbf{AB}$

(*c*) $\mathbf{A^2}$ (*d*) $\mathbf{A^T}$

(*e*) $\mathbf{B^{-1}}$ (*f*) $\mathbf{B^TA^T}$

(*g*) $\mathbf{A^2 + B^2 - AB}$

(*h*) determinant of **A**, determinant of **B** and determinant of **AB**.

P2.38 Use MATLAB to define the following matrices:

$$\mathbf{A} = \begin{bmatrix} 2 & 1 \\ 0 & 5 \\ 7 & 4 \end{bmatrix}$$

$$\mathbf{B} = \begin{bmatrix} 5 & 3 \\ -2 & -4 \end{bmatrix}$$

$$\mathbf{C} = \begin{bmatrix} 2 & 3 \\ -5 & -2 \\ 0 & 3 \end{bmatrix}$$

$$\mathbf{D} = [1 \quad 2]$$

Compute matrices and determinants if they exist.

(*a*) $\mathbf{(AC^T)^{-1}}$ (*b*) $|\mathbf{B}|$

(*c*) $|\mathbf{AC^T}|$ (*d*) $\mathbf{(C^TA)^{-1}}$

P2.39 Consider the two matrices

$$\mathbf{A} = \begin{bmatrix} 1 & 0 & 1 \\ 2 & 3 & 4 \\ -1 & 6 & 7 \end{bmatrix} \quad \text{and} \quad \mathbf{B} = \begin{bmatrix} 7 & 4 & 2 \\ 3 & 5 & 6 \\ -1 & 2 & 1 \end{bmatrix}$$

Using MATLAB, determine the following:

(*a*) $\mathbf{A + B}$ (*b*) $\mathbf{AB}$

(*c*) $\mathbf{A^2}$ (*d*) $\mathbf{A^T}$

(*e*) $\mathbf{B^{-1}}$ (*f*) $\mathbf{B^TA^T}$

(*g*) $\mathbf{A^2 + B^2 - AB}$ (*h*) det **A**, det **B**, and det of **AB**.

P2.40 Find the inverse of the following matrices:

(*a*) $$\mathbf{A} = \begin{bmatrix} 3 & 2 & 1 \\ -1 & 5 & 4 \\ 5 & 7 & -9 \end{bmatrix}$$

(*b*) $$\mathbf{B} = \begin{bmatrix} 1 & 6 & 3 \\ -4 & -5 & 7 \\ 8 & 4 & 2 \end{bmatrix}$$

(*c*) $$\mathbf{C} = \begin{bmatrix} -1 & -2 & 5 \\ -4 & 7 & 2 \\ 7 & -8 & -1 \end{bmatrix}$$

P2.41 Determine the eigenvalues and eigenvectors of the following matrices using MATLAB.

$$A = \begin{bmatrix} 1 & -2 \\ 1 & 5 \end{bmatrix}, \quad B = \begin{bmatrix} 1 & 5 \\ -2 & 7 \end{bmatrix}$$

P2.42 If $A = \begin{bmatrix} 4 & 6 & 2 \\ 5 & 6 & 7 \\ 10 & 5 & 8 \end{bmatrix}$

Use MATLAB to determine the following:

(*a*) the three eigenvalues of **A**

(*b*) the eigenvectors of **A**

(*c*) Show that **AQ** = **Qd** where *Q* is the matrix containing the eigenvectors as columns and **d** is the matrix containing the corresponding eigenvalues on the main diagonal and zeros else where.

P2.43 Determine eigenvalues and eigenvector of **A** using MATLAB.

(*a*) $\mathbf{A} = \begin{bmatrix} 0.5 & -0.8 \\ 0.75 & 1.0 \end{bmatrix}$ (*b*) $\mathbf{A} = \begin{bmatrix} 8 & 3 \\ -3 & 4 \end{bmatrix}$

P2.44 Determine the eigenvalues and eigenvectors of the following matrices using MATLAB.

(*a*) $\mathbf{A} = \begin{bmatrix} 1 & -2 \\ 1 & 3 \end{bmatrix}$ (*b*) $\mathbf{A} = \begin{bmatrix} 1 & 5 \\ -2 & 4 \end{bmatrix}$

P2.45 Determine the eigenvalues and eigenvectors of *A* * *B* using MATLAB.

$$\mathrm{A} = \begin{bmatrix} 3 & -1 & 2 & 1 \\ 1 & 2 & 7 & 4 \\ 7 & -1 & 8 & 6 \\ 1 & -2 & 3 & 4 \end{bmatrix} \qquad \mathrm{B} = \begin{bmatrix} 1 & 2 & 5 & 7 \\ 2 & -1 & -2 & 4 \\ 3 & 2 & 5 & 1 \\ 4 & 1 & -3 & 6 \end{bmatrix}$$

P2.46 Determine the eigenvalues and eigenvectors of A and B using MATLAB

(*a*) $\mathbf{A} = \begin{bmatrix} 4 & 5 & -3 \\ -1 & 2 & 3 \\ 2 & 5 & 7 \end{bmatrix}$

$$\mathbf{B} = \begin{bmatrix} 1 & 2 & 3 \\ 8 & 9 & 6 \\ 5 & 3 & -1 \end{bmatrix}$$

P2.47 Determine the eigenvalues and eigenvectors of $A = a*b$ using MATLAB.

$$\mathbf{a} = \begin{bmatrix} 6 & -3 & 4 & 1 \\ 0 & 4 & 2 & 6 \\ 1 & 3 & 8 & 5 \\ 2 & 2 & 1 & 4 \end{bmatrix}$$

$$\mathbf{b} = \begin{bmatrix} 0 & 1 & 2 & 3 \\ 4 & 5 & 6 & -1 \\ 1 & 5 & 4 & 2 \\ 2 & -3 & 6 & 7 \end{bmatrix}$$

P2.48 Determine the values of *x*, *y*, and *z* for the following set of linear algebraic equations:

$$x_2 - 3x_3 = -7$$
$$2x_1 + 3x_2 - x_3 = 9$$
$$4x_1 + 5x_2 - 2x_3 = 15$$

P2.49 Determine the values of x, y, and z for the following set of linear algebraic equations:

$$2x - y = 10$$
$$-x + 2y - z = 0$$
$$-y + z = -50$$

P2.50 Solve the following set of equations using MATLAB.

(*a*)
$$2x_1 + x_2 + x_3 - x_4 = 12$$
$$x_1 + 5x_2 - 5x_3 + 6x_4 = 35$$
$$-7x_1 + 3x_2 - 7x_3 - 5x_4 = 7$$
$$x_1 - 5x_2 + 2x_3 + 7x_4 = 21$$

(*b*)
$$x_1 - x_2 + 3x_3 + 5x_4 = 7$$
$$2x_1 + x_2 - x_3 + x_4 = 6$$
$$-x_1 - x_2 - 2x_3 + 2x_4 = 5$$
$$x_1 + x_2 - x_3 + 5x_4 = 4$$

P2.51 Solve the following set of equations using MATLAB.

(*a*)
$$2x_1 + x_2 + x_3 - x_4 = 10$$
$$x_1 + 5x_2 - 5x_3 + 6x_4 = 25$$
$$-7x_1 + 3x_2 - 7x_3 - 5x_4 = 5$$
$$x_1 - 5x_2 + 2x_3 + 7x_4 = 11$$

(*b*)
$$x_1 - x_2 + 3x_3 + 5x_4 = 5$$
$$2x_1 + x_2 - x_3 + x_4 = 4$$
$$-x_1 - x_2 + 2x_3 + 2x_4 = 3$$
$$x_1 + x_2 - x_3 + 5x_4 = 1$$

P2.52 Solve the following set of equations using MATLAB.

(*a*)
$$x_1 + 2x_2 + 3x_3 + 5x_4 = 21$$
$$-2x_1 + 5x_2 + 7x_3 - 9x_4 = 17$$
$$5x_1 + 7x_2 + 2x_3 - 5x_4 = 23$$
$$-x_1 - 3x_2 - 7x_3 + 7x_4 = 26$$

(*b*)
$$x_1 + 2x_2 + 3x_3 + 4x_4 = 9$$
$$2x_1 - 2x_2 - x_3 + x_4 = -5$$
$$x_1 - 3x_2 + 4x_3 - 4x_4 = 7$$
$$2x_1 + 2x_2 - 3x_3 + 4x_4 = -6$$

P2.53 Determine the inverse of the following matrix using MATLAB.

$$A = \begin{bmatrix} 3s & 2 & 0 \\ 7s & -s & -5 \\ 3 & 0 & -3s \end{bmatrix}$$

P2.54 Expand the following function F(s) into partial fractions with MATLAB:

$$F(s) = \frac{5s^3 + 7s^2 + 8s + 30}{s^4 + 15s^3 + 62s^2 + 85s + 25}$$

P2.55 Determine the Laplace transform of the following time functions using MATLAB.

(*a*) $f(t) = u(t + 9)$

(*b*) $f(t) = e^{5t}$

(*c*) $f(t) = (5t + 7)$

(*d*) $f(t) = 5u(t) + 8e^{7t} - 12e^{-8t}$

(*e*) $f(t) = e^{-t} + 9t^3 - 7t^{-2} + 8$

(*f*) $f(t) = 7t^4 + 5t^2 - e^{-7t}$

(*g*) $f(t) = 9\ u\ t + 5e^{-3t}$

P2.56 Determine the inverse Laplace transform of the following rotational function using MATLAB.

$$F(s) = \frac{7}{s^2 + 5s + 6} = \frac{7}{(s+2)(s+3)}$$

P2.57 Determine the inverse transform of the following function having complex poles

$$F(s) = \frac{15}{(s^3 + 5s^2 + 11s + 10)}$$

P2.58 Determine the inverse Laplace transform of the following functions using MATLAB:

(*a*) $F(s) = \dfrac{s}{s(s+2)(s+3)(s+5)}$

(*b*) $F(s) = \dfrac{1}{s^2(s+7)}$

(*c*) $F(s) = \dfrac{5s+9}{(s^3 + 8s + 5)}$

(*d*) $F(s) = \dfrac{s - 28}{s(s^2 + 9s + 33)}$

CHAPTER 3

MATLAB TUTORIAL

3.1 CONTROL SYSTEMS

The application of MATLAB to the analysis and design of control systems, engineering mechanics (statics and dynamics), mechanical vibration analysis, electrical circuits, and numerical methods is presented in this chapter with a number of illustrative examples. The MATLAB computational approach to the transient response analysis, steps response, impulse response, ramp response, and response to the simple inputs are presented. Plotting root loci, Bode diagrams, polar plots, Nyquist plot, Nichols plot, and state space method are obtained using MATLAB. Extensive worked examples are included with a significant number of exercise problems to guide the student to understand and as an aid for learning about the analysis and design of control systems, engineering mechanics, vibration analysis of mechanical systems, electrical circuits, and numerical methods using MATLAB.

3.2 TRANSIENT RESPONSE ANALYSIS

When the numerator and denominator of a closed-loop transfer function are known, the commands **step (num, den)**, **step (num, den, t)** in MATLAB can be used to generate plots of unit-step responses. Here, *t* is the user specified time.

3.3 RESPONSE TO INITIAL CONDITION

3.3.1 Case 1: State Space Approach

Consider a system defined in state-space given by

$$\dot{\mathbf{x}} = \mathbf{A}\mathbf{x} \tag{3.1}$$

$$\mathbf{x}\,(\mathbf{0}) = \mathbf{x}_o$$

Assuming that there is no external input acting on the system, the response **x (t)** knowing the initial condition x (0) and that **x** is an n -vector, is obtained as follows:

Taking Laplace transform of both sides of Eq. (3.1), we obtain

$$s\,\mathbf{x}\,(s) - \mathbf{x}\,(0) = \mathbf{AX}\,(s) \tag{3.2}$$

Equation (3.2) can be rearranged as

$$s\,\mathbf{x}\,(s) = \mathbf{AX}\,(s) + \mathbf{x}\,(0) \tag{3.3}$$

Taking inverse Laplace transform of Eq. (3.3), we get

$$\dot{\mathbf{x}} = A\,\mathbf{x} + \mathbf{x}\,(0)\,\delta\,(t) \tag{3.4}$$

Defining $\dot{\mathbf{z}} = \mathbf{x}$, Eq. (3.4) can be written as

$$\ddot{\mathbf{z}} = \mathbf{A}\,\dot{\mathbf{z}} + \mathbf{x}\,(0)\,\delta\,(t) \tag{3.5}$$

Integrating Eq. (3.5), we obtain

$$\dot{\mathbf{z}} = A\mathbf{z} + \mathbf{x}\,(0)\,1(t) = \mathbf{A}\mathbf{z} + \mathbf{B}u \tag{3.6}$$

where $\mathbf{B} = \mathbf{x}\,(0)$ and $u = 1(t)$

Noting that $\dot{\mathbf{z}} = \mathbf{x}$ and $x\,(t) = \dot{\mathbf{z}}\,(t)$, we have

$$\mathbf{x} = \dot{\mathbf{z}} = A\mathbf{z} + \mathbf{B}u \tag{3.7}$$

The response to initial condition is obtained by solving Eqs. (3.6) and (3.7).

The corresponding MATLAB command used to obtain the response curves are given as follows:

```
[x, z, t] = step (A, B, A, B);
x1 = [1 0 0 ...0] * x';
x2 = [1 0 0 ...0] * x';
   .   .
   .   .
   .   .
xn = [0 0 0 ...1] * x';
plot (t, x1, x2,..., t, xn)
```

3.3.2 Case 2: State Space Approach

Consider the system defined in state space by

$$\dot{\mathbf{x}} = \mathbf{A}\mathbf{x} \qquad \mathbf{x}\,(0) = \mathbf{x}_o \tag{3.8}$$

$$\mathbf{y} = \mathbf{C}\mathbf{x} \tag{3.9}$$

where $\mathbf{x}$ is an n vector and $\mathbf{y}$ is an m vector.

By defining $\dot{\mathbf{z}} = \mathbf{x}$ (3.10)

we obtain

$$\dot{\mathbf{z}} = \mathbf{A}\mathbf{z} + \mathbf{x}\,(0)\,1\,(t) = \mathbf{A}\mathbf{z} + \mathbf{B}u \tag{3.11}$$

where

$$\mathbf{B} = \mathbf{x}\,(0) \quad \text{and} \quad u = 1\,(t) \tag{3.12}$$

Since $\mathbf{x} = \dot{\mathbf{z}}$, Eq. (3.9) becomes

$$\mathbf{y} = \mathbf{C}\dot{\mathbf{z}} \tag{3.13}$$

From Eqs. (3.11) and (3.13), we obtain

$$\mathbf{y} = \mathbf{C}\,(\mathbf{A}\mathbf{z} + \mathbf{B}\,u) = \mathbf{C}\mathbf{A}\mathbf{z} + \mathbf{C}\mathbf{B}\,u \tag{3.14}$$

The response of the system is obtained from the Eqs. (3.11) and (3.14) to a given initial condition.

The following MATLAB commands may be used to obtain the response curves:

```
[y, z, t] = step (A, B, C*A , C*B);
y1 = [1  0  0 ...0] * y' ;
y2 = [0  1  0 ...0] * y' ;          (3.15)
   .   .
   .   .
```

$$ym = [0\ \ 0\ \ 0 \ldots 1] * y' ;$$

plot $(t\ , y_{1,}\ t\ , y_2\ ,\ldots\ldots\ldots, t, ym)$

3.4 SECOND ORDER SYSTEMS

The standard form of a second-order system is defined by

$$G(s) = \frac{{\omega_n}^2}{s^2 + 2\xi\omega_n s + \omega_n^2} \tag{3.16}$$

where ξ is the damping ratio of the system and ω_n is the undamped natural frequency of the system.

The dynamic behavior of the second order system is then described in terms of two parameters ξ and ω_n.

If $0 < \xi < 1$, the closed loop poles are complex conjugates and lie in the left-half s plane. The system is called underdamped, and the transient response is oscillatory, If $\xi = 0$, the transient response does not die out. If $\xi = 1$, the system is called critically damped. Overdamped system correspond to $\xi = 1$.

Given ω_n and ξ, then the MATLAB command

printsys (num, den)

or

printsys(num, den, s)

prints the num/den as a ratio of polynomials in s.

The unit-step response of the transfer-function system using MATLAB is obtained with the use of step-response commands with left-hand arguments.

c = step (num, den, t)

or

[y , x , t] = step (num , den ,t)

3.5 ROOT LOCUS PLOTS

Consider the system equation

$$1 + \frac{K(s+z_1)\,(s+z_2)\ldots(s+z_n)}{(s+p_1)\,(s+p_2)\ldots(s+p_n)} = 0 \tag{3.17}$$

Equation (3.17) can be written as

$$1 + K\frac{num}{den} = 0 \tag{3.18}$$

where *num* is the numerator of the polynomial and *den* is the denominator polynomial, and K is the gain (K>0). The vector K contains all the gain values for which the closed loop poles are to be computed.

The root loci is plotted by using the MATLAB command

rlocus (num, den)

The gain vector K is supplied by the user.

The matrix r and gain vector K are obtained by the following MATLAB commands:

[r, k] = rlocus (num, den)
[r, k] = rlocus (num, den,k)
[r, k] = rlocus (A, B, C, D)
[r, k] = rlocus (A, B, C, D, K) (3.19)
[r, k] = rlocus (sys)

In Eqs. (3.19), r has length K rows and length [den – 1] columns containing the complex root locations.

For plotting the root loci, the MATLAB command plot (r, ' ') is used.

The following MATLAB command are used for plotting the root loci with mark '0' or 'x':

r = rlocus (num, den)
plot (r, '0') or plot (r, 'x')

MATLAB provides its own set of gain values used to compute a root locus plot. It also uses the automatic axis scaling features of the plot command.

3.6 BODE DIAGRAMS

Bode diagrams are rectangular plots. Bode diagrams are also known as logarithmic plot and consist of two graphs: the first one is a plot of the logarithmic of the magnitude of a sinusoidal transfer function, the second one is a plot of the phase angle. Both these graphs are plotted against the frequency on a logarithmic scale.

The MATLAB command "bode" obtains the magnitudes and phase angles of the frequency response of continuous-time, linear, time-invariant systems.

The MATLAB bode commands commonly used are:

Bode(num, den)
bode(num, den, W)
bode(A, B, C, D) (3.20)
bode(A, B, C, D, W)
bode(sys)

where W is the frequency vector.

MATLAB bode commands with left hand arguments commonly used are:

[mag, phase, w] = bode (num, den)
[mag, phase, w] = bode (num, den, w)
[mag, phase, w] = bode (A, B, C, D)
[mag, phase, w] = bode (A, B, C,D, w) (3.21)
[mag, phase, w] = bode (A, B, C, D, iu, w)
[mag, phase, w] = bode (sys)

The MATLAB commands given in Eq. (3.21) returns the frequency response of the system in matrices *mag*, *phase*, and w. The plot is not drawn on the screen. The matrices *mag*, *phase* provide the magnitudes and phase angles of frequency response of the system, computed at the specified frequency points.

The magnitude may be converted into decibles using the MATLAB statement

magdB = 20 * log 10 (mag) (3.22)

In MATLAB , the following command

logspace (d1 , d2) (3.23)

or

logspace(d1, d2, n). logspace(d1, d2) (3.24)

are used to specify the frequency range that will generate a vector of 50 points logarithmically equally spaced between decades 10^{d1} and 10^{d2}

The MATLAB command

w = logspace (– 1, 2) (3.25)

may be used to generate 50 points between 0.1 and 100 rad/sec.

Similarly, the MATLAB command

logspace (d1, d2, n) (3.26)

generates n points logarthimatically equally spaced between 10^{d1} and 10^{d2} whereby the n points include both the endpoints.

3.7 NYQUIST PLOTS

Nyquist plots are also used in the frequency-response representation of linear, time invariant, continuous-time feedback control systems. Nyquist plots are polar plots.

The MATLAB command

nyquist (num, den) (3.27)

Draw the Nyquist plot of the transfer function

$$G(s) = \frac{num(s)}{den(s)} \tag{3.28}$$

where *num* and *den* contain the polynomial coefficients in descending powers of *s*. The other MATLAB command used for drawing Nyquist plots are:

nyquist (num ,den , w)
nyquist (A, B, C, D)
nyquist (A, B, C, D, w)
nyquist (A, B, C, D,iu, w) (3.29)
nyquist (sys)

where *w* is the frequency vector.

The MATLAB command involving the user-specified vector *w* in Eq. (3.29) computes the frequency response at the specified frequency points.

The following MATLAB commands

[re , im ,w] = nyquist (num , den)
[re , im ,w] = nyquist (num , den ,w)
[re , im ,w] = nyquist (A, B, C, D)
[re , im ,w] = nyquist (A, B, C, D, w) (3.30)
[re , im ,w] = nyquist (A, B, C, D, iu, w)
[re , im ,w] = nyquist (sys)

are used to obtain the frequency response of the system in the matrices *Re*, *im*, and *w*. The plot is not drawn on the screen. The matrices *Re* and *im* contain the real and imaginary parts of the frequency response of the system, computed at the frequency points specified in the vector *w*.

3.8 NICHOLS CHART

The chart consisting of the M and N loci in the log magnitude versus phase diagram is called the Nichols chart. The G (jw) locus drawn on the Nichols chart gives both the gain characteristics and phase characteristics of the closed loop transfer function at the same time. The Nichols chart contains curves of constant closed loop magnitude and phase angle. The Nichols chart is symmetric about the 180^0 axis. The M loci are centered about the critical point (0 dB, – 180). The Nichols chart is useful in determining the frequency response of the closed loop from that of the open loop. The Nichols chart is produced by using the MATLAB command nichols(num, den). The command ngrid creates the dotted lines that allow reading closed-loop gain and phase from the Nichols chart. In order to customize the axes of the Nichols chart, the MATLAB command ***axis*** is used.

3.9 GAIN MARGIN, PHASE MARGIN, PHASE CROSSOVER FREQUENCY, AND GAIN CROSSOVER FREQUENCY

The MATLAB command

```
[Gm, pm, wcp, wcg] = margin (sys)                                   (3.31)
```

can be used to obtain the gain margin, phase margin, phase crossover frequency, and gain crossover frequency.

In Equation (3.31), Gm is the gain margin, pm is the phase margin, wcp is the phase-crossover frequency, and wcg is the gain crossover frequency.

The following MATLAB command is commonly used for obtaining the resonant peak and resonant frequency :

```
[ mag, phase, w] = bode (num, den, w)
                  or
[mag, phase, w] = bode (sys, w)
[Mp, k] = max (mag)                                                 (3.32)
resonant peak = 20 * log 10 (Mp)
resonant frequency = w (k)
```

The following lines are used in MATLAB program to obtain bandwidth :

```
n = 1
while 20 * log 10 (mag (n)) > – 3
n = n + 1
end                                                                 (3.33)
bandwidth = w (n)
```

3.10 TRANSFORMATION OF SYSTEM MODELS

In this section, we consider two cases of transformation of system models.

1. Transformation of system model from transfer function to state space
2. Transformation of system model from state space to transfer function

3.10.1 Transformation of System Model from Transfer Function to State Space

The closed-loops transfer function can be written as

$$\frac{Y(s)}{U(s)} = \frac{\textit{numerator of polynominal in s}}{\textit{denominator of polynominal in s}} = \frac{num}{den} \tag{3.34}$$

The state-space representation is obtained by the MATLAB command

[A, B, C, D] = tf2ss (num, den) (3.35)

3.10.2 Transformation of System Model from State Space to Transfer Function

The transfer function from state-space equations is obtained by using the MATLAB command:

[num, den] = ss2tf (A, B, C, D, iu) (3.36)

where *iu* corresponds to the system with more than one input. *iu* is either 1, 2, or 3, where 1 implies input u_1, 2 implies input u_2, and 3 implies input u_3.

For system with only one input, the MATLAB command

[num, den] = ss2tf (A, B, C, D,) (3.37)

or

[num, den] = ss2tf (A, B, C, D, 1) (3.38)

may be used.

3.11 BODE DIAGRAMS OF SYSTEMS DEFINED IN STATE SPACE

Let the control system defined in State Space be

$$\begin{aligned}\dot{\mathbf{x}} &= \mathbf{Ax} + \mathbf{Bu}\\ \mathbf{y} &= \mathbf{Cx} + \mathbf{Du}\end{aligned} \tag{3.39}$$

where

A = state matrix (n × n matrix)
B = control matrix (n × r matrix)
C = output matrix (m × n matrix)
D = output matrix (m × n matrix)
u = control vector (r – vector)
x = state vector (n – vector)
y = output vector (m – vector)

The MATLAB command **bode[A, B, C, D]** may be used to obtain the Bode diagram of this system. In fact, the command **bode[A, B, C, D]** gives a series of Bode plots, one for each input of the system, with the frequency range automatically determined.

If we use the scalar iu as an index into the inputs of the control system that specifies which input is to be used for the Bode plot, then the MATLAB command **Bode[A, B, C, D iu]** produces the Bode plots from the input iu to all the outputs ($y_1, y_2, \ldots\ldots, y_m$) of the system with the frequency range automatically determined.

If the system has three inputs, then $\mathbf{u} = \begin{bmatrix} u_1 \\ u_2 \\ u_3 \end{bmatrix}$

For a system with only one input u , then the MATLAB command

Bode[A, B, C, D] (3.40)

or

Bode [A, B, C, D, 1] can be used (3.41)

3.12 NYQUIST PLOTS OF A SYSTEM DEFINED IN STATE SPACE

Consider the system defined in state space given by Equation (3.39). Nyquist plots of the system defined in Eq. (3.39) may be obtained by using the MATLAB command

nyquist (A, B, C, D) (3.42)

The MATLAB command given by Eq. (3.42) produces a series of Nyquist plots one corresponding to each input and output combination of the system, with the frequency range automatically determined.

If we use the scalar *iu* as an index to the inputs of the control system that specifies which input is to be used for the Nyquist plot, then the MATLAB command nyquist (A, B, C, D, *iu*, *w*) produces Nyquist plots from the input to all the outputs (y_1, y_2, ..., y_m) of the system with the frequency range automatically determined.

The MATLAB command

nyquist (A, B, C, D, iu, w) (3.43)

considers the user-supplied frequency vector *w*. The vector *w* specifies the frequency at which the frequency response should be determined.

3.13 TRANSIENT-RESPONSE ANALYSIS IN STATE SPACE

In this section, we present the transient-response analysis of systems in state-space using MATLAB. Specifically, we present the step response, impulse, ramp response, and responses to other forms of simple inputs.

3.13.1 Unit Step Response

For a control system defined in a state space form as in Eq. (3.39), the MATLAB command

step (A, B, C, D) (3.44)

will generate plots of unit step responses, with the time vector automatically determined provided t is not explicitly provided in the step commands.

The MATLAB command step (sys) may also be used to obtain the unit-step response of a system.

The command

step (sys) (3.45)

can be used where the system is defined by

sys = tf (num, den) (3.46)

or

sys = ss (A, B, C, D) (3.47)

If the following MATLAB step commands with left hand arguments are used then no plot is shown on the screen.

[y, x , t] = step [num, den, t]

[y, x, t] = step (A, B, C, D, iu) (3.48)

[y, x, t] = step (A, B, C, D, iu ,t)

Hence, in order to obtain the response curves, plot commands should be used. The matrices x, and y contain the state response of the system and the output respectively, computed at the time points t. In Eq. (3.48), iu is a scalar index of the inputs of the system, which specifies the input to be used for the response, and t is the user specified time. The step command in Eq. (3.48) can be used to obtain a series of step response plots , one for each input and output combination of

$$\dot{\mathbf{x}} = \mathbf{Ax} + \mathbf{Bu}$$
$$\mathbf{y} = \mathbf{Cx} + \mathbf{Du} \tag{3.49}$$

when the system involves multiple inputs and multiple outputs.

3.13.2 Impulse Response

The following MATLAB commands may be used to obtain the unit impulse response of a control system:

impulse (num, den) (3.50)

impulse(A, B, C, D) (3.51)

[y, x, t] = impulse (num, den) (3.52)

[y, x, t] = impulse (num, den ,t) (3.53)

[y, x, t] = impulse (A, B,C, D) (3.54)

[y, x, t] = impulse (A, B, C, D, iu) (3.55)

[y, x, t] = impulse (A, B, C, D, iu, t) (3.56)

The command in Eq. (3.50) impulse (num, den) shows the plots of the unit impulse response on the monitor (screen). The command in Eq. (3.51) , impulse (A, B, C, D) produces a series of unit impulse-response plots one for each input and output combination of the system defined in Eq. (3.39) with the time vector automatically obtained. The vector t in Eqs. (3.53) and (3.56) is the user supplied time vector, which specifies the times at which the impulse response is to be obtained. The scalar iu in Eqs. (3.55) and (3.56) is an index into the inputs of the system and specifies which input is to be used for the impulse response. The matrices x and y in Eqs. (3.52) to (3.56) contain the state responses of the system and the output respectively, evaluated at the time points t.

3.13.3 Unit Ramp Response

Consider the system described in state space as

$$\dot{\mathbf{x}} = \mathbf{Ax} + \mathbf{B}u$$
$$\mathbf{y} = \mathbf{Cx} + \mathbf{D}u \tag{3.57}$$

where u is the unit-ramp function.

When all the initial conditions are zeros, the unit ramp response is the integral of the unit step response. Therefore, the unit ramp response is given by

$$Z = \int_0^t y\, dt \tag{3.58}$$

or
$$\dot{\mathbf{z}} = \mathbf{y} = \mathbf{x}_1 \tag{3.59}$$

Defining

$$\mathbf{z} = \mathbf{x}_3 \tag{3.60}$$

Equation (10.59) can be written as

$$\dot{\mathbf{x}}_3 = \mathbf{x}_1 \tag{3.61}$$

Combining Eqs. (3.57) and (3.61), we can write

$$\begin{aligned}\dot{\mathbf{x}} &= \mathbf{AAx} + \mathbf{BB}u \\ \mathbf{z} &= \mathbf{CCx} + \mathbf{DD}u\end{aligned} \tag{3.62}$$

The MATLAB command

$$[z, x, t] = \text{step}\ (AA, BB, CC, DD) \tag{3.63}$$

can be used to obtain the unit-ramp response curve $z(t)$.

3.13.4 Response to Arbitrary Input

The response to an arbitrary input can be obtained by using the following MATLAB commands :

$$\text{lsim (num, den, } t) \tag{3.64}$$

$$\text{lsim } (A, B, C, D, u, t) \tag{3.65}$$

$$y = \text{lsim (num, den, } r, t) \tag{3.66}$$

$$y = \text{lsim } (A, B, C, D, u, t) \tag{3.67}$$

The MATLAB commands in Eqs. (3.64) to (3.67) will generate the response to input time function r or u.

3.14 RESPONSE TO INITIAL CONDITION IN STATE SPACE

Consider the system defined in state space by

$$\dot{\mathbf{x}} = \mathbf{Ax} + \mathbf{B}u, \quad \mathbf{x}(0) = \mathbf{x}_0 \tag{3.68}$$

$$y = \mathbf{cx} + \mathbf{D}u \tag{3.69}$$

The MATLAB command

$$\text{initial } (A, B, C, D, [\text{initial condition}], t) \tag{3.70}$$

may be used to provide the response to the initial condition.

3.15 EXAMPLE PROBLEMS AND SOLUTIONS

3.15.1 Control Systems

Example E3.1. *Reduce the system shown in Fig.E3.1 to a single transfer function, T(s) = C(s)/R(s) using MATLAB. The transfer functions are given as*

$$G_1(s) = \frac{1}{(s+7)}$$

$$G_2(s) = \frac{1}{(s^2+6s+5)}$$

$$G_3(s) = \frac{1}{(s+8)}$$

$$G_4(s) = \frac{1}{s}$$

$$G_5(s) = \frac{7}{(s+3)}$$

$$G_6(s) = \frac{1}{(s^2+7s+5)}$$

$$G_7(s) = \frac{5}{(s+5)}$$

$$G_8(s) = \frac{1}{(s+9)}$$

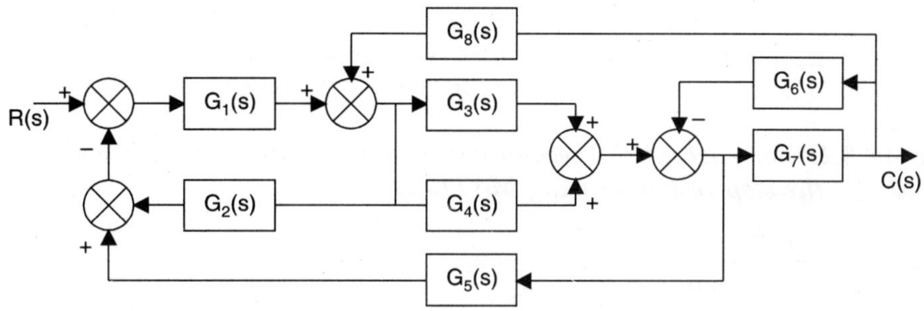

Fig. E3.1

The transfer functions are given as:

G1 (s) = 1/(s + 7)

G2 (s) = 1/(s^2 + 3s + 5)

G3 (s) = 1/(s + 8)

G4 (s) = 1/s

G5 (s) = 7/(s + 3)

G6 (s) = 1/(s^2 + 7s + 5)

G7 (s) = 5/(s + 5)

G8 (s) = 1/(s + 9)

Solution:

% MATLAB Program

```
G1 = tf ( [0 0 1], [0 1 7]);
G2 = tf ( [0 0 1], [1 6 5]);
G3 = tf ( [0 0 1], [0 1 8]);
G4 = tf ( [0 0 1], [0 1 0]);
G5 = tf ( [0 0 7], [0 1 3]);
G6 = tf ( [0 0 1], [1 7 5]);
G7 = tf ( [0 0 5], [0 1 5]);
G8 = tf ( [0 0 1], [0 1 9]);
G9 = tf ( [0 0 1], [0 0 1]);
```

```
T1 = append (G1, G2, G3, G4, G5, G6, G7, G8, G9);
Q = [ 1 -2 -5 9 ]
      2 1 8 0
      3 1 8 0
      4 1 8 0
      5 3 4 -6
      6 7 0 0
      7 3 4 -6
      8 7 0 0];
Inputs = 9;
Outputs = 7;
Ts = connect (T1, Q, Inputs, Outputs);
T = Tf (Ts) computer response
Transfer function:
10 s^7 + 290 s^6 + 3350 s^5 + 1.98e004 s^4 + 6.369e004 s^3 + 1.089e005 s^2
+ 8.895e004 s + 2.7e004 s^10 + 45 s^9 + 866 s^8 + 9305 s^7 + 6.116e004 s^6 +
2.533e005 s^5 + 6.57e005 s^4 + 1.027e006 s^3 + 8.909e005 s^2 + 3.626e005 s +
4.2e004
```

Example 3.2. *For each of the second order systems below, find ξ, ωn, Ts, Tp, Tr, % overshoot, and plot the step response using MATLAB.*

(a) $T(s) = \dfrac{130}{s^2 + 15s + 130}$

(b) $T(s) = \dfrac{0.045}{s^2 + 0.025s + 0.045}$

(c) $T(s) = \dfrac{10^8}{s^2 + 1.325 \times 10^3 s + 10^8}$

Solution:

(*a*)

```
>> clf
>> numa=130;
>> dena=[1 15 130];
>> Ta=tf(numa, dena)
Transfer function:
      130
..................
s^2 + 15 s + 130
>> omegana=sqrt (dena(3))
omegana =
11.4018
>> zetaa=dena(2)/(2*omegana)
zetaa =
0.6578
>> Tsa=4/(zetaa*omegana)
```

```
Tsa =
0.5333
>> Tpa=pi/(omegana*sqrt(1-zetaa^2))
Tpa =
0.3658
>> Tra=(1.76*zetaa^3 - .417*zetaa^2 + 1.039*zetaa + 1)/omegana
Tra =
0.1758
>> percenta=exp(-zetaa*pi/sqrt(1-zetaa^2))*100
percenta =
6.4335
>> subplot(221)
>> step(Ta)
>> title('(a)')
>> '(b)'
ans =
```

(*b*)

```
>> numb=.045;
>> denb=[1 .025 .045];
>> Tb=tf(numb,denb)
Transfer function:
        0.045
.........
s^2 + 0.025s + 0.045
>> omeganb=sqrt(denb(3))
omeganb =
0.2121
>> zetab=denb(2)/(2*omeganb)
zetab =
0.0589
>> Tsb=4/(zetab*omeganb)
Tsb =
320
>> Tpb=pi/(omeganb*sqrt(1-zetab^2))
Tpb =
14.8354
>> Trb=(1.76*zetab^3 - .417*zetab^2 + 1.039*zetab + 1)/omeganb
Trb =
4.9975
>> percentb=exp(-zetab*pi/sqrt(1-zetab^2))*100
percentb =
83.0737
>> subplot(222)
>> step(Tb)
>> title('(b)')
```

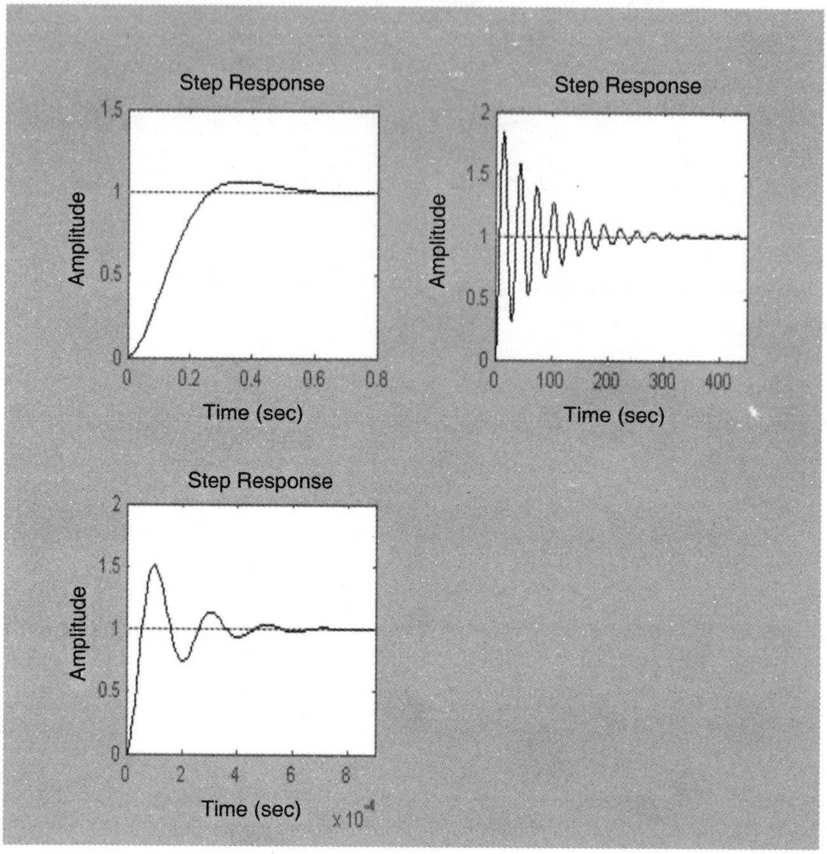

Fig. E3.2

```
>> '(c)'
ans =
(c)
>> numc=10E8;
>> denc=[1 1.325*10E3 10E8];
>> Tc=tf(numc,denc)
Transfer function:
1e009
.........
s^2 + 13250 s + 1e009
>> omeganc=sqrt(denc(3))
omeganc =
3.1623e+004
>> zetac=denc(2)/(2*omeganc)
zetac =
0.2095
>> Tsc = 4/(zetac*omeganc)
```

```
Tsc =
6.0377e-004
>> Tpc =pi/(omeganc*sqrt (1-zetac^2))
Tpc =
1.0160e-004
>> Trc = (1.76*zetac^3 - .417*zetac^2 + 1.039*zetac + 1)/omeganc
Trc =
3.8439e-005
>> percentc =exp(-zetac*pi/sqrt(1-zetac^2))*100
percentc =
51.0123
>> subplot (223)
>> step (Tc)
>> title ('(c)')
```

Example 3.3. *Determine the pole locations for the system shown below using MATLAB.*

$$\frac{C(s)}{R(s)} = \frac{s^3 - 6s^2 + 7s + 15}{s^5 + s^4 - 5s^3 - 9s^2 + 11s - 12}$$

Solution:

>> *%MATLAB Program*

```
>> den= [1 1 -5 -9 11 -12];
>> A=roots (den)
A =
- 2.1586 + 1.2396i
- 2.1586 - 1.2396i
2.3339
0.4917 + 0.7669i
0.4917 - 0.7669i
```

Example 3.4. *Determine the pole locations for the unity feedback system shown below using MATLAB.*

$$G(s) = \frac{150}{(s+5)(s+7)(s+9)(s+11)}$$

Solution:

>> *%MATLAB Program*

```
>> numg =150
numg =
150
>> deng =poly ([-5 -7 -9 -11]);
>> 'G(s)'
ans =
G(s)
>> G=tf (numg, deng)
```

Transfer function:

```
      150
.....................
s^4 + 32s^3 + 374s^2 + 1888s + 3465
>> 'Poles of G(s)'
ans =
Poles of G(s)
>> pole(G)
ans =
- 11.0000
- 9.0000
- 7.0000
- 5.0000
>> 'T(s)'
ans =
T(s)
>> T=feedback (G, 1)
```

Transfer function:

```
    150
....................
s^4 + 32s^3 + 374s^2 + 1888s + 3615
>> pole (T)
ans =
- 10.9673 + 1.9506i
- 10.9673 - 1.9506i
- 5.0327 + 1.9506i
- 5.0327 - 1.9506i
```

Example 3.5. *A plant to be controlled is described by a transfer function*

$$G(s) = \frac{s+7}{s^2 + 9s + 30}$$

Obtain the root locus plot using MATLAB.

Solution:

>> *%MATLAB Program*

```
>> clf
>> num= [1 7];
>> den= [1 9 30];
>> rlocus (num, den);
```

Computer response is shown in Fig. E3.5.

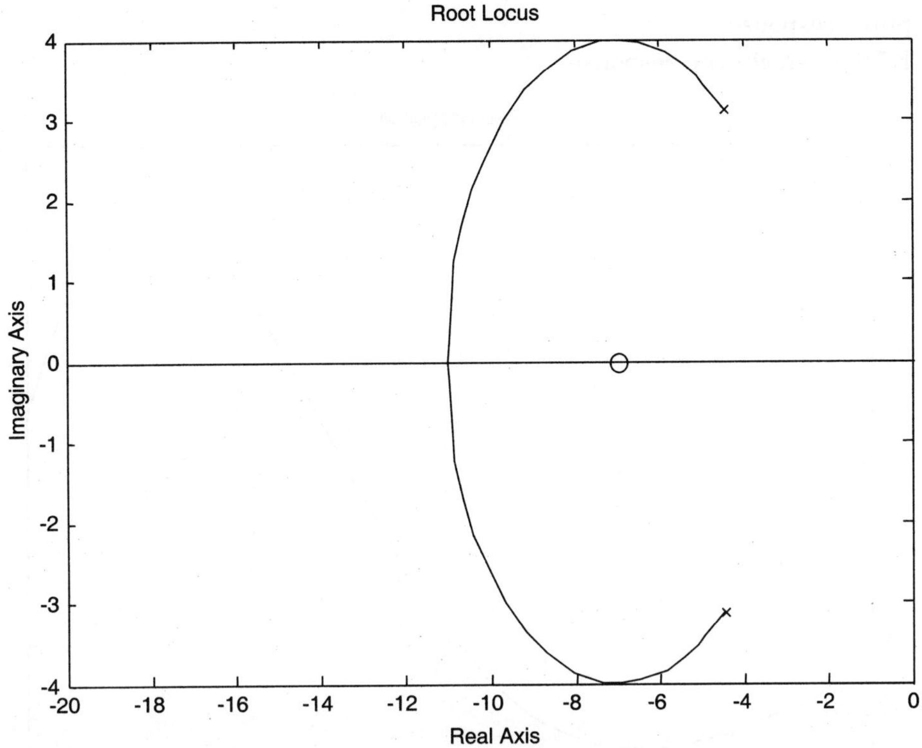

Fig. E3.5

Example 3.6. *For the unity feedback system shown in Fig. E3.6, G(s) is given as*

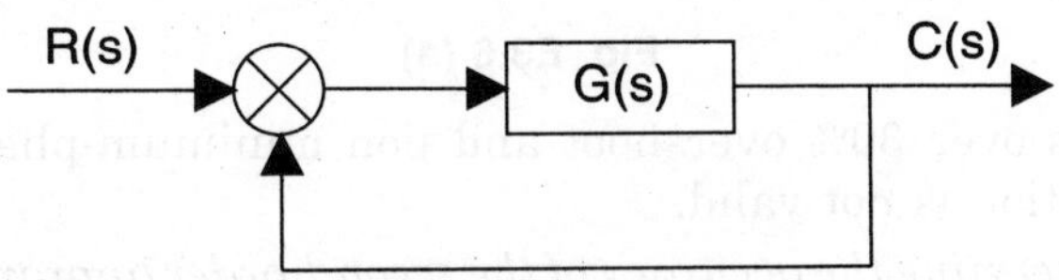

Fig. E3.6

$$G(s) = \frac{30\,(s^2 - 5s + 3)}{(s+1)\,(s+2)\,(s+4)\,(s+5)}$$

Determine the closed-loop step response using MATLAB.

Solution:

>> *%MATLAB Program*

```
>>numg=40*[1 -5 7];
>> deng =poly ([-1 -2 -4 -5]);
>> G=tf (numg, deng);
>> T=feedback (G, 1)
```

Transfer function:

```
40s^2 - 200s + 280
........................
```

```
s^4 + 12s^3 + 89s^2 - 122s + 320
>> step (T)
```

Computer response:

Fig. E3.6 (a) shows the response.

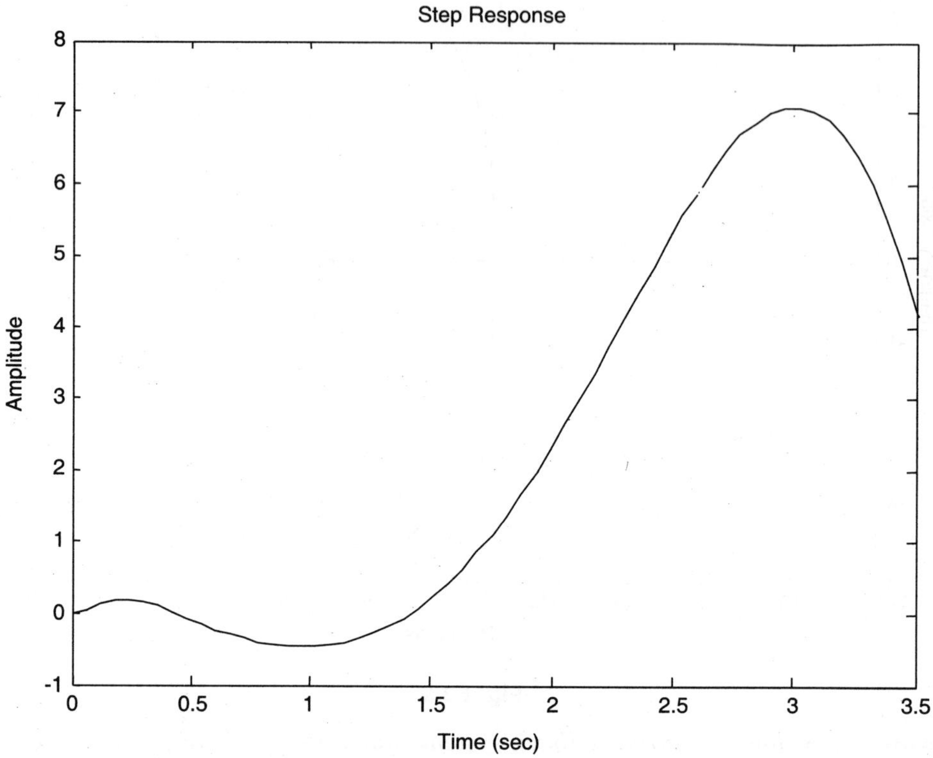

Fig. E3.6 (a)

Simulation shows over 30% overshoot and non minimum-phase behavior. Hence the second-order approximation is not valid.

Example 3.7. *Determine the accuracy of the second-order approximation using MATLAB to simulate the unity feedback system shown in Fig. E3.7 where*

$$G(s) = \frac{12\,(s^2 + 3s + 9)}{(s^2 + 3s + 9)\,(s + 1)\,(s + 5)}$$

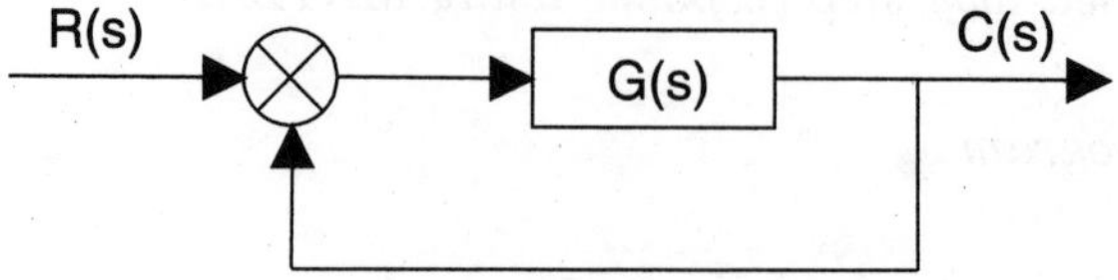

Fig. E3.7

Solution:

>> *%MATLAB Program*

```
>> numg=12*[1 3 9];
>> deng =conv([1 3 9], poly([-1 -5]));
>> G=tf(numg, deng);
>> T=feedback(G, 1);
>> step(T)
```

Computer response [see Fig E3.7 (*a*)].

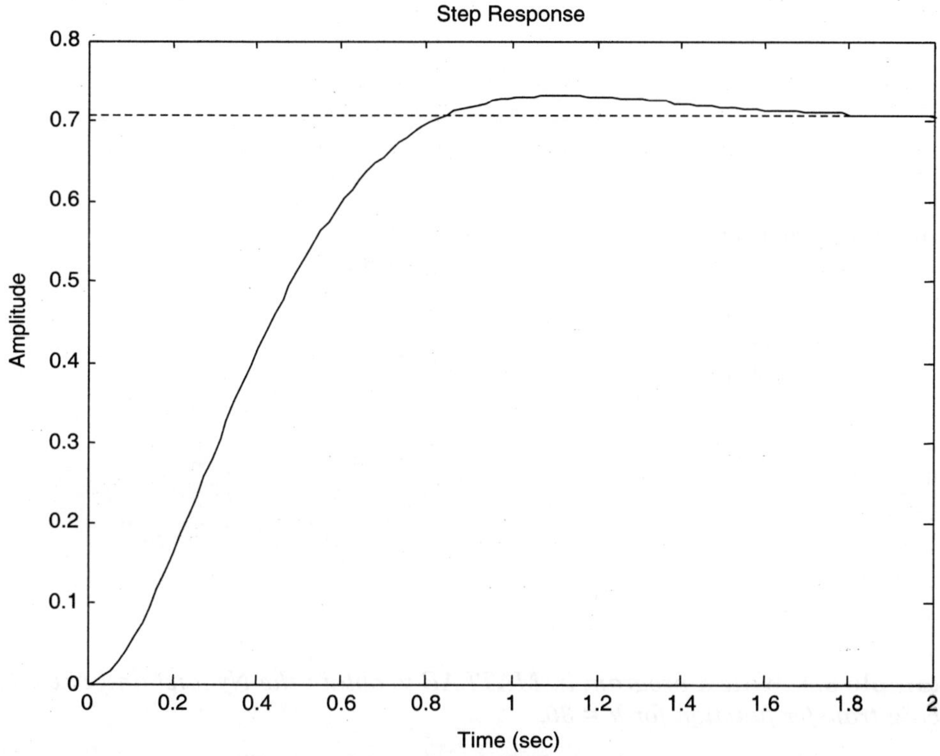

Fig. E3.7 (*a*)

Example 3.8. *For the unity feedback system shown in Fig. E3.8 with*

$$G(s) = \frac{K(s+1)}{s(s+1)(s+5)(s+6)}$$

determine the range of K for stability using MATLAB.

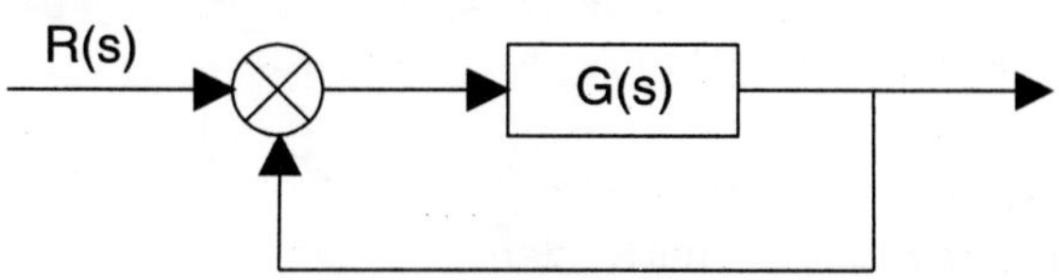

Fig. E3.8

Solution:

>> *%MATLAB Program*

```
>> K = [0:0.2:200];
>> for i = 1: length (K);
>> deng = poly ([0 -1 -5 -6]);
>> dent = deng+ [0 0 0 K (i) K (i)];
>> R=roots (dent);
>> A = real(R);
>> B = max (A);
>> if B>0
>> R
>> K = K (i)
>> break
>> end
>> end
```

Computer response:

```
R =
-10.0000
-0.5000 + 4.4441i
-0.5000 - 4.4441i
-1.0000
A =
-10.0000
-0.5000
-0.5000
-1.0000
B =
-0.5000
```

Example 3.9. *Write a program in MATLAB to obtain the Nyquist and Nichols plots for the following transfer function for k = 30.*

$$G(s) = \frac{k(s+1)(s+3+7i)(s+3-7i)}{(s+1)(s+3)(s+5)(s+3+7i)(s+3-7i)}$$

Solution:

>> *%MATLAB Program*

```
>> %Simple Nyquist and Nichols plots
>> clf
>> z= [-1 -3+7*i -3-7*i];
>> p= [-1 -3 -5 -3+7*i -3-7*i];
>> k=30;
>> [num, den] =zp2tf (z', p', k);
>> subplot (211), nyquist (num, den)
>> subplot (212), Nichols (num, den)
>> ngrid
>> axis ([50 360 -40 30])
```

Computer response: The Nyquist and Nichols plots are shown in Fig. E3.9.

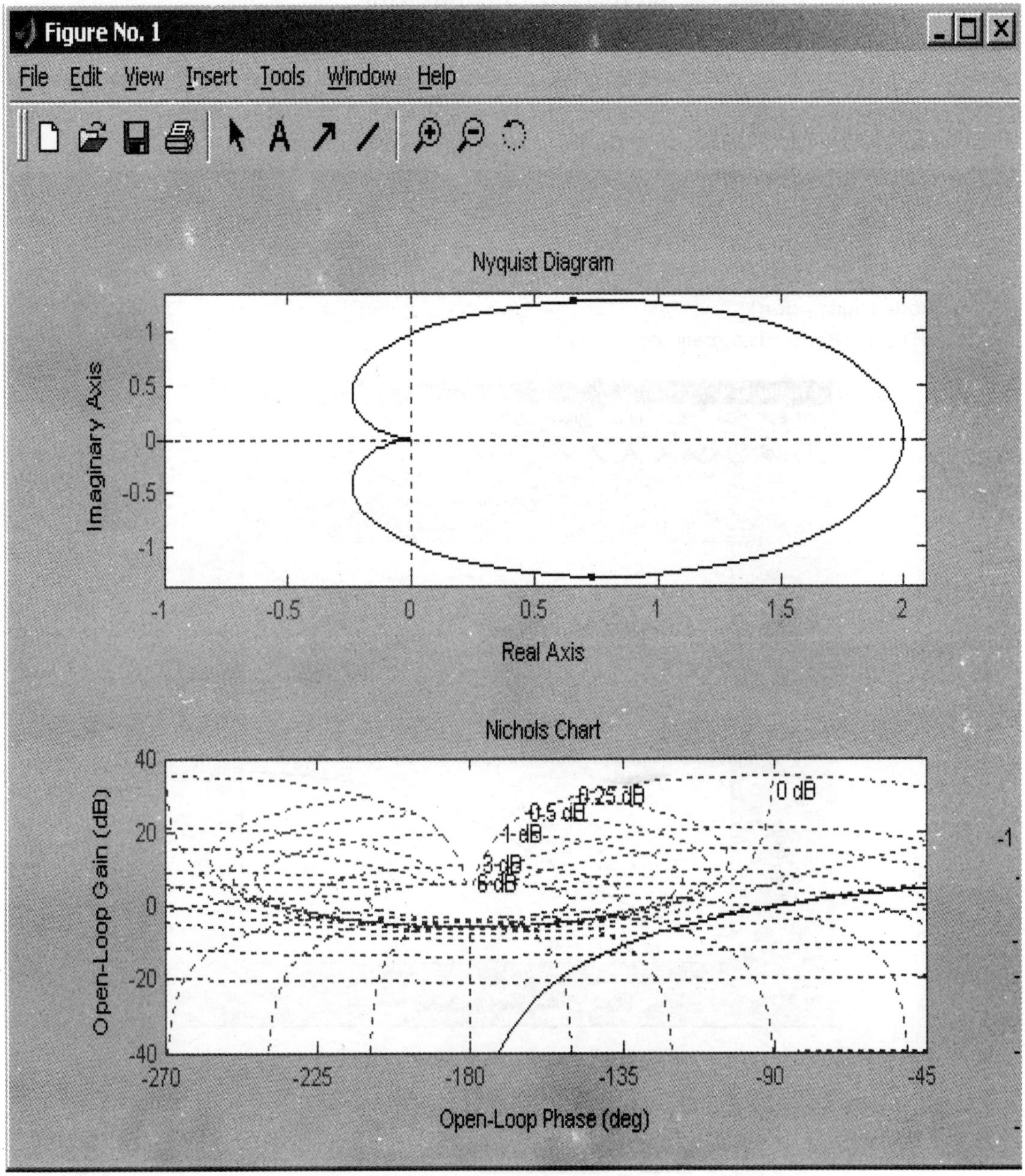

Fig. E3.9

Example 3.10. *A PID controller is given by*

$$G_C(s) = 29.125\ \frac{(s+0.57)^2}{s}$$

Draw a Bode diagram of the controller using MATLAB.

Solution:

$$G_C(s) = \frac{29.125(s^2 + 1.14s + 0.3249)}{s}$$

$$= \frac{29.125s^2 + 33.2025s + 9.4627)}{s}$$

The following MATLAB program produces the Bode diagram

>> *%MATLAB Program*

```
>> %Bode diagram
>> num = [29.125 33.2025 9.4627];
>> den = [0 1 0];
>> bode(num, den)
>> title('Bode diagram of G(s)')
```

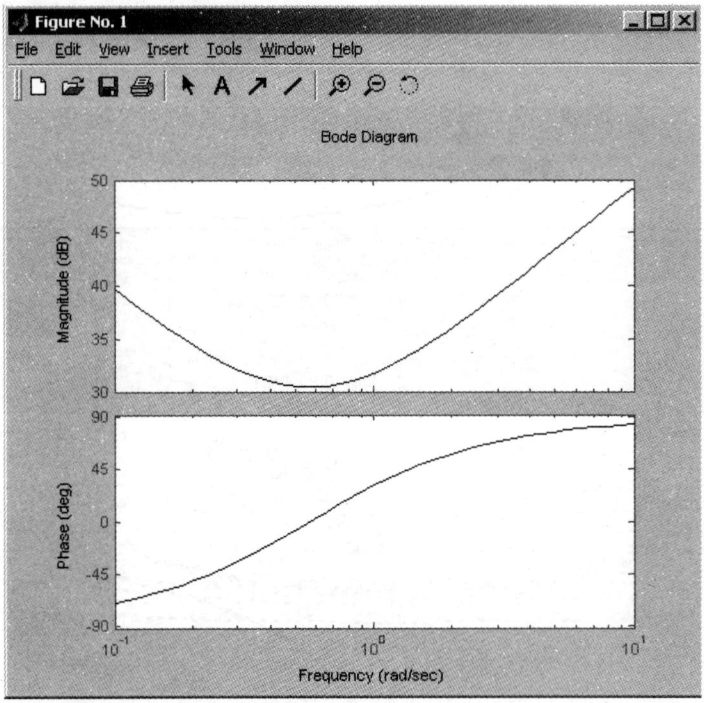

Fig. E3.10 Bode diagram of G(s).

Example 3.11. *For the closed-loop system defined by*

$$\frac{C(s)}{R(s)} = \frac{1}{s^2 + 2\zeta s + 1}$$

(a) plot the unit-step response curves c (t) for ζ *=0, 0.1, 0.2, 0.4, 0.5, 0.6, 0.8, and1.0.* ω_n *is normalized to 1.*

(b) plot a three dimensional plot of (a).

Solution:

```
>> %Two-dimensional plot and three-dimensional plot of unit-step
>> %response curves for the standard second-order system with wn = 1
```

```
>> %and zeta=0, 0.1, 0.2, 0.4, 0.5, 0.6, 0.8, and 1.0
>> t=0:0.2:10;
>> zeta = [0 0.1 0.2 0.4 0.5 0.6 0.8 1.0];
>> for n=1:8;
>> num = [0 0 1];
>> den = [1 2*zeta (n) 1];
>> [y (1:51, n), x, t] = step (num, den, t);
>> end
>> %Two-dimensional diagram with the command plot (t, y)
>> plot (t, y)
```

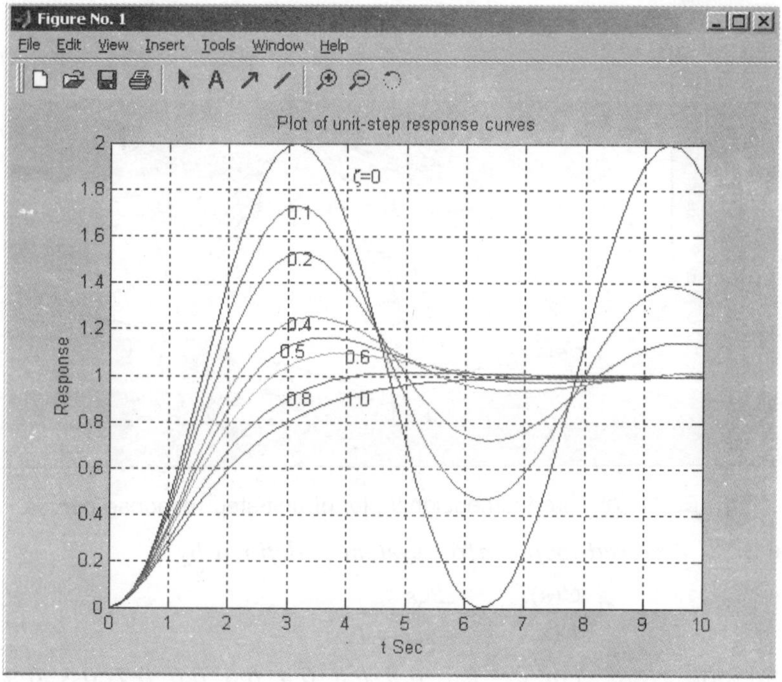

Fig. E3.11 (*a*) Plot of unit-step response curves.

```
>> grid
>> title ('Plot of unit-step response curves')
>> xlabel ('t Sec')
>> ylabel ('Response')
>> text (4.1, 1.86,'\zeta=0')
>> text (3.0, 1.7,'0.1')
>> text (3.0, 1.5,'0.2')
>> text (3.0, 1.22,'0.4')
>> text (2.9, 1.1,'0.5')
>> text (4.0, 1.08,'0.6')
>> text (3.0, 0.9,'0.8')
>> text (4.0, 0.9,'1.0')
>> %For three dimensional plot, we use the command mesh (t, eta, y')
>> mesh (t, eta, y')
```

```
>> title ('Three-dimensional plot of unit-step response curves')
>> xlabel ('t Sec')
>> ylabel ('\zeta')
>> zlabel ('Response')
```

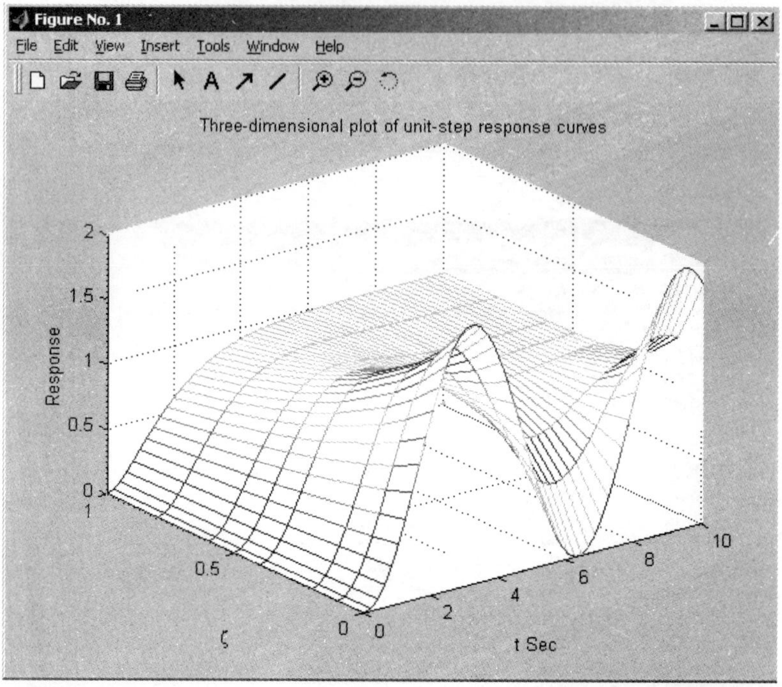

Fig. E3.11 (*b*) Three-dimensional plot of unit-step response curves.

Example 3.12. *A closed-loop control system is defined by*

$$\frac{C(s)}{R(s)} = \frac{2\zeta s}{s^2 + 2\zeta s + 1}$$

where ζ is the damping ratio. For ζ = 0.1, 0.2, 0.3, 0.4, 0.5, 0.6, 0.7, 0.8, 0.9, and 1.0, using MATLAB, plot

(*a*) *a two-dimensional diagram of unit-impulse response curves*

(*b*) *a three-dimensional plot of the response curves.*

Solution:

A MATLAB program that produces a two-dimensional diagram of unit-impulse response curves and a three-dimensional plot of the response curves is given below:

```
>> %To plot a two-dimensional diagram
>> t=0:0.2:10;
>> zeta = [0.1 0.2 0.3 0.4 0.5 0.6 0.7 0.8 0.9 1.0];
>> for n = 1:10;
>> num = [0 2*zeta (n) 1];
>> den = [1 2*zeta (n) 1];
>> [y(1:51, n), x, t] = impulse(num, den, t);
>> end
```

```
>> plot (t, y)
>> grid
>> title ('Plot of unit-impulse response curves')
>> xlabel ('t Sec')
>> ylabel ('Response')
>> text (2.0, 0.85,'0.1')
>> text (1.5, 0.75,'0.2')
>> text (1.5, 0.6,'0.3')
>> text (1.5, 0.5,'0.4')
>> text (1.5, 0.38,'0.5')
>> text (1.5, 0.25,'0.6')
>> text (1.7, 0.12,'0.7')
>> text (2.0,-0.1,'0.8')
>> text (1.5, 0.0,'0.9')
>> text (.5, 1.5,'1.0')
>> %Three-dimensional plot
>> mesh (t, eta, y')
>> title ('Three-dimensional plot')
>> xlabel ('t Sec')
>> ylabel ('\zeta')
>> zlabel ('Response')
```

The two-dimensional diagram and three-dimensional diagram produced by this MATLAB program are shown in Figs. E3.12 (*a*) and (*b*) respectively.

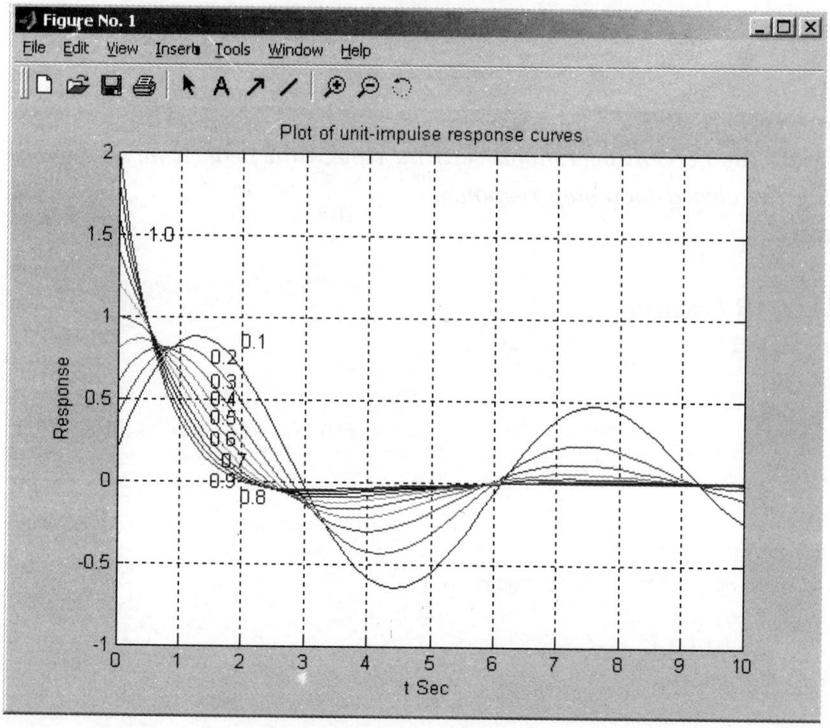

Fig. E3.12 (a) Two-dimensional plot.

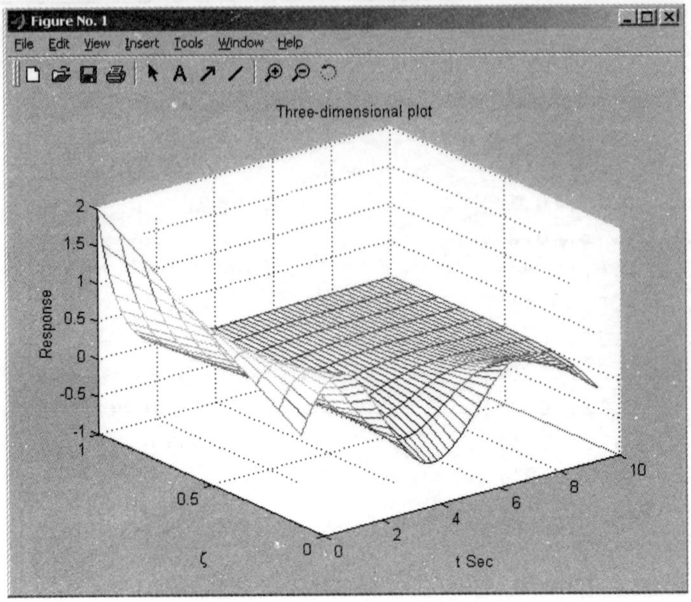

Fig. E3.12 (*b*) Three-dimensional plot.

Example 3.13. *For the system shown in Fig. E3.13 write a program in MATLAB that will use an open-loop transfer function G(s):*

$$G(s) = \frac{50\,(s+1)}{s(s+3)\,(s+5)}$$

$$G(s) = \frac{25\,(s+1)\,(s+7)}{s(s+2)\,(s+4)\,(s+8)}$$

(*a*) *Obtain a Bode plot*

(*b*) *Estimate the percent overshoot, settling time, and peak time*

(*c*) *Obtain the closed-loop step response.*

Solution:

(*a*)

```
>> %MATLAB Program
>> G=zpk ([-1], [0 -3 -5], 50)
>> G=tf (G)
>> bode (G)
>> title ('System 1')
>> %title ('System 1')
>> pause
>> %Find phase margin
>> [Gm, Pm, Wcg, Wcp] = margin (G);
>> w=1:.01:20;
>> [M, P, w] = bode (G, w);
>> %Find bandwidth
```

```
>> for k=1:1: length(M);
>> if 20*log10(M (k)) +7<=0;
>> 'Mag'
>> 20*log10(M(k))
>> 'BW'
>> wBW =w(k)
>> break
>> end
>> end
>> %Find damping ratio, percent overshoot, settling time, and peak time
>> for z=0:.01:10
>> Pt=atan (2*z/(sqrt(-2*z^2+sqrt (1+4*z^4))))*(180/pi);
>> if (Pm-Pt) <=0
>> z;
>> Po=exp (-z*pi/sqrt(1-z^2));
>> Ts= (4/(wBW*z))*sqrt((1-2*z^2) +sqrt(4*z^4-4*z^2+2));
>> Tp = (pi/(wBW*sqrt(1-z^2)))*sqrt((1-2*z^2) +sqrt(4*z^4-4*z^2+2));
>> fprintf ('Bandwidth=%g', wBW)
>> fprintf ('Phase margin=%g', Pm)
>> fprintf (', Damping ratio=%g', z)
>> fprintf (', Percent overshoot=%g', Po*100)
>> fprintf (', Settling time=%g', Ts)
>> fprintf (', Peak time=%g', Tp)
>> break
>> end
>> end
>> T=feedback (G, 1);
>> step (T)
>> title ('Step response system 1')
>>%title ('Step response system 1')
Computer response:
Zero/pole/gain:
  50 (s+1)
---------
s(s+3)(s+5)
Transfer function:
    50s + 50
------------
s^3 + 8 s^2 + 15s
```

The Bode plot is shown in Fig. E3.13 (*a*)

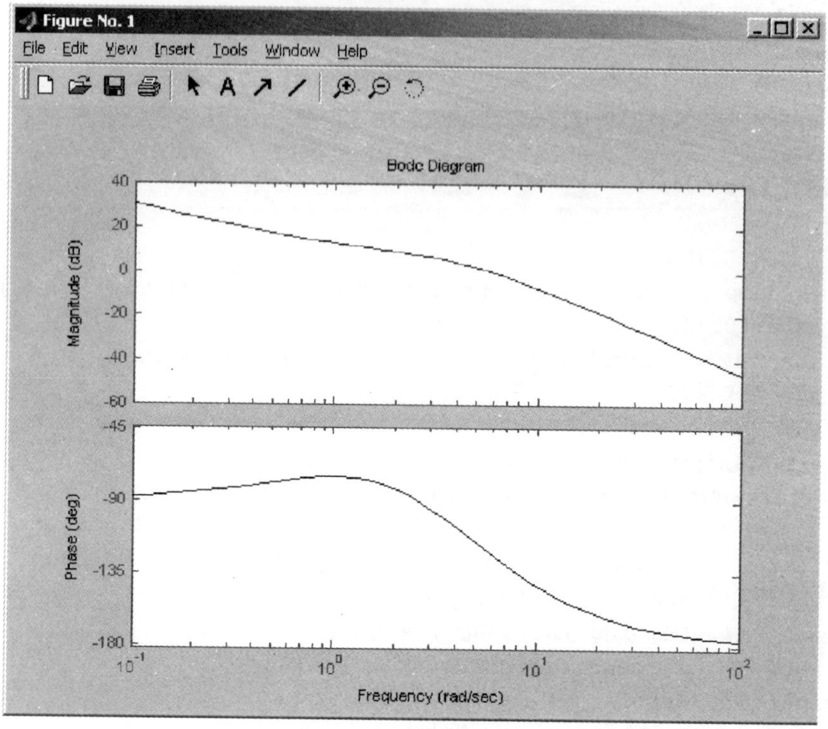

Fig. E3.13 (*a*)

```
ans =
Mag
ans =
-3.0032
ans =
BW
wBW =
9.7900
Bandwidth=9.79Phase margin=53.892, Damping ratio=0.59, Percent
overshoot=10.0693, Settling time=0.804303, Peak time=0.461606
```

The step response is shown in Fig. E3.13 (*b*)

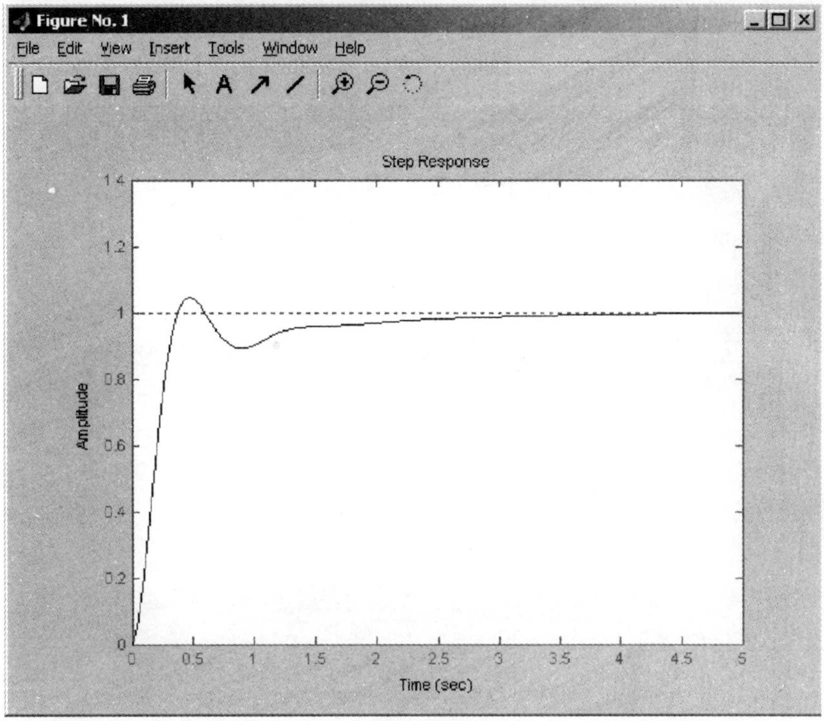

Fig. E3.13 (*b*)

(*b*) Likewise, for this problem

```
>> G=zpk ([-1 -7], [0 -2 -4 -8], 25)
>> G=tf (G)
```

The following Bode plot and step response are obtained [see Figs. E3.13 (*c*) and (*d*).

```
Zero/pole/gain:
25 (s+1) (s+7)
__________

s (s+2) (s+4) (s+8)
Transfer function:
25 s^2 + 200 s + 175
________________

s^4 + 14 s^3 + 56 s^2 + 64 s
ans =
Mag
ans =
-7.0110
ans =
BW
wBW =
6.5500
    Bandwidth=6.55, Phase margin=63.1105, Damping ratio=0.67, Percent
overshoot=5.86969, Settling time=0.959175, Peak time=0.679904
```

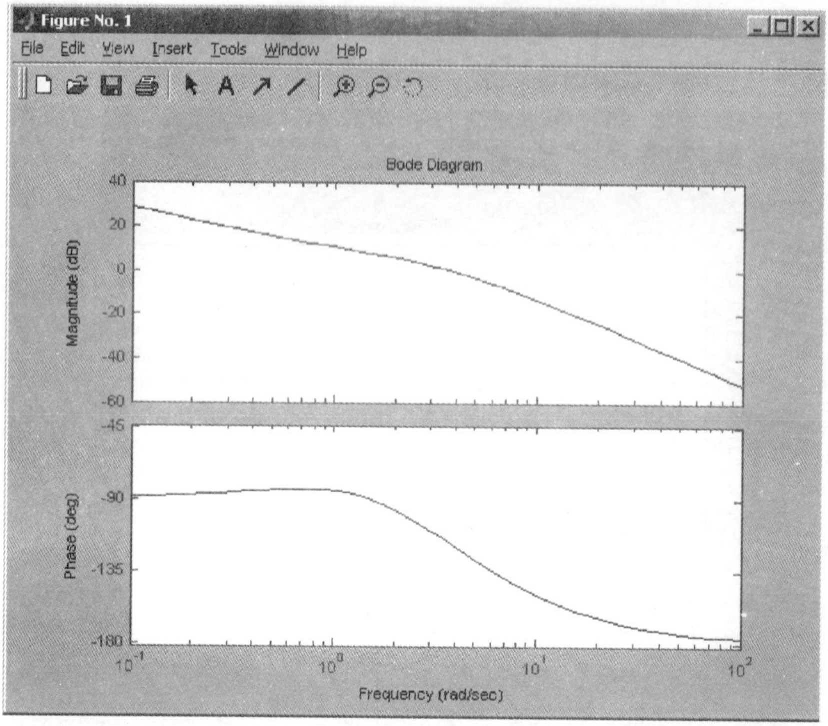

Fig. E3.13 (c)

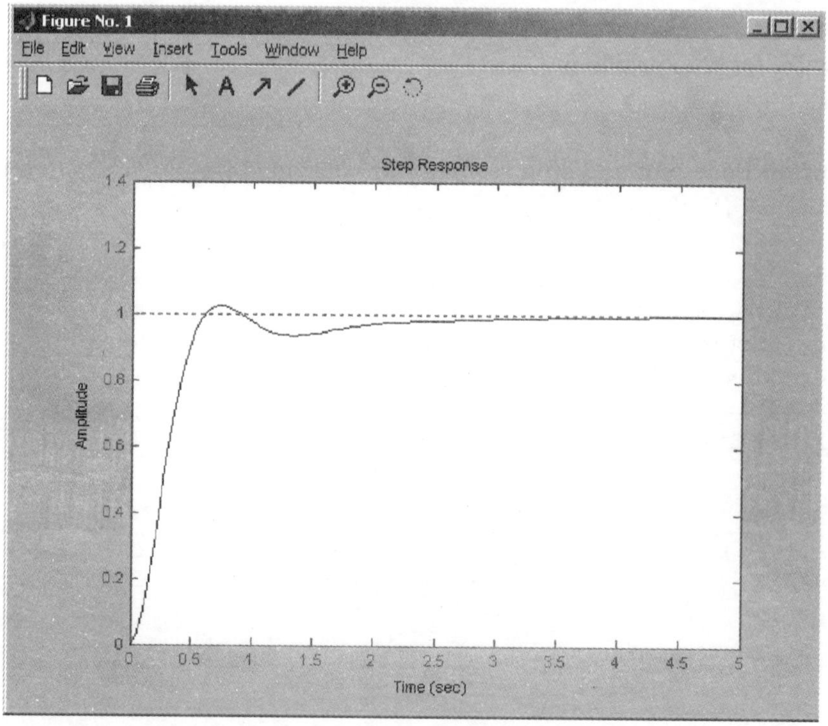

Fig. E3.13 (*d*)

Example 3.14. *Write a program in MATLAB for a unity-feedback system with*

$$G(s) = \frac{K(s+7)}{(s^2+3s+52)(s^2+2s+35)}$$

(a) plot the Nyquist diagram

(b) display the real-axis crossing value and frequency.

Solution:

>> *%MATLAB Program*

```
>> numg= [1 7]
>> deng=conv ([1 3 52], [1 2 35]);
>> G=tf (numg, deng)
>> 'G(s)'
>> Gap=zpk (G)
>> inquest (G)
>> axis ([-3e-3, 4e-3,-5e-3, 5e-3])
>> w=0:0.1:100;
>> [re, im] =nyquis t (G, w);
>> for i=1:1: length (w)
>> M (i) =abs (re (i) +j*im (i));
>> A (i) =atan2 (im (i), re (i))*(180/pi);
>> if 180-abs (A (i)) <=1;
>> re (i);
>> im (i);
>> K=1/abs (re (i));
>> fprintf ('\nw =%g', w (i))
>> fprintf (', Re=%g', re (i))
>> fprintf (', Im =%g', im (i))
>> fprintf (', M=%g', M (i))
>> fprintf (', K=%g', K)
>> Gm=20*log10 (1/M (i));
>> fprintf (', Gm=&G', Gm)
>> break
>> end
>> end
Computer response:
numg=
1    7
Transfer function:
  s + 7
-----------------
s^4 + 5 s^3 + 93 s^2 + 209 s + 1820
ans =
G(s)
Zero/pole/gain:
(s+7)
-----------------
(s^2 + 2s + 35) (s^2 + 3s + 52)
```

The Nyquist plot is shown in Fig. E3.14.

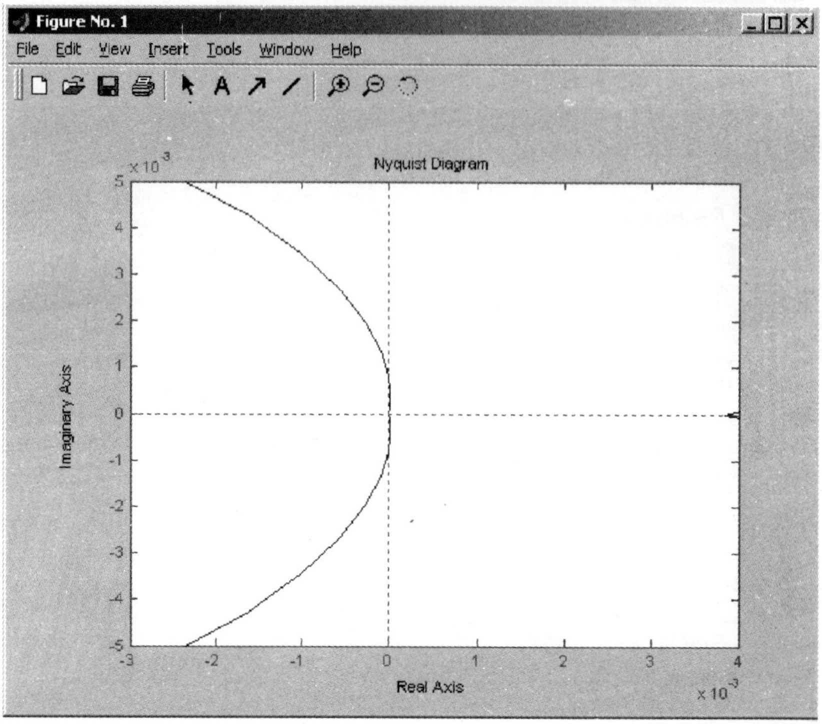

Fig. E3.14

Example 3.15. *Determine the unit-ramp response of the following system using MATLAB and lsim command*

$$\frac{C(s)}{R(s)} = \frac{1}{3s^2 + 2s + 1}$$

Solution:

>> *%MATLAB Program*

```
>> %Unit-ramp response
>> num=[0 0 1];
>> den=[3 2 1];
>> t=0:0.1:10;
>> r=t;
>> y=lsim(num,den,r,t);
>> plot(t,r,'-',t,y,'o')
>> grid
>> title('Unit-ramp response')
>> xlabel('t Sec')
>> ylabel('Unit-ramp input and output')
>> text(1.0,4.0,'Unit-ramp input')
>> text(5.0,2.0,'Output')
```

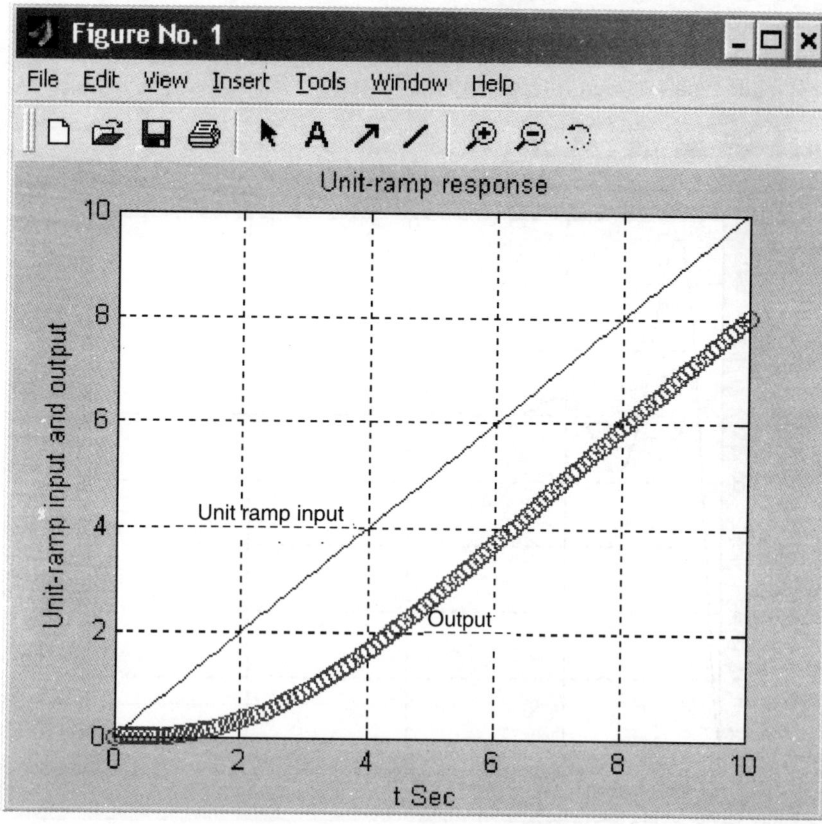

Fig. E3.15 Unit-ramp response.

Example 3.16. *A higher-order system is defined by*

$$\frac{C(s)}{R(s)} = \frac{7s^2 + 16s + 10}{s^4 + 5s^3 + 11s^2 + 16s + 10}$$

(a) plot the unit-step response curve of the system using MATLAB

(b) obtain the rise time, peak time, maximum overshoot, and settling time using MATLAB.

Solution:

```
>> %Unit-step response curve
>> num=[0 0 7 16 10];
>> den=[1 5 11 16 10];
>> t=0:0.02:20;
>> [y,x,t]=step(num,den,t);
>> plot(t,y)
>> grid
>> title('Unit-step response')
>> xlabel('t Sec')
>> ylabel('Output y(t)')
>> %Response to rise from 10% to 90% of its final value
>> r1=1;while y(r1)<0.1,r1=r1+1;end
```

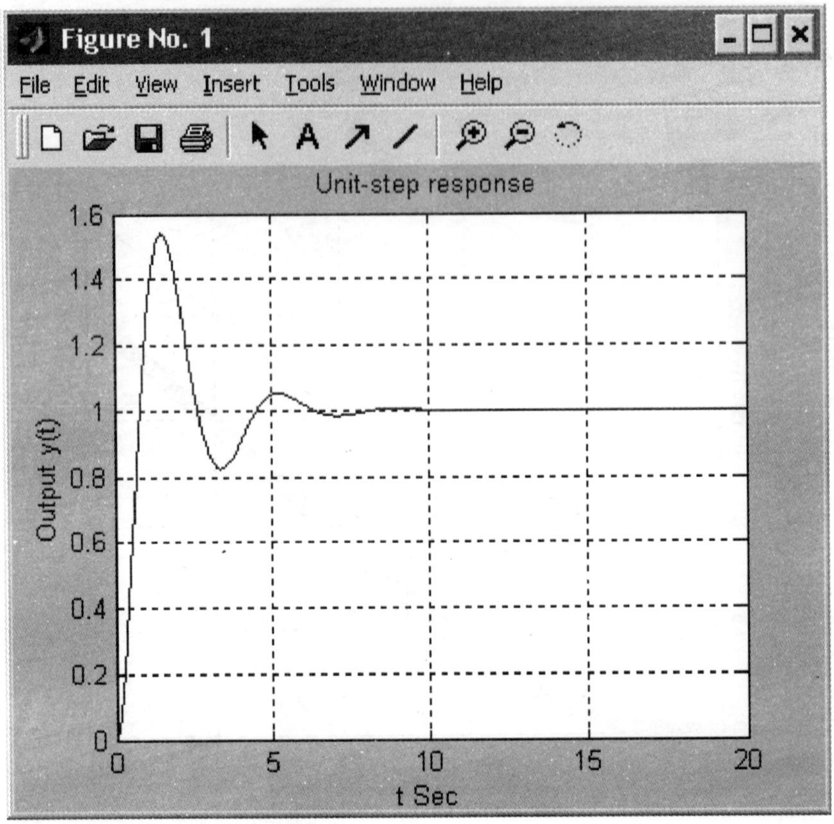

Fig. E3.16 Unit-step response.

```
>> r2=1;while y(r2)<0.9,r2=r2+1;end
>> rise_time=(r2-r1)*0.02
rise_time =
0.5400
>> [ymax,tp]=max(y);
>> peak_time=(tp-1)*0.02
peak_time =
1.5200
>> max_overshoot=ymax-1
max_overshoot =
0.5397
>> s=1001;while y(s)>0.98 & y(s)<1.02;s=s-1;end
>> settling_time=(s-1)*0.02
settling_time =
6.0200
```

Example 3.17. *Obtain the unit-ramp response of the following closed-loop control system whose closed-loop transfer function is given by*

$$\frac{C(s)}{R(s)} = \frac{s+12}{s^3+5s^2+8s+12}$$

Determine also the response of the system when the input is given by

$$r = e^{-0.7t}$$

Solution:

```
>> %Unit-ramp response - lsim command
>> num=[0 0 1 12];
>> den=[1 5 8 12];
>> t=0:0.1:10;
>> r=t;
>> y=lsim(num,den,r,t);
>> plot(t,r,'-',t,y,'o')
>> grid
>> title('Unit-ramp response')
>> xlabel('t Sec')
>> ylabel('Output')
>> text(3.0,6.5,'Unit-ramp input')
>> text(6.2,4.5,'Output')
```

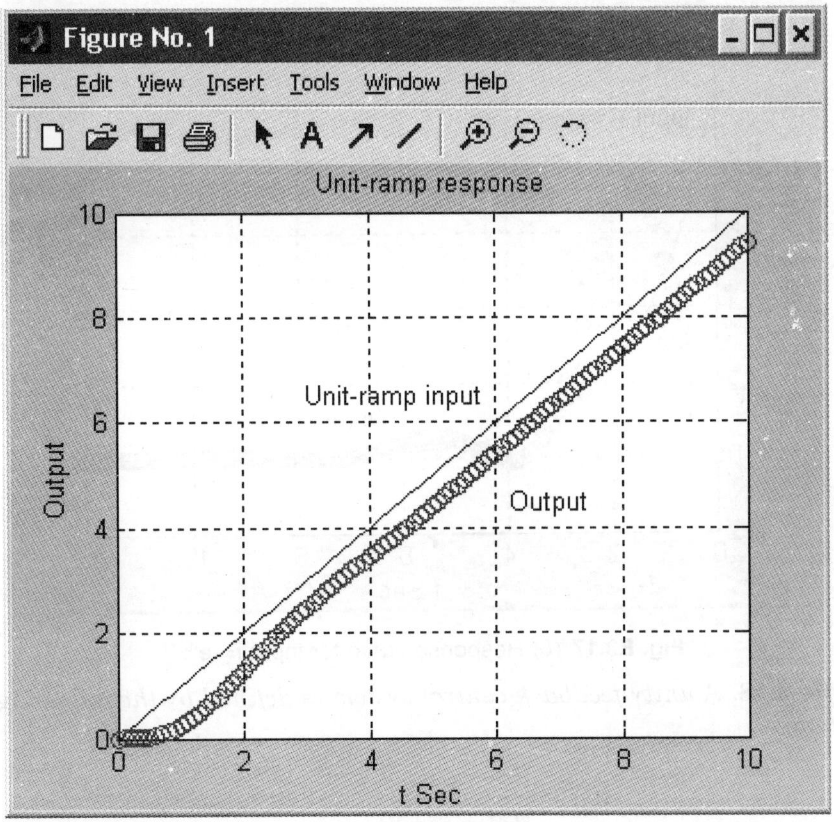

Fig. E3.17 (*a*) Unit-ramp response curve.

```
>> %Input r1=exp(-0.7t)
>> num=[0 0 1 12];
>> den=[1 5 8 12];
```

```
>> t=0:0.1:12;
>> r1=exp(-0.7*t);
>> y1=lsim(num,den,r1,t);
>> plot(t,r1,'-',t,y1,'o')
>> grid
>> title('Response to input r1=exp(-0.7t)')
>> xlabel('t Sec')
>> ylabel('Input and output')
>> text(0.5,0.9,'Input r1=exp(-0.7t)')
>> text(6.3,0.1,'Output')
```

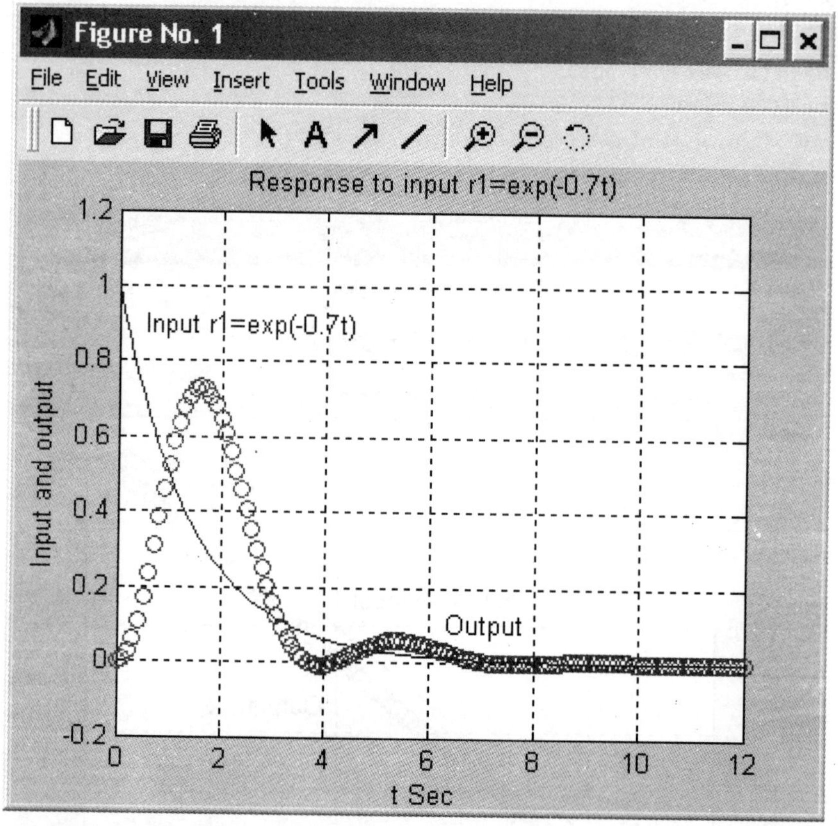

Fig. E3.17 (*b*) Response curve for input r= $e^{-0.7t}$.

Example 3.18. *A unity-feedback control system is defined by the following feedforward transfer function*

$$G(s) = \frac{K}{s(s^2 + 7s + 9)}$$

(*a*) *determine the location of the closed-loop poles, if the value of gain is equal to 3*

(*b*) *plot the root loci for the system using MATLAB.*

Solution:

>> %MATLAB Program to find the closed-loop poles

```
>> p=[1 7 9 3];
>> roots(p)
ans =
-5.4495
-1.0000
-0.5505
```

>> %MATLAB Program to plot the root-loci

```
>> num=[0 0 0 1];
>> den=[1 5 9 0];
>> rlocus(num,den);
>> axis('square')
>> grid
>> title('Root-locus plot of G(s)')
```

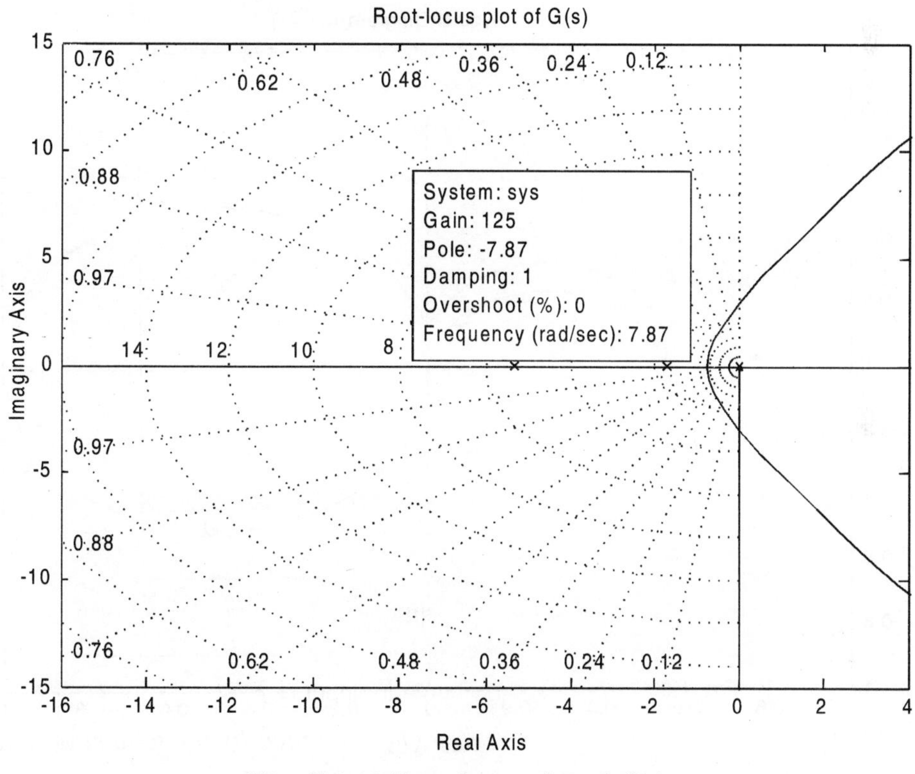

Fig. E3.18 Root-locus plot of $G(s)$.

Example 3.19. *The open-loop transfer function of a unity-feedback control system is given by*

$$G(s) = \frac{1}{s^3 + 0.4s^2 + 7s + 1}$$

*(a) draw a Nyquist plot of G(**s**) using MATLAB*

(b) determine the stability of the system.

Solution:

>> % Open-loop poles

```
>> p=[1 0.4 7 1];
>> roots(p)
ans =
-0.1282 + 2.6357i
-0.1282 - 2.6357i
-0.1436
```

>> % Nyquist plot

```
>> num=[0 0 0 1];
>> den=[1 0.4 7 1];
>> nyquist(num,den)
>> v=[-3 3 -2 2];axis(v);axis('square')
>> grid
>> title('Nyquist plot of G(s)')
```

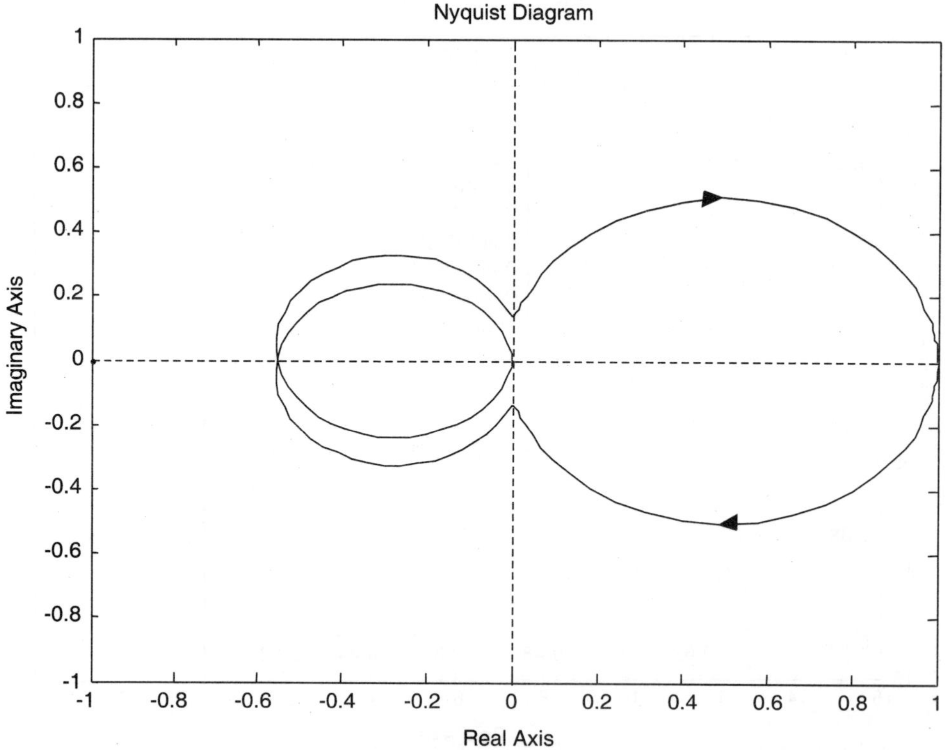

Fig. E3.19 Nyquist plot of G(s).

There are two open-loop poles in the right half *s* plane and no encirclement of the critical point, the closed-loop system is unstable.

Example 3.20. *For the closed-loop control system shown in Fig. E3.20, obtain the range of gain K for stability and plot a root-locus diagram for the system.*

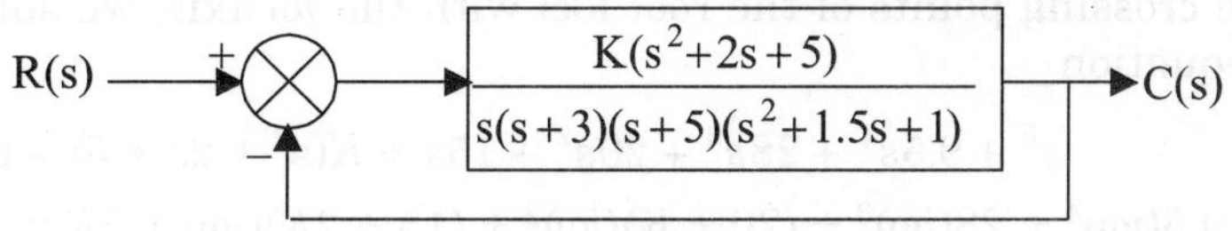

Fig. E3.20

Solution:

The range of gain K for stability is obtained by first plotting the root loci and then finding critical points (for stability) on the root loci. The open-loop transfer function $G(s)$ is

$$G(s) = \frac{K(s^2 + 2s + 5)}{s(s+3)\,(s+5)\,(s^2 + 1.5s + 1)}$$

$$= \frac{K(s^2 + 2s + 5)}{s^5 + 9.5s^4 + 28s^3 + 15s}$$

A MATLAB program to generate a plot of the root loci for the system is given below. The resulting root-locus plot is shown in Fig. E3.20(*a*).

% MATLAB program

```
num = [0   0     0    1    2     5];
den = [1   9.5   28   20   15    0];
rlocus(num,den)
v = [-8    2     -5   5];  axis(v);  axis('square')
grid
title('Root-Locus  Plot')
```

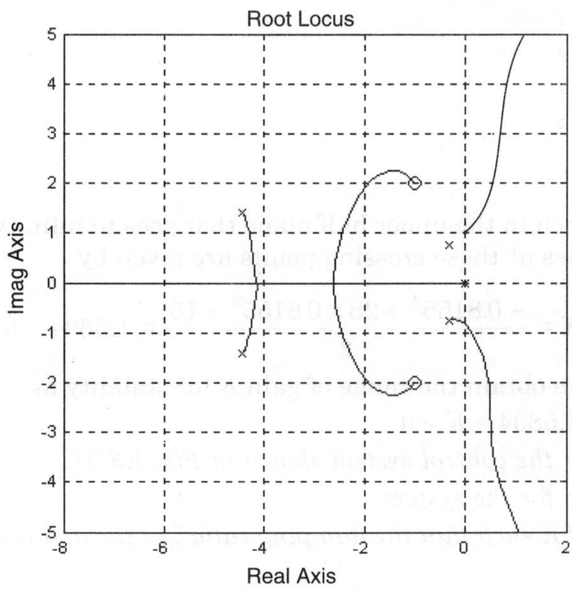

Fig. E3.20 (*a*)

From Fig. E3.20 (*a*), we notice that the system is conditionally stable. All critical points for stability lie on the $j\omega$ axis.

To obtain the crossing points of the root loci with the $j\omega$ axis, we substitute $s = jw$ into the characteristic equation

$$s^5 + 9.5s^4 + 28s^3 + 20s^2 + 15s + K(s^2 + 2s + 5) = 0$$

or $$(j\omega)^5 + 9.5(j\omega)^4 + 28(j\omega)^3 + (20 + K)(j\omega)^2 + (15 + 2K)(j\omega) + 5K = 0$$

or $$[9.5\omega^4 - (20 + K)\,\omega^2 + 5K] + j[\omega^5 - 28\omega^3 + (15 + 2K)\,\omega] = 0$$

Equating the real part and imaginary part equal to zero, respectively, we get

$$9.5\omega^4 - (20 + K)\,\omega^2 + 5K = 0 \qquad (1)$$

$$\omega^5 - 28\omega^3 + (15 + 2K)\,\omega = 0 \qquad (2)$$

Equation.(2) can be written as

$$\omega = 0$$

or $$\omega^4 - 28\omega^2 + 15 + 2K = 0 \qquad (3)$$

$$K = \frac{-\omega^4 + 28\omega^2 - 15}{2} \qquad (4)$$

Substituting Eq.(4) into Eq.(1), we obtain

$$9.5\omega^4 - [20 + ½(-\omega^4 + 28\omega^2 - 15)]\,\omega^2 - 2.5\omega^4 + 70\omega^2 - 37.5 = 0$$

or $$0.5\omega^6 - 2\omega^4 + 57.5\omega^2 - 37.5 = 0$$

The roots of the above equation can be obtained by MATLAB program given below.

% MATLAB program

```
a = [0.5  0     -2    0     57.5  0     -37.5];
roots(a)
```

MATLAB output:

```
ans =
-2.4786 + 2.1157i
-2.4786 - 2.1157i
 2.4786 + 2.1157i
 2.4786 - 2.1157i
 0.8155
-0.8155
```

The root-locus branch in the upper half plane that goes to infinity crosses the jw axis at $\omega = 0.8155$. The gain values at these crossing points are given by

$$K = \frac{-0.8155^4 + 28 \times 0.8155^2 - 15}{2} = 1.5894 \quad \text{for } w = 0.8155$$

For this K value, we obtain the range of gain K for stability as

$$1.5894 > K > 0$$

Example 3.21. *For the control system shown in Fig. E3.21:*

(a) plot the root loci for the system

(b) find the value of K such that the damping ratio ζ of the dominant closed-loop poles is 0.6

(c) obtain all closed-loop poles

(d) plot the unit-step respond curve using MATLAB.

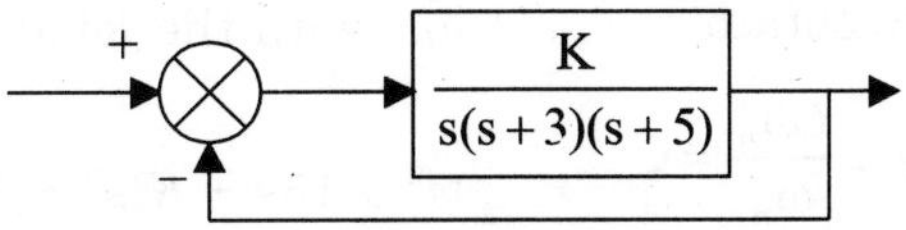

Fig. E3.21

Solution:

(*a*) The MATLAB program given below generates a root-locus plot for the given system. The resulting plot is shown in Fig. E3.21 (*a*).

% MATLAB program

```
num = [0  0     0    1];
den = [1  8     15   0];
rlocus(num,den)
v = [-6   4     -5   5]; axis(v); axis('square')
grid
title('Root-Locus Plot')
```

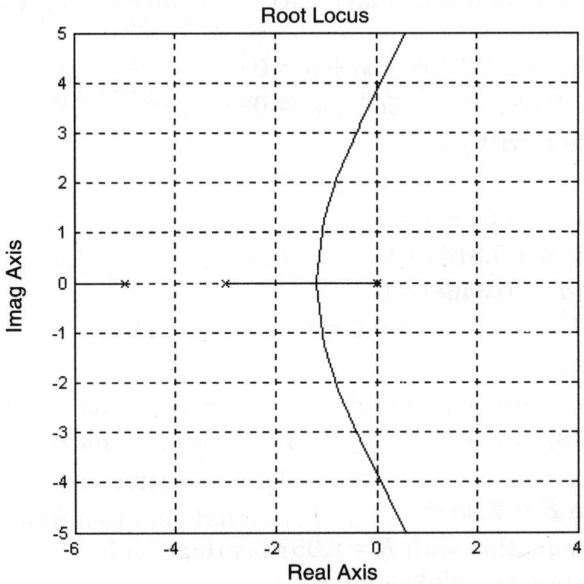

Fig. E3.21 (*a*)

(*b*) We note that the constant ζ points $(0 < \zeta < 1)$ lie on a straight line having angle θ from the $j\omega$ axis as shown in Fig. E3.17 (*b*).

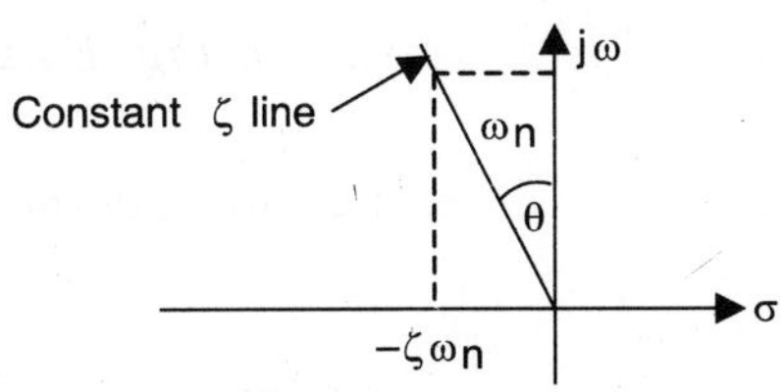

Fig. E3.21 (*b*)

From Fig. E3.21 (*b*), we obtain

$$\sin\theta = \frac{\zeta\omega_n}{\omega_n} = \zeta$$

Also that $\zeta = 0.6$ line can be defined by

$$s = -\,0.75a + ja$$

where a is a variable $(0 < a < \infty)$. To obtain the value of K such that the damping ratio ζ of the dominant closed-loop poles is 0.6 we determine the intersection of the line $s = -\,0.75a + ja$ and the root locus. The intersection point can be obtained by solving the following simultaneous equations for a.

$$s = -\,0.75a + ja \qquad (1)$$

$$s(s + 3)(s + 5) + K = 0 \qquad (2)$$

From Eqs. (1) and (2), we obtain

$$(-\,0.75a + ja)(-\,0.75a + ja + 3)(-0.75a + ja + 5) + K = 0$$

or
$$(1.8281a^3 - 2.1875a^2 - 3a + K) + j(0.6875a^3 - 7.5a^2 + 15a) = 0$$

Equating the real part and imaginary part of the above equation to zero, respectively, we obtain

$$1.8281a^3 - 2.1875a^2 - 3a + K = 0 \qquad (3)$$

$$0.6875a^3 - 7.5a^2 + 4a = 0 \qquad (4)$$

Equation (4) can be rewritten as

$$a = 0$$

or
$$0.6875a^2 - 7.5a + 4 = 0$$

or
$$a^2 - 10.90991a + 5.8182 = 0$$

or
$$(a - 0.5623)(a - 10.3468) = 0$$

Therefore
$$a = 0.5323 \text{ or } a = 10.3468$$

From Eq.(3) we obtain

$$K = -1.8281a^3 + 2.1875a^2 + 3a = 2.0535 \qquad \text{for } a = 0.5626$$

$$K = -1.8281a^3 + 2.1875a^2 + 3a = -1759.74 \qquad \text{for } a = 10.3468$$

Since the K value is positive for $a = 0.5623$ and negative for $a = -10.3468$, we select $a = 0.5623$. The required gain K is 2.0535.

The characteristic equation with $K = 2.0535$ is then

$$s(s + 3)(s + 5) + 2.0535 = 0$$

or
$$s^3 + 8s^2 + 15s + 2.0535 = 0$$

(*c*) The closed-loop poles can be obtained by the following MATLAB program.

% MATLAB Program

```
p = [1    8    15    2.0535];
roots(p)
ans =
-5.1817
-2.6699
-0.1484
```

Hence, the closed-loop poles are located at

$$s = -\,5.1817,\ s = -\,2.6699,\ s = -\,4.1565$$

(*d*) The unit-step response of the system for $K = 2.0535$ can be obtained from the following MATLAB program. The resulting unit-step response curve is shown in Fig. E3.21(C).

% MATLAB program

```
num = [0  0    0    2.0535];
den = [1  8    15   2.0535];
step(num,den)
grid
title('Unit-Step Response')
xlabel('t Sec')
ylabel('Output')
```

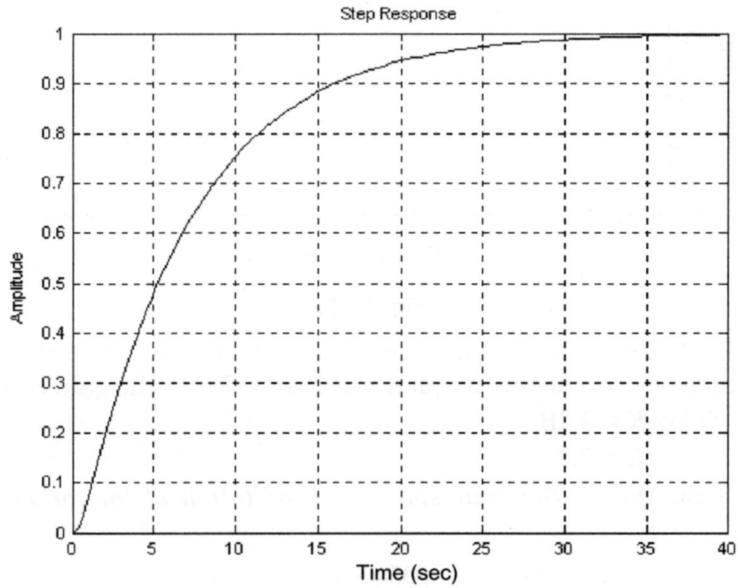

Fig. E3.21 (c)

Example 3.22. *The open-loop transfer function of a unity-feedback control system is given by*

$$G(s) = \frac{K}{s(s^2 + s + 5)}$$

(a) determine the value of gain K such that the phase margin is 50º

(b) find the gain margin for the gain K obtained in (a).

Solution:

$$G(s) = \frac{K}{s(s^2 + s + 5)}$$

The undamped natural frequency is $\sqrt{5}$ rad/s and the damping ratio of $0.1.\sqrt{5}$ from the denominator.

Let the frequency corresponding to the angle of – 130° (Phase Margin of 50) be ω_1 and therefore

$$\angle G(j\omega_1) = -130°$$

The Bode diagram is shown in Fig. E3.36 from MATLAB program.

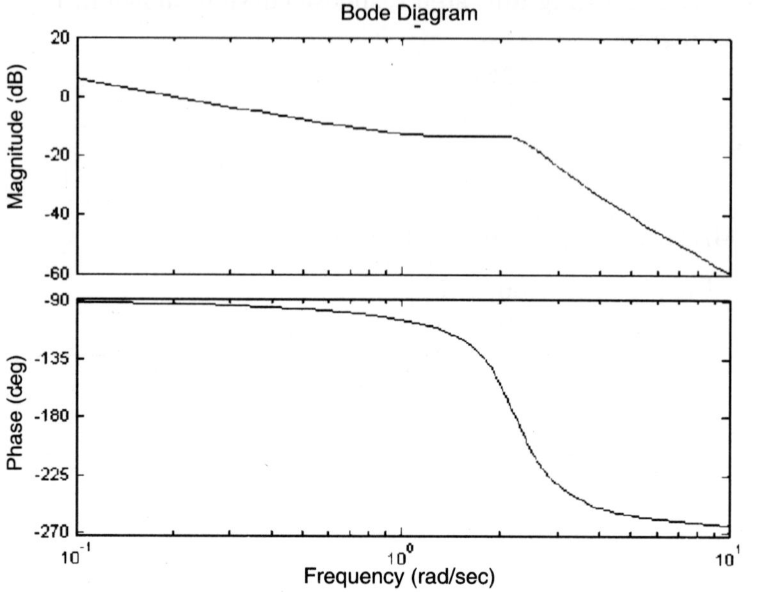

Fig. E3.22

From Fig. E3.22, the required phase margin of 50° and occurs at the frequency ω=1.06 rad/s. The magnitude of G(*j*ω) at this frequency is then – 7 dB. The gain *K* must then satisfy

$$20\log K = 7 \text{ dB}$$

or

$$K = 2.23$$

Example 3.23. *Obtain the state-space representation of the following system using MATLAB.*

$$\frac{C(s)}{R(s)} = \frac{35s + 7}{s^3 + 5s^2 + 36s + 7}$$

Solution:

A MATLAB program to obtain a state-space representation of this system is given below.

% MATLAB program

```
>> num=[0 0 35 7];
>> den=[1 5 36 7];
>> g=tf(num,den)
Transfer function:
  35 s + 7

_____________

s^3 + 5 s^2 + 36 s + 7
>>  [A,B,C,D]=tf2ss(num,den)
A =
-5    -36     -7
 1      0      0
 0      1      0
```

```
B =
1
0
0
C =
0    35    7
D =
0
```

From the MATLAB output we obtain the following state space equations:

$$\begin{bmatrix} \dot{x}_1 \\ \dot{x}_2 \\ \dot{x}_3 \end{bmatrix} = \begin{bmatrix} -5 & -36 & -7 \\ 1 & 0 & 0 \\ 0 & 1 & 0 \end{bmatrix} \begin{bmatrix} x_1 \\ x_2 \\ x_3 \end{bmatrix} + \begin{bmatrix} 1 \\ 0 \\ 0 \end{bmatrix} u$$

$$y = [0 \quad 35 \quad 7] \begin{bmatrix} x_1 \\ x_2 \\ x_3 \end{bmatrix} + [0]u$$

Example 3.24. *Find the transfer function for the following system using MATLAB.*

$$\begin{bmatrix} \dot{x}_1 \\ \dot{x}_2 \\ \dot{x}_3 \end{bmatrix} = \begin{bmatrix} 0 & 1 & 0 \\ -5 & -2 & 0 \\ 0 & 2 & -6 \end{bmatrix} \begin{bmatrix} x_1 \\ x_2 \\ x_3 \end{bmatrix} + \begin{bmatrix} 0 & 0 \\ 3 & -1 \\ 5 & 0 \end{bmatrix} u$$

$$y = \begin{bmatrix} 1 & 0 & 0 \\ 0 & 0 & 0 \end{bmatrix} \begin{bmatrix} x_1 \\ x_3 \\ x_3 \end{bmatrix}$$

Solution:

The transfer function matrix is given by

$$\mathbf{G(s) = C[sI - A]^{-1} B}$$

where

$$\mathbf{A} = \begin{bmatrix} 0 & 1 & 0 \\ -5 & -2 & 0 \\ 0 & 2 & -6 \end{bmatrix} \quad \mathbf{B} = \begin{bmatrix} 0 & 0 \\ 3 & -1 \\ 5 & 0 \end{bmatrix} \quad \mathbf{C} = \begin{bmatrix} 1 & 0 & 0 \\ 0 & 0 & 1 \end{bmatrix}$$

Hence

$$G(s) = \begin{bmatrix} 1 & 0 & 0 \\ 0 & 0 & 1 \end{bmatrix} \begin{bmatrix} s & -1 & 0 \\ 5 & s+2 & 0 \\ 0 & -2 & s+6 \end{bmatrix} \begin{bmatrix} 0 & 0 \\ 3 & -1 \\ 5 & 0 \end{bmatrix}$$

>> *%MATLAB Program*

```
>> syms s
>> C=[1 0 0;0 0 1];
>> M=[s -1 0;5 s+2 0;0 -2 s+6];
>> B=[0 0;3 -1;5 0];
>> C*inv(M)*B
ans =
[ 3/(s^2+2*s+5),  -1/(s^2+2*s+5)]
[  6*s/(s^3+8*s^2+17*s+30)+5/(s+6),  -2*s/(s^3+8*s^2+17*s+30)]
```

Example 3.25. *Determine the transfer function G(s) = Y(s)/R(s), for the following system representation in state space form.*

$$\dot{x} = \begin{bmatrix} 0 & 3 & 7 & 0 \\ 0 & 0 & 1 & 0 \\ 0 & 0 & 0 & 1 \\ -5 & -6 & 9 & 5 \end{bmatrix} x + \begin{bmatrix} 0 \\ 5 \\ 7 \\ 2 \end{bmatrix} r$$

$$y = [1 \quad 3 \quad 6 \quad 5]x$$

Solution:

```
A=[0 3 5 0;0 0 1 0;0 0 0 1;-5 -6 8 5];
B=[0;5;7;2];
C=[1 3 7 5];
D=0;
statespace=ss(A,B,C,D)
 a =
              x1          x2          x3          x4
 x1            0           3           5           0
 x2            0           0           1           0
 x3            0           0           0           1
 x4           -5          -6           8           5
 b =
            u1
x1          0
x2          5
x3          7
x4          2
c =
            x1          x2          x3          x4
y1          1           3           7           5
 d =
            u1
y1          0
```

Continuous-time model.

```
[A,B,C,D]=tf2ss(num,den);
G=tf(num,den)
```

Transfer function:

```
s^4 + 3 s^3 + 10 s^2 + 5 s + 6
-------------------------------
  s^5 + 7 s^4 + 8 s^3 + 6 s^2
```

Example 3.26. *Determine the transfer function and poles of the system represented in state space as follows using MATLAB.*

$$\dot{x} = \begin{bmatrix} 9 & -3 & -1 \\ -3 & 2 & 0 \\ 6 & 8 & -2 \end{bmatrix} + \begin{bmatrix} 1 \\ 2 \\ 3 \end{bmatrix} u(t)$$

$$y = [2 \quad 9 \quad -12]x \ ; \ x(0) = \begin{bmatrix} 0 \\ 0 \\ 0 \end{bmatrix}$$

Solution:

% MATLAB Program

```
>> A = [8 -3 4;-7 1 0;3 4 -7]
A =
 8     -3      4
-7      1      0
 3      4     -7
>> B=[1;3;8]
B =
1
3
8
>> C=[1 7 -2}
C =
1      7     -2
>> D=0
D =
0
>>  [numg,deng]=ss2tf(A,B,C,D,1)
numg =
1.0e+003  *
0      0.0060     0.0730     -2.8770
deng =
1.0000    -2.0000   -88.0000    33.0000
>> G=tf(numg,deng)
```

Transfer function:

```
   6 s^2 + 73 s - 2877
-------------
s^3 - 2 s^2 - 88 s + 33
>> poles=roots(deng)
poles =
10.2620
-8.6344
   0.3724
```

Example 3.27. *A control system is defined by*

$$\begin{bmatrix} \dot{x}_1 \\ \dot{x}_2 \end{bmatrix} = \begin{bmatrix} 0 & 1 \\ -25 & -9 \end{bmatrix} \begin{bmatrix} x_1 \\ x_2 \end{bmatrix} + \begin{bmatrix} 1 & 1 \\ 0 & 1 \end{bmatrix} \begin{bmatrix} u_1 \\ u_1 \end{bmatrix}$$

$$\begin{bmatrix} y_1 \\ y_2 \end{bmatrix} = \begin{bmatrix} 1 & 0 \\ 0 & 1 \end{bmatrix} \begin{bmatrix} x_1 \\ x_2 \end{bmatrix}$$

Plot the four sets of Bode diagrams for the system [two for input1, and two for input 2] using MATLAB.

Solution:

There are 4 sets of Bode diagrams (2 for input1 and 2 for input2)

```
>> %Bode Diagrams
>> A=[0 1;-25 -9];
>> B=[1 1;0 1];
>> C=[1 0;0 1];
>> D=[0 0;0 0];
>> bode(A,B,C,D)
```

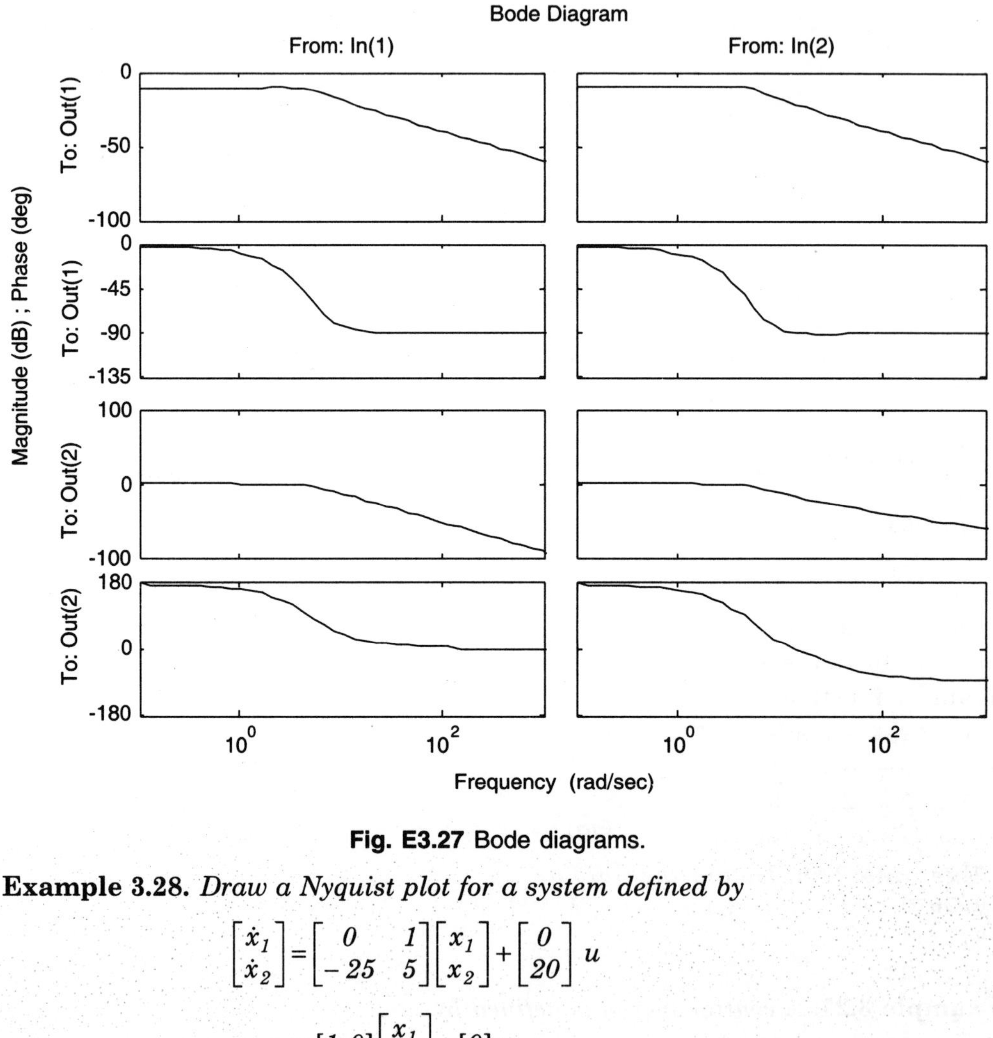

Fig. E3.27 Bode diagrams.

Example 3.28. *Draw a Nyquist plot for a system defined by*

$$\begin{bmatrix} \dot{x}_1 \\ \dot{x}_2 \end{bmatrix} = \begin{bmatrix} 0 & 1 \\ -25 & 5 \end{bmatrix} \begin{bmatrix} x_1 \\ x_2 \end{bmatrix} + \begin{bmatrix} 0 \\ 20 \end{bmatrix} u$$

$$y = [1 \;\; 0] \begin{bmatrix} x_1 \\ x_2 \end{bmatrix} + [0]u$$

using MATLAB.

Solution:

Since the system has a single input u and a single output y, a Nyquist plot can be obtained by using the command nyquist (A, B, C, D) or nyquist (A, B, C, D, 1).

```
>> %MATLAB Program
>> A=[0 1;-25 5];
>> B=[0;20];
>> C=[1 0];
>> D=[0];
>> nyquist(A,B,C,D)
>> grid
>> title('Nyquist plot')
```

The Nyquist plot is shown in Figure E3.28.

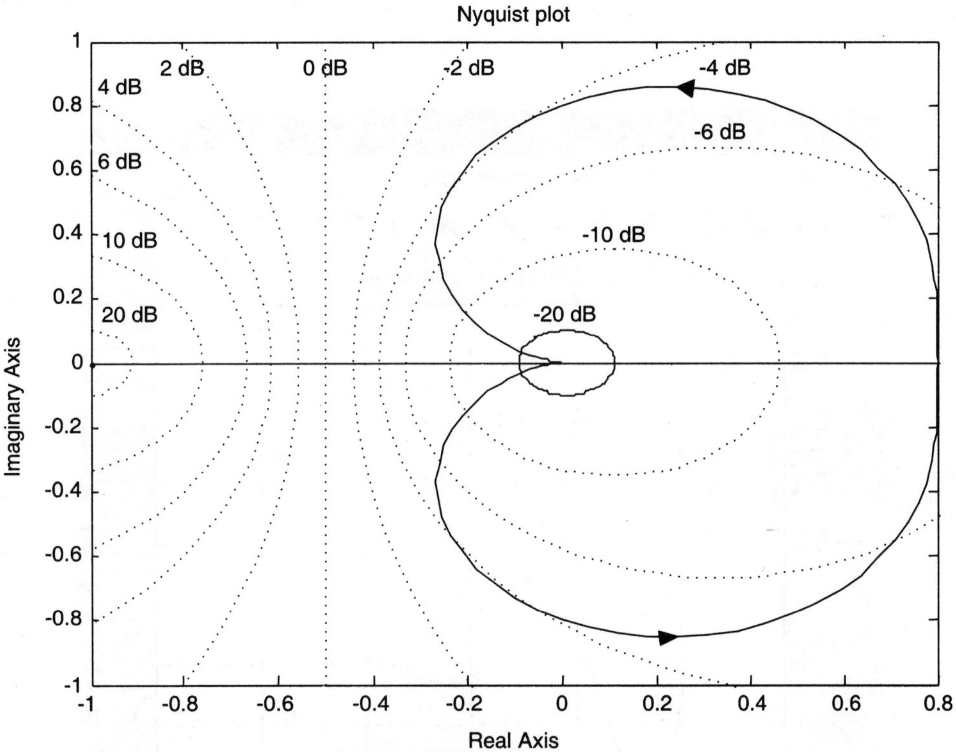

Fig. E3.28 Nyquist plot.

Example 3.29. *Obtain the unit-step response, unit-ramp response, and unit-impulse response of the following system using MATLAB*

$$\begin{bmatrix} \dot{x}_1 \\ \dot{x}_2 \end{bmatrix} = \begin{bmatrix} -1 & -1.5 \\ 2 & 0 \end{bmatrix} \begin{bmatrix} x_1 \\ x_2 \end{bmatrix} + \begin{bmatrix} 1.5 \\ 0 \end{bmatrix} u$$

$$y = [1 \quad 0] \begin{bmatrix} x_1 \\ x_2 \end{bmatrix}$$

where u is the input and y is the output.

Solution:

>> %Unit-step response

```
>> A=[-1 -1.5;2 0];
>> B=[1.5;0];
>> C=[1 0];
>> D=[0];
>>  [y,x,t]=step(A,B,C,D);
>> plot(t,y)
>> grid
>> title('Unit-step response')
>> xlabel('t Sec')
>> ylabel('Output')
```

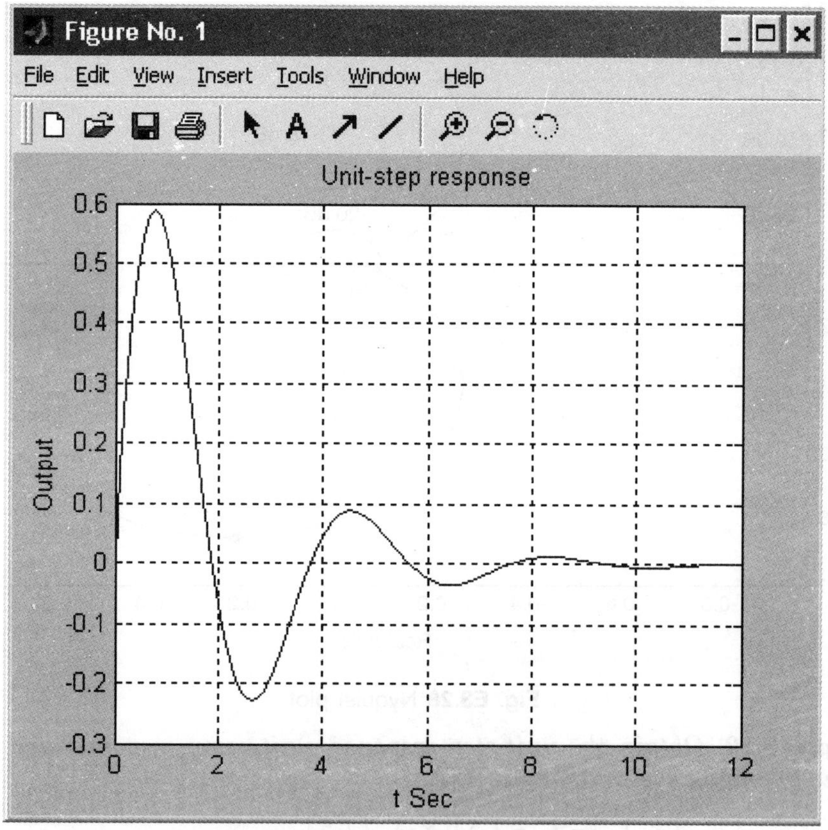

Fig. E3.29 (*a*) Unit-step response.

>> %Unit-ramp response

```
>> A= [-1 -1.5; 2 0];
>> B= [1.5; 0];
>> C= [1 0];
>> D= [0];
>> % New enlarged state and output equations
>> AA= [A zeros (2, 1); C 0];
```

```
>> BB= [B; 0];
>> CC= [0 1];
>> DD= [0];
>> [z, x, t] =step (AA, BB, CC, DD);
>> x3= [0 0 1]*x'; plot (t, x3, t, t,'-')
>> grid
>> title ('Unit-ramp response')
>> xlabel ('t Sec')
>> ylabel ('Output and unit-ramp input')
>> text (12, 1.2,'Output')
```

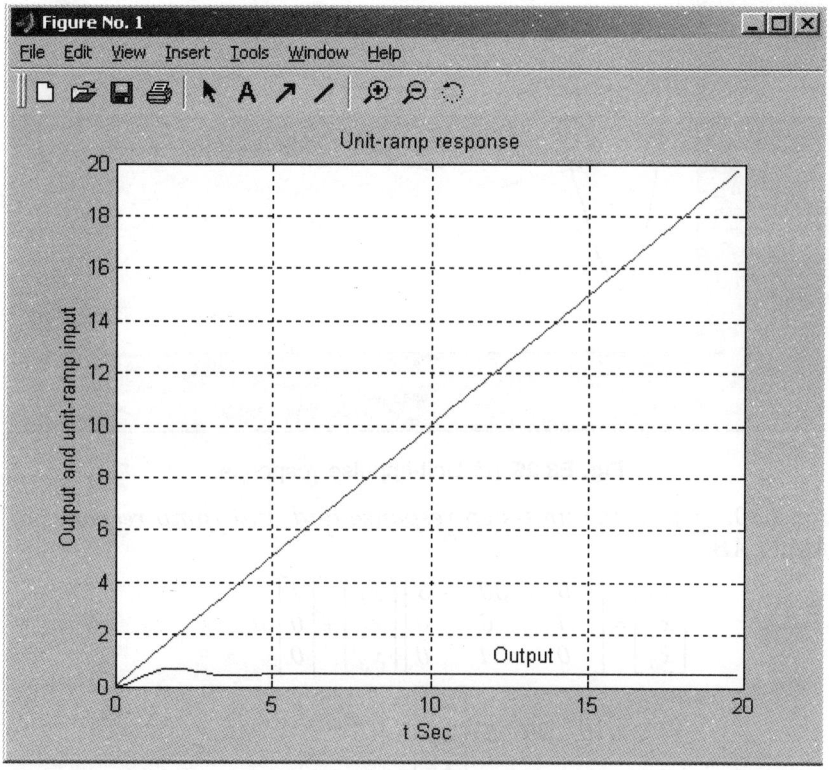

Fig. E3.29 (*b*) Unit-ramp response.

>> *%Unit-impulse response*

```
>> A= [-1 -1.5; 2 0];
>> B= [1.5; 0];
>> C= [1 0];
>> D= [0];
>> impulse (A, B, C, D)
```

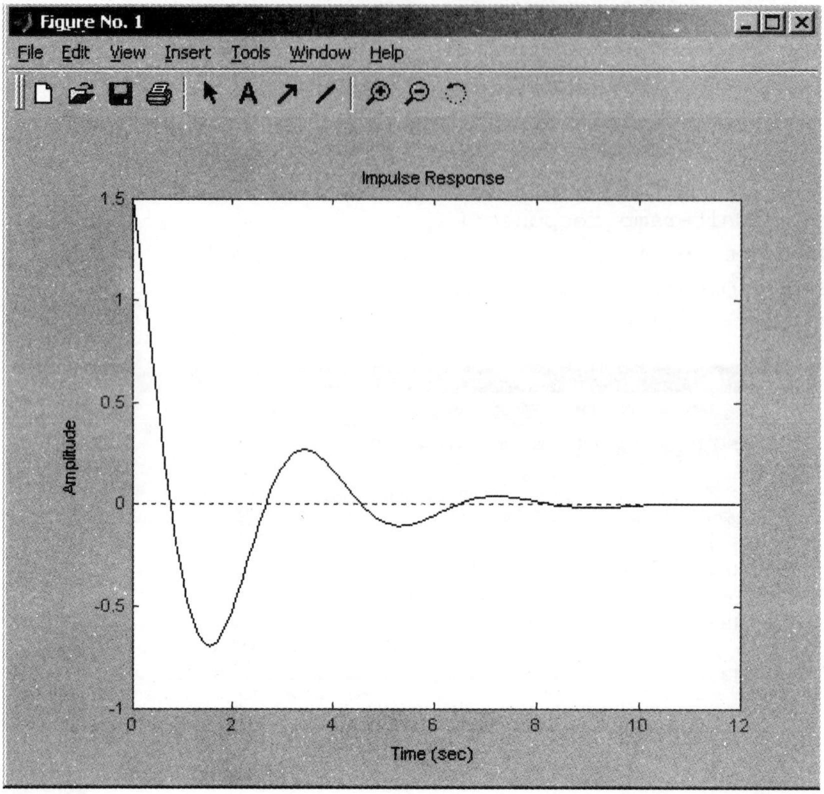

Fig. E3.29 (*c*) Unit-impulse response.

Example 3.30. *Obtain the unit-step response and unit-ramp response of the following system using MATLAB.*

$$\begin{bmatrix} \dot{x}_1 \\ \dot{x}_2 \\ \dot{x}_3 \end{bmatrix} = \begin{bmatrix} -5 & -30 & -5 \\ 1 & 0 & 0 \\ 0 & 1 & 0 \end{bmatrix} \begin{bmatrix} x_1 \\ x_2 \\ x_3 \end{bmatrix} + \begin{bmatrix} 1 \\ 0 \\ 0 \end{bmatrix} u$$

$$y = [0 \quad 20 \quad 5] \begin{bmatrix} x_1 \\ x_2 \\ x_3 \end{bmatrix} + [0]\, u$$

Solution:

>> *%MATLAB Program*

```
>> A= [-5 -30 -5; 1 0 0; 0 1 0];
>> B= [1; 0; 0];
>> C= [0 20 5];
>> D= [0];
>> [y, x, t] =step (A, B, C, D);
>> plot (t, y)
>> grid
>> title ('Unit-response')
>> xlabel ('t Sec')
>> ylabel ('Output y (t)')
```

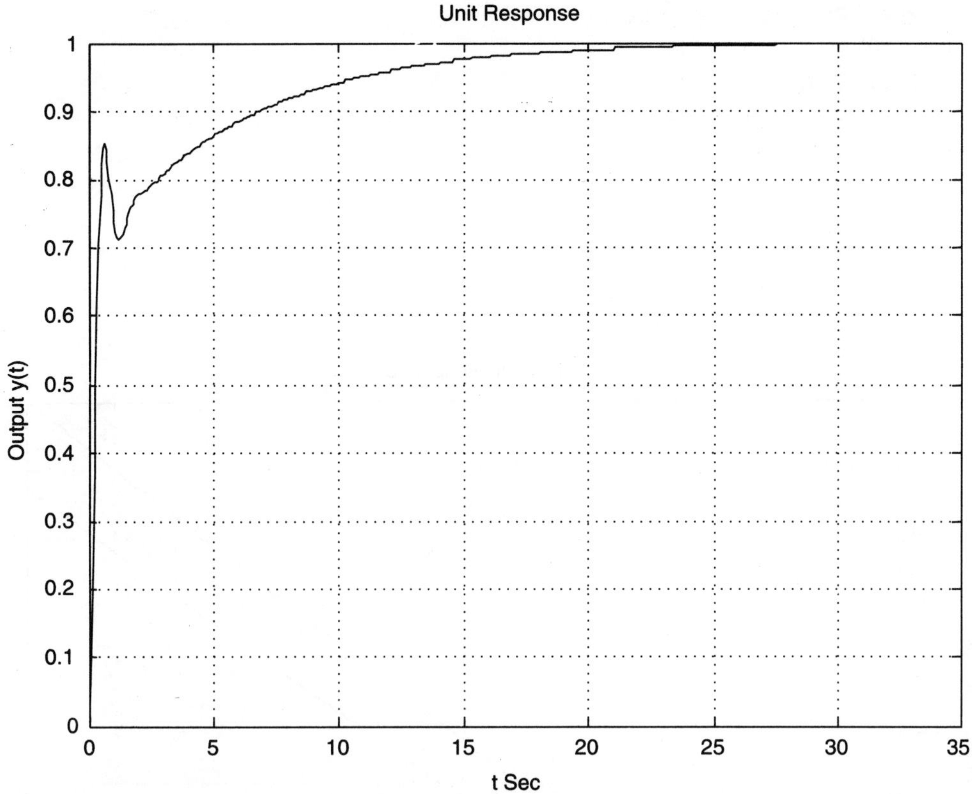

Fig. E3.30 (*a*) Unit-step response.

Unit-ramp response:

$$AA = \begin{bmatrix} -5 & -30 & -5 & 0 \\ 1 & 0 & 0 & 0 \\ 0 & 1 & 0 & 0 \\ 0 & 25 & 5 & 0 \end{bmatrix} = \begin{bmatrix} & A & & 0 \\ & & & 0 \\ & & & 0 \\ 0 & 25 & 5 & 0 \end{bmatrix} = \text{A zeros (2, 1); C 0]}$$

$$BB = \begin{bmatrix} 1 \\ 0 \\ 0 \\ 0 \end{bmatrix} = \begin{bmatrix} B \\ \\ 0 \end{bmatrix}$$

$$CC = [0 \quad 25 \quad 5 \quad 0] = [C \quad 0]$$

>> *%MATLAB Program*

```
>> A= [-5 -30 -5; 1 0 0; 0 1 0];
>> B= [1; 0; 0];
>> C= [0 25 5];
>> D= [0];
>> AA= [A zeros (3, 1); C 0];
>> BB= [B; 0];
```

```
>> CC=[C 0];
>> DD= [0];
>> t=0:0.01:5;
>> [z, x, t] =step (AA, BB, CC, DD, 1, t);
>> P= [0 0 0 1]*x';
>> plot (t, P, t, t)
>> grid
>> title ('Unit-ramp response')
>> xlabel ('t Sec')
>> ylabel ('Input and output')
```

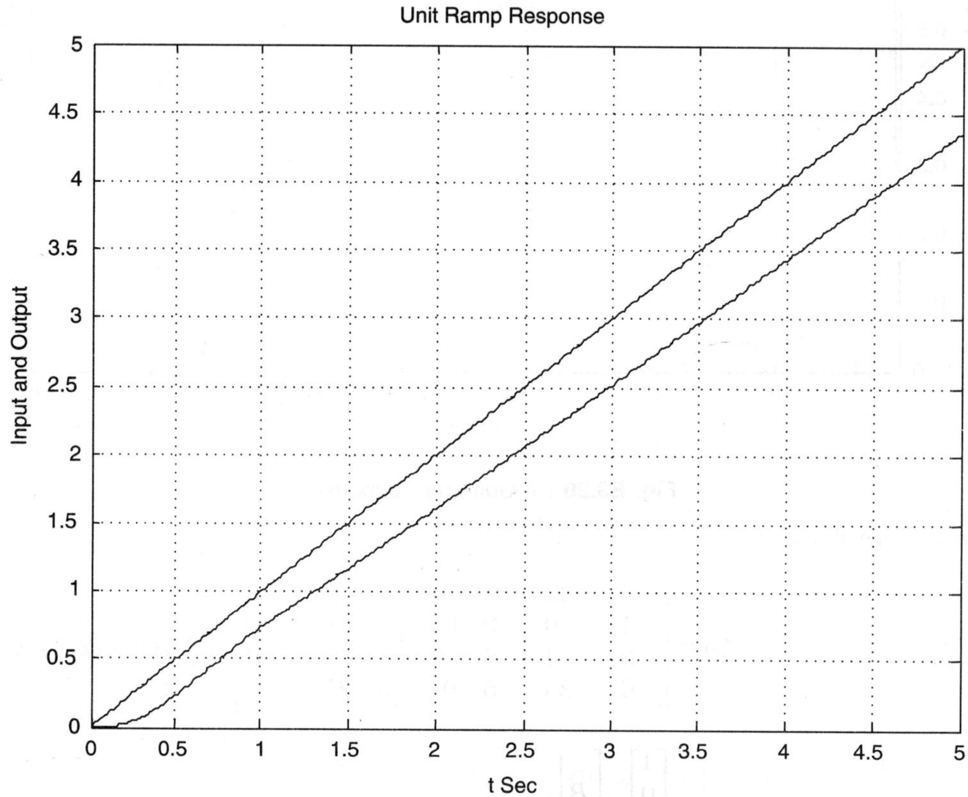

Fig. E3.30 (*b*) Unit-ramp response.

Example 3.31. *Consider the system*

$$\begin{bmatrix} \dot{x}_1 \\ \dot{x}_2 \\ \dot{x}_3 \end{bmatrix} = \begin{bmatrix} 3 & 0 & 0 \\ 0 & 1 & 0 \\ 0 & 3 & 2 \end{bmatrix} \begin{bmatrix} x_1 \\ x_2 \\ x_3 \end{bmatrix}$$

The output is given by

$$y = [1\ 1\ 1] \begin{bmatrix} x_1 \\ x_2 \\ x_3 \end{bmatrix}$$

(a) determine the observability of the system using MATLAB

(b) show that the system is completely observable if the output is given by

$$\begin{bmatrix} y_1 \\ y_2 \end{bmatrix} = \begin{bmatrix} 1 & 1 & 1 \\ 1 & 3 & 2 \end{bmatrix} \begin{bmatrix} x_1 \\ x_2 \\ x_3 \end{bmatrix}$$

using MATLAB.

Solution:

>> *%MATLAB Program*

```
>> A= [3 0 0; 0 1 0; 0 3 2];
>> C= [1 1 1];
>> rank ([C' A'*C' A'^2*C'])
ans =
3
>> A= [3 0 0; 0 1 0; 0 3 2];
>> C= [1 1 1; 1 3 2];
>> rank ([C' A'*C' A'^2*C'])
ans =
3
```

From the above, we observe that the system is observable and controllable.

Example 3.32. *Consider the following state equation and output equation*

$$\begin{bmatrix} \dot{x}_1 \\ \dot{x}_2 \\ \dot{x}_3 \end{bmatrix} = \begin{bmatrix} -1 & -3 & -2 \\ 0 & -2 & 1 \\ 1 & 0 & -1 \end{bmatrix} \begin{bmatrix} x_1 \\ x_2 \\ x_3 \end{bmatrix} + \begin{bmatrix} 3 \\ 0 \\ 1 \end{bmatrix} u$$

$$y = [1\ 1\ 0] \begin{bmatrix} x_1 \\ x_2 \\ x_3 \end{bmatrix}$$

Determine if the system is completely state controllable and completely observable using MATLAB.

Solution:

The controllability and observability of the system can be obtained by examining the rank condition of

$$[\mathbf{B\ AB\ A^2B}] \text{ and } [\mathbf{C'\ A'\ C'\ (A')^2\ C'}]$$

>> *%MATLAB Program*

```
>> A= [-1 -3 -2; 0 -2 1; 1 0 -1];
>> B= [3; 0; 1];
>> C= [1 1 0];
>> D= [0];
>> rank ([B A*B A^2*B])
ans =
3
>> rank ([C' A'*C' A'^2*C'])
ans =
3
```

We observe the rank of $[B\ AB\ A^2B]$ is 3 and the rank of **[C′ A′*C′ (A′)²* C′]** is 3, the system is completely state controllable and observable.

Example 3.33. *Determine the eigenvalues of the following system using MATLAB.*

$$\dot{x} = \begin{bmatrix} 0 & 2 & 0 \\ 0 & 2 & -9 \\ -2 & 2 & 5 \end{bmatrix} x + \begin{bmatrix} 0 \\ 0 \\ 2 \end{bmatrix} r$$

$$y = [0\ 0\ 1]\ x$$

Solution:

```
>> A = [0 2 0; 0 2 -7; -2 2 5]; %Define the matrix above
>> eig (A) %Calculate the eigenvalues of matrix A.
ans =
2.0000
2.5000 + 3.4278i
2.5000 - 3.4278i
```

3.15.2 Engineering Mechanics

MATLAB has an excellent collection of commands and functions that are useful for solving engineering mechanics problems. The problems presented in this chapter are basic and are normally presented in introductory mechanics courses. The application of MATLAB to the problems in the analysis and design of engineering mechanics is presented in this chapter with a number of illustrative examples. These examples cover different topics of Mechanics. Each solution begins with a problem formulation followed by a corresponding MATLAB code with simple explanations wherever possible. The background required for solving Mechanics problems is the knowledge of differentiation, integration, trigonometry along with basic physical laws. Table-3.1 summarizes some of the important mathematical relations often used for computations in statics and dynamics.

Table 3.1 Some Mathematical Expressions

S. No.	*Mathematical entity*
1	If $ax^2 + bx + c = 0$, then $x = \dfrac{-b \pm \sqrt{b^2 - 4ac}}{2a}$ is root
2	If $\sin\theta = A$ and $\cos\theta = B$ then (*i*) $A^2 + B^2 = 1$, (*ii*) $\sin 2\theta = 2AB$ (*iii*) $\cos 2\theta = B^2 - A^2$ (*iv*) $\tan\theta = \dfrac{A}{B}$ (*v*) $\sin(\theta \pm \phi) = \sin\theta\cos\phi \pm \cos\theta\sin\phi$. In a triangle: (*a*) **sine rule** $a/\sin(\theta) = b/\sin(\phi)$, where a is side opposite to angle ϕ and b is side opposite to angle θ.
	(*b*) **Cosine rule**: $c^2 = a^2 + b^2 - 2ab\cos(\theta)$, where θ is angle opposite to side c.
3	Important derivatives (*i*) $\dfrac{d}{dx}(u^n) = nu^{n-1}\dfrac{du}{dx}$ (*ii*) $\dfrac{d}{dx}(uv) = u\dfrac{dv}{dx} + v\dfrac{du}{dx}$ (*iii*) $\dfrac{d}{dx}\left(\dfrac{u}{v}\right) = \dfrac{v\dfrac{du}{dx} - u\dfrac{dv}{dx}}{v^2}$ (*iv*) $\dfrac{d}{dx}(\sin\theta) = \cos\theta\dfrac{d\theta}{dx}$ (*v*) $\dfrac{d}{dx}(\cos\theta) = -\sin\theta\dfrac{d\theta}{dx}$ (*vi*) $\dfrac{d}{dx}(\tan\theta) = \sec^2\theta\dfrac{d\theta}{dx}$

4	Important integrals (here c is a constant) (i) $\int x^p dx = \frac{x^{p+1}}{p+1} + c, p \neq -1$ (ii) $\int \frac{dx}{p+qx} = \frac{1}{q}\log_e (p+qx) + c$ (iii) $\int \frac{dx}{p+qx^2} = \frac{1}{2\sqrt{-pq}} \log_e \frac{p+x\sqrt{-pq}}{p-x\sqrt{-pq}} + c$, where $pq < 0$ (iv) $\int \sin x\, dx = -\cos x + c$, (v) $\int \cos x\, dx = \sin x + c$, (vi) $\int e^{ax} dx = \frac{1}{a} e^{ax} + c$

In the computational approach the first step is to understand the theory necessary to solve the problem. The second step is to develop an algorithm that outlines how the problem input can be used along with the theory to generate the required output. In the final step, this algorithm must be translated into correct MATLAB statements. It must be noted that MATLAB solutions need not be unique. In all the examples solved in this chapter, a script is written in the form of '.m' files and the program is executed by typing the *m* file name with extension at the command prompt. It is emphasized to develop each program with different commands to learn MATLAB with perfection.

3.15.2.1 Statics

Problems in statics are presented on the following common topics: forces on a particle, rigid-bodies and equivalent forces, equilibrium of rigid bodies, truss analysis, beams, friction and distributed forces (centroids).

ES 3.1. *Write a MATLAB software program to determine the magnitude and direction of the resultant of n coplanar forces applied at point A in Fig. ES3.1. Use this software to solve the following:*

$F_1 = 20$ kN, $F_2 = 40$ kN, $F_3 = 200$ kN, $a_1 = 40°$, $a_2 = 25°$ and $a_3 = 58°$

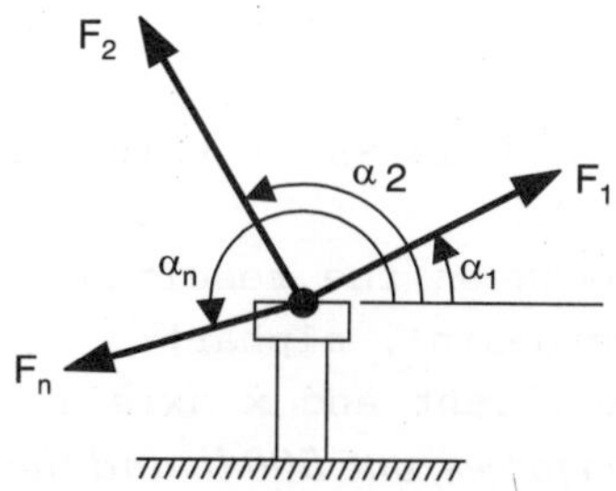

Fig. ES3.1

Solution:

We know that

$$R_x = \sum_{i=1}^{n} F_i \cos \alpha_i, \quad R_y = \sum_{i=1}^{n} F_i \sin \alpha_i \tag{E.1}$$

Therefore $$R = \sqrt{R_x^2 + R_y^2} \tag{E.2}$$

and $$\alpha_R = \tan^{-1} \frac{R_y}{R_x} \tag{E.3}$$

Let α^*_R be the value defined by Eq. (E.3) and such that $-90° \leq \alpha^*_R \leq 90°$. Then, we have

If $R_x \geq 0$ and $R_y \geq 0$: $\alpha_R = \alpha^*_R$ (E.4)

If $R_x \geq 0$ and $R_y < 0$: $\alpha_R = 360° + \alpha^*_R$ (E.5)

If $R_x < 0$: $\alpha_R = 180° + \alpha^*_R$ (E.6)

MATLAB Solution:

```
>> % Example ES3.1
>> n=3;
>> alpha= [40 25 58];
>> alpha1=alpha;
>> force= [20 40 200];
>> sumx=0;
>> sumy=0;
>> for i=1: n
sumx=sumx + force (i)*cos (alpha (i));
sumy=sumy + force (i)*sin (alpha (i));
end
>> r=sqrt (sumx^2+sumy^2);
>> alphar = atan2 (sumy, sumx);
>> alphar=alphar*180/pi;
>> if alphar < 0;
alphar = alphar + 360;
end
>> fprintf ('The resultant R is %4.1f kN\n', r);
The resultant R is 214.1 kN
>> fprintf ('The angle between the resultant
and x axis is % 4.1f degrees\n', alphar);
The angle between the resultant and x axis is 76.5 degrees
```

ES 3.2. *Fig. ES3.2 shows two forces, one 500 lb and the other P applied by cables on each side of the obstruction A in order to remove the spike. Write a MATLAB program to determine*

(a) the magnitude of P necessary such that the resultant T is directed along the spike

(b) the magnitude of T

(c) Plot P and T as a function of d. (Range of d between 1 and 15 inches).

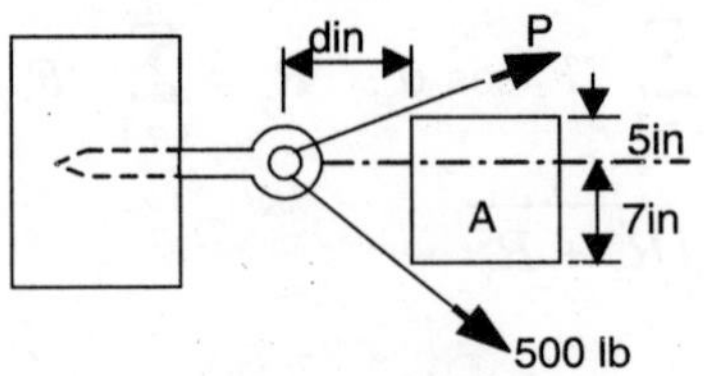

Fig. ES3.2

Solution:

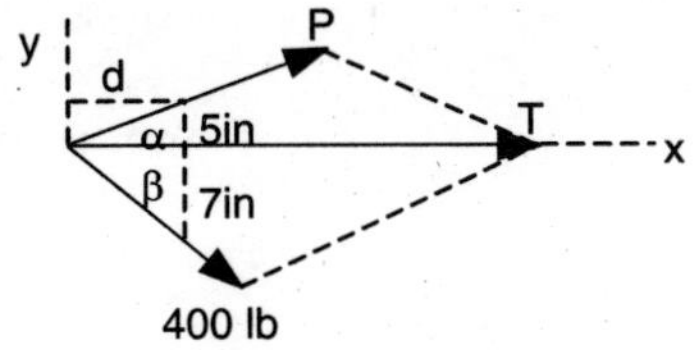

Fig. ES3.2 (*a*)

From equilibrium equations, we have

$$T = \Sigma F_x = P\cos\alpha + 500\cos\beta$$

$$0 = \Sigma F_x = P\sin\alpha - 500\cos\beta$$

Solving the above two equations, we obtain

$$P = \frac{500\sin\beta}{\sin\alpha}$$

and

$$T = 500(\sin\beta\cot\alpha + \cos\beta)$$

From Fig. ES3.2 (*a*), we have

$$\alpha = \tan^{-1}(5/d) \quad \text{and} \quad \beta = \tan^{-1}(7/d)$$

Fig. ES3.2 (*b*) shows the plot of P and T as a function of d.

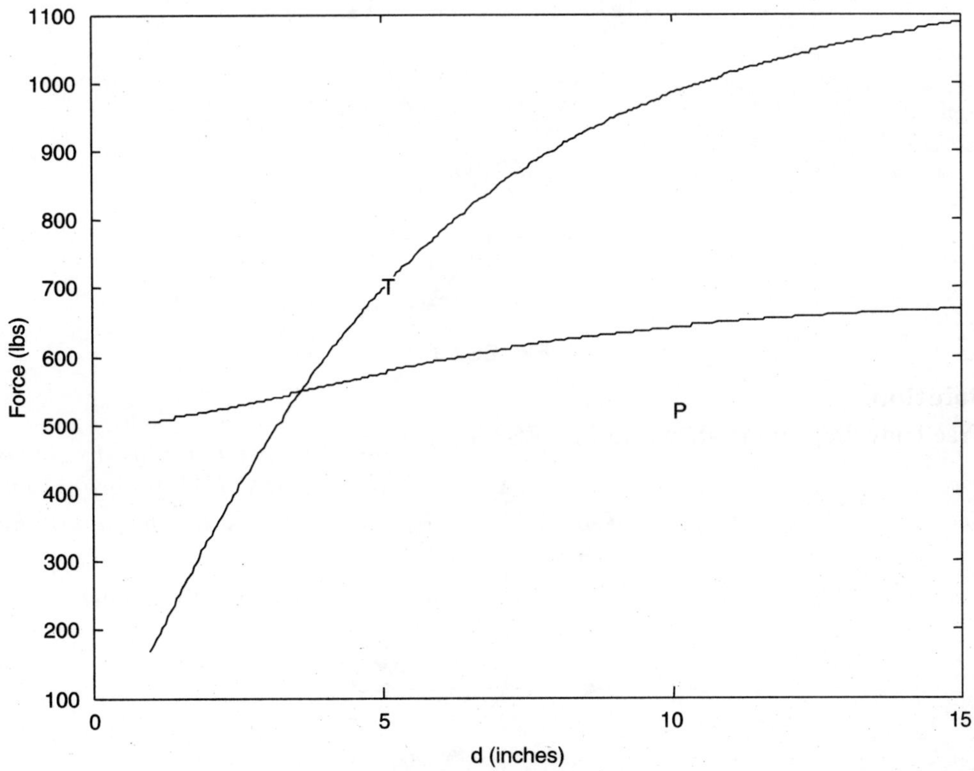

Fig. ES3.2 (*b*)

```
>> % Example ES3.2
>> % Range of d
>> d=1:0.05:15;
>> % Define alpha
>> alpha = atan (5./d);
>> %Define beta
>> beta = atan(7./d);
>> % Compute force P
>> P = 500*sin(beta)./sin(alpha);
>> % Define force T
>> T=500*(sin(beta).*cot(alpha) +cos(beta));
>> plot (d, P, d, T)
>> xlabel ('d (inches)');
>> ylabel ('Force (lbs)');
```

ES 3.3. *Fig. ES3.3 shows the two cables MO and NO tied together at O and the loadings are also shown. The magnitude of F is 150 lbs.*

(*a*) *Derive the expressions relating the tension in each cable as a function of* α.

(*b*) *Write a MATLAB program to plot the tension in each cable for* $0 \le \alpha \le 90°$.

(*c*) *Determine the smallest value of* α *for which both cables are in tension.*

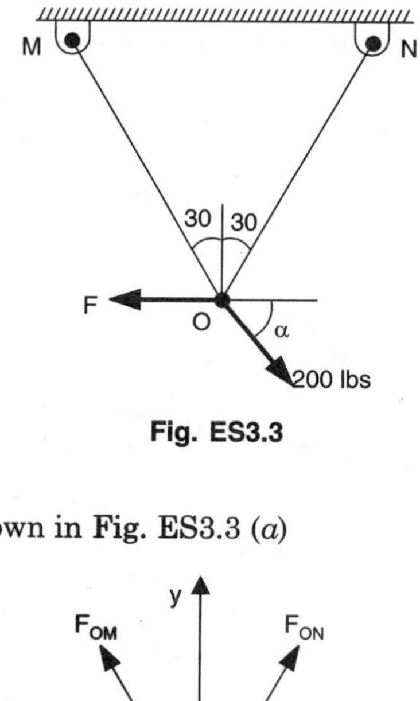

Fig. ES3.3

Solution:

Free body diagram is shown in **Fig. ES3.3** (*a*)

Fig. ES 3.3 (*a*)

(*a*) $$\Sigma F_x = 0 \Rightarrow (F_{ON} - F_{OM}) \sin 30 - F + 200 \cos \theta = 0$$

or $$(F_{ON} - F_{OM}) = 2F - 400 \cos \theta$$

$$(F_{ON} - F_{OM}) = 300 - 400 \cos \theta \quad (1)$$

$$\Sigma F_y = 0 \Rightarrow (F_{OM} + F_{ON}) \cos 30 - 200 \sin \theta = 0$$

or $$(F_{OM} + F_{ON}) = 200 \times \frac{2}{\sqrt{3}} .\sin \theta = 230.94 \sin \theta \quad (2)$$

Solving (1) and (2)

$$F_{ON} = 150 - 200 \cos \theta + 115.47 \sin \theta$$

$$F_{OM} = 115.47 \sin \theta + 200 \cos \theta - 150$$

(*b*) *MATLAB Program:*

```
%
Theta= [0:.1:90];
theta=Theta*pi/180;
fon =150-200*cos(theta) +115.47*sin(theta);
fom=-150+115.47*sin(theta) +200*cos(theta);
ton=abs (fon);
tom=abs (fom);
[ton_min, i]=min (ton);
[tom_min, j]=min (tom);
Ang1_min=Theta (i);
Ang2_min=Theta (j);
plot (Theta, fon, Theta, fom);
%
legend ('Fon','Fom', 2)
%
xlabel ('Theta (degrees)')
ylabel ('Cable Tension (lb)')
grid on
fprintf ('(c)\n')
fprintf ('Smallest value of Theta for which the tensions are positive
is from %g to %g degrees\n', Ang1_min, Ang2_min)
```

The output is shown in Fig. ES3.3 (*b*).

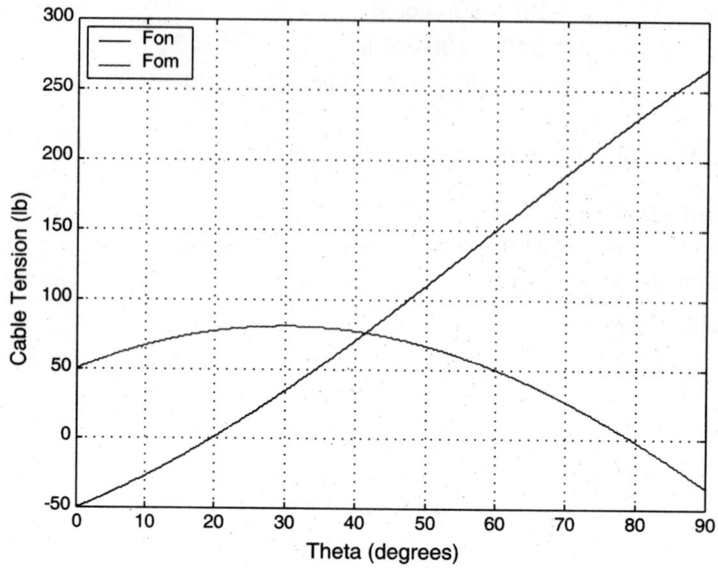

Fig. ES 3.3 (*b*)

(*c*) The smallest value of theta for which the tensions are positive is from 19.5 to 79.5 degrees.

ES 3.4. *Fig.ES3.4 shows the location of the center of gravity of a 5000 lb truck for the unladen condition. The location of the added load W_L is at a distance of x inches behind the rear axle. Write a MATLAB program and plot W_L as a function of x for x ranging from 0 to 60 inches.*

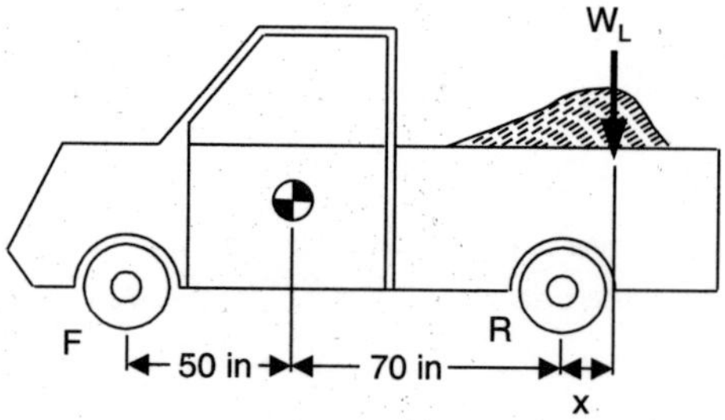

Fig. ES3.4

Solution:

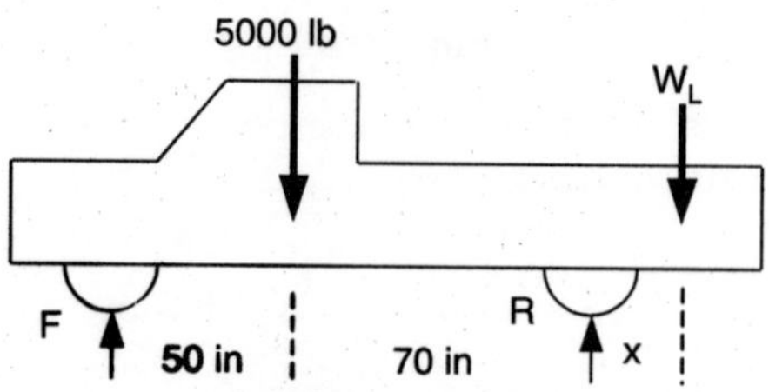

Fig. ES3.4 (a)

Fig.ES.3.4 (*a*) shows the free body diagram. The equilibrium equations can be written as

$$\Sigma M_R = 0 = 5000(70) - N(120) - W_L x = 0$$
$$\Sigma F_y = 0 = N + N - 5000 - W_L = 0$$

Solving the above equations for W_L, we obtain

$$W_L = \frac{5000}{(60 + x)}$$

The plot of W_L as a function of x is shown in Fig.ES3.4 (*b*).

MATLAB Solution:

```
>> % Example ES3.4
>> % Define the range of x for the plot
>> x= (0:0.05:60);
>> % Define Wl
>> Wl=50000./(60+x);
>> plot(x, Wl)
>> % Labels
>> xlabel ('x (inches)')
>> ylabel ('Load Weight (lbs)')
```

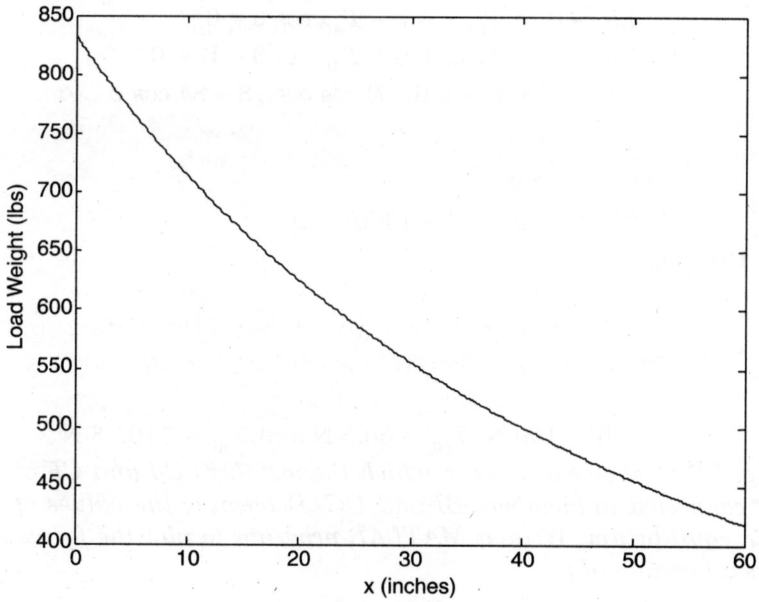

Fig. ES3.4 (*b*)

ES 3.5. *Fig. ES3.5 shows a container of weight W suspended from ring C to which cable CB of length 6 m and spring AC are attached.*

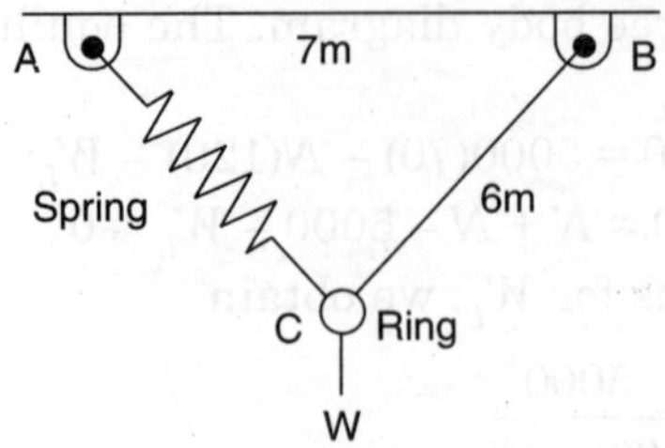

Fig. ES3.5

Write a MATLAB program to determine the tension in the cable when W = 150 N. Given: The spring constant as 120 N/m, and its unstretched length as 4 m.

Solution:

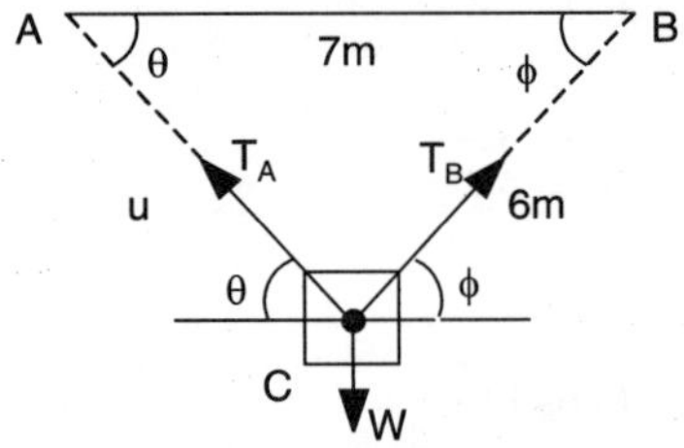

$$\Sigma F_x = 0 \Rightarrow T_{BC}.\cos\phi - T_{AC}\cos\theta = 0 \quad (1)$$

$$\Sigma F_y = 0 \Rightarrow T_{BC}.\sin\phi + T_{AC}\sin\theta - W = 0 \quad (2)$$

Geometry: $$u^2 = 6^2 + 7^2 - 2(6)(7)\cos\phi = 78 - 84\cos\phi \quad (3)$$

$$\frac{u}{\sin\phi} = \frac{6}{\sin\theta} \quad (4)$$

Stiffness: $$T_{AC} = k(u - u_0) = 120(u - 4) \quad (5)$$

MATLAB Program

```
syms b c p t u;
[b c p t u]=solve('b*cos(p)-c*cos(t)=0','b*sin(p)+c*sin(t)
-150=0','u^2=78-84*cos(p)','u/sin(p)=6/sin(t)','c=120*(u-4)')
```

Output is

$$W = 150\text{ N},\ T_{BC} = 60.5\text{ N and } T_{AC} = 110.53\text{ N}$$

ES 3.6. *Fig. ES3.6 shows a truss in which the members CJ and CF of the loaded truss cross but are not connected to members BI and DG. Determine the values of α for which the truss cannot be in equilibrium. Write a MATLAB program to plot the forces in members BC, JC, IC, and IG as a function of α.*

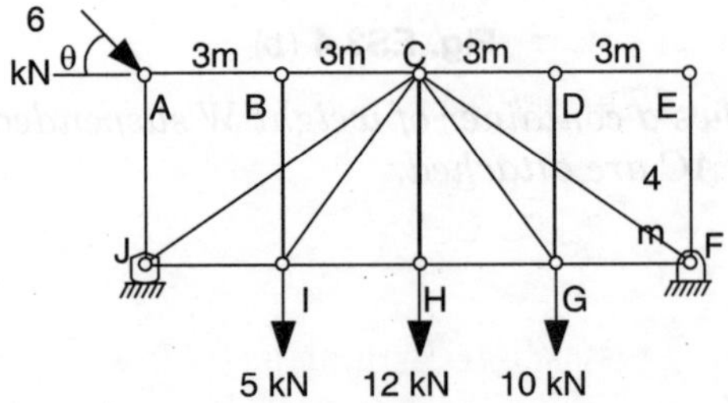

Fig. ES3.6

Solution:

Free body diagram is shown in Fig. ES3.6 (*a*).

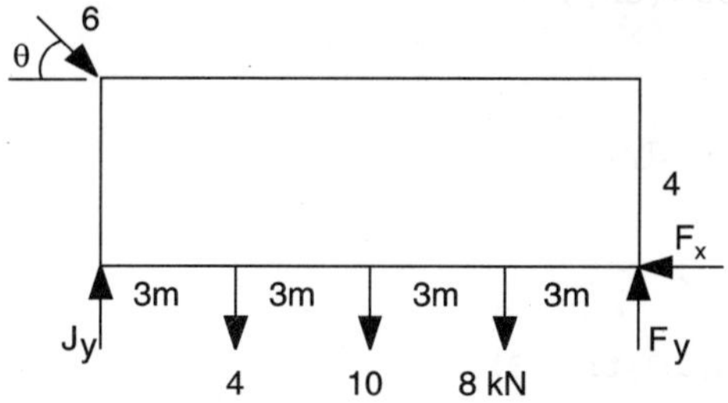

Fig. ES 3.6 (*a*)

First we determine the reaction force at *J* from a free-body diagram for the entire truss.

$$\circlearrowleft \Sigma M_F = 0 = 6 \sin \theta(12) - 6 \cos \theta(4) + 4(9) + 10(6) + 8(3) - J_y (12)$$

$$J_y = 10 + 6 \sin \theta - 2 \cos \theta$$

Note that the rocker can only exert an upward force at A. Thus, to be in equilibrium, J_y must be positive. Since sin θ and cos θ vary between plus and minus one, the above equation indicates that J_y will be positive for all θ. Thus, the truss will be in equilibrium for all values of θ.

To obtain the required forces we now consider free-body diagrams for joints *A*, *J* and *I*. Note that each member is assumed to be in tension. Thus, positive answers will imply tension and negative answers compression. Also note the order in which the joints are analyzed. In each case there are only two unknown forces.

Joint A

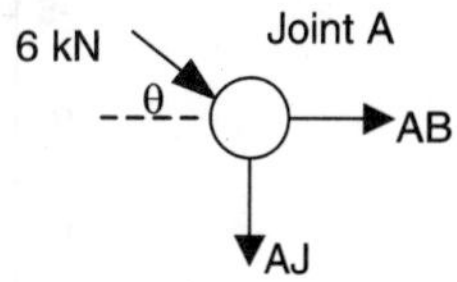

$$\Sigma F_x = 0 = AB + 6 \cos \theta; \quad \Sigma F_y = 0 = -6 \sin \theta - AJ$$

$$AJ = -6 \sin \theta \qquad AB = -6 \cos \theta$$

Note that $AB = BC$ since *BI* is a zero-force member.

Joint J

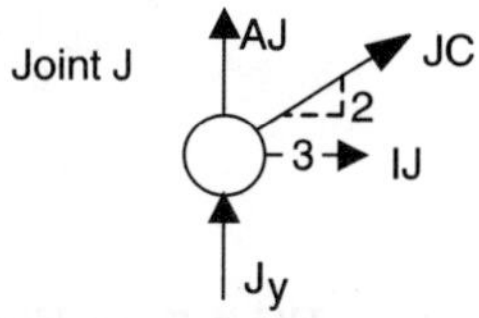

$$\Sigma F_x = 0 = IJ + \frac{3}{\sqrt{13}} JC; \; \Sigma F_y = 0 = AJ + J_y + \frac{2}{\sqrt{13}} JC$$

$$JC = -\frac{\sqrt{13}}{2} (AJ + J_y) = -\frac{\sqrt{13}}{2} (10 - 2 \cos\theta)$$

$$IJ = -\frac{3}{\sqrt{13}} JC = 15 - 3 \cos \theta$$

Joint I

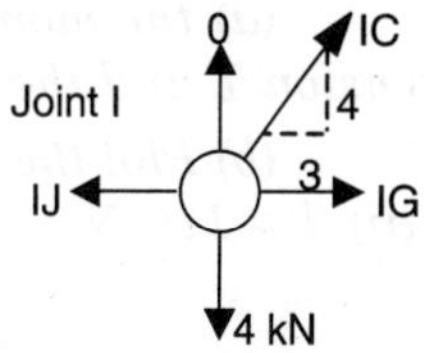

$$\Sigma F_x = 0 = GI - IJ + \frac{3}{5} IC; \; \Sigma F_y = 0 = \frac{4}{5} IC - 4$$

$$IC = 5 \text{ kN} \; IG = IJ - \frac{3}{5} 5 = 12 - 3 \cos \theta$$

Note in the above that substitutions have been made in order to express each force explicitly in terms of θ. This has been done primarily for completeness. The program below shows that the explicit substitution is not necessary. The computer will substitute automatically provided each force is expressed in terms of functions that have been previously defined.

MATLAB Program

```
th=0:0.05:2*pi;
Jy=10+6*sin(th)-2*cos(th);
AJ=-6*sin(th);
BC=-6*cos(th);
JC=sqrt (13)/2*(AJ + Jy);
IJ=-3/sqrt (13)*JC;
IC=5;
IG=IJ-3;
plot(th,BC,th,JC,th,IC,th,IG)
legend ('BC','JC','IC','IG')
xlabel ('theta (rads)')
ylabel ('Force (kN)')
```

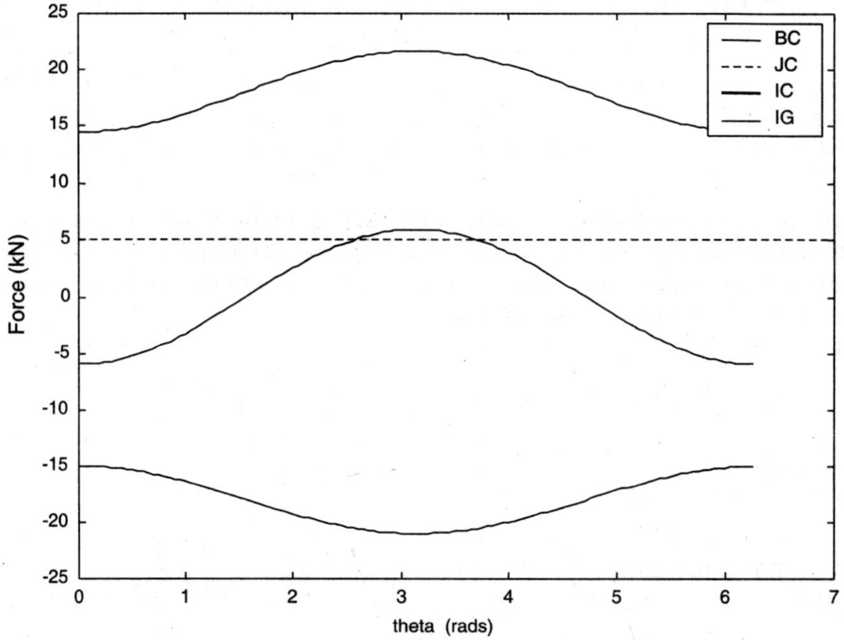

Fig. ES 3.6 (*b*)

ES 3.7. *In Fig. ES3.7 rod CB is held by a cord AC which has a tension T. Write a MATLAB program to determine,*

(a) the moment about B of the force exerted by the cord at point C as a function of the tension T and the distance d.

(b) Plot the moment about B for 300 mm ≤ d ≤ 1000 mm when (i) T = 60 N, (ii) T = 80 N, (iii) T = 110 N.

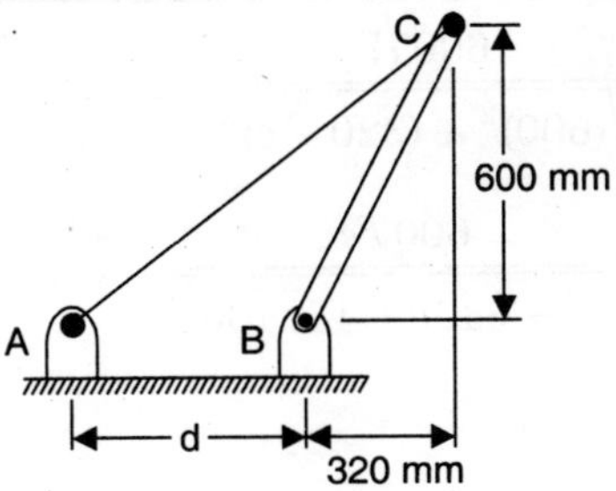

Fig. ES3.7

Solution:

Free body diagram is shown in Fig. ES3.7 (*a*).

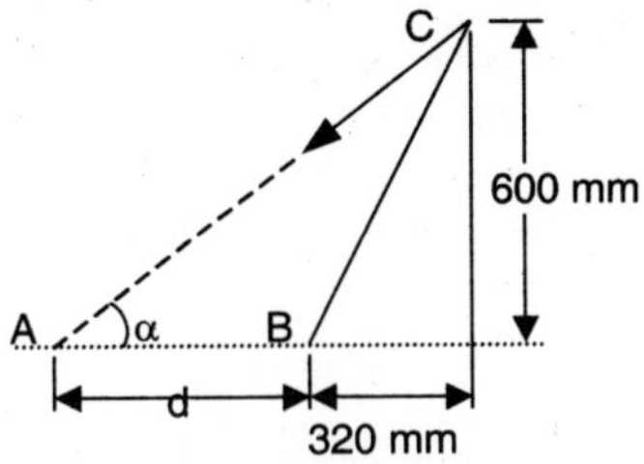

Fig. ES3.7 (*a*)

The length of *AC* is from the figure,

$$AC = \sqrt{(600)^2 + (320 + c)^2}$$

For the angle α we have

$$\cos \alpha = \frac{320 + c}{AC}$$

$$\sin \alpha = \frac{600}{AC}$$

The tension **T** is given by

$$\mathbf{T} = -\,T \cos \alpha \mathbf{I} - T \sin \mathbf{j}$$

and the position vector from point *B* as

$$\mathbf{r} = 320\mathbf{i} + 600\mathbf{j}$$

The moment about point *B* is

$$\mathbf{M}_B = \mathbf{r} \times \mathbf{T} = (320\mathbf{i} + 600\mathbf{j}) \times (-\cos \alpha\mathbf{i} - \sin \alpha\mathbf{j})\ T$$

$$= \mathrm{T}\ (600 \cos \alpha - 320 \sin \alpha)\mathbf{k}$$

The magnitude of the moment is

$$M_B = \frac{600Tc}{\sqrt{c^2 + 640c + 462400}}$$

$$M_B = \frac{T}{AC}\ [600(320 + c) - 320(600)]$$

$$M_B = \frac{600Tc}{\sqrt{(600)^2 + (320+c)^2}}$$

or

$$M_B = \frac{600Tc}{\sqrt{c^2 + 640c + 4624000}}$$

MATLAB Program

```
c=[320:1:960];
Mb1=600*60*c./(sqrt(c.^2+640*c+462400));
Mb2=600*80*c./(sqrt(c.^2+640*c+462400));
Mb3=600*110*c./(sqrt(c.^2+640*c+462400));
%
plot(c,Mb1,c,Mb2,c,Mb3)
legend('T=60 N','T=80 N','T=110 N',2)
%
xlabel('c(mm)')
ylabel('Moment(N-mm)')
grid on
```

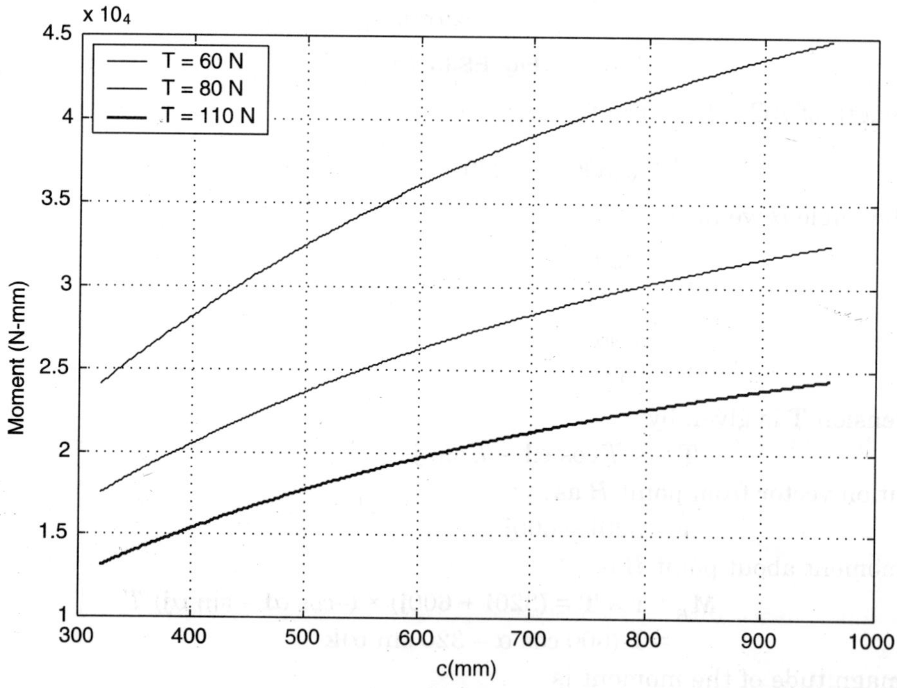

Fig. ES3.7 (*b*)

ES 3.8. *Fig. ES 3.8 shows a frame in which the structural members support the 5 kN load. The load may be applied at any angle α (– 90° to + 90°). The pins at A and B need to be designed to support the maximum force transmitted to them. Write a MATLAB program to plot the forces at A and B as a function of α and find their maximum values and corresponding angles α.*

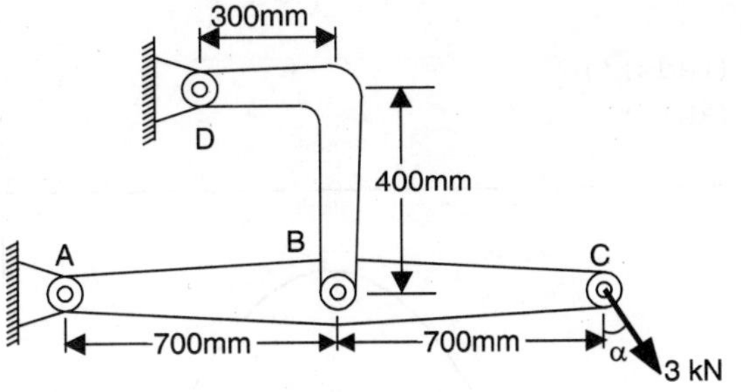

Fig. ES3.8

Solution:

Free body diagram of *ABC* is shown in Fig. ES3.8 (*a*)

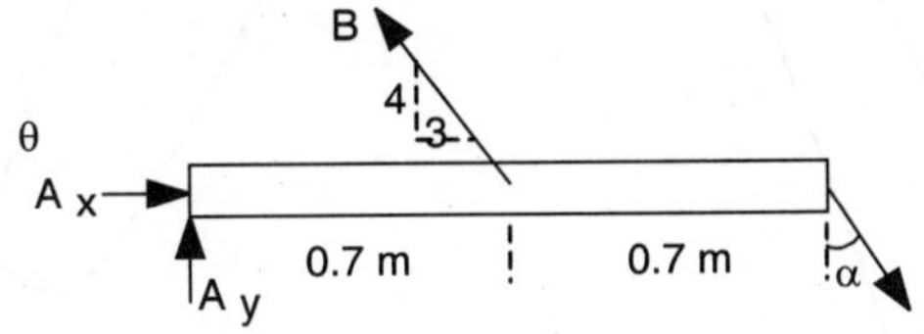

Fig. ES3.8 (*a*)

Note that member *BD* is a two-force member, thus the direction of the force **B** is from *B* to *D*. The equilibrium equations are,

$$\circlearrowleft \Sigma M_A = \frac{4}{5} B(0.6) - 5 \cos \alpha (1.4) = 0$$

$$\Sigma F_x = A_x - \frac{3}{5} B + 5 \sin \alpha = 0; \quad \Sigma F_y = A_y + \frac{4}{5} B - 5 \cos \alpha = 0$$

Solving these equations yields,

$$B = 12.5 \cos \alpha,\ A_x = 0.6\,B - 5 \cos\alpha \quad \text{and} \quad A_y = 5 \cos\alpha - 0.8\,B$$

$$A = \sqrt{A_x^2 + A_y^2}$$

Substitution and simplification yields

$$A = \sqrt{81.25 \cos^2 \alpha + 25 \sin^2 \alpha - 75 \cos \alpha \sin \alpha}$$

Maximum value of *A* is obtained from MATLAB program while maximum value of *B* is 12.5 N at $\alpha = 0$.

The maximum value of *A* and the corresponding angle α will be found in the MATLAB program.

MATLAB program

```
al=-pi/2:0.01:pi/2;
B=12.5*cos(al);
Ax=0.6*B-5*sin(al);
Ay=5*cos(al)-0.8*B;
A=sqrt(Ax.^2+Ay.^2);
plot(al,B,al,A)
```

```
legend('A','B')
xlabel('alpha   (rads)')
ylabel('Force   (kN)')
```

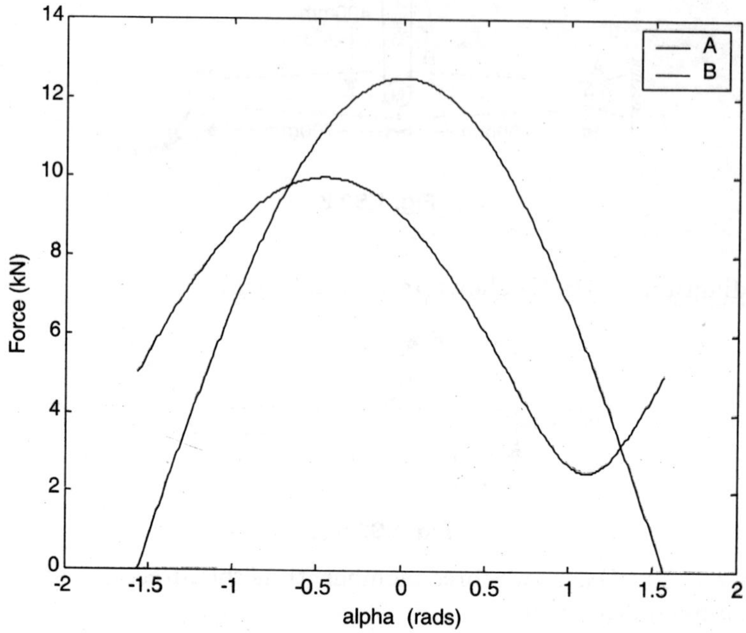

Fig. ES3.8 (*a*)

ES 3.9. *The coefficient of friction, μ, can be determined by μ = F/mg where F is the measured force (N), m is the mass (kg) and g = acceleration due to gravity (9.81 m/s²). The following table gives the experimental data. Determine*

(*a*) *the coefficient of friction in each test*

(*b*) *the average from all tests.*

Test #	*1*	*2*	*3*	*4*	*5*	*6*	*7*
Mass m (kg)	2	4	5	*10*	*20*	*50*	*100*
Force F (N)	*12.4*	*23.2*	*30.5*	*60.8*	*116.5*	*293.8*	*597.3*

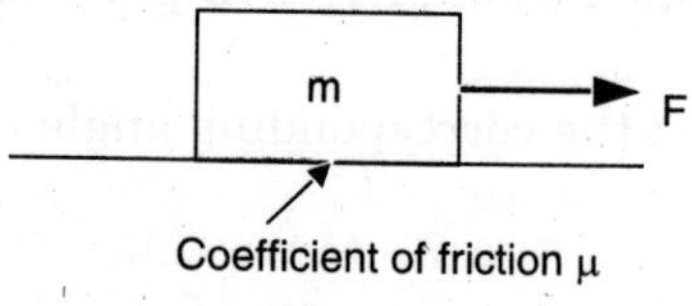

Fig. ES3.9

Solution:

```
>> %Example E3.9
>> % Force and mass values in vector form
>> F=[12.4 23.2 30.5 60.8 116.5 293.8 597.3];
>> m=[2 4 5 10 20 50 100];
```

```
>> % Coefficient of friction, mu
>> mu=F./(m*9.81)
mu =
0.6320    0.5912    0.6218    0.6198    0.5938    0.5990    0.6089
>> averagemu=mean(mu)
averagemu =
0.6095
```

ES 3.10. *Fig. ES3.10 shows a uniform quarter-circular member of mass m lying in the vertical plane and hinged at A and supported against the vertical wall by a small roller at B.*

(a) Derive expressions for the shear force V, the compressive force C, and the bending moment M due to the weight of the member.

(b) Write a MATLAB program to plot the non-dimensional forces and moment (V/mg, C/mg and M/mgR) as a function of the section orientation angle α.

(c) Determine the maximum bending moment and its location angle α.

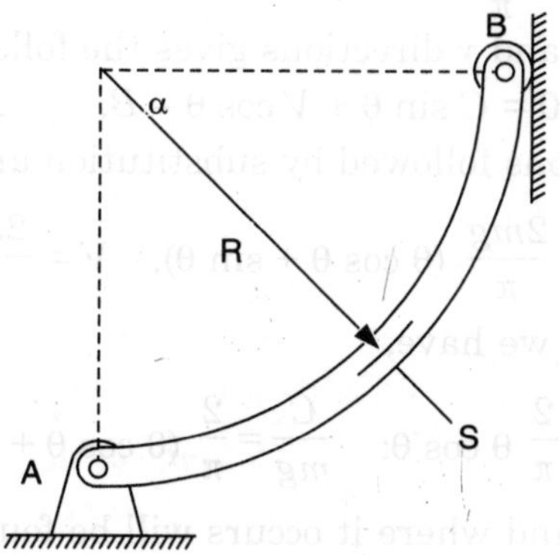

Fig. ES3.10

Solution:

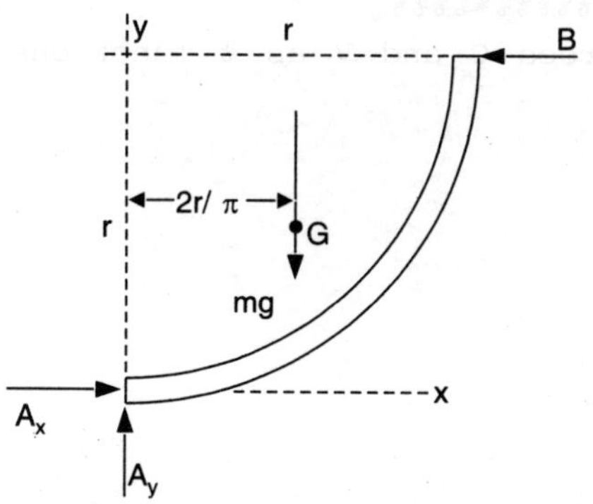

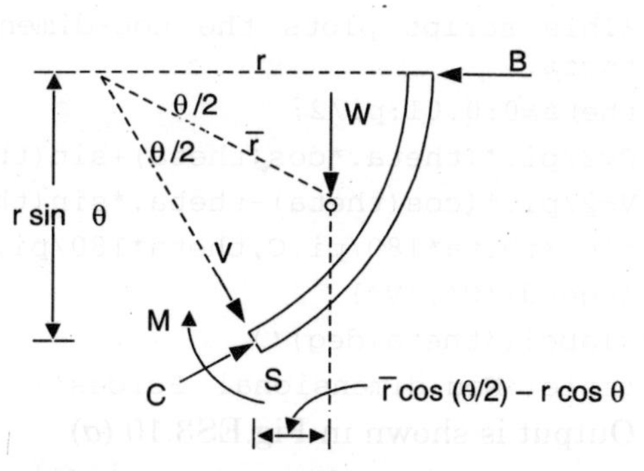

First consider the free-body diagram for the entire member. The centroid for a quarter-circular arc is $2r/\pi$ (see the figure above). All we need from this free-body diagram is the force B, which we can find by summing moments about A.

$$\circlearrowleft \Sigma M_A = 0 = Br - mg\left(\frac{2r}{\pi}\right) \qquad B = \frac{2mg}{k}$$

Now we construct another free-body diagram by cutting through at an arbitrary angle θ as shown to the figure above. From Table D/3, the centroid $\bar{r}$ is,

$$\bar{r} = \frac{r\sin(\theta/2)}{\theta/2} = \frac{2r\sin(\theta/2)}{\theta}$$

The weight W of the section in the diagram will be mg (the total weight of the member) times the ratio of the length of the section ($r\theta$) to the total length of the member ($r\pi/2$). Thus,

$$W = \frac{r\theta}{r\pi/2}\, mg = \frac{2\theta}{\pi}\, mg$$

Now we sum moment about S. For sake of clarity, the moment arms for W and B are shown in the diagram.

$$\Sigma M_S = 0 = M + W(\bar{r}\cos(\theta/2) - r\cos\theta) - Br\sin\theta$$

Substitution and simplification yields,

$$M = \frac{2mgr}{\pi}(\theta\cos\theta + \sin\theta - 2\sin(\theta/2) = \frac{2mgr}{\pi}\theta\cos\theta$$

Summing forces in the x and y directions gives the following two equations.

$$\Sigma F_x = 0 = C\sin\theta + V\cos\theta - B, \qquad \Sigma F_y = 0 = C\cos\theta - V\sin\theta - W$$

Solving these two equations followed by substitution and simplification yields

$$C = \frac{2mg}{\pi}(\theta\cos\theta + \sin\theta), \quad V = \frac{2mg}{\pi}(\cos\theta - \theta\sin\theta)$$

In non-dimensional form we have,

$$\frac{M}{mgr} = \frac{2}{\pi}\theta\cos\theta; \quad \frac{C}{mg} = \frac{2}{\pi}(\theta\cos\theta + \sin\theta); \quad \frac{V}{mg} = \frac{2}{\pi}(\cos\theta - \theta\sin\theta)$$

The maximum moment and where it occurs will be found in the worksheet below. The results are

$$M_{max} = 0.357\ mgr \text{ at } \theta = 0.860 \text{ radians } (49.3^\circ).$$

MATLAB program

```
%%%%%%%%%%%%%%%%%%Script #1%%%%%%%%%%%%%%%%%%%
%This script plots the non-dimensional forces C and V as % functions of
theta.
theta=0:0.01:pi/2;
C=2/pi.*(theta.*cos(theta)+sin(theta));
V=2/pi.*(cos(theta)-theta.*sin(theta));
plot(theta*180/pi,C,theta*180/pi,V)
legend('C','V')
xlabel('theta(deg)')
title('Non-dimensional Forces')
```

Output is shown in Fig.ES3.10 (*a*)

```
% This script plots the non-dimensional bending moment
% as a function of theta
theta=0:0.01:pi/2;
M=2/pi*theta.*cos(theta);
plot(theta*180/pi,M)
xlabel('theta  (deg)')
```

```
Title('Non-dimensional  Bending  Moment')
```

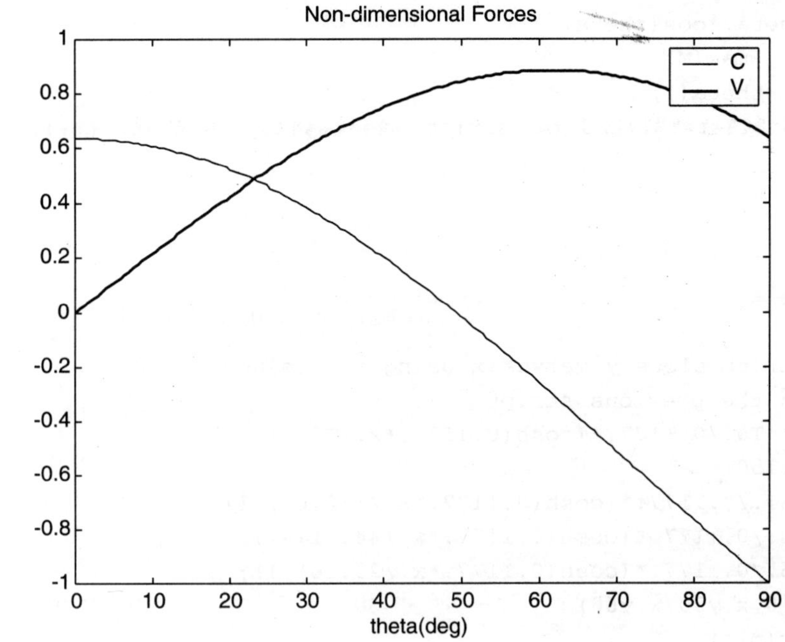

Fig. ES3.10 (*a*)

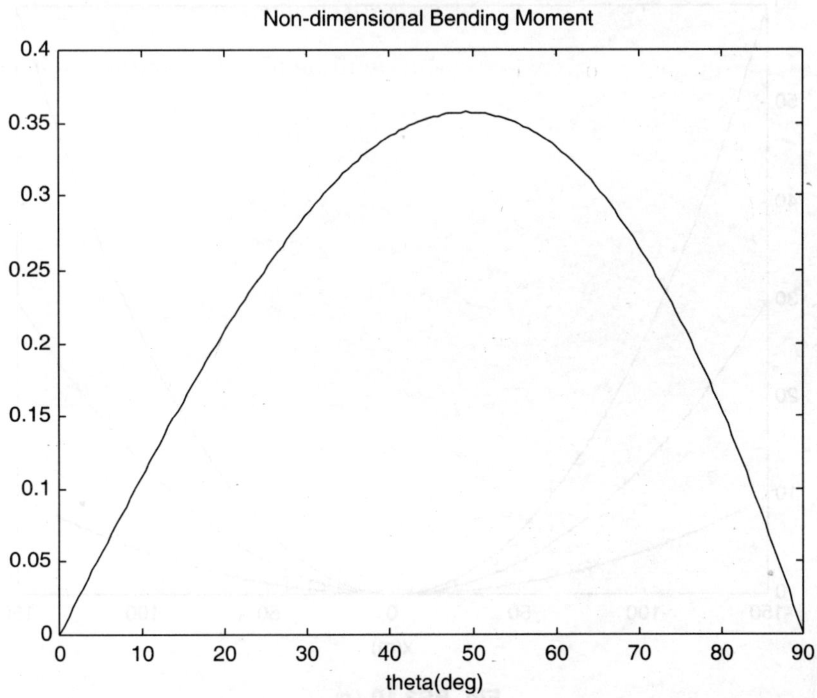

Fig. ES3.10 (*b*)

```
% This program finds the maximum moment
% using MATLAB's max function
theta=0:0.01:pi/2;
M=2/pi*theta.*cos(theta);
[M_max, i]=max(M)
theta_max=theta(i)
%%%%%%%%%%%%%%%%%%%%End of script %%%%%%%%%%%%%%%%%%%%%%%%%%%%%%%
M_max =
0.3572
i =
87
theta_max =
0.8600
% This script plots y versus x using the values of T0
%found in the previous script
y=inline('T0./0.1177.*(cosh(0.1177.*x./T0)-1)');
x=-150:1:150;
y10=132.61./0.1177.*(cosh(0.1177.*x./132.61)-1);
y30=44.71./0.1177.*(cosh(0.1177.*x./44.71)-1);
y60=23.16./0.1177.*(cosh(0.1177.*x./23.16)-1);
plot(x,y10,x,y30,x,y60)
xlabel('x(m)')
ylabel('y(m)')
```

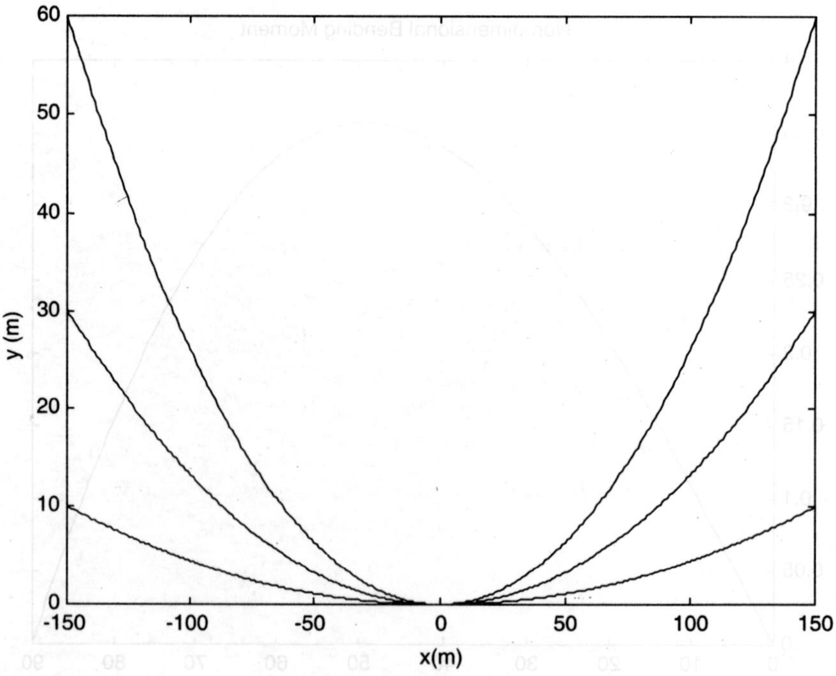

Fig. ES3.10 (*c*)

ES 3.11. *Fig. ES3.11 shows a cylindrical silo where*

H = height of the cylindrical portion

r = radius of the cylindrical silo

R = radius of the spherical cap roof

V = volume of the silo

Write a MATLAB program to compute

(a) the height for given values of r, R, and V

(b) the surface area of the silo, S

Use the program to determine the height and surface area of a silo, given r = 32 ft, R = 50 ft and V = 125,000 ft³.

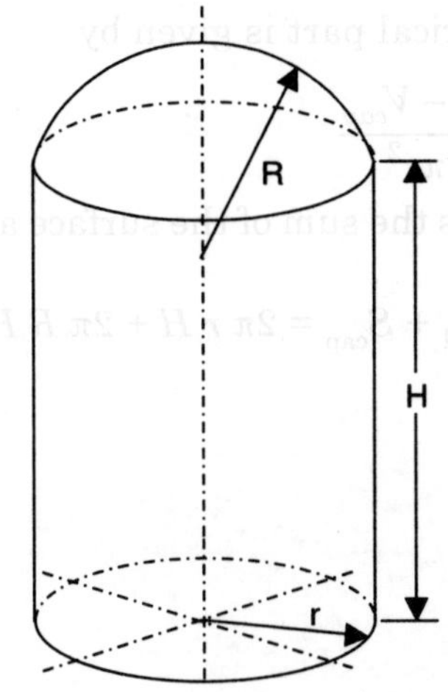

Fig. ES3.11

Solution:

The total volume of the silo is the sum of the volumes of the cylindrical part and the spherical cap.

$$V_{total} = V_{cyl} + V_{cap}$$

$$= \pi r^2 H + \frac{1}{3} rh^2 (3R - h)$$

where $$h = R - R\cos\theta = R(1 - \cos\theta)$$

and $$r = R\sin\theta \quad \text{or} \quad \theta = \sin^{-1}(r/R)$$

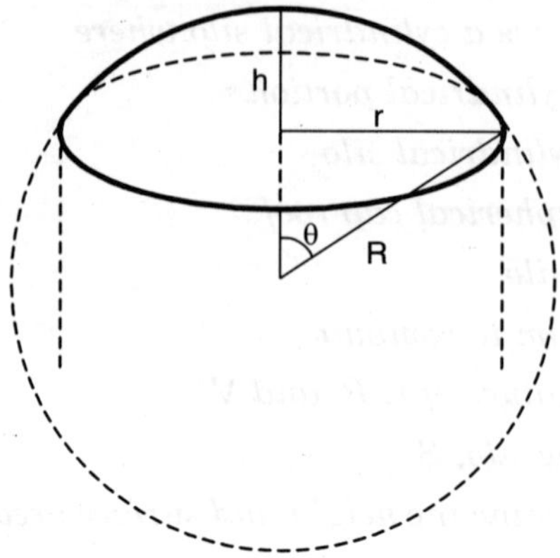

Fig. ES3.11 (*a*)

The height, H, of the cylindrical part is given by

$$H = \frac{V - V_{cap}}{\pi r^2}$$

The surface area of the silo is the sum of the surface areas of the cylindrical part and the spherical cap.

$$S = S_{cyl} + S_{cap} = 2\pi\, r\, H + 2\pi\, R\, H$$

>> %MATLAB Solution:

```
>> r=32;
>> R=50;
>> V=125000;
>> Theta=asin(r/R)
Theta =
    0.6945
>> h = R*(1-cos(Theta))
h =
   11.5813
>> Vcap=pi*h^2*(3*R-h)/3
Vcap =
  1.9442e+004
>> H=(V-Vcap)/(pi*r^2);
>> S=2*pi*(r*H+R*h);
>> fprintf('The height H = %f ft.',H)
The height H = 32.812738 ft.
>> fprintf('The surface area of the silo S = %f square ft.',S)
The surface area of the silo s = 10235.750764 square ft.
```

ES 3.12. *Fig ES3.12 shows a cable loaded uniformly along the horizontal, which has a mass of 15 kg per meter of its own length and supports its own weight only.*

(*a*) *Determine the tension at mid length, the maximum tension and the total length of the cable for H = 10, 20, 30, 40, 50, 60, 70 and 80 meters.*

(*b*) *Write a MATLAB program to plot y as a function of x for the above values of H.*

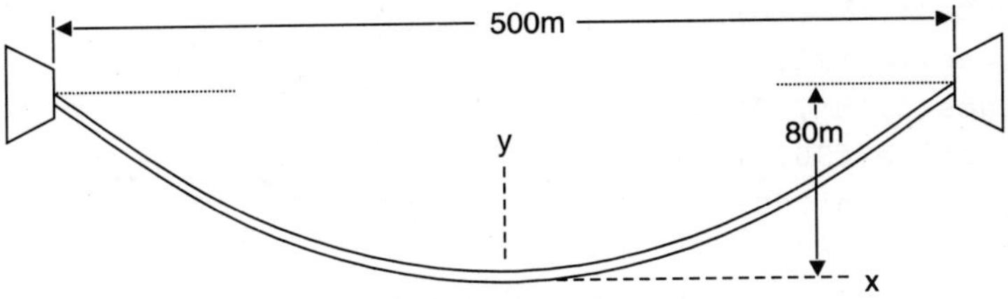

Fig. ES3.12

Solution:

For a uniformly distributed load, we have a catenary shape for the cable. The curve assumed by the cable ($y(x)$) is given by

$$y = \frac{T_0}{\mu}\left(\cosh\frac{\mu x}{T_0} - 1\right)$$

where $\mu = (15)\,(9.81)\,(10^{-3}) = 0.1472$ kN/m. To find the tension at mid length (T_0), we substitute $x = 250$ m and $y = h$, giving

$$h = \frac{T_0}{\mu}\left(\cosh\frac{250\mu}{T_0} - 1\right)$$

To use the solve function in MATLAB, we first need to re-write the above in terms of a function f whose root (zero) provides the solution to the original equation.

$$f = h - \frac{T_0}{0.1472}\left(\cosh\frac{250 \times 0.1472}{T_0} - 1\right)$$

The roots of this equation will be obtained for the three specified values of h. With T_0 known we can easily find the maximum cable tension T_{max} (with $y = h$) and the total length of the cable ($L_c = 2s$) (with $x = 250$ m).

$$T_{max} = T_0 + \mu h$$

$$L_c = 2s = \frac{2T_0}{\mu}\sinh\left(\frac{250\mu}{T_0}\right)$$

The results are:

$h(m)$	$T_0\,(kN)$	$T_{max}\,(kN)$	$L_c\,(m)$
10	132.6	133.8	300.9
30	44.7	48.2	307.9
60	23.2	30.2	329.9

MATLAB program

```
% This script finds the roots (T0) of function f
%(see problem formulation) for h= 10,30,and
% 60 meters. the maximum tension and total cable
%length is then found for the same values of h.
f='h-T0/0.1472*(cosh(0.1472*250/T0)-1)'
```

```
f10=subs(f,'h',10);
T0_10=solve(f10,'T0')
f30=subs(f,'h',30);
T0_30=solve(f30,'T0')
f60=subs(f,'h',60);
T0_60=solve(f60,'T0')
% Now we find the maximum cable tension.
Tm=inline('T0+0.1472*h')
Tmax_10=Tm(T0_10,10)
Tmax_30=Tm(T0_30,30)
Tmax_60=Tm(T0_60,60)
% Next we determine the total length of the cable.
Lc=inline('2*T0/0.1472*(sinh(0.1472*250/T0))')
Lc_10=Lc(T0_10)
Lc_30=Lc(T0_30)
Lc_60=Lc(T0_60)
```

Output is:

```
f =
h-T0/0.1472*(cosh(0.1472*250/T0)-1)
T0_10 =
460.24512430858055194348798961110
T0_30 =
154.06375896035471416316009559463
T0_60 =
78.095825828643598721536606035324
Tm =
Inline function:
Tm(T0,h) = T0+0.1472*h
Tmax_10 =
461.71712430858055194348798961110
Tmax_30 =
158.47975896035471416316009559463
Tmax_60 =
86.927825828643598721536606035324
Lc =
Inline function:
Lc(T0) = 2*T0/0.1472*(sinh(0.1472*250/T0))
Lc_10 =
500.53293571245060777586927743822
Lc_30 =
504.76817584749182092133632282053
Lc_60 =
518.71022894786080143160489959183
```

ES 3.13. *Fig. ES3.13 shows three flat blocks positioned on an incline oriented at angle α.*

(a) Write a MATLAB program to plot the maximum value of P (if no slipping occurs) versus α. Assume only positive values of P and indicate the regions over which

(i) the 40 kg block slides alone,

(ii) the 40 kg and 30 kg blocks slide together.

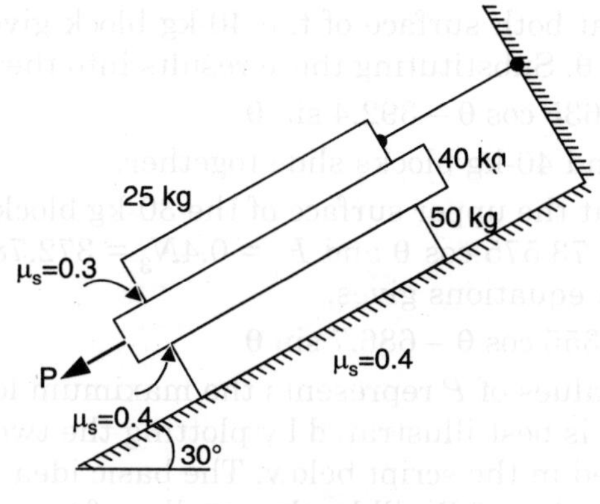

Fig. ES3.13

Solution:

The free-body diagrams for the three blocks are shown in figure ES3.13 (a) below. To obtain the required plot it will be convenient to use a slightly different approach than that used in the sample problem in your text. We start by writing down the equilibrium equations without making any assumptions about where sliding occurs.

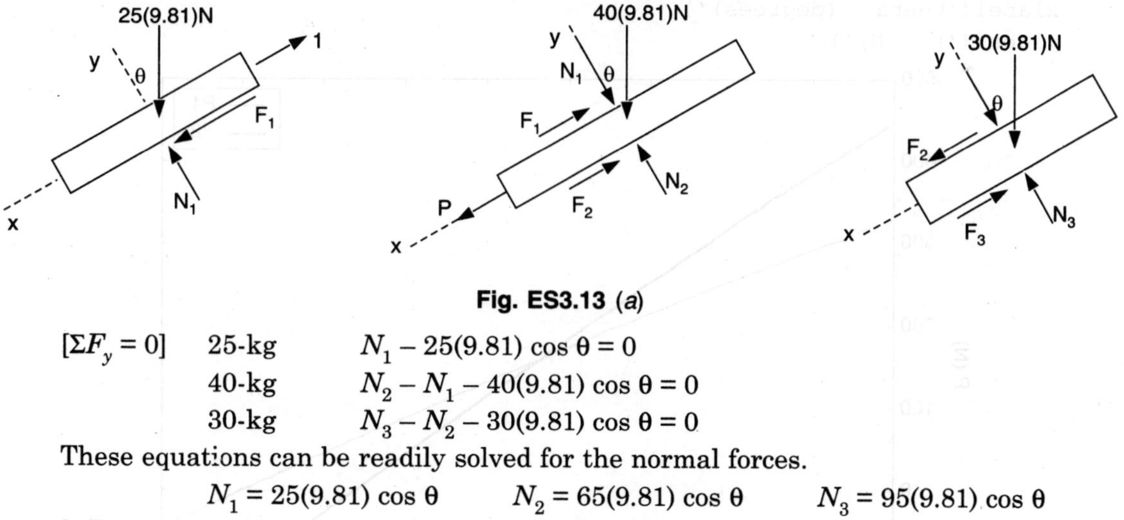

Fig. ES3.13 (*a*)

$[\Sigma F_y = 0]$ 25-kg $N_1 - 25(9.81)\cos\theta = 0$

40-kg $N_2 - N_1 - 40(9.81)\cos\theta = 0$

30-kg $N_3 - N_2 - 30(9.81)\cos\theta = 0$

These equations can be readily solved for the normal forces.

$N_1 = 25(9.81)\cos\theta \qquad N_2 = 65(9.81)\cos\theta \qquad N_3 = 95(9.81)\cos\theta$

$[\Sigma F_x = 0]$ 40-kg $P - F_1 - F_2 + 40(9.81)\sin\theta = 0$

30-kg $F_2 - F_3 + 30(9.81)\sin\theta = 0$

Now we have two equations with four unknowns, P, F_1, F_2 and F_3. Note that we have not written the equation for the summation of forces in the x-direction for the 25-kg block. The reason is that this equation introduces an additional unknown (T) that we are not interested in determining.

The next step is to make assumptions about which block(s) slide. As will be seen, either of the two possible assumptions about impending motion will reduce two of the friction forces to functions of θ only. This will result in two equations that may be solved for P and the remaining friction force. The forces calculated will be designated P_1 or P_2 to distinguish the two cases for impending slip.

Case-1: Only the 40-kg block slips.

Impending slippage at both surface of the 40-kg block gives $F_1 = 0.3N_1 = 73.575 \cos\theta$ and $F_2 = 0.4N_2 = 255.06 \cos\theta$. Substituting these results into the equilibrium equations yields

$$P_1 = 328.635 \cos\theta - 392.4 \sin\theta$$

Case-2: The 30 and 40-kg blocks slide together.

Impending slippage at the upper surface of the 30-kg block and lower surface of the 40-kg block gives $F_1 = 0.3N_1 = 73.575 \cos\theta$ and $F_3 = 0.4N_3 = 372.78 \cos\theta$. Substitution of these results into the equilibrium equations gives,

$$P_2 = 446.355 \cos\theta - 686.7 \sin\theta$$

Which of these two values of P represents the maximum load that can be applied without slippage on any surface is best illustrated by plotting the two expressions as a function of θ. This plot will be generated in the script below. The basic idea is that at any specified angle θ, the critical or maximum value of P will be the smaller of two values calculated.

MATLAB Program:

```
theta=0:0.01:pi/4;
P1=328.635*cos(theta)-392.4*sin(theta);
P2=446.335*cos(theta)-686.7*sin(theta);
plot(theta*180/pi, P1, theta*180/pi, P2)
Legend('P1','P2')
xlabel('theta  (degrees)')
ylabel('P  (N)')
```

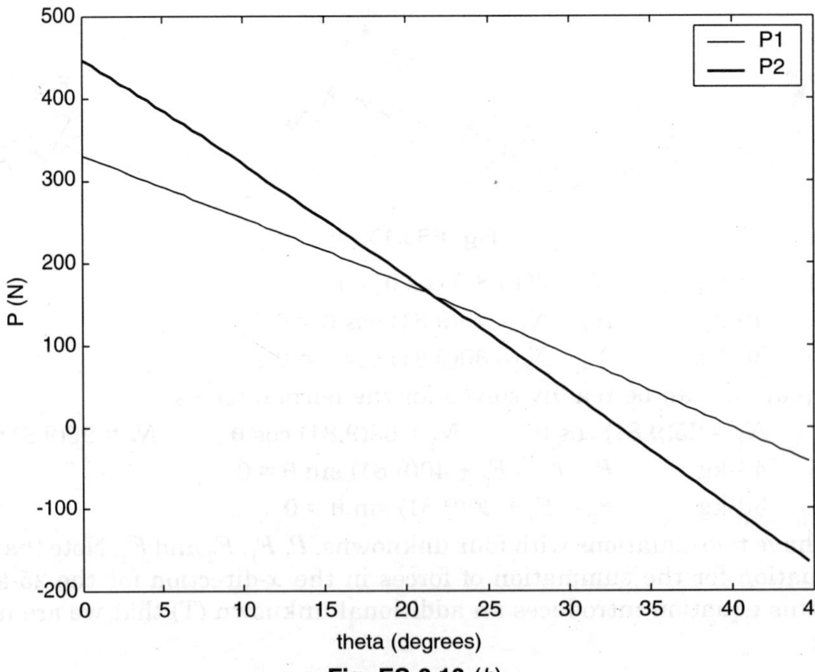

Fig. ES 3.13 (*b*)

Fig. 3.13 (*b*) shows P_1 and P_2 plotted as a function of θ. For each θ, the critical or maximum value of P will be the smaller of two values calculated. By setting $P_1 = P_2$ we find that the two curves intersect at $\theta = 0.503$ rads (28.8°). Thus, for $\theta \le 28.8°$ P controls and the 50-kg block slides by itself while for $\theta \le 28.8°$ P_2 controls and the 40 and 50-kg blocks slide together.

ES 3.14. *Fig. ES3.14 shows a slender rod AB of weight W attached to blocks at A and B which move freely in the guides.*

(a) Derive a relationship between α, W, L and k that need to be satisfied for the rod to be in equilibrium. The spring constant is k, W = 20 lb and L = 50 in.

(b) Write a MATLAB program to compute and plot k as a function of α for 15° ≤ α ≤ 40°.

(c) Determine the two values of α corresponding to equilibrium when k = 0.8 lb/in.

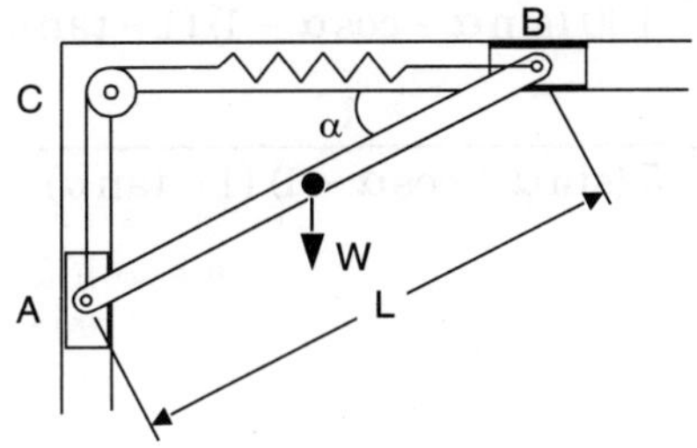

Fig. ES3.14

Solution:

Refer to the free-body diagram shown in Fig.ES3.14 (*a*).

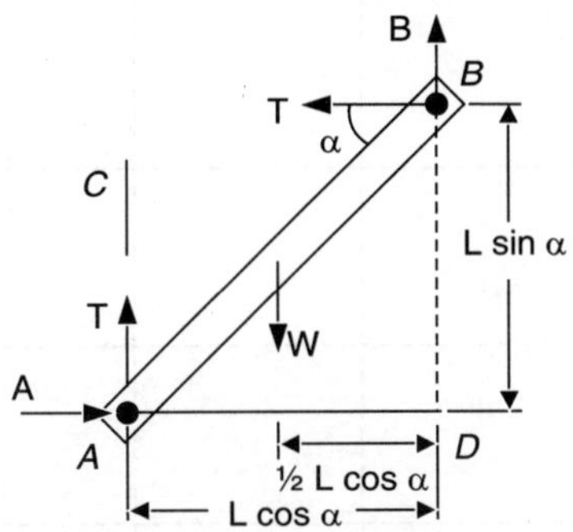

Fig. ES3.14 (*a*)

If x equals the elongation of the spring, then

$$x = AC + BC - L$$
$$= L(\sin\alpha + \cos\alpha - 1)$$

The tension in the spring is given by

$$T = kx = kL(\sin\alpha + \cos\alpha - 1)$$

The moment about point D gives

$$\Sigma M_D = 0: \qquad T(L\sin\alpha) - T(L\cos\alpha) + W\left(\frac{1}{2}L\cos\alpha\right) = 0$$

$$T(\sin\alpha - \cos\alpha) + \frac{W}{2}\cos\alpha = 0$$

Substituting for T gives

$$kL(\sin\alpha + \cos\alpha - 1)\,(\sin\alpha - \cos\alpha) + \frac{W}{2}\cos\alpha = 0$$

$$(\sin\alpha + \cos\alpha - 1)\,(\tan\alpha - 1) + \frac{W}{2}\cos\alpha = 0$$

and
$$\frac{W}{2kL} = (\sin\alpha + \cos\alpha - 1)\,(1 - \tan\alpha)$$

For $W = 20$ lb and $L = 50$ in.

$$k = \frac{20}{100\,(\sin\alpha + \cos\alpha - 1)\,(1 - \tan\alpha)}$$

$$= \frac{1}{5\,(\sin\alpha + \cos\alpha - 1)\,(1 - \tan\alpha)}$$

MATLAB program

```
%
syms pt real
W=20;
L=50;
K=0.8;
rhs=W/(2*K*L);
```

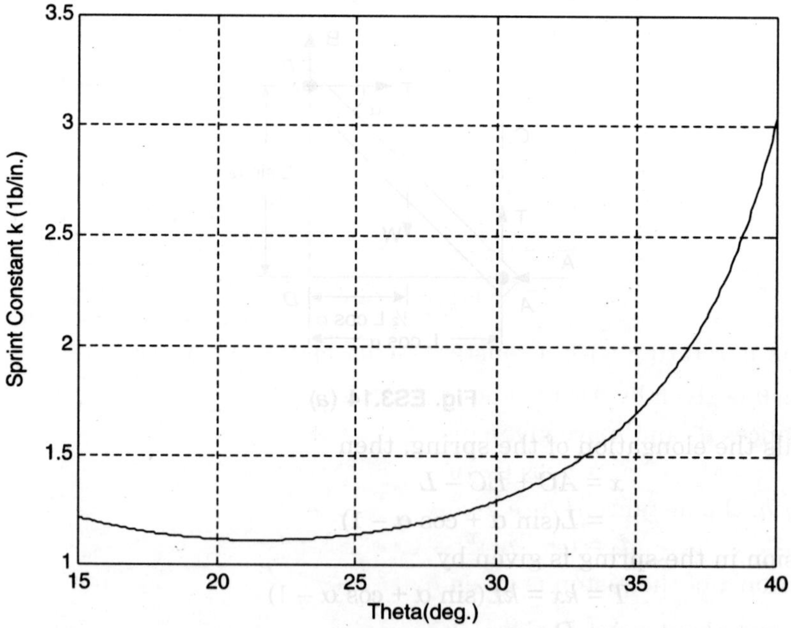

Fig. ES3.14 (*b*)

```
Tha=[15:.1:40];
t=Tha*pi/180;
k=1./(5*(sin(t)+cos(t)-1).*(1-tan(t)));
```

```
plot(Tha,k)
xlabel('Theta(deg.)')
ylabel('Spring Constant k (lb/in.)')
grid on
eq1=(sin(pt)+cos(pt)-1)*(1-tan(pt))-rhs;
[pt]=solve(eq1);
Ang=real(double(pt));
Theta=Ang*180/pi;
fprintf('The calculated angles for Theta are:%8.2fdegrees\n',Theta)
```

The calculated angles for Theta are: – 132.18 degrees

The calculated angles for Theta are: 21.09 degrees

The calculated angles for Theta are: 21.09 degrees

ES 3.15. *(Slider crank mechanism). Fig.ES3.15 shows the slider crank mechanism. Write a MATLAB program that calculates and plots the position, velocity and acceleration of the piston for one full revolution of the crank. Assume that the crank is rotating at a constant speed of 550 rpm. Given*

radius of crank = 125 mm

radius of crank shaft = 250 mm.

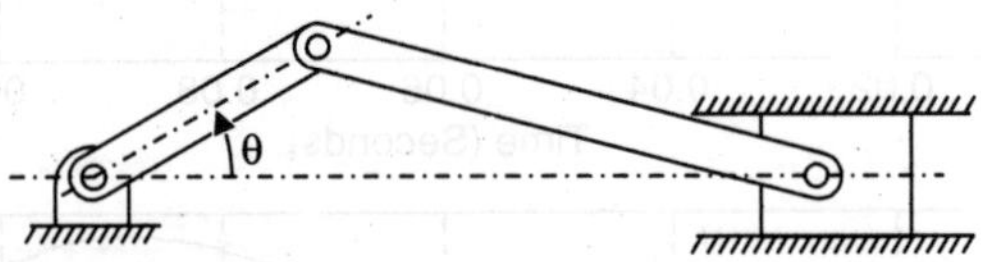

Fig. ES3.15

Solution:

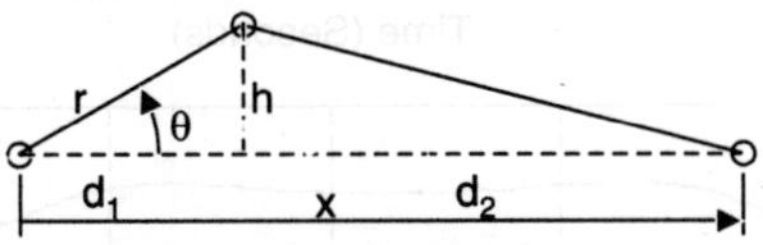

Fig. ES.3.15 (a)

The crank is rotating with a constant angular velocity $\dot{\theta}$. If $\theta = 0°$ when $t = 0$, then at time t the angle θ is given by $\theta = \dot{\theta}t$, and that $\ddot{\theta} = 0$ at all times.

The distances d_1 and h are given by

$$d_1 = r\cos\theta \quad \text{and} \quad h = r\sin\theta$$

Knowing h, the distance d_2 are given by

$$d_2 = (c^2 - h^2)^{1/2} = (c^2 - r^2\sin^2\theta)^{1/2}$$

The position x of the piston is given by

$$x = d_1 + d_2 = r\cos\theta + (c^2 - r^2\sin^2\theta)^{1/2}$$

The velocity of the piston is given by

$$\dot{x} = -r\dot{\theta}\sin\theta - \frac{r^2}{2(c^2 - r^2\sin^2\theta)^{1/2}}$$

The acceleration of the piston is given by

$$\ddot{x} = -r\dot{\theta}^2\cos\theta - \frac{4r^2\dot{\theta}^2\cos 2\theta(c^2 - r^2\sin^2\theta) + (r^2\,\dot{\theta}\sin 2\theta)^2}{4(c^2 - r^2\sin\theta)^{3/2}}$$

>> % MATLAB Solution:

```
>> % Define TD, r, and c
>> TD=550*2*pi/60; % Radians/sec
>> tf=2*pi/TD;
>> r=0.125; % meters
>> c=0.250; % meters
>> t=linspace(0,tf,200); % Create a vector with 200 elements
>> TH=TD*t; % Compute Theta for each t
>> d2s=c^2-r^2*sin(TH).^2; % d2 squared
>> x=r*cos(TH)+sqrt(d2s); % Calculate x for each Theta
>> xd=-r*TD*sin(TH)-(r^2*TD*sin(2*TH))./(2*sqrt(d2s)); % Velocity
```

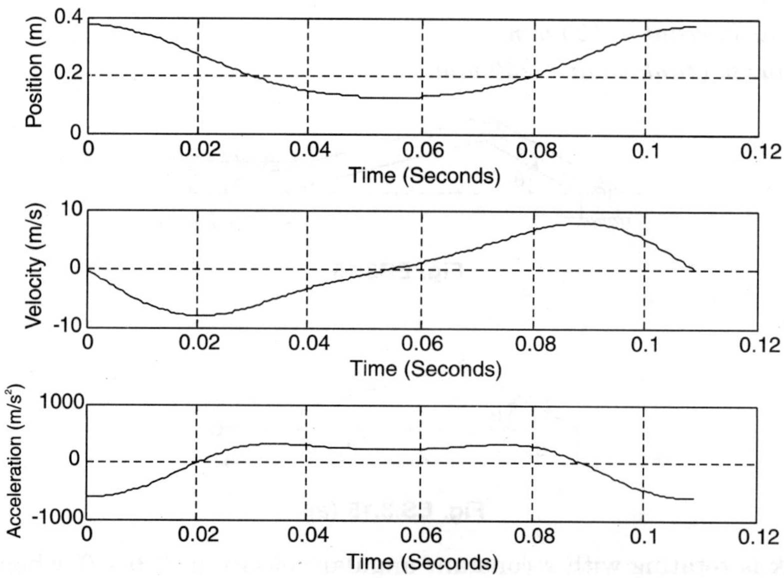

Fig. ES3.15 *(b)*

```
% Acceleration
>>xdd=-r*TD^2*cos(TH)-(4*r^2*TD^2*cos(2*TH).*d2s+(r^2*sin(2*TH)*TD).^2)./
(4*d2s.^(3/2));
>> subplot(3,1,1)
>> plot(t, x) % Plot x versus t
>> grid
>> xlabel('Time (Seconds)')
>> ylabel(' Position (m)')
>> subplot(3,1,2)
>> plot(t, xd) % Plot Velocity vs. t
>> grid
```

```
>> xlabel(' Time (Seconds)')
>> ylabel('Velocity (m/s)')
>> subplot(3,1,3)
>> plot(t, xdd) % Plot Acceleration Vs. t
>> grid
>> xlabel('Time (Seconds)')
>> ylabel('Acceleration (m/s^2)')
```

ES 3.16. *In Fig.ES3.16, the horizontal position of the rectangular block is adjusted by the wedge under the action of the force P. The wedge angle is α. μ_1 and μ_2 are the coefficient of the static friction at the two wedge surfaces and between the block and the horizontal surfaces respectively.*

(a) Obtain a general expression for P (the least force required to move the block) in terms of α, μ_1 and μ_2

(b) Write a MATLAB program to plot P as a function of μ_1 for α = 15, 20, and 25°. μ_2 = 0.5

(c) For α = 10°, plot P as a function of μ_1 for μ_2 = 0.2, 0.4, 0.6 and 0.8.

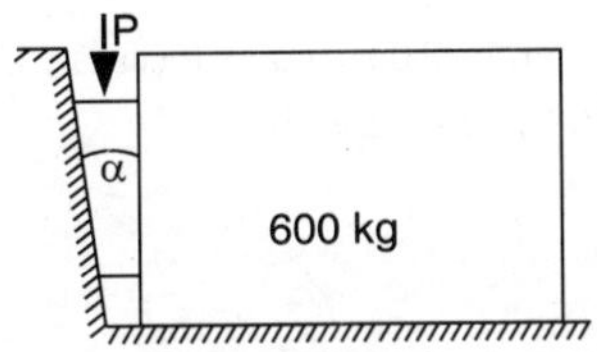

Fig. ES3.16

Solution:

Free body diagram of wedge and block is shown in Fig. ES3.16 (*a*)

Fig. ES3.16 (a)

(*a*) First we write the equilibrium equations from the free-body diagrams for the wedge and for the block.

For the Block

$$\Sigma F_x = 0 = N_2 - \mu_2 N_3 \qquad \Sigma F_y = 0 = N_3 - mg - \mu_1 N_2$$

These two equations are readily solved for N_2 and N_3.

$$N = \frac{\mu_2 mg}{1-\mu_1\mu_2} \qquad N_3 = \frac{mg}{1-\mu_1\mu_2}$$

For the Wedge

$$\Sigma F_x = 0 = N_1 \cos\alpha - \mu_1 N_1 \sin\alpha - N_2$$

$$\Sigma F_y = 0 = N_1 \sin\alpha + \mu_1 N_1 \cos\alpha + \mu_1 N_2 - P$$

After substituting for N_2 we solve the first equation for N_1. Substituting this result into the second yields an expression for P.

$$P = \frac{\mu_2 mg\,(2\mu_1 + (1-\mu_1^2)\,(\tan\alpha)}{(1-\mu_1 \tan\alpha)\,(1-\mu_1\mu_2)}$$

This result is used to generate the plots for parts (*b*) and (*c*) I the worksheet below.

MATLAB program:

```
% This script produces the plot for part(b)
% Note the conversion factor on theta in the following
% function. This allows us to use degrees rather than
% radians when we use P. Also note the division by 1000,
% converting tp kN.
P=inline('.5*600*9.81*(tan(th*pi/180)*(1-mu1^2)+2*mu1)/(1-mu1*tan(th*pi/
180))/(1-mu1*.6)/1000');
P=vectorize(P);
mu1=0:0.01:0.8;
plot(mu1,P(mu1,15),mu1,P(mu1,20),mu1,P(mu1,25))
legend('mu_1=15','mu_1=20','mu_1=25')
xlabel('mu_1')
ylabel('P(kN)')
```

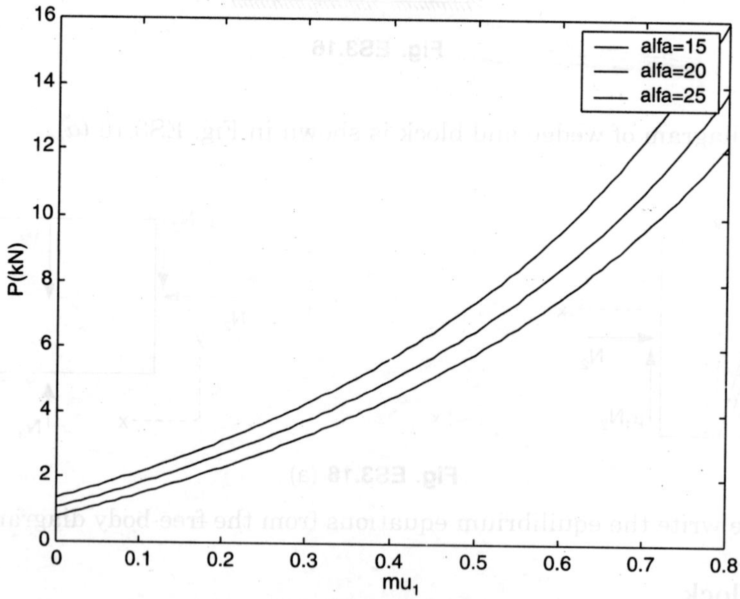

Fig. ES 3.16 (*b*)

```
% This script produces the plot for part(c)
% Note that P is a function of mu1 and mu2in this case.
% Also note that the division by 1000, converting to kN.
P=inline('mu2*600*9.81*(tan(10*pi/180)*(1-mu1^2)+2*mu1)/(1-mu1*tan(10*pi/
180))/(1-mu1*mu2)/1000');
P=vectorize(P);
```

```
mu1=0:0.01:0.8;
plot(mu1,P(mu1,.2),mu1,P(mu1,.4),mu1,P(mu1,.6),mu1,P(mu1,.8))
legend('mu_2=0.2','mu_2=0.4','mu_2=0.6','mu_2=0.8')
xlabel('mu_1')
ylabel('P(kN)')
```

The output is shown in Fig. ES 3.16 (*c*).

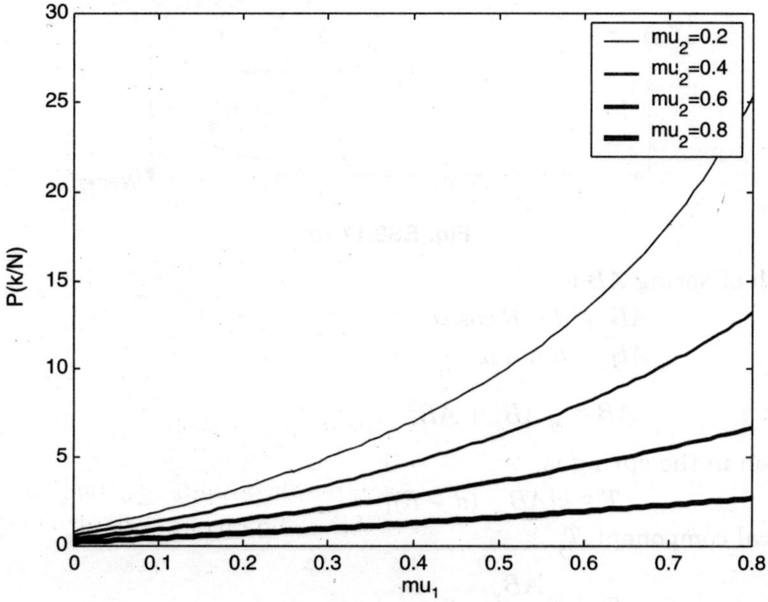

Fig. ES 3.16 (*c*)

ES 3.17. *In Fig. ES3.17, the spring is unstretched when $\alpha = 0$ and k is the spring constant. Write a MATLAB program to compute and plot the mass m corresponding to equilibrium as a function of α for values of α from 0 to 90°. Find the value of α corresponding to equilibrium m = 2.5 kg. Given R = 210 mm, d = 50 mm and k = 1.2 kN/m.*

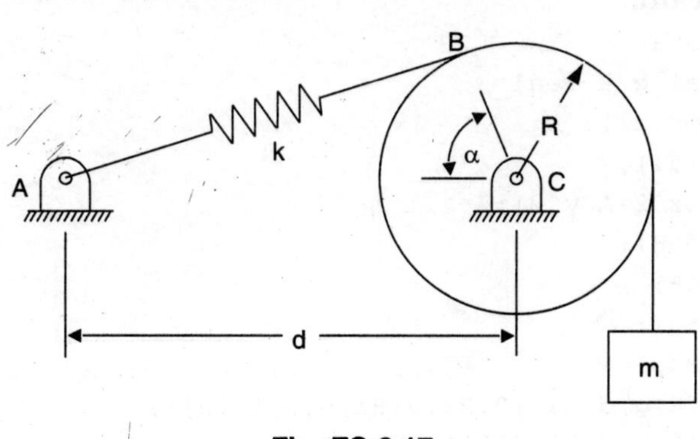

Fig. ES 3.17

Solution:

The free body diagram is shown in Fig. ES3.17 (*a*)

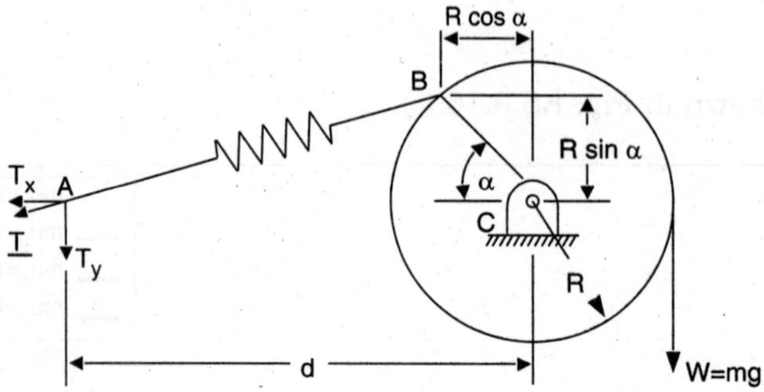

Fig. ES3.17 (*a*)

The length of spring *AB* is

$$AB_x = d - R\cos\alpha$$
$$AB_y = R\sin\alpha$$
$$AB = \sqrt{AB_x^2 + AB_y^2}$$

The tension in the spring is

$$T = k[AB - (d - R)]$$

The vertical component T_y

$$T_y = T\frac{AB_y}{AB}$$

Summing moments about point *C* gives

$$\Sigma M_C = 0: \quad T_y d - mg\,R = 0 \quad m = T_y\frac{d}{Rg}$$

MATLAB program:

```
% Input Data
syms d R al g k real
ABx=d-R*cos(al);
ABy=R*sin(al);
AB=sqrt(ABx^2+ABy^2);
T=k*(AB-(d-R));
Ty=T*ABy/AB;
m=Ty*d/(R*g);
u=Ty*d-2*R*g;
u=subs(u,{R,g,d,k},{0.21,9.81,0.5,1200});
m=subs(m,{R,g,d,k},{0.21,9.81,0.5,1200});
th=solve(u,al);
Angle=th*180/pi;
Alpha=double(Angle)
td=[0:5:90];
ang=td*pi/180;
```

```
yd=subs(m,al,ang);
%
plot(td,yd)
xlabel('Alpha(deg.)')
ylabel('Mass m (kg)')
grid on
```

Output is:

```
Alpha =
1.0e + 002*
- 0.0957 + 0.1850i
1.7683
- 0.0957 - 0.1850i
0.2271
```

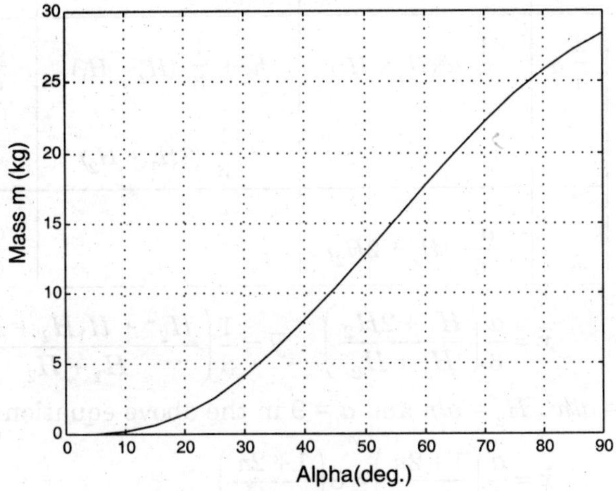

Fig. ES3.17 (*b*)

ES 3.18. *Fig. ES3.18 shows a trapezoid.*

(a) Obtain the x and y coordinates of the centroid when $H_1 = a/n^2$ *and* $H_2 = a/n$.

(b) Write a MATLAB program to plot the values of $\bar{x}$ *and* $\bar{y}$ *for* $1 \le n \le 5$ *and* $a = 9$ *inches.*

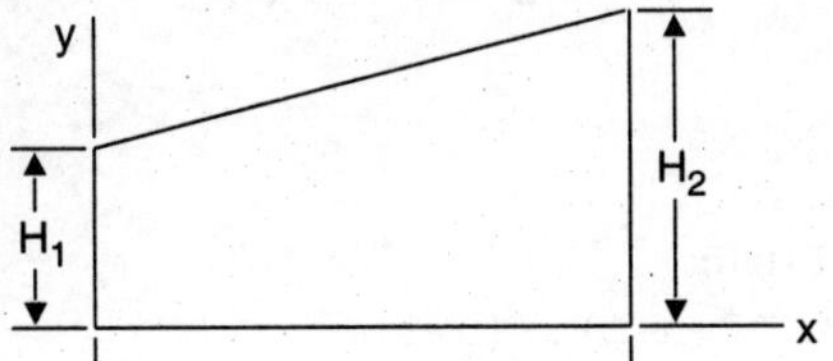

Fig. ES3.18

Solution:

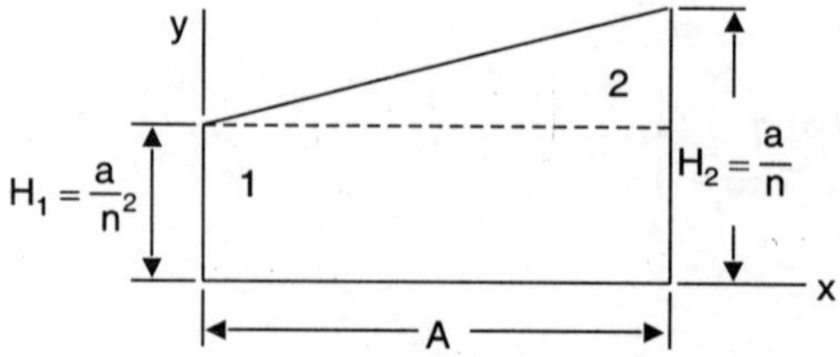

Fig. ES3.18 (a)

	Area	$\bar{x}$	$\overline{Ax}$	$\bar{y}$	$\overline{Ay}$
1.	aH_1	$\frac{a}{2}$	$\frac{1}{2}\ a^2H_1$	$\frac{1}{2}\ H_1$	$\frac{1}{2}\ aH_1^{\ 2}$
2.	$\frac{1}{2}\ a\ (H_2 - H_1)$	$\frac{2}{3}\ a$	$\frac{1}{3}\ a^2(H_2 - H_1)$	$h_1 + \frac{1}{3}\ (H_2 - H_1)$ $= \frac{1}{3}\ (2H_1 + H_2)$	$\frac{a}{6}\ (H_2^{\ 2} + H_1H_2 - 2H_1^{\ 2})$
Σ	$\frac{a}{2}\ (H_1 + H_2)$		$\frac{a^2}{6}\ (H_1 + 2H_2)$		$\frac{a}{6}\ (H_2^{\ 2} + H_1H_2 - 2H_1^{\ 2})$

Hence $$\bar{x} = \frac{a}{3}\left(\frac{H_1 + 2H_2}{H_1 + H_2}\right) \quad \bar{y} = \frac{1}{3}\left(\frac{H_2^{\ 2} + H_1H_2 + H_1^{\ 2}}{H_1 + H_2}\right)$$

Substituting $H_1 = a/n^2$, $H_2 = a/n$ and $a = 9$ in the above equations yields

$$\bar{x} = \frac{a}{3}\left(\frac{1+2n}{1+n}\right) = 3\left(\frac{1+2n}{1+n}\right)$$

$$\bar{y} = \frac{a}{3n^2}\left(\frac{n^2+n+1}{1+n}\right) = \frac{3}{n^2\left(\frac{n^2+n+1}{1+n}\right)}$$

The plot for $1 \le n \le 5$ is shown in Fig. ES3.18 (*b*).

MATLAB Solution:

```
>> % Example ES3.18
>> n=[1:.1:5];
>> xbar=3*(1+2*n)./(1+n);
>> ybar=(3./n.^2).*(n.^2+n+1)./(n+1);
>> xlabel('n')
>> ylabel('Centroid (in.)')
>> grid on
>> plot(n,xbar,ybar)
>> text(3,5,'xbar')
>> text(3,1.4,'ybar')
```

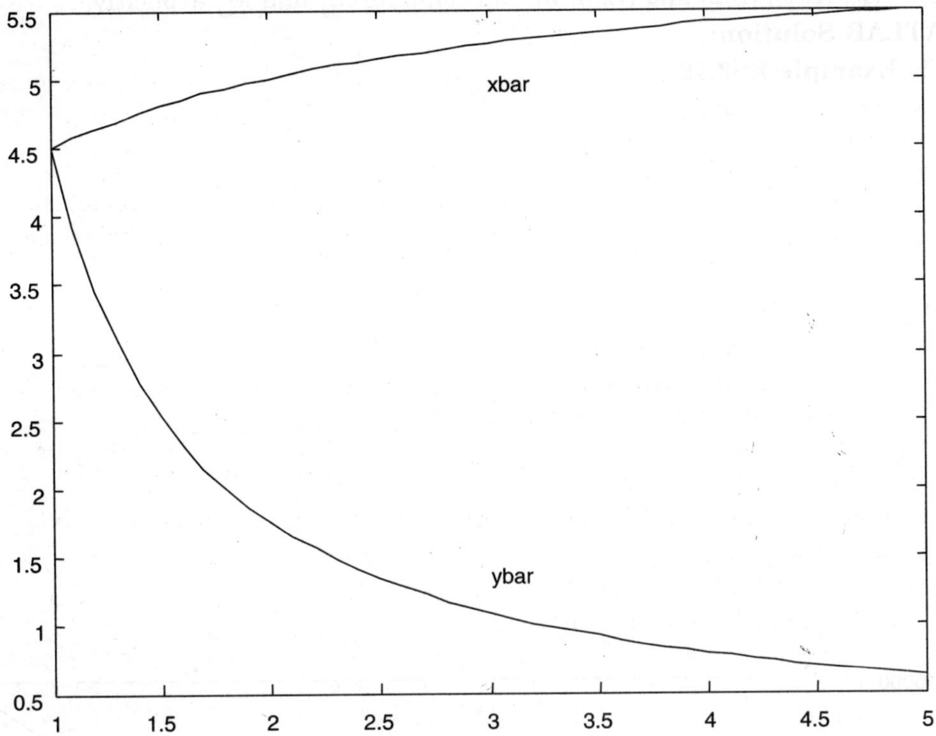

Fig. ES3.18 *(b)*

ES 3.19. *Fig. ES3.19 shows a large turnbuckle which supports a cable tension of 12,000 lb. The mean diameter of the two 1.25 in. screws is 1.15 in. and has five square threads per inch. Both screws have single threads.*

(a) Determine the moments M_T and M_L that must be applied to the body of the turnbuckle in order to tighten and loosen it respectively.

(b) Write a MATLAB program to plot the moments MT and ML as functions of μ for $0 \leq \mu \leq 1$, where μ is the coefficient of friction for the threads.

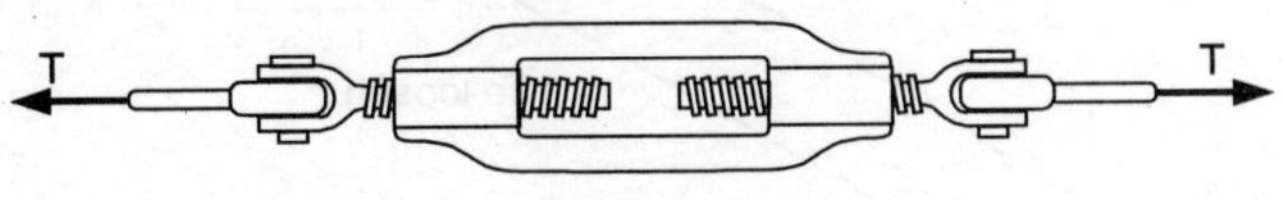

Fig. ES3.19

Solution:

M_T and M_L can be obtained as

$$M_T = 2Tr \tan(\phi + \alpha) \qquad \text{(E.1)}$$

and

$$M_L = 2T\, r \tan(\phi + \alpha) \qquad \text{(E.2)}$$

where $T = 12{,}000$ lb and the lead $L = 1/5$ in/rev.

The mean radius is given by $r = 1.15/2 = 0.575$ in.

We also have

$$\alpha = \tan^{-1}\left(\frac{L}{2\pi r}\right) \quad \text{and} \quad \phi = \tan^{-1}(\mu) \qquad \text{(E.3)}$$

From Eqs. (E.1), (E.2) and (E.3), we can compute M_L and M_T explicitly.

MATLAB Solution:

>> % Example ES3.19

```
>> % Input T,L, and r
>> T=12000;
>> L=1/5;
>> r=1.15/2;
>> alpha = atan(L/2/pi/r);
>> mu = 0:0.01:1;
>> phi = atan(mu);
>> MT = 2*T*r*tan(alpha + phi);
>> ML=2*T*r*tan(phi-alpha);
>> % Place a horizontal line at x = 0
>> M=0*mu;
>> plot(mu,MT,mu,ML,mu,M)
>> xlabel('Coefficient of friction')
>> ylabel('Moment (lb-in)')
>> text(0.5,5000,'To loosen')
>> text(0.3,8000,'To tighten')
```

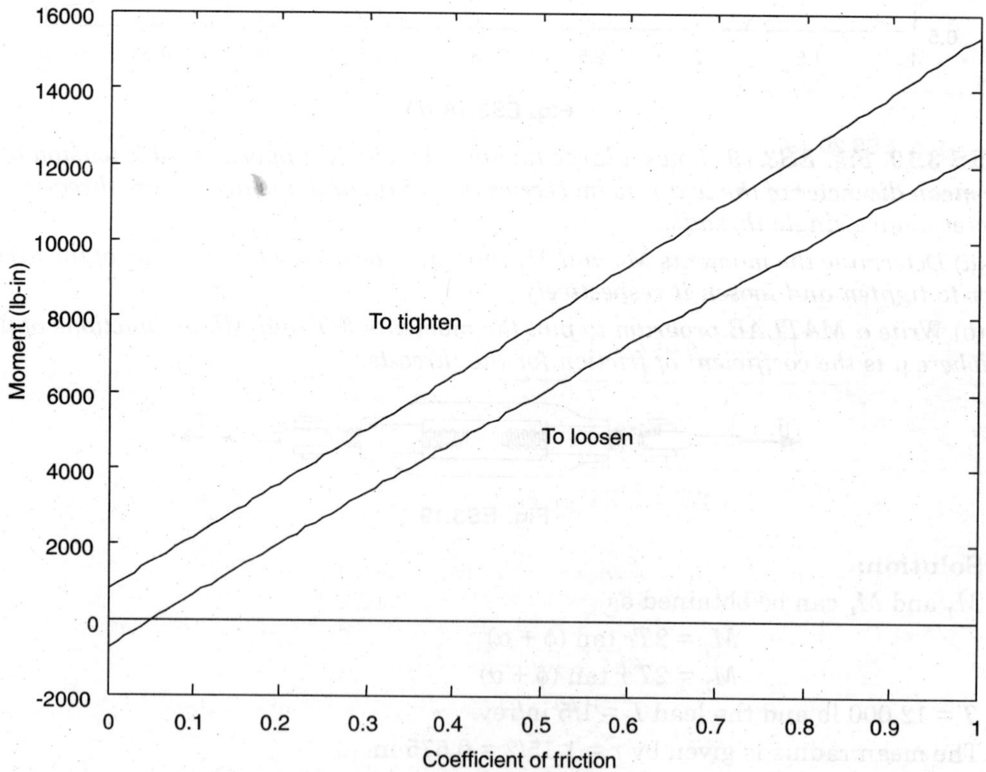

Fig. ES3.19 *(a)*

The plot of moments *MT* and *ML* as functions of μ for $0 \le \mu \le 1$ is shown in Fig. ES3.19 (*a*).

ES 3.20. *Determine the y coordinate of the centroid of the body shown in Fig. ES3.20 when $h = nb$ and $h = n^2b$. Write a MATLAB program to plot $\bar{y}$ as a function of n for both cases using $1 \le n \le 12$ and (i) $b = 5$ inches, (ii) $b = 7$ inches and (iii) $b = 9$ inches.*

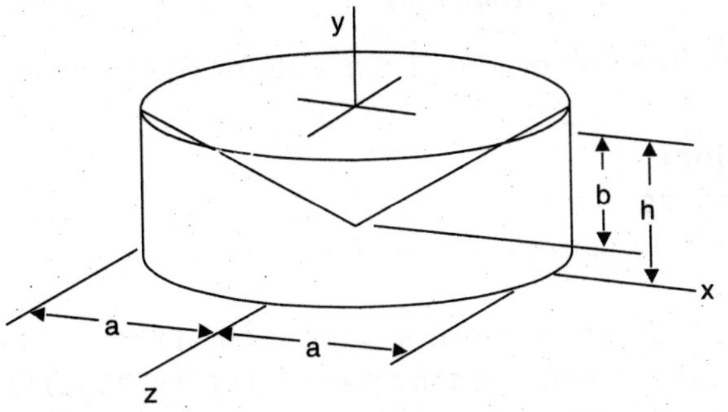

Fig. ES3.20

Solution:

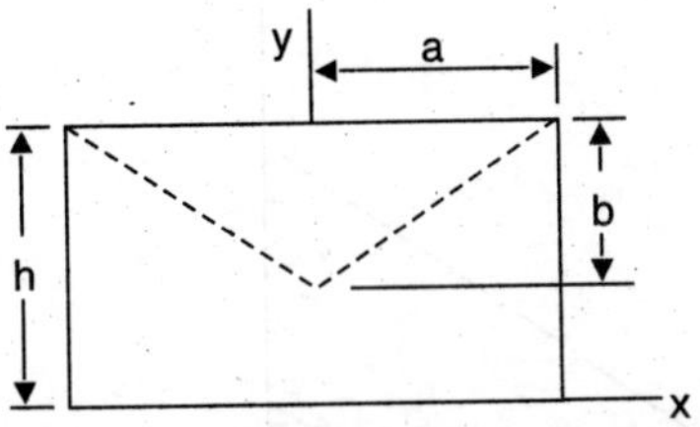

Fig. ES3.20 (*a*)

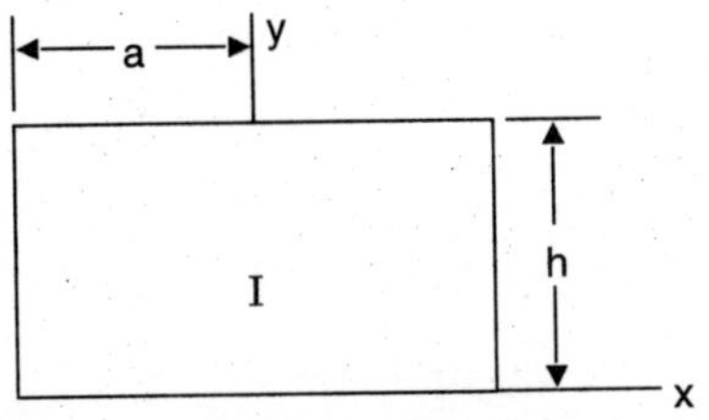

Fig. ES3.20 (*b*)

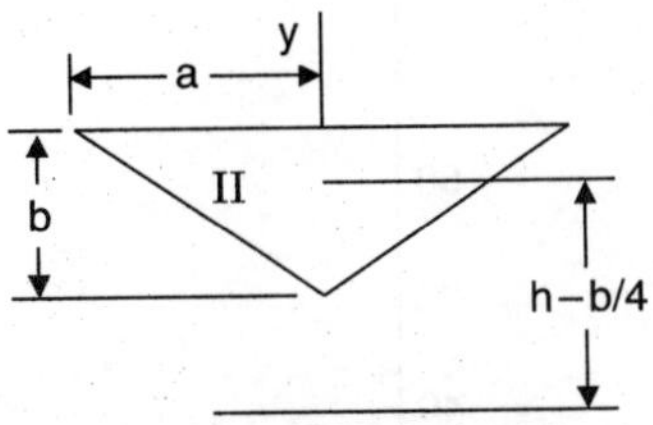

Fig. ES3.20 (*c*)

Based on the Fig. ES3.20 (*a*), the following table is constructed.

	V	$\bar{y}$	$\overline{yV}$
I	$\pi a^2 h$	$\frac{1}{2} h$	$\frac{1}{2} \pi a^2 h^2$
II	$-\frac{1}{3} \pi a^2 b$	$h - \frac{1}{4} b$	$-\frac{\pi}{3} a^2 bh + \frac{\pi}{12} a^2 b^2$

$\bar{y}\, \Sigma V = \Sigma \bar{y}\, V$ gives

$$\bar{y}\left(\pi a^2 h - \frac{\pi}{3} a^2 b\right) = \frac{\pi}{12} a^2 b^2 + \frac{\pi}{2} a^2 h^2 - \frac{\pi}{3} a^2 bh$$

and
$$\bar{y} = \frac{6h^2 - 4bh + b^2}{4(3h - b)}$$

For $h = nb$

$$\bar{y} = \frac{b(6n^2 - 4n + 1)}{4(3n - 1)}$$

For $h = n^2b$

$$\bar{y} = \frac{b(6n^2 - 4n + 1)}{4(3n^2 - b)}$$

The plot of $\bar{y}$ as a function of n for $1 \le n \le 12$ for the cases is shown in Fig. ES3.20 (*d*) and (*e*).

MATLAB Solution:

```
>> % Example ES3.20
>> % Use symbolic notation in MATLAB
>> syms b, n
>> ybarn=inline('b*(6*n.^2-4*n.+1)/(4*(3*n-1))');
>> ybarn2=inline('b*(6*n.^4-4*n.^2+1)/(4*(3*n.^2-1))');
>> ybarn=vectorize(ybarn);
>> ybarn2=vectorize(ybarn2);
>> n=[1:1:12];
```

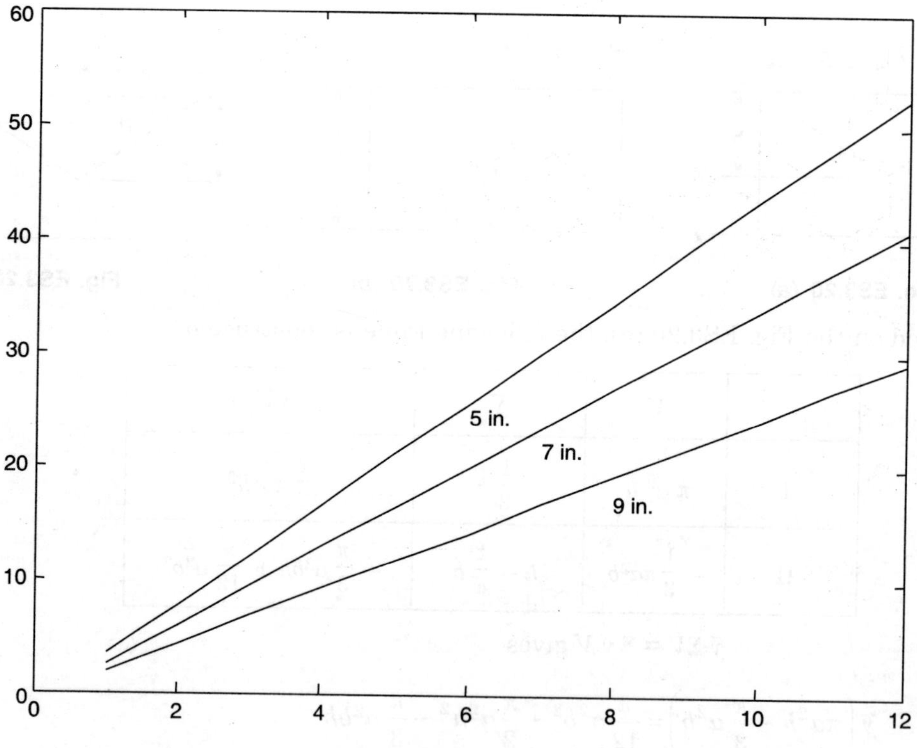

Fig. ES3.20 (*d*)

```
>> subplot(2,1,1)
>> plot(n, ybarn(5,n),n, ybarn(7,n),n, ybarn(9,n))
>> xlabel('n')
>> ylabel('Centroid   (in)')
>> grid on
>> subplot(2,1,2)
```

```
>> plot(n,ybarn2(5,n),n,ybarn2(7,n),n,ybarn2(9,n))
>> xlabel('n')
>> ylabel('Centroid  (in)')
>> grid on
>> text(6,25,'5 in.')
>> text(7,22,'7 in.')
>> text(8,11,'9 in.')
>> text(4,100,'5 in.')
>> text(10,300,'7 in.')
>> text(8,100,'9 in.')
```

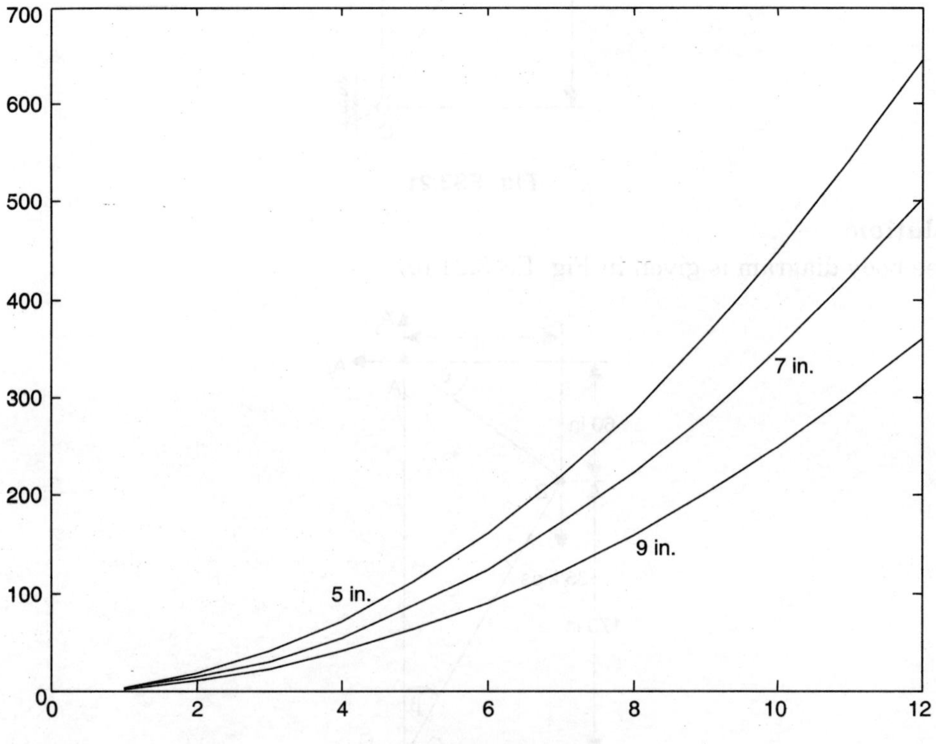

Fig. ES3.20 (*e*)

ES 3.21. *For the system shown in Fig. ES3.21, obtain the force in each member of the truss as a function of d. Write a MATLAB program to plot the force in each member for 30 in ≤ d ≤ 250 inches.*

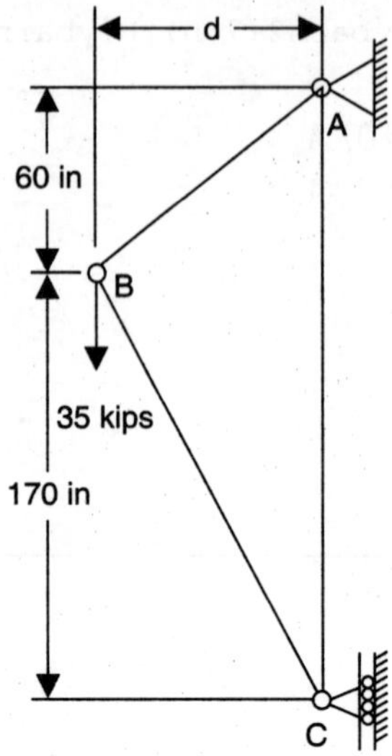

Fig. ES3.21

Solution:

Free body diagram is given in Fig. ES 3.21 (*a*)

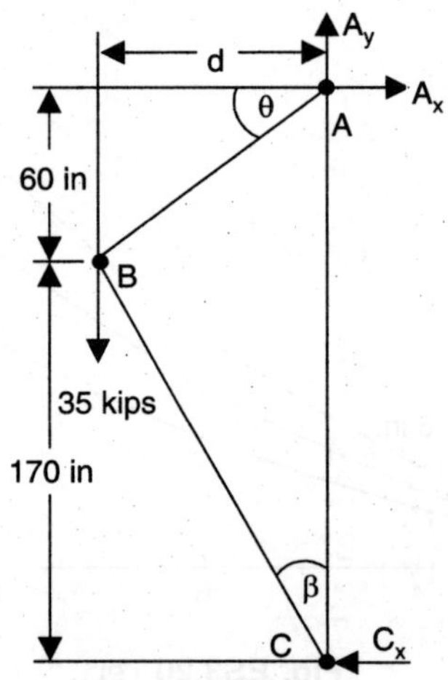

Fig. ES 3.21 (*a*)

The length of *AB* and *BC* is given by

$$AB = \sqrt{d^2 + (60)^2}$$

$$BC = \sqrt{d^2 + (172)^2}$$

and the angles θ and β by

$$\sin\theta = \frac{60}{AB}, \quad \cos\theta = \frac{d}{AB}$$

$$\sin\beta = \frac{d}{BC}, \quad \cos\beta = \frac{170}{BC}$$

Reactions:

$$\Sigma F_x = 0: \quad A_x - C_x = 0$$
$$\Sigma F_y = 0: \quad A_y - 35000 = 0, A_y = 35000$$
$$\Sigma M_C = 0: \quad a(35000) - 230A_x = 0$$

Thus $$A_x = \frac{35000}{230} d = 152.17\, d = C_x$$

Joint *A* and *C*:

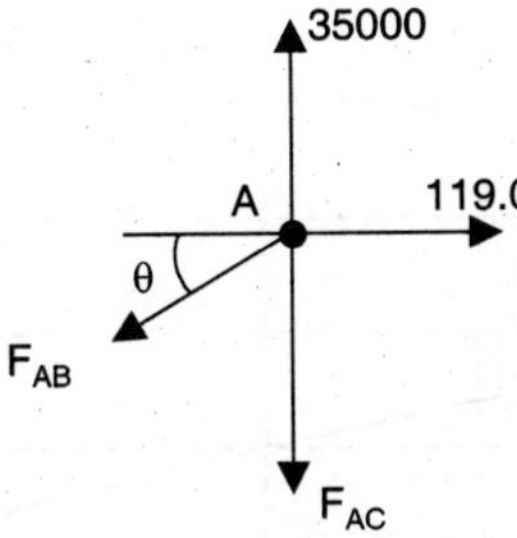

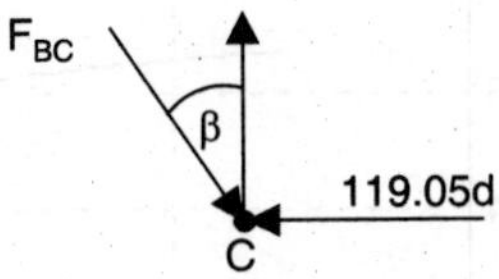

Fog *A*:

$$\Sigma F_x = 0: \quad 152.17d - F_{AB} \cos \theta = 0$$
$$F_{AB} = \frac{AB}{d} (152.17d)$$
$$F_{AB} = 152.17\, AB\ lb(T)$$
$$\Sigma F_y = 0: \quad 35000 - F_{AC} - F_{AB} \sin \theta = 0$$
$$F_{AC} = 35000 - 152.17\, AB \left(\frac{60}{AB}\right)$$
$$F_{AC} = 25868.9\ lb(T)$$

For *C*:

$$\Sigma F_y = 0: \quad 25869 - F_{BC} \cos \beta = 0$$
$$F_{BC} = \frac{BC}{170} 25868.9$$
$$F_{BC} = 152.17BC\ lb(C)$$

MATLAB program:

```
%
d=30:1:250;
AB=sqrt(d.^2+60^2);
BC=sqrt(d.^2+170^2);
Fab=152.17*AB;
Fac=25868.9+0*d;
Fbc=-152.17*BC;
plot(d,Fab,d,Fac,d,Fbc)
xlabel('d(in)')
ylabel('Member Force(lb)')
legend('FAb','FAC','FBC',2)
grid on
```

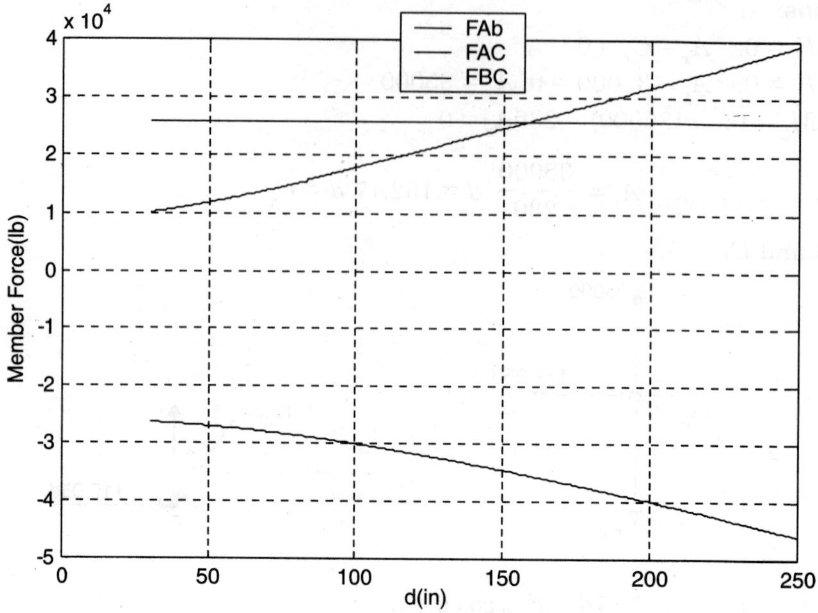

Fig. ES3.21 (*b*)

ES 3.22. *Fig. ES3.22 shows the loading on a semicircular member. The radius of the semicircular member is 25 inches. Write a MATLAB program to plot the internal forces, namely, the axial forces, shearing force and bending moment as functions of* α *for* $0 < \alpha < 90°$.

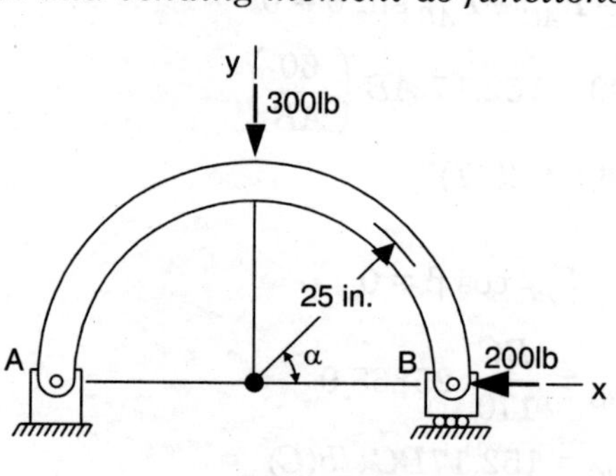

Fig. ES3.22

Solution:

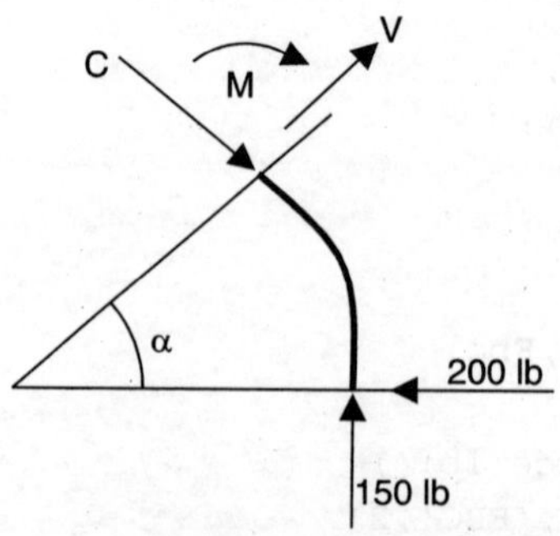

Fig. ES3.22 (*a*)

For $0 \le \theta \le 90°$:

$$C = 200 \sin \alpha + 150 \cos \alpha$$
$$V = 200 \cos \alpha - 150 \sin \alpha$$
$$M = -200\, r \sin \alpha + 100r\,(1 - \cos \alpha)$$

Plot these values as a function of θ for $0 \le \theta \le 90°$ with $r = 25$ in.

MATLAB Solution:

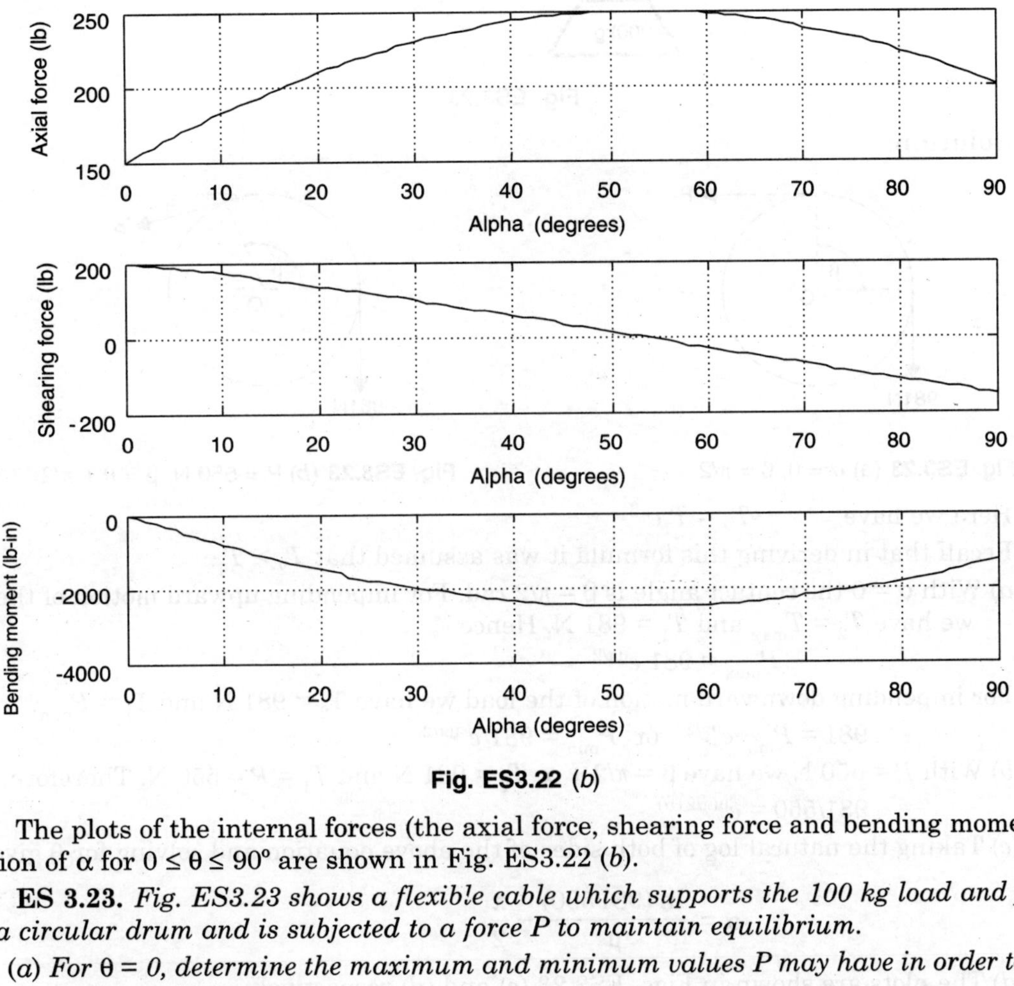

Fig. ES3.22 (*b*)

The plots of the internal forces (the axial force, shearing force and bending moment) as function of α for $0 \le \theta \le 90°$ are shown in Fig. ES3.22 (*b*).

ES 3.23. *Fig. ES3.23 shows a flexible cable which supports the 100 kg load and passes over a circular drum and is subjected to a force P to maintain equilibrium.*

(*a*) *For θ = 0, determine the maximum and minimum values P may have in order to raise or lower the load.*

(*b*) *Write a MATLAB program to plot P_{max} and P_{min} versus μ for $0 \le \mu \le 1$, where μ is the coefficient of static friction between the cable and the fixed drum.*

(*c*) *For P = 550 N, determine the minimum value for which the angle θ may have before the load begins to slip.*

(*d*) *Plot θ_{min} versus μ for $0 \le \mu \le 1$.*

Limit to vertical (θ) axis to be between – 60° and 360°.

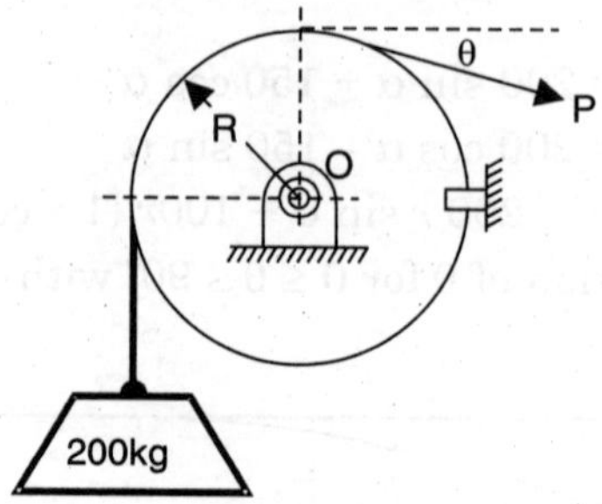

Fig. ES3.23

Solution:

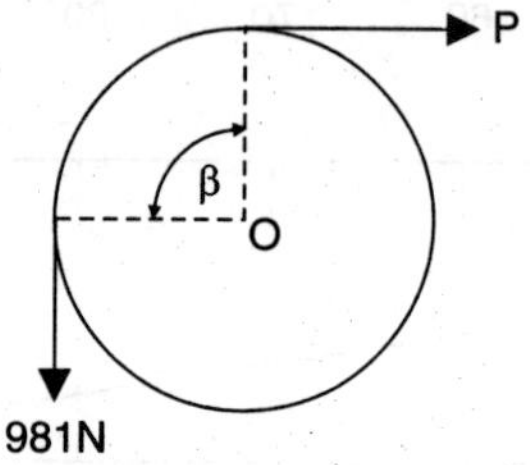

Fig. ES3.23 (*a*) $\alpha = 0$, $\beta = \pi/2$

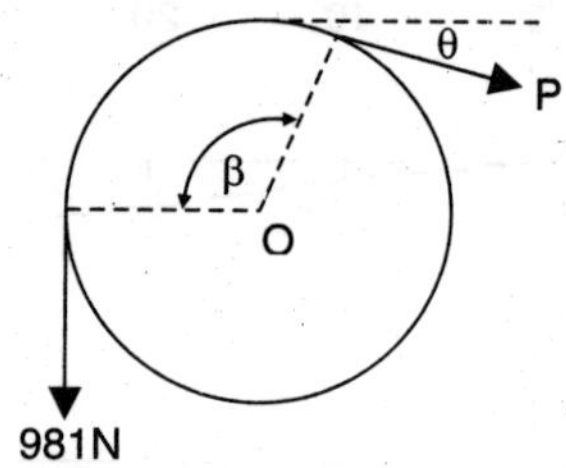

Fig. ES3.23 (*b*) P = 550 N, $\beta = \theta + \pi/2$

Here we have $T_2 = T_1 e^{\mu\beta}$

Recall that in deriving this formula it was assumed that $T_2 > T_1$.

(*a*) With $\alpha = 0$ the contact angle is $\beta = \pi/2$ rad. For impending upward motion of the load we have $T_2 = T_{max}$ and $T_1 = 981$ N. Hence

$$P_{max} = 981\ e^{\mu\pi/2}$$

For impending downward motion of the load we have $T_2 = 981$ N and $T_1 = P_{min}$.

$$981 = P_{min}\ e^{\mu\pi/2} \quad \text{or} \quad P_{min} = 981\ e^{-\mu\pi/2}$$

(*b*) With $P = 550$ N we have $\beta = \pi/2 + \alpha$, $T_2 = 981$ N and $T_1 = P = 550$ N. Therefore,

$$981/550 = e^{\mu(\pi/2+\alpha)}$$

(*c*) Taking the natural log of both sides of the above equation and solving for θ gives,

$$\alpha = \frac{\ln(981/500)}{\mu} - \frac{\pi}{2}$$

(*d*) The plots are shown in Figs. ES3.23 (*c*) and (*d*) respectively.

MATLAB Solution:

```
>> % Example ES3.23
>> % Plots of Pmax and P min versus coefficient of friction
>> mu=0:0.005:1;
>> pmax=981*exp(mu*pi/2)/1000;
>> pmin=981*exp(-mu*pi/2)/1000;
>> plot(mu,pmax,mu,pmin)
>> grid on
>> xlabel('Coefficient of friction')
>> ylabel('Force P (kN)')
```

```
>> text(0.7,2.5,'Pmax')
>> text(0.6,0.5,'Pmin')
```

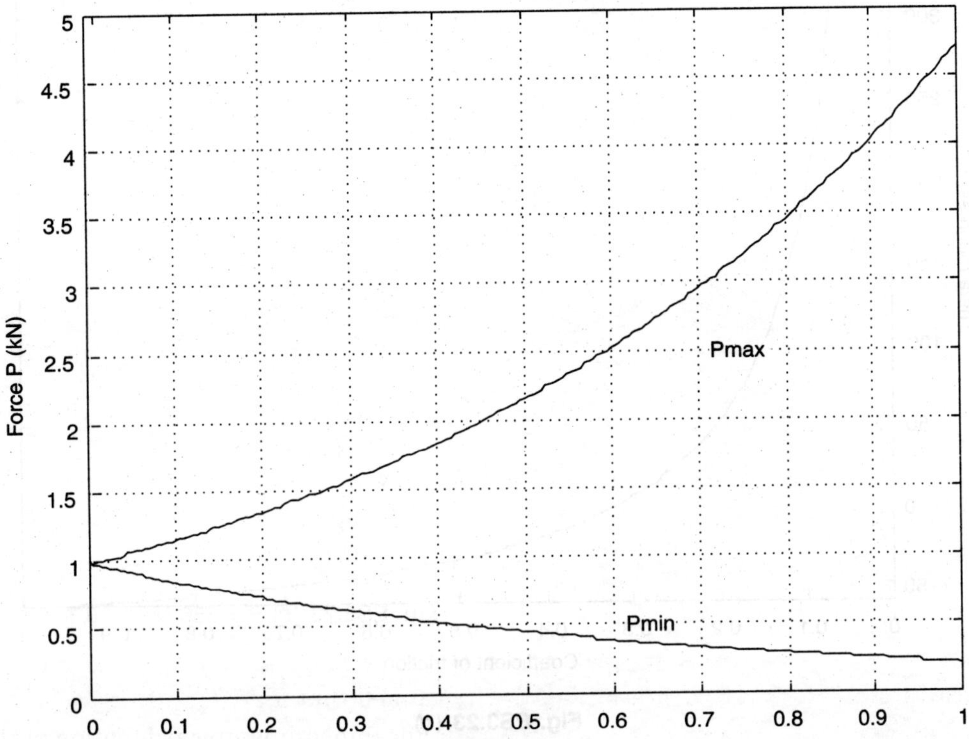

Fig. ES3.23 (*c*)

>> % Example ES3.23

```
>> % Plot of minimum angle theta versus mu
>> mu=0:0.005:1;
>> thet=log(981/550)./mu-pi/2
>> plot(mu,thet*180/pi)
>> axis([1 1 -60 360])
>> xlabel('Coefficient of friction')
>> ylabel('Theta (degrees)')
```

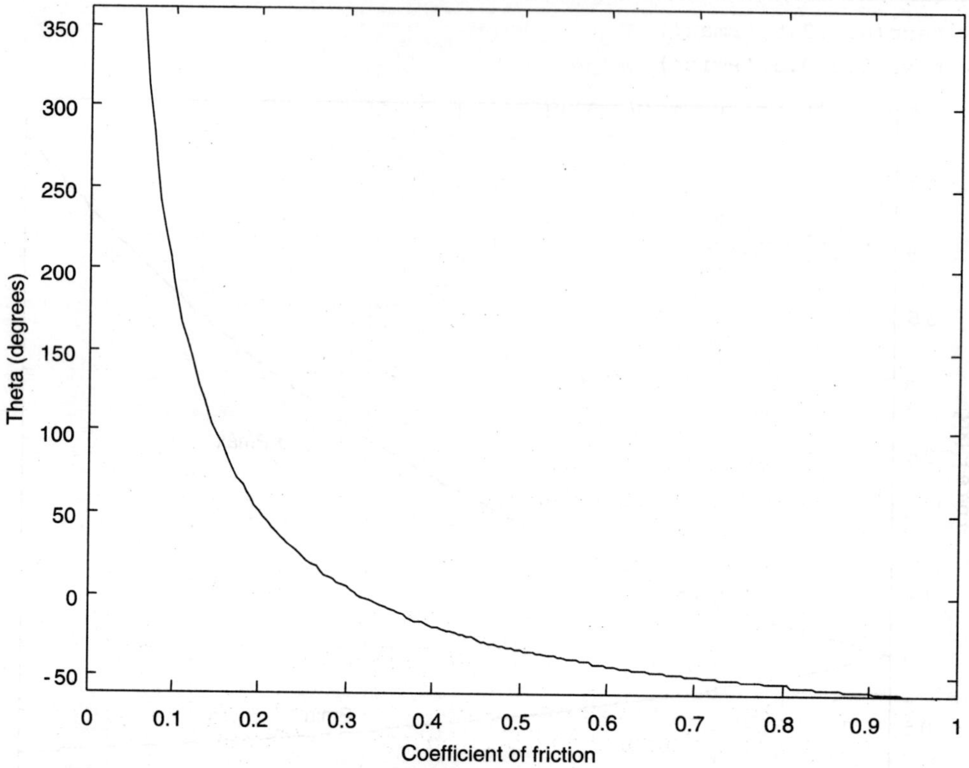

Fig. ES3.23 (*d*)

ES 3.24. *Fig. ES3.24 shows a curved beam member AB which is a parabola and the vertex is at A and $l = nH$. Write a MATLAB program to determine,*

(*a*) *The internal forces and bending moment at an arbitrary point C*

(*b*) *Plot the internal forces and bending moment as a function of x for $0.1l \le x \le 0.9l$.*

Given: The magnitude of the vertical load $F = 1.5$ kN, $l = 500$ mm and $n = 2, 3, 4$.

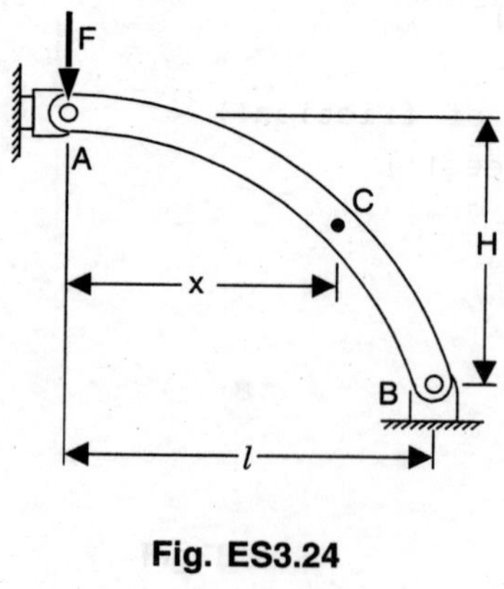

Fig. ES3.24

Solution:

Free body diagram is given in Fig. ES 3.24 (*a*).

Summing forces and moments yields

$$\Sigma F_y = 0: \quad B_y - 1.5 = 0$$

$$B_y - 1.5 \text{ kN}$$

$$\Sigma M_B = 0: 1.5(L) - A(H) = 0$$

$$A = 1.2\,\frac{l}{H} = 1.5\text{n kN}$$

For a parabola $y = kx^2$, we have

$$h = kl^2 = k(nH)^2$$

and
$$k = \frac{1}{Hn^2}$$

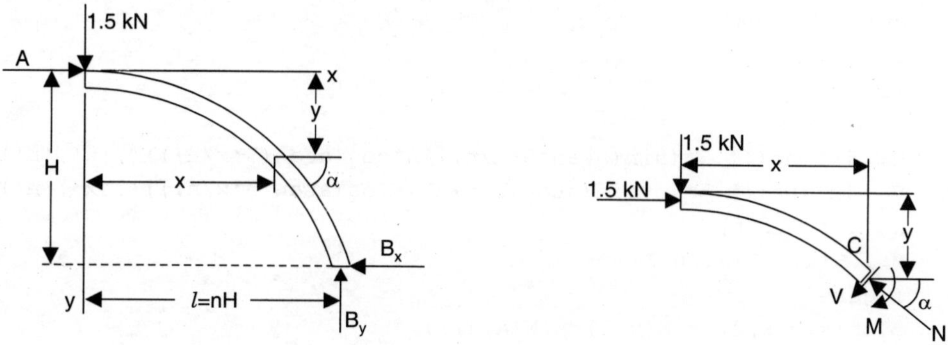

Fig. ES 3.24 (*a*)

Arbitrary Section *AJ:*

At an arbitrary point *C*, we have

$$\text{slope} = \frac{dy}{dx} = \frac{2x}{Hn^2}$$

and
$$\tan\alpha = \frac{2x}{Hn^2}$$

$$\alpha = \tan^{-1}\left(\frac{2x}{Hn^2}\right)$$

At an arbitrary section *AC:*

$$\Sigma M_J = 0: \quad 1.5(x) - 1.5n(y) - M = 0$$

$$M = 1.5(x - ny)$$

$$= 1.5\left(x - \frac{nx^2}{Hn^2}\right)$$

$$M = 1.5x\left(1 - \frac{x}{Hn}\right)$$

Summing forces gives:

$$\Sigma F_N = 0: \quad N - 1.5(\sin\alpha) - 1.5n(\cos\alpha) = 0$$

$$N = 1.5(\sin\alpha - n\cos\alpha)$$

$$\Sigma F_V = 0: \quad -V - 1.5(\cos\alpha) + 1.5n(\sin\alpha) = 0$$

$$V = 1.5(n\sin\alpha - \cos\alpha)$$

With $l = 0.5$ m $= Hn$

$$M = 1.5x\left(1 - \frac{x}{0.5}\right)$$

where
$$\alpha = \tan^{-1}\left(\frac{2x}{ln}\right) = \tan^{-1}\left(\frac{x}{0.25n}\right)$$

MATLAB program:

```
%
syms n t real
L=0.5;
x1=0.1*L;
xf=0.9*L;
N= inline('1.5*(sin(atan(5*x/n))+n*cos(atan(5*x/n)))','x','n');
V= inline('1.5*(n*sin(atan(5*x/n))-cos(atan(5*x/n)))','x','n');
x=[x1:.01:xf];
M=1.5*x.*(1-x./0.5);
figure(1)
plot(x,N(x,2),x,N(x,3),x,N(x,4));
xlabel('x(m)')
ylabel('Normal  Force(kN)')
legend('n=2','n=3','n=4',4)
grid on
figure(2)
plot(x,V(x,2),x,V(x,3),x,V(x,4));
xlabel('x(m)')
ylabel('Shear  Force(kN)')
legend('n=2','n=3','n=4',2)
grid on
figure(3)
plot(x,M)
xlabel('x(m)')
ylabel('Moment(kN-m)')
grid on
```

Output is

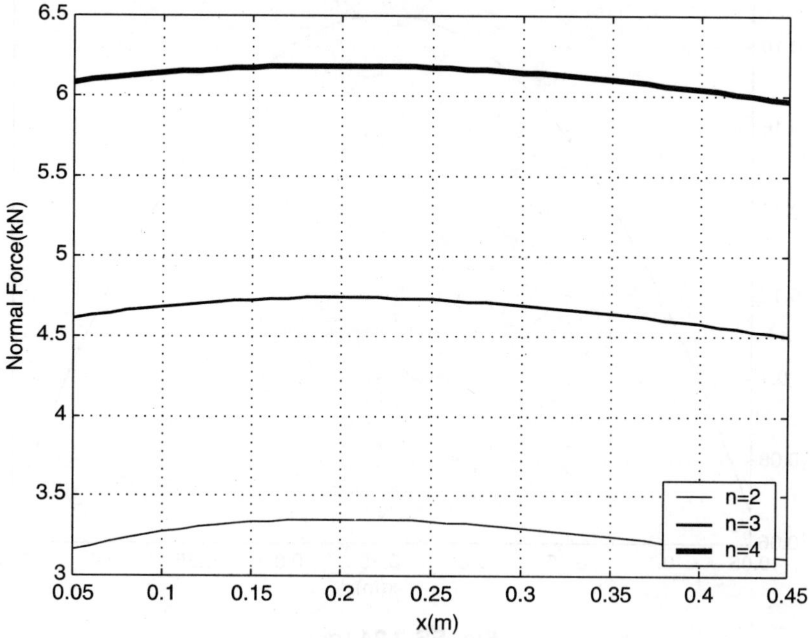

Fig. ES3.24 (*b*)

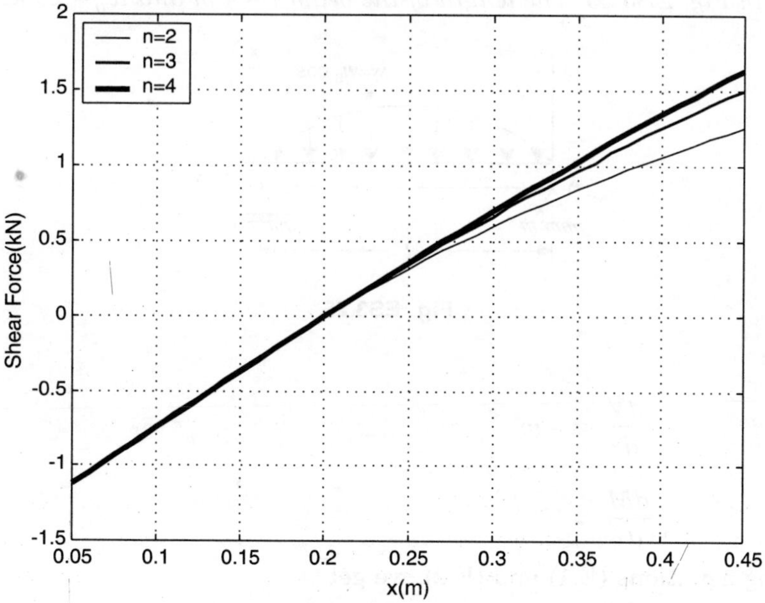

Fig. ES3.24 (*c*)

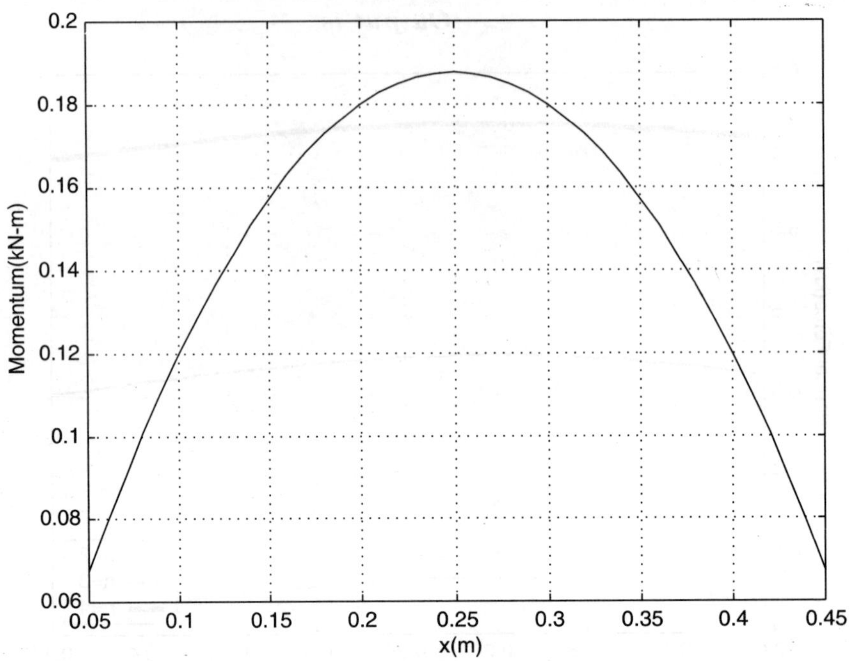

Fig. ES 3.24 (*d*)

ES 3.25. *Write a MATLAB program to plot the shear and bending moment diagrams for the beam shown in Fig. ES3.25. The length of the beam $l = 4$ m and $w_0 = 20$ kN/m.*

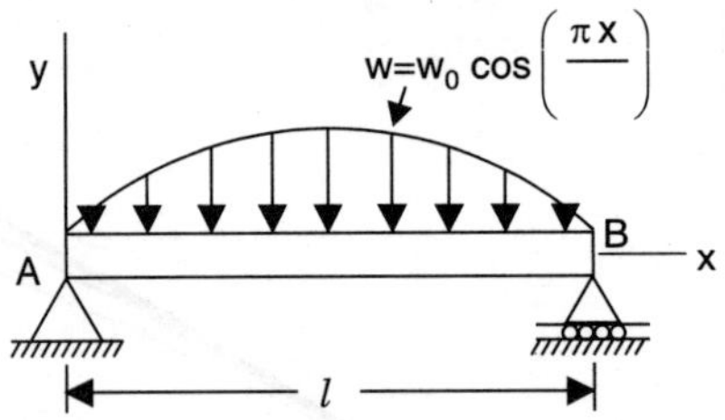

Fig. ES3.25

Solution:

$$\frac{dV}{dx} = -w \tag{E.1}$$

$$\frac{dM}{dx} = V \tag{E.2}$$

Integrating equations (E.1) and (E.2), we get

$$V = -\int w dx = \int w_0 \sin \frac{\pi x}{l} dx = w_0 \frac{L}{\pi} \cos \frac{\pi x}{l} + C_1$$

$$M = \int V dx = \int \left(w_0 \frac{l}{\pi} \cos \frac{\pi x}{l} + C_1 \right) dx = w_0 \left(\frac{l}{\pi} \right)^2 \sin \frac{\pi x}{l} + C_1 x + C_2$$

The boundary conditions are written as

At $x = 0$, $$M = 0 = w_0 \left(\frac{l}{\pi}\right)^2 \sin\frac{\pi(0)}{l} + C_1(0) + C_2, \text{ which implies that } C_2 = 0.$$

At $x = l$, $$M = 0 = w_0 \left(\frac{A}{\pi}\right)^2 \sin\frac{\pi x}{A} + C_1, \text{ which implies that } C_1 = 0.$$

Hence $$V = w_0 \frac{A}{\pi} \cos\frac{\pi x}{A}, \quad M = w_0 \left(\frac{A}{\pi}\right)^2 \sin\frac{\pi x}{A}$$

MATLAB Solution:

```
>> % Example ES3.25
>> % Input w0 and l
>> w0=20;
>> l = 4;
>> x = [0:0.1:4];
>> v1  = w0*(l/pi)*cos(pi*x/l);
>> m1= w0*(l/pi)^2*sin(pi*x/l);
>> plot(x,v1)
>> xlabel('x(m)')
>> ylabel('Shear force (kN)')
>> grid on
>> plot(x,m1)
>> xlabel('x,(m)')
>> ylabel('Bending moment (kN-m)')
```

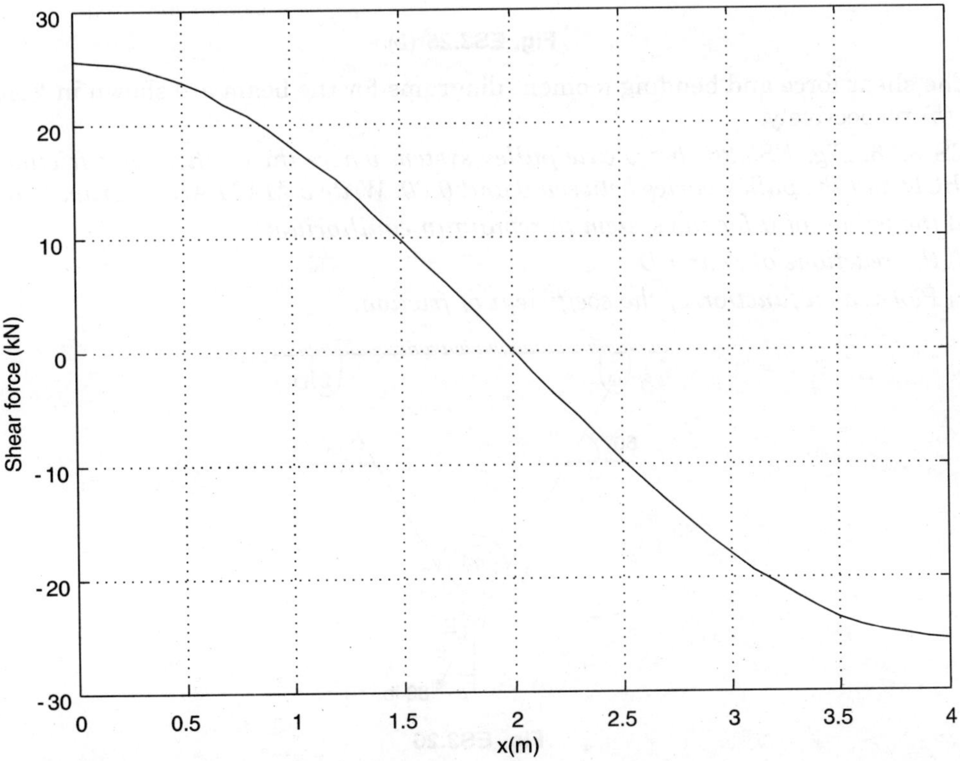

Fig. ES3.25 (*a*)

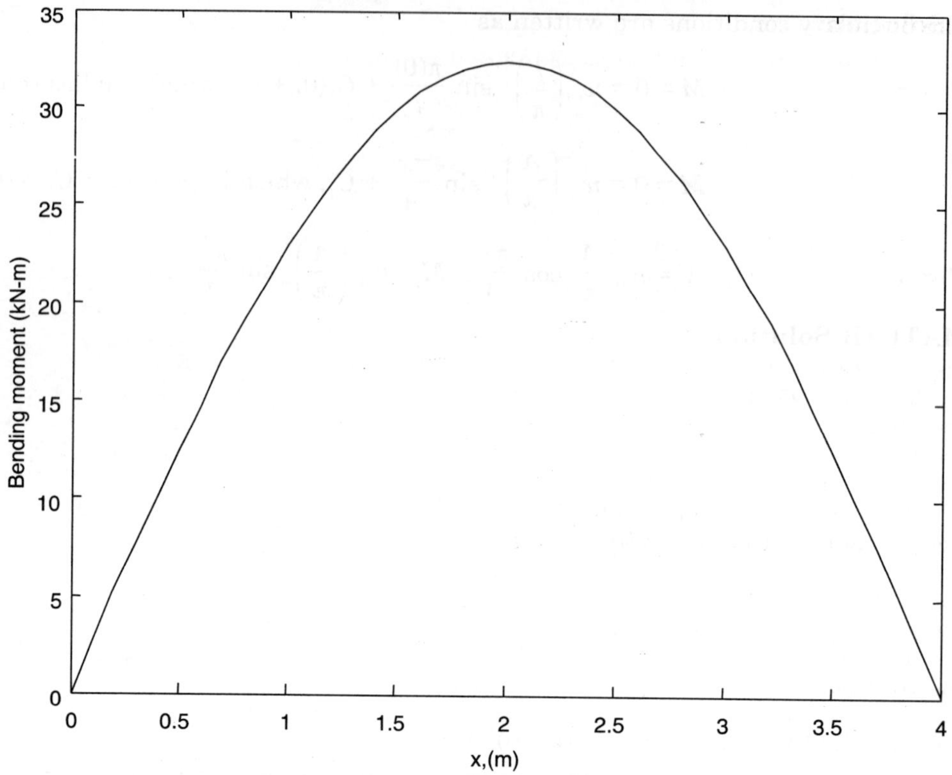

Fig. ES3.25 (*b*)

The shear force and bending moment diagrams for the beam are shown in Figs. ES3.25 (*a*) and (*b*) respectively.

ES 3.26. *Fig. ES3.26 shows axle pulley system where the coefficient of friction between cable ABCD and the pulley varies between 0 and 0.60. Write a MATLAB program to determine,*

(*a*) *the values of* α *for the system to remain in equilibrium*

(*b*) *the reactions at A and D*

(*c*) *Plot* α *as a function of the coefficient of friction.*

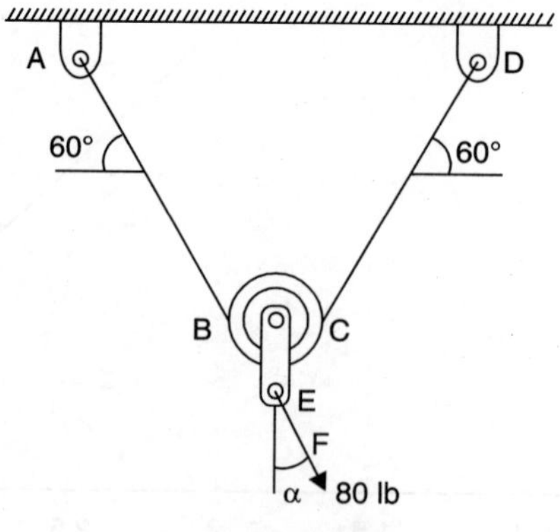

Fig. ES3.26

Solution:

Free body diagram and force triangle are given in Figs. ES3.26 (*a*) and (*b*)

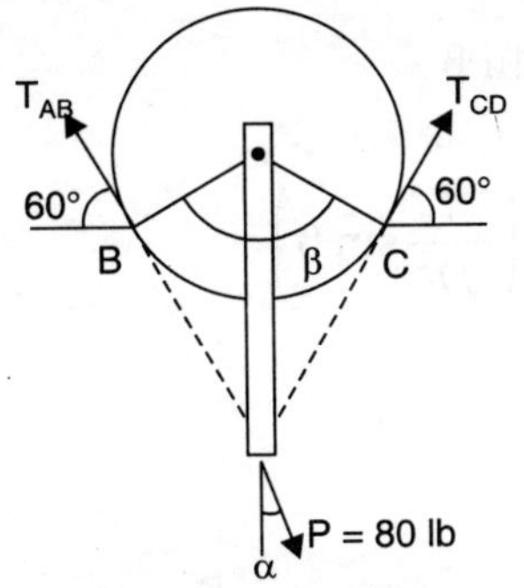

Fig. ES3.26 (*a*)

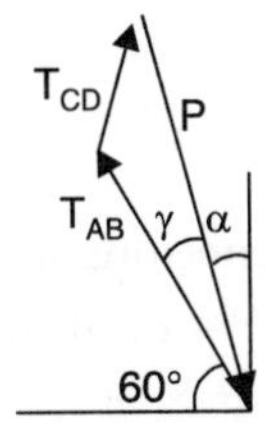

Fig. ES3.26 (*b*)

Since the 80 lb force tends to rotate the pulley counterclockwise, the cable tends to slip relative to the pulley clockwise and we have

$$T_1 = T_{CD},\ T_2 = T_{AB},\ \mu_s = \text{static friction}$$

$$\beta = 120° = \frac{2\pi}{3} \text{ rad}$$

$$\frac{T_2}{T_1} = e^{\mu_s \beta}, \frac{T_{AB}}{T_{CD}} = e^{\frac{2\pi}{3}\mu_s}$$

$$T_{AB} = e^{\frac{2\pi}{3}\mu_s}\ T_{CD} \tag{1}$$

From the force triangle we have using the law of cosines

$$P^2 = T_{AB}^2 + T_{CD}^2 - 2T_{AB}\,T_{CD}\cos\beta$$

$$= \left(e^{\frac{2\pi}{3}\mu_s}\ T_{CD}\right) + T_{CD}^2 - 2\left(e^{\frac{2\pi}{3}\mu_s}\ T_{CD}\right)T_{CD}\left(-\frac{1}{2}\right)$$

$$= \left[\left(e^{\frac{2\pi}{3}\mu_s}\right)^2 + 1 + e^{\frac{2\pi}{3}\mu_s}\right]T_{CD}^2$$

$$P^2 = FT_{CD}^2$$

where
$$F = \left[\left(e^{\frac{2\pi}{3}\mu_s}\right)^2 + 1 + e^{\frac{2\pi}{3}\mu_s}\right]$$

Hence, we have that

$$T_{CD} = \frac{1}{\sqrt{F}}\ P \tag{2}$$

(*a*) The corresponding values of α for the system to remain in equilibrium. Using the law of sines we have

$$\frac{\sin\gamma}{T_{CD}} = \frac{\sin\beta}{P}, \quad \sin\gamma = \frac{T_{CD}}{P}\sin\beta$$

$$\sin\gamma = \frac{1}{\sqrt{F}}\sin\beta, \quad \gamma = \sin^{-1}\left(\frac{1}{\sqrt{F}}\sin\beta\right)$$

$$\alpha = 90^\circ - (60 + \gamma)$$

(*b*) The reactions at *A* and *D* are as follows:
Making $P = 80$ lb in Eq.(2), then

$$D = T_{CD} = \frac{1}{\sqrt{F}}\,(80)\text{ lb}$$

and from Eq. (1)

$$A = T_{AB} = e^{\frac{2\pi}{3}\mu_s}\left(\frac{1}{\sqrt{F}}\right)(50)\text{ lb}$$

MATLAB program:

```
    %
mu=0;
%Output  Headings
fprintf('Friction   Angle\n')
fprintf('\n')
ang=120.*pi/180.;
N=1;
while mu<0.60
    Y(N)=mu;
    a=exp(pi*ang*mu);
    E=sqrt(a^2+1.+a);
    ooe=1./E;
    sgamma=ooe*sin(ang);
    Gamma=asin(sgamma);
    alpha(N)=pi/2.-(1.0472+Gamma);
    X(N)=alpha(N)*180/pi;
    D(N)=ooe*80.;
    A(N)=a*D(N);
    fprintf('%5.3f      %5.3f\n',Y(N),X(N))
    mu=mu+.05;
    N=N+1;
end%while
fprintf('\n')
fprintf('\n')
fprintf('Friction      Reac.A      Reac.D\n')
fprintf('\n')
```

```
for I=1:N-1
    fprintf('%5.3f      %5.3f      %5.3f\n',Y(I),A(I),D(I))
end
figure(1)
plot(Y,X)
xlabel('Coefficient of Friction')
ylabel('Angle alpha')
grid on
figure(2)
plot(Y,D,Y,A)
xlabel('Coefficient of Friction')
ylabel('Reactions')
legend('Reaction at A','Reaction at D',2)
grid on
```

Output is:

```
>> es28
Friction       Angle
0.000         -0.000
0.050          5.377
0.100         10.391
0.150         14.780
0.200         18.425
0.250         21.332
0.300         23.582
0.350         25.286
0.400         26.558
0.450         27.497
0.500         28.186
0.550         28.687
Friction     Reac.A          Reac.D
0.000        46.188          46.188
0.050        53.481          38.488
0.100        59.860          31.001
0.150        65.068          24.251
0.200        69.105          18.535
0.250        72.125          13.922
0.300        74.336          10.326
0.350        75.934           7.591
0.400        77.083           5.546
0.450        77.907           4.033
0.500        78.498           2.925
0.550        78.921           2.116
```

Also following Figs. ES3.26 (*c*) and (*d*) are obtained.

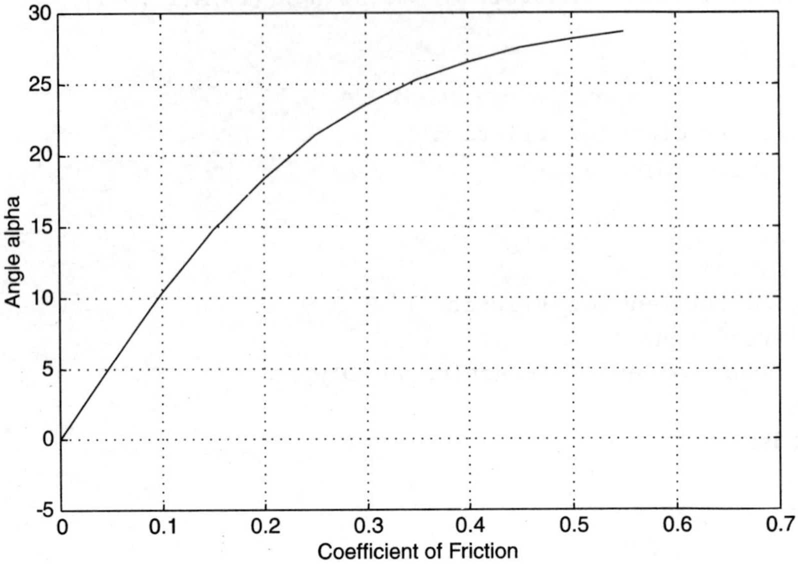

Fig. ES 3.26 (*c*)

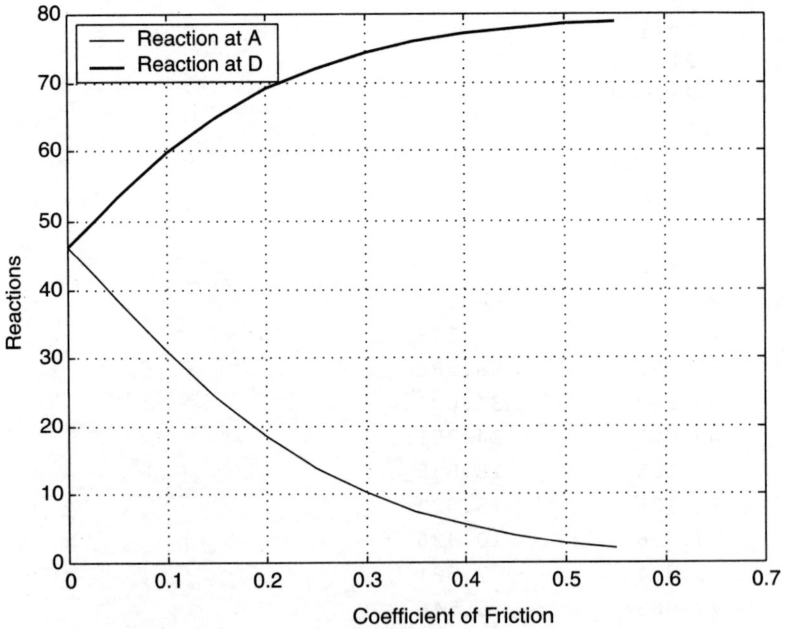

Fig. ES 3.26 (*d*)

ES 3.27. *Fig. ES3.27 shows the end of a trough which is filled with water in the line AA′. Determine,*

(*a*) *the depth D of the point of application of the resultant of the hydrostatic forces acting on the panel (center of pressure)* (*b*) *plot the depth of the center of pressure D as a function of n. Given B = 250 mm, H = 500 mm and A = nB, where n varies from 1 to 10.*

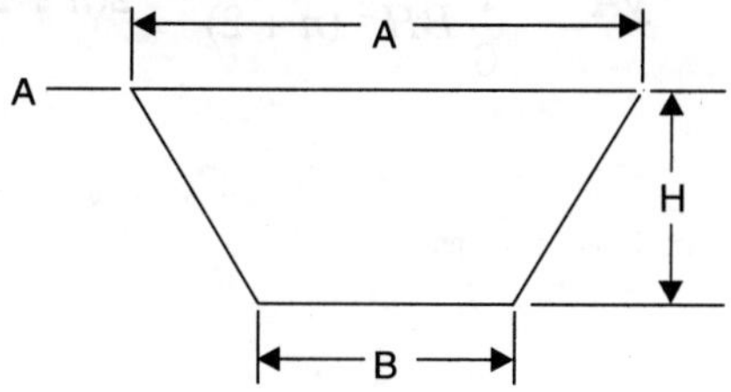

Fig. ES3.27

Solution:

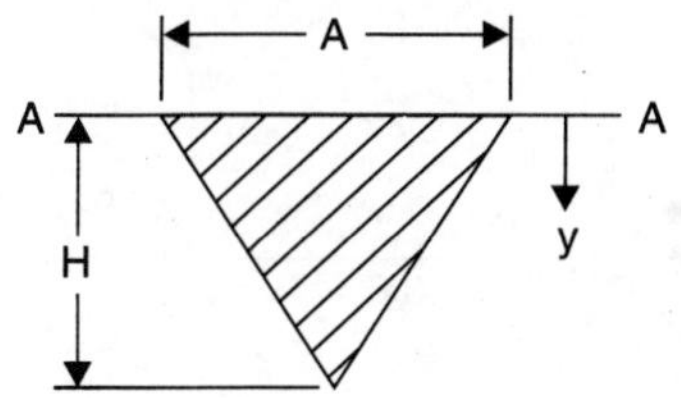

Fig. ES3.27 (*a*)

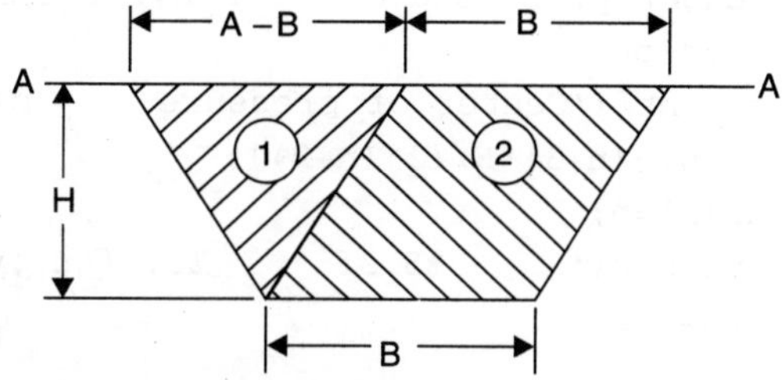

Fig. ES3.27 (*b*)

Let y_p = distance of center of pressure from AA'. Then

$$Ry_p = M_{AA'}$$

and
$$y_p = \frac{M_{AA'}}{R} = \frac{\gamma\int y^2 dA}{\gamma\int y dA} = \frac{I_{AA'}}{\bar{y}A}$$

Dividing the trapezoid as shown in Fig.ES3.29 (*a*), we have

$$I_{AA'} = \frac{1}{12}(A - B)\,H^3 + \frac{1}{3}\,BH^3$$

and since
$$A = nB$$

$$I_{AA'} = \frac{1}{12}\,(nB - B)\,H^3 + \frac{1}{3}\,BH^3 = \frac{1}{12}\,BH^3\,(n + 3)$$

$$\bar{y}\,A = yA + yA = \frac{1}{3}\,H\left[\frac{1}{2}(A - B)H\right] + \frac{1}{12}\,H(BH)$$

$$= \frac{1}{3}\,H\left[\frac{1}{2}(nB - B)\,H\right] + \frac{1}{12}\,H(BH)$$

$$= \frac{1}{6}\,BH^2\,(n - 1) + BH^2$$

$$= \frac{1}{6}\,BH^2\,(n + 2)$$

Hence $$y_p = \frac{I_{AA'}}{\bar{y}A} = \frac{\frac{1}{12}BH^3\,(n+3)}{\frac{1}{6}BH^2\,(n+2)} = \frac{n+3}{2(n+2)}\,H$$

MATLAB Solution:

```
>> % Example ES3.27
>> % Input data of H and B in mm
>> H=500;
>> B=250;
>> fprintf('H=%g mm and B=%gmm\n',H,B);
H=500 mm and B=250mm
>> fprintf('n');
Center pf pressure      n/n>> for n=1:10
grid on
fprintf('Center of pressure      n/n')
yp(n)=(n+3)*H/(2*(n+2));
nn(n)=n;
fprintf('      %5.3f       %5.3f\n',yp(n),nn(n))
end%for
```

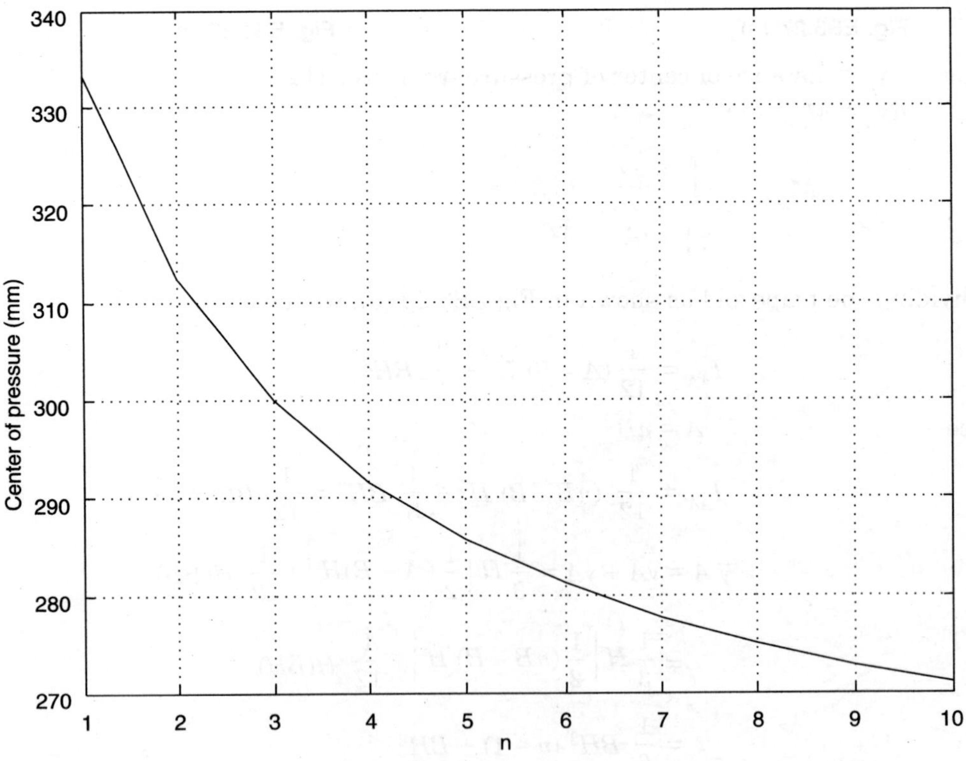

Fig. ES3.27 *(c)*

```
Center of pressure          n
   333.333              1.000
   312.500              2.000
   300.000              3.000
   291.667              4.000
   285.714              5.000
   281.250              6.000
   277.778              7.000
   275.000              8.000
   272.727              9.000
   270.833             10.000
>> plot(nn,yp)
>> xlabel('n')
>> ylabel('Center of pressure (mm)')
>> grid on
```

ES 3.28. *For the area shown in Fig. ES3.28, write a MATLAB program to plot I_{xy} as a function of n for values of N from 1 to 12 knowing that A = 100 mm and B = 80 mm.*

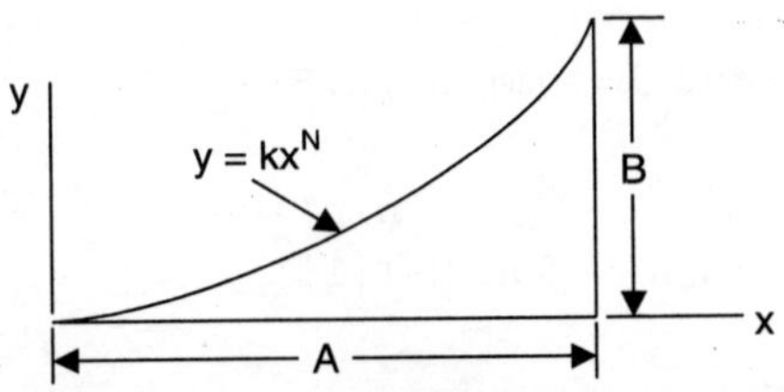

Fig. ES3.28

Solution:

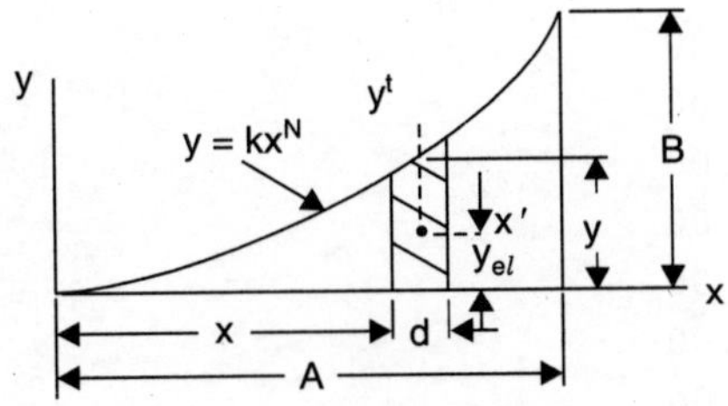

Fig. ES3.28 *(a)*

When $x = A$, $B = ka^N$

Therefore, $$k = \frac{B}{A''}$$

and $$y = \frac{B}{A^N}\ x^N$$

Also $$dI_{xy} = \overline{dI}_{x'y'} + \overline{x}_{el}\overline{y}_{el}\ dA$$

but due to symmetry

$$\overline{dI}_{x'y'} = 0$$

Since $\bar{x}_{el} = x$, $\bar{y}_{el} = 1/2y$ and $dA = ydx$, we have

$$I_{xy} = \int dI_{xy} = \int_0^a x\left(\frac{1}{2}y\right)ydx = \frac{1}{2}\int_0^A x\left(\frac{B}{A^N}x^N\right)^2 dx$$

$$I_{xy} = \frac{1}{2}\frac{B^N}{A^{2N}}\int_0^A x^{2n+1}\,dx = \frac{1}{2}\frac{B^N}{A^{2N}}\frac{A^{2N+2}}{2N+2}$$

$$I_{xy} = \frac{A^2B^2}{4(N+1)}$$

MATLAB Solution:

```
>> % Example ES3.28
>> % Input A and B [in mm]
>> A=100;
>> B=80;
>> fprintf('A=%g mm, and B=%gmm\n',A,B);
A=100 mm, and B=80mm
>> fprintf('I(xy)                    n/n')
I(xy)                    n/n>> for N-1:15
>> for N=1:15
Ixy(N)=A^2*B^2/(4.*(N+1));
NN(N)=N;
fprintf('%5.2e            %5.2f\n',Ixy(N),NN(N))
end%for
8.00e+006        1.00
5.33e+006        2.00
4.00e+006        3.00
3.20e+006        4.00
2.67e+006        5.00
2.29e+006        6.00
2.00e+006        7.00
1.78e+006        8.00
1.60e+006        9.00
1.45e+006       10.00
1.33e+006       11.00
1.23e+006       12.00
1.14e+006       13.00
1.07e+006       14.00
1.00e+006       15.00
>> plot(NN, Ixy)
>> xlabel('N')
>> ylabel('Computed moment of inertia I(xy)(mm^4)')
>> grid on
```

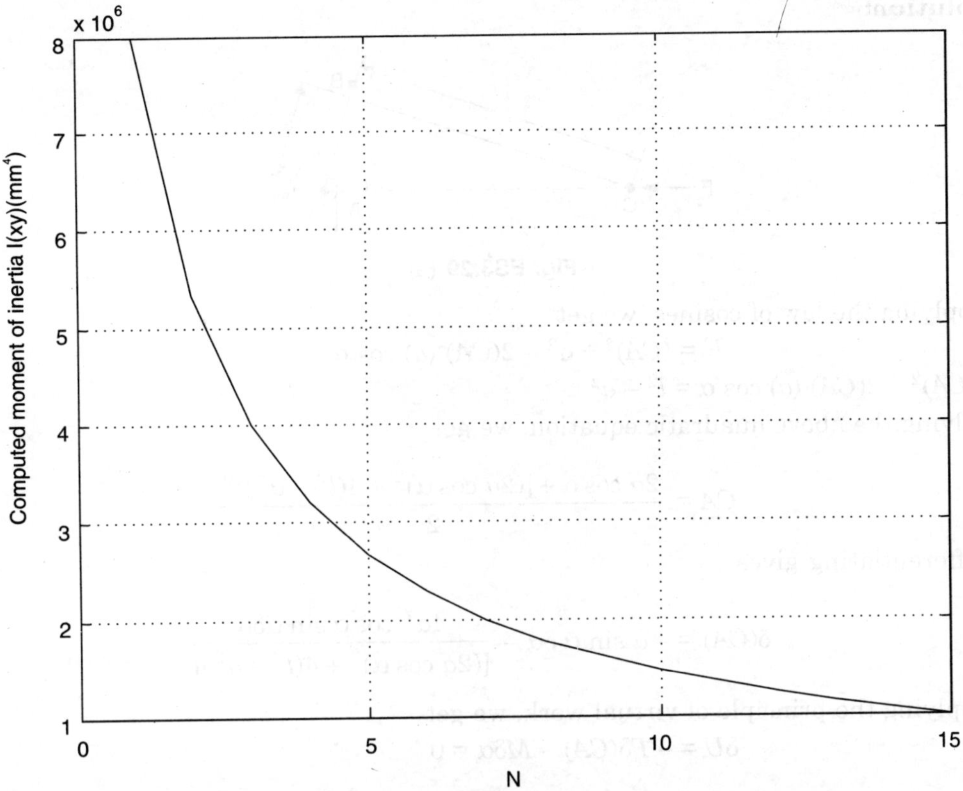

Fig. ES3.28 (*b*)

ES 3.29. *Fig. ES3.29 shows a crank shaft mechanism where a couple M is applied to the crank AB to maintain the equilibrium of the system. The force applied to the system is F. Write a MATLAB program to plot the ratio of M/F as a function of crank angle α from 0 to 180 degrees. Given α = 50 mm and l = 150 mm, determine*

(*a*) *the value of crank angle α for which the ratio M/F is maximum and the corresponding value of M/F.*

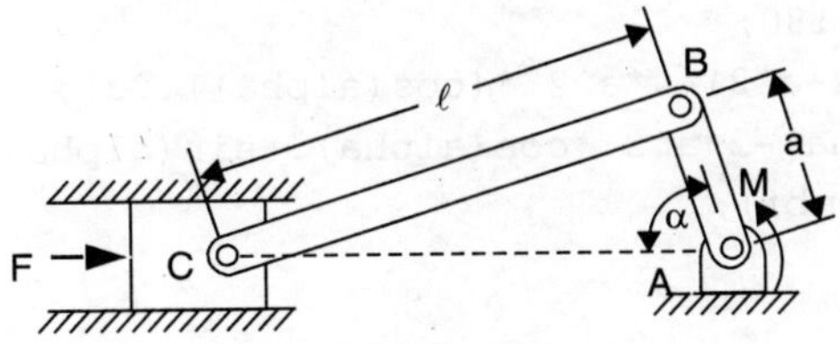

Fig. ES3.29

Solution:

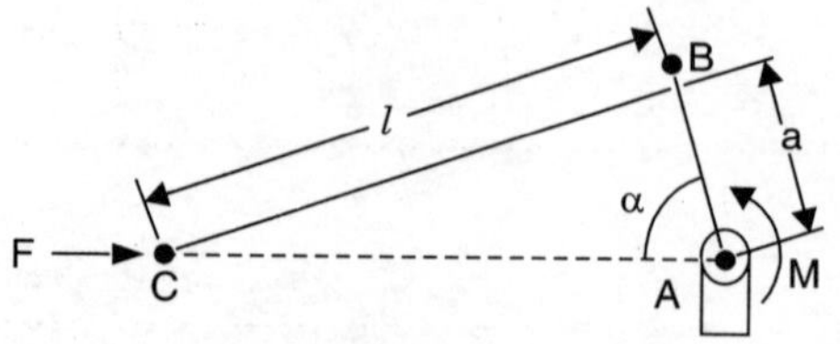

Fig. ES3.29 (*a*)

Applying the law of cosines, we get

$$l^2 = (CA)^2 + a^2 - 2(CA)\,(a)\cos\alpha$$

or $(CA)^2 - 2(CA)\,(a)\cos\alpha = l^2 - a^2$

Solving the above quadratic equation, we get

$$CA = \frac{2a\cos\alpha + [(2a\cos\alpha)^2 + 4(l^2 - a^2)]^{1/2}}{2}$$

Differentiating gives

$$\delta(CA) = -\,a\sin\alpha\,\delta\alpha - \frac{4a^2\cos\alpha\sin\alpha\delta\alpha}{[(2a\cos\alpha)^2 + 4(l^2 - a^2)]}$$

Applying the principle of virtual work, we get

$$\delta U = -\,P\delta(CA) - M\delta\alpha = 0$$

or

$$\frac{M}{P} = -\,\delta(CA) = a\sin\alpha + \frac{4a^2\cos\alpha\sin\alpha}{[(2a\cos\alpha)^2 + 4(l^2 - a^2)]}$$

MATLAB Solution:

```
>> % Example ES3.29
>> % Program inputs: a and l
>> a = 0.050;
>> l = 0.150;
>> alpha = Ho:1:180;
>> alpha=alpha*pi/180;
>> den=sqrt(4*(l^2-a^2)+4*a^2.*(cos(alpha)).^2);
>> mbf=a.*sin(alpha)-2*a^2.*cos(alpha).*sin(alpha)./den;
>> [mbfmax,i]=max(mbf);
>> vt=alpha(i);
>> mbfMax=mbfmax
>> fprintf('a) Alpha when M/F is a maximum is = %3.0f degrees and M/F =
%6.4f(m)\n',vt,mbfmax)
(a) Alpha when M/F is a maximum is =  93 degrees and M/F = 0.0501(m)>>
vt=alpha(i);
>> plot(alpha,mbf)
>> grid on
>> xlabel('Angle Alpha (Degrees)')
>> ylabel('M/F ( meters)')
>>
```

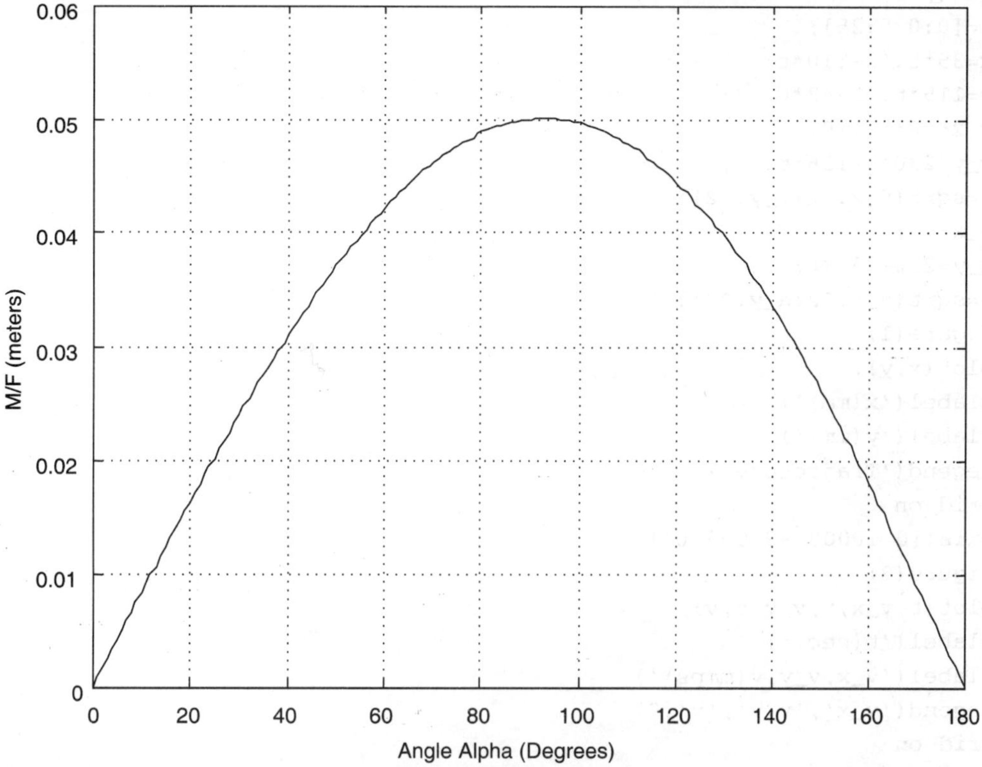

Fig. ES3.29 (*b*)

3.15.2.2 Dynamics

ED 3.1. *The motion of a particle is defined by the equation*

$$x = 35\,t^2 - 110\,t$$

and

$$y = 115\,t^2 - 42\,t^3$$

where x and y = displacement of the particle (mm)

t = time (sec.)

For the time interval $0 \le t \le 25$ *s, write a MATLAB program to plot*

(*a*) *the path of the particle in the xy plane,*

(*b*) *the components of the velocity* v_x *and* v_y *and the magnitude of the velocity v,*

(*c*) *the components of the acceleration* a_x *and* a_y *and the magnitude of the acceleration.*

Solution:

Given $x = 35\,t^2 - 110\,t$ $\qquad y = 115\,t^2 - 42\,t^3$

$$v_x = \dot{x} = 70\,t - 110 \qquad v_y = \dot{y} = 230\,t - 126\,t^2$$

$$a_x = \ddot{x} = 70 \qquad a_y = \ddot{y} = 230 - 252\,t$$

MATLAB solution:

(*a*), (*b*) and (*c*): The plots are shown in Figs. ED3.1 (*a*), (*b*) and (*c*) respectively.

Program is as follows:

```
t=[0:0.5:25];
x=35*t.^2-110*t;
y=115*t.^2-42*t.^3;
v_x=70*t-110;
v_y=230*t-126*t.^2;
v=sqrt(v_x.^2+v_y.^2);
a_x=70;
a_y=230-252*t;
a=sqrt(a_x.^2+a_y.^2);
figure(1)
plot(x,y);
xlabel('x(mm)')
ylabel('y(mm)')
legend('Trajectory')
grid on
axis([0 10000 -2.5e5 0])
figure(2)
plot(t,v_x,t,v_y,t,v)
xlabel('t(sec)')
ylabel('v_x,v_y,v(mmps)')
legend('v_x','v_y','v',2)
grid on
figure(3)
plot(t,a_x,t,a_y,t,a)
xlabel('t(sec)')
ylabel('a_x,a_y,a(mmps^2)')
legend('a_x','a_y','a',2)
grid on
```

Program output:

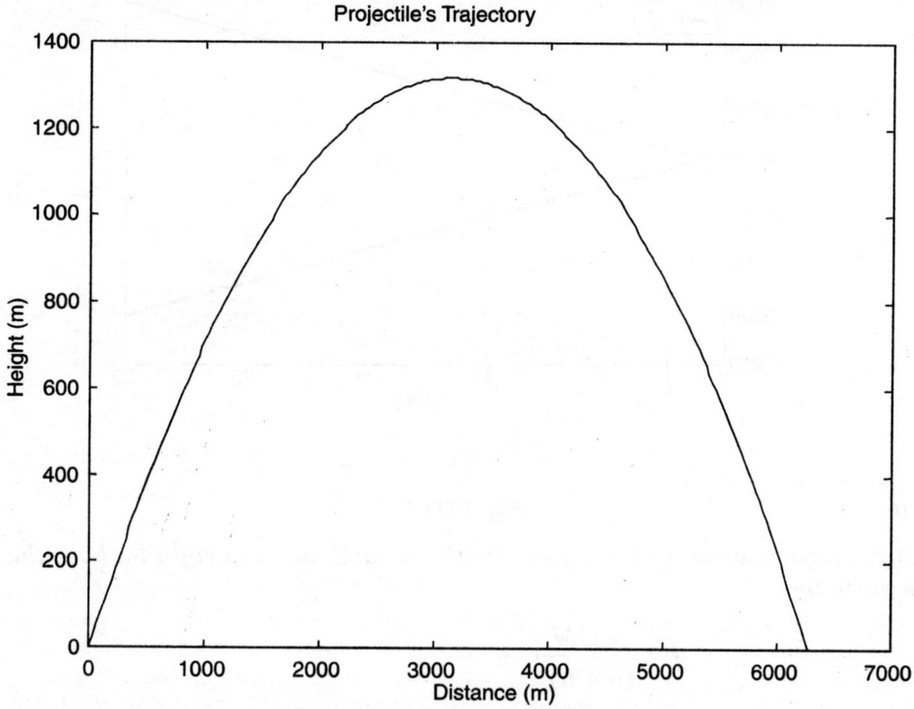

(*a*)

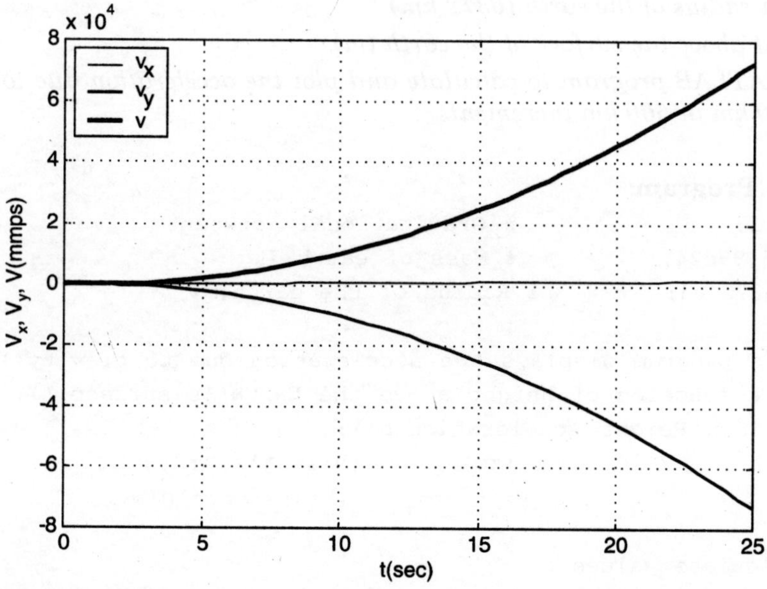

(*b*)

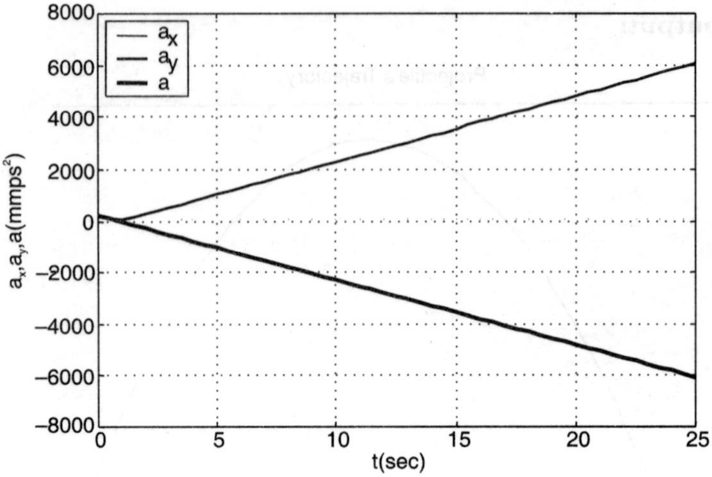

(c)

Fig. ED3.1

ED 3.2. *The acceleration due to gravity due to earth at any height h above the surface of the earth is given by*

$$g = \frac{-GM}{(R+h)^2}\ m/s^2$$

where G = *the gravitational constant* (6.67×10^{-11} Nm^2/kg^2)

M = *mass of the earth* (5.99×10^{24} *kg*)

R = *mean radius of the earth (6372 km)*

h = *height above the surface of the earth (m).*

Write a MATLAB program to calculate and plot the acceleration due to earth's gravity for $0 \le h \le 45{,}000$ *km in 500 km increments.*

Solution:

MATLAB Program:

```
G=6.67E-11;                  % Gravitational constant
m_earth=5.99e24;             % Mass of earth (kg)
r_earth=6372.e3;            % Radius of the earth(m)

disp('This program displays the acceleration due to gravity');
disp('as a function of height above the Earth''s surface:');
fprintf('\n\n Height Acceleration\n');
fprintf(       '            (km)       (m/sec**2) \n');
fprintf(       '            =======    ============\n');

% Now calculate values
jj=0;
for ii=0:500000:45000000
```

```
% Increment counter
   jj=jj+1;

   %Get height
   height(jj)=ii;

   % Calculate acceleration
    grav(jj)=-G*m_earth/(r_earth+height(jj))^2;

   % Write out results
   fprintf('       %7d       %8.3f\n',height(jj)/1000,grav(jj));
end

% Now plot the acceleration due to gravity vs height
figure(1);
plot(height/1000,grav,'LineWidth',2);
title('\bfAcceleration due to gravity vs height');
xlabel('\bfHeight above Earth''s surface (km)');
ylabel('\bfAcceleration (m/s^{2})');
```

Program output:

This program displays the acceleration due to gravity as a function of height above the Earth's surface

```
     Height          Acceleration
     (km)         (m/sec**2)
     =======      ============
           0            -9.840
         500            -8.460
        1000            -7.352
        1500            -6.447
        2000            -5.700
        2500            -5.076
        3000            -4.549
        3500            -4.100
        4000            -3.714
        4500            -3.380
        5000            -3.089
        5500            -2.835
        6000            -2.610
        6500            -2.411
        7000            -2.234
        7500            -2.076
        8000            -1.934
        8500            -1.806
        9000            -1.691
        9500            -1.586
       10000            -1.491
```

```
10500       -1.404
11000       -1.324
11500       -1.251
12000       -1.184
12500       -1.122
13000       -1.065
13500       -1.012
14000       -0.963
14500       -0.917
15000       -0.875
15500       -0.835
16000       -0.798
16500       -0.764
17000       -0.731
17500       -0.701
18000       -0.673
18500       -0.646
19000       -0.621
19500       -0.597
20000       -0.574
20500       -0.553
21000       -0.533
21500       -0.514
22000       -0.496
22500       -0.479
23000       -0.463
23500       -0.448
24000       -0.433
24500       -0.419
25000       -0.406
25500       -0.393
26000       -0.381
26500       -0.370
27000       -0.359
27500       -0.348
28000       -0.338
28500       -0.329
29000       -0.319
29500       -0.310
30000       -0.302
30500       -0.294
31000       -0.286
31500       -0.279
32000       -0.271
32500       -0.264
33000       -0.258
33500       -0.251
34000       -0.245
34500       -0.239
35000       -0.233
```

```
35500         -0.228
36000         -0.223
36500         -0.217
37000         -0.212
37500         -0.208
38000         -0.203
38500         -0.198
39000         -0.194
39500         -0.190
40000         -0.186
40500         -0.182
41000         -0.178
41500         -0.174
42000         -0.171
42500         -0.167
43000         -0.164
43500         -0.161
44000         -0.157
44500         -0.154
45000         -0.151
```

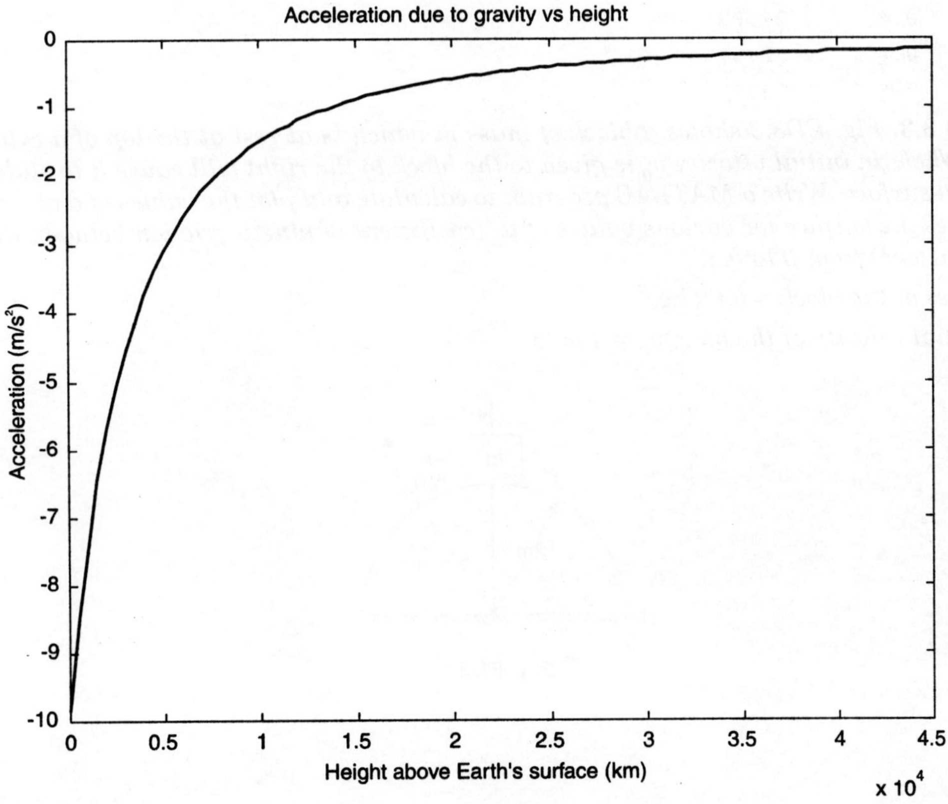

Fig. ED3.2

```
t0=0;tf=0.275;x0=[4,0]';
g=9.8;r0=2.0;mk=0.2;
[t,x]=ode45('f3',t0,tf,x0,mk,g,r0);
eq1=g*cos(x(:,2))-x(:,1).^2/r0;
[y,k]=min(abs(eq1));
theta=x(k,2)*180/pi;
fprintf('%4.2f\n',theta);

function vdot = f3(t,x,mk,g,r0)
mk=0.2;
g=9.8;
r0=2;
vdot = [g*(sin(x(2))-mk*cos(x(2)))+mk*x(1)^2/r0;x(1)/r0];
```

Ouput is

```
    mk        theta
     0        20.41
   0.1        20.34
   0.2        21.11
   0.3        21.87
   0.4        22.61
   0.5        23.35
```

ED 3.3. *Fig. ED3.3 shows a block of mass m which is at rest at the top of a cylindrical surface. When an initial velocity v_0 is given to the block to the right will cause it to slide on the cylindrical surface. Write a MATLAB program to calculate and plot the values of a at which the block leaves the surface for various values of μ_k (coefficient of kinetic friction between the block and the surface) from 0 to 0.5.*

Mass of the block = 0.60 kg.

Initial velocity of the bloc, v_0 = 4 m/s.

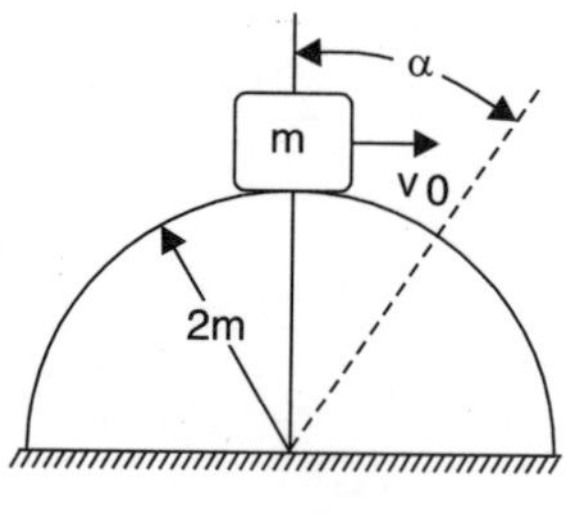

Fig. ED3.3

Solution:

The freebody diagram of block is given in Fig. ED33 (a)

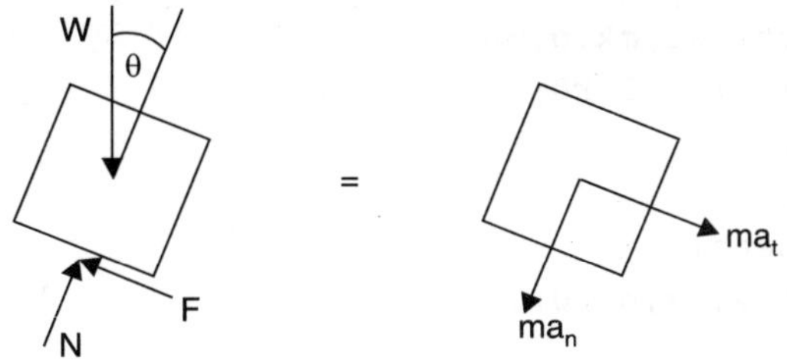

Fig. ED3.3 (a)

Summing forces in the normal direction gives

$$\Sigma F_n = ma_n:$$

$$W \cos \alpha - Nm \frac{V^2}{r}$$

or
$$N = m\left(g \cos \alpha - \frac{V^2}{r}\right)$$

For sliding we have that

$$F = \mu_k N = \mu_k m\left(g \cos \alpha - \frac{V^2}{r}\right).$$

Summing forces in the tangential direction gives

$$\Sigma F_t = ma_t:$$

$$W \sin \alpha - F = ma_t$$

or
$$a_t = g \sin \alpha - \frac{1}{m} F$$

Substituting for F in the above expression yields

$$a_t = g \sin \alpha - \frac{1}{m} \mu_k\, m\left(g \cos \alpha \frac{v^2}{r}\right) = g\,(\sin \alpha - \mu_k \cos \alpha) + \mu_k \frac{v^2}{r}$$

Now $\quad a_t = \dfrac{dv}{dt}$, thus

$$\frac{dv}{dt} = g(\sin \alpha - \mu_k \cos \alpha) + \mu_k \frac{v^2}{\rho} \quad \text{(E.1)}$$

Also, we have that

$$v = r\dot{\theta} \quad \text{or} \quad \frac{d\alpha}{dt} = \frac{1}{r} v \quad \text{(E.2)}$$

Thus, differential equations (E.1) and (E.2) define the motion of the block. As the block leaves the surface, N $\Rightarrow$ 0. Hence,

$$g \cos \alpha - \frac{v^2}{r} = 0$$

defines the value of α at which the block leaves the surface.

The program for this problem is given as follows:

```
t0=0;tf=0.275;x0=[4,0]';
g=9.8;r0=2.0;mk=0.2;
[t,x]=ode45('f3',t0,tf,x0,mk,g,r0);
eq1=g*cos(x(:,2))-x(:,1).^2/r0;
[y,k]=min(abs(eq1));
theta=x(k,2)*180/pi;
fprintf('%4.2f\n',theta);
function vdot = f3(t,x,mk,g,r0)
mk=0.2;
g=9.8;
r0=2;
vdot = [g*(sin(x(2))-mk*cos(x(2)))+mk*x(1)^2/r0;x(1)/r0];
Ouput is
          mk        theta
           1        20.41
         1.1        20.34
         1.2        21.11
         1.3        21.87
         1.4        22.61
         1.5        23.35
```

ED 3.4. *Fig ED3.4 shows the motion of a particle. The initial velocity and the angle at which the projectile is fired are known. Write a MATLAB program to calculate and plot the maximum height and distance. Use the program to calculate and plot the trajectory of a projectile that is fired at a velocity of 250 m/s at an angle of 40°.*

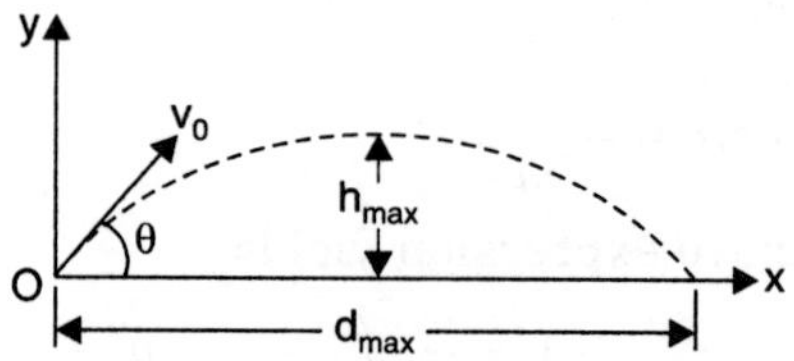

Fig. ED3.4

Solution:

Components of velocity $v_{0x} = v_0 \cdot \cos\theta$ and $v_{0y} = v_0 \sin\theta$.

Height $$h_{max} = \frac{{v_0}^2 y}{2g}$$

and corresponding time

$$t_{max} = \frac{v_0 y}{g}$$

But to draw trajectory as a function of distance (x)

$$y = v_{0y}t - \frac{1}{2}gt^2 \quad \text{and} \quad x = v_{0x}t$$

Open the file "trajectory.*m*" and type the following

```
function[hmax,dmax]=trajectory(v0,theta)
%trajectory calculates the max height and distance of a projectile, and %
makes a plot of the trajectory.
% Input arguments are:
%v0:initial velocity in(m/s).
%theta:angle in degrees.
%Output arguments are:
%hmax:maximum height in (m).
%dmax:maximum distance in(m).
%The function creates also a plot of the trajectory.
g=9.81;
v0x=v0*cos(theta*pi/180);
v0y=v0*sin(theta*pi/180);
```

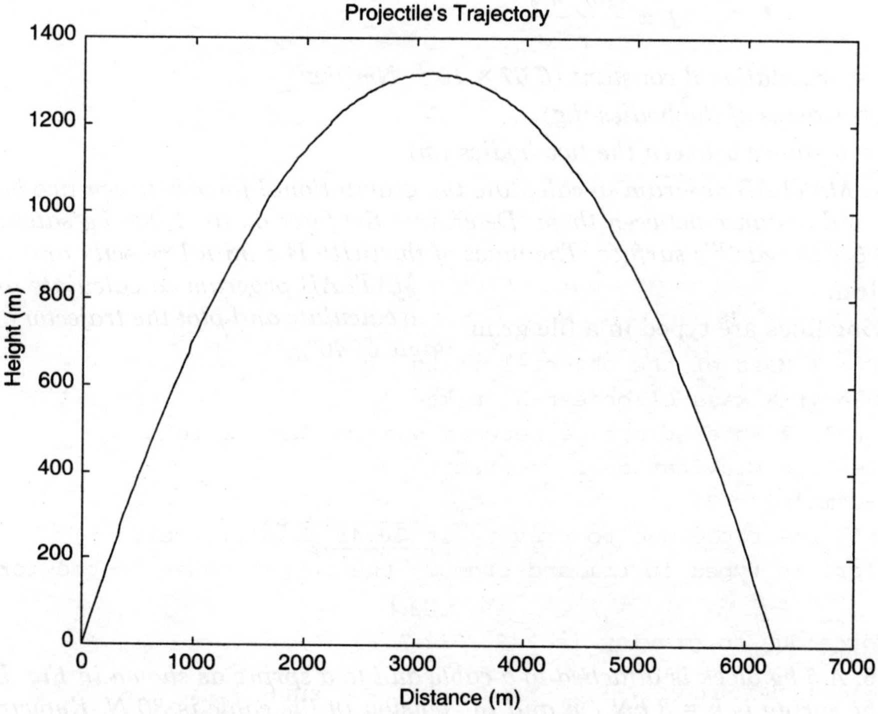

Fig. ED 3.4 (*a*)

```
thmax=v0y/g;
hmax=v0y^2/(2*g);
ttot=2*thmax;
dmax=v0x*ttot;
%Creating a trajectory plot
tplot=linspace(0,ttot,200);
x=v0x*tplot;
y=v0y*tplot-0.5*g*tplot.^2;
```

```
plot(x,y)
xlabel('DISTANCE(m)')
ylabel('HEIGHT(m)')
title('PROJECTILE'S TRAJECTORY')
```

To execute it type the following at the MATLAB command prompt

```
>> [h d] = trajectory(250,40)
```

Then the output comes as:

```
h =
      1.3162e+003
d =
      6.2743e+003
```

This is followed by the Fig. ED 3.4(*a*).

ED 3.5. *The gravitational force F between two bodies of masses m_1 and m_2 is given by*

$$F = \frac{Gm_1 m_2}{r^2}$$

where G = *gravitational constant* (6.67×10^{-11} Nm^2/kg^2)

m_1, m_2 = *masses of the bodies (kg)*

r = *distance between the two bodies (m).*

Write a MATLAB program to calculate the gravitational force between two bodies given their masses and distance between them. Determine the force on an 1,000 kg satellite in orbit 40,000 km above the earth's surface. The mass of the earth is 5.98×10^{24} kg.

Solution:

Following lines are typed in a file gr.m

```
m1=1000; % Mass of the object-1 in kg
m2=5.98e24; % Mass of object-2 in kg
r=40000e3; % Enter distance between the two bodies in m
G=6.67e-11;% Gravitational constant
force=G*m1*m2/r^2;
fprintf('The force due to gravity is %0.4f N.\n',force);
```

Once 'gr' is typed in command prompt, the output comes in the form:

```
The force due to gravity is 249.2912 N.
```

ED 3.6. *A 5 kg block is attached to a cable and to a spring as shown in Fig. ED3.6. The constant of the spring is k = 3 kN./m and the tension in the cable is 30 N. Knowing that the cable is cut, (a) derive an expression for the velocity of the block as a function of its displacement x, (b) determine the maximum displacement x_m and the maximum speed v_m, (c) plot the speed of the block as a function of s for $0 \le s \le s_m$.*

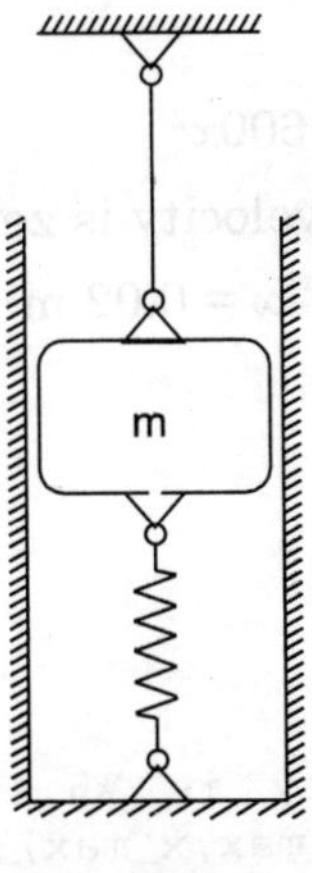

Fig. ED3.6

Solution:

Free body diagram of the block is shown in Fig. ED3.6 (*a*)

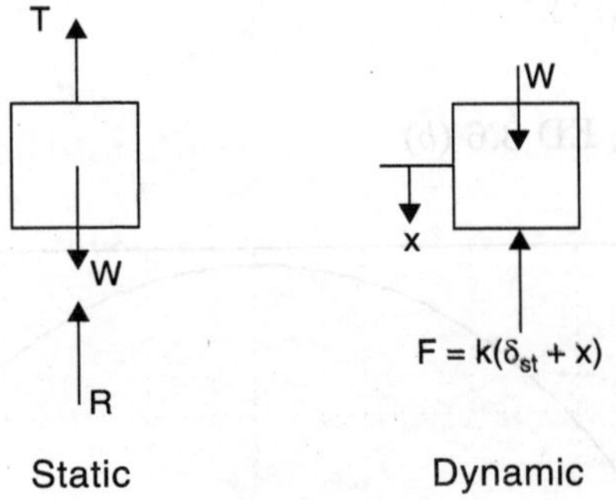

Fig. ED3.6 (*a*)

For the static case we have

$T = 30\ N$, $W = mg = 50\ N$, $R = 20$ N

Thus,

$$R = k\delta_{st} \qquad \delta_{st} = \frac{R}{k}$$

$$\delta_{st} = \frac{20}{3000} = 0.006667 \text{ m}$$

Using the principle of work and energy we have

$$T_1 + U_{1\to 2} = T_2$$

or

$$\int_0^x [W - k(\delta_{st} + x)]dx = \frac{1}{2}mv^2$$

Substituting yields

$$\int_0^x [50 - 3000(0.006667 + x)]dx = 2.5v^2$$

$$\Rightarrow \quad 50x - 20x - 1500x^2 = 2.5v^2$$

$$v^2 = 12x - 600x^2$$

At maximum displacement, the velocity is zero.

$$\therefore \quad 12x - 600x^2 = 0 \quad \text{or} \quad x = 0.02 \text{ m.}$$

The program is

```
x=[0:.0001:0.02];
v=sqrt(12.*x-600.*x.^2);
[v_max,i]=max(v);
x_max=x(i);
fprintf('The maximum velocity is %5.4fm/s and the corresponding maximum displacement is %5.4fm\n',v_max,x_max);
figure(1)
plot(x,v)
xlabel('x,(m)')
ylabel('Velocity(m/s)')
grid on
```

Output is shown in Fig. ED 3.6 (*b*)

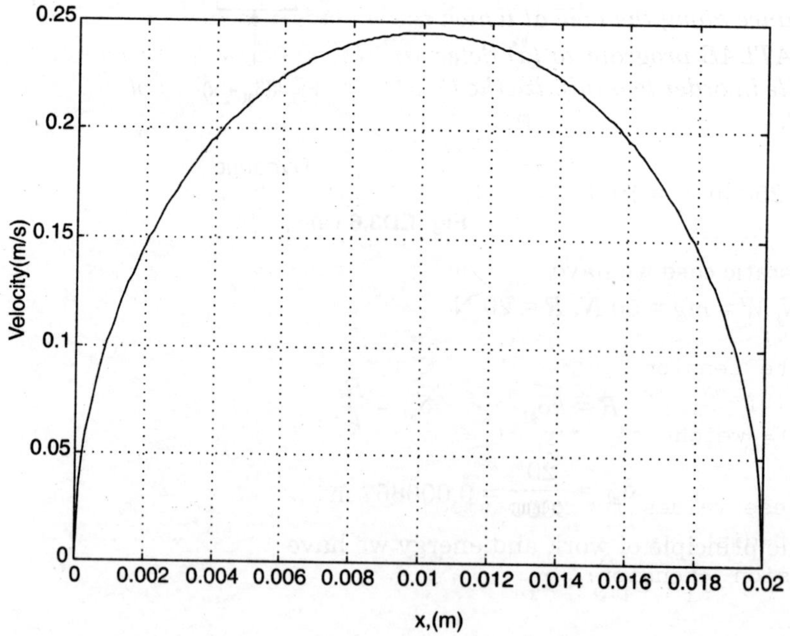

Fig. ED 3.6 (*b*)

ED 3.7. *Fig ED3.7 shows a weight W hung from the end of a horizontal pole of negligible weight.*

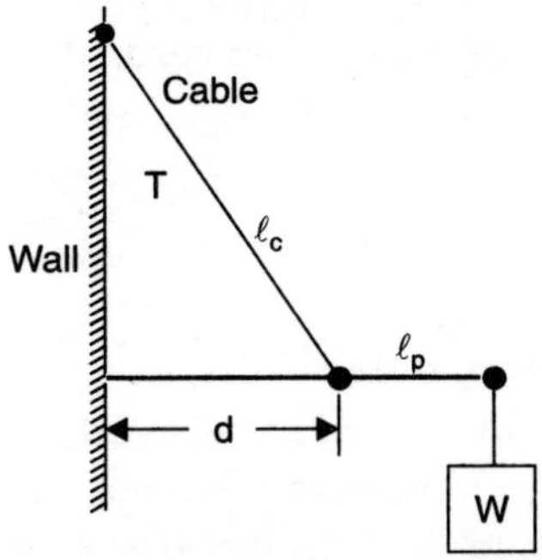

Fig. ED3.7

The pole is attached to the wall by a pivot and is supported by a cable attached to the wall at a higher point. The tension, T, in the cable is given by

$$T = \frac{wl_c l_p}{d\sqrt{l_p^{\,2} - d^2}}$$

where T = tension in the cable, W = weight of the object, l_c = length of the cable, l_p = length of the pole and d = distance along the pole at which the cable is attached.

Write a MATLAB program to (a) determine the distance (d) at which the cable can be attached to the pole in order to minimize the tension in the cable, (b) plot the tension in the cable as a function of d.

Solution:

Given $W = 250$ lb, $l_p = 10$ ft, $l_c = 10$ ft.

```
% Initial values
lc=10;
lp=10;
weight=250;
% Calculate tension
dist(1)=1.0;
tension(1)= weight *lc *lp /( dist(1)* sqrt(lp^2 - dist(1)^2) );

% Save these values for comparison
saved_dist=dist;
saved_tension =tension;

% Calculate for those values between 1 and 9 ft
for ii = 2:80
     % Calculate tension
     dist(ii) =ii/10 + 1;
     var=dist(ii);
     den=var* sqrt(lp^2 - var^2);
     tension(ii)= weight *lc*lp/den;
```

```
        % Find minimum tension point
        if tension(ii) < saved_tension
            saved_tension = tension(ii);
            saved_dist = dist(ii);

        end
    end

    % Display minimum tension point
    disp(['Minimum tension at 'num2str(saved_dist)]);
    disp(['Minimum tension = 'num2str(saved_tension)]);
    %plot tension vs distance
    figure(1);
    plot(dist,tension,'b-','LineWidth',2);
    grid on;
    title('\bfPlot of tension vs attachment distance');
    xlabel('\bfDistance');
    ylabel('\bfTension');
```

The output is given in Fig. ED 3.7 (*a*).

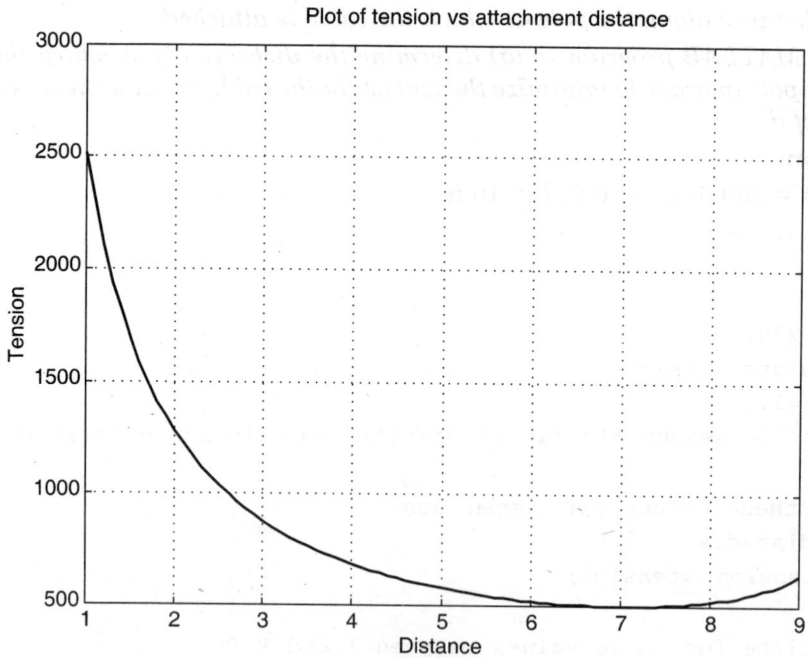

Fig. ED 3.7(*a*)

At the command prompt following output is obtained:

>>   Minimum tension at 7.1

Minimum tension = 500.0168

ED 3.8. *Fig. ED3.8 shows the 3-D projectile trajectory where a projectile is fired with an initial velocity of v_0 at an angle of θ relative to the ground. The projectile is aimed directly north. The projectile also moves in the western direction at a constant speed of 25 m/s. Write a MATLAB program to (a) determine and plot the trajectory of the projectile until it hits the ground, (b) plot the trajectory that the projectile would have had if the projectile does not move in the western direction. Given v_0 = 300 m/s and θ = 60°.*

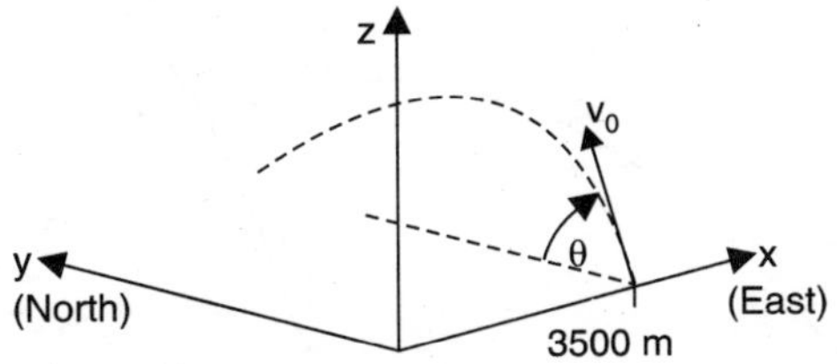

Fig. ED3.8

Solution:

x and y axes represent East and North directions. Consider vertical direction as z. As projectile is fired in north, the initial velocity v_0 can be resolved into y and z direction as:

$$v_{0y} = v_0.\cos\theta \quad \text{and} \quad v_{0z} = v_0.\sin\theta$$

Time taken by projectile to reach highest point ($v_z = 0$) is

$$t_h = \frac{v_{0z}}{g}$$

Total flying time, $\quad t_t = 2.t_h$

Due to wind, projectile has a constant velocity in the negative x direction, $v_x = -25$.

∴ Position of projectile is given by

$$x = x_0 + v_x t \quad \text{and} \quad y = y_0 + v_{oy} t$$

$$z = z_0 + v_{0z} t - \frac{1}{2} g t^2$$

Start with initial point (x_0, y_0, z_0) = (3000.0.0)

MATLAB program

```
v0=300;g=9.81;theta=60;
x0=3000;vx=-25;
v0z=v0*sin(theta*pi/180);
v0y=v0*cos(theta*pi/180);
t=2*v0z/g;
tplot=linspace(0,t,100); % CREATING A TIME VECTOR WITH 100 ELEMENTS
z=v0z*tplot-0.5*g*tplot.^2;
y=v0y*tplot;
x=x0+vx*tplot;
 % CALCULATING X,Y,Z COORDINATES OF THE PROJECTILE AT EACH TIME

xnowind(1:length(y))=x0; % CONSTANT X-COORDINATE WHEN NO WIND
plot3(x,y,z,'k-',xnowind,y,z,'k--') % TWO 3-D LINE PLOTS
grid on
```

```
xlabel('x(m)');
ylabel('y(m)');zlabel('z(m)')
```

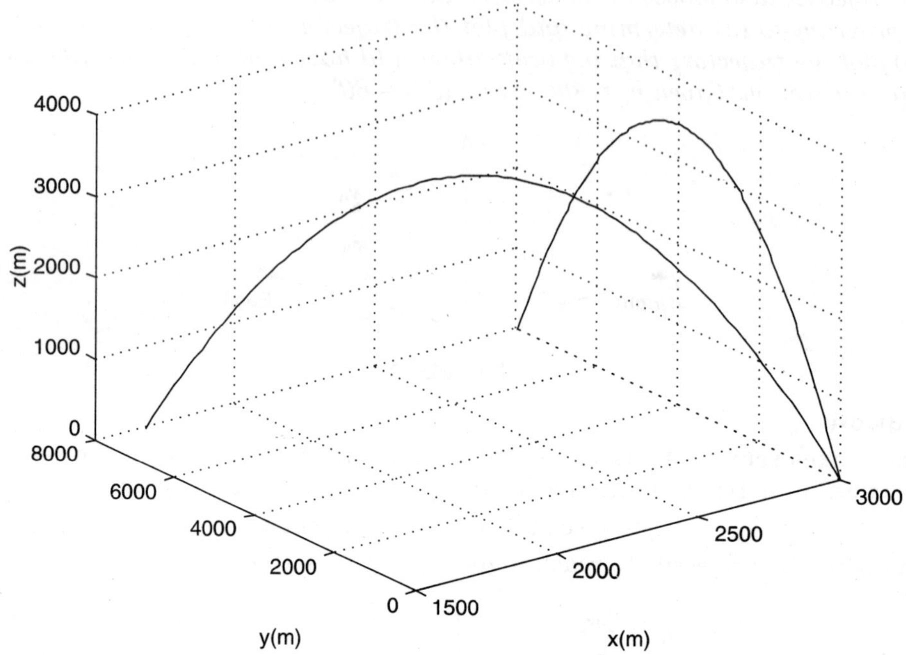

Fig. ED 3.8 (*a*)

ED 3.9. *Fig. ED3.9 shows a block B of mass m_B starts from rest and slides down an inclined plane of a wedge of mass m_A which is supported by a horizontal surface. (a) Obtain an expression for the speed of block B relative to wedge A. (b) The speed of wedge A, (c) Write a MATLAB program to plot the speed of B relative to A and the speed of A as function of s, where s is the distance travelled by the block B down the surface of the wedge for $0 \le s \le 1.0$ m.*

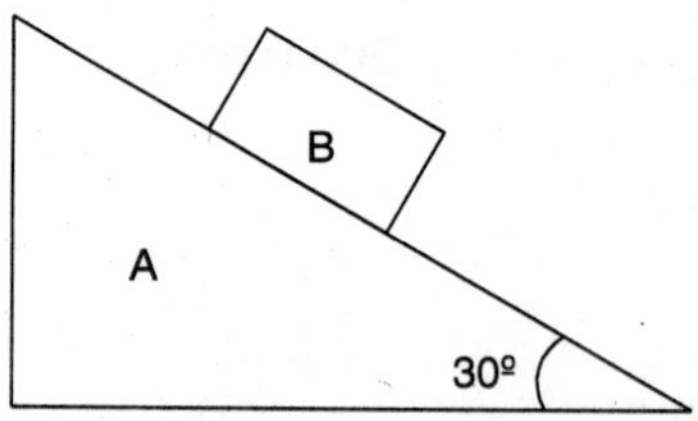

Fig. ED3.9

Solution:

Given: $m_B = 10$ kg and $m_A = 16$ kg. Velocity triangle is given in Fig. ED3.9 (*a*).

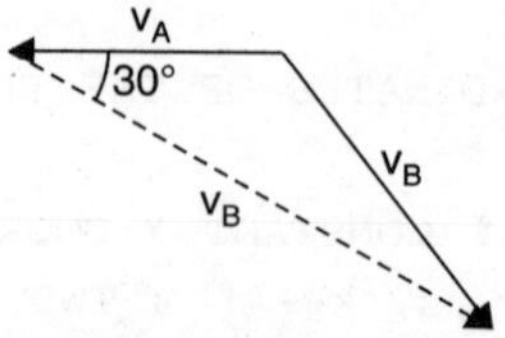

Fig. ED3.9 (*a*)

$$m_B\, v_{BA} \cos 30 - m_A\, v_A - m_B\, v_A = 0$$

Speed of block B relative to wedge A

$$v_{BA} = \frac{(m_A + m_B)\, v_A}{m_B \cos 30} = \frac{26 \times v_A}{10 \cos 30} = 3v_A$$

Applying conservation of energy rule

$$m_B gh = \frac{1}{2}\, m_A\, v_A^2 + \frac{1}{2}\, m_B\, v_B^2$$

where $\quad v_B^2 = v_A^2 + v_{BA}^2 - 2v_A\, v_{BA} \,.\, \cos 30$

$$= v_A^2 + 9v_A^2 - 6v_A^2 \cos 30 = 4.804v_A^2$$

$$\Rightarrow \quad 10 \times 9.81 \times s \cos 30 = \frac{1}{2} \times 16v_A^2 + \frac{1}{2} \times 4.804v_A^2 \quad \Rightarrow \quad v_A^2 = 2.654s.$$

$$v_A = 1.63\sqrt{s}$$

$$\therefore \quad v_{BA} = 3 \times 1.63\sqrt{s} = 4.89\sqrt{s}$$

MATLAB program

```
s=[0:0.01:1.0];
va=1.63*sqrt(s);
vb_a=4.89*sqrt(s);
figure(1)
plot(s,va,s,vb_a)
xlabel('s(m)')
ylabel('Velocity  (m/s)')
legend('Velocity of A','Velocity of B relative to A',2);
grid on
```

Output is shown in Fig. ED3.9 (*b*)

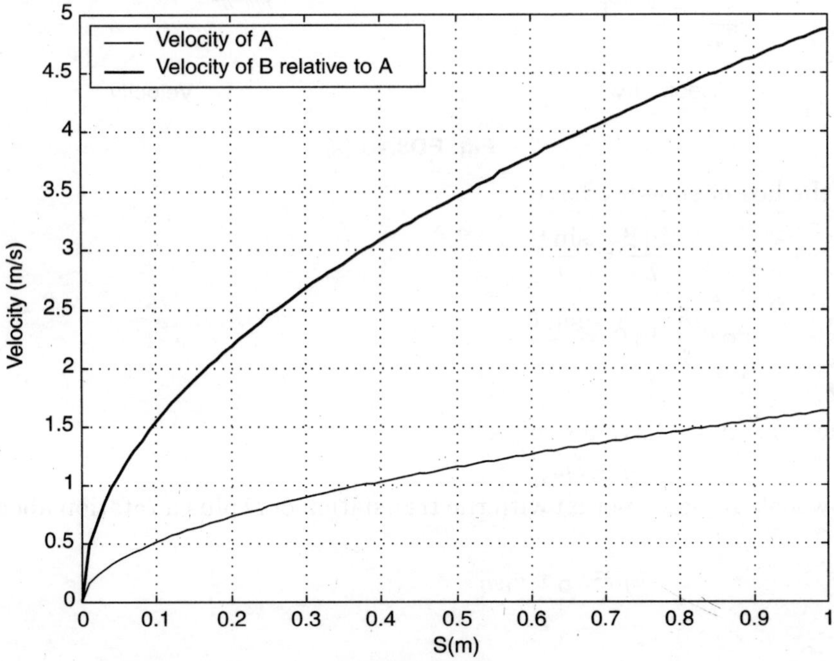

Fig. ED 3.9 (*b*)

ED 3.10. *Fig. ED3.10 shows an engine system where the crank AB rotates with a constant angular velocity ω_{AB} clockwise. Write a MATLAB program to determine the plot for values of θ from 0 to 180° (a) the angular velocity and angular acceleration of connecting rod BD, (b) the velocity and acceleration of piston P. The length of the connecting rod, l = 10 in. Crank length, AB = 3.5, ω_{AB} = 1000 rpm.*

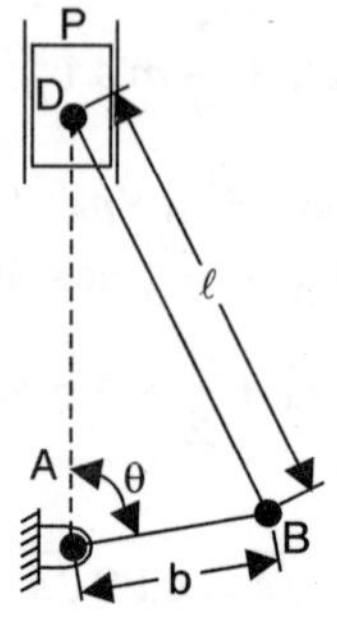

Fig. ED3.10

Solution:

Free body diagrams are shown in Fig. ED3.10 (*a*).

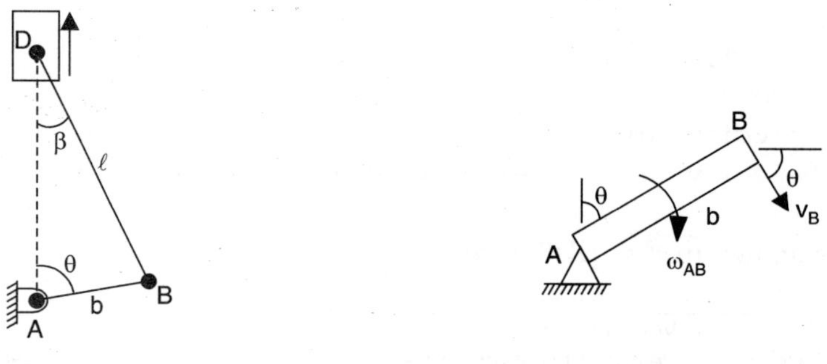

Geometry

Velocity

Fig. ED3.10 (*a*)

From the law of sines we have

$$\frac{\sin\beta}{b} = \frac{\sin\theta}{l}$$

$$\sin\beta \frac{b}{l}\sin\theta = \frac{\sin\theta}{n} \quad (1)$$

where $n = l/b$.

For the velocity we have for rod AB

$$v_B = b\omega_{AB} \quad (2)$$

We now look at plane motion with the translation of D plus a rotation about D. For B we have

$$v_B = v_D + v_{B/D}$$

Angular velocity of connecting rod

$$BD\left(\frac{d\beta}{dt} = \omega_{BD} = \frac{\omega \cos\theta}{\sqrt{n^2 - \sin^2\theta}}\right)$$

Angular acceleration

$$\alpha_{CD} = \frac{\omega^2 . \sin\theta\,(n^2 - 1)}{(n^2 - \sin^2\theta)^{3/2}}$$

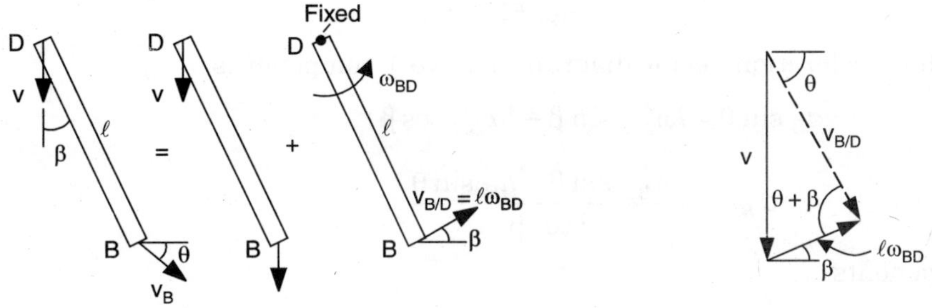

Fig. ED3.10 (*b*)

Using the law of sine gives

$$\frac{v_D}{\sin(\theta+\beta)} = \frac{v_B}{\sin(90-\beta)} = \frac{l\omega_{BD}}{\sin(90-\theta)}$$

$$v_D = v_B \frac{\sin(\theta+\beta)}{\sin(90-\beta)} = v_B \frac{\sin(\theta+\beta)}{\cos\beta} \tag{3}$$

$$\omega_{BD} = \frac{V_B \sin(90-\theta)}{l \sin(90-\beta)} = \frac{v_B \cos\theta}{l\cos\beta} \tag{4}$$

For the acceleration we have for rod AB

$$a_B = b\omega_{AB}^2 | \theta \tag{5}$$

and for rod BD:

$$(a_{D/B})_t = l\alpha BD$$

$$(a_{D/B})_n = l\omega_{BD}^2$$

Acceleration:

For the acceleration we again look at plane motion with the translation of D plus a rotation about D. For B we have

$$a_B = a_D + a_{B/D}$$

$$a_B = a_D + (a_{B/D})_t + (a_{B/D})_n$$

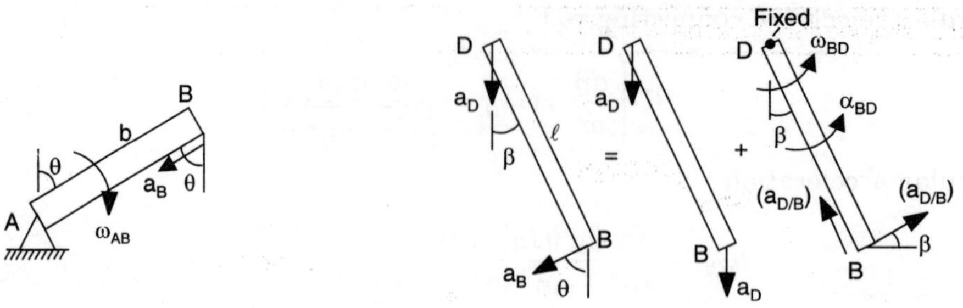

Fig. ED3.10 (*c*)

Using the acceleration vector diagram *w* have + components.

$$a_B \sin\theta = l\omega_{BD}^2 \sin\beta - l\alpha_{BD} \cos\beta$$

$$\alpha_{BD} = \frac{l\omega_{BD}^2 \sin\beta - a_B \sin\theta}{l\cos\beta} \tag{6}$$

+↓ components

$$a_B \cos\theta = a_D - l\omega_{BD}^2 \cos\beta - l\alpha_{BD} \sin\beta$$

$$a_D = a_B \cos\theta + l\omega_{BD}^2 \cos\beta + l\alpha_{BD} \sin\beta \tag{7}$$

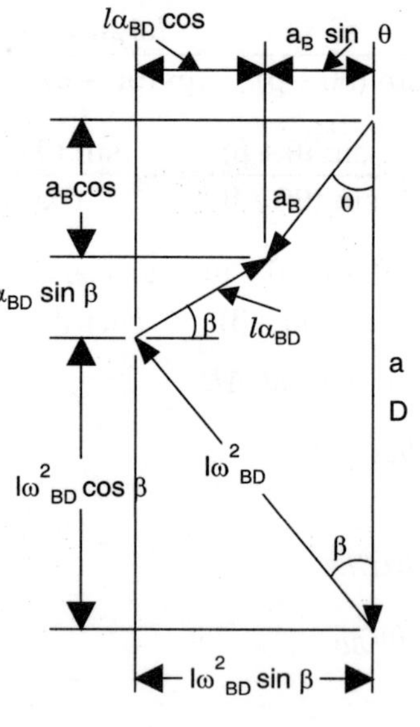

Fig. ED3.10 (*d*)

MATLAB program

```
%
L=10; % Length of connecting rod
b=3.5;% crank radius
%Velocity
omega_AB=1000*2*pi/60; % angular velocity
v_B=b*omega_AB; % linear velocity of crank pin
```

```
Theta=[0:5:180];
t=Theta*pi/180;
beta=asin((b/L)*sin(t));% inclination of connecting rod with axis of piston
v_D=v_B*sin(t+beta)./cos(beta);% velocity of piston in in/sec
v_D=v_D/12; % velocity in ft/sec
omega_BD=v_B*cos(t)./(L*cos(beta));% angular velocity of connecting rod
% acceleration
a_B=b*omega_AB^2; % acceleration of point B
alpha_BD=(L.*omega_BD.^2.*sin(beta)-a_B.*cos(t))./(L.*cos(beta));
a_D=a_B.*cos(t)+L.*omega_BD.^2.*cos(beta)+L.*alpha_BD.*sin(beta);
a_D=a_D/12;
% Determine and plot values
z=[Theta;omega_BD;alpha_BD;v_D;a_D];
fprintf('Theta   Angular    Angular       Piston       Piston\n')
fprintf('     velocity   Acceleration    velocity    acceleration\n')
fprintf('(deg)    (rad/s)    (rad/s^2)    (ft/s)    (ft/s^2)\n')
fprintf('\n');
fprintf('%5.3f  %6.3f      %6.3f    %6.3f    %6.3f\n',z);
fprintf('\n');
%fprintf('The two values of theta for zero collar speed are %5.3f and %5.3f
degrees\n',a1,a2)
figure(1)
plot(Theta,omega_BD)
xlabel('Theta(degrees)')
ylabel('Angular Velocity Rod BD (rad/s)')
grid on
figure(2)
plot(Theta,alpha_BD)
xlabel('Theta (degrees)')
ylabel('Angular Acceleration Rod BD(rad/s^2)')
grid on
figure(3)
plot(Theta,v_D)
xlabel('Theta(degrees)')
ylabel('Piston Velocity (ft/s)')
grid on
figure(4)
plot(Theta,a_D)
xlabel('Theta(degrees)')
ylabel('Piston Acceleration D (ft/s^2)')
grid on
```

Output is as follows:

Theta	Angular velocity	Angular acceleration	Piston velocity	Piston acceleration
(deg)	(rad/s)	(rad/s^2)	(ft/s)	(ft/s^2)
0.000	36.652	–3838.179	0.000	4317.952

5.000	36.529	–3784.630	3.591	4201.587
10.000	36.162	–3707.245	7.135	4049.852
15.000	35.549	–3607.751	10.589	3865.943
20.000	34.691	–3487.725	13.907	3653.345
25.000	33.587	–3348.540	17.048	3415.809
30.000	32.239	–3191.321	19.973	3157.328
35.000	30.647	–3016.912	22.646	2882.120
40.000	28.816	–2825.851	25.035	2594.606
45.000	26.749	–2618.379	27.114	2299.362
50.000	24.455	–2394.461	28.861	2001.068
55.000	21.944	–2153.852	30.262	1704.410
60.000	19.231	–1896.200	31.309	1413.962
65.000	16.333	–1621.186	31.999	1134.026
70.000	13.274	–1328.702	32.339	868.446
75.000	10.080	–1019.047	32.342	620.416
80.000	6.780	–693.123	32.027	392.281
85.000	3.408	–352.596	31.417	185.390
90.000	0.000	–0.000	30.543	0.000
95.000	–3.408	361.239	29.437	–164.733
100.000	–6.780	726.881	28.132	–310.665
105.000	–10.080	1092.042	26.663	–440.487
110.000	–13.274	1451.434	25.063	–557.474
115.000	–16.333	1799.649	23.364	–665.186
120.000	–19.231	2131.457	21.594	–767.172
125.000	–21.944	2442.066	19.777	–866.685
130.000	–24.455	2727.333	17.934	–966.462
135.000	–26.749	2983.908	16.081	–1068.563
140.000	–28.816	3209.296	14.230	–1174.287
145.000	–30.647	3401.866	12.392	–1284.150
150.000	–32.239	3560.796	10.570	–1397.923
155.000	–33.587	3685.985	8.768	–1514.710
160.000	–34.691	3777.938	6.986	–1633.044
165.000	–35.549	3837.653	5.222	–1751.006
170.000	–36.162	3866.494	3.472	–1866.339
175.000	–36.529	3866.078	1.733	–1976.552
180.000	–36.652	3838.179	0.000	–2079.014

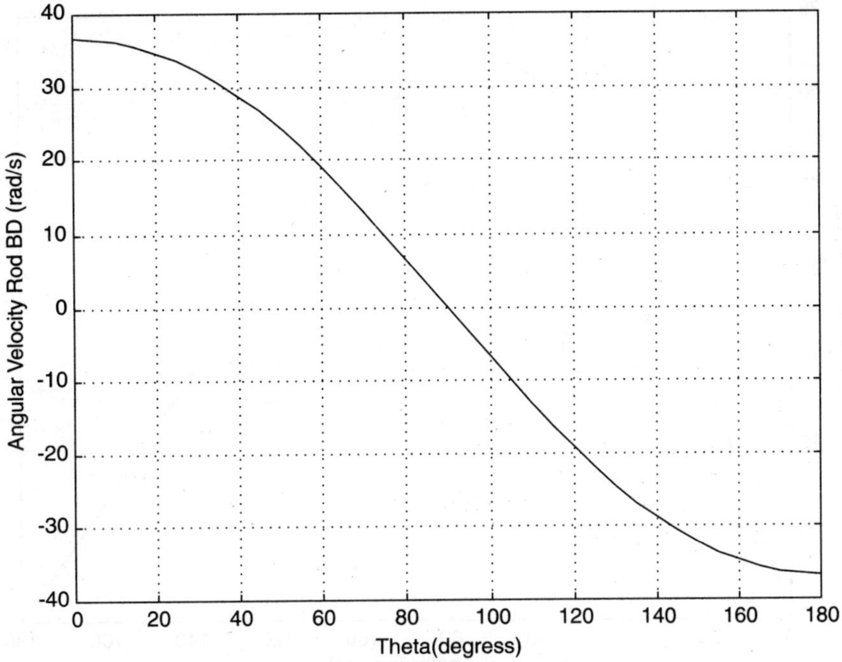

Angular velocity of connecting rod

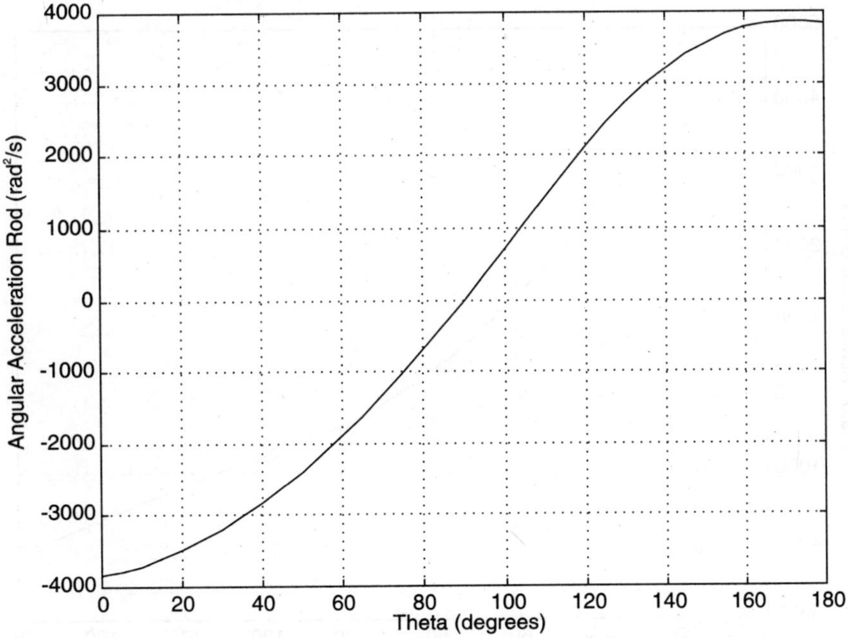

Angular acceleration of connecting rod

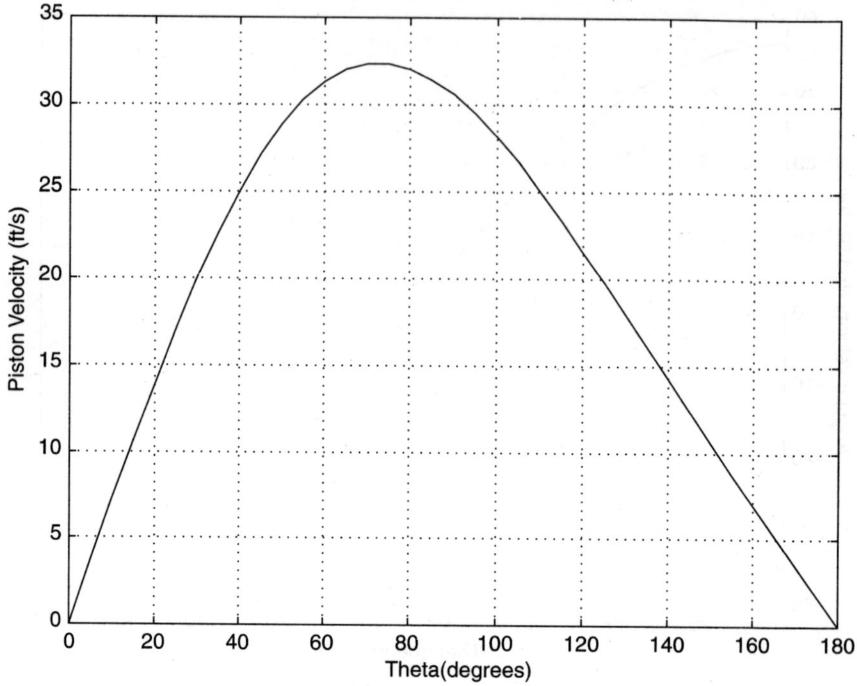

Piston velocity

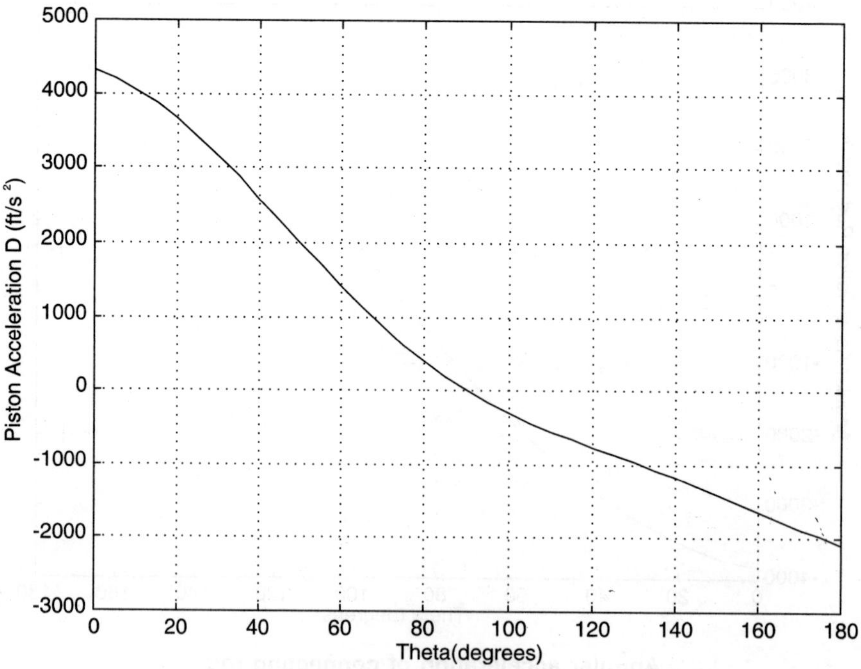

Piston acceleration

ED 3.11. *Fig. ED3.11 shows a safety bumber placed at the end of a racetrack to stop out-of-control vehicles. The force that the bumber applies to the vehicle is given by*

$$F = kv^3(x + 1)^3$$

where $k = 32$ *kg-s/m*^{5} *(a constant)*

x = *displacement of the front edge of the bumber*

v = *velocity of the front edge of the bumber.*

A vehicle of mass 2000 kg hits the bumper at a speed of 100 km/h. Write a MATLAB program to determine and plot the velocity of the vehicle as a function of x for $0 \le x \le 5$ *m. The mass of the vehicle = 2000 kg. Vehicle speed when it hit the bumber = 100 km/h.*

Fig. ED3.11

Solution:

The deceleration of the car once it hits the bumber can be calculated from Newton's second law of motion.

$$ma = -Kv^3(x + 1)^3$$

which can be solved for the acceleration a as a function of v and x:

$$a = -\frac{Kv^3\,(x+1)^3}{m}$$

The velocity as a function of x can be calculated by substituting the acceleration in the equation:

$$vdv = adx$$

which gives:

$$\frac{dv}{dx} = \frac{-Kv^2\,(x+1)^3}{m}$$

The last equation is a first order *ODE* that needs to be solved for the interval $0 \le x \le 3$ with the initial condition:

$$v = 90 \text{ km/h at } x = 0.$$

Numerical solution of differential equation is shown in the following program written in a script file (*m*-file), which should be executed by typing the file name in command prompt.

MATLAB program

```
global k m
k=32;m=2000;v0=100;
xspan=[0:0.2:5];% a vector that specifies the interval of the solution
v0mps=v0*1000/3600; % changing velocity in m/s
[x v]=ode45('bumper',xspan,v0mps)% Solving the ODE
plot(x,v)
xlabel('x(m)');ylabel('velocity(m/s)')
```

The function file with the differential equation named bumper.m is listed below:

```
function dvdx=bumper(x,v)
global k m
dvdx=-(k*v^2*(x+1)^3)/m;
```

The output of the above program is the vectors x and v as follows along with plot.

```
x =
         0
    0.2000
    0.4000
    0.6000
    0.8000
    1.0000
    1.2000
    1.4000
    1.6000
    1.8000
    2.0000
    2.2000
    2.4000
    2.6000
    2.8000
    3.0000
    3.2000
    3.4000
    3.6000
    3.8000
    4.0000
    4.2000
    4.4000
    4.6000
    4.8000
    5.0000

v =
   27.7778
   24.8173
   21.1176
   17.1814
   13.5109
   10.4127
    7.9665
    6.0685
    4.6826
    3.6212
    2.8257
    2.2268
    1.7734
    1.4268
    1.1588
    0.9505
    0.7855
    0.6552
    0.5501
    0.4651
```

```
0.3959
0.3389
0.2919
0.2526
0.2197
0.1920
```

Output is given in Fig.ED3.11 (a)

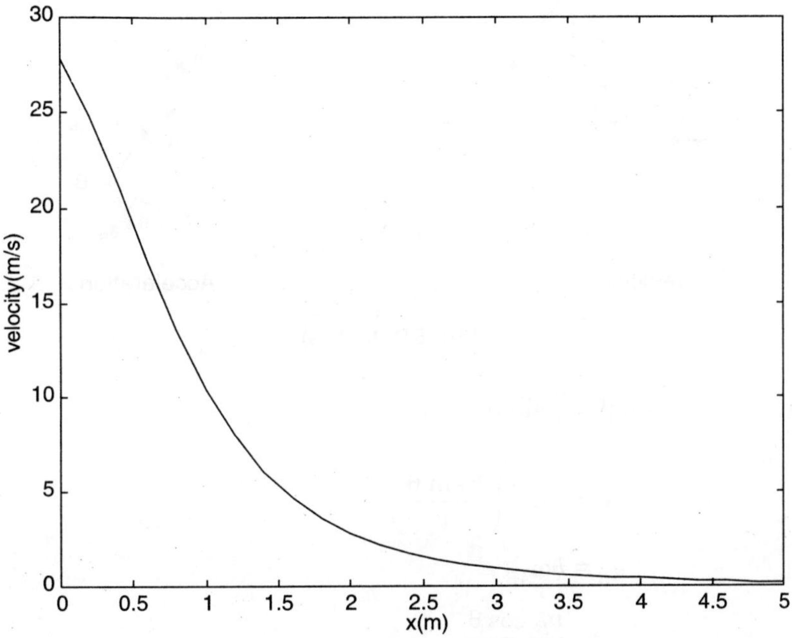

Fig. ED 3.11 *(a)*

ED 3.12. *Fig. ED3.12 shows an engine system where the connecting rod BD is attached to the piston P. The crank AB rotates with a constant angular velocity of ω_{AB} rpm clockwise with no force applied to the face of the piston. Write a MATLAB program to plot the horizontal and vertical components of the dynamic reactions exerted on the connecting rod at B and D for $0 \le \theta \le 180°$. Weight of the connecting rod BD = 5.5 lb. Weight of the piston, W_P = 6.3 lb.*

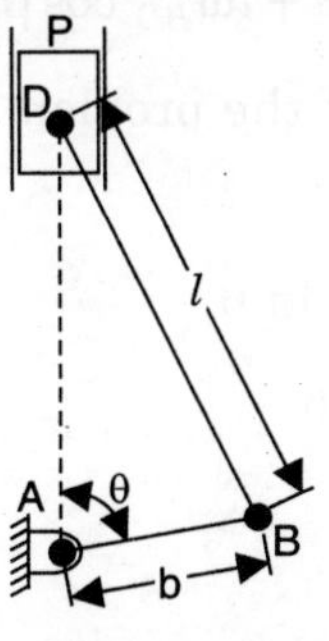

Fig. ED3.12

Solution:

Directions of velocity and accelerations is shown in Fig.ED3.12 (*a*).

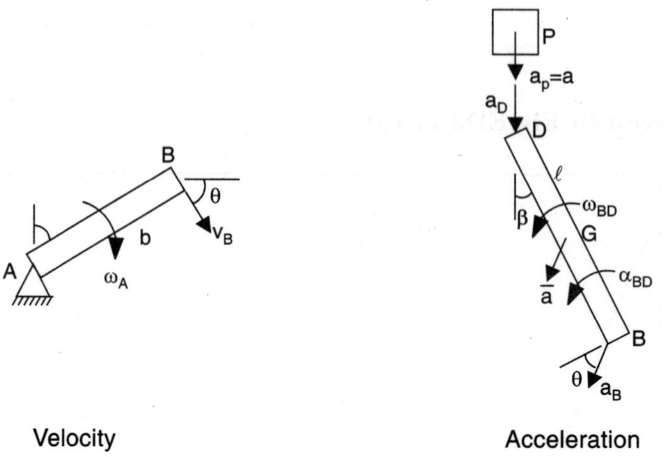

Fig. ED 3.12 (*a*)

We have $$\sin\beta = \frac{b}{l}\sin\theta$$

or $$\beta = \sin^{-1}\left(\frac{b\sin\theta}{l}\right) \quad (1)$$

$$v_B = b\omega_{AB} \quad (2)$$

$$\omega_{BD} = \frac{v_B\cos\theta}{l\cos\beta} \quad (3)$$

$$a_B = b\omega_{AB}^2 \quad (4)$$

$$\alpha_{BD} = \frac{l\omega_{BD}^2\sin\beta - a_B\sin\theta}{l\cos\beta} \quad (5)$$

$$a_D = a_B\cos\theta + l\omega_{BD}^2\cos\beta + l\alpha_{BD}\sin\beta \quad (6)$$

Now, we study the kinematics of the problem. First we must find $\bar{a}_x$ and $\bar{a}_y$. From the figure, we have

$$\overset{+}{\rightarrow}\ (a_B)_x = -\,a_B\sin\theta$$

$$+\downarrow (a_B)_y = a_B\cos\theta$$

Kinetics:

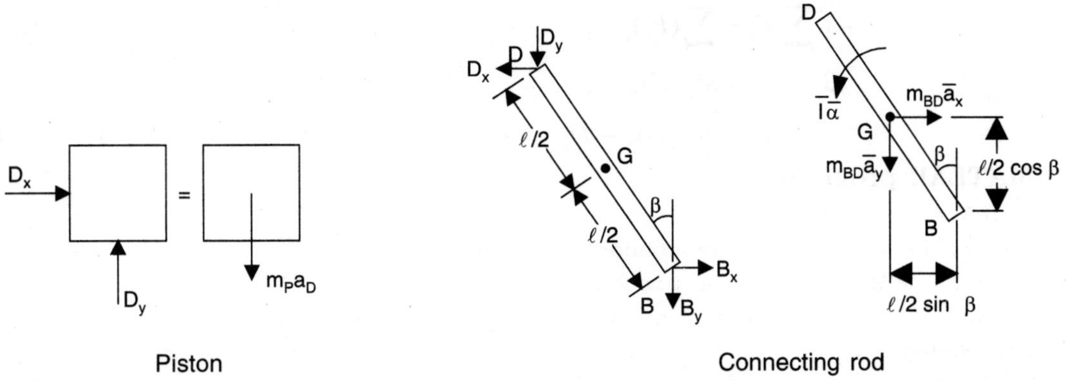

Fig. ED 3.12 (*b*)

Since G is at the middle of BD we have

$$\underset{\rightarrow}{+}\ \bar{a}_x = \frac{1}{2}(a_B)_x \tag{7}$$

$$+\downarrow \bar{a}_y = \frac{1}{2}[(a_B)_x + a_D] \tag{8}$$

For the piston we find

$$+\downarrow \sum F_y = \sum (F_y)_{eff} :$$

$$D_y = -\, m_P a_D \tag{9}$$

Since we are after the dynamic reactions, we shall omit the weight of the piston and connecting rod. For the connecting rod we have

$$\bar{I} = \frac{1}{12} m_{BD}\, l^2 \tag{10}$$

Summing moments about B yields

$$\sum M_B = \sum (M_B)_{eff} :$$

$$-D_x\, l \cos\beta - D_y\, l \sin\beta = -\,\bar{I}\alpha + m_{BD} a_x \left(\frac{1}{2}\cos\beta\right) - m_{BD}\left(\frac{1}{2}\sin\beta\right)$$

Dividing by l and solving for D_x gives

$$D_x = -\,D_y \frac{\sin\beta}{\cos\beta} + \frac{\bar{I}\alpha}{l\cos\beta} - \frac{m_{BD}}{2}\bar{a}_x + \frac{m_{BD}}{2}\bar{a}_y \frac{\sin\beta}{\cos\beta}$$

$$D_x = -\,D_y \tan\beta + \frac{\bar{I}\alpha}{l\cos\beta} - \frac{m_{BD}}{2}\bar{a}_x + \frac{m_{BD}}{2}\bar{a}_y \tan\beta \tag{11}$$

$$\underset{\rightarrow}{+} \sum F_x = \sum (F_x)_{eff} :$$

$$B_x - D_x = m_{BD}\, \bar{a}_x$$

$$B_x = m_{BD}\,\overline{a}_x + D_x \tag{12}$$

$$+\downarrow \sum F_x = \sum (F_x)_{eff}:$$

$$B_y + D_y = m_{BD}\,\overline{a}_y$$

$$B_y = m_{BD}\,\overline{a}_y - D_y \tag{13}$$

MATLAB program

```
g=386.4;
Wbd=5.5;% weight of the connecting rod
Wp=6.3; % Weight of the piston
mp=Wp/g; % mass of the piston
mbd=Wbd/g;
L=10; % Length of connecting rod
b=3.5; % crank radius
I_bar=(1/12)*mbd*L^2; % mass moment of inertia
omega_AB=1000*(2*pi)/60;
v_B=b*omega_AB;
Theta=[0:10:180];
t=Theta*pi/180;
beta=asin(b*sin(t)/L);
omega_BD=v_B*cos(t)./(L*cos(beta));
%acceleration
a_B=b*omega_AB^2;
alpha_BD=(L.*omega_BD.^2.*sin(beta)-a_B.*sin(t))./(L.*cos(beta));
a_D=a_B.*cos(t)+L.*omega_BD.^2.*cos(beta)+L.*alpha_BD.*sin(beta);
%
ax_bar=-0.5*a_B*sin(t);
ay_bar=0.5*a_B*cos(t)+0.5*a_D;
Dy=-mp*a_D;
Dx=-Dy.*tan(beta)+(I_bar*alpha_BD)./(L*cos(beta))-mbd*ax_bar./
2+mbd*ay_bar.*tan(beta)./2;
Bx=mbd*ax_bar+Dx;
By=mbd*ay_bar–Dy;
%
%Determine and plot values
z=[Theta;Bx;By;Dx;Dy];
fprintf('Theta     Bx        By Dx      Dy\n')
fprintf('(deg)  (lb)        (lb) (lb)    (lb)\n')
fprintf('\n');
fprintf('%5.0f   %8.3f      %8.3f %7.3f      %8.3f\n',z);
fprintf('\n');
figure(1)
plot(Theta,Bx,Theta,By)
xlabel('Theta(degrees)')
ylabel('Dynamic Reactions(lb)')
legend('Bx','By')
```

```
grid on
figure(2)
plot(Theta,Dx,Theta,Dy)
xlabel('Theta(degrees)')
ylabel('Dynamic  Reactions(lb)')
legend('Dx','Dy',2)
grid on
```

The output of the program is given below

Theta (deg)	Bx (lb)	By (lb)	Dx (lb)	Dy (lb)
0	0.000	-202.885	0.000	-844.817
10	38.553	-194.894	85.988	-823.271
20	66.317	-171.499	159.744	-759.882
30	74.055	-134.466	210.636	-658.451
40	55.678	-86.832	231.263	-525.450
50	9.741	-32.933	218.996	-370.047
60	-59.547	21.748	177.018	-203.789
70	-142.450	71.082	114.238	-39.655
80	-225.508	109.214	43.504	109.638
90	-294.892	131.753	-21.729	233.815
100	-340.144	136.812	-71.132	326.972
110	-356.607	125.439	-99.918	388.410
120	-345.479	101.213	-108.913	422.001
130	-311.888	69.226	-102.634	434.453
140	-262.328	34.916	-86.742	433.316
150	-202.697	3.172	-66.115	425.449
160	-137.391	-22.153	-43.964	416.219
170	-69.257	-38.378	-21.823	409.295
180	-0.000	-43.954	-0.000	406.764

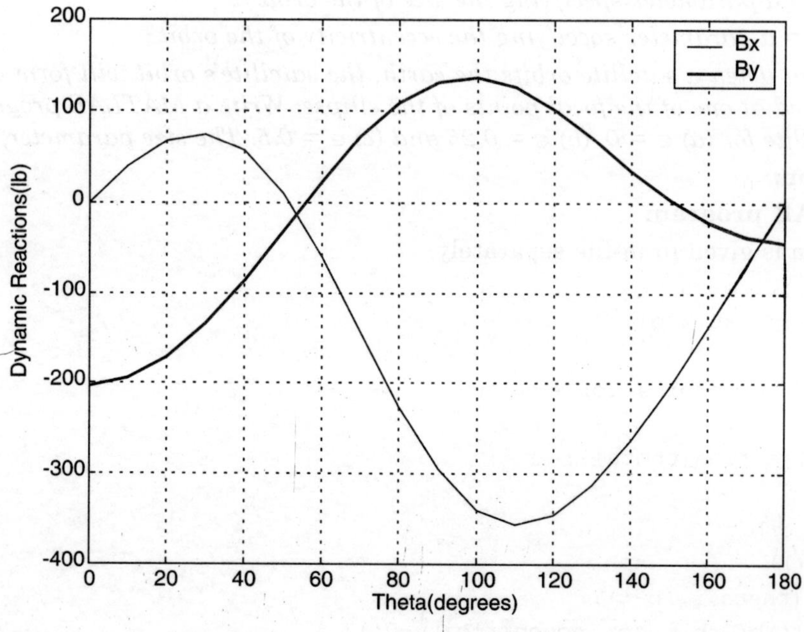

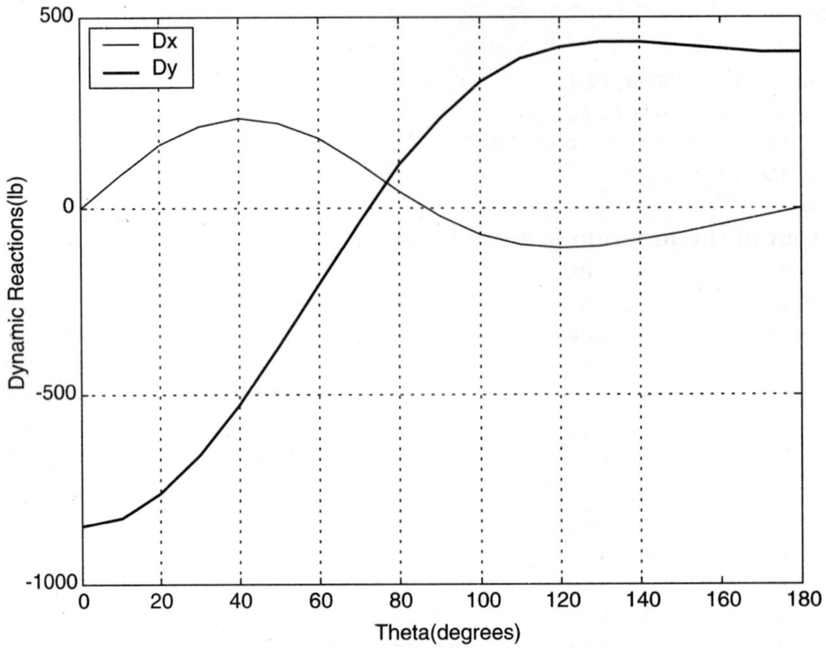

Fig. ED 3.12 (*c*)

ED 3.13. *The satellite's orbit in polar coordinates is given by*

$$r = \frac{p}{1 - \in \cos \theta}$$

where r = *distance of the satellite from the center of the earth*

θ = *angle of the satellite from the center of the earth*

p = *a parameter specifying the size of the orbit*

$\in$ = *a parameter specifying the eccentricity of the orbit*

Note that when a satellite orbits the earth, the satellite's orbit will form an ellipse with the earth located at one of the focal points of the ellipse. Write a MATLAB program to plot the orbit of a satellite for (*a*) $\in = 0$, (*b*) $\in = 0.25$ *and* (*c*) $\in = 0.5$. *The size parameter, p = 1100 km.*

Solution:

MATLAB program

Program is given in m-file separately.

```
% Initialize values
theta=0:pi/20:2*pi;
p=1100;
% Calculate orbits for e=0
e=0.0;
r=p./(1-e.*cos(theta));

% Plot orbit
figure(1)
polar (theta,r,'b-');
title ('\bfOrbit for eccentricity=0');
```

```
% Calculate orbits for e=0.25
e=0.25;
r=p./(1-e.*cos (theta));
%plot orbit
figure(2)
polar (theta,r,'b-');
title ('\bfOrbit for eccentricity =0.25');

%Calculate orbits for e=0.5
e=0.5;
r=p./(1-e.*cos(theta));

%Plot orbit
figure(3)
polar (theta,r,'b-');
title('\bfOrbit for eccentricity=0.5');
```

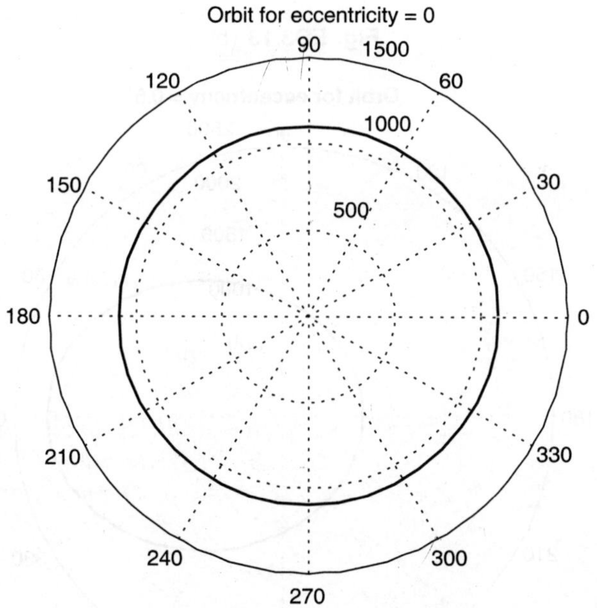

Fig. ED3.13 *(a)*

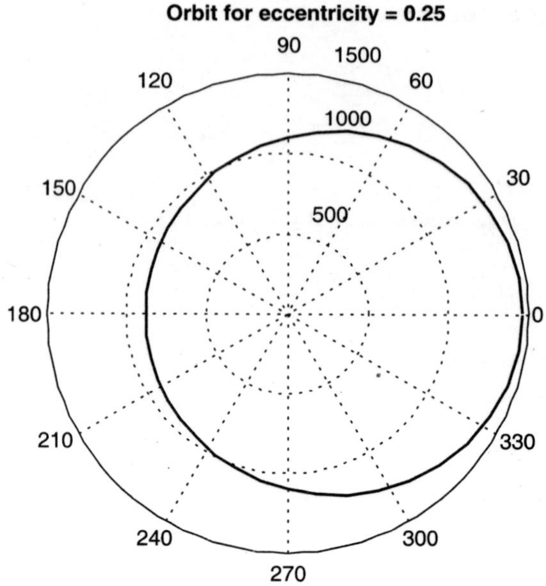

Fig. ED3.13 (*b*)

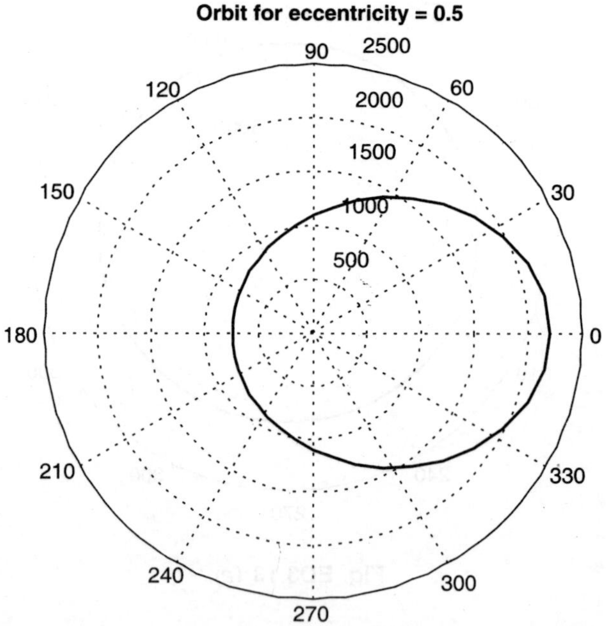

Fig. ED 3.13 (*c*)

ED 3.14. *A motor of weight W is supported by four springs with a spring constant of 400 lb/in each. The unbalance of the motor is found to be equivalent to a weight of 0.05 lb located 5 in. from the axis of rotation. If the motor is constrained to move vertically, write a MATLAB program to calculate and plot the amplitude of the vibration and the maximum acceleration of the motor for motor speeds from 0 to 1200 rpm. The weight of the motor W = 35 lb.*

Solution:

From the equation

$$x_m = \frac{P_m / k}{1-(\omega_f / \omega_n)^2} \tag{1}$$

where $P_m = mr\omega_f^{\,2}$

$$k = 4(350)\ (12) = 16{,}800 \text{ lb/ft}$$

$$P_m = \left(\frac{0.04}{g}\right)\left(\frac{5}{12}\right)\omega_f^2 = 0.0005176\,\omega_f^2$$

$$\omega_n^2 = \frac{k}{m} = \frac{kg}{W} = \frac{32.2(16{,}800)}{30} = 18{,}032 s^{-2}$$

Substituting values into (1)

$$x_m = \frac{(0.0005176\omega_f^2)/(16{,}800)}{1-\omega_f^2 / 18{,}032} \text{ ft} \tag{2}$$

$$a_m = \omega_f^2 x_m \text{ ft/s}^2 \tag{3}$$

where $\omega_f = (rpm)\left(\frac{2\pi}{60}\right)$

MATLAB program

```
W=35;
g=32.2;
mW=W/g; % Mass of motor in lb
ks=400*12; % one spring stiffness
m=0.05/g; % mass of unbalance
r=5/12; % radius of rotation
k=4*ks; % total stiffness of all springs
omega_n2=k/mW; % square of natural frequency
omega_f=[0:50:1200]; % variation of motor speeds
omeg=omega_f*2*pi/60;
Pm=m*r*omeg.^2;% Unbalance force due to rotation
xm=(Pm/k)./(1-omega_f.^2/omega_n2);
am=omega_f.^2.*xm; % acceleration in ft/s^2
amp=xm*12;% amplitude in inches
z=[omega_f;amp;am];
fprintf('omega_f(rpm)     Amplitude(in.)        Acceleration(ft/s^2)\n');
fprintf('\n');
fprintf('%5.0f            %f          %5.2f\n',z);
figure(1)
plot(omega_f,amp)
xlabel('Motor Speed (rpm)')
ylabel('Amplitude of Vibration(in.)')
grid on
```

```
figure(2)
plot(omega_f,am)
xlabel('Motor Speed (rpm)')
ylabel('Maximum Motor Acceleration(ft/s^2)')
grid on
```

Output is as follows:

omega_f(rpm)	Amplitude(in.)	Acceleration(ft/s^2)
0	0.000000	0.00
50	0.000013	0.00
100	0.000102	0.09
150	–0.000364	–0.68
200	–0.000140	–0.47
250	–0.000109	–0.57
300	–0.000097	–0.73
350	–0.000092	–0.93
400	–0.000088	–1.17
450	–0.000086	–1.45
500	–0.000084	–1.76
550	–0.000083	–2.10
600	–0.000082	–2.47
650	–0.000082	–2.88
700	–0.000081	–3.32
750	–0.000081	–3.79
800	–0.000081	–4.30
850	–0.000080	–4.83
900	–0.000080	–5.41
950	–0.000080	–6.01
1000	–0.000080	–6.64
1050	–0.000080	–7.31
1100	–0.000079	–8.02
1150	–0.000079	–8.75
1200	–0.000079	–9.52
1250	–0.000079	–10.32

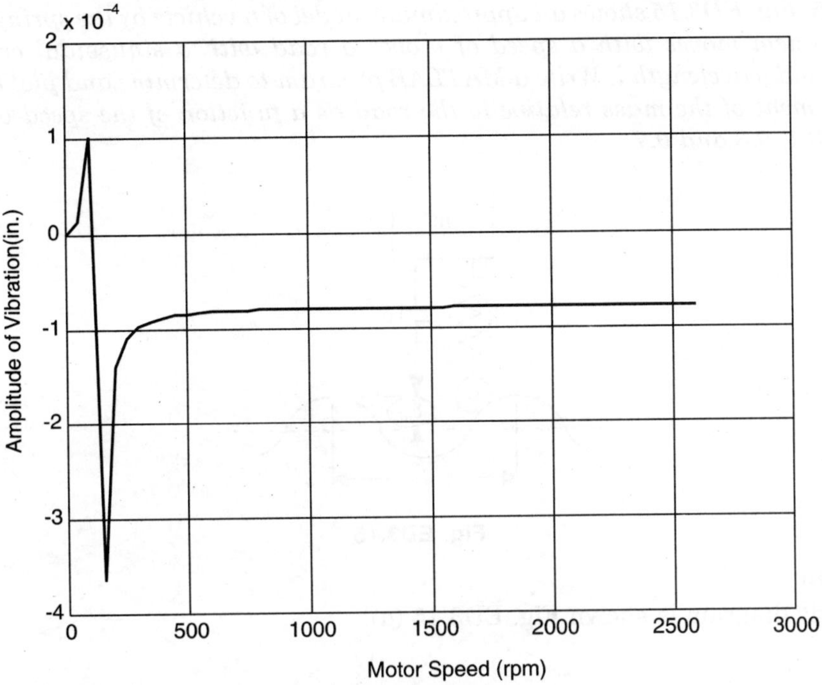

Fig. ED 3.14 *(a)*

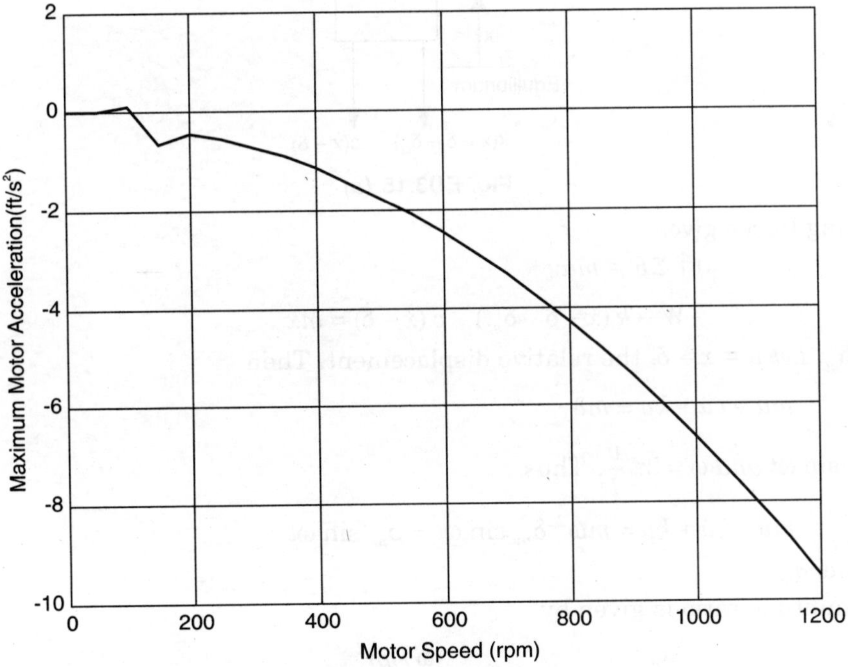

Fig. ED 3.14 *(b)*

ED 3.15. *Fig. ED3.15 shows an approximate model of a vehicle by the spring and dashpot system. The system moves with a speed of v over a road with a sinusoidal cross-section of amplitude d_m and wavelength l. Write a MATLAB program to determine and plot the amplitude of the displacement of the mass relative to the road as a function of the speed v for damping factors $c/c_c = 0.1$, 0.3 and 0.7.*

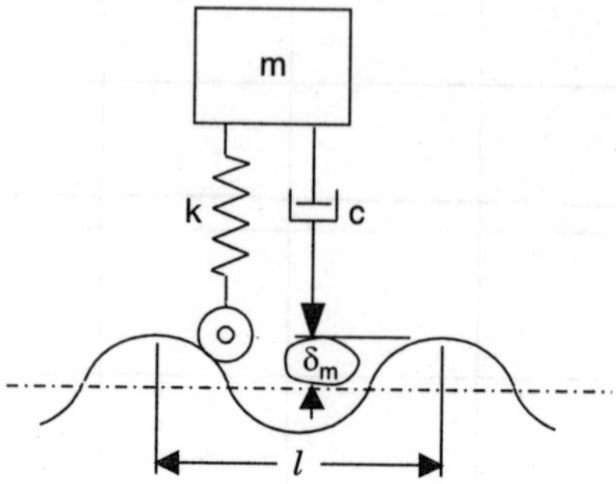

Fig. ED3.15

Solution:

Free body diagram is shown Fig. ED3.15 (*a*).

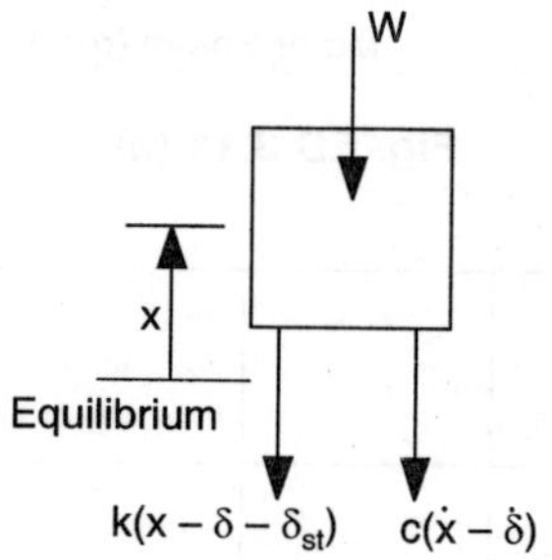

Fig. ED3.15 (*a*)

Summing forces gives

$$+\uparrow \Sigma F = ma:$$

$$-W - k(x - \delta - \delta_{st}) - c(\dot{x} - \dot{\delta}) = m\ddot{x}$$

where $W = k\delta_{st}$. Let $u = x - \delta$, the relative displacement. Then

$$m\ddot{u} + c\dot{u} + ku = m\ddot{\delta}$$

where $\delta = \delta_m \sin \omega t$ and $\omega = 2\pi \frac{v}{l}$. Thus

$$m\ddot{u} + c\dot{u} + ku = m\omega^2 \delta_m \sin \omega t = p_m . \sin \omega t$$

where $p_m = m\omega^2\partial_m$.

The amplitude ratio is given by

$$\frac{u_m}{\delta_m} = \frac{(\omega / p)^2}{\sqrt{[1 - (\omega / p)^2]^2 + [2(c / c_c)(\omega / p)]^2}}$$

where $p = \sqrt{k / m}$ and $\omega = 2\pi \frac{v}{l}$.

MATLAB program

```
i=0;
for cR=0.1:0.3:0.9;
i=i+1;
Op=[0:0.5:10];
den=sqrt((1-Op.^2).^2+(2*cR.*Op).^2);
umdm=Op.^2./den;
fprintf('For Damping Ratio=%5.3f\n',cR)
fprintf('Speed v      Relative Displacement\n');
fprintf('\n');
z=[Op;umdm];
fprintf('%5.2f                      %5.3f\n',z);
fprintf('\n')
plot(Op,umdm)
hold on % In same plot to draw another curve
xlabel('Speed (Omega/p)')
ylabel('Relative Displacement')
grid on
end
```

Output of the above program is

For Damping Ratio = 0.100

Speed v	Relative Displacement
0.00	0.000
0.50	0.330
1.00	5.000
1.50	1.750
2.00	1.322
2.50	1.185
3.00	1.122
3.50	1.087
4.00	1.065
4.50	1.051
5.00	1.041
5.50	1.033
6.00	1.028
6.50	1.024
7.00	1.020
7.50	1.018
8.00	1.016
8.50	1.014
9.00	1.012
9.50	1.011
10.00	1.010

For Damping Ratio=0.400

Speed v	Relative Displacement
0.00	0.000
0.50	0.294
1.00	1.250
1.50	1.298
2.00	1.176
2.50	1.112
3.00	1.078
3.50	1.057
4.00	1.043
4.50	1.034
5.00	1.027
5.50	1.023
6.00	1.019
6.50	1.016
7.00	1.014
7.50	1.012
8.00	1.011
8.50	1.009
9.00	1.008
9.50	1.008
10.00	1.007

For Damping Ratio = 0.700

Speed v	Relative Displacement
0.00	0.000
0.50	0.244
1.00	0.714
1.50	0.921
2.00	0.975
2.50	0.991
3.00	0.996
3.50	0.998
4.00	0.999
4.50	1.000
5.00	1.000
5.50	1.000
6.00	1.000
6.50	1.000
7.00	1.000
7.50	1.000
8.00	1.000
8.50	1.000
9.00	1.000
9.50	1.000
10.00	1.000

The output is shown in Fig. ED3.15 (*b*)

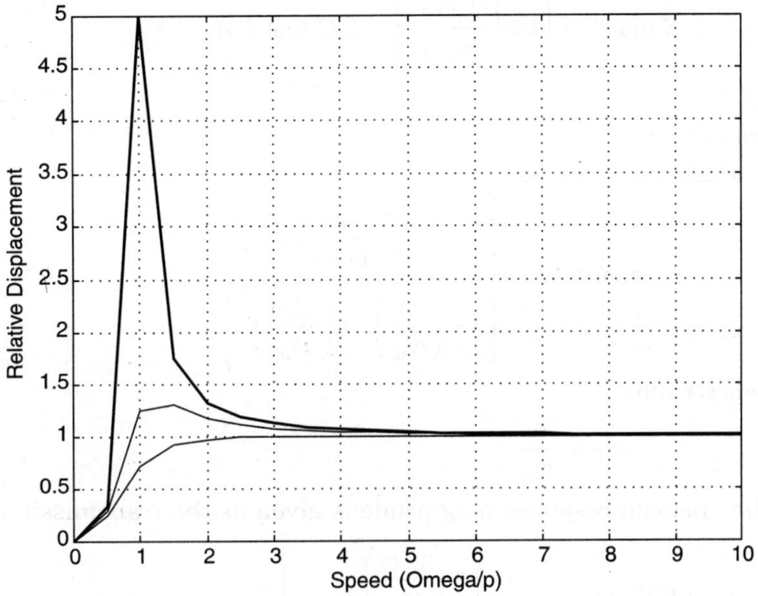

Fig. ED3.15 (*b*)

3.15.3 Mechanical Vibrations

EM3.1. *Write a MATLAB script for plotting*

(*a*) *the non-dimensional response magnitude for a system with harmonically moving base shown in Fig. EM3.1.*

(*b*) *the response phase angle for system with harmonically moving base.*

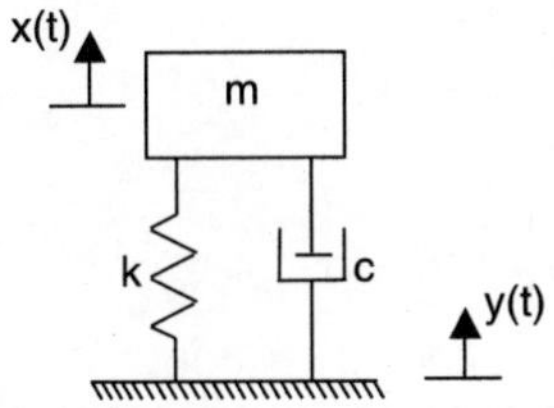

Fig. EM3.1 Single degree of freedom system with moving base.

Solution:

The magnitude of the frequency response is given as

$$| G(i\omega) | = \frac{1}{\left[\left[\left(1-\frac{\omega}{\omega_n}\right)^2\right]^2 + \left(2\zeta\frac{\omega}{\omega_n}\right)^2\right]^{1/2}}$$

The magnitude of $X(i\omega)$ is given as

$$|X(i\omega)| = \left[1+\left(\frac{2\zeta\omega}{\omega_n}\right)^2\right]^{1/2} |G(i\omega)| A$$

where $y(t) = \text{Re}\, A^{i\omega t}$

$x(t) = X(i\omega)\, ei^{\omega t}$

The phase angle ϕ is given as

$$\phi(\omega) = \tan^{-1}\left[\frac{2\zeta\left(\frac{\omega}{\omega_n}\right)^3}{1-\left(\frac{\omega}{\omega_n}\right)^2+\left(\frac{2\zeta\omega}{\omega_n}\right)^2}\right]$$

The frequency ratio

$$r = \frac{\omega}{\omega_n}$$

The non-dimensional response magnitude is given as the transmissibility

$$\frac{|X(i\omega)|}{A} = \left[\frac{1+\left(\frac{2\zeta\omega}{\omega_n}\right)^2}{1-\left(\frac{\omega}{\omega_n}\right)^2+\left(\frac{2\zeta\omega}{\omega_n}\right)^2}\right]$$

Based on these equations MATLAB script is written as follows:

```
zeta = [0.05; 0.1; 0.15; 0.25; 0.5; 1.25; 1.5]; % damping factors
r= [0:0.01:3]; %frequency ratio
for k=1: length (zeta)
     G(k,:)=sqrt((1+(2*zeta(k)*r).^2)./((1-r.^2).^2+(2*zeta(k)*r).^2));
     phi(k,:)=atan2(2*zeta(k)*r.^3,1-r.^2+(2*zeta(k)*r).^2);
end
figure (1)
plot(r, G)
xlabel ('\omega/\omega_n')
ylabel ('|x (i\omega)|/A')
grid
legend('\zeta_1=0.05','\zeta_2=0.1','\zeta_3=0.15','\zeta_4=0.25','\zeta_5=0.5','\zeta_6
=1.25','\zeta_7=1.5')
figure (2)
plot(r, phi)
xlabel ('\omega/\omega_n')
ylabel ('\phi (\omega)')
grid
ha=gca;
set (ha,'ytick',[0:pi/2:pi])
set(ha,'yticklabel',{[];'pi/2';'p'})
legend('\zeta_1=0.05','\zeta_2=0.1','\zeta_3=0.15','\zeta_4=0.25','\zeta_5=0.5','
\zeta_6=1.25','\zeta_7=1.5')
```

The output of this program is shown in Figs. EM3.1 (a) and (b).

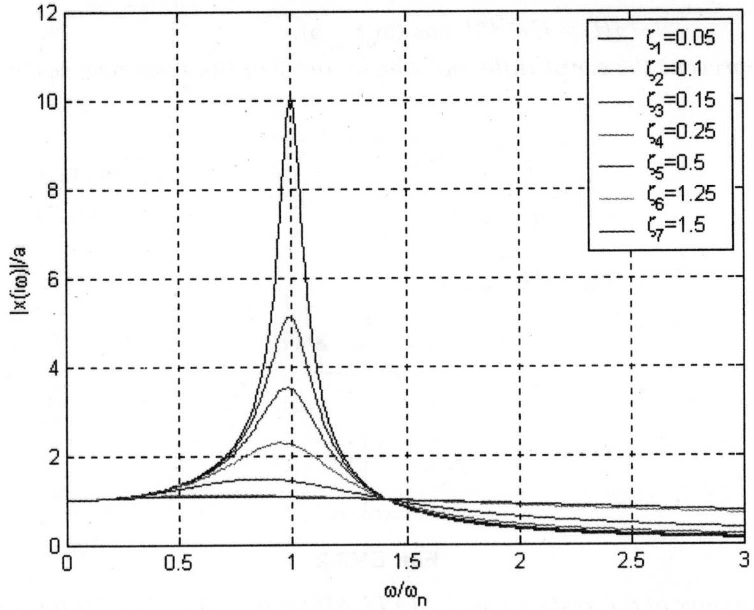

(a)

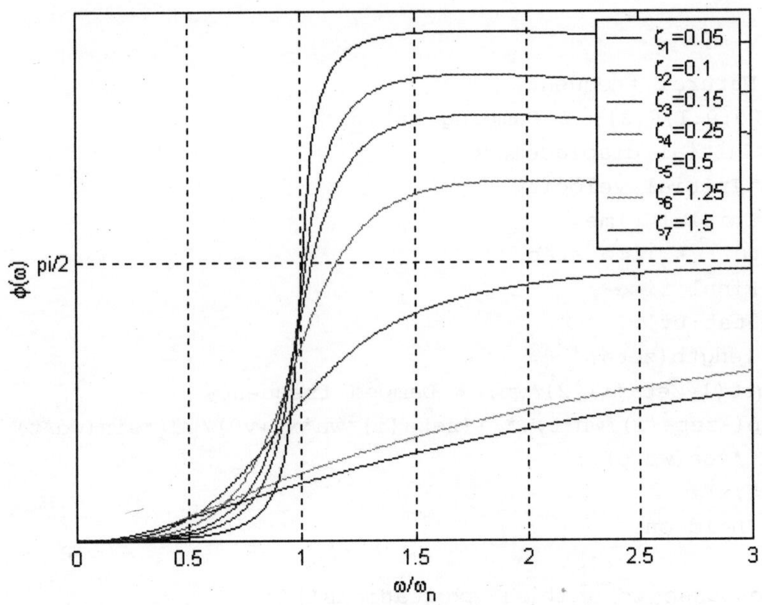

(b)

Fig. EM3.1

EM3.2. *An analytical expression for the response of an damped single degree of freedom system (Fig. EM3.2) to given initial displacement and velocity is given by*

$$x(t) = Ce^{-\zeta\omega_n t} \cos(\omega_d t - \phi)$$

where C and ϕ represent the amplitude and phase angle of the response, respectively having the values

$$C = \sqrt{x_0^{\,2} + \left(\frac{\zeta\omega_n x_0 + v_0}{\omega_d}\right)^2}, \; \phi = tan^{-1}\left(\frac{\zeta\omega_n x_0 + v_0}{\omega_d x_0}\right)$$

and

$$\omega_d = \sqrt{1-\zeta^2}\;\omega_n$$

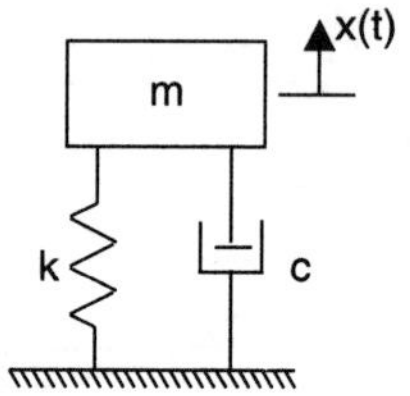

Fig. EM3.2

Plot the response of the system using MATLAB for $\omega_n = 5$ rad/s, $\zeta = 0.05, 0.1, 0.2$ subjected to the initial conditions $x(0) = 0$, $\dot{x}(0) = v_0 = 60$ cm/s.

Solution:

```
clear
clf
wn=5; % Natural frequency
zeta=[0.05;0.1;0.2]; % Damping ratio
x0=0; % Initial displacement
v0=60; % Initial velocity
t0=0; % Initial time
deltat=0.01; % Time step
tf=6; % Final time
t=[t0:deltat:tf];
for i=1:length(zeta),
    wd=sqrt(1-zeta(i)^2)*wn; % Damped frequency
     x=exp(-zeta(i)*wn*t).*(((zeta(i)*wn*x0+v0)/wd)*sin(wd*t)
      + x0*cos(wd*t));
    plot(t,x)
           hold on
end
title('Response to initial excitations')
xlabel('t[s]')
ylabel('x(t)')
grid
```

The output of this program is as follows:

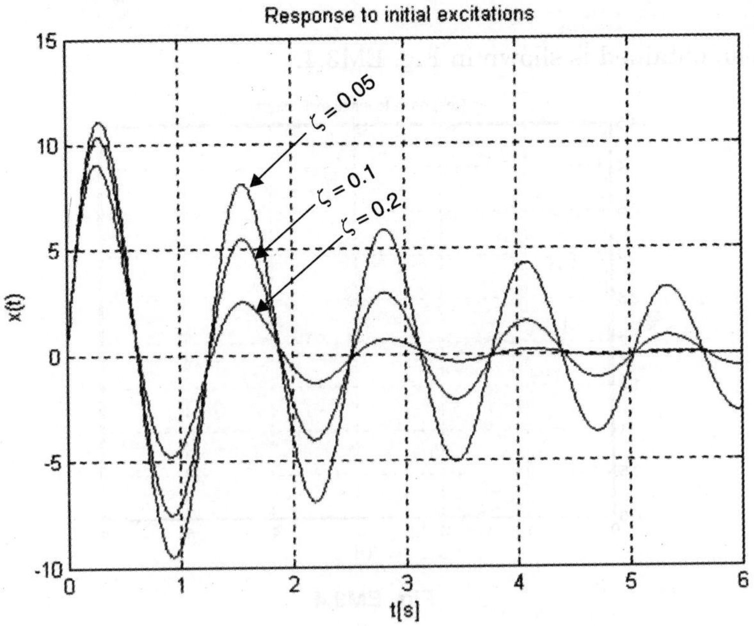

Fig. EM3.2 (*a*)

EM3.3. *Plot the response of the system in Problem EM3.2 using MATLAB for* $\omega_n = 5$ *rad/sec,* $\zeta = 1.3, 1.5, 2.0$ *subjected to the initial conditions* $x(0) = 0,\ \dot{x}(0) = v_0 = 60$ *cm/s.*

Solution:

Changing the program slightly, with zeta = [1.3, 1.5, 2.0] in EM3.2, we obtain Fig. EM3.3.

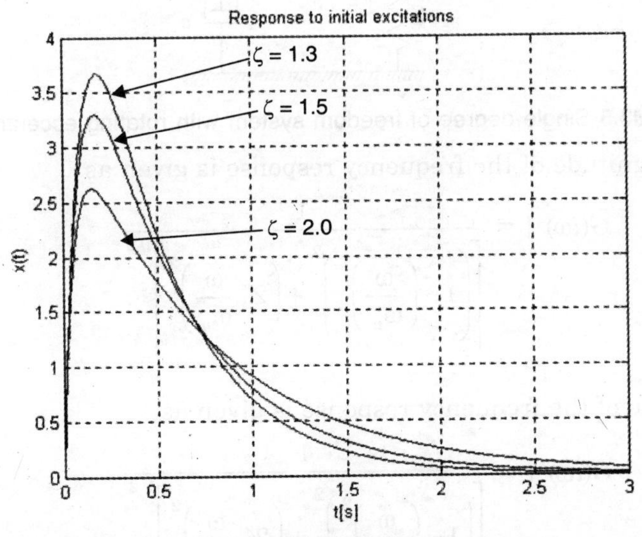

Fig. EM3.3

EM3.4. *Plot the response of the system in Problem EM3.2 using MATLAB for $\omega_n = 5$ rad/sec and $\zeta = 1.0$ subjected to the initial conditions $x(0) = 0$, $\dot{x}(0) = v_0 = 60$ cm/s.*

Solution:

The solution obtained is shown in Fig. EM3.4.

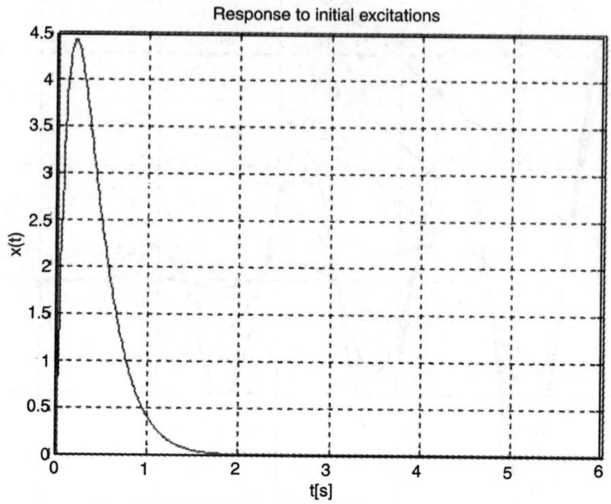

Fig. EM3.4

EM3.5. *Write MATLAB script for plotting the magnitude of the frequency response of a system with rotating unbalanced masses as shown in Fig.EM3.5.*

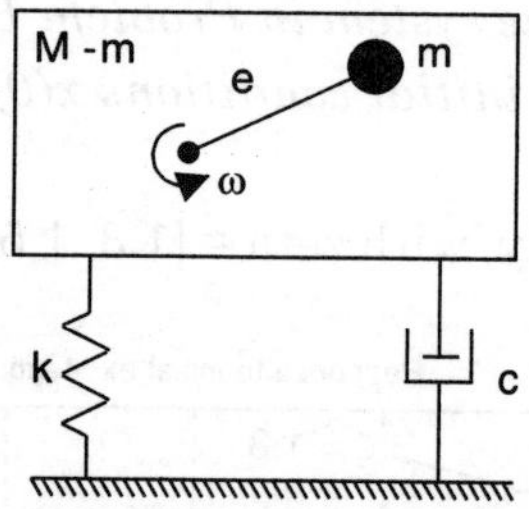

Fig. EM3.5 Single degree of freedom system with rotating eccentric mass.

Hint: The magnitude of the frequency response is given as

$$|G(i\omega)| = \frac{1}{\left[\left(1-\left(\frac{\omega}{\omega_n}\right)^2\right)^2 + \left(2\zeta\frac{\omega}{\omega_n}\right)^2\right]^{1/2}}$$

Solution:

The magnitude of the frequency response is given as

$$|G(i\omega)| = \frac{1}{\left[\left(1-\left(\frac{\omega}{\omega_n}\right)^2\right)^2 + \left(2\zeta\frac{\omega}{\omega_n}\right)^2\right]^{1/2}}$$

for ξ = 0.05, 0.01, 0.15, 0.20, 0.20, 0.25, 0.5, 0.75, 1.0, 1.25, 1.5.

$r = \omega/\omega n$ = 0 to 3 in steps of 0.01.

MATLAB Program:

```
zeta=[0.05;0.1;0.15;0.25;0.5;1;1.25;1.5]; % Damping factors
r=[0:0.01:3]; % Frequency ratios
for k=1:length(zeta),
    G=(r.^2)./sqrt((1-r.^2).^2+(2*zeta(k)*r).^2);
   plot(r,G)
   hold on
end
xlabel('\omega/\omega_n')
ylabel('({\omega/\omega_n})^2|G(I\omega)|')
grid
```

Fig. EM3.5 (*a*) shows the output of the program

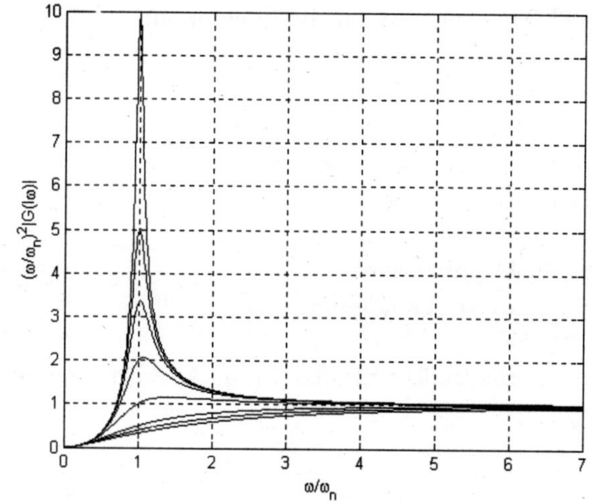

Fig. EM3.5 (*a*)

For showing legends on the curves, gtext command can be employed.

EM3.6. *A single degree of freedom spring-mass system subjected to coulomb damping is shown in Fig. EM3.6.*

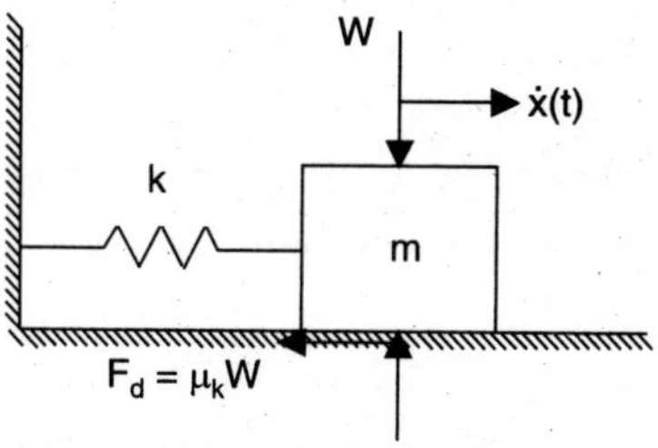

Fig. EM3.6

The parameters of the system have the values $m = 600$ *kg,* $k = 20 \times 10^4$ *N/m,* $\mu_s = 0.15$ *and* $m_k = 0.10$*. The initial conditions are* $x(0) = x_0 = 1.5$ *cm,* $\dot{x}(0) = 0$*. Plot the response x(t) versus t using MATLAB.*

The magnitude of the average response value f_d *is given as*

$$f_d = \frac{F_d}{k} = \frac{\mu_k mg}{k}$$

If n denotes the half-cycle just prior to the cessation of motion, then n is the smallest integer satisfying the inequality

$$x_0 - (2n - 1) f_d < \left(1 + \frac{\mu_s}{\mu_f}\right) f_d$$

where μ_s = *static coefficient of friction*

μ_k = *kinetic coefficient of friction*

Solution:

The following MATLAB program can be developed:

```
m=600; % Mass
k=200000; % Stiffness
mus=0.15; % Static friction coefficient
muk=0.10; % Kinetic friction coefficient
x0=1.5; % Initial displacement
t0=0;
deltat=0.005; % Time increment
wn=sqrt(k/m); % Natural frequency
fd=100*muk*m*9.81/k;
N=ceil(0.5*((x0-(1+mus/muk)*fd)/fd+1)); % Half cycles
t=[];
x=[];
if N>0
    for n=1:N,
        t1=[t0:deltat:t0+pi/wn];
        x1=(x0-(2*n-1)*fd)*cos(wn*t1)+fd*(-1)^(n+1);
        t=[t t1];
        x=[x x1];
        t0=t0+pi/wn;
    end
end
plot(t,x,t,fd*ones(length(t)),'—',t,-fd*ones(length(t)),'—')
title('Response to initial excitations')
xlabel('t[s]')
ylabel('x(t)[cm]')
grid
```

The output is shown in Fig. EM3.6 (*a*).

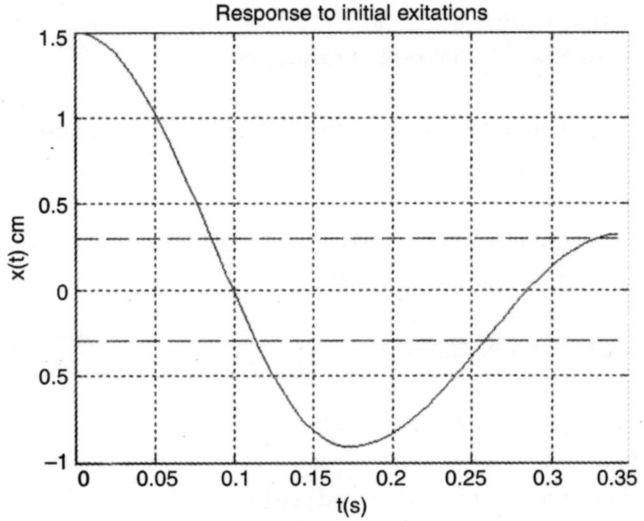

Fig. EM3.6 (*a*)

EM3.7. *Write a MATLAB script for obtaining the response of a viscosity damped single degree of freedom system to the force* $F(t) = F0\, e^{-\alpha t}\, u(t)$ *by means of the convolution integral. The pulse is rectangular as shown in Fig. EM3.7 with T = 0.1 seconds.*

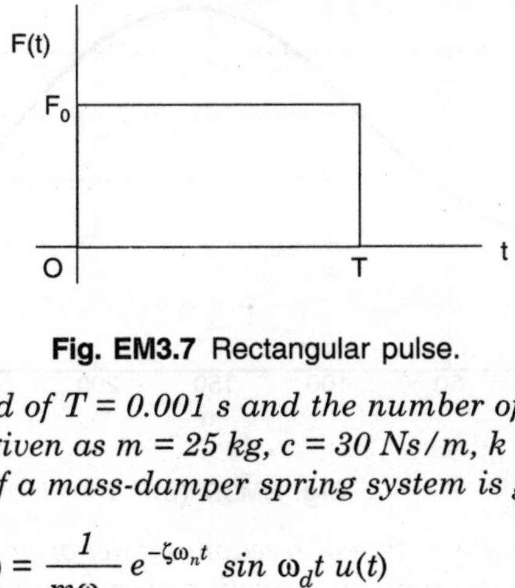

Fig. EM3.7 Rectangular pulse.

Use the sampling period of T = 0.001 s and the number of sampling times n = 300. The parameters of the system are given as m = 25 kg, c = 30 Ns/m, k = 6000 N/m, F0 = 300 N, and $\alpha = 1$. *The impulse response of a mass-damper spring system is given by*

$$g(t) = \frac{1}{m\omega_d} e^{-\zeta\omega_n t} \sin \omega_d t\, u(t)$$

Solution:

```
m=25; % mass
c=30;% damping
k=6000; % stiffness
F0=300; % Force amplitude
T=0.1;
wn=sqrt(k/m);% Natural frequency
```

```
zeta=c/(2*sqrt(m*k));%damping factor
Ts=0.001;% sampling period
N=301;% sampling times
wd=wn*sqrt(1-zeta^2);% damped frequency
for n=1:N,
  if n<=T/Ts+1; F(n)=F0; else F(n)=0; end     %force
end
n=[1:N];
g=Ts*exp(-(n-1)*zeta*wn*Ts).*sin((n-1)*wd*Ts)/(m*wd);
% discrete-time impulse response
c0=conv(F,g);%convolution sum
c=c0(1:N); % plot to N samples
n=[0:N-1];
axes('position',[0.1 0.2 0.8 0.7])
plot(n,c,'.')
title('Response to Rectangular pulse')
xlabel('n')
ylabel('x(n) m');
grid
```

The output is shown in Fig. EM3.7 (*a*)

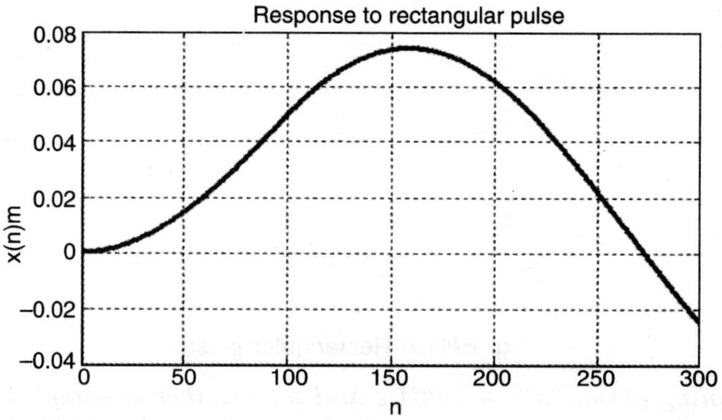

Fig. EM3.7 (*a*)

EM3.8. *A simplified single degree of freedom model of an automobile suspension system is shown in Fig. EM3.8. The automobile is traveling over a rough road at a constant horizontal speed when it encounters a bump in the road of the shape shown in Figs. EM3.8 (a), (b). The velocity of the automobile is 20 m/s, m = 1500 kg, k = 150,000 N/m, and* $\zeta = 0.10$. *Determine the response of the automobile.*

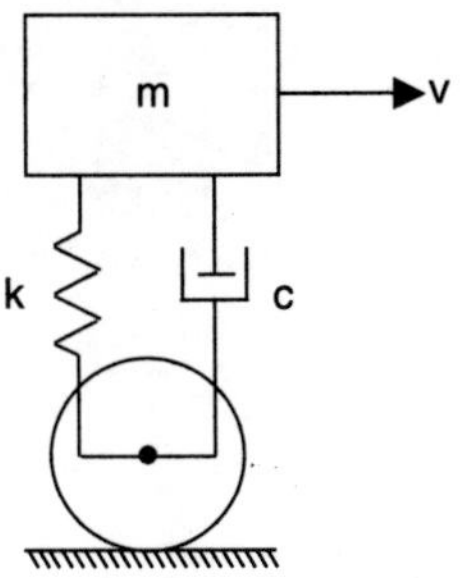

Fig. EM3.8 Simplified single degree of automobile model.

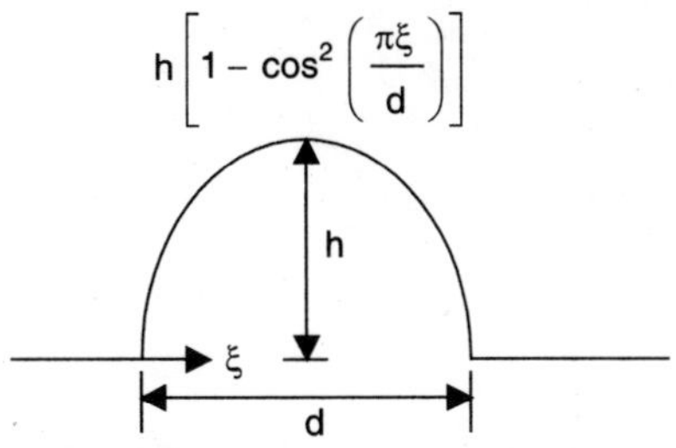

Fig. EM3.8 (*a*) Versed sine freedom pulse model for bump.

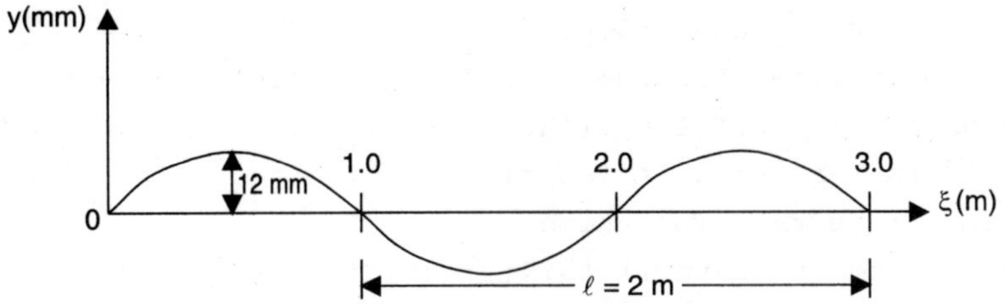

Fig. EM3.8 (*b*) Road contour.

$$y(\xi) = h\left[1-\cos^2\left(\frac{\pi\xi}{d}\right)\right][1-u(\xi-d)]$$

Here $h = 0.012$ *m,* $d = 1.0$ *m and for constants automobile speed,* $\xi = vt$. *The vertical displacement of the automobile wheels is given by*

$$y(t) = h\left[1-\cos^2\left(\frac{\pi v}{d}\right)\right]\left[1-u\left(t-\frac{d}{v}\right)\right]$$

The system response as per convolution integral is

$$x(t) = -\,m_{eq}\int_0^t [2\zeta\omega_n\,\dot{y}(\tau) + \omega_n^2 y(\tau)]\,h(t-\tau)\,d\tau$$

The wheel velocity becomes

$$\dot{y}(t) = 2\left(\frac{\pi v}{d}\right)\sin\left(\frac{2\pi v}{d}\,t\right)\left[1-u\left(t-\frac{d}{v}\right)\right]$$

Solution:

MATLAB program for this is given below:

```
% Simplified one-degree-of-freedom model of vehicle suspension system
% Vehicle encounters bump in road modelled as a versed sinusoidal pulse
% y(t)=h(1-(cos(pi*v*t/ t0))^2)*(u(t)-u(t-d/v))
%convolution integral is used to evaluate system response
syms t tau
% input parameters
digits(10)
```

```
format short e
m=1500;
k=150000;
zeta=0.10;
hb=0.012;
d=1.0;
v=20;
% system parameters and constants
omega_n=sqrt(k/m); % Natural frequency
omega_d=omega_n*sqrt(1-zeta^2); % damped natural frequency
c1=pi/d;
% wheel displacement and velocity
% MATLAB 'Heaviside' for the unit step function
y=hb*(1-cos(c1*v*t)^2)*(1-sym('Heaviside(t-0.04)'));
ydot=hb*c1*sin(2*c1*v*tau)*(1-sym('Heaviside(tau-0.04)'))
%convolution integral evaluation
h=exp(-zeta*omega_n*(t-tau)).*sin(omega_d*(t-tau))/(m*omega_d);
g1=-2*zeta*m*omega_n*ydot*h;
g2=-omega_n^2*m*y*h;
g1a=vpa(g1,5);
g2a=vpa(g2,5);
I1=int(g1a,tau,0,t);
I1a=vpa(I1,5);
I2=int(g2a,tau,0,t);
I2a=vpa(I2,5);
x1=I1a+I2a;
x=vpa(x1,5);
vel=diff(x);
acc=diff(vel);
time=linspace(0,0.3,50);
for i=1:50
  x1=subs(x,t,time(i));
  xa(i)=vpa(x1);
end
xp=double(xa);
plot(time,xp,'-');
grid;
xlabel('time(sec)')
ylabel('x(t)  [m]')
```

The output of this MATLAB program is given in Fig. EM3.8 (*c*)

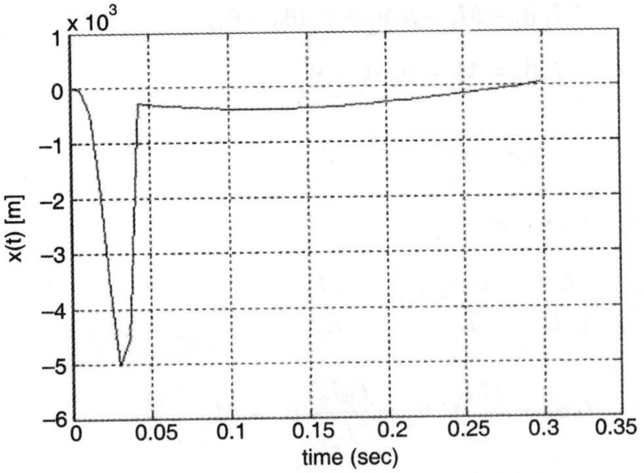

Fig. EM3.8 (*c*)

EM3.9. *Fig. EM3.9 shows two disks of mass polar moments of inertia I_1 and I_2 mounted on a circular shaft with torsional stiffnesses G_{J1} and G_{J2}. Neglect the mass of the shaft.*

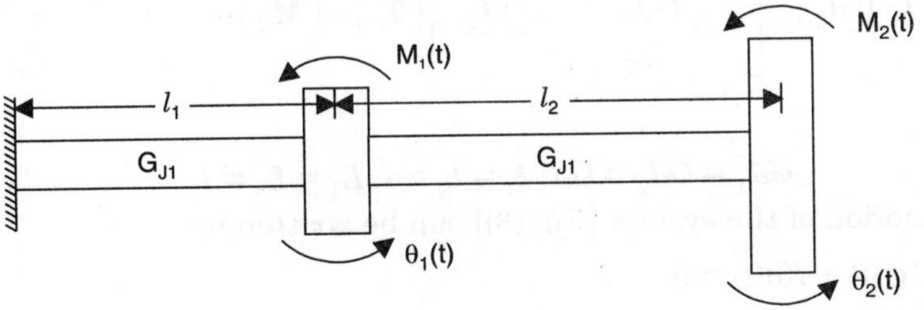

Fig. EM3.9

(*a*) *Obtain the differential equations of motion for the angular displacements of the disks*

(*b*) *Determine the natural frequencies and natural modes of the system if $I_1 = I_2 = I$, $G_{J1} = G_{J2} = G_J$, and $l_1 = l_2 = l$*

(*c*) *Obtain the response of the system to the torques $M_1(t) = 0$, and $M_2(t) = M_2e^{-\alpha t}$ in discrete time*

(*d*) *Obtain the response of the system to the torques $M_1(t) = 0$, and $M_2(t) = M_2e^{-\alpha t}$*

(*e*) *Obtain in discrete time the response of the system to the torques $M_1(t) = 0$, and $M_2(t) = M_2e^{-\alpha t}$ using MATLAB.*

Solution:

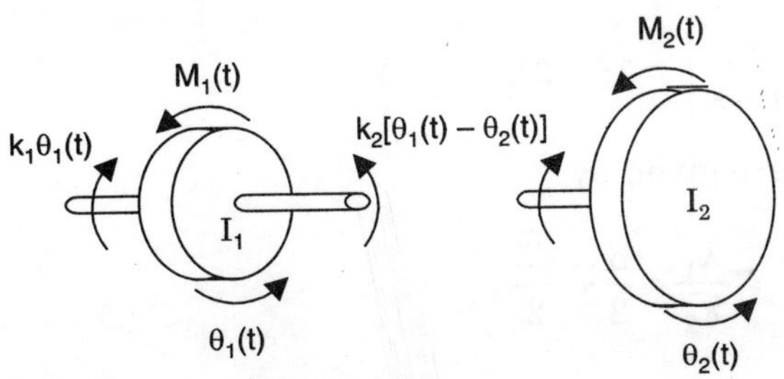

Fig. EM3.9 (*a*)

(*a*) The equations of motion are given by

$$I_1\ddot{\theta}_1 = M_1 - k_1\theta_1 + k_2(\theta_2 - \theta_1)$$

$$I_2\ddot{\theta}_2 = M_2 - k_2(\theta_2 - \theta_1) \tag{1}$$

where $$k_i = \frac{GJ_i}{L_i},\ i = 1, 2.$$

Rearranging Eq. (1), we get

$$I_1\ddot{\theta}_1 + \left(\frac{GJ_1}{L_1} + \frac{GJ_2}{L_2}\right)\theta_1 - \frac{GJ_2}{L_2}\theta_1 = M_1$$

$$I_2\ddot{\theta}_2 - \frac{GJ_2}{L_2}\theta_1 + \frac{GJ_2}{L_2}\theta_2 = M_2 \tag{2}$$

In matrix form, we can write

$$\begin{bmatrix} I_1 & 0 \\ 0 & I_2 \end{bmatrix}\begin{bmatrix} \ddot{\theta}_1 \\ \ddot{\theta}_2 \end{bmatrix} + \begin{bmatrix} \frac{GJ_1}{L_1} + \frac{GJ_2}{L_2} & -\frac{GJ_2}{L_2} \\ -\frac{GJ_2}{L_2} & \frac{GJ_2}{L_2} \end{bmatrix}\begin{bmatrix} \theta_1 \\ \theta_2 \end{bmatrix} = \begin{bmatrix} M_1 \\ M_2 \end{bmatrix} \tag{3}$$

(*b*) Denoting

$$GJ_1 = GJ_2 = GJ,\ I_1 = I_2 = I,\ L_1 = L_2 = L \tag{4}$$

the equations of motion of the system [Eq. (3)] can be written as

$$M\ddot{\theta}(t) + K\theta(t) = 0 \tag{5}$$

where $$M = I\begin{bmatrix} 1 & 0 \\ 0 & 1 \end{bmatrix},\ K = \frac{GJ}{L}\begin{bmatrix} 2 & -1 \\ -1 & 1 \end{bmatrix},\ \theta(t) = \begin{bmatrix} \theta_1(t) \\ \theta_2(t) \end{bmatrix} \tag{6}$$

are the mass matrix, stiffness matrix and configuration vector, respectively. The free vibration solution can be written as

$$\theta_i(t) = \Theta_i e^{i\omega t},\ i = 1, 2 \tag{7}$$

where ω is the frequency of oscillation and $\Theta = [\Theta_1\ \ \Theta_2]^T$ is a vector of constants, we have

$$\begin{bmatrix} 2 & -1 \\ -1 & 1 \end{bmatrix}\begin{bmatrix} \Theta_1 \\ \Theta_2 \end{bmatrix} = \lambda\begin{bmatrix} \Theta_1 \\ \Theta_2 \end{bmatrix},\ \lambda = \omega^2\frac{IL}{GJ} \tag{8}$$

The characteristic equation can be written as

$$\begin{vmatrix} 2-\lambda & -1 \\ -1 & 1-\lambda \end{vmatrix} = \lambda^2 - 3\lambda + 1 = 0 \tag{9}$$

The eigenvalues are given by

$$\frac{\lambda_1}{\lambda_2} = \frac{3}{2} \mp \frac{\sqrt{5}}{2} \tag{10}$$

The natural frequencies are given by

$$\omega_1 = 0.6180\ \sqrt{GJ/IL}\ ,\ \omega_2 = 1.6180\ \sqrt{GJ/IL} \tag{11}$$

Denote the modal vector corresponding to λ_1 by $\Theta_1 = [\Theta_{11}\ \ \Theta_{21}]^T$, the modal vector is from the matrix equation as

$$\begin{bmatrix} 2 & -1 \\ -1 & 1 \end{bmatrix}\begin{bmatrix} \Theta_{11} \\ \Theta_{21} \end{bmatrix} = \lambda_1 \begin{bmatrix} \Theta_{11} \\ \Theta_{21} \end{bmatrix} = \frac{3-\sqrt{5}}{2}\begin{bmatrix} \Theta_{11} \\ \Theta_{21} \end{bmatrix} \tag{12}$$

or

$$\Theta_1 = \Theta_{11}\begin{bmatrix} 1 \\ 1.6180 \end{bmatrix} \tag{13}$$

In a similar way by letting $\Theta_2 = [\Theta_{12}\ \ \Theta_{22}]^T$, we have

$$\begin{bmatrix} 2 & -1 \\ -1 & 1 \end{bmatrix}\begin{bmatrix} \Theta_{12} \\ \Theta_{22} \end{bmatrix} = \lambda_2 \begin{bmatrix} \Theta_{12} \\ \Theta_{22} \end{bmatrix} = \frac{3+\sqrt{5}}{2}\begin{bmatrix} \Theta_{12} \\ \Theta_{22} \end{bmatrix} \tag{14}$$

or

$$\Theta_2 = \Theta_{12}\begin{bmatrix} 1 \\ 1.6180 \end{bmatrix} \tag{15}$$

The modal vectors are shown below

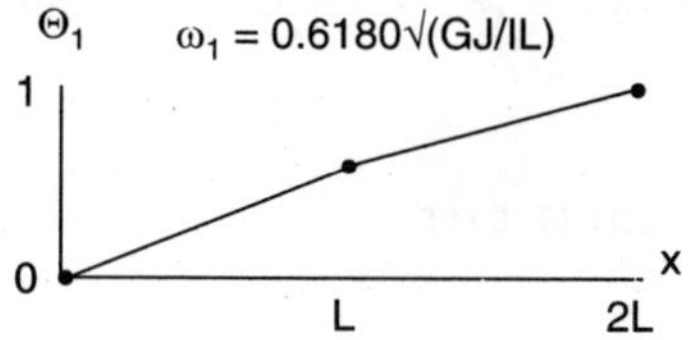

Fig. EM3.9 (*b*)

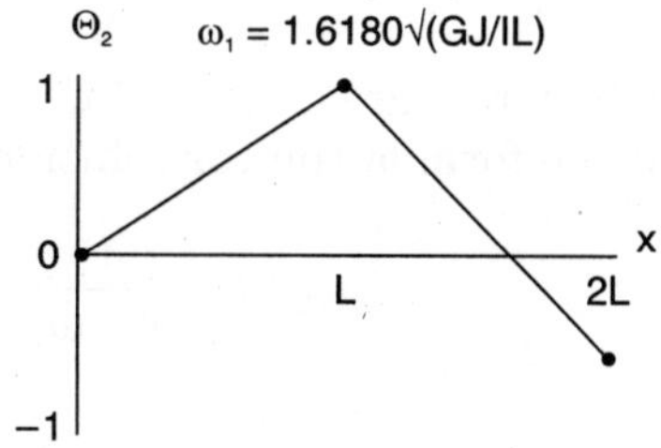

Fig. EM3.9 (*c*)

(*c*) The equations of motion are

$$M\ddot{\theta}(t) + K\theta(t) = M(t) \tag{16}$$

where

$$M = I\begin{bmatrix} 1 & 0 \\ 0 & 1 \end{bmatrix}, K = \frac{GJ}{L}\begin{bmatrix} 2 & -1 \\ -1 & 1 \end{bmatrix}, \theta(t) = \begin{bmatrix} \theta_1(t) \\ \theta_2(t) \end{bmatrix} M(t) = \begin{bmatrix} M_1(t) \\ M_2(t) \end{bmatrix} \tag{17}$$

The solution $\theta(t)$ is given by

$$\theta(t) = \eta_1(t)\ \Theta_1 + \eta_2(t)\ \Theta_2 \tag{18}$$

where $\eta_1(t)$ and $\eta_2(t)$ are modal coordinates and Θ_1 and Θ_2 are modal vectors. The modal equations can be written as

$$m'_{11}\ddot{\eta}_1(t) + m'_{11}\omega_1^2\eta_1(t) = N_1(t),\ m'_{22}\ddot{\eta}_2(t) + m'_{22}\omega_2^2\eta_2(t) = N_2(t) \tag{19}$$

where

$$\omega_1 = \sqrt{\frac{3-\sqrt{5}}{2}\frac{GJ}{IL}}, \omega_2 = \sqrt{\frac{3+\sqrt{5}}{2}\frac{GJ}{IL}} \tag{20}$$

The natural frequencies are given by Eq. (20).

$$\Theta_1 = \begin{bmatrix} 1 \\ \dfrac{1+\sqrt{5}}{2} \end{bmatrix}, \Theta_2 = \begin{bmatrix} 1 \\ \dfrac{1-\sqrt{5}}{2} \end{bmatrix} \tag{21}$$

The modal vectors are given by Eq. (21).

$$m'_{11} = \boldsymbol{\Theta}_1^T M \Theta_1 = \begin{bmatrix} 1 \\ \dfrac{1+\sqrt{5}}{2} \end{bmatrix}^T = \begin{bmatrix} I & 0 \\ 0 & I \end{bmatrix} \begin{bmatrix} 1 \\ \dfrac{1+\sqrt{5}}{2} \end{bmatrix} = \frac{5+\sqrt{5}}{2} I$$

$$m'_{22} = \boldsymbol{\Theta}_2^T M \Theta_2 = \begin{bmatrix} 1 \\ \dfrac{1-\sqrt{5}}{2} \end{bmatrix}^T = \begin{bmatrix} I & 0 \\ 0 & I \end{bmatrix} \begin{bmatrix} 1 \\ \dfrac{1-\sqrt{5}}{2} \end{bmatrix} = \frac{5-\sqrt{5}}{2} I \tag{22}$$

The modal mass coefficients are given by Eq. (22).

$$N_1(t) = \boldsymbol{\Theta}_1^T M(t) = \begin{bmatrix} 1 \\ \dfrac{1+\sqrt{5}}{2} \end{bmatrix}^T = \begin{bmatrix} 0 \\ M_2 e^{-\alpha t} \end{bmatrix} = \frac{1+\sqrt{5}}{2} M_2 e^{-\alpha t}$$

$$N_2(t) = \boldsymbol{\Theta}_2^T M(t) = \begin{bmatrix} 1 \\ \dfrac{1-\sqrt{5}}{2} \end{bmatrix}^T = \begin{bmatrix} 0 \\ M_2 e^{-\alpha t} \end{bmatrix} = \frac{1+\sqrt{5}}{2} M_2 e^{-\alpha t} \tag{23}$$

The modal forces are given by Eq. (23). The solutions $\eta_1(t)$ and $\eta_2(t)$ of the modal equations are written in the form of the convolution integrals

$$\eta_1(t) = \frac{1}{m'_{11}\,\omega_1} \int_0^t N_1(t-\tau) \sin \omega_1 \tau \, d\tau$$

$$= \frac{1+\sqrt{5}}{(5+\sqrt{5})(\alpha^2+\omega_1^2)} \frac{M_2}{I} \left[e^{-\alpha t} - \left(\cos \omega_1 t - \frac{\alpha}{\omega_1} \sin \omega_1 t \right) \right]$$

$$\eta_2(t) = \frac{1-\sqrt{5}}{(5-\sqrt{5})(\alpha^2+\omega_2^2)} \frac{M_2}{I} \left[e^{-\alpha t} - \left(\cos \omega_2 t - \frac{\alpha}{\omega_2} \sin \omega_2 t \right) \right] \tag{24}$$

where ω_1 and ω_2 are given above. Hence, the response can be written as

$$\theta_1(t) = \eta_1(t) + \eta_2(t), \quad \theta_2(t) = \frac{1+\sqrt{5}}{2} \eta_1(t) + \frac{1-\sqrt{5}}{2} \eta_2(t) \tag{25}$$

(*d*) The equations of motion are

$$M\ddot{\theta}(t) + K\theta(t) = M(t) \tag{26}$$

where

$$M = I\begin{bmatrix} 1 & 0 \\ 0 & 1 \end{bmatrix}, K = \frac{GJ}{L}\begin{bmatrix} 2 & -1 \\ -4 & 1 \end{bmatrix}, \theta(t) = \begin{bmatrix} \theta_1(t) \\ \theta_2(t) \end{bmatrix}, M(t) = \begin{bmatrix} 0 \\ M_2 e^{-\alpha t} \end{bmatrix} \tag{27}$$

Assuming a solution of the form

$$\theta(t) = \eta_1(t)\,\Theta_1 + \eta_2(t)\,\Theta_2 \tag{28}$$

in which $\eta_1(t)$ and $\eta_2(t)$ are modal coordinates and

$$\Theta_1 = \begin{bmatrix} 1 \\ \dfrac{1+\sqrt{5}}{2} \end{bmatrix}, \Theta_2 = \begin{bmatrix} 1 \\ \dfrac{1-\sqrt{5}}{2} \end{bmatrix} \tag{29}$$

are the modal vectors. The modal equations are given by

$$m'_{11}\ddot{\eta}_1(t) + m'_{11}\omega_1^2\eta_1(t) = N_1(t)$$

$$m'_{22}\ddot{\eta}_2(t) + m'_{22}\omega_2^2\eta_2(t) = N_2(t) \tag{30}$$

in which

$$\omega_1 = \sqrt{\frac{3-\sqrt{5}}{2}\frac{GJ}{IL}}, \omega_2 = \sqrt{\frac{3+\sqrt{5}}{2}\frac{GJ}{IL}} \tag{31}$$

are the natural frequencies

$$m'_{11} = \frac{5+\sqrt{5}}{2}I, m'_{22} = \frac{5-\sqrt{5}}{2}I \tag{32}$$

are modal mass coefficients.

The modal forces are given by

$$N_1(t) = \frac{1+\sqrt{5}}{2}M_2e^{-\alpha t}, N_2(t) = \frac{1-\sqrt{5}}{2}M_2e^{-\alpha t} \tag{33}$$

The response is given by

$$\theta(n) = \eta_1(n)\Theta_1 + \eta_2(n)\Theta_2, \quad n = 1, 2,... \tag{34}$$

where

$$\eta_1(n) = \sum_{k=0}^{n} N_1(k)g_1(n-k),$$

$$\eta_2(n) = \sum_{k=0}^{n} N_2(k)g_2(n-k), \quad n = 1, 2,... \tag{35}$$

are the discrete-time modal coordinates given in the form of convolution sums, in which the discrete time impulse responses are given by

$$g_i(n) = \frac{T}{m'_{ii}\omega_i}\sin n\omega_i T, \; i = 1, 2 \tag{36}$$

where T is the sampling period. The discrete-time response is given by

$$\theta(n) = T\sum_{k=0}^{n}\left\{\left[\frac{N_1(k)}{m'_{11}\omega_1}\sin(n-k)\omega_1 T\right]\Theta_1 + \left[\frac{N_2(k)}{m'_{22}\omega_2}\sin(n-k)\omega_2 T\right]\Theta_2\right\}$$

$$= \frac{TM_2}{1\sqrt{GJ/IL}}\sum_{k=0}^{n}\left\{e^{-\alpha kT}\sin(n-k)\,0.618034\sqrt{\frac{GJ}{IL}}\,T\begin{bmatrix}0.723607\\1.17082\end{bmatrix}\right.$$

$$\left. + e^{-\alpha kT}\sin(n-k)\,1.618034\sqrt{GJ/IL}\,T\begin{bmatrix}-0.276393\\0.17082\end{bmatrix}\right\} \tag{37}$$

Denoting $\sqrt{GJ/IL} = 1$, $M_2/I = 1$, $\alpha = 1$ and $T = 0.01$ s, the response is given by

$$\theta(n) = 0.01 \sum_{k=0}^{n} e^{-0.01k} \left\{ \sin 0.618034\,(n-k) \begin{bmatrix} 0.723607 \\ 1.170820 \end{bmatrix} + \sin 1.618034(n-k)T \begin{bmatrix} -0.276393 \\ 0.170820 \end{bmatrix} \right\} \quad (38)$$

The discrete-time response sequence is given by

$$\theta(0) = \begin{bmatrix} 0 \\ 0 \end{bmatrix}$$

$$\theta(1) = 0.01 \left\{ \sin 0.00618034 \begin{bmatrix} 0.723607 \\ 1.170820 \end{bmatrix} + \sin 0.0161803 \begin{bmatrix} -0.276393 \\ 0.170820 \end{bmatrix} \right\}$$

$$= \begin{bmatrix} 1.82291 \times 10^{-9} \\ 9.999 \times 10^{-5} \end{bmatrix}$$

$$\theta(2) = 0.01 \left\{ [\sin (0.00618034 \times 2) + e^{-0.01} \sin 0.00618034 \begin{bmatrix} 0.723607 \\ 1.170820 \end{bmatrix} + [\sin (0.0161803 \times 2) + e^{-0.01} \sin 0.0161803] \begin{bmatrix} -0.276393 \\ 0.170820 \end{bmatrix} \right\}$$

$$= \begin{bmatrix} 1.54497 \times 10^{-8} \\ 2.990 \times 10^{-4} \end{bmatrix} \quad (39)$$

(*e*) The response $\theta_i(n)$ (i = 1, 2) is plotted in Fig. EM3.9 (*d*) obtained from the following MATLAB program.

```
% Response of 2-degree of freedom system
clear
clf
I=1; % mass
k=1;%=GJ/L torsional stiffness
M=I*[1 0;0 1];% mass matrix
K=k*[2 -1;-1 1];%stiffness matrix
[u,W]=eig(K,M);% eigenvalue problem
% W= eigenvalues
u(:,1)=u(:,1)/max(u(:,1)); % normalization
u(:,2)=u(:,2)/max(u(:,2));
[w(1),I1]=min(max(W)); % relabeling of the eigenvalues
[w(2),I2]=max(max(W));
w(1)=sqrt(w(1)); % Lowest natural frequency
w(2)=sqrt(w(2)); % highest natural frequency
U(:,1)=u(:,I1); % relabelling of the eigenvectors
U(:,2)=u(:,I2);
m1=U(:,1)'*M*U(:,1); % mass quantities
m2=U(:,2)'*M*U(:,2);
```

```
T=0.01; % sampling period
N=2000; % sampling times
M2=1; % second disk torque amplitude
alpha=1;
n=[1:N];
N1=U(:,1)'*[zeros(1,N);M2*exp(-alpha*n*T)]; % modal forces
N2=U(:,2)'*[zeros(1,N);M2*exp(-alpha*n*T)];
g1=T*sin((n-1)*w(1)*T)/(m1*w(1)); %discrete time impulse responses
g2=T*sin((n-1)*w(2)*T)/(m2*w(2));
c1=conv(N1,g1); %convolution sum
c2=conv(N2,g2);
theta=U(:,1)*c1(1:N)+U(:,2)*c2(1:N); % N samples for plotting
n=[0:N-1];
axes('position',[0.1 0.2 0.8 0.7])
plot(n,theta(1,:),'.',n,theta(2,:),'.')
h=title('Response by the convolution sum');
set(h,'FontName', 'Times','FontSize',12)
h=xlabel('n')
set(h,'FontName','Times','FontSize',12)
h=ylabel('\theta_1(n),\theta_2(n)');
set(h,'FontName','Times','FontSize',12)
grid
```

Its output is shown in Fig. EM3.9 (*d*).

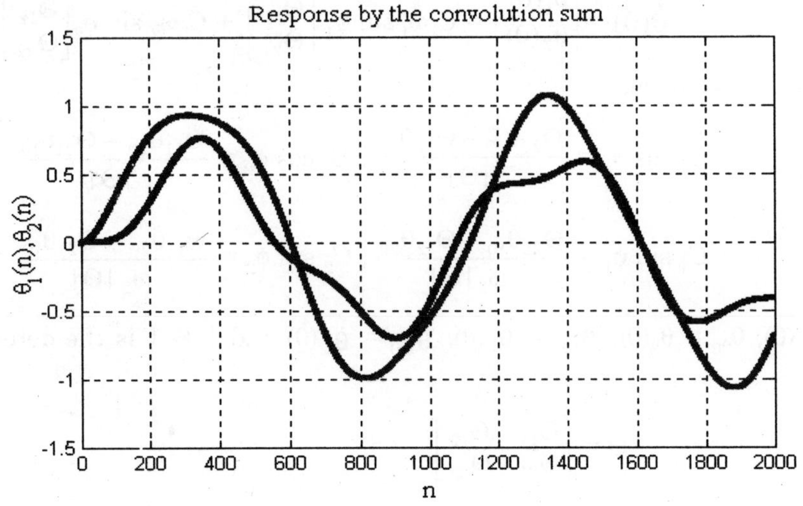

Fig. EM3.9 (*d*)

EM3.10. *Obtain the response of the system of Problem EM3.9 to the initial excitation $\theta_1(0) = 0$, $\theta_2(0) = 1.5$, $\dot{\theta}_1(0) = 1.8\sqrt{GJ/Il}$, and $\dot{\theta}_2(0) = 0$. Plot the response of the system using MATLAB.*

Solution:

The initial conditions are given as

$$\theta_1(0) = 0,\ \theta_2(0) = 1.5,\ \dot{\theta}_1(0) = 1.8\sqrt{\frac{GJ}{IL}},\ \dot{\theta}_2(0) = 0 \tag{1}$$

From Problem V3.9, we have

$$\omega_1 = 0.6180\sqrt{\frac{GJ}{IL}},\ \Theta_1 = \begin{bmatrix}\Theta_{11}\\ \Theta_{21}\end{bmatrix} = \Theta_{11}\begin{bmatrix}1\\ 1.6180\end{bmatrix}$$

$$\omega_2 = 1.6180\sqrt{\frac{GJ}{IL}},\ \Theta_2 = \begin{bmatrix}\Theta_{12}\\ \Theta_{22}\end{bmatrix} = \Theta_{12}\begin{bmatrix}1\\ 1.6180\end{bmatrix} \tag{2}$$

The response to the initial excitation is a superposition of the natural modes. Hence

$$\theta(t) = C_1\cos(\omega_1 t - \phi_1)\Theta_1 + C_2\cos(\omega_2 t - \phi_2)\Theta_2$$

or

$$\theta(t) = C_1(\cos\omega_1 t\cos\phi_1 + \sin\omega_1 t\sin\phi_1)\Theta_1 + C_2(\cos\omega_2 t\cos\phi_2 + \sin\omega_2 t\sin\phi_2)\Theta_2 \tag{3}$$

and

$$\dot{\boldsymbol{\theta}}(t) = C_1\omega_1(\sin\omega_1 t\cos\phi_1 - \cos\omega_1 t\sin\phi_1)\Theta_1 - C_2\omega_2(\sin\omega_2 t\cos\phi_2 - \cos\omega_2 t\sin\phi_2)\Theta_2$$

If $t = 0$, then Eq. (3) becomes

$$\theta(0) = \begin{bmatrix}\theta_1(0)\\ \theta_2(0)\end{bmatrix} = C_1\cos\phi_1\begin{bmatrix}\Theta_{11}\\ \Theta_{21}\end{bmatrix} + C_2\cos\phi_2\begin{bmatrix}\Theta_{12}\\ \Theta_{22}\end{bmatrix}$$

$$\dot{\boldsymbol{\theta}}(0) = \begin{bmatrix}\dot{\theta}_1(0)\\ \dot{\theta}_2(0)\end{bmatrix} = C_1\omega_1\sin\phi_1\begin{bmatrix}\Theta_{11}\\ \Theta_{21}\end{bmatrix} + C_2\omega_2\sin\phi_2\begin{bmatrix}\Theta_{12}\\ \Theta_{22}\end{bmatrix} \tag{4}$$

From Eq. (4) we have,

$$C_1\cos\phi_1 = \frac{\Theta_{22}\theta_{10} - \Theta_{12}\theta_{20}}{|\Theta|},\ C_2\cos\phi_2 = \frac{\Theta_{11}\theta_{20} - \Theta_{21}\theta_{10}}{|\Theta|}$$

$$C_1\sin\phi_1 = \frac{\Theta_{22}\dot{\theta}_{10} - \Theta_{12}\dot{\theta}_{20}}{\omega_1|\Theta|},\ C_2\sin\phi_2 = \frac{\Theta_{11}\dot{\theta}_{20} - \Theta_{21}\dot{\theta}_{10}}{\omega_2|\Theta|} \tag{5}$$

where $\theta_{10} = \theta_1(0)$, $\theta_{20} = \theta_2(0)$, $\dot{\theta}_{10} = \dot{\theta}_1(0)$, $\dot{\theta}_{20} = \dot{\theta}_2(0)$ and $|\Theta|$ is the determinant of the matrix

$$\Theta = \begin{bmatrix}\Theta_{11} & \Theta_{12}\\ \Theta_{21} & \Theta_{22}\end{bmatrix} \tag{6}$$

Letting $\Theta_{11} = \Theta_{12} = 1$,

$$C_1\cos\phi_1 = -\frac{1.5}{|\Theta|},\ C_2\cos\phi_2 = \frac{1.5}{|\Theta|},$$

$$C_1\sin\phi_1 = -\frac{1.8}{|\Theta|},\ C_2\sin\phi_2 = -\frac{1.8}{|\Theta|}$$

$$|\Theta| = \begin{vmatrix} 1 & 1 \\ 1.6180 & -0.6180 \end{vmatrix} = -2.2360 \tag{7}$$

The response to the given initial excitation is given by

$$\theta(t) = \begin{bmatrix} \theta_1(t) \\ \theta_2(t) \end{bmatrix} = \frac{1}{2.2360}\left[\left[(1.5\cos\omega_1 t + 1.8\cos\omega_1 t)\begin{bmatrix} 1 \\ 1.6180 \end{bmatrix}\right.\right.$$

$$\left.\left. + (-1.5\cos\omega_2 t + 1.8\sin\omega_2 t)\begin{bmatrix} 1 \\ -0.6180 \end{bmatrix}\right]\right]$$

or by components

$$_1(t) = 0.6708\left(\cos 0.6180\sqrt{\frac{GJ}{IL}}\,t - \cos 1.6180\sqrt{\frac{GJ}{IL}}\,t\right)$$

$$+ 0.8060\left(\sin 0.6180\sqrt{\frac{GJ}{IL}}\,t - \sin 1.6180\sqrt{\frac{GJ}{IL}}\,t\right)$$

$$\theta_2(t) = 1.0854\cos 0.6180\sqrt{\frac{GJ}{IL}}\,t + 1.3025\sin 0.6180\sqrt{\frac{GJ}{IL}}\,t$$

$$+ 0.4146\cos 1.6180\sqrt{\frac{GJ}{IL}}\,t - 0.4975\sin 106180\sqrt{\frac{GJ}{IL}}\,t \tag{8}$$

The MATLAB program is listed as follows:

```
% response of a two-degree-of freedom system to initial excitations
clear
clf
I=1; % inertia
k=1;%=GJ/L stiffness
M=I*[1 0;0 1];% mass
K=k*[2 -1;-1 1];%stiffness
[u,W]=eig(K,M);% eigenvalue problem
% W=matrix of eigenvalues
u(:,1)=u(:,1)/max(u(:,1)); % normalization
u(:,2)=u(:,2)/max(u(:,2));
[w(1),I1]=min(max(W)); % relabeling
[w(2),I2]=max(max(W));
w(1)=sqrt(w(1)); % lowest natural frequency
w(2)=sqrt(w(2)); % highest natural frequency
U(:,1)=u(:,I1); % relabelling
U(:,2)=u(:,I2);
x0=[0;2];% Initial displacement
v0=[2*sqrt(k/I);0]; % initial velocity
t=[0:0.1:20]; % initial time, time increment, final time
% displacement
x1=(((U(2,2)*x0(1)-U(1,2)*x0(2))*cos(w(1)*t)+(U(2,2)*v0(1)-
U(1,2)*v0(2))*sin(w(1)*t)/w(1))*U(1,1)+((U(1,1)*x0(2)-
```

```
U(2,1)*x0(1))*cos(w(2)*t)+(U(1,1)*v0(2)-U(2,1)*v0(1))*sin(w(2)*t)/
w(2))*U(1,2))/det(U);
x2=(((U(2,2)*x0(1)-U(1,2)*x0(2))*cos(w(1)*t)+(U(2,2)*v0(1)-
U(1,2)*v0(2))*sin(w(1)*t)/w(1))*U(2,1)+((U(1,1)*x0(2)-
U(2,1)*x0(1))*cos(w(2)*t)+(U(1,1)*v0(2)-U(2,1)*v0(1))*sin(w(2)*t)/
w(2))*U(2,2))/det(U);
axes('position',[0.2 0.3 0.6 0.5])
plot(t,x1,t,x2)
title('Response to initial excitation')
ylabel('\theta_1(t),\theta_2(t)')
xlabel('t[s]')
legend('\theta_1(t)','\theta_2(t)',1)
grid
```

The corresponding output obtained is shown in Fig. EM3.10

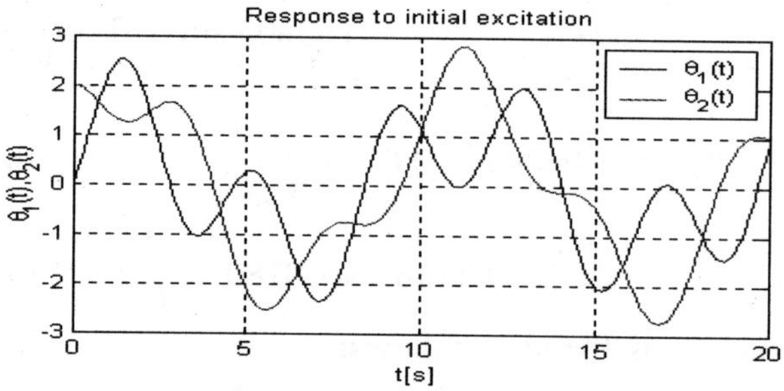

Fig. EM3.10

EM3.11. *A simplified model of an automobile suspension system is shown in Fig. EM3.11 as a two degree of freedom system. Write a MATLAB script to determine the natural frequencies of this model.*

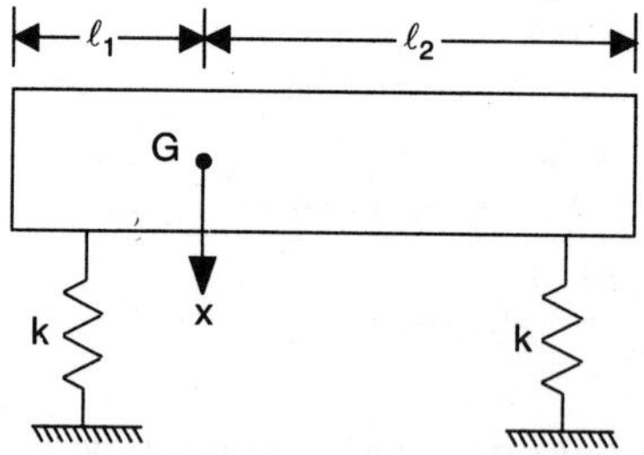

Fig. EM3.11 Simplified model of an automobile.

The differential equations governing the motion of the system are given as

$$\begin{bmatrix} m & 0 \\ 0 & I \end{bmatrix}\begin{bmatrix} \ddot{x} \\ \ddot{\theta} \end{bmatrix} + \begin{bmatrix} 2k & (l_2 - l_1)k \\ (l_2 - l_1)k & (l_2^2 - l_1^2)k \end{bmatrix}\begin{bmatrix} x \\ \theta \end{bmatrix} = \begin{bmatrix} 0 \\ 0 \end{bmatrix}$$

where x is the displacement of the mass center and θ is the angular rotation of the body from its horizontal position.

The parameters are given as

Automobile weight, W = 5000 lb

Centroidal moment of inertia, I = 400 slug-ft²
Spring stiffness, k = 2500 lb/ft

$l_1 = 3.4\ ft$

$l_2 = 4.6\ ft$

Solution:

The MATLAB program is given as follows:

```
%Two-degree-of-freedom system
W=input('Vehicle weight in lb');
I=input('Mass moment of inertia in slugs-ft^2')
k=input('Stiffness in lb/ft')
a=input('Distance from rear springs to cg in ft')
b=input('Distance from front springs to cg')
% mass matrix
g=32.2;
m=W/g;
M=[m,0;0,I];
% stiffness matrix
K=[2*k,(b-a)*k;(b-a)*k,(b^2+a^2)*k];
% eigenvalues and eigenvectors calculation
C=inv(M)*K;
[V,D]=eig(C);
om_1=sqrt(D(1,1));
om_2=sqrt(D(2,2));
X1=[V(1,1);V(2,1)];
X2=[V(1,2);V(2,2)];
% Output
disp('Vehicle weight in lb='); disp(W)
disp('moment of inertia in slugs-ft^2');disp(I)
disp('Stiffness in lb/ft='); disp(k)
disp('Distance from rear springs to cg in ft='); disp(a)
disp('Distance from front springs to cg in ft=');disp(b)
disp('Mass-matrix');disp(M)
disp('Stiffness-matrix');disp(K)
disp('Natural frequencies in rad/s=');
disp(om_1)
disp(om_2)
disp('Mode shape vectors'); disp(X1)
disp(X2)
```

The output of this program is as follows:

Vehicle weight in lb 5000

W=5000

Mass moment of inertia in slugs-ft^2 400

I = 400

```
Stiffness in lb/ft 2500
k = 2500
Distance from rear springs to cg gravity in ft 3.4
a = 3.4000
Distance from front springs to cg 4.6
b = 4.6000
Vehicle weight in lb= 5000
Moment of inertia in slugs-ft^2    400
Stiffness in lb/ft=2500
Distance from rear springs to cg in ft=3.4000
Distance from front springs to cg in ft=4.6000
Mass-matrix
      155.2795              0
             0       400.0000
Stiffness-matrix
  1.0e+004 *
    0.5000     0.3000
    0.3000     8.1800
Natural frequencies in rad/s=
    5.6003
   14.3296
Mode shape vectors
   -0.9991
    0.0433

   -0.1109
   -0.9938
```

EM3.12. *Determine the free-vibration response of a two degree of freedom system shown in Fig. EM3.12 with the initial conditions* $x_1(0) = 0$, $x_2(0) = 0.005$ *m*, $\dot{x}_1(0) = 0$, $\dot{x}_2(0) = 0$. *The parameters of the system are given as* $m = 30$ *kg*, $k = 20{,}000$ *N/m, and* $c = 150$ *N.s/m.*

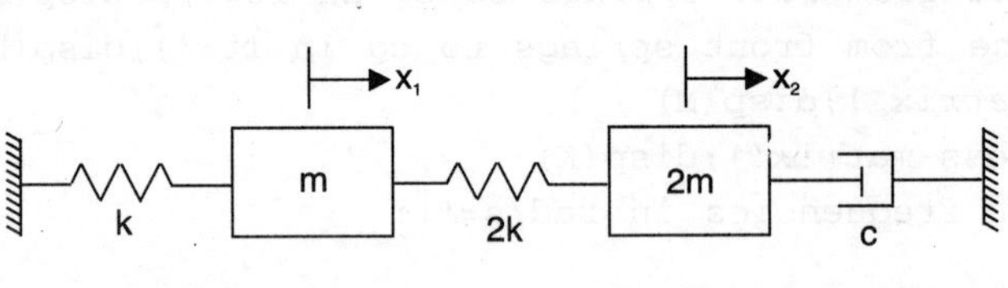

Fig. EM3.12 Two degree of freedom system.

The differential equations governing the motion of the system are

$$\begin{bmatrix} m & 0 \\ 0 & 2m \end{bmatrix}\begin{bmatrix} \ddot{x}_1 \\ \ddot{x}_2 \end{bmatrix} + \begin{bmatrix} 0 & 0 \\ 0 & 0 \end{bmatrix}\begin{bmatrix} \dot{x}_1 \\ \dot{x}_2 \end{bmatrix} + \begin{bmatrix} 3k & -2k \\ -2k & 2k \end{bmatrix}\begin{bmatrix} x_1 \\ x_2 \end{bmatrix} = \begin{bmatrix} 0 \\ 0 \end{bmatrix}$$

or $$M\dot{y} + Ky = 0$$

where $$M = \begin{bmatrix} 0 & M \\ M & c \end{bmatrix}; K = \begin{bmatrix} -M & 0 \\ 0 & K \end{bmatrix}; Y = \begin{bmatrix} \dot{x} \\ x \end{bmatrix}$$

The solution is assumed as

$$y = \phi e^{-\gamma t}$$

where γ *are the eigenvalues of* $M^{-1}K$ *and* ϕ *are the eigenvectors. The general solution is a linear combination over all solutions, that is,*

$$y = \sum_{j=1}^{4} c_j \phi_j e^{-\gamma_j t}$$

and application of initial conditions gives

$$y_0 = \sum_{j=1}^{4} c_j \phi_j = VC$$

and $$C = V^{-1} y_0$$

Solution:

The MATLAB program is given as follows:

```
m=30; % Mass
k=20000; % Stiffness
c=150; % Damping
% 4 x 4 matrices
disp('4 x 4 Mass matrix');
mt=[0,0,m,0;0,0,0,2*m;m,0,0,0;0,2*m,0,c];
disp('4 x 4 stiffness matrix');
kt=[-m,0,0,0;0,-2*m,0,0;0,0,3*k,-2*k;0,0,-2*k,2*k];
Z=inv(mt)*kt;
[V,D]=eig(Z);
disp('Eigenvalues');
V
disp('Initial  conditions');
x0=[0;0;0.005;0]
disp('Integration  constants');
S=inv(V)*x0
tk=linspace(0,2,101);
% Evaluation  of  time  dependent  response
% Recall  that  x1=y3  and  x2=y4
for  k=1:101
  t=tk(k);
  for  i=3:4
     x(k,i-2)=0;
     for  j=1:4
x(k,i-2)=x(k,i-2)+(real(S(j))*real(V(i,j))
imag(S(j))*imag(V(i,j)))*cos(imag(D(j,j))*t);
x(k,i-2)=x(k,i-2)+(imag(S(j))*real(V(i,j))
real(S(j))*imag(V(i,j)))*sin(imag(V(i,j))*t);
        x(k,i-2)=x(k,i-2)*exp(-real(D(j,j))*t);
     end
  end
end
```

```
plot(tk,x(:,1),'-',tk,x(:,2),':')
title('Solution of problem EM3.12')
xlabel('t[sec]')
ylabel('x(m)')
legend('x1(t)','x2(t)')
```

The output of this program is given below. See also Fig.EM3.12(a).

V =

–0.9390 – 0.9390	0.5886 – 0.0085i	0.5886 + 0.0085i	
0.3428 – 0.0185i	0.3428 + 0.0185i	0.8050	0.8050
0.0001 – 0.0188i	0.0001 + 0.0188i	–0.0026 + 0.0440i	–0.0026 – 0.0440i
0.0003 + 0.0069i	0.0003 – 0.0069i	–0.0044 + 0.0601i	–0.0044 – 0.0601i

Initial conditions

x0 =

0
0
0.0050
0

Integration constants

S =

– 0.0013 + 0.1048i
– 0.0013 – 0.1048i
– 0.0019 – 0.0119i
– 0.0019 + 0.0119i

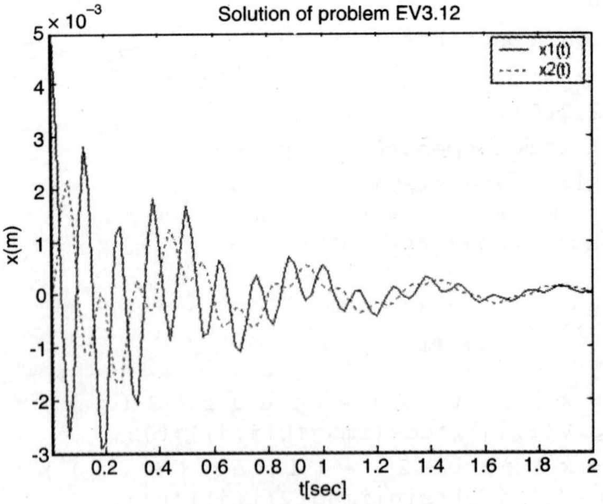

Fig. EM3.12 (*a*)

EM3.13. *For systems with arbitrary viscous-damping, the response must be obtained in the state-space, which implies the use of the transition-matrix. If the response is to be evaluated on a computer, then the state-equations must be transformed to discrete time. Determine the free-vibration response of a 2-degree-of-freedom damped system with initial conditions X(0) = {0, 0.01} and $\dot{X}$ (0)={0, 0}. Given*

$$[M] = \begin{bmatrix} 0 & 0 & 30 & 0 \\ 0 & 0 & 0 & 50 \\ 30 & 0 & 0 & 0 \\ 0 & 50 & 0 & 80 \end{bmatrix}$$

$$[K] = \begin{bmatrix} -40 & 0 & 0 & 0 \\ 0 & -50 & 0 & 0 \\ 0 & 0 & 35000 & -25000 \\ 0 & 0 & -25000 & 4000 \end{bmatrix}$$

Solution:

The solution is similar to the problem EM3.12, and the MATLAB program is written as follows:

```
mt=[0 0 30 0;0,0,0,50;30,0,0,0;0,50,0,80];
kt=[- 40,0,0,0;0,- 50,0,0;0,0,35000,- 25000;0,0,- 25000, 4000];
Z=inv(mt)*kt;
[V,D]=eig(Z);
disp('Eigenvalues')
DS=[D(1,1),D(2,2),D(3,3),D(4,4)]
disp('Eigenvectors')
V
x0=[0;0;0.01;0];
S=inv(V)*x0;
tk=linspace(0,2,101);
for k=1:101
t=tk(k);
for i=3:4
x(k,i-2)=0;
for j=1:4
x(k,i-2)=x(k,i-2)+(real(S(j))*real(V(i,j))-
imag(S(j))*imag(V(i,j)))*cos(imag(D(j,j))*t);
x(k,i-2)=x(k,i-2)+(imag(S(j))*real(V(i,j))-
imag(S(j))*imag(V(i,j)))*sin(imag(V(i,j))*t);
x(k,i-2)=x(k,i-2)*exp(-real(D(j,j))*t);
end
end
end
plot(tk,x(:,1),'-',tk,x(:,2),':')
title('Free Vibration response of damped system')
xlabel('t (sec)')
ylabel('x (m)')
legend('x1(t)','x2(t)')
```

```
The output obtained is given as follows:
Eigenvalues
DS =
Columns 1 through 2
1.1082e - 001 + 4.4615e + 001i  1.1082e - 001 - 4.4615e + 001i
Columns 3 through 4
- 1.5162e + 001                 1.6541e + 001
Eigenvectors
V =
Columns 1 through 2
 9.5465e-001                    9.5465e-001
- 2.9638e - 001 + 9.4404e - 003i - 2.9638e - 001 - 9.4404e-003i
- 6.3783e - 005 + 2.5677e - 002i - 6.3783e - 005 - 2.5677e-002i
- 1.9510e - 004 - 6.6437e - 003i - 1.9510e - 004 + 6.6437e-003i
Columns 3 through 4
4.6275e - 001       - 4.5475e - 001
8.8381e - 001       - 8.8839e - 001
3.6624e - 002         3.2991e - 002
5.8290e - 002         5.3710e - 002
```

Fig. EM3.13 shows the response in time-domain obtained from the output of the MATLAB program.

Note: Here $M = \begin{bmatrix} 25 & 0 \\ 0 & 50 \end{bmatrix}, C = \begin{bmatrix} 0 & 0 \\ 0 & 80 \end{bmatrix}$ and

$$K = \begin{bmatrix} 35000 & -25000 \\ -25000 & 3000 \end{bmatrix}$$ and state matrices are

respectively $$M^T = \begin{bmatrix} 0 & 0 & 25 & 0 \\ 0 & 0 & 0 & 50 \\ 25 & 0 & 0 & 0 \\ 0 & 50 & 0 & 80 \end{bmatrix}$$

and $$K^T = \begin{bmatrix} -30 & 0 & 0 & 0 \\ 0 & -50 & 0 & 0 \\ 0 & 0 & 35000 & -25000 \\ 0 & 0 & -25000 & 3000 \end{bmatrix}$$

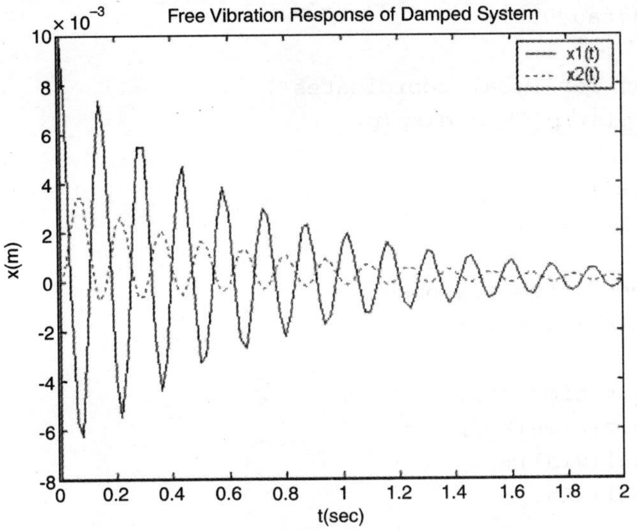

Fig. EM3.13

EM3.14. *In the Example EM3.13, if a force $F_0 exp(-at)$ acts on the system, find the forced vibration response using the MATLAB program. Given $F_0 = 60$.*

Solution:

Here first few steps are common as in free-vibration response problem.

```
syms t tau
m=25;
k=12500;
c=80;
F0=60;
alpha=1.5;
mt=[0,0,m,0;0,0,0,2*m;m,0,0,0;0,2*m,0,c];
kt=[-m,0,0,0;0,-2*m,0,0;0,0,3*k,-2*k;0,0,-2*k,2*k];
z=inv(mt)*kt;
[V,D]=eig(z);
L=conj(V)'*mt*V;
for j=1:4
ss=1/sqrt(L(j,j));
for i=1:4
P(i,j)=V(i,j)*ss;
end
end
F=[0;0;0;F0*exp(-alpha*tau)];
G=P'*F;
G=vpa(G);
%Convolution integral solution
for i=1:4
f(i)=G(i)*exp(-D(i,i)*(t-tau));
```

```
p(i)=int(f(i),tau,0,t);
end
disp('solution for modal coordinates')
p=[p(1);p(2);p(3);p(4)]; disp(p)
y=P*p;
disp('response')
disp('x1=y3, x2=y4 ')
y=vpa(y);
% Plotting the system response
time=linspace(0,1.5,101);
for k=1:101
x1a=subs(y(3),t,time(k));
x2a=subs(y(4),t,time(k));
x1b(k)=vpa(real(x1a));
x2b(k)=vpa(real(x2a));
end
x1=double(x1b);
x2=double(x2b);
plot(time,x1,'-',time,x2,':')
xlabel('t(seconds)')
ylabel('response(m)')
legend('x1(t)', 'x2(t)')
```

The output of the program is shown as the forced vibration response in Fig. EM3.14.

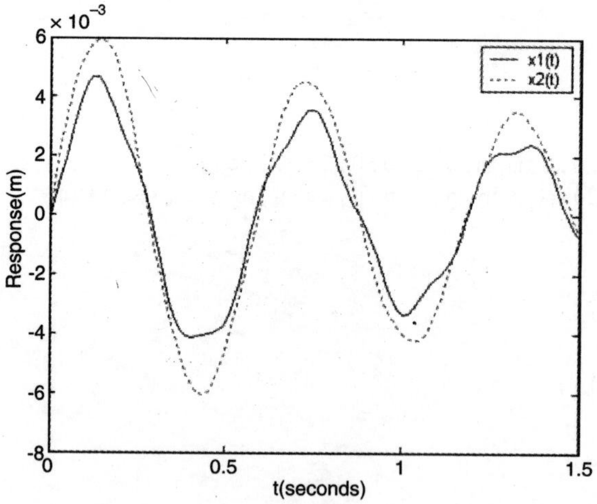

Fig. EM3.14

EM3.15. *Two gears A and B in mesh are mounted on two uniform circular shafts of equal stiffness* $\frac{GJ}{L}$. *If the gear A is subjected to a torque* $M_0 \cos \omega t$, *derive an expression for angular motion of B. Assume the radius ratio as:* $\frac{R_A}{R_B} = n$. *Here L is length of each shaft. Write a MATLAB script to plot the response.*

Solution:

Here the equation of motion is given by

$$I_{eq}\ \ddot{\theta} + k_{eq}.\theta_A = M_A = M_0 \cos\omega t \tag{1}$$

where the equivalent stiffness of the gears $k_{eq} = \dfrac{GJ}{L}(1 + n^2)$ and equivalent moment of inertia of gears $I_{eq} = I_A + n^2 I_B$.

Simplifying the above equation of motion we get:

$$\ddot{\theta}_A + \omega_n{}^2\,\theta_A = \frac{M_0}{I_A + n^2 I_B}\cos\omega t, \text{ and } \omega_n{}^2 = \frac{GJ(1+n^2)}{L(I_A + n^2\ I_B)} \tag{2}$$

Since $\theta_B = n\,\theta_A$, the solution is given by

$$\theta_B = \frac{M_0 Ln}{GJ(1+n^2)\,[1-(\omega/\omega_n)^2]}\cos\omega t$$

The MATLAB program to plot the values of amplitude of θ_B for various values of ω is given as follows:

```
M0=1;% amplitude of the moment
L=1; %  length of shaft
GJ=1;%  torsional stiffness
n=3;%  gear ratio
r=[0:0.01:3];% frequency ratio
thetab=(M0*L*n)./(GJ*(1+(n.^n))*(1-r.^2)); % amplitude
plot(r,thetab)
title('Response to torque')
ylabel('\theta_b')
xlabel('\omega/\omega_n')
grid
```

The output is shown in Fig. EM3.15.

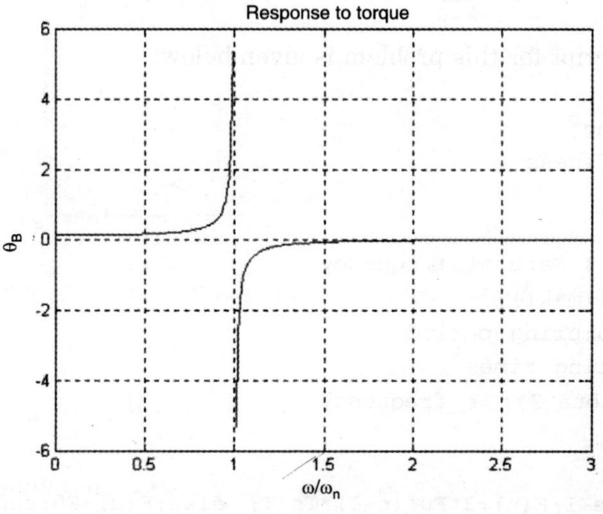

Fig. EM3.15

EM3.16. *Derive the response of a viscously damped single-degree of freedom system to the trapezoidal pulse shown in Fig. EM3.16. Plot response for system parameters, m = 15 kg, c = 25 NS/m and k = 5000 N/m. Use convolution sum*

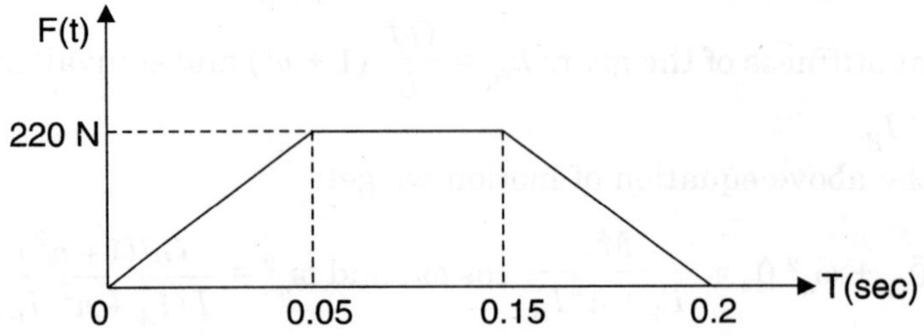

Fig. EM3.16

Solution:

The system is described by:

$$\ddot{x} + 2\xi\omega_n \dot{x} + \omega_n^2 x = \frac{F(t)}{m}$$

where

$$F(t) = \begin{cases} \dfrac{2F_0}{T}t, & 0 < t < \dfrac{T}{2} \\ F_0, & \dfrac{T}{2} < t < \dfrac{3T}{2} \\ 2F_0\left(2 - \dfrac{t}{T}\right), & \dfrac{3T}{2} < t < 2T \\ 0, & t > 2T \end{cases}$$

where $T = 0.2$ sec in Fig. EM3.16

The discrete time response by convolution sum is:

$$x(n) = \sum_{k=0}^{n} F(k)\, g(n-k)$$

The MATLAB script for this problem is given below:

```
m=15; % mass
c=25; % damping
k=5000; % stiffness
F0=220;
T=0.2;
wn=sqrt(k/m); % Natural frequency
zeta=c/(2*sqrt(m*k));
Ts=0.003; % Sampling period
N=201; % sampling times
wd=wn*sqrt(1-zeta^2); % frequency
% force
for n=1:N,
if n<=(T/2)/Ts+1;F(n)=2*F0*(n-1)*Ts/T;  else;F(n)=F0;end
```

```
if n>(3*T/2)/Ts+1;F(n)=2*F0*(2-(n-1)*Ts/T);end
if n>2*T/Ts+1;F(n)=0;end
end
n=[1:N];
g=Ts*exp(-(n-1)*zeta*wn*Ts).*sin((n-1)*wd*Ts)/(m*wd);
% discrete-time impulse response
c0=conv(F,g); % Convolution sum
c=c0(1:N); % plot to N samples
n=[0:N-1];
axes('position',[0.1 0.2 0.8 0.7])
plot(n,c,'.');
title('Response to the Trapezodial pulse');
xlabel('n')
ylabel('x(n) [m]')
grid
```

Output of this program is the Fig. EM3.16 (*a*)

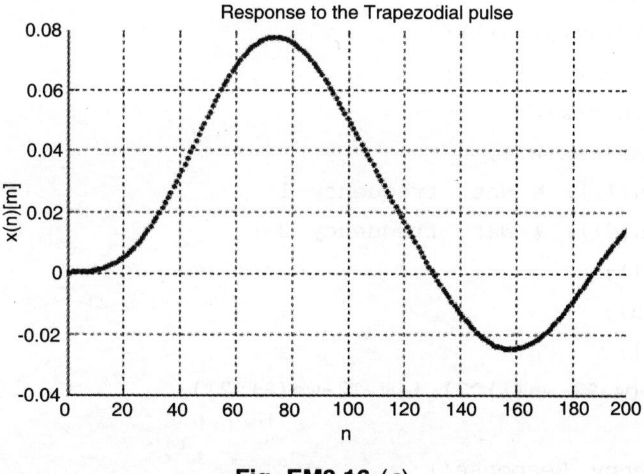

Fig. EM3.16 (*a*)

EM3.17. *A two storey building is undergoing a horizontal motion $y(t) = Y_0.\sin\omega t$. Derive expression for displacement of second floor. Write MATLAB script to plot the response. Assume appropriate values of stiffness and mass of the system.*

Equations of motion for building can be written as:

$$\begin{bmatrix}2 & 0\\0 & 2\end{bmatrix}\begin{Bmatrix}\ddot{x}_1\\\ddot{x}_2\end{Bmatrix}_{+a2}\begin{bmatrix}4 & -2\\-2 & 2\end{bmatrix}\begin{Bmatrix}x_1\\x_2\end{Bmatrix}_{=a2}\begin{Bmatrix}Y_0 \sin wt\\0\end{Bmatrix}$$

where $$\alpha^2 = \frac{12EI}{mH^3} = \frac{12}{m} = 6$$

Solving for steady-state response we get:

$$X1 = \frac{(\alpha^2 - \omega^2)\,\alpha^2}{(\omega^2 - \omega_1^2)(\omega^2 - \omega_2^2)}\, Y_0$$

$$X2 = \frac{\alpha^4}{(\omega^2 - \omega_1^2)(\omega^2 - \omega_2^2)}\, Y_0$$

These values are to be plotted against various values of ω.

Solution:

The MATLAB script for this problem is given as follows:

```
m=20;% mass
k=200; % k=12EI/H^3  stiffness
w0=k/m;
M=[m 0;0 m]; %mass matrix
K=[2*k -k;-k k]; % stiffness matrix
%eigenvalues
[u,W]=eig(K,M);
u(:,1)=u(:,1)/max(u(:,1));
u(:,2)=u(:,2)/max(u(:,2));
[wn(1),I1]=min(max(W));
[wn(2),I2]=max(max(W));
wn(1)=sqrt(wn(1)); % Nat. frequency 1
wn(2)=sqrt(wn(2)); % Nat. frequency 2
U(:,1)=u(:,I1);
U(:,2)=u(:,I2);
w=[0:0.002:6];
T2=(w0^2)./((w.^2-wn(1)^2).*(w.^2-wn(2)^2));
plot(w,T2)
title('Frequency Response')
ylabel('{\it{X}}_2(\omega)/{\it{Y}}_0')
xlabel('\omega')
axis([0 8 -5 5])
grid
```

The MATLAB output is shown in Fig. EM3.17.

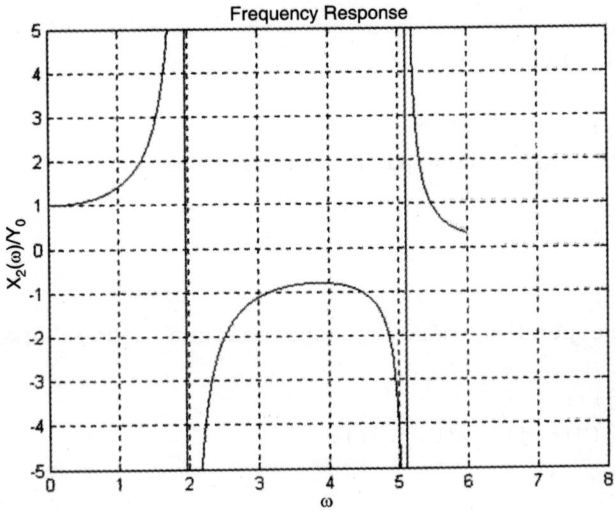

Fig. EM3.17

EM3.18. *A 3-degree of freedom system shown in Fig. EM3.18. Obtain the natural frequencies and mode shapes using MATLAB script. Assume k = m = 1.*

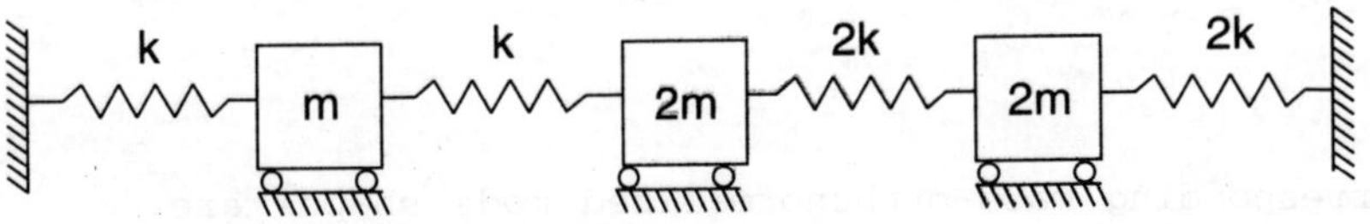

Fig. EM3.18

The equations of motion can be written as:

$$M\,\ddot{x}(t) + Kx(t) = 0$$

with $x(t) = [x1(t)\ x2(t)\ x3(t)]T$ *as the displacement vector*

$$M = \begin{bmatrix} m & 0 & 0 \\ 0 & 2m & 0 \\ 0 & 0 & 2m \end{bmatrix} \quad \text{and} \quad K = \begin{bmatrix} 2k & -k & 0 \\ -k & 3k & -2k \\ 0 & -2k & 4k \end{bmatrix}$$

as the mass and stiffness matrices.

Solution:

The MATLAB script for finding the natural frequencies and mode shapes is given as follows:

```
k=1; % stiffness
m=1; % mass
M=m*[1 0 0;0 2 0;0 0 2]; % mass matrix
K=k*[2 -1 0;-1 3 -2;0 -2 4]; % stiffness matrix
N=3;
R=chol(M); % Cholesky decomposition technique
L=R';
A=inv(L)*K*inv(L');
```

```
[x,W]=eig(A);
v=inv(L')*x;
for i=1:N,
w1(i)=sqrt(W(i,i));
end
[w,I]=sort(w1);
disp('The first three natural frequencies are')
disp(w(1))
disp(w(2))
disp(w(3))
n=[1:N];
disp('The corresponding mass-orthonormalized mode shapes are')
for j=1:N,
U(:,j)=v(:,I(j));
U(:,j)=U(:,j)/(U(:,j)'*M*U(:,j));
disp('mode-')
disp(j)
disp(U(:,j))
end
```

The outputs are as follows:

```
The first three natural frequencies are
0.7071
1.4142
1.7321
The corresponding mass-orthonormalized mode shapes are
mode-
1
- 0.3651
- 0.5477
- 0.3651
mode-
2
0.8165
- 0.0000
- 0.4082
mode-
3
- 0.4472
 0.4472
- 0.4472
```

EM3.19. *In Fig. EM3.18, if mass m is subjected to unit step function u(t), determine the response using modal analysis. Write a MATLAB script to plot the displacement response of all the masses.*

Solution:

The MATLAB program is given as follows:

```
M=[1 0 0;0 2 0;0 0 2]; % mass matrix
```

```
C=[0 0 0;0 0 0;0 0 0]; % damping matrix
K=[2 -1 0;-1 3 -2;0 -2 4]; % stiffness
A=[zeros(size(M)) eye(size(M));-inv(M)*K -inv(M)*C];
B=[zeros(size(M)); inv(M)];
TO=10; %RISE TIME OF FORCE
N=200;
T=0.1; % SAMPLING PERIOD
NO=TO/T;
phi=eye(size(A))+T*A+T^2*A^2/2+T^3*A^3/6;
gamma=inv(A)*(phi-eye(size(A)))*B;
x(:,1)=zeros(2*length(M),1);
for k=1:N,
f(k)=1;
F(:,k)=[1;0;0]*f(k); % Force is only applied to mass m
x(:,k+1)=phi*x(:,k)+gamma*F(:,k);
end
k=[0:N];
plot(k,x(1,:),'o',k,x(2,:),'s',k,x(3,:),'.')
title('system response for unit step at first mass EM3.19')
ylabel('x_1(k),x_2(k),x_3(k)')
xlabel('k')
legend('x_1(k)','x_2(k)','x_3(k)')
grid
```

The output obtained is shown in Fig. EM3.19.

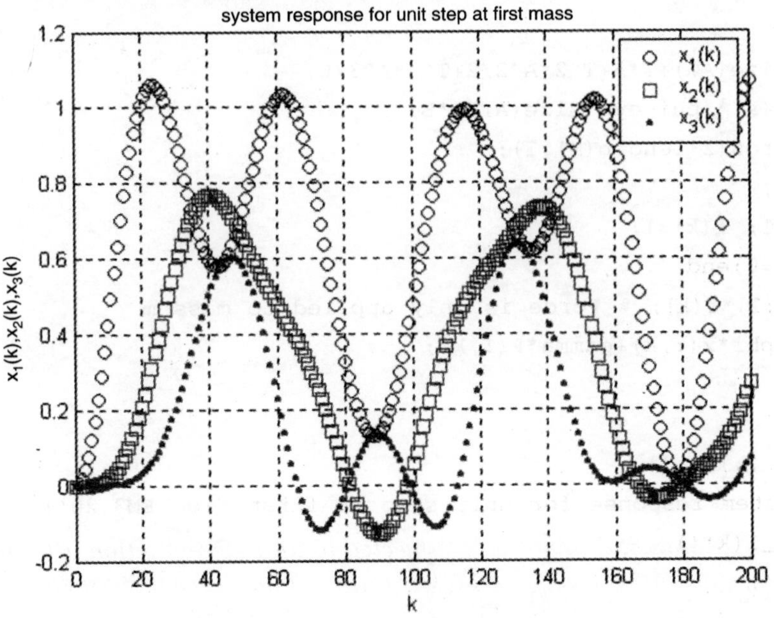

Fig. EM3.19

EM3.20. *A two-degree of freedom torsional system shown in Fig. EM3.20 and is subjected to a torque of unit pulse nature [u(t) – u(t – 4)] at the disc B.*

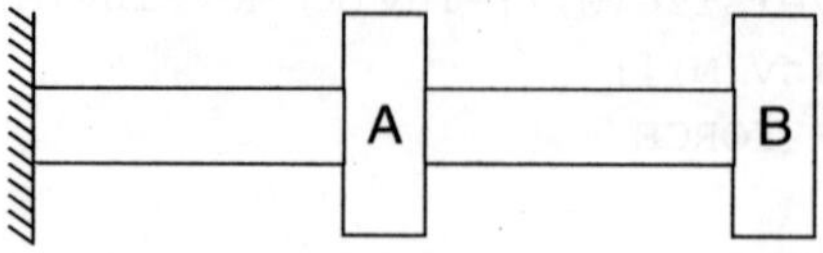

Fig. EM3.20

The mass, stiffness and damping matrices are

$$M = \begin{bmatrix} 3 & 0 \\ 0 & 5 \end{bmatrix}, K = \begin{bmatrix} 5 & -4 \\ -4 & 4 \end{bmatrix} \text{ and } C = \begin{bmatrix} 1.6 & -0.8 \\ -0.8 & 0.8 \end{bmatrix}$$

Plot the response of disc A using MATLAB.

Solution:

The MATLAB script is given as follows:

```
M=[1 0;0 2]; % mass matrix
C=[1.6 -0.8; -0.8 0.8]; % damping matrix
K=[5 -4;-4 4]; % stiffness matrix
A=[zeros(size(M)) eye(size(M));-inv(M)*K -inv(M)*C];
B=[zeros(size(M)); inv(M)];
TO=4; %RISE TIME OF FORCE
N=600;
T=0.1; % SAMPLING PERIOD
NO=TO/T;
phi=eye(size(A))+T*A+T^2*A^2/2+T^3*A^3/6;
gamma=inv(A)*(phi-eye(size(A)))*B;
x(:,1)=zeros(2*length(M),1);
for k=1:N,
if k<=NO+1; f(k)=1;
else;f(k)=0;end
F(:,k)=[0;1]*f(k); % Force is only applied to mass m
x(:,k+1)=phi*x(:,k)+gamma*F(:,k);
end
k=[0:N];
plot(k,x(1,:),'.')
title('system response for unit step at first disc EM3.20')
ylabel('x_1(k)')
xlabel('k')
grid
```

The output of this program is given in Fig. EM3.20 (*a*).

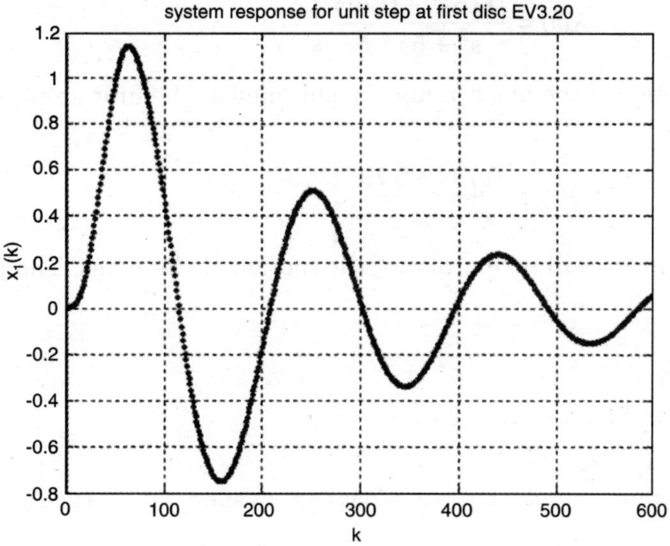

Fig. EM3.20 (*a*)

EM3.21. *For the single degree of freedom vibrating system shown in Fig. EM3.21, determine the motion of the mass subjected to the initial conditions x(0) = 0.15 m and $\dot{x}(0)$ = 0.04 m/s. Given m = 1 kg, c = 5 N-s/m, and k = 5 N/m.*

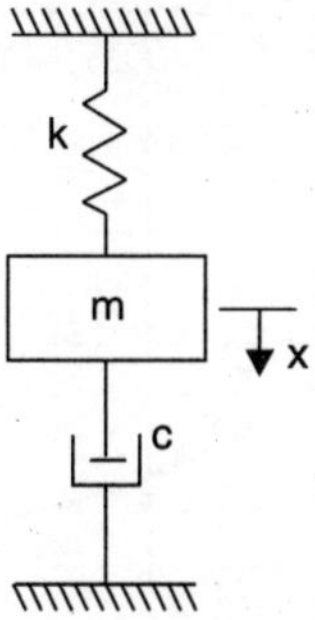

Fig. EM3.21

Solution:

The system equation is

$$m\ddot{x} + c\dot{x} + kx = 0$$

with the initial conditions $x(0) = 0.15$ m and $\dot{x}(0) = 0.04$ m/s. The Laplace transform of the system equation gives

$$m[s^2X(s) - sx(0) - \dot{x}(0)] + c[sX(s) - x(0)] + kX(s) = 0$$

or $$(ms^2 + cs + k)X(s) = mx(0)s + m\dot{x}(0) + cx(0)$$

Solving this last equation for $X(s)$ and substituting the given numerical values, we obtain

$$X(s) = \frac{mx\,(0)s + m\dot{x}(0) + cx(0)}{ms^2 + cs + k} = \frac{0.15s + 0.79}{s^2 + 5s + 5}$$

This equation can be written as

$$X(s) = \frac{0.15s^2 + 0.79s}{s^2 + 5s + 5}\frac{1}{s}$$

Hence the motion of the mass m may be obtained as the unit-step response of the following system:

$$G(s) = \frac{0.15s + 0.79}{s^2 + 5s + 5}$$

MATLAB program will give a plot of the motion of the mass. The plot is shown in Fig. EM3.21 (*a*).

```
num = [0.15    0.79    0];
den = [1    5    5];
step(num,den)
grid
title('Response of spring mass-damper system to initial condition')
```

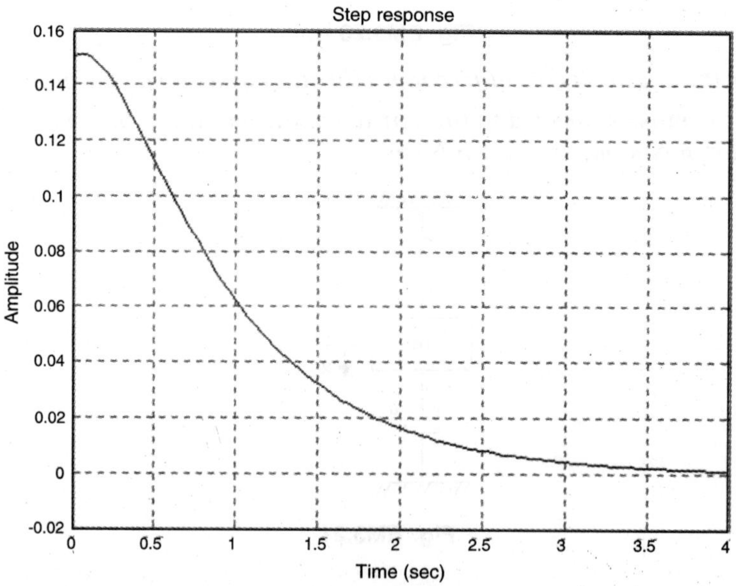

Fig. EM3.21 (*a*)

EM3.22. *For the vibrating single degree of freedom shown in Fig.EM3.22, determine the response of the system when 12 N of free (step input) is applied to the mass m and plot the response using MATLAB. Given that the system is at rest initially and the displacement x is measured from the equilibrium position. Assume that m = 2 kg, c = 10 N-s/m, and k = 80 N/m.*

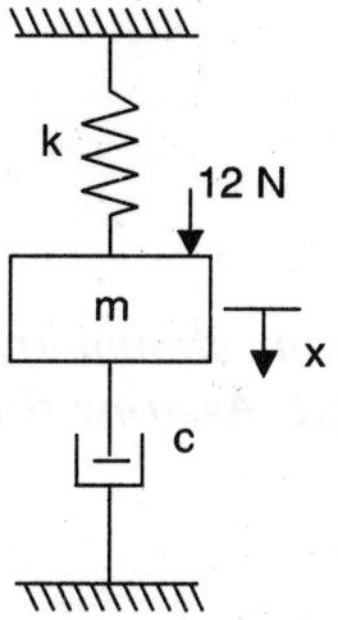

Fig. EM3.22

Solution:

The equation of motion for the system is

$$m\ddot{x} + c\dot{x} + kx = P$$

By substituting the numerical values into this last equation, we get

$$2\ddot{x} + 10\dot{x} + 80x = 12$$

By taking the Laplace transform of this last equation and substituting the initial conditions [$x(0) = 0$ and $\dot{x}(0) = 0$], the result is

$$(s^2 + 5s + 40)X(s) = \frac{6}{s}$$

Solving for $X(s)$, we obtain

$$X(s) = \frac{6}{s\,(s^2 + 5s + 40)}$$

The response exhibits damped vibrations.

MATLAB program is used to plot response curve, which is shown in Fig. EM3.22 (*a*).

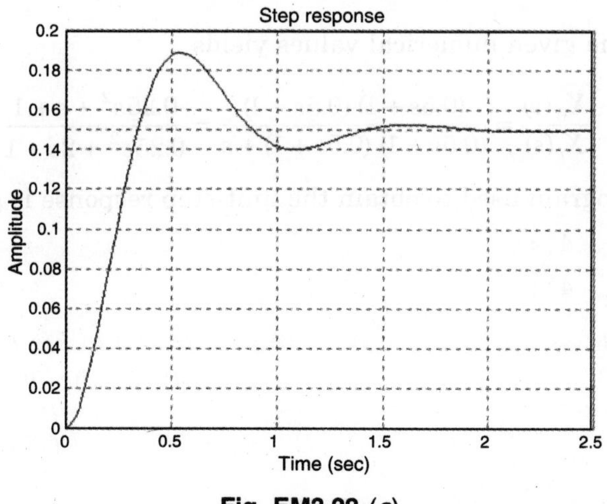

Fig. EM3.22 (*a*)

```
num = [0   0   6];
den = [1   5  40];
step(num,den)
grid
```

EM3.23. *For the mechanical system shown in Fig. EM3.23, obtain the response $x_0(t)$ when $x_i(t)$ is a unit step displacement input. Assume that $k_1 = 2$ N/m, $k_2 = 4$ N/m, $c_1 = 1$ N-s/m, and $c_2 = 2$ N-s/m.*

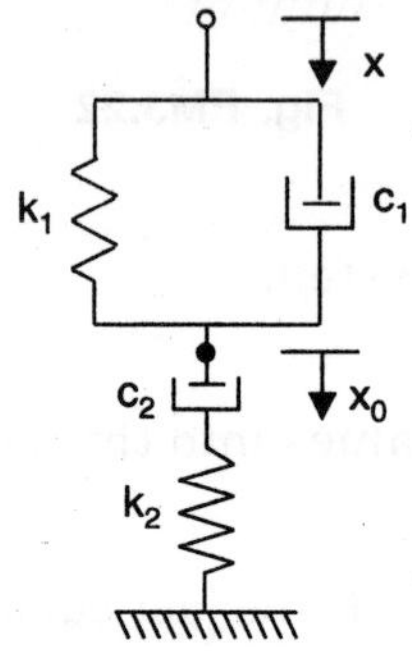

Fig. EM3.23

Solution:

The transfer function $X_0(s)/X_i(s)$ is given by

$$\frac{X_0(s)}{X_i(s)} = \frac{\left(\frac{c_1}{k_1}s+1\right)\left(\frac{c_2}{k_2}s+1\right)}{\left(\frac{c_1}{k_1}s+1\right)\left(\frac{c_2}{k_2}s+1\right)\frac{c_2}{k_1}s}$$

Substitution of the given numerical values yields

$$\frac{X_0(s)}{X_i(s)} = \frac{(0.5s+1)(0.5s+1)}{(0.5s+1)(0.5s+1)+s} = \frac{0.25s^2+s+1}{0.25s^2+2s+1} = \frac{s^2+4s+4}{s^2+8s+4}$$

The MATLAB program used to obtain the unit-step response is given below:

```
num = [1   4   4];
den = [1   8   4];
step(num,den)
grid
```

The output is shown as in Fig. EM3.23 (*a*)

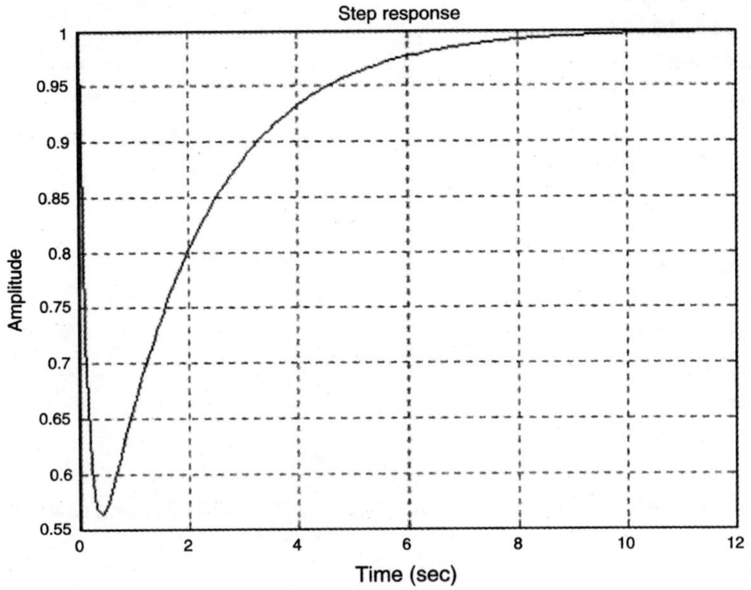

Fig. EM3.23 (*a*)

EM3.24. *The impulse response of a second-order system is given as*

$$\frac{C(s)}{R(s)} = \frac{\omega_n^2}{s^2 + 2\xi\omega_n s + \omega_n^2}$$

For a unit-impulse input R(s) = 1, and $\omega_n = 1$ *rad/sec, C(s) is given by*

$$C(s) = \frac{1}{s^2 + 2\xi s + 1}$$

Plot the ten unit-impulse response curves in one diagram using MATLAB for $\xi = 0.1, 0.2, 0.3, 0.4, 0.5, 0.6, 0.7, 0.8, 0.9,$ *and* 1.0.

Solution:

The MATLAB program:

```
num = [0    0    1];
den1 = [1   0.2   1];
t = 0:0.1:10;
impulse(num,den1,t);
text(2.2, 0.88, 'Zeta = 0.1')
hold
current plot held
den2 = [1    0.4    1]; den3 = [1    0.6    1]; den4 = [1    0.8    1];
den5 = [1    1    1]; den6 = [1    1.2    1]; den7=[1    1.4    1];
den8 = [1    1.6    1]; den9 = [1    1.8    1]; den10 = [1    2.0    1];
impulse(num,den2,t)
text(1.3,0.7,'0.3')
impulse(num,den3,t)
```

```
text(1.15,0.58,'0.5')
impulse(num,den4,t)
text(1.1,0.46,'0.7')
impulse(num,den5,t)
text(0.8,0.38,'1.0')
impulse(num,den6,t)
text(0.7,0.28,'1.0')
impulse(num,den7,t)
text(0.6,0.24,'1.0')
impulse(num,den8,t)
text(0.5,0.21,'1.0')
impulse(num,den9,t)
text(0.4,0.18,'1.0')
impulse(num,den10,t)
text(0.3,0.15,'1.0')
grid
title('Impulse-response curve for G(s) = 1/[s^2+2(zeta)s+1]')
hold
current plot released
```

The output is shown in Fig. EM3.24.

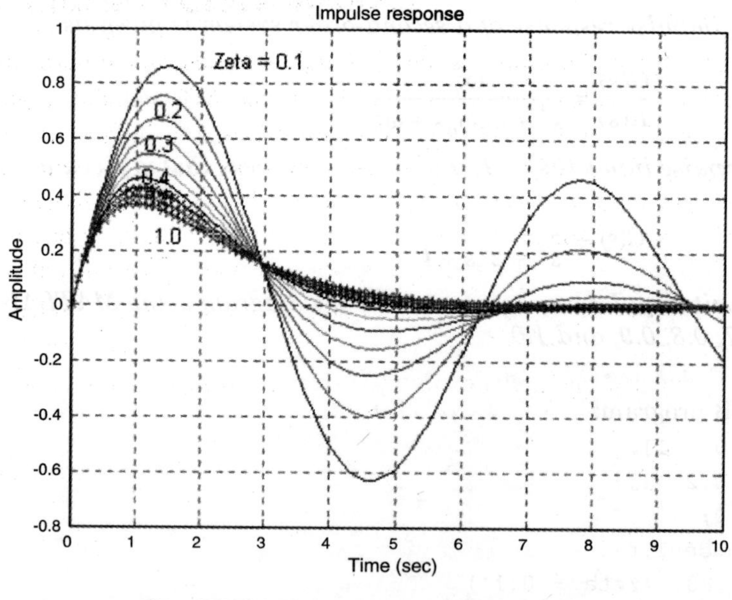

Fig. EM3.24 Unit-impulse response curves.

EM3.25. *For the mechanical system shown in Fig. EM3.25, assume that $m = 1$ kg, $m_1 = 2$ kg, $k_1 = 15$ N/m, and $k_2 = 60$ N/m. Determine the vibration when the initial conditions are given as: $x(0) = 0.23$ m, $\dot{x}(0) = 0$ m/s, $y(0) = 1$ m, $\dot{y}(0) = 0$ m/s. Write a MATLAB program to plot curves x(t) versus t and y(t) versus t for the initial conditions.*

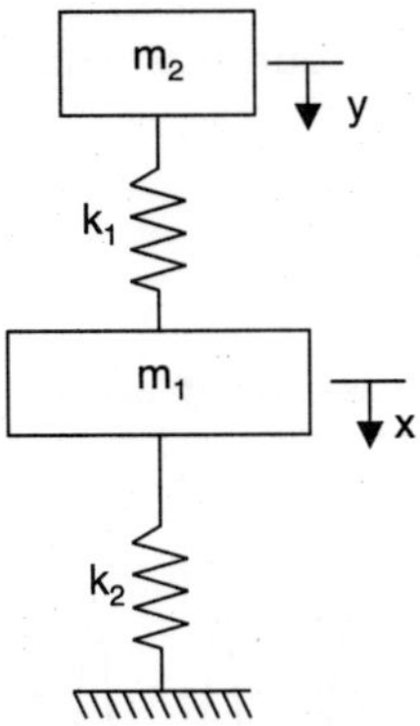

Fig. EM3.25

Solution:

The equations for the system are

$$(2s^2 + 75)X(s) = 2sx(0) + 15Y(s) \quad (1)$$

$$(s^2 + 15)Y(s) = sy(0) + 15X(s) \quad (2)$$

Solving we obtain

$$X(s) = \frac{2(s^2 + 15)\, sx\,(0) + 15sy(0)}{2s^4 + 105s^2 + 900} \quad (3)$$

For the initial conditions

$$x(0) = 0.23 \text{ m},\ \dot{x}(0) = 0 \text{ m/s},\ y(0) = 1 \text{ m},\ \dot{y}(0) = 0 \text{ m/s}$$

Equation (3) becomes as follows:

$$X(s) = \frac{0.46s^3 + 21.9s}{2s^4 + 105s^2 + 900}$$

$$= \frac{0.46s^4 + 21.9s^2}{2s^4 + 105s^2 + 900}\frac{1}{s} \quad (4)$$

By substituting Eq.(4) into Eq.(2) and solving for $Y(s)$, we obtain

$$Y(s) = \frac{1}{s^2 + 15}\, sy(0) + 15X(s)$$

Substituting $y(0) = 1$ into the last equation and simplifying, we get

$$Y(s) = \frac{1}{s^2 + 15}\, \frac{24s^5 + 105s^3 + 900s + 6.9s^3 + 328.5s}{2s^4 + 105s^2 + 900}$$

To obtain plot of $x(t)$ versus t, we may enter the following MATLAB program into the computer. The resulting plots are shown in Fig. EM3.25 (*a*). Likewise $y(t)$ versus t can be also plotted.

```
num1 = [0.46  0  21.9  0  0];
den = [2   0  105  0 900];        % see equation (4)
t=0:0.01:20;
x=step(num1,den,t)
```

```
plot(t,x)
title('Responses mass m1- x(t) due to initial conditions')
xlabel('t sec')
ylabel('x(t)')
grid
```

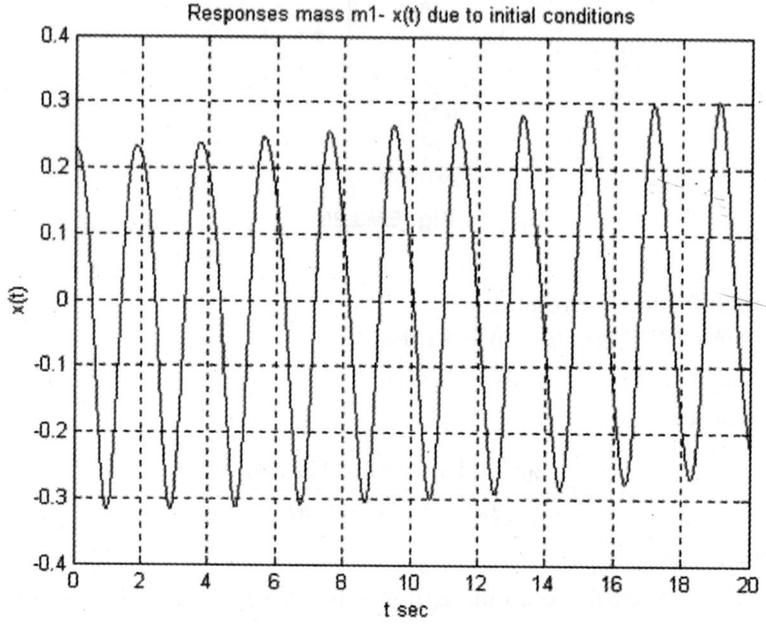

Fig. EM3.25 *(a)*

EM3.26. *For the mechanical system shown in Fig.EM3.25, assume that m = 1 kg, m_1 = 2 kg, k_1 = 15 N/m, and k_2 = 60 N/m. Determine the vibration when the initial conditions are given as: x(0) = 1.75 m, $\dot{x}(0)$ = 0 m/s, y(0) = – 1 m, $\dot{y}(0)$ = 0 m/s. Write a MATLAB program to plot curves x(t) versus t and y(t) versus t for the initial conditions.*

Solution:

$$X(s) = \frac{2\,(s^2 + 15)\,sx\,(0) + 15sy(0)}{2s^4 + 105s^2 + 900}$$

$$Y(s) = \frac{1}{s^2 + 15}\,sy(0) + 15X(s)$$

For the initial conditions

$$x(0) = 1.75 \text{ m},\ \dot{x}(0) = 0 \text{ m/s},\ y(0) = -1 \text{ m},\ \dot{y}(0) = 0 \text{ m/s}$$

we obtain the following expressions for X(s) and Y(s):

$$X(s) = \frac{3.5s^3 + 37.5s}{2s^4 + 105s^2 + 900} = \frac{3.5s^4 + 37.5s^2}{2s^4 + 105s^2 + 900}\,\frac{1}{s}$$

$$Y(s) = \frac{1}{s^2 + 15} - s + 15X(s)$$

A MATLAB program for obtaining plots of $x(t)$ versus t given below. The resulting plot is shown in Fig. EM3.26.

```
num1 = [3.5  0 37.5 0  0];
den = [2   0  105 0 900];
step(num1,den)
```

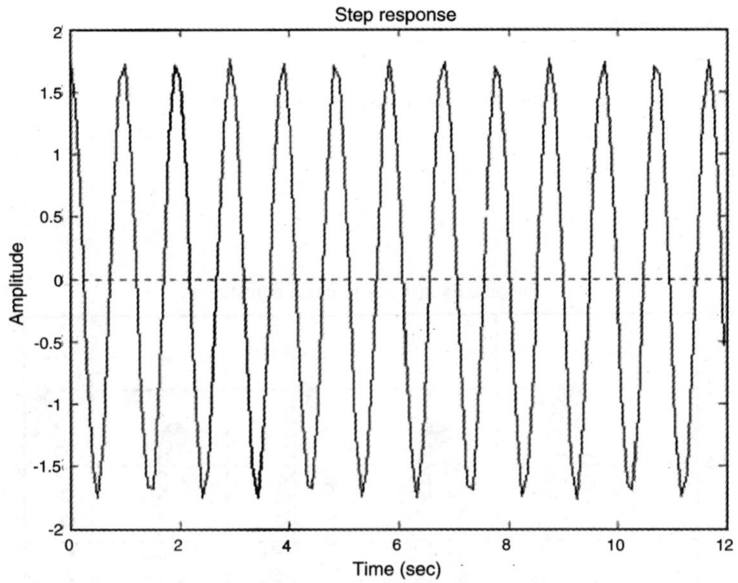

Fig. EM3.26 Plot of motion of mass m_1.

EM3.27. *For the mechanical system shown in Fig. EM3.25, assume that m = 1 kg, m_1 = 2 kg, k_1 = 15 N/m, and k_2 = 60 N/m. Determine the vibration when the initial conditions are given as: $x(0) = 0.5$ m, $\dot{x}(0) = 0$ m/s, $y(0) = -0.5$ m, $\dot{y}(0) = 0$ m/s. Write a MATLAB program to plot curves x(t) versus t and y(t) versus t for the initial conditions.*

Solution:

$$X(s) = \frac{2\,(s^2 + 15)\,sx(0) + 15sy(0)}{2s^4 + 105s^2 + 900}$$

$$Y(s) = \frac{1}{s^2 + 15}\,sy(0) + 15X(s)$$

For the initial conditions

$$x(0) = 0.5 \text{ m},\ \dot{x}(0) = 0 \text{ m/s},\ y(0) = -0.5 \text{ m},\ \dot{y}(0) = 0 \text{ m/s}$$

we obtain the following expressions for $X(s)$ and $Y(s)$:

$$X(s) = \frac{s^4 + 7.5s^2}{2s^4 + 105s^2 + 900}\,\frac{1}{s}$$

$$Y(s) = \frac{-0.5s}{s^2 + 15} + \frac{15X(s)}{s^2 + 15}$$

A MATLAB program for obtaining plots of $x(t)$ versus t given below. The resulting plots are shown in Fig. EM3.27. Likewise $y(t)$ can also be plotted

```
num1 = [1  0  7.5   0   0];
den = [2   0   105   0  900];
t = 0:0.02:5;
x = step(num1,den,t)
plot(t, x, 'o')
title('Responses x(t) due to initial conditions ')
xlabel('t sec')
ylabel('x(t)')
grid
```

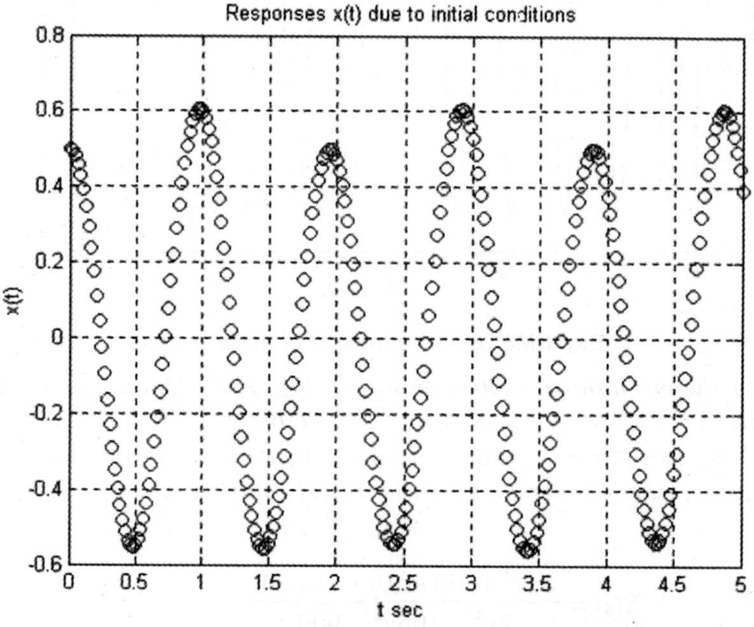

Fig. EM3.27

3.15.4 Electrical Circuits

EC 3.1. *Fig. EC3.1 shows an electrical circuit with resistors and voltage sources. Write a MATLAB program to determine the current in each resistor using the mesh current method based on Kirchhoff's voltage law. Given:* $V_1 = 22V$, $V_2 = 12V$, $V_3 = 44V$, $R_1 = 20\Omega$, $R_2 = 12\Omega$, $R_3 = 15\Omega$, $R_4 = 7\Omega$, $R_5 = 16\Omega$, $R_6 = 10\Omega$, $R_7 = 10\Omega$, $R_8 = 15\Omega$

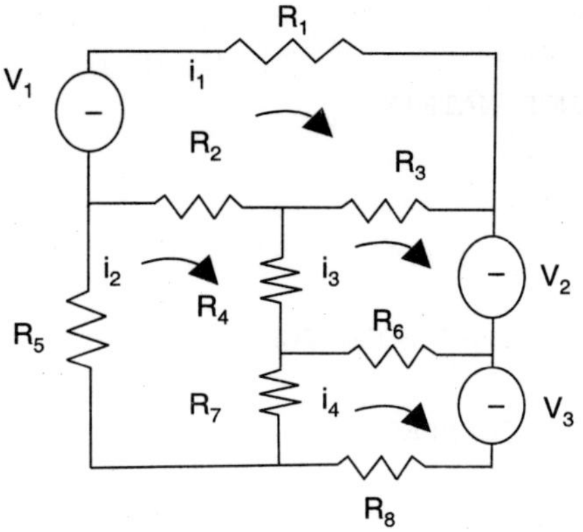

Fig. EC3.1

Solution:

Let i_1, i_2, i_3 and i_4 be the loop currents as shown in Figure.EC3.1.

According to Kirchhoff's voltage law: sum of voltage around closed circuit is zero.

Thus the loop equations can be written by taking in each loop clockwise direction as reference.

$$V_1 - R_1 i_1 - R_3(i_1 - i_3) - R_2(i_1 - i_2) = 0$$
$$-R_2(i_2 - i_1) - R_4(i_2 - i_3) - R_7(i_2 - i_4) - R_5 i_2 = 0$$
$$-R_3(i_3 - i_1) - V_2 - R_6(i_3 - i_4) - R_4(i_3 - i_2) = 0$$
$$V_3 - R_8 i_4 - R_7(i_4 - i_2) - R_6(i_4 - i_3) = 0$$

The equations can be written in matrix form as follows:

$$\begin{bmatrix} -(R_1+R_2+R_3) & R_2 & R_3 & 0 \\ R_2 & -(R_2+R_4+R_5+R_7) & R_4 & R_7 \\ R_3 & R_4 & -(R_3+R_4+R_6) & R_6 \\ 0 & R_7 & R_6 & -(R_6+R_7+R_8) \end{bmatrix} \begin{bmatrix} i_1 \\ i_2 \\ i_3 \\ i_4 \end{bmatrix} = \begin{Bmatrix} -V_1 \\ 0 \\ V_2 \\ -V_3 \end{Bmatrix}$$

MATLAB solution of this system of equations is given below:

MATLAB program:

```
%INITIALIZING THE VARIABLES
V1=22;
V2=12;
V3=44;
V=[-V1;0;V2;-V3] % CREATE THE VOLTAGE VECTOR
R1=20;
R2=12;
R3=15;
R4=7;
R5=16;
R6=10;
```

```
R7=10;
R8=15;
% CREATE THE RESISTANCE MATRIX
R=[-(R1+R2+R3) R2 R3 0;
R2 -(R2+R4+R5+R7) R4 R7;
R3 R4 -(R3+R4+R6) R6;
0 R7 R6 -(R6+R7+R8)];
% GET THE CURRENT VECTOR AS SOLUTION
I=inv(R)*V;
% ALLOT VALUES TO FOUR CURRENTS
i1=I(1)
i2=I(2)
i3=I(3)
i4=I(4)
```

The output obtained is as follows:

```
V =
  -22
   0
  12
  -44

i1 =
   0.8785

i2 =
   0.7154

i3 =
   0.7138

i4 =
   1.6655
```

The current in resistor R_2 = i1 –i2 = 0.1631 A
The current in resistor R_3 = i1 –i3 = 0.1647 A
The current in resistor R_4 = i2 –i3 = 0.0016 A
The current in resistor R_6 = i4 –i3 = 0.9517 A
The current in resistor R_7 = i4 –i2 = 0.9501 A

EC3.2. *Write a MALTAB program that computes the voltage across each resistor and the power dissipated in each resistor for the circuit shown in Fig. EC3.2 that has resistors connected in series.*

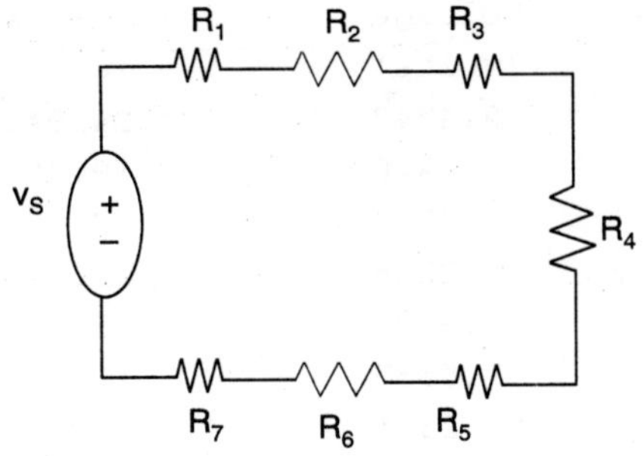

Fig. EC3.2

The voltage across each of the several resistors connected in series is given by the voltage divider rule $v_s\, v_s\, v_s\;\; v_n\, v_n = \dfrac{R_n}{R_{eq}}\; v_s$

where

v_n *and* R_n *are the voltage across resistor n and its resitance*
$R_{eq} = \Sigma R_n =$ *the equivalent resistance*
v_s *is the source voltage*

The power dissipated in each resistor is given by $P_n = \dfrac{R_n}{R_{eq}}\; v_s^{\,2}$

Solution:
MATLAB program is given for the following data:

V_s = 12V, R_1 = 10Ω, R_2 = 7Ω, R_3 = 6Ω, R_4 = 9Ω, R_5 = 4Ω, R_6 = 7.5Ω, R_7 = 10Ω

```
% THIS PROGRAM CALCULATES THE VOLTAGE ACROSS EACH RESISTOR
% IN A CIRCUIT THAT HAS RESISTORS CONNECTED IN SERIES
vs=input('Enter the source voltage');
rn=input('Enter values of resistors as elements in a row vector\n');
req=sum(rn);              % CALCULATING EQUIVALENT RESISTANCE
vn=rn*vs/req;             % APPLY VOLTAGE DIVIDE RULE
pn=rn*vs^2/req/2;         % CALCULATING POWER IN EACH CIRCUIT
i=vs/req;                 % CALCULATE CURRENT IN THE CIRCUIT
ptotal=vs*i;              % CALCULATE POWER IN THE CIRCUIT
table=[rn',vn',pn'];      % CREATE TABLE
disp('  Resistance  Voltage   Power') %DISPLAY HEADINGS FOR COLUMNS
disp('     (ohms)       (volts)  (watts)')
disp(table)               % DISPLAY THE VARIABLE 'TABLE'
fprintf('The curent in the circuit is %f Amp',i)
fprintf('\nThe total power dissipated in the circuit is %f Watts\n',ptotal)
```

MATLAB output:

```
Enter the source voltage 12
Enter values of resistors as elements in a row vector
[10 7 6 9 4 7.5 5]
```

```
      Resistance              Voltage              Power
          (ohms)              (volts)             (watts)
         10.0000               2.4742             14.8454
          7.0000               1.7320             10.3918
          6.0000               1.4845              8.9072
          9.0000               2.2268             13.3608
          4.0000               0.9897              5.9381
          7.5000               1.8557             11.1340
          5.0000               1.2371              7.4227
The curent in the circuit is 0.247423 Amp
The total power dissipated in the circuit is 2.969072 Watts
```

EC3.3. *Fig. EC3.3 shows a semiconductor diode and the current flowing through the diode is given by:*

$$i_D = i_0\left[exp\left(\frac{qv_d}{kT}\right) - 1\right]$$

where v_d = *the voltage across the diode (volts)*

i_0 = *the leakage current of the diode (amps)*

k = *Boltzmann's constant* (1.38×10^{-23} *Joule/K*)

q = *the charge of an electron* (1.6×10^{-19} *Coulombs*)

T = *Temperature in °K*

i_D = *the current flow through the diode (amps)*

(*a*) *write a MATLAB program to calculate the current flowing through this diode for all voltages from – 0.2V to + 0.25 V in 0.01 V steps.*

(*b*) *repeat the procedure in* (*a*) *for 70°F, 200 °F and 400 °F.*

(*c*) *plot the current as a function of applied voltage*

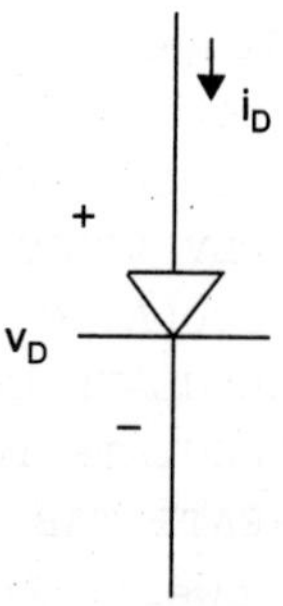

Fig. EC3.3

Solution:

MATLAB program for calculation of current flow in diode is given below:

```
% INITIAL VALUES
i0=2e-4;                    % leakage current in amp
k=1.38e-23;                      % Boltzmann constant (J/K)
q=1.602e-19;                     % charge of electron in C
vd=-0.2:0.01:0.25;               % diode voltage (V)
t_f=[75 200 400];                % temperature in F
```

```
for ii=1:length(t_f)
    t_k=(5/9)*(t_f(ii)-32)+273.15;          %convert temperature to kelvin
    id=i0.*(exp((q*vd)/(k*t_k))-1); %calculate diode current
    if ii ==1
    plot(vd,id,'-o');                        % plot lines in various ways
    hold on;
    elseif ii ==2
        plot(vd,id,'-');
    elseif ii==3
        plot(vd,id,':o');
        hold off;
    end
end
legend('75 deg F','200 deg F','400 deg F')
grid on;
title('\bf plot of diode voltage Vs diode current');
xlabel('v_{D}');
ylabel('i_{D}');
```

The output is shown in Fig. 3.3EA (*a*)

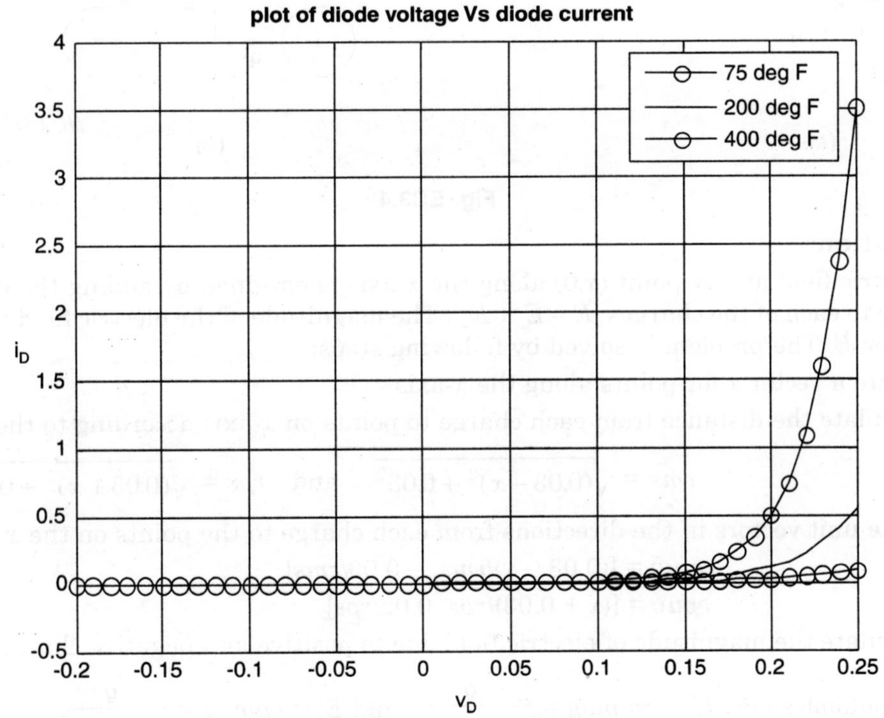

Fig. EC3.3 (*a*) MATLAB Output.

EC3.4. *Fig. EC3.4 (a) shows the electric field at a point due to a charge which is a vector E. The magnitude of E is given by Coulomb's law* $E = \dfrac{q}{4\pi\varepsilon_o r^2}$, *where q is magnitude of the charge, r is the distance between the charge and the point, and eo is the permittivity constant* ($8.8542 \times 10^{-12}\ C^2/Nm^2$). *The electric field E at any point is obtained by superposition of the electric field of each charge. An electric dipole with* $q = 12 \times 10^{-19}$ *C is created as shown in Fig. EC3.4 (b).*

Write a MATLAB program to determine and plot the magnitude of the electric field along the x-axis from x = – 8 cm to x = 8 cm

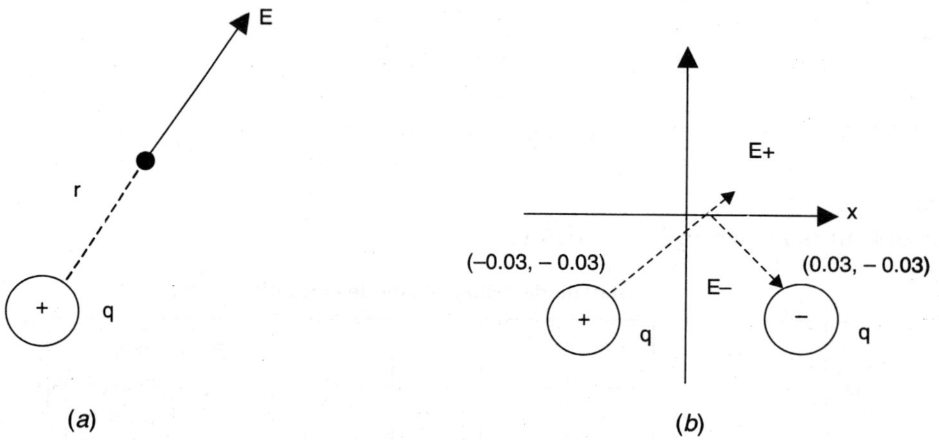

Fig. EC3.4

Solution:

Electric field at any point (*x*,0) along the *x*-axis is obtained by adding the electric field vectors due to each of the charges. $E = E_- + E_+$. The magnitude of the electric field is the length of the vector *E*. The problem is solved by following steps:

1. Create a vector *x* for points along the *x*-axis
2. Calculate the distance from each charge to points on *x*-axis according to the equations

$$rms = \sqrt{(0.03 - x)^2 + 0.03^2} \quad \text{and} \quad rps = \sqrt{(0.03 + x)^2 + 0.03^2}$$

3. Write unit vectors in the directions from each charge to the points on the *x*-axis as

emuv = [(0.03 – *x*)/*rms*, – 0.03/*rms*]

and *epuv* = [(*x* + 0.03)/*rps*, 0.02/*rps*]

4. Calculate the magnitude of electric field due to positive and negative charges according to Coulomb's law: $E\_ = emmag = \dfrac{q}{4\pi\varepsilon_o rms^2}$ and $E_+ = epmag = \dfrac{q}{4\pi\varepsilon_o rps^2}$
5. Calculate em and ep by multiplying the unit vectors by emmag and epmag
6. Calculate *E* as *e* = *em* + *ep*;
7. Find the magnitude of *e*
8. Plot e as a function of *x*.

MATLAB program for this is given below:

```
q=12e-9;
ep=8.8542e-12;
x=[-0.08:0.001:0.08]';                            %COLUMN VECTOR OF x
rms=(0.03-x).^2+0.03^2;rm=sqrt(rms);
rps=(0.03+x).^2+0.03^2;rp=sqrt(rps);
emuv=[((0.03-x)./rm),(-0.03./rm)];                     % Unit vector of em
epuv=[((0.03+x)./rp),(0.03./rp)];                      % Unit vector of ep
emmag=(q/(4*pi*ep))./rms;
epmag=(q/(4*pi*ep))./rps;
em=[emmag.*emuv(:,1),emmag.*emuv(:,2)];
% Multiplication of magnitude and uv
ep=[epmag.*epuv(:,1),epmag.*epuv(:,2)];
e=em+ep;
emag=sqrt(e(:,1).^2+e(:,2).^2);
plot(x,emag,'k');
xlabel('Position along the x-axis(m)')
ylabel('Magnitude of electric field (N/C)')
title('Electric field due to an electric dipole')
```

The output of the program is shown in Fig. EC3.4 (*c*)

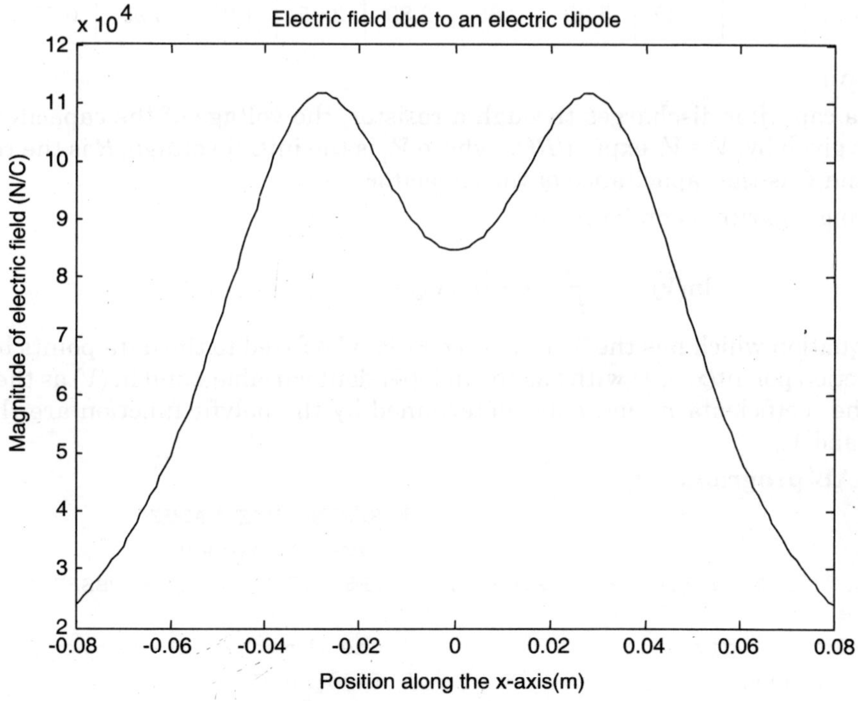

Fig. EC3.4 (*c*) MATLAB output.

Example EC3.5. *Figure EC3.5 shows a circuit to determine the electrical capacitance of an electrical capacitor.*

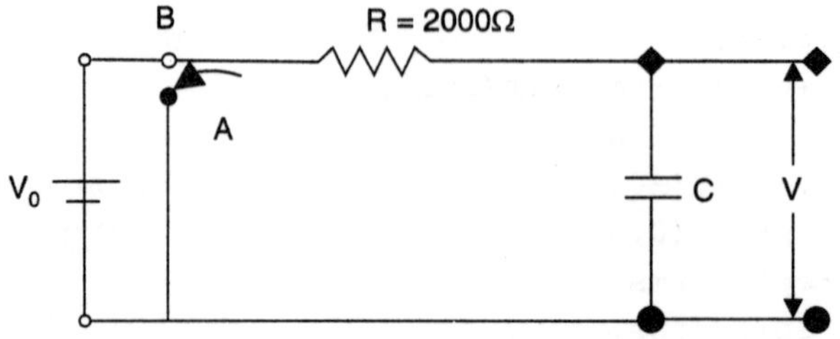

Fig. EC3.5

The process involves the following steps: first the switch is connected to B and the capacitor is charged, then the switch is switched to A and the capacitor discharges through the resistor. The voltage across the capacitor is measured as the capacitor discharges. The measurements obtained are given below in a table: (a) write a MATLAB program to plot the voltage as a function of time (b) determine the capacitance of the capacitor by fitting an experimental curve to the data points.

Table 3.1

t(*s*)	1	2	3	4	5	6	7	8	9	10
V(volts)	9.5	7.35	5.25	3.65	2.85	2.05	1.25	0.95	0.75	0.61

Solution:

When a capacitor discharges through a resistor, the voltage of the capacitor as a function of time is given by: $V = V_0 \exp(-t/RC)$, where V_0 is the initial voltage, R is the resistance of the resistor and C is the capacitance of the capacitor.

By taking logarithms on both sides

$$\ln(V) = -\frac{1}{RC}\, t + \ln (V_0)$$

This equation which has the form $y = mx + c$ can be fitted to the data points by using the MATLAB function polyfit(x, y, 1) with t as the independent variable x and ln (V) as the dependent variable y. The coefficients m and c are determined by the polyfit function are then used to determine C and V_0.

MATLAB program:

```
r=2000;                          % RESISTANCE VALUE
t=1:10;                          % time in seconds
v=[9.5 7.35 5.25 3.65 2.85 2.05 1.25 0.95 0.75 0.61]; % OBSERVED VALUES of
voltage
p=polyfit(t,log(v),1);           % one dimensional polynomial fit
c=-1/(r*p(1));                   % finding C
v0=exp(p(2));                    % finding V0
tplot=0:0.1:10;                  % choosing plotting coordinates
vplot=v0*exp(-tplot./(r*c));
disp('Capacitance');c
```

```
plot(t,v,'o',tplot,vplot)
xlabel('t(s)')
ylabel('voltage');
```

MATLAB output:

```
Capacitance
c =
     0.0016
```

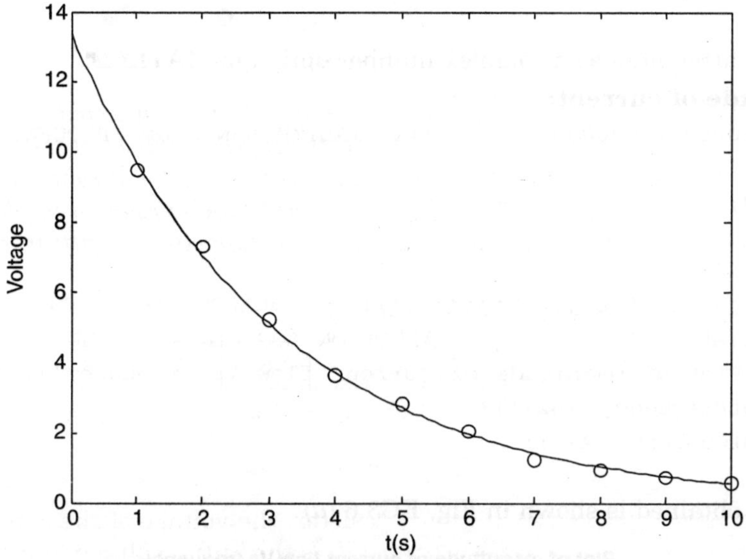

Fig. EC3.5 (*a*) MATLAB output.

Example EC3.6. *A series RLC circuit driven by a sinusoidal AC voltage source (120 ∠0° volts) is shown in Fig. EC3.6. The impedance of the inductor is given by ZL = j2πfL, where j =* $\sqrt{-1}$*, f is frequency of the voltage source (Hz) and L is the inductance (Henries). The impedance of the capacitor is given by* $\frac{-j}{2\pi fC}$*, where C is the capacitance (farads). The current I flowing is given by Kirchhoff's voltage law that is* $i = \frac{120\angle 0^\circ}{R + j2\pi fL - j/(2\pi fC)}$*. Write a MATLAB program to calculate and plot (a) the magnitude of the current as function of frequency for the range 100 kHz to 10 MHz. (b) the phase angle as a function of frequency for the range 100 kHz to 10 MHz. (c) the magnitude and phase angle of the current as a function of frequency on two subplots of a single figure.*

Given R = 120 Ω, L = 0.15 mH and C = 0.26 nF

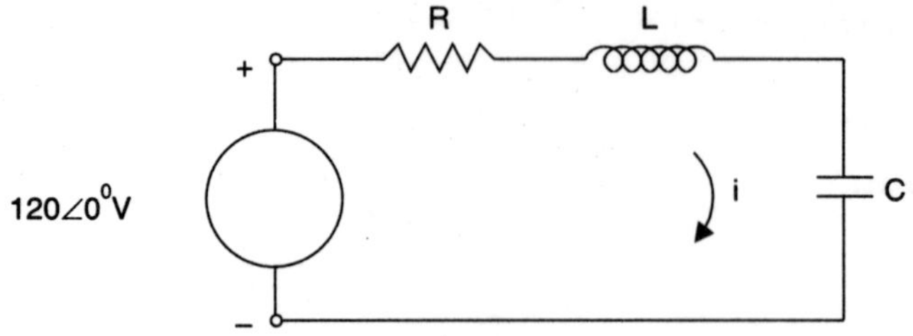

Fig. EC3.6

Solution:

This can be attempted as a complex number option in MATLAB.

(*a*) **Magnitude of current:**

```
f=100000:50000:10000000;        %INITIALIZE RANGE OF FREQUENCY
vs=120;
c=0.265e-9;
L=0.15e-3;
r=120;
i0=vs./(r+j*2*pi*f*L-j./(2*pi*f*c));       %CALCULATE OUTPUT CURRENT
semilogx(f,abs(i0));            %PLOT ON LOG-LINEAR SCALE
title('\bfPlot of magnitude of current flow Vs frequency');
xlabel('\bfFrequency  (Hz)');
ylabel('\bfCurrent  (A)');
grid on;
```

The output obtained is shown in Fig. EC3.6 (*a*)

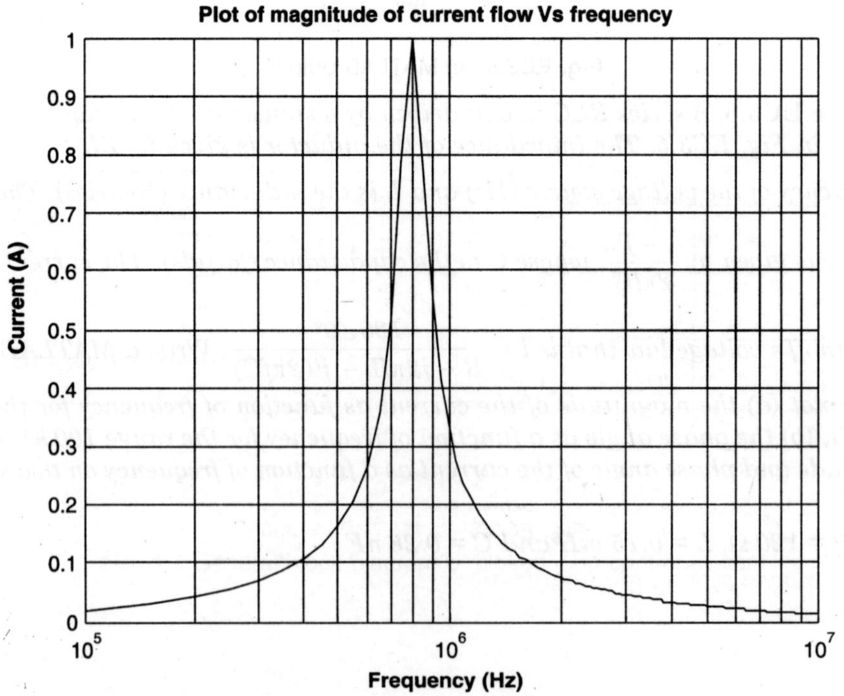

Fig. EC3.6 (*a*) MATLAB output.

(*b*) Phase angle

```
f=100000:50000:10000000;          %INITIALIZE RANGE OF FREQUENCY
vs=120;
c=0.265e-9;
L=0.15e-3;
r=120;
i0=vs./(r+j*2*pi*f*L-j./(2*pi*f*c));       %CALCULATE OUTPUT CURRENT
phase=angle(i0)*180/pi;
figure(1);
semilogx(f,phase,'LineWidth',2);           %PLOT ON LOG-LINEAR SCALE
title('\bfPlot of phase of current flow Vs frequency');
xlabel('\bfFrequency  (Hz)');
ylabel('\bfCurrent  (A)');
grid on;
```

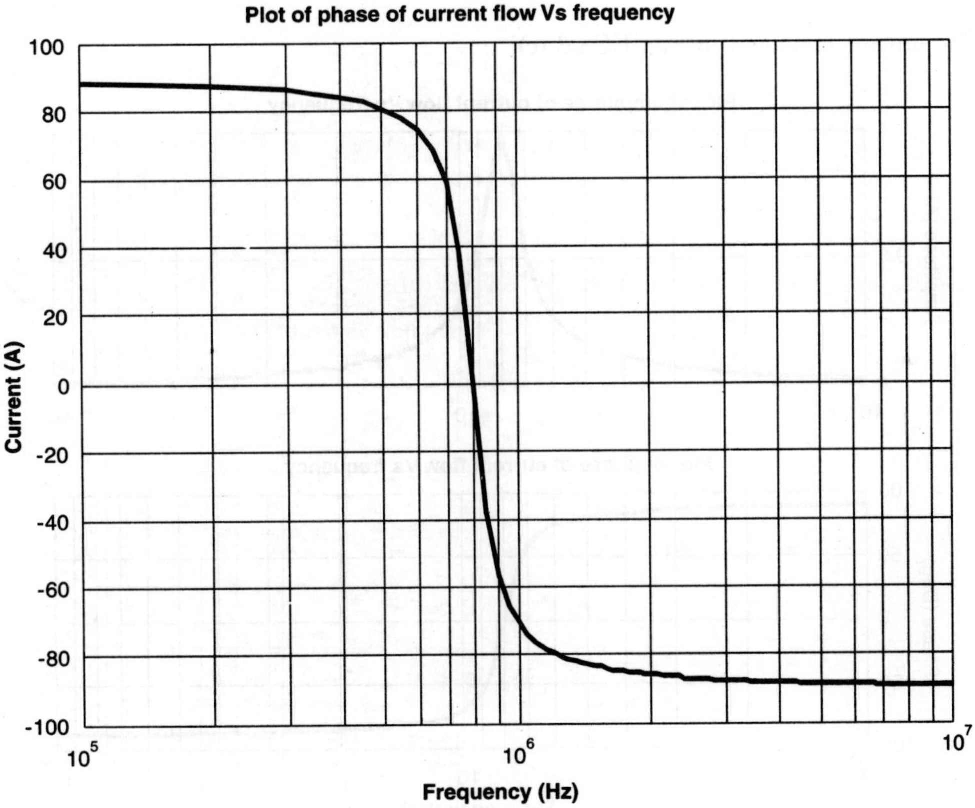

Fig. EC3.6 (*b*) MATLAB output.

(*c*) Both phase angle and magnitude on single plot

```
f=100000:50000:10000000;          %INITIALIZE RANGE OF FREQUENCY
vs=120;
c=0.265e-9;
L=0.15e-3;
```

```
r=120;
w=2*pi*f; %CALCULATE W
i0=vs./(r+j*2*pi*f*L-j./(2*pi*f*c));        %CALCULATE OUTPUT CURRENT
phase=angle(i0)*180/pi;                     %PHASE ANGLE IN DEGREES
figure(1);
subplot(2,1,1);                     %SUB-PLOT-1
semilogx(f,abs(i0),'Linewidth',2);          %MAGNITUDE
title('\bfPlot of amplitude of current flow Vs frequency');
ylabel('\bfAmplitude (A)');
grid on;
subplot(2,1,2);                     % SUB-PLOT-2
semilogx(f,phase,'Linewidth',2);            %PHASE
title('\bfPlot of phase of current flow Vs frequency');
xlabel('\bfFrequency (Hz)');
ylabel('\bfPhase (Deg)');
grid on;
```

The output is shown in Fig. EC3.6 (*c*)

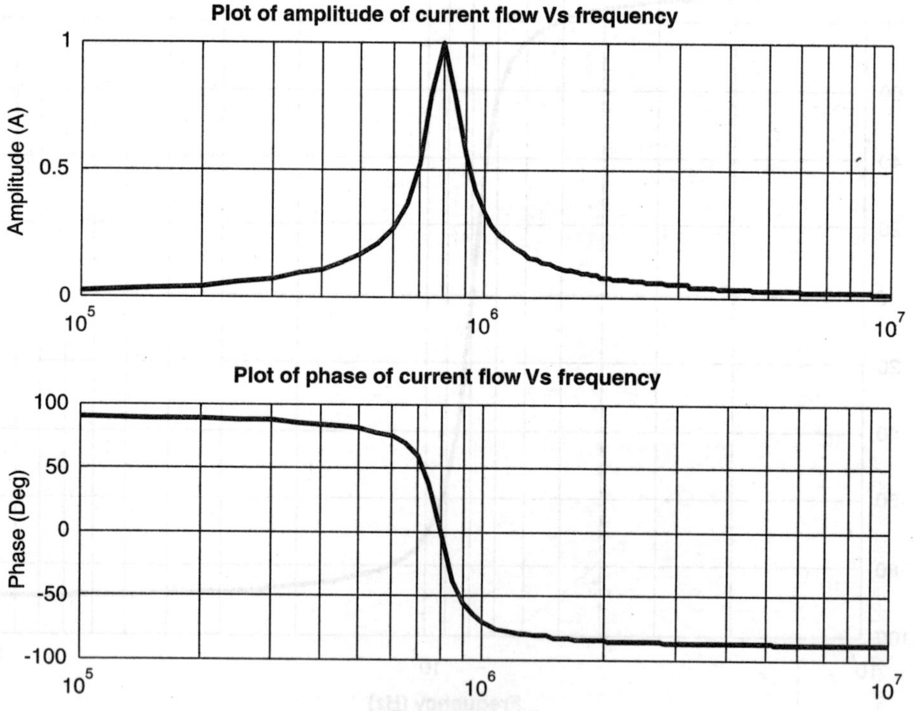

Fig. EC3.6 (*c*) MATLAB output.

Example EC3.7. *The Table below gives the viscosity* μ *of an oil at different temperatures. Write a MATLAB program to determine an equation that can be fitted to the data.*

T(°C)	– 20	0	20	40	60	80	100	120
$\mu \times 10^{-5}$(N-S/m²)	4.2	0.4	0.092	0.034	0.016	0.0077	0.0046	0.0033

Solution:

Usually viscosity varies with absolute temperature exponentially. To have a best fit of the given points, a curved exponential figure is suitable, whose equations can be written as:

$$\mu = \exp(aT^2 + bT + c)$$

Taking logarithms and simplifying

$$\log \mu = aT^2 + bT + c.$$

Here the constants can be obtained from the polyfit function ployfit(T, log μ, 2), then finally μ is obtained from the exponential relation and plotted as a function of temperature in Kelvin.

MATLAB program for this application is shown below:

```
TC=-20:20:120;          %TEMPERATURE RANGE IN DEGREE CENTRIGRADE
mu=[4.2 0.4 0.092 0.034 0.016 0.0077 0.0046 0.0033]; %GIVEN VISCOSITIES
TK=TC+273;       % TEMPERATURE IN KELVIN
p=polyfit(TK,log(mu),2)
% POLYNOMIAL FITTING WITH LOG(MU) AND TK, SECOND ORDER
Tplot=273+[-20:120];% DEFINING TK AS AN ARRAY
muplot=exp(p(1)*Tplot.^2+p(2)*Tplot+p(3));%CORRESPONDING MU ARRAY
semilogy(TK,mu,'o',Tplot,muplot,'-')%PLOTTING ON SEMI-LOG SCALE
xlabel('\bfTemperature K');
ylabel('\bfViscosity in N-S/meter square');
```

Fig. EC3.7 shows the MATLAB output.

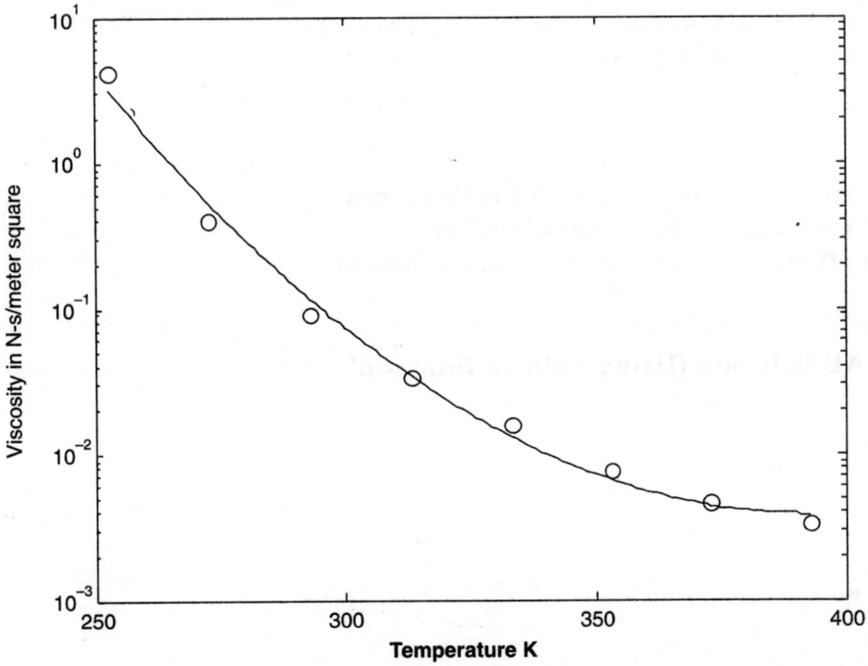

Fig. EC3.7 MATLAB output.

3.15.5 Numerical Methods

Example EN3.1. *Use Gaussian elimination scheme to solve the set of equations.*

$$2x_1 + x_2 - 3x_3 = 11$$
$$4x_1 - 2x_2 + 3x_3 = 8$$
$$-2x_1 + 2x_2 - x_3 = -6$$

Solution:

```
>> %Prob EN 3.1
>> A=[2 1 -3;4 -2 3;-2 2 -1];
>> b=[11;8;-6];
>> format long
>>  [x,det]=gauss(A,b)
x =
     3
    -1
    -2
det = -22
function [x,det] = gauss(A,b)
% Solves A*x = b by Gauss elimination & finds det(A)
% USAGE: [x,det] = gauss(A,b)
if size(b,2) > 1; b = b'; end    % b column vector
n = length(b);
for k = 1:n-1 % Elimination
for i= k+1:n
if A(i,k) ~= 0
lambda = A(i,k)/A(k,k);
A(i,k+1:n) = A(i,k+1:n) - lambda*A(k,k+1:n);
b(i)= b(i) - lambda*b(k);
end
end
end
if nargout == 2; det = prod(diag(A)); end
for k = n:-1:1  % Back substitution
b(k) = (b(k) - A(k,k+1:n)*b(k+1:n))/A(k,k);
end
x = b;
```

MATLAB Solution [Using built-in function]:

```
>> A=[2 1 -3;4 -2 3;-2 2 -1];
>> B=[11; 8; -6];
>> x=inv(A)*B
>> x=A\B
x =
     3
    -1
    -2
>> x=inv(A)*B
```

```
x =
     3.0000
    -1.0000
    -2.0000
```

Example EN3.2. *Using Gaussian elimination method, solve the system of linear equations:*

$$\begin{aligned} 2x_1 + x_2 + x_3 - x_4 &= 10 \\ x_1 + 5x_2 - 5x_3 + 6x_4 &= 25 \\ -7x_1 + 3x_2 - 7x_3 - 5x_4 &= 5 \\ x_1 - 5x_2 + 2x_3 + 7x_4 &= 11 \end{aligned}$$

Solution:

```
      EDU>> Run_pr_EN3.2
A =
         2         1         1        - 1
         1         5       - 5          6
       - 7         3       - 7        - 5
         1       - 5         2          7
b =
      10
      25
       5
      11

x =
     25.3587
    -19.6143
    -28.9058
     -7.8027

det =
   -223.0000

x =
     25.3587
    -19.6143
    -28.9058
     -7.8027
```

Run_pr_EN3.2.m

```
A=[2 1 1 -1;1 5 -5 6; -7 3 -7 -5; 1 -5 2 7];
b=[10;25;5;11];
[x,det]=gauss(A,b)
x=A\b
```

gauss.m

```
function [x,det] = gauss(A,b)
```

```
% Solves A*x = b
if size(b,2) > 1; b = b'; end
n = length(b);
for k = 1:n-1
for i= k+1:n
if A(i,k) ~= 0
lambda = A(i,k)/A(k,k);
A(i,k+1:n) = A(i,k+1:n) - lambda*A(k,k+1:n);
b(i)= b(i) - lambda*b(k);
end
end
end
if nargout == 2; det = prod(diag(A)); end
for k = n:-1:1
b(k) = (b(k) - A(k,k+1:n)*b(k+1:n))/A(k,k);
end
x = b;
```

MATLAB Solution [Using built-in function]:

```
>> A=[2 1 1 -1;1 5 -5 6;-7 3 -7 -5;1 -5 2 7];
>> B=[10;25;5;11];
>> x=A\B

x =
   25.3587
  -19.6143
  -28.9058
   -7.8027

>> x=inv(A)*B
x =
   25.3587
  -19.6143
  -28.9058
   -7.8027
```

Example EN3.3. *Tridiagonalize the given matrix* $[A] = \begin{bmatrix} 4 & 3 & 2 & 1 \\ 3 & 8 & 6 & 12 \\ 2 & 6 & 8 & 3 \\ 1 & 2 & 3 & 4 \end{bmatrix}$ *by Householder's method.*

Solution:

The following program is used.

```
%Input - A is an nxn symmetric matrix
A = [4 3 2 1; 3 8 6 12;2 6 8 3; 1 2 3 4];
%Output - T is a tridiagonal matrix
```

```
[n,n]=size(A);
for k=1:n-2
   s=norm(A(k+1:n,k));
   if (A(k+1,k)<0)
      s=-s;
   end
   r=sqrt(2*s*(A(k+1,k)+s));
   W(1:k)=zeros(1,k);
   W(k+1)=(A(k+1,k)+s)/r;
   W(k+2:n)=A(k+2:n,k)'/r;
   V(1:k)=zeros(1,k);
   V(k+1:n)=A(k+1:n,k+1:n)*W(k+1:n)';
   c=W(k+1:n)*V(k+1:n)';
   Q(1:k)=zeros(1,k);
   Q(k+1:n)=V(k+1:n)-c*W(k+1:n);
   A(k+2:n,k)=zeros(n-k-1,1);
   A(k,k+2:n)=zeros(1,n-k-1);
   A(k+1,k)=-s;
   A(k,k+1)=-s;
  A(k+1:n,k+1:n)=A(k+1:n,k+1:n)-2*W(k+1:n)'*Q(k+1:n)-2*Q(k+1:n)'*W(k+1:n);
end
T=A;
fprintf('Matrix in tridiagonal form is\n');
disp(T)
Matrix in tridiagonal form is
     4.0000    -3.7417         0         0
    -3.7417    14.5714    2.6245         0
          0     2.6245    2.6878   -0.7622
          0          0   -0.7622    2.7407
```

Example EN3.4. *Use the QR factorisation method with Householder transformation to calculate the eigen values and the corresponding eigenvectors of the matrix [K] where*

$$[K] = \begin{bmatrix} 5 & -4 & 1 & 0 \\ -4 & 6 & -4 & 1 \\ 1 & -4 & 6 & -4 \\ 0 & 1 & -4 & 5 \end{bmatrix}$$

MATLAB Solution [Using built-in function]:

```
>> A=[5 -4 1 0;-4 6 -4 1;1 -4 6 -4;0 1 -4 5];
>> kmax=200;
>>  [eigs,A]=eig_QR(A,kmax)
eigs =
   13.0902
    6.8541
    1.9098
    0.1459
```

```
A =
       13.0902      0.0000    0.0000     -0.0000
       -0.0000      6.8541    0.0000      0.0000
        0.0000     -0.0000    1.9098      0.0000
             0           0   -0.0000      0.1459
>> A=[5 -4 1 0;-4 6 -4 1;1 -4 6 -4;0 1 -4 5];
>> [Q,d]=eig(A)
Q =
   -0.3717    -0.6015    0.6015    -0.3717
   -0.6015    -0.3717   -0.3717     0.6015
   -0.6015     0.3717   -0.3717    -0.6015
   -0.3717     0.6015    0.6015     0.3717
d =
     0.1459         0         0          0
          0    1.9098         0          0
          0         0    6.8541          0
          0         0         0    13.0902
function [eigs,A]=eig_QR(A,kmax)
%Find eigenvalues by using QR factorization
if margin<2, kmax=200; end
for k=1:kmax
[Q,R]=qr(A); %A=Q*R; R=Q'*A=Q^-1*A
A=R*Q; %A=Q^-1*A*Q
end
eigs=diag(A);
function [Q,R]=qr_my(A)
%QR factorization
N= size(A,1); R=A; Q=eye(N);
for k=1:N-1
H=Householder(R(:,k),k);
R=H*R;
Q=Q*H;
end
function H=Householder(x,k)
%Householder transform to zero out tail part starting from k+1
H=eye(N)-2*w*w'; %Householder matrix
N= length(x);
w=zeros(N,1);
w(k)=(x(k)+g)/c; w(k+1:N)=x(k+1:N)/c;%Eq.(P8.4-10)
tmp= sum(x(k+1:N).^2);
c=sqrt((x(k)+g)^2 +tmp);
g=sqrt(x(k)^2+tmp);
```

Example EN 3.5. *Using Cholesky's method of solution, solve the following linear equations.*

$$x_1 + x_2 + x_3 = 7$$
$$3x_1 + 3x_2 + 4x_3 = 23$$
$$2x_1 + x_2 + x_3 = 10$$

MATLAB Solution [Using built-in function]:

Cholesky's method:

```
>> A=[1 1 1;3 3 4;2 1 1];
>> B=[7;23;10];
>> [L,U]=lu(A)
L =
     0.3333    -0.0000     1.0000
     1.0000          0          0
     0.6667     1.0000          0
U =
     3.0000     3.0000     4.0000
          0    -1.0000    -1.6667
          0          0    -0.3333
>> L*U
ans =
     1.0000     1.0000     1.0000
     3.0000     3.0000     4.0000
     2.0000     1.0000     1.0000
>> d=L\B
d =
   23.0000
   -5.3333
   -0.6667
>> x=U\d
x =
     3.0000
     2.0000
     2.0000
```

Check with MATLAB [Using built-in function]:

```
>> A=[1 1 1;3 3 4;2 1 1];
>> B=[7;23;10];
>> x=A\B
x =
     3.0000
     2.0000
     2.0000
>> x=inv(A)*B
x =
     3.0000
     2.0000
     2.0000
```

Example EN3.6. *Solve the system of equations by Cholesky's factorization method.*

$$12x_1 - 6x_2 - 6x_3 - 1.5x_4 = 1$$
$$-6x_1 + 4x_2 + 3x_3 + 0.5x_4 = 2$$
$$-6x_1 + 3x_2 + 6x_3 + 1.5x_4 = 3$$
$$-1.5x_1 + 0.5x_2 + 1.5x_3 + x_4 = 4$$

MATLAB Solution [Using built-in function]:

Cholesky's method:

```
>> A=[12 -6 -6 -1.5;-6 4 3 0.5;-6 3 6 1.5;-1.5 0.5 1.5 1];
>> B=[1;2;3;4];
>> [L,U]=lu(A)
L =
    1.0000         0          0          0
   -0.5000    1.0000          0          0
   -0.5000         0     1.0000          0
   -0.1250   -0.2500     0.2500     1.0000
U =
   12.0000   -6.0000    -6.0000    -1.5000
         0    1.0000          0    -0.2500
         0         0     3.0000     0.7500
         0         0          0     0.5625
>> L*U
ans =
   12.0000   -6.0000    -6.0000    -1.5000
   -6.0000    4.0000     3.0000     0.5000
   -6.0000    3.0000     6.0000     1.5000
   -1.5000    0.5000     1.5000     1.0000
>> d=L\B
d =
    1.0000
    2.5000
    3.5000
    3.8750
>> x=U\d
x =
    2.7778
    4.2222
   -0.5556
    6.8889
```

MATLAB Solution [Using built-in function]:

```
>> A=[12 -6 -6 - 1.5;-6 4 3 0.5;-6 3 6 1.5;-1.5 0.5 1.5 1];
>> B=[1;2;3;4];
>> x=A\B
x =
    2.7778
    4.2222
```

```
   -0.5556
    6.8889
>> x=inv(A)*B
x =
    2.7778
    4.2222
   -0.5556
    6.8889
```

Example EN3.7. *Solve the set of equations given in Problem EN3.3. Use Jacobi method.*

Solution:

```
>> A=[3 1 -1;4 -10 1;2 1 5]; >> b=[-2 3 4]';
>> [x,k]=jacobi(A,b,[0 0 0]',1.e-10)
Jacobi iteration has converged in 38 iterations.
x =
   -0.2462
   -0.3026
    0.9590
k =
    38
function [x, k, diff] = jacobi(A,b,x0,tol,kmax)
% Jacobi iteration on the system Ax=b.
if margin<3, x0=zeros(size(b));, end
if margin<4, tol=1e-10;, end
if margin<5, kmax=100;, end
if min(abs(diag(A)))<eps
error('Coefficient matrix has zero diagonal entries, iteration cannot be performed.\r')
end
[n m]=size(A);
xold=x0;
k=1;, diff=[];
while k<=kmax
    xnew=b;
    for i=1:n
        for j=1:n
            if j~=i
                xnew(i)=xnew(i)-A(i,j)*xold(j);
            end
        end
        xnew(i)=xnew(i)/A(i,i);
    end
    diff(k)=norm(xnew-xold,2);
    if diff(k)<tol
          fprintf('Jacobi iteration has converged in %d iterations.\r', k)
            x=xnew;
            return
```

```
        end
        k=k+1;, xold=xnew;
    end
    fprintf('Jacobi iteration failed to converge.\r')
    x=xnew;
```

Example EN3.8. *Find the solution to the equations using Jacobi method with initial values [0 0 0 0].*

$$\begin{bmatrix} 4 & 2 & 0 & 0 \\ 2 & 8 & 2 & 0 \\ 0 & 2 & 8 & 2 \\ 0 & 0 & 2 & 4 \end{bmatrix} \begin{Bmatrix} x_1 \\ x_2 \\ x_3 \\ x_4 \end{Bmatrix} = \begin{Bmatrix} 4 \\ 0 \\ 0 \\ 0 \end{Bmatrix}$$

Solution:

```
>> A=[4 2 0 0;2 8 2 0;0 2 8 2;0 0 2 4];
>> b=[4;0;0;0];
>> x0=[0 0 0 0]';
>> [x,k]=jacobi(A,b,x0,1.e-10)
Jacobi iteration has converged in 34 iterations.
x =
    1.1556
   -0.3111
    0.0889
   -0.0444
k =
    34
function [x, k, diff] = jacobi(A,b,x0,tol,kmax)
% Jacobi iteration on the system Ax=b.
if margin<3, x0=zeros(size(b));, end
if margin<4, tol=1e-10;, end
if margin<5, kmax=100;, end
if min(abs(diag(A)))<eps
error('Coefficient matrix has zero diagonal entries, iteration cannot be performed.\r')
end
[n m]=size(A);
xold=x0;
k=1;, diff=[];
while k<=kmax
    xnew=b;
    for i=1:n
        for j=1:n
            if j~=i
                xnew(i)=xnew(i)-A(i,j)*xold(j);
            end
        end
```

```
            xnew(i)=xnew(i)/A(i,i);
        end
        diff(k)=norm(xnew-xold,2);
        if diff(k)<tol
              fprintf('Jacobi iteration has converged in %d iterations.\r', k)
               x=xnew;
               return
        end
        k=k+1;, xold=xnew;
end
fprintf('Jacobi iteration failed to converge.\r')
x=xnew;
```

Example EN3.9. *Find the solution to the equations using Gauss-Seidel method.*

$$\begin{bmatrix} 4 & 2 & 0 & 0 \\ 2 & 8 & 2 & 0 \\ 0 & 2 & 8 & 2 \\ 0 & 0 & 2 & 4 \end{bmatrix} \begin{Bmatrix} x_1 \\ x_2 \\ x_3 \\ x_4 \end{Bmatrix} = \begin{Bmatrix} 4 \\ 0 \\ 0 \\ 0 \end{Bmatrix}$$

Solution:

MATLAB Solution [Using built-in function]:

```
>> A=[4 2 0 0;2 8 2 0;0 2 8 2;0 0 2 4];
>> B=[4;0;0;0];
>> x=A\B
x =
    1.1556
   -0.3111
    0.0889
   -0.0444
>> x=inv(A)*B
x =
    1.1556
   -0.3111
    0.0889
   -0.0444
>> A=[4 2 0 0;2 8 2 0;0 2 8 2;0 0 2 4];
>> b=[4;0;0;0];
>> tol=1.0e-9;
>> format long;
>> xguess=[1 -0.5 0.1 -0.2];
>>  [x,iter]=GAUSSD(A,b,xguess,tol)
x =
  1.15555555568491  -0.31111111117579  0.08888888892123  -0.04444444446061
iter =
    16
```

```
function [Y,iter] = GAUSSD(A,r,yguess,tol)
```

```
% GAUSSD will iteratively solve Ay=r
n = length(r);  Y = yguess;  dy = ones(n,1); iter = 0;
while norm(dy)/norm(Y) > tol
for i = 1:n
if abs(A(i,i))<100*eps;error('zero pivot found');end
dy(i) = r(i)/A(i,i);
for j = 1:n
dy(i) = dy(i) - A(i,j)*Y(j)/A(i,i);
end
Y(i) = Y(i) + dy(i);
end
iter = iter + 1;
if iter>1000; error('not converging in 1000 steps');end
end
```

Example EN3.10. *Solve the system of equations given by [A]{x} = {b} using Gauss-Seidel method. The matrices [A] and {b} are given below.*

$$[A] = \begin{bmatrix} 4 & 2 & 0 & 0 \\ 2 & 8 & 2 & 0 \\ 0 & 2 & 8 & 2 \\ 0 & 0 & 2 & 4 \end{bmatrix} \{b\} = \begin{Bmatrix} 4 \\ 0 \\ 0 \\ 14 \end{Bmatrix}$$

Solution:

MATLAB Solution [Using built-in function]:

```
>> A=[4 2 0 0;2 8 2 0;0 2 8 2;0 0 2 4];
>> B=[4;0;0;14];
>> x=A\B
x =
     1.0000
        0
    -1.0000
     4.0000
>> x=inv(A)*B
x =
    1.0000
   -0.0000
   -1.0000
    4.0000
>> A=[4 2 0 0;2 8 2 0;0 2 8 2;0 0 2 4];
>> b=[4;0;0;14];
>> xguess=[1 -0.5 0.1 -0.2];
>> tol=1.0e-9;
>> format long;
>>  [x,iter]=GAUSSD(A,b,xguess,tol)
x =
  1.00000000075670  -0.00000000037835  -0.99999999981083  3.99999999990541
```

```
iter =
     17
function [Y,iter] = GAUSSD(A,r,yguess,tol)
% GAUSSD will iteratively solve Ay=r
n = length(r);   Y = yguess;   dy = ones(n,1); iter = 0;
while norm(dy)/norm(Y) > tol
for i = 1:n
if abs(A(i,i))<100*eps;error('zero pivot found');end
dy(i) = r(i)/A(i,i);
for j = 1:n
dy(i) = dy(i) - A(i,j)*Y(j)/A(i,i);
end
Y(i) = Y(i) + dy(i);
end
iter = iter + 1;
if iter>1000; error('not converging in 1000 steps');end
end
```

Example EN 3.11. *Solve the system of equations given below by Householder's factorization method:*

$$\begin{bmatrix} 4 & -1 & 0 & 0 \\ -1 & 4 & -1 & 0 \\ 0 & -1 & 4 & -1 \\ 0 & 0 & -1 & 4 \end{bmatrix} \begin{Bmatrix} x_1 \\ x_2 \\ x_3 \\ x_4 \end{Bmatrix} = \begin{Bmatrix} 1 \\ 0 \\ 0 \\ 0 \end{Bmatrix}$$

Solution:

```
  %Prob 4.25
>> A=[4 -1 0 0;-1 4 -1 0;0 -1 4 -1;0 0 -1 4];
>> b=[1;0;0;0];
>> householder(A)
ans =
      4        1        0        0
     -2        4       -1        0
      0        2        4        1
      0        0        1        4
>> [L,U]=lu(A)
L =
      1.0000         0             0             0
     -0.2500    1.0000             0             0
           0   -0.2667        1.0000             0
           0         0       -0.2679        1.0000
U =
      4.0000   -1.0000             0             0
           0    3.7500       -1.0000             0
           0         0        3.7333       -1.0000
           0         0             0        3.7321
>> L*U
```

```
ans =
     4    -1     0     0
    -1     4    -1     0
     0    -1     4    -1
     0     0    -1     4
>> A=[4 -1 0 0;-1 4 -1 0;0 -1 4 -1;0 0 -1 4];
>> d=L\b
d =
    1.0000
    0.2500
    0.0667
    0.0179
>> x=U\d
x =
    0.2679
    0.0718
    0.0191
    0.0048
```

MATLAB Solution [Using built-in function]:

```
>> A=[4 -1 0 0;-1 4 -1 0;0 -1 4 -1;0 0 -1 4];
>> B=[1;0;0;0];
>> x=A\B
x =
    0.2679
    0.0718
    0.0191
    0.0048
>> x=inv(A)*B
x =
    0.2679
    0.0718
    0.0191
    0.0048
```

```
function A=householder(A)
% Householder reduction of A to tridiagonal form A=[c\d\c]
%Extract c and d by d=diag(A), c=diag(A,1)
%Usage: A=householder(A)
n=size(A,1);
for k=1:n-2
u=A(k+1:n,k);
uMag=sqrt(dot(u,u));
if u(1)<0;uMag=-uMag; end
u(1)=u(1)+uMag;
A(k+1:n,k)=u;
H=dot(u,u)/2;
```

```
v=A(k+1:n,k+1:n)*u/H;
g=dot(u,v)/(2*H);
v=v-g*u;
A(k+1:n,k+1:n)=A(k+1:n,k+1:n)-v*u'-u*v';
A(k,k+1)=-uMag;
d=diag(A);
c=diag(A,1);
end
```

Example EN3.12. *Solve the system of equations given below by Householder's factorization method:*

$$\begin{bmatrix} 4 & 2 & 1 & 1 \\ 2 & 10 & 2 & 1 \\ 1 & 2 & 4 & 2 \\ 1 & 2 & 4 & 8 \end{bmatrix} \begin{Bmatrix} x_1 \\ x_2 \\ x_3 \\ x_4 \end{Bmatrix} = \begin{Bmatrix} 10 \\ 20 \\ 30 \\ 40 \end{Bmatrix}$$

MATLAB Solution:

>> %Prob EN3.12

```
>> A=[4 2 1 1;2 10 2 1;1 2 4 2;1 2 4 8];
% Householder reduction of A to tridiagonal form A=[c\d\c]
%Extract c and  by d=diag(A) and c=diag(A,1)
>> householder(A)
ans =
    4.0000    -2.4495     1.0000     1.0000
    4.4495    12.4082    -1.8257    -2.6330
    1.0000     2.6422     2.4400     0.3427
    1.0000    -1.6330     2.3427     7.1517
>> b=[10;20;30;40];
>> A=[4 2 1 1;2 10 2 1;1 2 4 2;1 2 4 8];
>> [L,U]=lu(A)
L =
    1.0000         0          0          0
    0.5000    1.0000          0          0
    0.2500    0.1667     1.0000          0
    0.2500    0.1667     1.0000     1.0000
U =
    4.0000    2.0000     1.0000     1.0000
         0    9.0000     1.5000     0.5000
         0         0     3.5000     1.6667
         0         0          0     6.0000
>> L*U
ans =
     4     2     1     1
     2    10     2     1
     1     2     4     2
     1     2     4     8
>> d=L\b
```

```
d =
    10
    15
    25
    10
>> x=U\d
x =
    0.2381
    0.5159
    6.3492
    1.6667
```

```
function A=householder(A)
% Householder reduction of A to tridiagonal form A=[c\d\c]
%Extract c and d by d=diag(A), c=diag(A,1)
%Usage: A=householder(A)
n=size(A,1);
for k=1:n-2
u=A(k+1:n,k);
uMag=sqrt(dot(u,u));
if u(1)<0;uMag=-uMag; end
u(1)=u(1)+uMag;
A(k+1:n,k)=u;
H=dot(u,u)/2;
v=A(k+1:n,k+1:n)*u/H;
g=dot(u,v)/(2*H);
v=v-g*u;
A(k+1:n,k+1:n)=A(k+1:n,k+1:n)-v*u'-u*v';
A(k,k+1)=-uMag;
d=diag(A);
c=diag(A,1);
end
```

MATLAB Solution [Using built-in function]:

```
>> A=[4 2 1 1;2 10 2 1;1 2 4 2;1 2 4 8];
>> B=[10;20;30;40];
>> x=A\B
x =
    0.2381
    0.5159
    6.3492
    1.6667
>> x=inv(A)*B
x =
    0.2381
    0.5159
    6.3492
    1.6667
```

Example EN 3.13. *Minimize the following function using Newton's method.*

$$f(x, y) = x_1^2 - x_1x_2 - 5x_1 + x_2^2 - x_2$$

Use initial guesses, x = 0 and y = 0.

Solution:

```
>> fn=inline('x(1)^2-x(1)*x(2)-5*x(1)+x(2)^2-x(2)','x');
>> gn=inline('[2*x(1)-x(2)-5   -x(1)+2*x(2)-1]','x');
>> x0=[0  0];TolX=1e-4;TolFun=1e-6;MaxIter=100;
>> [x0,g0,xx]=newtons(gn,x0,TolX,MaxIter)
x0 =
    3.6667    2.3333
g0 =
  1.0e-015 *
         0   -0.4441
xx =
    3.6667    2.3333
    3.6667    2.3333
function g= jacob(f,x,h,varargin)
%Jacobian of f(x)
if margin<3,  h=.0001; end
N= length(x); h2= 2*h; %h12=12*h;
x=x(:).'; I= eye(N);
for n=1:N
f1=feval(f,x+I(n,:)*h,varargin{:});
f2=feval(f,x-I(n,:)*h,varargin{:});
f3=feval(f,x+I(n,:)*h2,varargin{:});
f4=feval(f,x-I(n,:)*h2,varargin{:});
f12=(f1-f2)/h2;
f12=(8*(f1-f2)-f3+f4)/h12;
g(:,n)=f12(:);
end
if  sum(sum(isnan(g)))==0&rank(g)<N
format short e
fprintf('At x=%12.6e, Jacobian singular with J=',x);
disp(g); format short;
end
function  [x,fx,xx]=  newtons(f,x0,TolX,MaxIter,varargin)
%newtons.m  to solve a set of nonlinear eqs
%input:  f   = a 1st-order vector ftn equivalent to a set of equations
% x0  = the initial guess of the solution
% TolX = the upper limit of |x(k)-x(k-1)|
% MaxIter= the maximum # of iteration
% Output: x   = the point which the algorithm has reached
% fx  = f(x(last))
% xx  = the history of x
h=1e-5; TolFun=eps; EPS=1e-6;
fx=feval(f,x0,varargin{:});
```

```
Nf=length(fx); Nx=length(x0);
if Nf~=Nx, error('Incompatible dimensions of f and x0!'); end
if margin<4, MaxIter=100; end
if margin<3, TolX=EPS; end
xx(1,:)=x0(:).';
%fx0= norm(fx);
for k=1: MaxIter
J=jacob(f,xx(k,:),h,varargin{:});
if rank(J)<Nx
k=k-1; fprintf('Warning: Jacobian singular! with det(J)=%12.6e\n',det(J));
break;
else
dx= -J\fx(:); %-[dfdx]^-1*fx;
end
xx(k+1,:)= xx(k,:)+dx.';
fx= feval(f,xx(k+1,:),varargin{:}); fxn=norm(fx);
% if fxn<fx0,  break;  end
%end
if fxn<TolFun|norm(dx)<TolX, break; end
%fx0= fxn;
end
x= xx(k+1,:);
if k==MaxIter
fprintf('Do not depend on this, though the best in %d iterations\n',MaxIter)
end
```

Example EN3.14. *Minimize the two-variable objective function*

$$f(x_1,x_2) = x_1^2 - 2x_1x_2 - 4x_1 + x_2^2 - x_2$$

Use initial values: (0, 0).

Solution:

```
>> fn=inline('10*x(1)^2-10*x(1)*x(2)+3*x(2)^2+2*x(1)','x');
>> gn=inline('[20*x(1)-10*x(2)+2    -10*x(1)+6*x(2)]','x');
>> x0=[0 0];TolX=1e-4;TolFun=1e-6;MaxIter=50;
>> [x0,g0,xx]=newtons(gn,x0,TolX,MaxIter)
x0 =
   -0.6000   -1.0000
g0 =
     0     0
xx =
         0         0
   -0.6000   -1.0000
   -0.6000   -1.0000
```

```
function g= jacob(f,x,h,varargin)
%Jacobian of f(x)
if margin<3,  h=.0001; end
N= length(x); h2= 2*h; %h12=12*h;
x=x(:).'; I= eye(N);
```

```
for n=1:N
f1=feval(f,x+I(n,:)*h,varargin{:});
f2=feval(f,x-I(n,:)*h,varargin{:});
f3=feval(f,x+I(n,:)*h2,varargin{:});
f4=feval(f,x-I(n,:)*h2,varargin{:});
f12=(f1-f2)/h2;
f12=(8*(f1-f2)-f3+f4)/h12;
g(:,n)=f12(:);
end
if sum(sum(isnan(g)))==0&rank(g)<N
format short e
fprintf('At x=%12.6e, Jacobian singular with J=',x);
disp(g); format short;
end
```

```
function [x,fx,xx]= newtons(f,x0,TolX,MaxIter,varargin)
%newtons.m to solve a set of nonlinear eqs
%input: f = a 1st-order vector ftn equivalent to a set of equations
% x0 = the initial guess of the solution
% TolX = the upper limit of |x(k)-x(k-1)|
% MaxIter= the maximum # of iteration
% Output: x = the point which the algorithm has reached
% fx = f(x(last))
% xx = the history of x
h=1e-5; TolFun=eps; EPS=1e-6;
fx=feval(f,x0,varargin{:});
Nf=length(fx); Nx=length(x0);
if Nf~=Nx, error('Incompatible dimensions of f and x0!'); end
if margin<4, MaxIter=100; end
if margin<3, TolX=EPS; end
xx(1,:)=x0(:).';
%fx0= norm(fx);
for k=1: MaxIter
J=jacob(f,xx(k,:),h,varargin{:});
if rank(J)<Nx
k=k-1; fprintf('Warning: Jacobian singular! with det(J)=%12.6e\n',det(J));
break;
else
dx= -J\fx(:); %-[dfdx]^-1*fx;
end
xx(k+1,:)= xx(k,:)+dx.';
fx= feval(f,xx(k+1,:),varargin{:}); fxn=norm(fx);
% if fxn<fx0, break; end
%end
if fxn<TolFun|norm(dx)<TolX, break; end
%fx0= fxn;
end
```

```
x= xx(k+1,:);
if k==MaxIter
fprintf('Do not depend on this, though the best in %d iterations\n',MaxIter)
end
```

Example EN3.15. *Fit a polynomial by quadratic approximation and determine the values of X at which F(X) is minimum.*

X	*F(X)*
1	*8*
2	*3*
3	*17*

Solution:

X	F(X)
1	8
2	3
3	17

Let $F(X) = a_0 + a_1X + a_2X^2$. From the given data, we have

$$a_0 + a_1 + a_2 = 8$$
$$a_0 + 2a_1 + 4a_2 = 3$$
$$a_0 + 3a_1 + 9a_2 = 17$$

Solving above equations using MATLAB give

```
>> A=[1 1 1;1 2 4;1 3 9];
>> b=[8;3;17];
>> x=A\b
x =
   32.0000
  - 33.5000
    9.5000
>> x=inv(A)*b
x =
   32.0000
  - 33.5000
    9.5000
a0    = 32
a1    = -33.5
a2    = 9.5
>> x=[1 2 3];y=[8 3 17]; a2a1a0=polyfit(x,y,2)
a2a1a0 =
    9.50000000000001 - 33.50000000000002   32.00000000000003
```

and the function is

$F(X) = 32 - 33.5X + 9.5X^2$

For finding minimum of the function, we find the first derivative of the function

$$\frac{dF(X)}{dX} = -33.5 + 19X = 0 \text{ ; for a minima or maxima}$$

Thus, $X = 1.763$

To check for minima or maxima, we find second derivative of the function

$$\frac{d^2F(X)}{dX^2} = 19 > 0; \text{ thus it's a minimum.}$$

The minimum value of $F(X)$ at X is

$$F(X = 1.763) = 2.462$$

Example EN3.16. *Fit a polynomial by quadratic approximation and determine the values of X at which F(X) is minimum.*

X	$F(X)$
1	– 7
2	5
3	14

Solution:

X	$F(X)$
1	– 7
2	5
3	14

Let $F(X) = a_0 + a_1X + a_2X^2$. From the given data, we have

$$a_0 + a_1 + a_2 = -7$$
$$a_0 + 2a_1 + 4a_2 = 5$$
$$a_0 + 3a_1 + 9a_2 = 14$$

Solving above equations give

$$a_0 = -22$$
$$a_1 = 16.5$$
$$a_2 = -1.5$$

```
>> x=[1 2 3];y=[- 7 5 14];a2a1a0=polyfit(x,y,2)
a2a1a0 =
   - 1.50000000000000   16.50000000000002 -22.00000000000003
```

and the function is

$$F(X) = -22 + 16.5X - 1.5X^2$$

For finding minimum of the function, we find the first derivative of the function

$$\frac{dF(X)}{dX} = 16.5 - 3X = 0 \text{ ; for a minimum or maximum}$$

thus, $X = 5.5$

To check for minima or maxima, we find second derivative of the function

$$\frac{d^2F(X)}{dX^2} = -3 < 0; \text{ thus it's a maximum.}$$

Thus, for the given function there is no absolute minimum. The function attains its minimum values at $\pm\infty$.

MATLAB Solution:

```
% Problem 3.2
clc
clear
disp('Fit a Polynomial by Quadratic Aproximation')
x1=1;
x2=2;
x3=3;
f1=-7;
f2=5;
f3=14;
fx1=[1 x1 (x1)];
fx2=[1 x2 2*x2];
fx3=[1 x3 3*x3];
a=[fx1;fx2;fx3]
b=[f1 ;f2; f3]
poly_values=inv(a)*b
%solve for min
disp('(derivative) d(f(x))/dx= -3*x+16.5')
disp('Thus minimum is x=-16.5/3 ~ 5.5')
disp('Second derivative d2(f(x))/d2x=-3<0')
disp('Thus no absolute minimum')
```

Example EN 3.17. *Use Powell's method to find the minimum of the function*

$$f = 120(y - x^2)^2 + (1 - x)^2$$

Start with (– 1, 1).

Solution:

```
function y = fex3_17(X)
y = 120*(X(2)-X(1)^2)^2+(1-X(1))^2;
>> global X FUNC
>> FUNC = @fex3_17;
>> X=[-1.0,1.0];
>> [xMin,fMin,numCycles]=powell
xMin =
    1.0000
    1.0000
fMin =
  4.8369e-021
numCycles =
    12
function [xMin,fMin,nCyc] = powell(h,tol)
% Powell's method
% h   = initial search increment  = 0.1
% tol = error tolerance = 1.0e-6
% X = starting point
% FUNC = function that returns f
```

```
% xMin = minimum point
% fMin = minimum value of f
% nCyc = number of cycles to convergence
global X FUNC V
if margin < 2; tol = 1.0e-6; end
if margin < 1; h = 0.1; end
if size(X,2) > 1; X = X'; end
n = length(X);
df = zeros(n,1);
u = eye(n);
for j = 1:40        % 40 cycles
xOld = X;
fOld = feval(FUNC,xOld);
for i = 1:n
V = u(1:n,i);
[a,b] = goldBracket(@fLine,0.0,h);
[s,fMin] = goldSearch(@fLine,a,b);
df(i) = fOld - fMin;
fOld = fMin;
X = X + s*V;
end
V = X - xOld;
[a,b] = goldBracket(@fLine,0.0,h);
[s,fMin] = goldSearch(@fLine,a,b);
X = X + s*V;
% convergence criterion
if sqrt(dot(X-xOld,X-xOld)/n) < tol
xMin = X; nCyc = j; return
end
iMax = 1; dfMax = df(1);
for i = 2:n
if df(i) > dfMax
iMax = i; dfMax = df(i);
end
end
for i = iMax:n-1
u(1:n,i) = u(1:n,i+1);
end
u(1:n,n) = V;
end
error('No converge')
function z = fLine(s)
global X FUNC V
z = feval(FUNC,X+s*V);
```

Example EN3.18. *Use Powell's method to find the minimum of the function*

$$f(x) = 9x_1^2 + 4x_2^2 - 8x_1x_2 + 3x_1$$

Start with $x_0 = [0\ \ 0]^T$.

Solution:

```
function y = fex3_18(X)
y = 9*X(1)^2+4*X(2)^2-8*X(1)*X(2)+3*X(1);
>> global X FUNC
>> FUNC = @fex3_18;
>> X=[- 1.0,1.0];
>>  [xMin,fMin,numCycles]=powell
xMin =
   - 0.3000
   - 0.3000
fMin =
   - 0.4500
numCycles =
     2
function [xMin,fMin,nCyc] = powell(h,tol)
% Powell's method
% h    = initial search increment  = 0.1
% tol = error tolerance  = 1.0e-6
% X = starting point
% FUNC = function that returns f
% xMin = minimum point
% fMin = miminum value of f
% nCyc = number of cycles to convergence
global X FUNC V
if margin < 2; tol = 1.0e-6; end
if margin < 1; h = 0.1; end
if size(X,2) > 1; X = X'; end
n = length(X);
df = zeros(n,1);
u = eye(n);
for j = 1:40        % 40 cycles
xOld = X;
fOld = feval(FUNC,xOld);
for i = 1:n
V = u(1:n,i);
[a,b] = goldBracket(@fLine,0.0,h);
[s,fMin] = goldSearch(@fLine,a,b);
df(i) = fOld - fMin;
fOld = fMin;
X = X + s*V;
end
V = X - xOld;
[a,b] = goldBracket(@fLine,0.0,h);
[s,fMin] = goldSearch(@fLine,a,b);
X = X + s*V;
```

```
% convergence criterion
if sqrt(dot(X-xOld,X-xOld)/n) < tol
xMin = X; nCyc = j; return
end
iMax = 1; dfMax = df(1);
for i = 2:n
if df(i) > dfMax
iMax = i; dfMax = df(i);
end
end
for i = iMax:n-1
u(1:n,i) = u(1:n,i+1);
end
u(1:n,n) = V;
end
error('No converge')
function z = fLine(s)
global X FUNC V
z = feval(FUNC,X+s*V);
```

Example EN3.19. *Use Fletcher-Reeves method to locate the minimum of function*

$$F(x) = 10x_1^2 + 3x_2^2 - 10x_1x_2 + 2x_1.$$

Start with $[0 \quad 0.05]^T$.

Solution:

```
Global X FUNC DFUNC V
>> FUNC=@fex3_19;DFUNC=@dfex3_19;X=[0,0.5];
>> [xMin,fMin,nCyc]=fletcherReeves
xMin =
   - 0.6000
   - 1.0000
fMin =
   - 0.6000
nCyc =
     3
function [xMin,fMin,nCyc] = fletcherReeves(h,tol)
% Fletcher-Reeves method
% h    = initial search increment = 0.1
% tol = error tolerance = 1.0e-6
% X      = starting point
% FUNC  = handle of function that returns f
% DFUNC = handle of function that returns grad(f)
% xMin = minimum point
% fMin = miminum value of f
% nCyc = # of cycles to convergence
global X FUNC DFUNC V
if margin < 2; tol = 1.0e-6; end
if margin < 1; h = 0.1; end
```

```
if size(X,2) > 1; X = X'; end
n = length(X);
g0 = -feval(DFUNC,X);
V = g0;
for i = 1:50
[a,b] = goldBracket(@fLine,0.0,h);
[s,fMin] = goldSearch(@fLine,a,b);
X = X + s*V;
g1 = -feval(DFUNC,X);
if sqrt(dot(g1,g1)) <= tol
xMin = X; nCyc = i; return
end
gamma = dot((g1 - g0),g1)/dot(g0,g0);
V = g1 + gamma*V;
g0 = g1;
end
error('Method did not converge')
function z = fLine(s)
global X FUNC V
z = feval(FUNC,X+s*V);
function [a,b] = goldBracket(func,x1,h)
% Brackets the minimum point of f(x)
% func   =  returns f(x)
% x1     = starting value of x
% h      = initial step size
% a, b = limits on x
c = 1.618033989;
f1 = feval(func,x1);
x2 = x1 + h; f2 = feval(func,x2);
if f2 > f1
h = -h;
x2 = x1 + h; f2 = feval(func,x2);
if f2 > f1
a = x2; b = x1 - h; return
end
end
% Search loop
for i = 1:100
h = c*h;
x3 = x2 + h; f3 = feval(func,x3);
if f3 > f2
a = x1; b = x3; return
end
x1 = x2; f1 = f2; x2 = x3; f2 = f3;
end
error('Failed to find minimum')
```

```
function [xMin,fMin] = goldSearch(func,a,b,tol)
% Golden section search method
% func   = function that returns f(x)
% a, b   = limits of the interval for the minimum
% tol    = error tol = 1.0e-6
% fMin = min of f(x)
% xMin = x at min point
if margin < 4; tol = 1.0e-6; end
nIter = ceil(-2.078087*log(tol/abs(b-a)));
R = 0.618033989;
C = 1.0 - R;
x1 = R*a + C*b;
x2 = C*a + R*b;
f1 = feval(func,x1);
f2 = feval(func,x2);
for i =1:nIter
if f1 > f2
a = x1; x1 = x2; f1 = f2;
x2 = C*a + R*b;
f2 = feval(func,x2);
else
b = x2; x2 = x1; f2 = f1;
x1 = R*a + C*b;
f1 = feval(func,x1);
end
end
if f1 < f2; fMin = f1; xMin = x1;
else; fMin = f2; xMin = x2;
end
```

Example EN3.20. *Use Fletcher-Reeves method to find the minimum of function in Problem EN3.18.*

Start with $[0 \;\; 0.05]^T$.

Solution:

```
Global X FUNC DFUNC V
X=[0,0];
>> FUNC=@fex3_20;DFUNC=@dfex3_20;X=[0,0];
>> [xMin,fMin,nCyc]=fletcherReeves
xMin =
   - 0.3000
   - 0.3000
fMin =
   - 0.4500
nCyc =
     3
function [xMin,fMin,nCyc] = fletcherReeves(h,tol)
```

```
% Fletcher-Reeves method
% h    = initial search increment = 0.1
% tol = error tolerance = 1.0e-6
% X      = starting point
% FUNC  = handle of function that returns f
% DFUNC = handle of function that returns grad(f)
% xMin = minimum point
% fMin = miminum value of f
% nCyc = # of cycles to convergence
global X FUNC DFUNC V
if margin < 2; tol = 1.0e-6; end
if margin < 1; h = 0.1; end
if size(X,2) > 1; X = X'; end
n = length(X);
g0 = -feval(DFUNC,X);
V = g0;
for i = 1:50
[a,b] = goldBracket(@fLine,0.0,h);
[s,fMin] = goldSearch(@fLine,a,b);
X = X + s*V;
g1 = -feval(DFUNC,X);
if sqrt(dot(g1,g1)) <= tol
xMin = X; nCyc = i; return
end
gamma = dot((g1 - g0),g1)/dot(g0,g0);
V = g1 + gamma*V;
g0 = g1;
end
error('Method did not converge')
function z = fLine(s)
global X FUNC V
z = feval(FUNC,X+s*V);
```

function [a,b] = goldBracket(func,x1,h)

```
% Brackets the minimum point of f(x)
% func    =  returns f(x)
% x1      = starting value of x
% h       = initial step size
% a, b = limits on x
c = 1.618033989;
f1 = feval(func,x1);
x2 = x1 + h; f2 = feval(func,x2);
if f2 > f1
h = -h;
x2 = x1 + h; f2 = feval(func,x2);
if f2 > f1
```

```
a = x2; b = x1 - h; return
end
end
% Search loop
for i = 1:100
h = c*h;
x3 = x2 + h; f3 = feval(func,x3);
if f3 > f2
a = x1; b = x3; return
end
x1 = x2; f1 = f2; x2 = x3; f2 = f3;
end
error('Failed to find minimum')
function [xMin,fMin] = goldSearch(func,a,b,tol)
% Golden section search method
% func   = function that returns f(x)
% a, b   = limits of the interval for the minimum
% tol    = error tol = 1.0e-6
% fMin = min of f(x)
% xMin = x at min point
if margin < 4; tol = 1.0e-6; end
nIter = ceil(-2.078087*log(tol/abs(b-a)));
R = 0.618033989;
C = 1.0 - R;
x1 = R*a + C*b;
x2 = C*a + R*b;
f1 = feval(func,x1);
f2 = feval(func,x2);
for i =1:nIter
if f1 > f2
a = x1; x1 = x2; f1 = f2;
x2 = C*a + R*b;
f2 = feval(func,x2);
else
b = x2; x2 = x1; f2 = f1;
x1 = R*a + C*b;
f1 = feval(func,x1);
end
end
if f1 < f2; fMin = f1; xMin = x1;
else; fMin = f2; xMin = x2;
end
```

Example EN3.21. *Minimize the following function f(x) by the penalty function method:*

$$f(x) = [(x_1 + 2)^2 + 5(x_2 - 2)^2]\,[(x_1 - 1.5)^2 + 0.5(x_2 - 0.5)^2]$$

$$\text{Subject to} \begin{bmatrix} -x_1 \\ -x_2 \\ 3x_1 - x_1x_2 + 4x_2 - 6 \\ 2x_1 + x_2 - 5 \\ 3x_1 - 4x_2^2 - 4x_2 \end{bmatrix} \le \begin{bmatrix} 0 \\ 0 \\ 0 \\ 0 \\ 0 \end{bmatrix}$$

Solution:

```
>> % Example EN3.21
>> % Penalty function method
>> clear,clf
>> f='f321p';
>> x0=[0.4 0.5];
>> TolX=1e-5;TolFun=1e-9;alpha0=1;
>> TolX=1e-5;
>> MaxIter=100;
>> [x0_Nelder,f0_Nelder]=opt_Nelder(f,x0,TolX,TolFun,MaxIter) % Nelder method
x0_Nelder =
    1.4423    0.6540
f0_Nelder =
    0.3176
>> [fc_Nelder,f0_Nelder,c0_Nelder]=f321p(x0_Nelder) % Its results
fc_Nelder =
    0.3176
f0_Nelder =
    0.3176
c0_Nelder =
   - 1.4423
   - 0.6540
   - 0.0002
   - 1.4613
   - 0.0002
>> [x0_s,f0_s]=fminsearch(f,x0) %MATLAB built-in fminsearch()
x0_s =
    1.4421    0.6540
f0_s =
    0.3178
>> [fc_s,f0_s,c0_s]=f321p(x0_s) % its results
fc_s =
    0.3178
f0_s =
    0.3178
c0_s =
   - 1.4421
   - 0.6540
   - 0.0009
```

```
   - 1.4618
   - 0.0001
>> [x0_u,f0_u]=fminunc(f,x0) % MATLAB built-in fminunc
Warning: Gradient must be provided for trust-region method;
  using line-search method instead.
> In fminunc at 243
Maximum number of function evaluations exceeded;
 increase options.MaxFunEvals
x0_u =
    1.4311    0.6599
f0_u =
    0.3637
>> [fc_u,f0_u,c0_u]=f321p(x0_u) % its results
fc_u =
    0.3637
f0_u =
    0.3637
c0_u =
   - 1.4311
   - 0.6599
   - 0.0114
   - 1.4778
   - 0.0880
function [fc,f,c]=f321p(x)
f=((x(1)+2)^2+5*(x(2)-2)^2)*((x(1)-1.5)^2+0.5*(x(2)-0.5)^2);
c=[-x(1); -x(2); 3*x(1)-x(1)*x(2)+4*x(2)-6;2*x(1)+x(2)-5;
    3*x(1)-4*x(2)^2-4*x(2)]; % Constraint vector
v=[1 1 1 1 1];e=[1 1 1 1 1]';% Weighting coefficient vector
fc=f+v*((c>0).*exp(e.*c)); %  New objective function
function [xo,fo]=opt_Nelder(f,x0,TolX,TolFun,MaxIter)
N=length(x0);
if N==1 %for 1-dimensional case
[xo,fo]=opt_quad(f,x0,TolX,TolFun); return
end
S= eye(N);
for i=1:N %repeat the procedure for each subplane
i1=i+1; if i1>N, i1=1; end
abc=[x0; x0+S(i,:); x0+S(i1,:)]; %each directional subplane
fabc=[feval(f,abc(1,:)); feval(f,abc(2,:)); feval(f,abc(3,:))];
[x0,fo]=Nelder0(f,abc,fabc,TolX,TolFun,MaxIter);
if N<3, break; end %No repetition needed for a 2-dimensional case
end
xo=x0;
function [xo,fo]=opt_quad(f,x0,TolX,TolFun,MaxIter)
%search for the minimum of f(x) by quadratic approximation method
if length(x0)>2, x012=x0(1:3);
```

```
else
if length(x0)==2, a=x0(1); b=x0(2);
else a=x0-10; b=x0+10;
end
x012= [a (a+b)/2 b];
end
f012= f(x012);
[xo,fo]=opt_quad0(f,x012,f012,TolX,TolFun,MaxIter);
function [xo,fo]=opt_quad0(f,x012,f012,TolX,TolFun,k)
x0= x012(1); x1= x012(2); x2= x012(3);
f0= f012(1); f1= f012(2); f2= f012(3);
nd= [f0-f2 f1-f0 f2-f1]*[x1*x1 x2*x2 x0*x0; x1 x2 x0]';
x3= nd(1)/2/nd(2); f3=feval(f,x3); %Eq.(7.1-4)
if k<=0|abs(x3-x1)<TolX|abs(f3-f1)<TolFun
xo=x3; fo=f3;
if k==0, fprintf('Just the best in given # of iterations'), end
else
if x3<x1
if f3<f1, x012=[x0 x3 x1]; f012= [f0 f3 f1];
else x012=[x3 x1 x2]; f012= [f3 f1 f2];
end
else
if f3<=f1, x012=[x1 x3 x2]; f012= [f1 f3 f2];
else x012=[x0 x1 x3]; f012= [f0 f1 f3];
end
end
[xo,fo]=opt_quad0(f,x012,f012,TolX,TolFun,k-1);
end
```

Example EN3.22. *Minimize* $f(x) = x_1^2 + x_2^2 - 6x_1 - 8x_2 + 10$

Subject to $4x_1^2 + x_2^2 \le 0$

$3x_1 + 5x_2 \le 0$

Using penalty function method.

Solution:

```
% Example EN3.22
% Penalty function method
 clear,clf
 f='f322p';
 x0=[0.4 0.5];
 TolX=1e-5;TolFun=1e-9;alpha0=1;
 TolX=1e-5; MaxIter=100;
 [x0_Nelder,f0_Nelder]=opt_Nelder(f,x0,TolX,TolFun,MaxIter) % Nelder method
 [fc_Nelder,f0_Nelder,c0_Nelder]=f322p(x0_Nelder) % Its results
 [x0_s,f0_s]=fminsearch(f,x0) %MATLAB built-in fminsearch()
 [fc_s,f0_s,c0_s]=f322p(x0_s) % its results
 [x0_u,f0_u]=fminunc(f,x0) % MATLAB built-in fminunc
```

```
 [fc_u,f0_u,c0_u]=f322p(x0_u) % its results
>> [x0_Nelder,f0_Nelder]=opt_Nelder(f,x0,TolX,TolFun,MaxIter) % Nelder method
x0_Nelder =
    0.4412 - 0.2647
f0_Nelder =
    9.7353
>> [fc_Nelder,f0_Nelder,c0_Nelder]=f322p(x0_Nelder) % Its results
fc_Nelder =
    9.7353
f0_Nelder =
    9.7353
c0_Nelder =
   - 0.7085
   - 0.0000
>> [x0_s,f0_s]=fminsearch(f,x0) %MATLAB built-in fminsearch()
x0_s =
    0.4412   -0.2647
f0_s =
    9.7353
>> [fc_s,f0_s,c0_s]=f322p(x0_s) % its results
fc_s =
    9.7353
f0_s =
    9.7353
c0_s =
   - 0.7085
   - 0.0000
>> [x0_u,f0_u]=fminunc(f,x0) % MATLAB built-in fminunc
Warning: Gradient must be provided for trust-region method;
  using line-search method instead.
> In fminunc at 243
Optimization terminated: relative infinity-norm of gradient less than
options.TolFun.
x0_u =
    0.4867   -0.1888
f0_u =
  10.5382
>> [fc_u,f0_u,c0_u]=f322p(x0_u) % its results
fc_u =
   10.5382
f0_u =
    8.8627
c0_u =
   - 0.9119
    0.5161
function [fc,f,c]=f322p(x)
```

```
f=(x(1)^2+x(2)^2-6*x(1)-8*x(2)+10);
c=[-4*x(1)^2+x(2)^2;  3*x(1)+5*x(2)]; % Constraint vector
v=[1 1];e=[1 1 ]';% Weighting coefficient vector
fc=f+v*((c>0).*exp(e.*c)); %  New objective function
function [xo,fo]=opt_quad(f,x0,TolX,TolFun,MaxIter)
%search for the minimum of f(x) by quadratic approximation method
if length(x0)>2, x012=x0(1:3);
else
if length(x0)==2, a=x0(1); b=x0(2);
else a=x0-10; b=x0+10;
end
x012= [a (a+b)/2 b];
end
f012= f(x012);
[xo,fo]=opt_quad0(f,x012,f012,TolX,TolFun,MaxIter);
function [xo,fo]=opt_quad0(f,x012,f012,TolX,TolFun,k)
x0= x012(1); x1= x012(2); x2= x012(3);
f0= f012(1); f1= f012(2); f2= f012(3);
nd= [f0-f2 f1-f0 f2-f1]*[x1*x1 x2*x2 x0*x0; x1 x2 x0]';
x3= nd(1)/2/nd(2); f3=feval(f,x3); %Eq.(7.1-4)
if k<=0|abs(x3-x1)<TolX|abs(f3-f1)<TolFun
xo=x3;  fo=f3;
if k==0, fprintf('Just the best in given # of iterations'), end
else
if x3<x1
if f3<f1, x012=[x0 x3 x1]; f012= [f0 f3 f1];
else  x012=[x3 x1 x2]; f012= [f3 f1 f2];
end
else
if f3<=f1, x012=[x1 x3 x2]; f012= [f1 f3 f2];
else  x012=[x0 x1 x3]; f012= [f0 f1 f3];
end
end
[xo,fo]=opt_quad0(f,x012,f012,TolX,TolFun,k-1);
end
```

Example EN3.23. *Find the minimum point of the following objective function f(x) using quadratic approximation method.*

$$f(x) = \frac{(x_2^{\,2} - 5)^2}{8} - 1$$

Solution:

```
>> clear,clf
>> f323=inline('(x.*x-5).^2/8-1','x');
>> a=0;b=3;TolX=1e-6;TolFun=1e-9;MaxIter=100;
>> [x0q,f0q]=opt_quad(f323,[a,b],TolX,TolFun,MaxIter)
x0q =
    2.2361
```

```
f0q =
   - 1.0000
>> % x0q= minimum point and f0q = its function value in above
>> [x0q,f0q]=fminbnd(f323,a,b) % MATLAB built-in function
x0q =
    2.2361
f0q =
   - 1.0000
```

```
function [xo,fo]=opt_quad(f,x0,TolX,TolFun,MaxIter)
%search for the minimum of f(x) by quadratic approximation method
if length(x0)>2, x012=x0(1:3);
else
if length(x0)==2, a=x0(1); b=x0(2);
else a=x0-10; b=x0+10;
end
x012= [a (a+b)/2 b];
end
f012= f(x012);
[xo,fo]=opt_quad0(f,x012,f012,TolX,TolFun,MaxIter);
```

```
function [xo,fo]=opt_quad0(f,x012,f012,TolX,TolFun,k)
x0= x012(1); x1= x012(2); x2= x012(3);
f0= f012(1); f1= f012(2); f2= f012(3);
nd= [f0-f2 f1-f0 f2-f1]*[x1*x1 x2*x2 x0*x0; x1 x2 x0]';
x3= nd(1)/2/nd(2); f3=feval(f,x3); %Eq.(7.1-4)
if k<=0|abs(x3-x1)<TolX|abs(f3-f1)<TolFun
xo=x3; fo=f3;
if k==0, fprintf('Just the best in given # of iterations'), end
else
if x3<x1
if f3<f1, x012=[x0 x3 x1]; f012= [f0 f3 f1];
else x012=[x3 x1 x2]; f012= [f3 f1 f2];
end
else
if f3<=f1, x012=[x1 x3 x2]; f012= [f1 f3 f2];
else x012=[x0 x1 x3]; f012= [f0 f1 f3];
end
end
[xo,fo]=opt_quad0(f,x012,f012,TolX,TolFun,k-1);
end
```

Example EN3.24. *Find the minimum point of the following objective function f(x) using quadratic approximation method.*

$$f(x) = \frac{(x^2+3)^2}{10} + e^x - 7$$

Solution:

```
>> clear,clf
>>  f324=inline('(x.*x+3).^2/10+exp(x)-7','x');
>>  a=0;b=3;TolX=1e-6;TolFun=1e-9;MaxIter=100;
>>  [x0q,f0q]=opt_quad(f324,[a,b],TolX,TolFun,MaxIter)
x0q =
   - 0.4793
f0q =
   - 5.3377
>> % x0q= minimum point and f0q = its function value
>> [x0q,f0q]=fminbnd(f324,a,b) % MATLAB built-in function
x0q =
  5.6302e-005
f0q =
   - 5.0999
function  [xo,fo]=opt_quad(f,x0,TolX,TolFun,MaxIter)
%search for the minimum of f(x) by quadratic approximation method
if length(x0)>2, x012=x0(1:3);
else
if length(x0)==2, a=x0(1); b=x0(2);
else a=x0-10; b=x0+10;
end
x012= [a (a+b)/2 b];
end
f012= f(x012);
[xo,fo]=opt_quad0(f,x012,f012,TolX,TolFun,MaxIter);
function  [xo,fo]=opt_quad0(f,x012,f012,TolX,TolFun,k)
x0= x012(1); x1= x012(2); x2= x012(3);
f0= f012(1); f1= f012(2); f2= f012(3);
nd= [f0-f2 f1-f0 f2-f1]*[x1*x1 x2*x2 x0*x0; x1 x2 x0]';
x3= nd(1)/2/nd(2); f3=feval(f,x3); %Eq.(7.1-4)
if  k<=0|abs(x3-x1)<TolX|abs(f3-f1)<TolFun
xo=x3;   fo=f3;
if k==0, fprintf('Just the best in given # of iterations'), end
else
if x3<x1
if f3<f1, x012=[x0 x3 x1]; f012= [f0 f3 f1];
else  x012=[x3 x1 x2]; f012= [f3 f1 f2];
end
else
if f3<=f1, x012=[x1 x3 x2]; f012= [f1 f3 f2];
else  x012=[x0 x1 x3]; f012= [f0 f1 f3];
end
end
[xo,fo]=opt_quad0(f,x012,f012,TolX,TolFun,k-1);
end
```

Example EN3.25. *Use the threshold Jacobi iteration method to calculate the eigen pairs of the matrix [K] where*

$$[K] = \begin{bmatrix} 5 & -4 & 1 & 0 \\ -4 & 6 & -4 & 1 \\ 1 & -4 & 6 & -4 \\ 0 & 1 & -4 & 5 \end{bmatrix}$$

Solution:

```
>> A=[5 -4 1 0;-4 6 -4 1;1 -4 6 -4;0 1 -4 5];
>> tol=1.0e-9;
>>  [eVal,eVecs]=jacobi(A,tol)
eVal =
      0.1459
     13.0902
      6.8541
      1.9098
eVecs =
      0.3717     - 0.3717     - 0.6015     - 0.6015
      0.6015       0.6015       0.3717     - 0.3717
      0.6015     - 0.6015       0.3717       0.3717
      0.3717       0.3717     - 0.6015       0.6015
function [eVals,eVecs] = jacobi(A,tol)
% Jacobi method for finding the eigenvalues and
% eigenvectors of a symmetric matrix A.
% USAGE: [eVals,eVecs] = jacobi(A,tol)
% tol = error tolerance (1.0e-9)
if margin < 2; tol = 1.0e-9; end
n = size(A,1);
maxRot = 5*(n^2);          % Limit number of rotations
P = eye(n);                % Initialize rotation matrix
for i = 1:maxRot           % Begin Jacobi rotations
[Amax,k,L] = maxElem(A);
if Amax < tol;
eVals = diag(A); eVecs = P;
return
end
[A,P] = rotate(A,P,k,L);
end
error('Too many Jacobi rotations')
function [Amax,k,L] = maxElem(A)
% Finds Amax = A(k,L) (largest off-diag. elem. of A).
n = size(A,1);
Amax = 0;
for i = 1:n-1
for j = i+1:n
if abs(A(i,j)) >= Amax
```

```
Amax = abs(A(i,j));
k = i; L = j;
end
end
end
function [A,P] = rotate(A,P,k,L)
% Zeros A(k,L) by a Jacobi rotation and updates
% transformation matrix P.
n = size(A,1);
diff = A(L,L) - A(k,k);
if abs(A(k,L)) < abs(diff)*1.0e-36
t = A(k,L);
else
phi = diff/(2*A(k,L));
t = 1/(abs(phi) + sqrt(phi^2 + 1));
if phi < 0; t = -t; end;
end
c = 1/sqrt(t^2 + 1); s = t*c;
tau = s/(1 + c);
temp = A(k,L); A(k,L) = 0;
A(k,k) = A(k,k) - t*temp;
A(L,L) = A(L,L) + t*temp;
for i = 1:k-1      % For i < k
temp = A(i,k);
A(i,k) = temp -s*(A(i,L) + tau*temp);
A(i,L) = A(i,L) + s*(temp - tau*A(i,L));
end
for i = k+1:L-1    % For k < i < L
temp = A(k,i);
A(k,i) = temp - s*(A(i,L) + tau*A(k,i));
A(i,L) = A(i,L) + s*(temp - tau*A(i,L));
end
for i = L+1:n       % For i > L
temp = A(k,i);
A(k,i) = temp - s*(A(L,i) + tau*temp);
A(L,i) = A(L,i) + s*(temp - tau*A(L,i));
end
for i = 1:n      % Update matrix
temp = P(i,k);
P(i,k) = temp - s*(P(i,L) + tau*P(i,k));
P(i,L) = P(i,L) + s*(temp - tau*P(i,L));
end
```

MATLAB Solution:

```
>> A=[5 -4 1 0;-4 6 -4 1;1 -4 6 -4;0 1 -4 5];
>> eig(A)
```

```
ans =
      0.1459
      1.9098
      6.8541
     13.0902
>> [Q,d]=eig(A)
Q =
   - 0.3717    - 0.6015      0.6015    - 0.3717
   - 0.6015    - 0.3717    - 0.3717      0.6015
   - 0.6015      0.3717    - 0.3717    - 0.6015
   - 0.3717      0.6015      0.6015      0.3717
d =
     0.1459         0           0           0
          0    1.9098           0           0
          0         0      6.8541           0
          0         0           0     13.0902
```

Example EN 3.26. *Use Jacobi method to calculate the eigen pairs of the matrix A.*

$$A = \begin{bmatrix} 11 & 2 & 3 & 1 & 4 \\ 2 & 9 & 3 & 5 & 2 \\ 3 & 3 & 15 & 4 & 3 \\ 1 & 5 & 4 & 12 & 4 \\ 4 & 2 & 3 & 4 & 17 \end{bmatrix}$$

Solution:

```
>> A=[11 2 3 1 4;2 9 3 5 2;3 3 15 4 3;1 5 4  12 4;4 2 3 4 17];
>> tol=1.0e-9;
>> [eVal,eVecs]=jacobi(A,tol)
eVal =
      8.6636
      4.8739
     10.9368
     26.0252
     13.5005
eVecs =
      0.7291    - 0.2673      0.5058      0.3274      0.1842
      0.4139      0.7414    - 0.3188      0.3260    - 0.2665
    - 0.4299      0.0502      0.5208      0.4983    - 0.5414
      0.0696    - 0.5949    - 0.6029      0.4475    - 0.2784
    - 0.3278      0.1497    - 0.0884      0.5813      0.7241
function [eVals,eVecs] = jacobi(A,tol)
% Jacobi method for finding the eigenvalues and
% eigenvectors of a symmetric matrix A.
% USAGE: [eVals,eVecs] = jacobi(A,tol)
% tol = error tolerance (1.0e-9)
if margin < 2; tol = 1.0e-9; end
n = size(A,1);
```

```
maxRot = 5*(n^2);          % Limit number of rotations
P = eye(n);                % Initialize rotation matrix
for i = 1:maxRot           % Begin Jacobi rotations
[Amax,k,L] = maxElem(A);
if Amax < tol;
eVals = diag(A); eVecs = P;
return
end
[A,P] = rotate(A,P,k,L);
end
error('Too many Jacobi rotations')
function [Amax,k,L] = maxElem(A)
% Finds Amax = A(k,L) (largest off-diag. elem. of A).
n = size(A,1);
Amax = 0;
for i = 1:n-1
for j = i+1:n
if abs(A(i,j)) >= Amax
Amax = abs(A(i,j));
k = i; L = j;
end
end
end
function [A,P] = rotate(A,P,k,L)
% Zeros A(k,L) by a Jacobi rotation and updates
% transformation matrix P.
n = size(A,1);
diff = A(L,L) - A(k,k);
if abs(A(k,L)) < abs(diff)*1.0e-36
t = A(k,L);
else
phi = diff/(2*A(k,L));
t = 1/(abs(phi) + sqrt(phi^2 + 1));
if phi < 0; t = -t; end;
end
c = 1/sqrt(t^2 + 1); s = t*c;
tau = s/(1 + c);
temp = A(k,L); A(k,L) = 0;
A(k,k) = A(k,k) - t*temp;
A(L,L) = A(L,L) + t*temp;
for i = 1:k-1              % For i < k
temp = A(i,k);
A(i,k) = temp -s*(A(i,L) + tau*temp);
A(i,L) = A(i,L) + s*(temp - tau*A(i,L));
end
```

```
for i = k+1:L-1     % For k < i < L
temp = A(k,i);
A(k,i) = temp - s*(A(i,L) + tau*A(k,i));
A(i,L) = A(i,L) + s*(temp - tau*A(i,L));
end
for i = L+1:n        % For i > L
temp = A(k,i);
A(k,i) = temp - s*(A(L,i) + tau*temp);
A(L,i) = A(L,i) + s*(temp - tau*A(L,i));
end
for i = 1:n        % Update matrix
temp = P(i,k);
P(i,k) = temp - s*(P(i,L) + tau*P(i,k));
P(i,L) = P(i,L) + s*(temp - tau*P(i,L));
end
>> eig(A)
ans =
      4.8739
      8.6636
     10.9368
     13.5005
     26.0252
>> [Q,d]=eig(A)
Q =
      - 2673     - 0.7291     0.5058    - 0.1842     0.3274
      0.7414     - 0.4139   - 0.3188      0.2665     0.3260
      0.0502       0.4299     0.5208      0.5414     0.4983
    - 0.5949     - 0.0696   - 0.6029      0.2784     0.4475
      0.1497       0.3278   - 0.0884    - 0.7241     0.5813
d =
     4.8739         0          0           0          0
        0        8.6636        0           0          0
        0           0       10.9368        0          0
        0           0          0        13.5005       0
        0           0          0           0       26.0252
```

Example EN3.27. *Use the MATLAB fminbnd function to find the maximum of $f(x) = 8 \sin x - \frac{x^2}{14}$ with the interval $x_l = 0$ and $x_u = 7$.*

Solution:

First we create an .m file to hold the function

```
function f=fx(x)
f=-(8*sin(x)-x^2/14)
The negative sign is for minimization.
>> x=fminbnd('fx', 0,7)
```

```
The result from MATLAB program is
f =
   - 7.8268
x =
    1.5432
```

Example EN3.28. *Use the MATLAB fminsearch function to find the maximum of f(x, y) = $3xy + 8x - x^2 - 9y^2$, using initial guesses, x = – 1 and y = 1.*

Solution:

Create an *m* file to hold the function

```
function f=fxy(x)
f=-(3*x(1)*x(2)+8*x(1)-x(1)^2-9*x(2)^2)
Invoke the fminsearch function with
>> x=fminsearch('fxy',[-1,1])
f = -21.3333
x = 5.3334    0.8889
```

Examples EN3.29 to EN3.34 are based on Tutorials and Demos from MATLAB (The Mathworks, Inc.)

Example EN3.29. *This is an unconstrained minimization example. Find the set of values* $[x_1, x_2]$ *that solve* $\min_x f(x) = e^{x_1}(4x_1^2 + 2x_2^2 + 4x_1x_2 + 2x_2 + 0.9)$.

Starting guess: $x^0 = [-1, 1]$.

MATLAB Solution [Using built-in function]:

```
Write an m-file objfun.m
function f=objfun(x)
f=exp(x(1))*(4*x(1)^2+2*x(2)^2+4*x(1)*x(2)+2*x(2)+0.9);
Write one of the unconstrained optimization routines:
>> x0=[-1,1];  % Starting guess at the solution
>> options=optimset('LargeScale','off');
>> [x,fval,exitflag,output]=fminunc(@objfun,x0,options)
Optimization terminated: relative infinity-norm of gradient less than
options.TolFun.
x =
    0.5000   - 1.0000
fval =
   3.6609e-016
exitflag =
    1
output =
          iterations: 8
           funcCount: 66
            stepsize: 1
      firstorderopt: 7.3704e-008
           algorithm: 'medium-scale: Quasi-Newton line search'
message: 'Optimization terminated: relative infinity-norm of gradient less
than options.TolFun.'
```

Example EN3.30. *This is a nonlinear inequality constrained example. Find x that solves*

$$\min_x f(x) = e^{x_1}(4x_1^2 + 2x_2^2 + 4x_1x_2 + 2x_2 + 0.9).$$

Subject to the constraints

$$x_1x_2 - x_1 - x_2 \le -2.0$$
$$x_1x_2 \ge -10$$

Starting guess: $x^0 = [-1, 1]$.

Solution:

The constraints are written in the form $c(x) \le 0$.

$$x_1x_2 - x_1 - x_2 + 2 \le 0$$
$$-x_1x_2 - 10 \le 0$$

MATLAB Solution [Using built-in function]:

Write an m-file objfun.m

```
function f=objfun(x)
f=exp(x(1))*(4*x(1)^2+2*x(2)^2+4*x(1)*x(2)+2*x(2)+0.9);
```

Write an m-file confun.m for the constraints:

```
function [c,ceq]=confun(x)
%Nonlinear inequality constraints
c=[2+x(1)*x(2)-x(1)  -x(2);-x(1)*x(2)-10];
%Nonlinear inequality constraints
ceq=[];
```

Invoke constrained optimization routine:

```
>> x0=[-1,1];   % Initial guess at the solution
>> options=optimset('LargeScale','off');
>> [x,fval]=fmincon(@objfun,x0,[],[],[],[],[],[],@confun,options)
Optimization terminated: first-order optimality measure less
 than options.TolFun and maximum constraint violation is less
 than options.TolCon.
Active inequalities (to within options.TolCon = 1e-006):
  lower        upper       ineqlin    ineqnonlin
                                          1

                                                     2

x = - 9.5474     1.0474
fval = 0.0236
>> [c,ceq]=confun(x)
c =
 1.0e-007 *
   -0.9035
    0.9035
ceq =
    []
```

Example EN3.31. *This is a constrained optimization example: inequalities and bounds.*

Minimize $f(x) = e^{x_1}(4x_1^2 + 2x_2^2 + 4x_1x_2 + 0.9)$

Subject to $2 + x_1x_2 - x_1 - x_2 \le 0$

$$-x_1x_2 \le 0$$
$$x_1 \ge 0$$
$$x_2 \ge 0$$

Initial values: $x^0 = [-1, 1]$.

MATLAB Solution [Using built-in function]:

```
function f=objfun(x)
f=exp(x(1))*(4*x(1)^2+2*x(2)^2+4*x(1)*x(2)+2*x(2)+0.9);
function [c, ceq]=confun(x)
%Nonlinear inequality constraints
c=[2+x(1)*x(2)-x(1)-x(2);-x(1)*x(2)-10];
%Nonlinear inequality constraints
ceq=[];
>> x0=[-1,1]; % Initial guess at the solution
>> lb=[0,0];    % Lower bounds
>> ub=[];     % Upper bounds
>>  options=optimset('LargeScale','off');
>>  [x,fval]=fmincon(@objfun,x0,[],[],[],[],lb,ub,@confun,options)
Optimization terminated: first-order optimality measure less
than options.TolFun and maximum constraint violation is less
than options.TolCon.
Active inequalities (to within options.TolCon = 1e-006):
lower        upper      ineqlin    ineqnonlin
    1                                   1
x = 0       1.5000
fval  =  8.5000
>>  [c,ceq]=confun(x)
c  =  0  -10
ceq  =  []
```

Example EN 3.32. *This is a constrained example with gradients.*

Minimize $f(x) = e^{x_1}(4x_1^2 + 2x_2^2 + 4x_1x_2 + 2x_2 + 0.9)$

Subject to $2 + x_1x_2 - x_1 - x_2 \le 0$

$$-x_1x_2 \le 0$$

Initial values: $x^0 = [-1, 1]$.

MATLAB Solution [Using built-in function]:

Write an m-file for the objective function and gradient:

The objective function and its gradient are defined in the m-file *objgrad.m* as follows:

```
function  [f,G]=objfungrad(x)
f=exp  (x(1))*(4*x(1)^2+2*x(2)^2+4*x(1)*x(2)+2*x(2)+0.9);
t=exp(x(1))*(4*x(1)^2+2*x(2)^2+4*x(1)*x(2)+2*x(2)+0.9);
G=[t+exp(x(1))*(8*x(1)+4*x(2)),
     exp(x(1))*(4*x(1)+4*x(2)+2)];
```

Write an m-file for the nonlinear constraints and the gradients of the nonlinear constraints:

The constraints and their partial derivatives are contained in the m-file *confungrad.m:*

```
function[c,ceq,dc,dceq]=confungrad(x)
c=[2+x(1)*x(2)-x(1)-x(2);
    -x(1)*x(2)-10];
dc=[x(2)-1,-x(2);
    x(1)-1,-x(1)];
ceq=[];
dceq=[];
```

Invoke constrained optimization routine:

```
Define a guess at the solution:
>> x0=[-1,1];
>> options=optimset('LargeScale','off');
>> options=optimset(options,'GradObj','on','GradConstr','on');
>> lb=[],ub=[], % no lower or upper bounds
>> [x,fval]=fmincon(@objfungrad,x0,[],[],[],[],lb,ub,@confungrad,options)
Optimization terminated: first-order optimality measure less
 than options.TolFun and maximum constraint violation is less
 than options.TolCon.
Active inequalities (to within options.TolCon = 1e-006):
  lower      upper     ineqlin   ineqnonlin
                                    1
                                    2
x =
   - 9.5474    1.0474
fval =
    0.0236
>> [c,ceq]=confungrad(x)   % check the constraint values at x
c =
  1.0e-007 *
   - 0.9032
    0.9032
ceq =
     []
```

Example EN3.33.

Minimize $f(x) = e^{x_1} (4x_1^2 + 2x_2^2 + 4x_1x_2 + 2x_2 + 0.9)$

Subject to $x_1^2 + x_2^2 = 1$

$-x_1x_2 \geq -10$

Initial values: $x^0 = [-1, 1]$.

MATLAB Solution [Using built-in function]:

The objective function and its gradient re defined in the m-file *objfun.m* as follows:

```
function f=objfun(x)
f=exp(x(1))*(4*x(1)^2+2*x(2)^2+4*x(1)*x(2)+2*x(2)+0.9);
The m-file confuneq.m contains the equality and inequality constraints:
function [c,ceq]=confuneq(x)
c=-x(1)*x(2)-10;
```

```
ceq=x(1)^2+x(2)-1;
```

Define a guess at the solution:

```
>> x0=[-1,1];
```

We will use the option as below:

```
>> options=optimset('LargeScale','off');
```

Call the optimization algorithm:.

```
>>[x,fval,exitflag,output]=fmincon(@objfun,x0,[],[],[],[],[],[],@confuneq,options);
Optimization terminated: first-order optimality measure less than
options.TolFun
 and maximum constraint violation is less than options.TolCon.
No active inequalities.
>> x
x =
   - 0.7488    0.4393
```

The function value at the solution is

```
>> fval
fval =
    1.4621
```

The constraint values at the solution are

```
>> [c,ceq]=confuneq(x)
c =
   - 9.6710
ceq =
  6.4376e-009
```

The total number of function evaluations was

```
>> output.funcCount
ans =
    21
```

Changing the default termination tolerances

Consider the original unconstrained problem solved first

Minimize f(x)= exp(x(1))*(4*x(1)^2+2*x(2)^2+4*x(1)*x(2)+2*x(2)+0.9

This time we will solve it more accurately by overriding the default termination criteria (options.TolX and options.TolFun).

Create an anonymous function of the objective to be minimized:

```
>> fun=@(x) exp(x(1))*(4*x(1)^2+2*x(2)^2+4*x(1)*x(2)+2*x(2)+0.9);
>> fun
fun =
    @(x) exp(x(1))*(4*x(1)^2+2*x(2)^2+4*x(1)*x(2)+2*x(2)+0.9)
```

Make a guess at the solution as

```
>> x0=[-1,1];
```

Set the optimization options: turn off the large-scale algorithms (the default):

```
>> options=optimset('LargeScale','off');
```

Override the default termination criteria:

```
% Termination tolerances on X and f.
>> options=optimset(options,'TolX',1e-3,'TolFun',1e-3);
```

```
Call the optimization algorithm:
>> [x,fval,exitflag,output]=fminunc(fun,x0,options);
Optimization terminated: relative infinity-norm of gradient less than
options.TolFun.
The optimizer has found a solution at
>> x
x =
    0.5246    - 1.0248
The function value at the solution is
>> fval
fval =
   - 0.1669
The total number of function evaluations was
>> output.funcCount
ans =
    48
Set optimization options: turn off the large scale algorithms (the default):
>> options=optimset(options,'Display','iter');
>> [x,fval,exitflag,output]=fminunc(fun,x0,options);
                                                              First-order
   Iteration   Func-count        f(x)        Step-size     optimality
       0           3          1.80261          0.736
       1           9          1.68824        0.377066          0.295
       2          30       - 0.124807         21.3836          0.964
       3          36       - 0.160866        0.239618          0.158
       4          45       - 0.166915       0.0531349         0.0219
       5          48       - 0.166933               1        0.00114
Optimization terminated: relative infinity-norm of gradient less than
options.TolFun.
```

Example EN3.34. *Optimal Fit of a Non-linear Function.*

This is a demonstration of the optimal fitting of a non-linear function to a set of data. It uses FMINSEARCH, an implementation of the Nelder-Mead simplex (direct search) algorithm, to minimize a nonlinear function of several variables.

MATLAB Solution [Using built-in function]:

```
First, create some sample data and plot it.
>> t = (0:.1:2)';
y = [6 4 3 2 1.8990 1.5 1.2 1.1 1.03
         0.8 0.68 0.61 0.59 0.39 0.39 0.54 0.34 0.13   0.2 0.17 0.26]';
plot(t,y,'ro'); hold on; h = plot(t,y,'b'); hold off;
title('Input data'); ylim([0 6])
The goal is to fit the following function with two linear parameters and two
nonlinear parameters to the data:
y = C(1)*exp(-lambda(1)*t) + C(2)*exp(-lambda(2)*t)
To fit this function, we've to create a function FITFUN. Given the nonlinear
parameter (lambda) and the data (t and y), FITFUN calculates the error in
the fit for this equation and updates the line (h).
type fitfun
```

```
function err = fitfun(lambda,t,y)
%FITFUN Used by FITDEMO.
%   FITFUN(lambda,t,y) returns the error between the data and the values
%   computed by the current function of lambda.
%   FITFUN assumes a function of the form
%     y =  c(1)*exp(-lambda(1)*t) + ... + c(n)*exp(-lambda(n)*t)
%   with n linear parameters and n nonlinear parameters.
 A = zeros(length(t),length(lambda));
for j = 1:length(lambda)
   A(:,j) = exp(-lambda(j)*t);
end
c = A\y;
z = A*c;
err = norm(z-y);
```

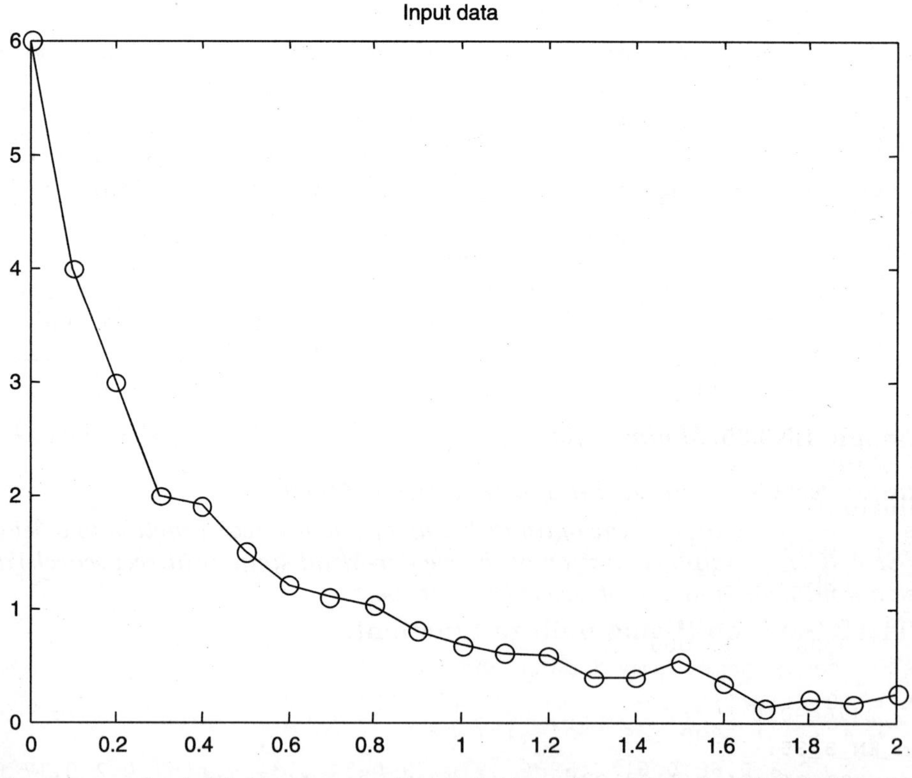

Make a guess for initial estimate of lambda (start) and invoke FMINSEARCH. It minimizes the error returned from FITFUN by adjusting lambda. It returns the final value of lambda. Use an output function to plot intermediate fits.

```
start = [1;0];
% We use an anonymous function to pass additional parameters t, y, h to the
% output function.
>> start = [1;0];
>> outputFcn = @(x,optimvalues,state) fitoutputfun(x,optimvalues,state,t,y,h);
options = optimset('OutputFcn',outputFcn,'TolX',0.1);
```

```
estimated_lambda  =  fminsearch(@(x)fitfun(x,t,y),start,options)
estimated_lambda  =
    7.0269
    1.3427
```

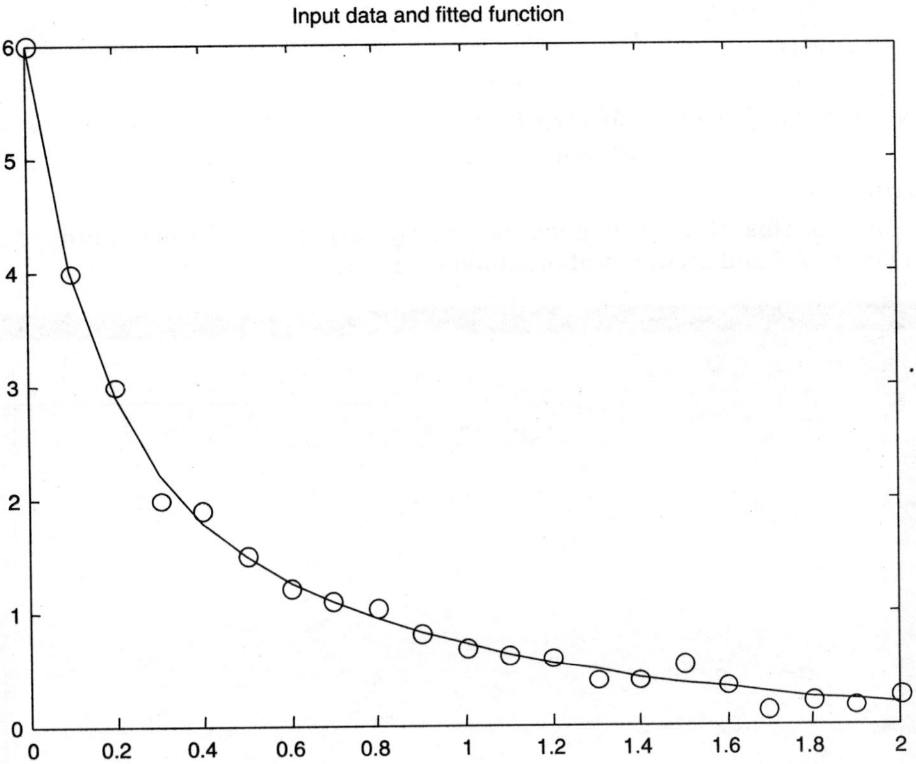

Example EN3.35. *Minimize* $f(x_1, x_2) = 90(x_2 - x_1^2)^2 + (1 - x_1)^2$ *using the starting point* $x_1 = -1.1, x_2 = 1.0.$

Solution:

```
The minimum = 0 at x = 1, y = 1
EDU>> Run_EN_3_35
x =
   1.0000    1.0000
fval =
  7.2476e-010
```

Run_EN_3_35:

```
y=@(x)90.*(x(2)-x(1)^2)^2  +  (1.-x(1))^2;
[x,fval]  =  fminsearch(y,[-1.1,1.0])
```

Example EN3.36. *Minimize* $f = 0.60 - 0.70/(1. + x^2) - 0.5*\tan^{-1}(1/x)$.

Solution:

The minimum value of the function is f = -0.8286 at x = 0.1001

```
EDU>> Run_EN_3_36
f =
    @(x)0.60  -  0.70/(1.  +  x.^2)  -  0.5*atan(1/x)
```

```
x =
     0.1001
f =
   -0.8286
```

Run_EN_3_36:

```
f = @(x)0.60 - 0.70/(1. + x.^2) - 0.5*atan(1/x)
x = fminbnd(f, .1,1.0)
f = 0.60 - 0.70/(1. + x.^2) - 0.5*atan(1/x)
```

Example EN3.37. *Use the MATLAB fminbnd function to find the maximum of*

$$y = (3^*(sin(\text{x})) - (x^2/12)$$

Solution:

The graph of this function is given below. By inspection, the maximum value of the function is close to 2.8 and occurs at approximately $x = 1.5$.

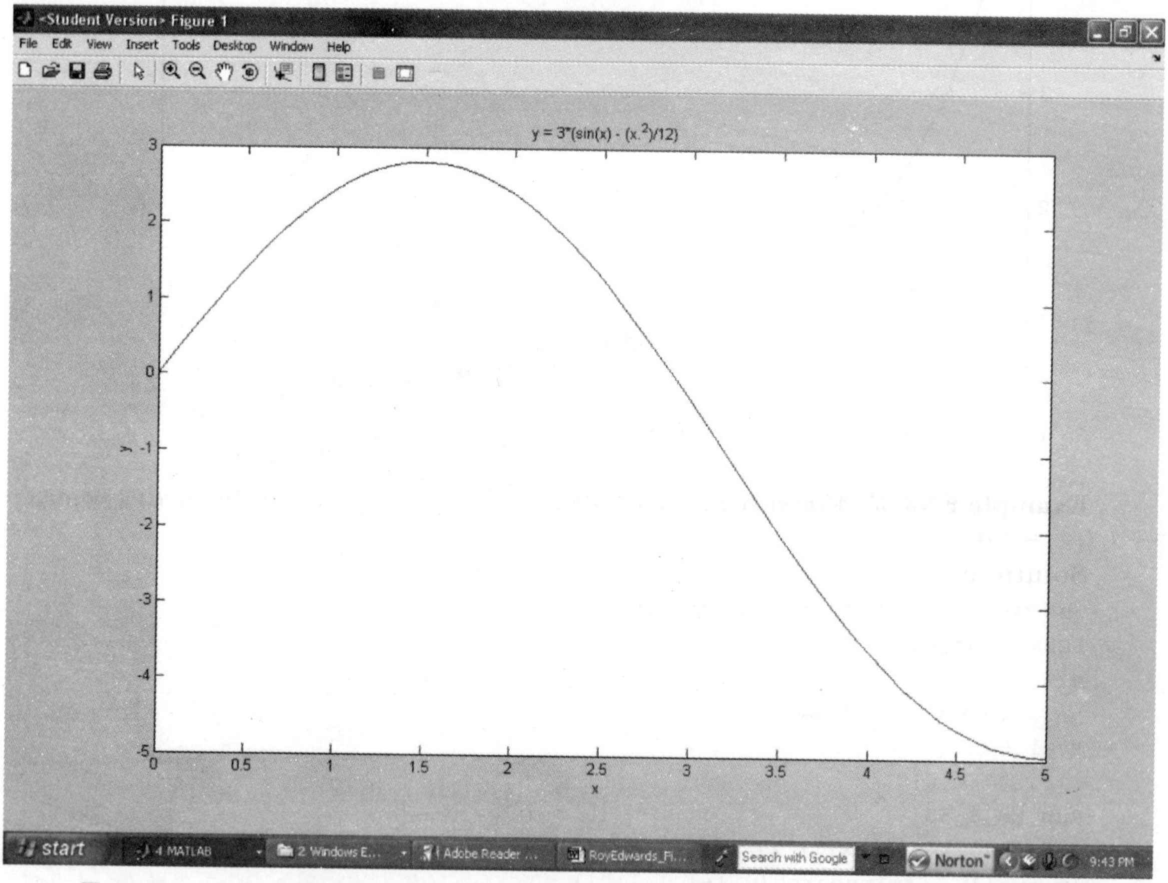

To compute the exact value of the *maximum* of the function, calculate the negative of the function (*y*). Use fminbnd on this new function (fun) to calculate the minimum value. The maximum of the original function occurs at the same x as the minimum of the new function. The maximum value of the original function is the negative of the new function.

```
EDU>> Run_EN_3_37
fun =
     @(x)(-1.0)* (3*(sin(x)) - (x.^2)/12)
x =
     1.4880
fun =
   - 2.8052
```

Therefore, the **maximum** of the original function, f, is

y = 2.8052 at

x = 1.4880

Run_EN_3_37:

```
fun = @(x)(-1.0)* (3*(sin(x)) - (x.^2)/12)
x = fminbnd(fun, 0,5.0)
fun = -1.0 * (3*(sin(x)) - (x.^2)/12)
```

Example EN3.38. *Using the MATLAB function fminsearch, maximize the function*

$$f(x_1, x_2) = 3*x_1*x_2 + 5*x_1 - x_1^2 - 3*x_2^2 \text{ with initial guess}$$

$$x_1 = -1.0, x_2 = 1.0$$

Solution:

The minimum of $-f$ occurs at $x = 10.0$, $y = 5.0$ and the minimum has a value of -25.0.

The maximum of the original function, f, occurs at $x = 10.0$, $y = 5.0$ and has a value of 25.0.

```
EDU>> Run_EN_3_38
x =
   10.0000    5.0000

fval =
   - 25.0000
```

Run_EN_3_38:

```
y=@(x)-1.0 * (3.*x(1).*x(2)+5.*x(1) -x(1).^2 - 3.*(x(2).^2));
[x,fval] = fminsearch(y,[-1.0,1.0])
```

Example EN3.39. *Use Powell's method to find the minimum of the function:*

$$f = 120(y - x^2)^2 + (1 - x)^2 \text{ starting with } (-1, 1)$$

Solution:

The minimum = 0 at $x = 1$, $y = 1$

```
EDU>> Run_EN_3_39
y =
   4.8369e-021

xMin =
     1.0000
     1.0000
fMin =
   4.8369e-021 (which is 0)
nCyc =
     12
```

```
Run_EN_3_39:
global X FUNC
FUNC = @fpr3_15;
X=[-1.0;1.0];
[xMin,fMin,nCyc]=powell
fpr3_39
function y = fpr3_15(X)
y=120.*(X(2)-X(1)^2)^2 + (1. - X(1))^2
```

ADDITIONAL EXAMPLES ON NUMERICAL METHODS

Example EN3.40. *Use the method of Gaussian elimination to solve the following system of linear equations:*

$$x_1 + x_2 + x_3 - x_4 = 2$$
$$4x_1 + 3x_2 + x_3 + x_4 = 11$$
$$x_1 - x_2 - x_3 + 2x_4 = 0$$
$$2x_1 + x_2 + 2x_3 - 2x_4 = 2$$

Solution:

Gaussian elimination method eliminates (makes zero) all coefficients below the main diagonal of the two-dimensional array. It does so by adding multiples of some equations to others in a symmetric way. Elimination makes the array of new coefficients have an upper triangular form since the lower triangular coefficients are all zero. Upper triangular equations can be solved by back substitution. Back substitution first solves last equation which has only one unknown $x(N) = b(N)/A(N,N)$.

Thus it is a two phase procedure.

(1) Forward elimination (Upper triangularization): First reduce the coefficients of first column of A below main diagonal to zero using first row. Then do the same for the second column using second row.

(2) Back Substitution: In this step, starting from the last equation, each of the unknowns is found.

Pitfalls of the method:

There are two pitfalls of Gauss Elimination method:

Division by zero: It is possible that division by zero may occur during forward elimination steps.

Round-off error: Gauss Elimination Method is prone to round-off errors.

```
The following MATLAB code is written for this problem
%%%%%%%%%%%%%%%%%%%%%%%%%%%%%%%%
%   This will perform Gaussian elmination
%   on the matrix that you pass to it.
%   i.e., given A and b it can be used to find x,
%         Ax = b
%
%   A - matrix for the left hand side.
%   b - vector for the right hand side
% This performs Gaussian elminination to find x.
```

```
% MATRIX DEFINITION
A=[1 1 1 -1;4 3 1 1;1 -1 -1 2;2 1 2 -2];
b=[2;11;0;2];
%%%%%%%%%%%%%%%%%%%%%%%%%%%%%%%
N = max(size(A))
% Perform Gaussian Elimination
 for j=2:N,
     for i=j:N,
        m = A(i,j-1)/A(j-1,j-1);
        A(i,:) = A(i,:) - A(j-1,:)*m;
        b(i) = b(i) - m*b(j-1);
     end
 end
disp('Upper triangular form of given matrix is=')
disp(A)
disp('b=')
disp(b)
% BACK-SUBSTITUTION
% Perform back substitution
 x = zeros(N,1);
 x(N) = b(N)/A(N,N);
 for j=N-1:-1:1,
   x(j)=(b(j)-A(j,j+1:N)*x(j+1:N))/A(j,j);
 end
disp('final solution is');
disp(x);
%%%%%%%%%%%%%%%%%%%%%%%%%%%%%%%
Output appears like this:
N
= 4
Upper triangular form of given matrix is=
    1.0000        1.0000        1.0000     - 1.0000
      0         - 1.0000      - 3.0000       5.0000
      0             0           4.0000     - 7.0000
      0             0             0          0.2500
b =
    2
    3
  - 8
    1
final solution is
  - 1
    2
    5
    4
......................
```

Check with MATLAB built-in function:

```
>> A=[1 1 1 -1;4 3 1 1;1 -1 -1 2;2 1 2 -2];
>> b=[2;11;0;2];
>> x=A\b
x =
  - 1.0000
    2.0000
    5.0000
    4.0000
```

Example EN3.41. *Solve the following system of equations using Cholesky's factorization.*

$$x_1 + x_2 + x_3 - x_4 = 2$$
$$x_1 - x_2 - x_3 + 2x_4 = 0$$
$$4x_1 + 4x_2 + x_3 + x_4 = 11$$
$$2x_1 + x_2 + 2x_3 - 2x_4 = 2$$

Solution:

Cholesky's factorization is basically applicable to only symmetric positive definite matrices.

Here original matrix [A] is decomposed as follows:

1. Form [A] = $L.L^T$ where L is lower triangular matrix
2. Forward substitution to solve $Ly = b$ for y
3. Back substitution to solve $L^Tx = y$ for x

For non-symmetric matrix a LU decomposition scheme can be employed using MATLAB function 'lu(A)'.

Complete MATLAB program is given below to solve the problem.

```
%%%%%%%%%%%%%%%%%%%%%%%
A=[1 1 1 -1;1 -1 -1 2;4 4 1 1;2 1 2 -2];
b=[2;0;11;2];
[L,U]=lu(A);
% solution of y
y=L\b;
%final solution x
x=U\y;
fprintf('Solution of the equations is\n');
disp(x)
%%%%%%%%%%%%%%%%%%%%%%%
```

Output is as follows:

Solution of the equations is

```
    1.0000
    2.0000
  - 1.0000
    0.0000
```

Check with MATLAB built-in function:

```
>> A=[1 1 1 -1;1 -1 -1 2;4 4 1 1;2 1 2 -2];
b=[2;0;11;2];
>> x=A\b
```

```
x =
        1.0000
        2.0000
      - 1.0000
        0.0000
```

Example EN3.42. *Using the Gauss-Seidel method, solve the system of equations given below:*

$$x + 2y + z = 0$$
$$3x + y - z = 0$$
$$x - y + 4z = 3$$

Solution:

The **Gauss-Seidel method** is a technique used to solve a linear system of equations. In solving equations $AX = b$, first the matrix A is written as: $A = D + L + U$ where the matrices D, L, and U represent the diagonal, negative strictly lower triangular, and negative strictly upper triangular parts of the coefficient matrix A.

Then the solution is given for every iteration counter k as:

$X^{(k+1)} = (D+ L)^{-1}(- U\, X^{(k)} + b)$ Gauss-Seidel Method

$X^{(k+1)} = D^{-1}(- (L+U)\, X^{(k)} + b)$ Jacobi Method

Disadvantages:

1. The matrix (D+L) is not always invertible. One **must** check that the values on the diagonal are non-zero before starting the iteration because it can lead to unpredictable results.
2. The spectral radius of the matrix $(D + L)^{-1} * \text{U}$ must have a modulus < 1 for the iteration to work correctly. Think of the spectral radius (the largest value of the set of eigenvalue modules) as being the greatest scalar factor that can be applied to a vector. If it is greater than 1, iteration on the corresponding eigenvector will result in an infinite limit.

Complete MATLAB program for solving given system of equations is given below:

```
% The display consists of a table of x-values with iterates of  x1, x2, ...,
xn
% in each column.
A=[1 2 1;3 1 -1;1 -1 4];b=[0;0;3];
X0=zeros(size(b)); % starting vector
tole=1e-6;kstop=30;% error tolerance and max. iterations
[n,n]=size(A);
P=tril(A);% lower triangular form
k=0;r=b-A*X0;
r0=norm(r);er=norm(r);
X=X0;
[L,U]=lu(P);
fprintf('iter#\tX(1)\t\tX(2)\n');
while er>tole & k<kstop
     fprintf('%d\t%f\t%f\n',k,X(1),X(2));
   k=k+1;
   dx=L\r;
   dx=U\dx;
```

```
    X=X+dx;
    r=b-A*X;
    er=norm(r)/r0;
    erp(k)=norm(r)/r0;
end
X
plot(erp,'-p');
grid on;
xlabel('Iteration #');
ylabel('normalized error');
```

Output of the program is as follows:

Final solution is X=

1.0e+024 *

– 1.7510

5.5007

1.8129

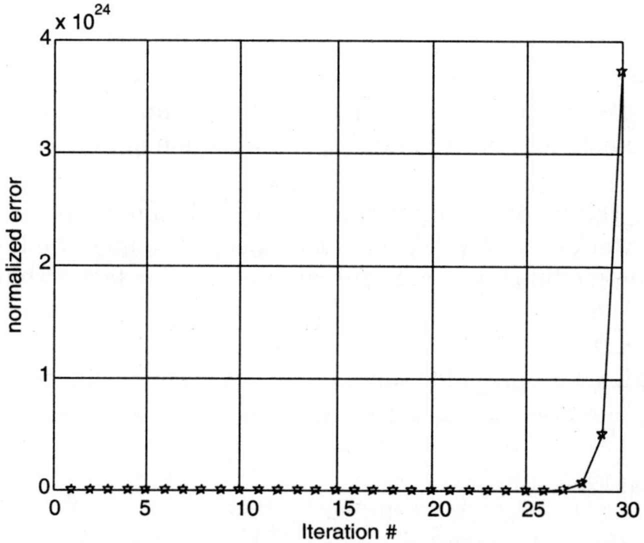

Fig. EN3.42

Here the method fails due to reason (2) given above.

Check with MATLAB built-in function:

```
>> A=[1 2 1;3 1 -1;1 -1 4];b=[0;0;3];
>> x=A\b
x =
      0.3333
    - 0.4444
      0.5556
```

Example EN3.43. *Using the Gauss-Seidel method, solve the system of equations given further:*

$$4x - y + z = 10$$
$$-x + 4y - 2z = -2$$
$$x - 2y + 4z = 5$$

Solution:

MATLAB program for this problem is given below:

```
A=[4 -1 1;-1 4 -2;1 -2 4];
b=[10;-2;5];
X0=zeros(size(b)); % starting vector
tole=1e-6;kstop=30;% error tolerance and max. iterations
[n,n]=size(A);
P=tril(A);% lower triangular form
k=0;r=b-A*X0;
r0=norm(r);er=norm(r);
X=X0;
[L,U]=lu(P);
fprintf('iter#\tX(1)\t\tX(2)\n');
while er>tole & k<kstop
    fprintf('%d\t%f\t%f\n',k,X(1),X(2));
   k=k+1;
   dx=L\r;
   dx=U\dx;
   X=X+dx;
   r=b-A*X;
   er=norm(r)/r0;
   erp(k)=norm(r)/r0;
end
X
plot(erp,'-p');
grid on;
xlabel('Iteration #');
ylabel('normalized error');
```

Output of the program is as follows:

iter#	X(1)	X(2)
0	0.000000	0.000000
1	2.500000	0.125000
2	2.359375	0.433594
3	2.389160	0.535767
4	2.403793	0.561245
5	2.407893	0.566810
6	2.408845	0.567927
7	2.409044	0.568137
8	2.409082	0.568174

```
The final solution is
    X =
       2.4091
       0.5682
       0.9318
```

The variation of error in each cycle is shown in Fig. EN3.43

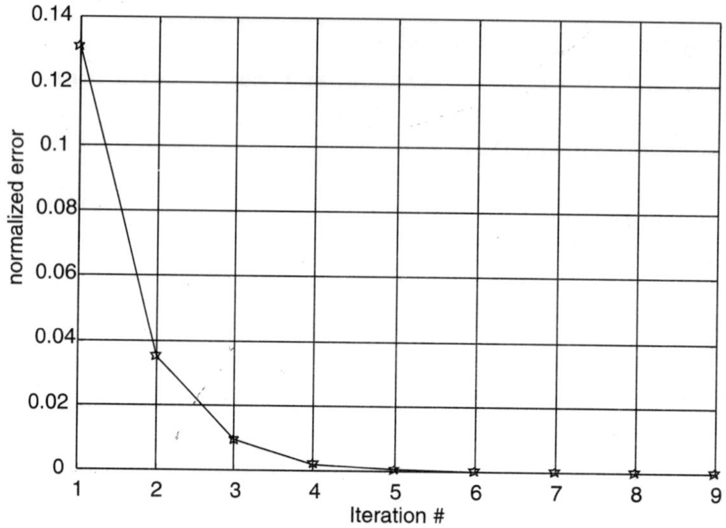

Fig. EN3.43 MATLAB output.

Check with MATLAB built-in function:

```
>> A=[4 -1 1;-1 4 -2;1 -2 4];b=[10;-2;5];
>> x=A\b
x =
    2.4091
    0.5682
    0.9318
```

Example EN3.44. *Find the minimum of the function $f = 90(y - x^2)^2 + (1 - x)^2$ with Powell's method starting at the point (– 1, 1).*

Solution:

Let $\vec{X}_0$ be an initial guess at the location of the minimum of the function $z = f(\vec{X}) = f(x_1, x_2,..., x_n)$. Assume that the partial derivatives of the function are not available. An intuitively appealing approach to approximating a minimum of the function f is to generate the next approximation $\vec{X}_1$ by proceeding successively to a minimum of f along each of the n standard base vectors. This process can be called the taxi-cab method and generates the sequence of points

$$\vec{X}_0 = \vec{P}_0, \vec{P}_1, \vec{P}_2, ..., \vec{P}_n = \vec{X}_1$$

Along each standard base vector $\vec{E}_k = (0, ..., 0, l_k, 0, ..., 0)$ the function f is a function of one variable, and the minimization of f might be accomplished by the application of either the golden ratio or Fibonacci searches if f is unimodal in this search direction. The iteration is then repeated to generate a sequence of points $\left\{\vec{X}_k\right\}_{k=0}^{\infty}$. Unfortunately, the method is, in general, inefficient due to the geometry of multivariable functions

Choose $N = 2$ and select two direction vectors $U_1 = \begin{bmatrix}1\\0\end{bmatrix}$ and $U_2 = \begin{bmatrix}0\\1\end{bmatrix}$. Start with point $P_0 = X = \begin{bmatrix}-1\\1\end{bmatrix}$ and construct $P_1 = P_0 + \lambda_1 U_1 = \begin{bmatrix}-1+\lambda_1\\1\end{bmatrix}$ by moving in the directions U_1 by optimal length λ_1. Substituting P_1 in f, a one-dimensional objective function is formed in λ_1, which is solved for minimum. Then λ_1 is substituted in P_1 and new point P_2 is evaluated as $P_2 = P_1 + \lambda_2 U_2$ by moving in the directions U_2 by optimal length λ_2 in the similar manner.

Then change the new search directions as $U_1 = \begin{bmatrix}0\\1\end{bmatrix}$ and $U_2 = P_2 - P_0$. The process is repeated for several iterations to get the best unconstrained optimum point X_n. For obtaining the optimum lengths λ_i an unconstrained one-dimensional problem is to be solved. The method is illustrated in Fig. EN3.44 (*a*).

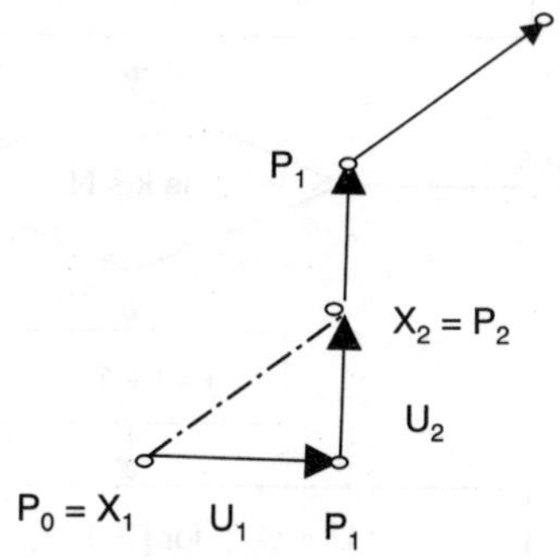

Fig. EN3.44 (*a*)

The solution by other method (Nedler and Meed' Simplex) is as follows with MATLAB readymade function:

Outputs with the function 'obj.m' given below is as follows:

```
function f =obj(x)
f=90*(x(2)-x(1)^2)^2+(1-x(1))^2;
>> [x fval]=fminsearch('obj',[-1,1]);
>> x
x=1.0000          1.0000
>> fval
fval=0.000
```

The complete flowchart of the method is shown in Fig. EN3.44 (*b*).

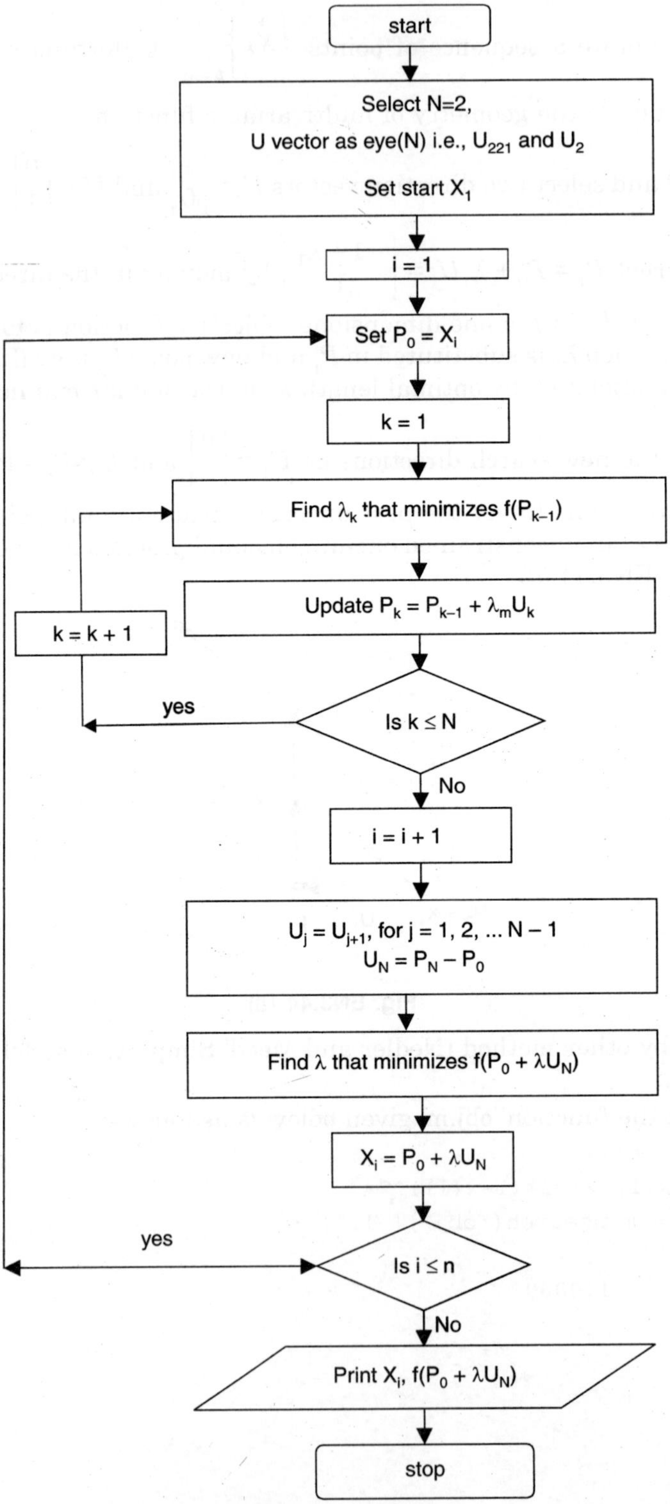

Fig. EN3.44 (*b*)

The program is as follows:

```
%%%%%%%%%%%%%%%%%%%%%%%%%%%%%%
global  X V
X=[-1;1];    %STARTING POINT
N=length(X);% number of design variables
df=zeros(N,1);% decreases of f stored here
u=eye(N);  % columns of u store search directions V
n=30;  % Number of cycles
for j=1:n
  xold=X;
  fold=90*(xold(2)-xold(1)^2)^2+(1-xold(1))^2;
  % FIRST N line searches record the decrease of f
  for i=1:N
   V=u(1:N,i);
   [s fmin]=fminbnd('fline',0,10);
   %Golden section search built in function
   df(i)=fold-fmin;
   fold=fmin;
   X=X+s*V;
  end
  % LAST LINE SEARCH IN THE CYCLE
  V=X-xold;
  [s fmin]=fminbnd('fline',0,10);
  X=X+s*V;
  % IDENTIFY BIGGEST DECREASE OF F
  % AND UPDATE SEARCH DIRECTIONS
  imax=1; dfmax=df(1);
  for i=2:N
    if df(i)>dfmax    imax=i; dfmax=df(i); end
  end
  for i=imax:N-1
    u(1:N,i)=u(1:N,i+1);
  end
  u(1:N,N)=V;
end
fprintf('Optimum point after %d cycles is\n',n);
fprintf('%f\n%f\n',X(1),X(2));
%%%%%%%%%%%%%%%%%%%%%%%%%%%%%%%%%%
The function with file name 'fline.m' used is as follows:
%%%%%%%%%%%%%%%%%%%%%%
function z=fline(s)
global X V
b1=V(1);
b2=V(2);
a1=X(1);a2=X(2);
```

```
p1=a1+s*b1;p2=a2+s*b2;
z=90*(p2-p1^2)^2+(1-p1)^2;
%%%%%%%%%%%%%%%%%%%%%%%%%%%%%%%%%%%%%%%%
```

The output is as follows:

```
Optimum point after 30 cycles is
1.003135
1.000263
```

Example EN3.45. *Use the Fletcher-Reeves method to locate the minimum of* $f(x) = 9x_1^2 + 3x_2^2 - 8x_1x_2 + 2x_1$. *Start with* $X_o = [0\ 0]^T$.

Solution:

This method is also called conjugate gradient method. It is a modification of Cauchy's method of gradient decent algorithm. The complete algorithm is given below:

Given x^0 perform the following steps:

1. Compute $\nabla f(x^0)$ and set $u^1 = -\nabla f(x^0)$.
2. For $i = 1, 2, \ldots, n$ do:

   2.1 set $x^i = x^{i-1} + \lambda_i u^i$ and λ_i such that

   $$f(x^{i-1} + \lambda_i u^i) = \min_{\lambda} f(x^{i-1} + \lambda u^i) \text{ (line search)},$$

   2.2 compute $\nabla f(x^i)$,

   2.3 if convergence criteria satisfied, then STOP and $x^* \cong x^i$, else go to step 2.4.

   2.4 if $1 \le i \le n - 1$, $u^{i+1} = -\nabla f(x^i) + \beta_i u^i$
3. Set $x^0 = x^n$ and go to Step 2 (restart).

Here
$$\beta_i = \frac{|\nabla f_i|^2}{|\nabla f_{i-1}|^2}.$$

In this problem
$$\nabla f = \begin{bmatrix} 18x_1 - 8x_2 + 2 \\ 6x_2 - 8x_1 \end{bmatrix} \text{ and } X^0 = \begin{bmatrix} 0 \\ 0 \end{bmatrix}$$

Iteration-1

Thus
$$U^1 = -\nabla f(X^0) = \begin{bmatrix} 2 \\ 0 \end{bmatrix}.$$

For $i = 1$, find $X^1 = X^0 + \lambda_1 U^1 = \begin{bmatrix} 2\lambda_1 \\ 0 \end{bmatrix}$ and substituting in function f and find the minimum value of f and corresponding λ_1 and X^1.

Compute $\nabla f(X^1)$

Iteration-2

$$U^2 = -\nabla f_1 + \frac{|\nabla f_1|^2}{|\nabla f_0|^2} U^1$$

$X^2 = X^1 + \lambda_2 U^2$ and perform line search to find the minimum of $f(X^2) = f(\lambda_2)$

Again compute $\nabla f(X^2)$. Stop if it is equal to zero otherwise continue.

Complete program is given below:

```
%%%%%%% CONJUGATE GRADIENT METHOD%%%%%%%%%%
global X V
X=[0;0];
N=length(X);
g0=-feval('df',X);
V=g0;
n=50;
for i=1:n
   [s fmin]=fminbnd('fline',-10,10);
  X=X+s*V
   g1=-feval('df',X);
   gamma=(norm(g1)/norm(g0))^2;
  V=g1+gamma*V;
  g0=g1;
end
fprintf('Minimum point for this function is \n');
fprintf('%f\n%f\n',X(1),X(2));
%%%%%%%%%%
This requires the following function files
%%%%%%%LINEAR FUNCTION OF s %%%%%%%%
function z=fline(s)
global V X
p1=X(1)+s*V(1);
p2=X(2)+s*V(2);
z=9*p1^2+3*p2^2-8*p1*p2+2*p1;
%%%%%%GRADIENT FUNCTION %%%%%%%%%%%%
function y=df(X)
y=[18*X(1)-8*X(2)+2;6*X(2)-8*X(1)];
%%%%%%%%%%%%%%%%%%
```

The output is as follows:

Minimum point for this function is

```
- 0.272727
- 0.363636
Exact output obtained from the function fminsearch('fxy',[-1,1]) is
- 0.2727
- 0.3637
```

Example EN3.46. *Use the Jacobi method to determine the eigenvalues and eigenvectors of the following matrix*

$$A = \begin{bmatrix} 11 & 2 & 8 \\ 2 & 2 & -10 \\ 9 & -10 & 5 \end{bmatrix}$$

Solution:

A solution is guaranteed for all real symmetric matrices when Jacobi's method is used. This limitation is not severe since many practical problems of applied mathematics and engineering involve symmetric matrices. From a theoretical viewpoint, the method embodies techniques that are found in more sophisticated algorithms. For instructive purposes, it is worthwhile to investigate the details of Jacobi's method.

Start with the real symmetric matrix A. Then construct the sequence of orthogonal matrices R_1, R_2, R_3,....R_n as follows:

$$\mathbf{D_0 = A}$$

and $$D_j = R_j^T D_j R_j \text{ for } j = 1, 2, \ldots .$$

It is possible to construct the sequence $\{R_j\}$ so that

$$\lim_{j \to \infty} \mathbf{D_j = D} = \text{diag}\,(\lambda_1, \lambda_2, \ldots, \lambda_n)$$

In practice we will stop when the off-diagonal elements are close to zero. Then we will have $D_m \approx D$.

The complete program is shown below:

```
%%%%%%%%%%%%%%%%%%%%%%%%%%%%%%%%%%%%%%%%%%%%
A=[11 2 8;2 2 -10;9 -10 5];
%Output - V is the nxn matrix of eigenvectors
%        - D is the diagonal nxn matrix of eigenvalues
D=A;
[n,n]=size(A);
V=eye(n);
%Calculate row p and column q of the off-diagonal element
%of greatest magnitude in A
[m1 p]=max(abs(D-diag(diag(D))));
[m2 q]=max(m1);
p=p(q);
i=1;
while(i<10)
%Zero out Dpq and Dqp
t=D(p,q)/(D(q,q)-D(p,p));
c=1/sqrt(t^2+1);
s=c*t;
R=[c s;-s c];
D([p q],:)=R'*D([p q],:);
D(:,[p q])=D(:,[p q])*R;
V(:,[p q])=V(:,[p q])*R;
[m1 p]=max(abs(D-diag(diag(D))));
[m2 q]=max(m1);
p=p(q);
i=i+1;
end
D=diag(diag(D))
fprintf('final eigenvalues are %f\t%f\t%f\n',D(1,1),D(2,2),D(3,3));
%%%%%%%%%%%%%%%%%%%%%%%%%%%%%%%%%%%%%%%%%%%%
```

The output of the program is as follows:

```
D =
        18.4278              0              0
              0        - 9.2213              0
              0              0         8.7934
Final eigenvalues are 18.427839     -9.221279 8.793440
```

Check with MATLAB built-in function:

```
>> A=[11 2 8;2 2 -10;9 -10 5];
>> [Q,D]=eig(A)
Q =
      0.3319     - 0.6460      - 0.6277
    - 0.6599       0.3380      - 0.6966
    - 0.6741     - 0.6845        0.3475
D =
    - 9.2213          0            0
           0    18.4308            0
           0          0       8.7905
```

Example EN3.47. *Transform the matrix* $A = \begin{bmatrix} 4 & 3 & 2 & 1 \\ 3 & 4 & 1 & 2 \\ 2 & 1 & 4 & 3 \\ 1 & 2 & 3 & 4 \end{bmatrix}$ *into tridiagonal form using Householder reduction. Also determine the transformation matrix.*

Solution:

Suppose that A is a symmetric $n \times n$ matrix.

Start with $A_0 = A$

Construct the sequence $\mathbf{P}_1, \mathbf{P}_2, ..., \mathbf{P}_{n-1}$ of Householder matrices, so that

$$\mathbf{A_k} = \mathbf{P_k}\, \mathbf{A_{k-1}}\, \mathbf{P_k} \quad \text{for } k = 1, 2, ..., n-2$$

where A_k has zeros below the subdiagonal in columns 1, 2, ..., k. Then A_{n-2} is a symmetric tridiagonal matrix that is similar to A. This process is called Householder's method.

```
%Input - A is an nxn symmetric matrix
A=[4 3 2 1;3 4 1 2;2 1 4 3;1 2 3 4];
%Output - T is a tridiagonal matrix
[n,n]=size(A);
for k=1:n-2
s=norm(A(k+1:n,k));
if (A(k+1,k)<0)
s=-s;
end
r=sqrt(2*s*(A(k+1,k)+s));
W(1:k)=zeros(1,k);
W(k+1)=(A(k+1,k)+s)/r;
W(k+2:n)=A(k+2:n,k)'/r;
V(1:k)=zeros(1,k);
V(k+1:n)=A(k+1:n,k+1:n)*W(k+1:n)';
```

```
c=W(k+1:n)*V(k+1:n)';
Q(1:k)=zeros(1,k);
Q(k+1:n)=V(k+1:n)-c*W(k+1:n);
A(k+2:n,k)=zeros(n-k-1,1);
A(k,k+2:n)=zeros(1,n-k-1);
A(k+1,k)=-s;
A(k,k+1)=-s;
A(k+1:n,k+1:n)=A(k+1:n,k+1:n) ...
  - 2*W(k+1:n)'*Q(k+1:n) - 2*Q(k+1:n)'*W(k+1:n);
end
T=A;
fprintf('Matrix in tridiagonal form is\n');
disp(T);
```

The output of the program is given below:

Matrix in tridiagonal form is

4.0000	– 3.7417	0	0
– 3.7417	6.5714	2.7180	0
0	2.7180	3.0529	1.2403
0	0	1.2403	2.3757

Example EN3.48. *Use the Sturm sequence property to find the interval of the smallest eigenvalue of* $A = \begin{bmatrix} 2 & -1 & 0 & 0 \\ -1 & 2 & -2 & 0 \\ 0 & -2 & 2 & -1 \\ 0 & 0 & -1 & 2 \end{bmatrix}$

Solution:

The sequence $\{f_k(a)\}$ and $\{f_k(b)\}$ can be used to determine the number of roots of $f_n(\lambda)$, which are contained in $[a, b]$.

The sequence $\{f_0, f_1, \ldots, f_n\}$ forms a *Sturm sequence* of polynomials, and such sequences have special properties. Given a point b, calculate

$$\{f_0,(b), f_1(b), \ldots, f_n(b)\}^1$$

and observe the signs of these quantities. If some $f_j(\lambda) = 0$, then choose the sign of $f_j(\lambda)$ to be opposite to that of $f_{j-1}(\lambda)$. It can be shown that $f_j(\lambda) = 0 \;\Rightarrow\; f_{j-1}(\lambda) \neq 0$. Having obtained a sequence of signs from (λ) let $s(\lambda)$ denote the number of agreements of sign between consecutive members of the sign sequences.

The MATLAB program for this is presented below:

```
%%%%Given the diagonals c and d of A = [c\d\c] and the value of λ, this
%%%%function returns the sturm sequence P0(λ), P1(λ), P2(λ),.......Pn(λ). Note
%%%%that Pn(λ)=|A − λI|
A=[2 -1 0 0;-1 2 -2 0;0 -2 2 -1;0 0 -1 2];
d=diag(A)';
c=[-1 -2 -1];
lam=input('Enter guess value lambda\n');
n=length(d)+1;
p=ones(n,1);
p(2)=d(1)-lam;
for i=2:n-1
```

```
    p(i+1)=(d(i)-lam)*p(i)-(c(i-1)^2)*p1(i-1);
end
fprintf('sturm sequence p(%f) is\n',lam);
disp(p);
%%% number of sign changes in the sturm sequence p is
%%% number eigenvalues of matrix A that are smaller than
%%% lam_min
n=length(p);
oldsign=1;
num_eval=0;
for i=2:n
    psign=sign(p(i));
    if psign==0  psign=-oldsign;
    end
    if psign*oldsign <0
       num_eval=num_eval+1;
    end
    oldsign=psign;
end
fprintf('Number of eigenvalues less than lamda=%f are %d\n',lam,num_eval);
-------------------------------------------------------
Output of the program is as follows (in two runs)
>>Enter guess value lambda
0
sturm sequence p(0.000000) is
     1
     2
     4
     0
   - 4
Number of eigenvalues less than lamda = 0.000000 are 1
» sturm
Enter guess value lambda
-1
sturm sequence p(- 1.000000) is
     1
     3
     9
    19
    53
Number of eigenvalues less than lamda = -1.000000 are 0
Hence lowest eigen value lies between 0 and -1.
```

Example EN3.49. *Use Gaussian elimination scheme to solve the set of equations:*

$$2x_1 + x_2 - 3x_3 = 11$$
$$4x_1 - 2x_2 + 3x_3 = 8$$
$$-2x_1 + 2x_2 - x_3 = -6$$

Solution:

Writing the equation in the form $[A]X = B$ and applying forward elimination and back-substitution we obtain

$$U = \begin{bmatrix} 2 & 1 & -3 \\ 0 & -4 & 9 \\ 0 & 0 & 2.75 \end{bmatrix} \text{ and right hand side } = \begin{bmatrix} 11 \\ -14 \\ -5 \end{bmatrix}$$

Finally, the solution from back substitution becomes $X = \begin{bmatrix} 3 \\ -1 \\ -2 \end{bmatrix}$

The complete MATLAB program is given below:

```
%%%%%%%%%%%%%%%%%%%%%%%%%%%%%%
%          Ax = b
%    A - matrix for the left hand side.
%    b - vector for the right hand side
% This performs Gaussian elminiation to find x.
% MATRIX DEFINITION
A=[2 1 -3;4 -2 3;-2 2 -1];
b=[11;8;-6];
N = max(size(A));
% Perform Gaussian (FORWARD) elimination
 for j=2:N,
     for i=j:N,
        m = A(i,j-1)/A(j-1,j-1);
        A(i,:) = A(i,:) - A(j-1,:)*m;
        b(i) = b(i) - m*b(j-1);
     end
 end
disp('Upper triangular form of given matrix is=')
disp(A);
disp('b=');
disp(b);
% BACK-SUBSTITUTION
% Perform back substitution
 x = zeros(N,1);
 x(N) = b(N)/A(N,N);
 for j=N-1:-1:1,
    x(j)=(b(j)-A(j,j+1:N)*x(j+1:N))/A(j,j);
 end
disp('final solution is');disp(x);
%%%%%%%%%%%%%%%%%%%%%%%%%%%%%%%%%%%%%
```

Output is as follows:

The final solution is

```
      3
```

```
    - 1
    - 2
```

Check with MATLAB built-in function:

```
>> A=[2 1 -3;4 -2 3;-2 2 -1];
b=[11;8;-6];
>> x=A\b
x =
      3
     -1
     -2
```

Example EN3.50. *Solve the system of linear equations by Gaussian elimination method:*

$$6x_1 + 3x_2 + 6x_3 = 30$$
$$2x_1 + 3x_2 + 3x_3 = 17$$
$$x_1 + 2x_2 + 2x_3 = 11$$

Solution:

Writing the equation in the form $[A]X = B$ and apply forward elimination and back-substitution

The complete MATLAB program and output are given below:

```
%   A - matrix for the left hand side.
%   b - vector for the right hand side
% This performs Gaussian elminiation to find x.
% MATRIX DEFINITION
A=[6 3 6;2 3 3;1 2 2];
b=[30;17;11];
%%%%%%%%%%%%%%%%%%%%%%%%%%%%%%%%
N = max(size(A))
% Perform Gaussian Elimination
 for j=2:N,
     for i=j:N,
        m = A(i,j-1)/A(j-1,j-1);
        A(i,:) = A(i,:) - A(j-1,:)*m;
        b(i) = b(i) - m*b(j-1);
     end
 end
disp('Upper triangular form of given matrix is=')
disp(A)
disp('b=')
disp(b)
% BACK-SUBSTITUTION
% Perform back substitution
 x = zeros(N,1);
 x(N) = b(N)/A(N,N);
 for j=N-1:-1:1,
   x(j)=(b(j)-A(j,j+1:N)*x(j+1:N))/A(j,j);
 end
```

```
disp('final solution is');
disp(x);
disp('matlab solution is');
x=inv(A)*b
```

OUPTUT is given below:

```
------------------------------------------------
N =
   3
Upper triangular form of given matrix is=
        6.0000        3.0000        6.0000
             0        2.0000        1.0000
             0             0        0.2500
b=
       30.0000
        7.0000
        0.7500
```

The final solution is

```
          1
          2
          3
```

MATLAB solution is

```
x =
     1
     2
     3
```

Check with MATLAB built-in function:

```
>> A=[6 3 6;2 3 3;1 2 2];
b=[30;17;11];
>>
>> x=A\b
x =
     1
     2
     3
```

Example EN3.51. *Using Cholesky's method solve the following linear equations:*

$$x_1 + x_2 + x_3 = 7$$
$$3x_1 + 3x_2 + 4x_3 = 23$$
$$2x_1 + x_2 + x_3 = 10$$

Solution:

The complete program and output are given below:

```
%%%%%%%%%%%%%%%%%%%%%%%%
A=[1 1 1;3 3 4;2 1 1];
b=[7;23;10];
[L,U]=lu(A);
% solution of y
```

```
y=L\b;
%final solution x
x=U\y;
fprintf('Solution of the equations is\n');
disp(x)
%%%%%%%%%%%%%%%%%%%%%%
```

Solution of the equations is

```
    3.0000
    2.0000
    2.0000
```

Check with MATLAB built-in function:

```
>> A=[1 1 1;3 3 4;2 1 1];
b=[7;23;10];
>> x=A\b
x =
    3.0000
    2.0000
    2.0000
```

Example EN3.52. *Solve the system of equations by Cholesky's factorization method:*

$$12x_1 - 6x_2 - 6x_3 - 1.5x_4 = 1$$
$$-6x_1 + 4x_2 + 3x_3 + 0.5x_4 = 2$$
$$-6x_1 + 3x_2 + 6x_3 + 1.5x_4 = 3$$
$$-1.5x_1 + 0.5x_2 + 1.5x_3 + x_4 = 4$$

Solution:

Here the matrix [*A*] is symmetric.

Program and the output are given below:

```
%%%%%%%%%%%%%%%%%%%%%%
A=[12 -6 -6 -1.5;-6 4 3 0.5;-6 3 6 1.5;-1.5 0.5 1.5 1];
b=[1;2;3;4];
[L,U]=lu(A);
% solution of y
y=L\b;
%final solution x
x=U\y;
fprintf('Solution of the equations is\n');
disp(x)
%%%%%%%%%%%%%%%%%%%%%%
```

The solution of the equations is

```
     2.7778
     4.2222
   - 0.5556
     6.8889
```

Check with MATLAB built-in function:

```
>> A=[12 -6 -6 -1.5;-6 4 3 0.5;-6 3 6 1.5;-1.5 0.5 1.5 1];
b=[1;2;3;4];
```

```
>> x=A\b
x =
     2.7778
     4.2222
   - 0.5556
     6.8889
```

Example EN3.53. *Find the solution to the equations using Gauss-Seidel method*

$$\begin{bmatrix} 4 & 2 & 0 & 0 \\ 2 & 8 & 2 & 0 \\ 0 & 2 & 8 & 2 \\ 0 & 0 & 2 & 4 \end{bmatrix} \begin{Bmatrix} x_1 \\ x_2 \\ x_3 \\ x_4 \end{Bmatrix} = \begin{Bmatrix} 4 \\ 0 \\ 0 \\ 0 \end{Bmatrix}$$

Solution:

The complete MATLAB program is given below with outputs of the program

```
%%%%%%%%%%%%%%%%
A=[4 2 0 0;2 8 2 0;0 2 8 2;0 0 2 4];b=[4;0;0;0];
X0=zeros(size(b)); % starting vector
tole=1e-6;kstop=30;% error tolerance and max. iterations
[n,n]=size(A);
P=tril(A);% lower triangular form
k=0;r=b-A*X0;
r0=norm(r);er=norm(r);
X=X0;
[L,U]=lu(P);
fprintf('iter#\tX(1)\t\tX(2)\t\tX(3)\t\tX(4)\n');
while er>tole & k<kstop
     fprintf('%d\t%f\t%f\t%f\t%f\n',k,X(1),X(2),X(3),X(4));
   k=k+1;
   dx=L\r;
   dx=U\dx;
   X=X+dx;
   r=b-A*X;
   er=norm(r)/r0;
   erp(k)=norm(r)/r0;
end
X
%%%%%%%%%%%%%%%%
```

The output as follows:

```
X =
       1.1556
     - 0.3111
       0.0889
     - 0.0444
```

Check with MATLAB built-in function:

```
>> A=[4 2 0 0;2 8 2 0;0 2 8 2;0 0 2 4];b=[4;0;0;0];
```

```
>> x=A\b
x =
     1.1556
   - 0.3111
     0.0889
   - 0.0444
```

Example EN3.54. *Solve the system of equations given by [A]{X} = {b} using Gauss-Seidel method. The matrices [A] and {b} are given below:*

$$[A] = \begin{bmatrix} 4 & 2 & 0 & 0 \\ 2 & 8 & 2 & 0 \\ 0 & 2 & 8 & 2 \\ 0 & 0 & 2 & 4 \end{bmatrix} \quad \text{and} \quad \{b\} = \begin{Bmatrix} 4 \\ 0 \\ 0 \\ 14 \end{Bmatrix}$$

Solution:

The following program and output are presented.

```
%%%%%%%%%%%%%%%%%%%%%%%%%%%%%%%%
A=[4 2 0 0;2 8 2 0;0 2 8 2;0 0 2 4];b=[4;0;0;14];
X0=zeros(size(b)); % starting vector
tole=1e-6;kstop=30;% error tolerance and max. iterations
[n,n]=size(A);
P=tril(A);% lower triangular form
k=0;r=b-A*X0;
r0=norm(r);er=norm(r);
X=X0;
[L,U]=lu(P);
fprintf('iter#\tX(1)\t\tX(2)\t\tX(3)\t\tX(4)\n');
while er>tole & k<kstop
     fprintf('%d\t%f\t%f\t%f\t%f\n',k,X(1),X(2),X(3),X(4));
    k=k+1;
    dx=L\r;
    dx=U\dx;
    X=X+dx;
    r=b-A*X;
    er=norm(r)/r0;
    erp(k)=norm(r)/r0;
end
X
%%%%%%%%%%%%%%%%%%%%%%%%%%%%%%%%
```

The output is as follows:

```
X =
     1.0000
   - 0.0000
   - 1.0000
     4.0000
```

Check with MATLAB built-in function:

```
>> A=[4 2 0 0;2 8 2 0;0 2 8 2;0 0 2 4];b=[4;0;0;14];
>> x=A\b
x =
        1.0000
          0
      - 1.0000
        4.0000
```

Example EN3.55. *Find the minimum of the function* $f(X) = 4x_1^2 + 3x_2^2 - 5x_1x_2 - 8x_1$ *starting from (0, 0) using Powell's method.*

Solution:

The process is shown in Fig. EN3.55.

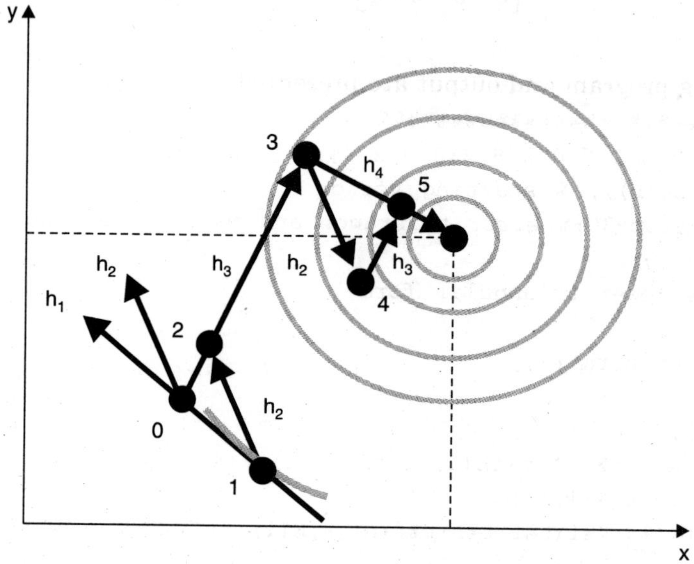

Fig. EN3.55

Complete program is as follows:

```
global  X V
X=[-1;1];    %STARTING POINT
N=length(X);% number of design variables
df=zeros(N,1);% decreases of f stored here
u=eye(N);  % columns of u store search directions V
n=30;  % Number of cycles
for j=1:n
  xold=X;
   fold=4*xold(1)^2+3*xold(2)^2-5*xold(1)*xold(2)-8*xold(1);
  % FIRST N line searches record the decrease of f
  for i=1:N
   V=u(1:N,i);
   [s fmin]=fminbnd('fline',0,10);
   %Golden section search built in function
```

```
    df(i)=fold-fmin;
    fold=fmin;
    X=X+s*V;
  end
  % LAST LINE SEARCH IN THE CYCLE
  V=X-xold;
  [s fmin]=fminbnd('fline',0,10);
  X=X+s*V;
  % IDENTIFY BIGGEST DECREASE OF F
  % AND UPDATE SEARCH DIRECTIONS
  imax=1; dfmax=df(1);
  for i=2:N
    if df(i)>dfmax    imax=i; dfmax=df(i); end
  end
  for i=imax:N-1
    u(1:N,i)=u(1:N,i+1);
  end
  u(1:N,N)=V;
end
fprintf('Optimum point after %d cycles is\n',n);
fprintf('%f\n%f\n',X(1),X(2));
%%%%%%%%%%%%%%%%%%%%%%%%%%%%%
Function is
%%%%%%%%%%%%%%%%%%%%%%%%%%%%
function z=fline(s)
global X V
b1=V(1);
b2=V(2);
a1=X(1);a2=X(2);
p1=a1+s*b1;p2=a2+s*b2;
z=4*p1^2+3*p2^2-5*p1*p2-8*p1;
%%%%%%%%%%%%%%%%%%%%%%%%%%%
```

Optimum point after 30 cycles is

```
2.091345
1.739733
```

Actual output from MATLAB function fminsearch('*fxy*', [0 0]) is

```
2.0870
1.7392
```

Example EN3.56. *Find the minimum value of the function* $f(X) = 7x_1^2 + 5x_2^2 - 8x_1x_2 - 5x_1$ *starting from (0, 0) using Powell's method.*

Solution:

The following program is used

```
%%%%%%%%%%%%%%%%%%%%%%%%%%%%%%%%%%%
global  X V
X=[0;0];     %STARTING POINT
N=length(X);% number of design variables
```

```
df=zeros(N,1);% decreases of f stored here
u=eye(N);  % columns of u store search directions V
n=30;  % Number of cycles
for j=1:n
  xold=X;
   fold=7*xold(1)^2+5*xold(2)^2-8*xold(1)*xold(2)-5*xold(1);
  % FIRST N line searches record the decrease of f
  for i=1:N
   V=u(1:N,i);
    [s fmin]=fminbnd('fline',0,10);
    %Golden section search built in function
    df(i)=fold-fmin;
    fold=fmin;
   X=X+s*V;
  end
  % LAST LINE SEARCH IN THE CYCLE
  V=X-xold;
   [s fmin]=fminbnd('fline',0,10);
  X=X+s*V;
  % IDENTIFY BIGGEST DECREASE OF F
  % AND UPDATE SEARCH DIRECTIONS
  imax=1; dfmax=df(1);
  for i=2:N
    if df(i)>dfmax    imax=i; dfmax=df(i); end
  end
  for i=imax:N-1
    u(1:N,i)=u(1:N,i+1);
  end
  u(1:N,N)=V;
end
fprintf('Optimum point after %d cycles is\n',n);
fprintf('%f\n%f\n',X(1),X(2));
%%%%%%%%%%%%%%%%%%%%%%%%%%%%%%%%%%%%%%%
Function used is as follows:
%%%%%%%%%%%%%%%%%%%%%%%%%%%%%%%%%%%%%%%
function z=fline(s)
global X V
a1=X(1);a2=X(2);
b1=V(1);
b2=V(2);
p1=a1+s*b1;p2=a2+s*b2;
z=7*p1^2+5*p2^2-8*p1*p2-5*p1;
```

Optimum point after 30 cycles is

```
0.657916
0.528062
```

The actual output from MATLAB fminsearch function is

```
0.6579
0.5263
```

Example EN3.57. *Repeat problem EN3.55 using Fletcher-Reeves method.*

Solution:

Given $f(x_1,x_2) = 4x_1^2 + 3x_2^2 - 5x_1x_2 - 8x_1$

Here gradient $df = [8x_1 - 5x_2 - 8;6x_2 - 5x_1]^T$

Complete program is given below:

```
%%%%%% CONJUGATE GRADIENT METHOD%%%%%%%%%%
global X V
X=[0;0];
N=length(X);
g0=-feval('df',X);
V=g0;
n=50;
for i=1:n
   [s fmin]=fminbnd('fline',-10,10);
  X=X+s*V
  g1=-feval('df',X);
   gamma=(norm(g1)/norm(g0))^2;
  V=g1+gamma*V;
  g0=g1;
end
fprintf('Minimum point for this function is \n');
fprintf('%f\n%f\n',X(1),X(2));
%%%%%%%%%%
```

This requires the following function files

```
%%%%%%%LINEAR FUNCTION OF s %%%%%%%%
function z=fline(s)
global V X
p1=X(1)+s*V(1);
p2=X(2)+s*V(2);
z=4*p1^2+3*p2^2-5*p1*p2-8*p1;
%%%%%%GRADIENT FUNCTION %%%%%%%%%%%%
function y=df(X)
y=[8*X(1)-5*X(2)-8;6*X(2)-5*X(1)];
%%%%%%%%%%%%%%%%%%
```

The output is as follows:

```
Minimum point for this function is
2.086957
1.739130
```

The actual output is $[2.087\ 1.7392]^T$

Example EN3.58. *Repeat problem EN3.56 using Fletcher-Reeves method.*

Solution:

Given $f(x_1, x_2) = 7x_1^2 + 5x_2^2 - 8x_1x_2 = 5x_1$

Here gradient $df = [14x_1 - 8x_2 - 5; 10x_2 - 8x_1]^T$

Complete program is as follows:

```
%%%%%%%%%%%%%%%
global X V
X=[0;0];
N=length(X);
g0=-feval('df',X);
V=g0;
n=50;
for i=1:n
   [s fmin]=fminbnd('fline',-10,10);
   X=X+s*V
   g1=-feval('df',X);
   gamma=(norm(g1)/norm(g0))^2;
   V=g1+gamma*V;
   g0=g1;
end
fprintf('Minimum point for this function is \n');
fprintf('%f\n%f\n',X(1),X(2));
%%%%%%%%%%%%%%%%%%
Functions are
%%%%%%%%%%%%%
function z=fline(s)
global V X
p1=X(1)+s*V(1);
p2=X(2)+s*V(2);
z=7*p1^2+5*p2^2-8*p1*p2-5*p1;
%%%%%%GRADIENT FUNCTION %%%%%%%%%%%
function y=df(X)
y=[14*X(1)-8*X(2)-5; 10*X(2)-8*X(1)];
%%%%%%%%%%%%%%%%%%
```

The output is as follows:

Minimum point for this function is

```
0.657895
0.526316
```

The actual output from MATLAB function is $[0.6579\ 0.5263]^T$.

Example EN3.59. *Use the Jacobi's method to determine the eigenvalues and eigenvectors of matrix* $[A] = \begin{bmatrix} 4 & -1 & -2 \\ -1 & 3 & 3 \\ -2 & 3 & 1 \end{bmatrix}$.

Solution:

The complete computer program is given below:

```
A=[4 -1 -2;-1 3 3;-2 3 1];
%Output - V is the nxn matrix of eigenvectors
%       - D is the diagonal nxn matrix of eigenvalues
D=A;
[n,n]=size(A);
V=eye(n);
%Calculate row p and column q of the off-diagonal element
%of greatest magnitude in A
[m1 p]=max(abs(D-diag(diag(D))));
[m2 q]=max(m1);
p=p(q);
i=1;
while(i<20)
    %Zero out Dpq and Dqp
    t=D(p,q)/(D(q,q)-D(p,p));
    c=1/sqrt(t^2+1);
    s=c*t;
    R=[c s;-s c];
    D([p q],:)=R'*D([p q],:);
    D(:,[p q])=D(:,[p q])*R;
    V(:,[p q])=V(:,[p q])*R;
     [m1 p]=max(abs(D-diag(diag(D))));
     [m2 q]=max(m1);
    p=p(q);
    i=i+1;
end
D=diag(diag(D))
fprintf('Final eigenvalues are %f\t%f\t%f\n',D(1,1),D(2,2),D(3,3));
```

The output is as follows:

```
D =
    2.6916         0          0
         0    6.6956          0
         0         0   - 1.3872
```

The final eigenvalues are 2.691611, 6.695589 and – 1.387200

Check with MATLAB built-in function:

```
>> A=[4 -1 -2;-1 3 3;-2 3 1];
>> [Q,D]=eig(A)
Q =
    0.2114    0.7636   - 0.6102
  - 0.5184    0.6168     0.5923
    0.8286    0.1911     0.5262
```

```
D =
  - 1.3872          0          0
        0      2.6916          0
        0           0     6.6956
```

Example EN3.60. *Find the eigenvalues and eigenvectors of* $[A] = \begin{bmatrix} 6 & -2 & 1 & -1 \\ -2 & 4 & -2 & 1 \\ 1 & -2 & 4 & -2 \\ -1 & 1 & -2 & 4 \end{bmatrix}$ *with the Jacobi's method.*

Solution:

The following MALTAB program is used for this:

```
A=[6 -2 1 -1;-2 4 -2 1;1 -2 4 -2;-1 1 -2 4];
%Output - V is the nxn matrix of eigenvectors
%       - D is the diagonal nxn matrix of eigenvalues
D=A;
[n,n]=size(A);
V=eye(n);
%Calculate row p and column q of the off-diagonal element
%of greatest magnitude in A
[m1 p]=max(abs(D-diag(diag(D))));
[m2 q]=max(m1);
p=p(q);
i=1;
while(i<100)
   %Zero out Dpq and Dqp
   t=D(p,q)/(D(q,q)-D(p,p));
   c=1/sqrt(t^2+1);
   s=c*t;
   R=[c s;-s c];
   D([p q],:)=R'*D([p q],:);
   D(:,[p q])=D(:,[p q])*R;
   V(:,[p q])=V(:,[p q])*R;
   [m1 p]=max(abs(D-diag(diag(D))));
   [m2 q]=max(m1);
   p=p(q);
   i=i+1;
end
D=diag(diag(D))
fprintf('Eigenvectors are\n')
disp(V)
```

The output is as follows:

```
D =
   9.1025         0         0         0
        0    1.5186         0         0
```

```
         0              0           4.5880            0
         0              0              0          2.7910
```

The eigenvectors are

```
V =
     0.6043       0.1788      - 0.7250      - 0.2778
   - 0.5006       0.5421      - 0.0252      - 0.6744
     0.4721       0.7046        0.4915        0.1976
   - 0.4016       0.4215      - 0.4818        0.6550
```

Check with MATLAB built-in function:

```
>> A=[6 -2 1 -1;-2 4 -2 1;1 -2 4 -2;-1 1 -2 4];
>> [Q,D]=eig(A)
Q =
   - 0.1788     - 0.2778        0.7250      - 0.6043
   - 0.5421     - 0.6744        0.0252        0.5006
   - 0.7046       0.1976      - 0.4915      - 0.4721
   - 0.4215       0.6550        0.4818        0.4016

D =
     1.5186          0             0             0
         0       2.7910            0             0
         0          0           4.5880           0
         0          0              0         9.1025
```

Example EN3.61. *Transform the matrix* $[A] = \begin{bmatrix} 7 & 2 & 3 & -1 \\ 2 & 8 & 5 & 1 \\ 3 & 5 & 12 & 9 \\ -1 & 1 & 9 & 7 \end{bmatrix}$ *into tridiagonal form using Householder reduction.*

Solution:

The following program is used.

```
%Input - A is an nxn symmetric matrix
A=[7 2 3 -1;2 8 5 1;3 5 12 9;-1 1 9 7];
%Output - T is a tridiagonal matrix
[n,n]=size(A);
for k=1:n-2
   s=norm(A(k+1:n,k));
   if (A(k+1,k)<0)
      s=-s;
   end
   r=sqrt(2*s*(A(k+1,k)+s));
   W(1:k)=zeros(1,k);
   W(k+1)=(A(k+1,k)+s)/r;
   W(k+2:n)=A(k+2:n,k)'/r;
   V(1:k)=zeros(1,k);
   V(k+1:n)=A(k+1:n,k+1:n)*W(k+1:n)';
   c=W(k+1:n)*V(k+1:n)';
```

```
        Q(1:k)=zeros(1,k);
        Q(k+1:n)=V(k+1:n)-c*W(k+1:n);
        A(k+2:n,k)=zeros(n-k-1,1);
        A(k,k+2:n)=zeros(1,n-k-1);
        A(k+1,k)=-s;
        A(k,k+1)=-s;
        A(k+1:n,k+1:n)=A(k+1:n,k+1:n)-2*W(k+1:n)'*Q(k+1:n)-2*Q(k+1:n)'*W(k+1:n);
end
T=A;
fprintf('Matrix in tridiagonal form is\n');
disp(T)
```

The output is as follows:

The matrix in tridiagonal form is

```
    7.0000    - 3.7417         0          0
  - 3.7417     10.6429     9.1309         0
         0      9.1309    10.5942    4.7716
         0           0     4.7716    5.7629
```

Example EN3.62. *Determine the transformation matrix in Problem PN3.13 above.*

Solution:

```
>> disp(V)
V =
         0      0.7071   - 0.4082         0
         0      0.7071     0.4082         0
  - 7.6079           0     0.8165         0
 - 10.7820           0          0    1.0000
```

3.16 SUMMARY

The classical methods of control systems engineering using MATLAB including the root locus analysis and design, Routh Hurwitz stability analysis, frequency response methods of analysis, Bode, Nyquist, and Nichols plots, steady-state error analysis, second order systems approximations, phase and gain margin and bandwidth, state space variable method, and controllability and observability are covered in this chapter. With this foundation of basic application of MATLAB, the Chapter provides opportunities to explore advanced topics in control systems engineering. With this foundation of basic application of MATLAB, this Chapter also provides opportunities to explore advanced topics in engineering: mechanics, vibration analysis, electrical circuits, and numerical methods. Extensive worked examples are included with a significant number of exercise problems to guide the student to understand and as an aid for learning about the analysis and design of control systems, engineering mechanics, vibration analysis of mechanical systems, electrical circuits, and numerical methods using MATLAB.

PROBLEMS

P3.1 [Reduction of multiple subsystems]

Reduce the system shown in Fig. P3.1 to a single transfer function, $T(s) = C(s)/R(s)$ using MATLAB. The transfer functions are given as

$G_1(s) = 1/(s + 3)$
$G_2(s) = 1/(s^2 + 3s + 5)$
$G_3(s) = 1/(s + 7)$
$G_4(s) = 1/s$
$G_5(s) = 7/(s + 5)$
$G_6(s) = 1/(s^2 + 3s + 5)$
$G_7(s) = 5/(s + 6)$
$G_8(s) = 1/(s + 8)$

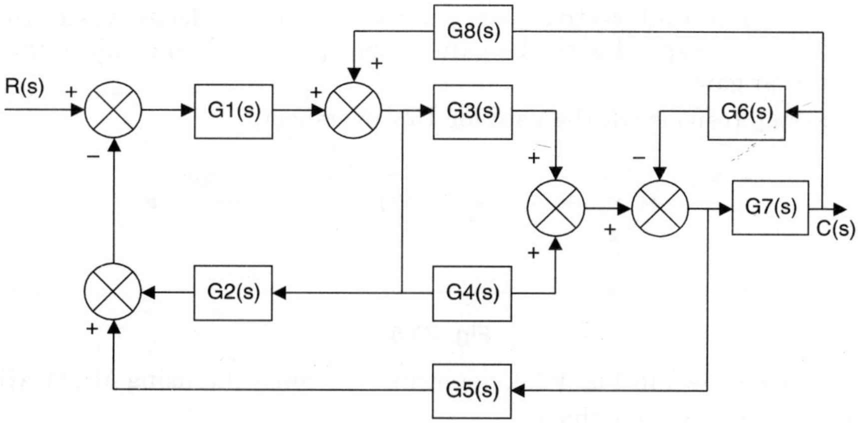

Fig. P3.1

P3.2 Obtain the unit-step response plot for the unity-feedback control system whose open loop transfer function is

$$G(s) = \frac{8}{s(s+1)(s+3)}$$

using MATLAB. Determine also the rise time, peak time, maximum overshoot, and settling time in the unit-step response plot.

P3.3 Obtain the unit-acceleration response curve of the unity-feedback control system whose open loop transfer function is given by

$$G(s) = \frac{8(s+1)}{s^2(s+3)}$$

using MATLAB. The unit-acceleration input is defined by

$$r(t) = \frac{1}{2}t^2 \; (t \geq 0)$$

P3.4 The feed forward transfer function $G(s)$ of a unity-feedback system is given by

$$G(s) = \frac{k(s+3)^2}{(s^2+5)(s+4)^2}$$

Plot the root loci for the system using MATLAB.

P3.5 For the unity feedback shown in Fig. P3.5, where

$$G(s) = \frac{k}{s(s+3)(s+4)(s+5)}$$

Obtain the following:

(*a*) display a root locus and pause

(*b*) draw a close-up of the root locus where the axes go from –2 to 0 on the real axis and – 2 to 2 on the imaginary axis

(*c*) overlay the 15% overshoot line on the close-up root locus

(*d*) allow you to select interactively the point where the root locus crosses the 15% overshoot line, and respond with the gain at that point as well as all of the closed-loop poles at that gain

(*e*) find the step response at the gain for 15% overshoot.

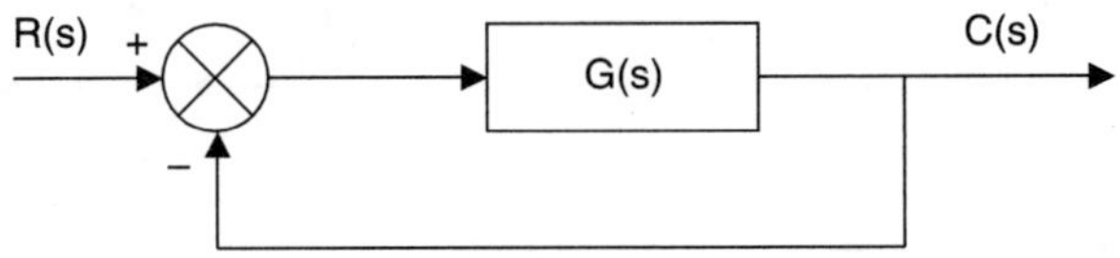

Fig. P3.5

P3.6 For the system shown in Fig. P3.6, determine the following using MATLAB

(*a*) display a root locus and phase

(*b*) display a close-up of the root locus where the axes go from –2 to 2 on the real axis and –2 to 2 on the imaginary axis

(*c*) overlay the 0.707 damping ratio line on the close-up root locus

(*d*) obtain the step response at the gain for 0.707 damping ratio

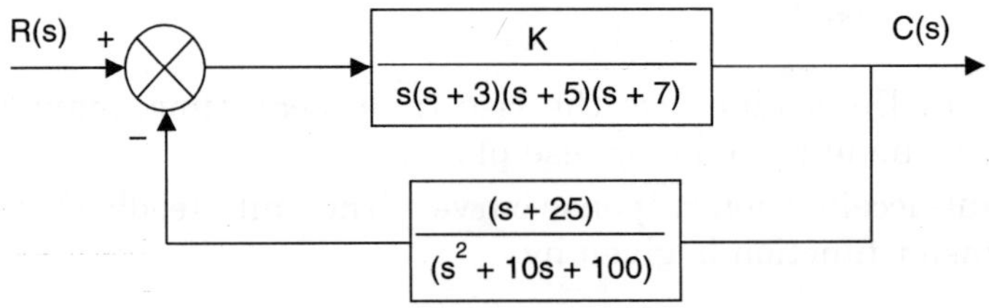

Fig. P3.6

P3.7 Write a program in MATLAB to obtain a Bode plot for the transfer function

$$G(s) = \frac{(5s^3 + 51s^2 + 20s + 400)}{(s^4 + 12s^3 + 60s^2 + 300s + 250)}$$

P3.8 Write a program in MATLAB for the unity feedback system with $G(s) = K/[s(s + 7)(s + 15)]$ so that the value of gain K can be input. Display the Bode plots of a system for the input value of K. Determine and display the gain and phase margin for the input value of K.

P3.9 Write a program in MATLAB for the system shown in Fig. P3.9 so that the value of K can be input ($K = 40$).

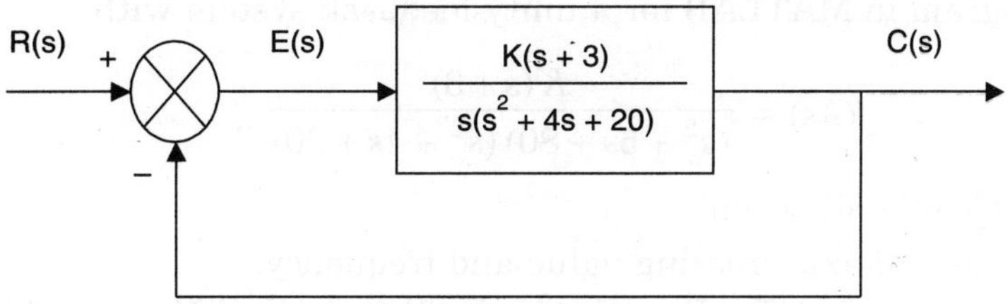

Fig. P3.9

(*a*) Display the closed-loop magnitude and phase frequency response for unity feedback system with an open-loop transfer function, *KG*(*s*).

(*b*) Determine and display the peak magnitude, frequency of the peak magnitude, and bandwidth for the closed-loop frequency response for the input value of *K*.

P3.10 Write a program in MATLAB for a unity feedback system with the forward-path transfer function given by

$$G(s) = \frac{7(s+3)}{s(s^2+4s+12)}$$

(*a*) Draw a Nichols plot of an open-loop transfer function

(*b*) The user can read the Nichols plot display and enter the value of M_p

(*c*) Obtain the closed-loop magnitude and phase plots.

(*d*) Display the expected values of percent overshoot, settling time, and peak time

(*e*) Plot the closed-loop step response.

P3.11 For the system shown in Fig. P3.11, write a program in MATLAB that will use an open-loop transfer function *G*(*s*).

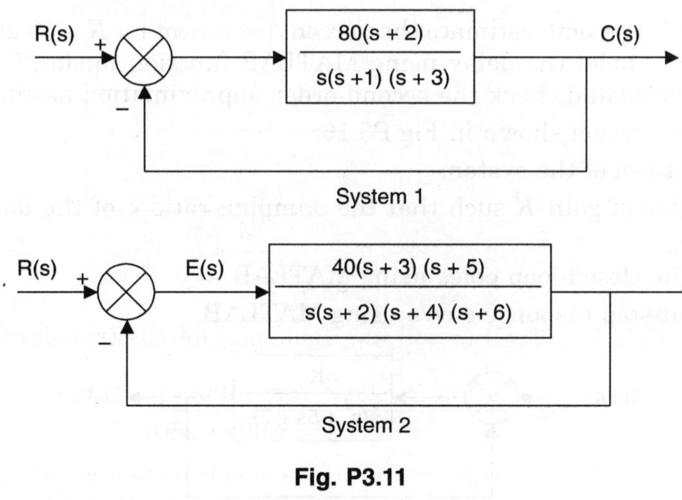

Fig. P3.11

(*a*) Obtain a Bode plot

(*b*) Estimate the percent overshoot, settling time, and peak time

(*c*) Obtain the closed-loop step response.

P3.12 Write a program in MATLAB for a unity-feedback system with

$$G(s) = \frac{K(s+3)}{(s^2+5s+80)(s^2+4s+20)}$$

(*a*) Plot the Nyquist diagram

(*b*) Display the real-axis crossing value and frequency.

P3.13 Write a program in MATLAB to obtain the Nyquist and Nichols plots for the following transfer function for $k = 30$.

$$G(s) = \frac{k(s+1)(s+2+5i)(s+2-5i)}{(s+2)(s+5)(s+7)(s+2+7i)(s+2-7i)}$$

P3.14 Write a program in MATLAB for a unity feedback system with the forward-path transfer function given by

$$G(s) = \frac{7(s+3)}{s(s^2+4s+12)}$$

(*a*) Draw a Nichols plot of an open-loop transfer function

(*b*) The user can read the Nichols plot display and enter the value of M_p

(*c*) Obtain the closed-loop magnitude and phase plots.

(*d*) Display the expected values of percent overshoot, settling time, and peak time

(*e*) Plot the closed-loop step response.

P3.15 For a unit feedback system with the forward-path transfer function

$$G(s) = \frac{K}{s(s+3)(s+10)}$$

and a delay of 0.5 second, estimate the percent overshoot for $K = 40$ using a second-order approximation. Model the delay using MATLAB function mode (T, n). Determine the unit step response and check the second-order approximation assumption mode.

P3.16 For the control system shown in Fig.P3.16:

(*a*) plot the root loci of the system

(*b*) find the value of gain K such that the damping ratio ξ of the dominant closed-loop poles is 0.5

(*c*) obtain all the closed-loop poles using MATLAB

(*d*) plot the unit-step response curve using MATLAB.

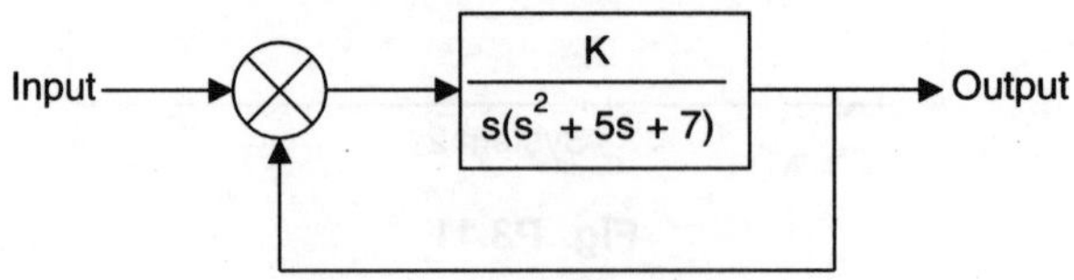

Fig. P3.16

P3.17 Fig. P3.17 shows a position control system with velocity feedback. What is the response $c(t)$ to the unit step input ?

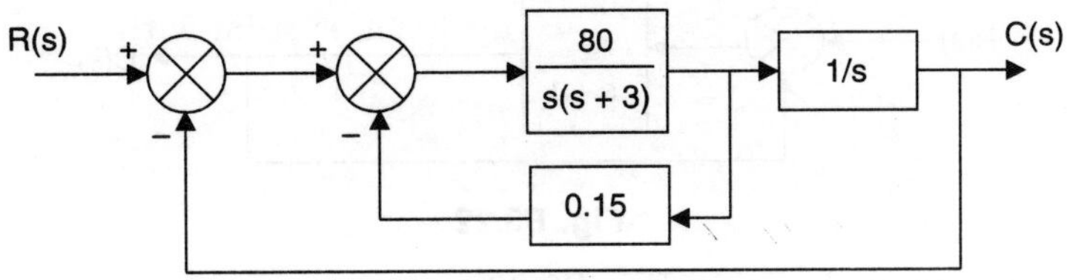

Fig. P3.17

P3.18 The open-loop transfer function G(s)H(s) of a control system is

$$G(s)H(s) = \frac{K}{s(s+0.5)(s^2+0.5s+8)} = \frac{K}{s^4+s^3+8.25s^2+4s}$$

Plot the root loci for the system using MATLAB.

P3.19 Design a compensator for the system shown in Fig.P3.19 such that the dominant closed-loop poles are located at $s = -1 \pm j\sqrt{3}$.

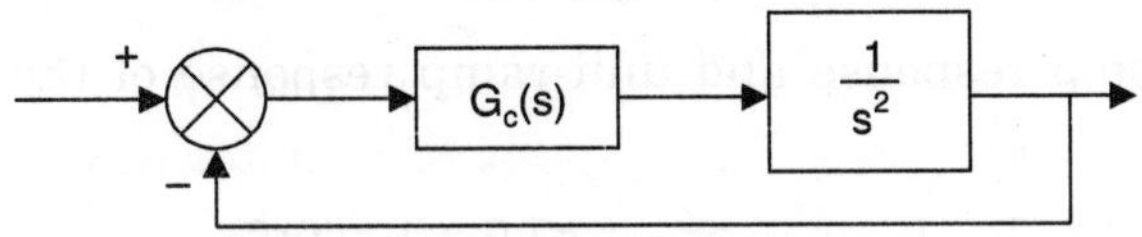

Fig. P3.19

P3.20 For the control system shown in Fig. 3.20:

(*a*) design a PID control $G_c(s)$ such that the dominant closed-loop poles located at s = – 1 ± *j*1.

(*b*) select a = 0.6 for the PID controller and find the values of *K* and *b*

(*c*) root-locus plot using MATLAB.

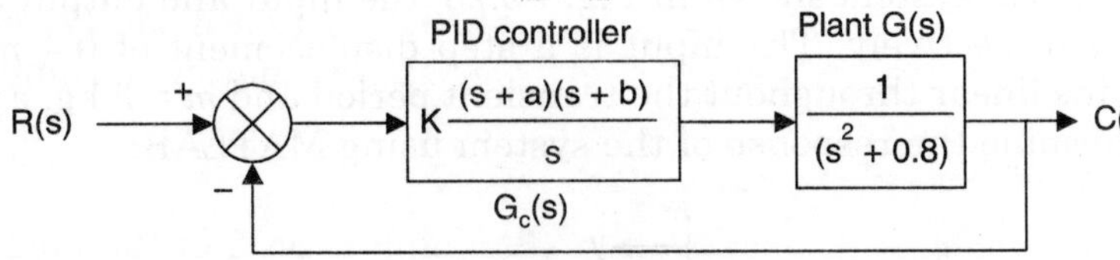

Fig. P3.20

P3.21 Draw a Bode diagram of the open-loop transfer function *G*(*s*) of the closed-loop system shown in Fig. P3.21 and obtain the phase margin and gain margin.

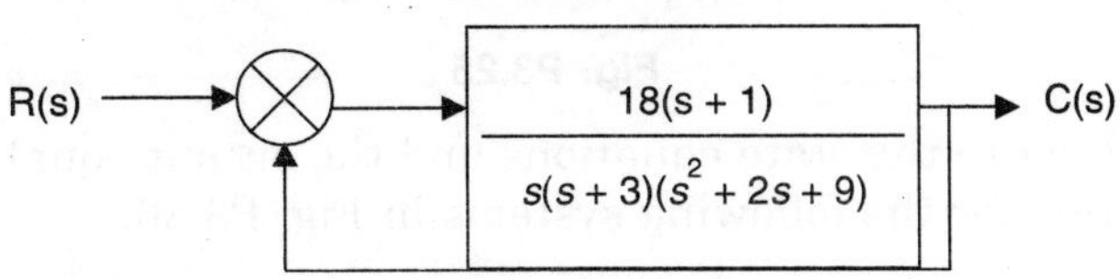

Fig. P3.21

P3.22 A block diagram of a process control system is shown in Fig. P3.22. Find the range of gain for stability.

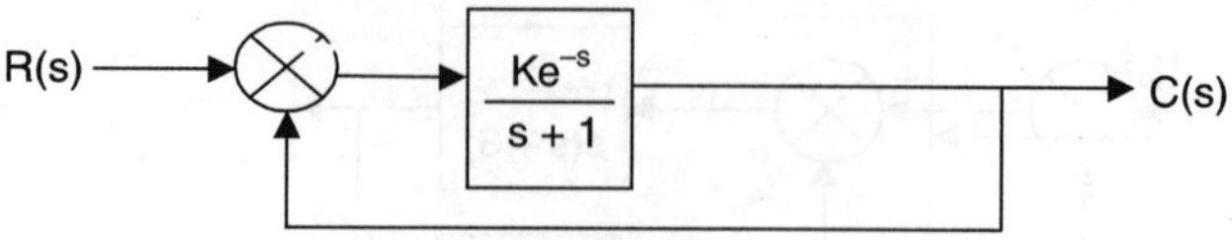

Fig. P3.22

P3.23 For the control system shown in Fig. P3.23:

(*a*) draw a Bode diagram of the open-loop transfer function

(*b*) find the value of the gain K such that the phase margin is 50°

(*c*) find the gain margin of the system with the gain obtained in (*b*).

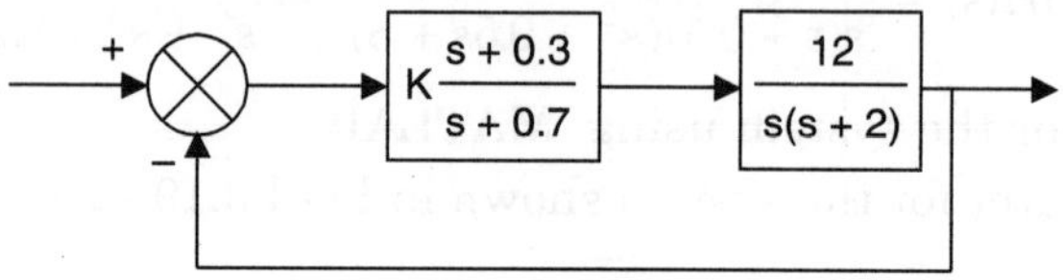

Fig. P3.23

P3.24 Obtain the unit-step response and unit-ramp response of the following system using MATLAB.

$$\begin{bmatrix} \dot{x}_1 \\ \dot{x}_2 \\ \dot{x}_3 \end{bmatrix} = \begin{bmatrix} -5 & -25 & -5 \\ 1 & 0 & 0 \\ 0 & 1 & 0 \end{bmatrix} \begin{bmatrix} x_1 \\ x_2 \\ x_3 \end{bmatrix} + \begin{bmatrix} 1 \\ 0 \\ 0 \end{bmatrix} u$$

$$y = [0 \quad 25 \quad 5] \begin{bmatrix} x_1 \\ x_2 \\ x_3 \end{bmatrix} + [0]u$$

P3.25 For the mechanical system shown in Fig. P3.25, the input and output are the displacement x and y respectively. The input is a step displacement of 0.4 m. Assuming the system remains linear throughout the transient period and m = 3 kg, c = 3 N-s/m, and k = 1 N/m, determine the response of the system using MATLAB.

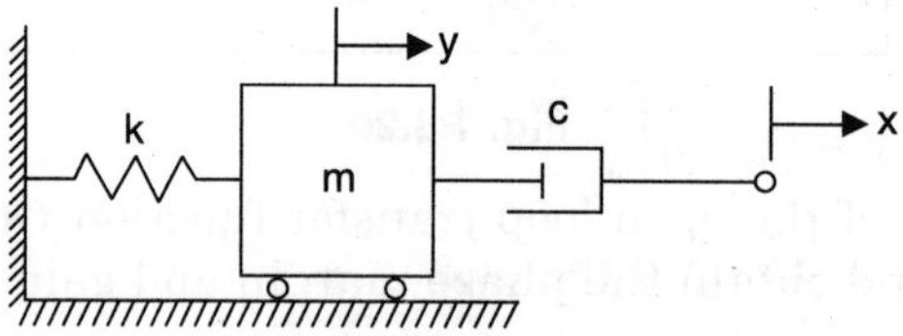

Fig. P3.25

P3.26 Using MATLAB, write the state equations and the output equation for the phase-variable representation for the following systems in Fig. P3.26.

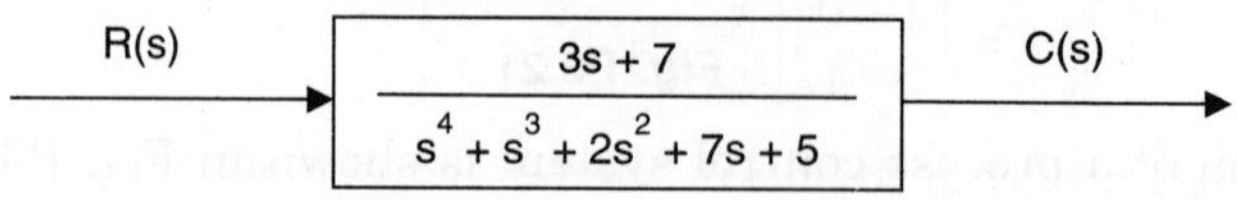

(*a*)

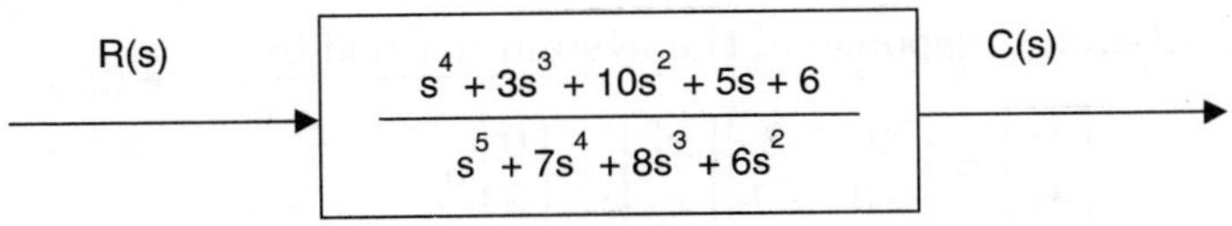

(*b*)

Fig. P3.26

P3.27 Determine the transfer function and poles of the system represented in state space as following using MATLAB.

$$\dot{x} = \begin{bmatrix} 9 & -5 & 2 \\ -4 & 1 & 0 \\ 3 & 5 & -7 \end{bmatrix} x + \begin{bmatrix} 2 \\ 5 \\ 7 \end{bmatrix} u(t)$$

$$y = [1 \quad 7 \quad -2]x; \quad x(0) = \begin{bmatrix} 0 \\ 0 \\ 0 \end{bmatrix}$$

P3.28 Obtain the root locus diagram of a system defined in state space using MATLAB. The system equations are

$$\dot{x} = Ax + Bu \quad \text{and} \quad y = Cx + Du \quad \text{and} \quad u = r - y$$

where r is the input and y is the output.

The matrices **A, B, C,** and **D** are:

$$A = \begin{bmatrix} 0 & 1 & 0 \\ 0 & 0 & 1 \\ -150 & -50 & -15 \end{bmatrix} \qquad B = \begin{bmatrix} 0 \\ 1 \\ -15 \end{bmatrix}$$

$$C = [1 \; 0 \; 0] \qquad D = [0]$$

P3.29 Obtain the Bode diagram of the following system using MATLAB.

$$\begin{bmatrix} \dot{x}_1 \\ \dot{x}_2 \end{bmatrix} = \begin{bmatrix} 0 & 1 \\ -30 & 7 \end{bmatrix} \begin{bmatrix} x_1 \\ x_2 \end{bmatrix} + \begin{bmatrix} 0 \\ 30 \end{bmatrix} u$$

$$y = [1 \quad 0] \begin{bmatrix} x_1 \\ x_2 \end{bmatrix}$$

The input of the system is u and the output is y.

P3.30 A control system is defined by

$$\begin{bmatrix} \dot{x}_1 \\ \dot{x}_2 \end{bmatrix} = \begin{bmatrix} -1 & -2 \\ 7.5 & 0 \end{bmatrix} \begin{bmatrix} x_1 \\ x_2 \end{bmatrix} + \begin{bmatrix} 1 & 1 \\ 1 & 0 \end{bmatrix} \begin{bmatrix} u_1 \\ u_2 \end{bmatrix}$$

$$\begin{bmatrix} y_1 \\ y_2 \end{bmatrix} = \begin{bmatrix} 1 & 0 \\ 0 & 1 \end{bmatrix} \begin{bmatrix} x_1 \\ x_2 \end{bmatrix} + \begin{bmatrix} 0 & 0 \\ 0 & 0 \end{bmatrix} \begin{bmatrix} u_1 \\ u_2 \end{bmatrix}$$

Write a MATLAB program to obtain the following plots:

(*a*) two Nyquist plots for the input u_1 in one diagram

(*b*) two Nyquist plots for the input u_2 in one diagram.

P3.31 Obtain the unit-ramp response of the system defined by

$$\begin{bmatrix} \dot{x}_1 \\ \dot{x}_2 \end{bmatrix} = \begin{bmatrix} 0 & 2 \\ -3 & -1 \end{bmatrix} \begin{bmatrix} x_1 \\ x_2 \end{bmatrix} + \begin{bmatrix} 0 \\ 2 \end{bmatrix} u$$

$$y = [1 \quad 0] \begin{bmatrix} x_1 \\ x_2 \end{bmatrix}$$

where u is the unit-ramp input. Use lsim command to obtain the response.

P3.32 Obtain the response curves $y(t)$ using MATLAB for the following system.

$$\begin{bmatrix} \dot{x}_1 \\ \dot{x}_2 \end{bmatrix} = \begin{bmatrix} -1 & 1 \\ -1 & 0 \end{bmatrix} \begin{bmatrix} x_1 \\ x_2 \end{bmatrix} + \begin{bmatrix} 0 \\ 2 \end{bmatrix} u$$

$$y = [1 \quad 0] \begin{bmatrix} x_1 \\ x_2 \end{bmatrix}$$

The input u is given by:

(*a*) u = unit-step input

(*b*) $u = e^{-t}$

The initial state $\mathbf{x}(0) = 0$

P3.33 Plot the step response using MATLAB for the following system represented in state space, where $u(t)$ is the unit step.

$$\dot{x} = \begin{bmatrix} -3 & 2 & 0 \\ 0 & -7 & 1 \\ 0 & 0 & -4 \end{bmatrix} x + \begin{bmatrix} 0 \\ 1 \\ 1 \end{bmatrix} u(t)$$

$$y = [0 \quad 1 \quad 1]x; \; x(0) = \begin{bmatrix} 0 \\ 0 \\ 0 \end{bmatrix}$$

P3.34 Diagonalize the following system using MATLAB.

$$\dot{x} = \begin{bmatrix} -10 & -5 & 7 \\ 15 & 4 & -12 \\ -8 & -3 & -6 \end{bmatrix} x + \begin{bmatrix} 1 \\ 2 \\ 3 \end{bmatrix} r$$

$$y = [1 \quad -2 \quad 3]x$$

P3.35 Determine the unit-ramp response of the system defined by

$$\begin{bmatrix} \dot{x}_1 \\ \dot{x}_2 \end{bmatrix} = \begin{bmatrix} 0 & 2 \\ -3 & -3 \end{bmatrix} \begin{bmatrix} x_1 \\ x_2 \end{bmatrix} + \begin{bmatrix} 0 \\ 2 \end{bmatrix} u$$

$$y = [1 \quad 0] \begin{bmatrix} x_1 \\ x_2 \end{bmatrix}$$

using MATLAB where u is the unit-ramp input. Use lsim command in MATLAB.

P3.36 Obtain the unit-impulse response of the following system using MATLAB

$$\begin{bmatrix} \dot{x}_1 \\ \dot{x}_2 \end{bmatrix} = \begin{bmatrix} 0 & 1 \\ -1 & -2 \end{bmatrix} \begin{bmatrix} x_1 \\ x_2 \end{bmatrix} + \begin{bmatrix} 0 \\ 1 \end{bmatrix} u$$

$$y = [1 \ 1] \begin{bmatrix} x_1 \\ x_2 \end{bmatrix} + [0]\, u$$

P3.37 A control system is defined by

$$\begin{bmatrix} \dot{x}_1 \\ \dot{x}_2 \\ \dot{x}_3 \end{bmatrix} = \begin{bmatrix} -1 & -3 & -3 \\ 0 & -2 & 1 \\ 2 & 0 & -1 \end{bmatrix} \begin{bmatrix} x_1 \\ x_2 \\ x_3 \end{bmatrix} + \begin{bmatrix} 3 \\ 0 \\ 1 \end{bmatrix} u$$

$$y = [1 \quad 2 \quad 0] \begin{bmatrix} x_1 \\ x_2 \\ x_3 \end{bmatrix}$$

Determine the controllability and observability of the system using MATLAB.

P3.38 Determine the eigenvalues of the following system using MATLAB.

$$\dot{x} = \begin{bmatrix} 0 & 1 & 0 \\ 0 & 1 & -5 \\ -2 & 1 & 3 \end{bmatrix} x + \begin{bmatrix} 0 \\ 0 \\ 1 \end{bmatrix} u$$

$$y = [0 \quad 0 \quad 1]x$$

P3.39 For the following path of a unity feedback system in state space representation, determine if the closed-loop system is stable using the Routh-Hurwitz criterion and MATLAB.

$$\dot{x} = \begin{bmatrix} 0 & 1 & 0 \\ 0 & 1 & 5 \\ -3 & -4 & -5 \end{bmatrix} + \begin{bmatrix} 0 \\ 0 \\ 1 \end{bmatrix} u$$

$$y = [0 \quad 1 \quad 1]x$$

P3.40 Consider the differential equation system given by

$$\ddot{y} + 4\dot{y} + 3y = 0 \qquad y(0) = 0.2 \qquad \dot{y}(0) = 0.1$$

Find the state-space equation for the system. Also, obtain the response $y(t)$ of the system subject to the given initial conditions using MATLAB.

STATICS

PS3.1 Determine the resultant or equivalent force applied to the bracket for the system shown in Fig. PS3.1.

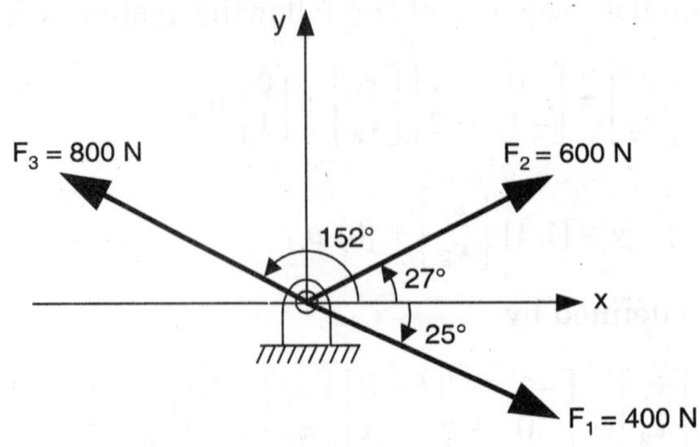

Fig. PS3.1

PS3.2 Figure PS3.2 shows a load W supported by two cables AC and BC.

(*a*) Determine the tension in AC and BC as a function of W and γ.

(*b*) Write a MATLAB program to plot the tensions T_{AC} and T_{BC} for values of γ ranging from $\gamma_1 = \beta - 90°$ to $\beta_2 = 90°$.

(*i*) $\alpha = 30°$, $\beta = 80°$, W = 1.5 kN

(*ii*) $\alpha = 60°$, $\beta = 40°$, W = 2.5 kN

(*iii*) $\alpha = 35°$, $\beta = 65°$, W = 1.2 kN

(*c*) Determine from the plots the value of γ for which the tension in the cables T_{AC} and T_{BC} are as small as possible and the respective values of T_{AC} and T_{BC}.

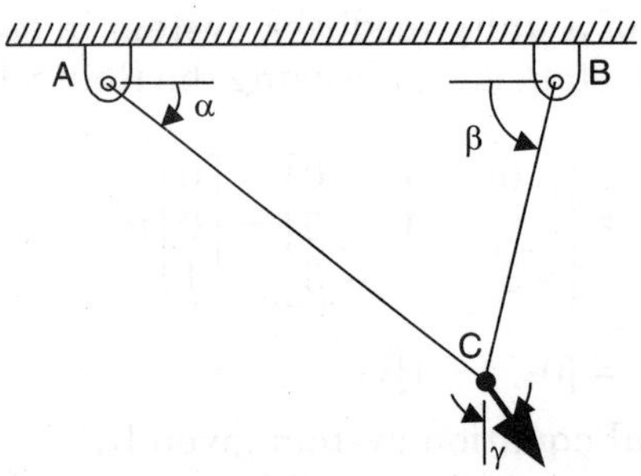

Fig. PS3.2

PS3.3 Figure PS3.3 shows an aluminium thin-walled sphere used as a marker buoy. Write a MATLAB program to determine the height H between the top of the buoy and the surface of the water. Given

the radius of the sphere = 70 cm
wall thickness = 15 mm
the density of aluminium = 2700 kg/m^3
the density of water in the ocean = 1000 kg/m^3.

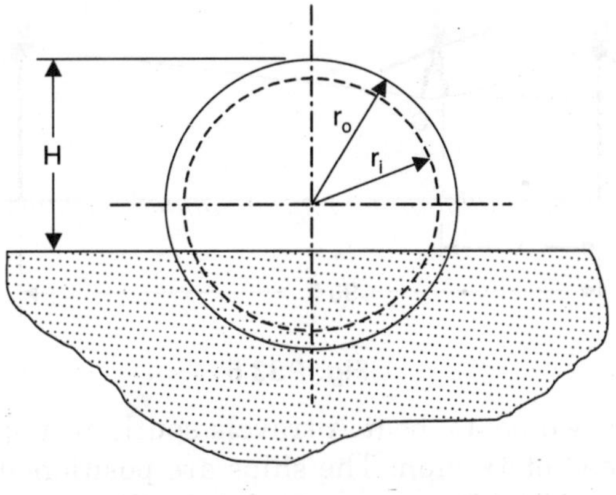

Fig. PS3.3

PS3.4 Two cars A and B are approaching an intersection. At time $t = 0$ car A is 500 ft south of the intersection traveling north at a constant speed of 60 mph. At the same time car B is 250 ft west of the intersection traveling east at a speed of 30 mph and accelerating at 5 ft/s^2. Determine the positions of the cars A and B, the distance between them, and the speed of car A relative to car B every second for the next 20 seconds.

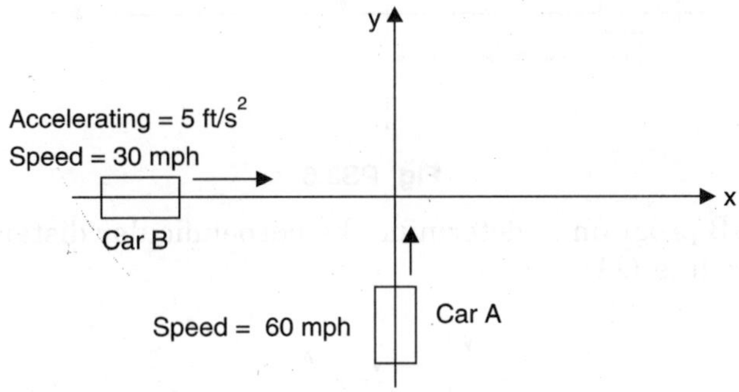

Fig. PS3.4

PS3.5 Figure PS3.5 shows an acrobat walking on a tight rope attached to supports at A and B. The friction between his shoes and the rope is high enough to prevent him from slipping. The weight of the rope and elastic deformation may be neglected. Write a MATLAB program to determine the deflection y and the tension in AC and BC of the rope for values of x ranging from 0.2 ft to 50 ft using 0.5 ft increments. Determine also

(*a*) the maximum deflection of the rope

(*b*) the maximum tension in the rope

(*c*) the minimum values of the tension in portions AC and BC of the rope.

The length of the tight rope = 90.6 ft

The combined weight of the acrobat and his balancing pole is 210 lbs.

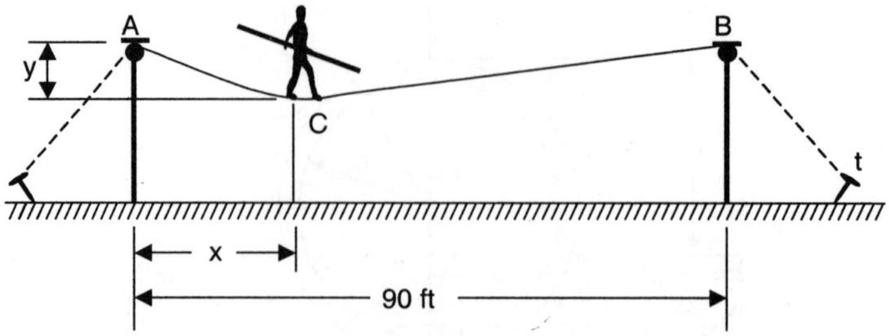

Fig. PS3.5

PS3.6 Figure PS3.6 shows two boats: boat A travels south at a speed of 10 mph, and boat *B* travels east at a speed of 19 mph. The ships are positioned at 8 am are also shown in Fig.PS3.19. Write a MATLAB program to plot the distance between the ships as a function of time for the next 5 hours.

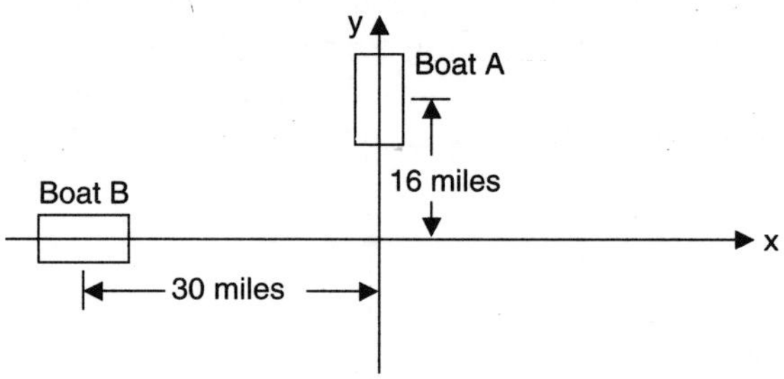

Fig. PS3.6

PS3.7 Write a MATLAB program to determine the perpendicular distance between the line of a force *F* and the line *OA*.

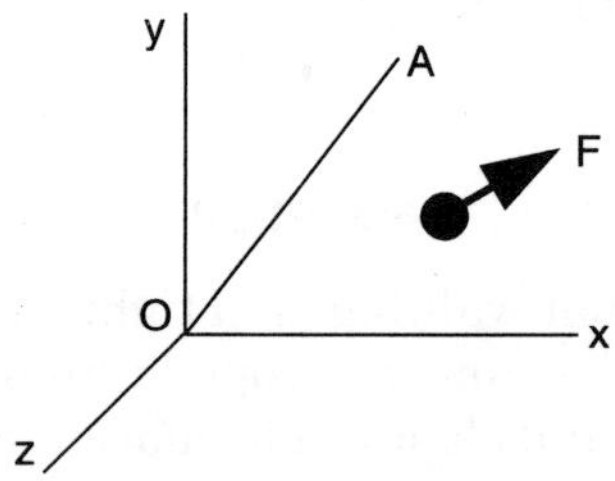

Fig. PS3.7

PS3.8 A tension T is applied to the cable attached to the top *A* of the rigid mast and secured to the ground at *B* as shown in Fig.PS3.8.

(*a*) Obtain a general expression for M_O (the moment of *T* about the base *O*) as a function of x_B and z_B

(*b*) Write a MATLAB program to plot the magnitude of M_O and its components about the *x* and *z* axes.

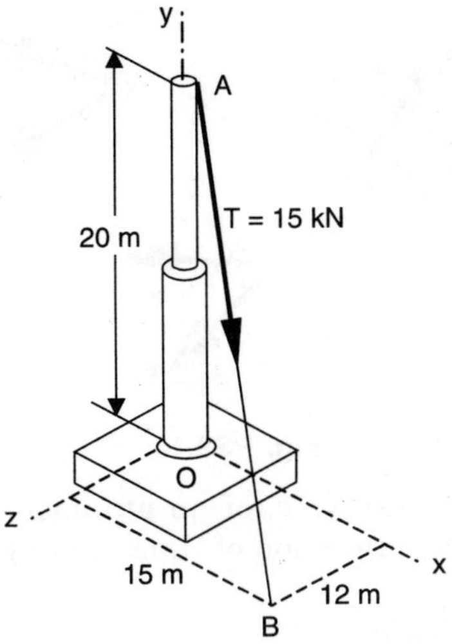

Fig. PS3.8

PS3.9 Figure PS3.9 shows a cone shaped foam cup designed to have a volume of 300 cm^3. The volume, V, and the surface area, S, of the foam are given by

$$V = \frac{1}{3}\pi R^2 H$$

and

$$S = \pi R\sqrt{R^2 + H^2}$$

where R = radius of the base

H = height of the cone

Determine the radius and surface area of the foam for cups with heights of 4, 5, 6, 7, 8, 9 and 10 cm.

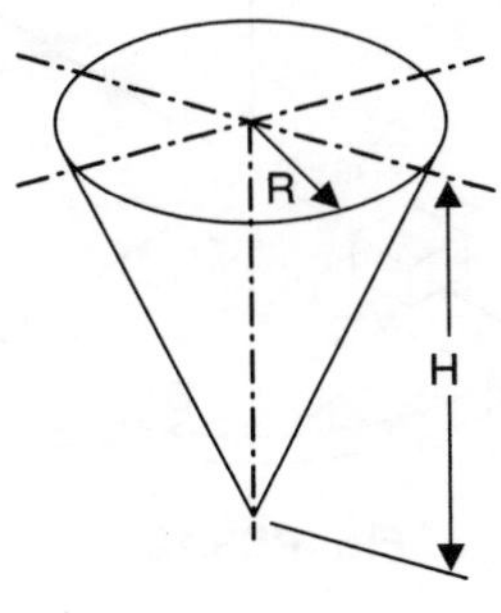

Fig. PS3.9

PS3.10 Write a MATLAB program to determine the single equivalent force R and the point where the line of action of R intersects the yz plane for the system shown in Fig.PS3.10.

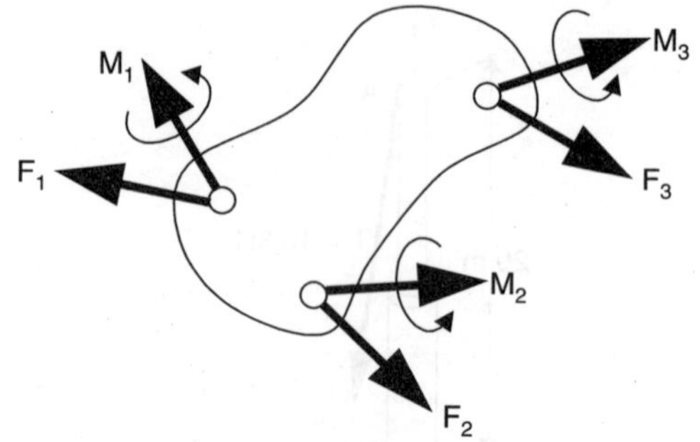

Fig. PS3.10

PS3.11 The distance, d, and the velocity, v, of an automobile that accelerates from rest at constant acceleration, a, as a function of time, t, are given by

$$d = \frac{1}{2} a t^2$$

and $$v = a t$$

Write a MATLAB program to determine d and v for the first 15 seconds for an automobile with an acceleration of $a = 1.6$ m/s^2.

PS3.12 Figure PS3.12 shows a uniform circular plate supported by three vertical wires that are equally spaced around the edge. A small block E is placed on the plate at D and is then slowly moved along the diameter CD until it reaches C. Write a MATLAB program to plot the tension in the wires A and C as a function of d, where d is the distance of the block from D. Find the value of d for which the tension in wires A and C is minimum. The radius of the circular plate = 310 mm. Mass of the circular plate = 30 kg. Mass of the small block = 3.5 kg.

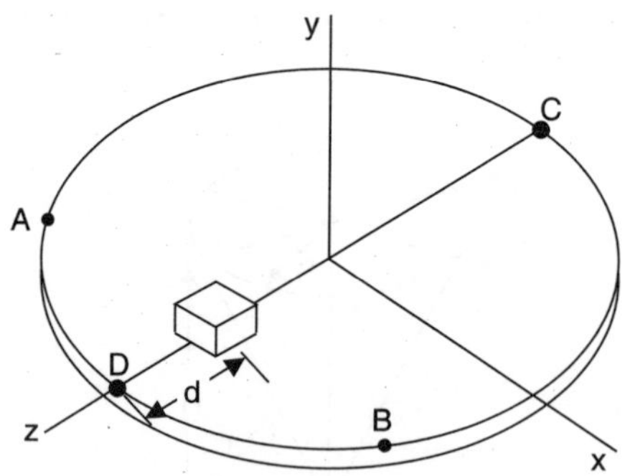

Fig. PS3.12

PS3.13 Figure PS3.13 shows a truss supporting a ramp (shown as a dashed line). The ramp extends from a fixed approach level at joint F to a fixed exit level at joint J. The loads shown in Fig. PS3.13 represent the weight of the ramp.

(*a*) Determine the forces in members CD, BH, and IH as a function of α and l.

(*b*) Write a MATLAB program to plot the non-dimensional forces CD/l, BH/l, and IH/l as a function of α for α between 0° and 45°.

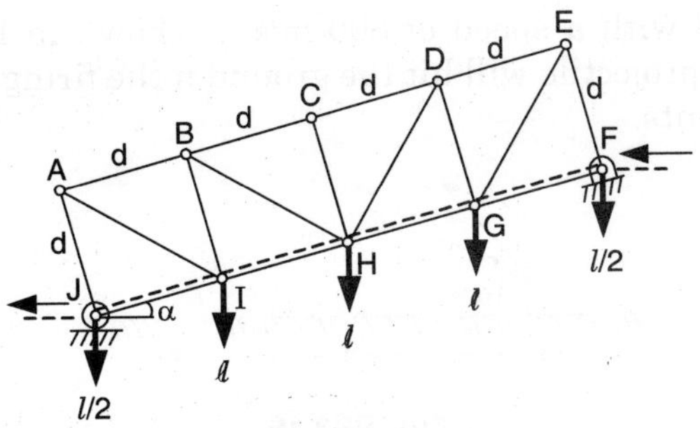

Fig. PS3.13

PS3.14 For the truss shown in Fig. PS3.14, the forces in the seven members are obtained by solving simultaneously the following seven equations.

$$F_1 \sin 30^\circ = -3000$$
$$F_1 \cos 30^\circ + F_2 = 0$$
$$F_3 + F_1 \sin 30^\circ = 0$$
$$F_4 - F_1 \cos 30^\circ = 0$$
$$-F_3 - F_4 \sin 30^\circ = 4000$$
$$F_6 + F_5 \cos 30^\circ - F_2 = 0$$
$$F_5 \sin 30^\circ + F_7 = 0$$

Determine the forces in the members F_1 to F_7 by writing the above equations in matrix form and solve them using MATLAB.

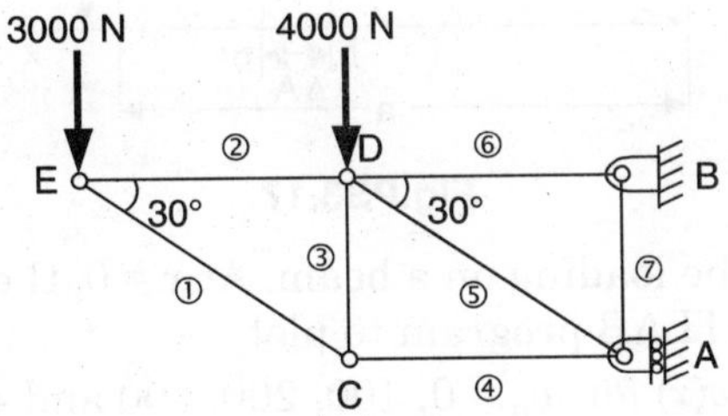

Fig. PS3.14

PS3.15 Determine the distance H for which the centroid of the shaded area in Fig. PS3.15 is as high above BB' as possible. Write a MATLAB program to plot the ratio H/B as a function of k for $0.1 \le k \le 0.9$.

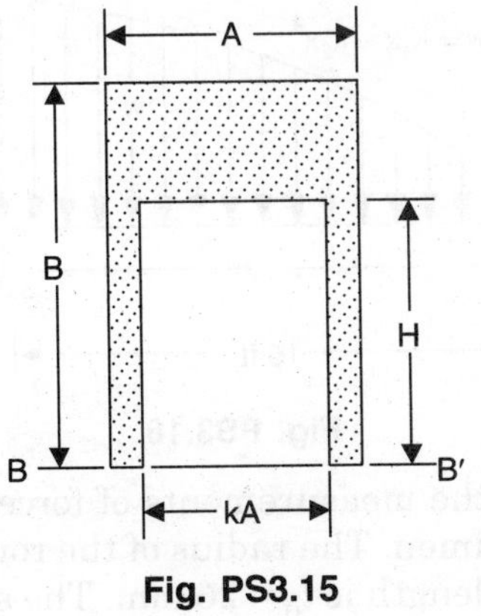

Fig. PS3.15

PS3.16 A projectile is fired with a speed of 800 m/s as shown in Fig.PS3.16. Determine the distance d that the projectile will hit the ground if the firing angle α is changed from 2° to 88° in 2° increments.

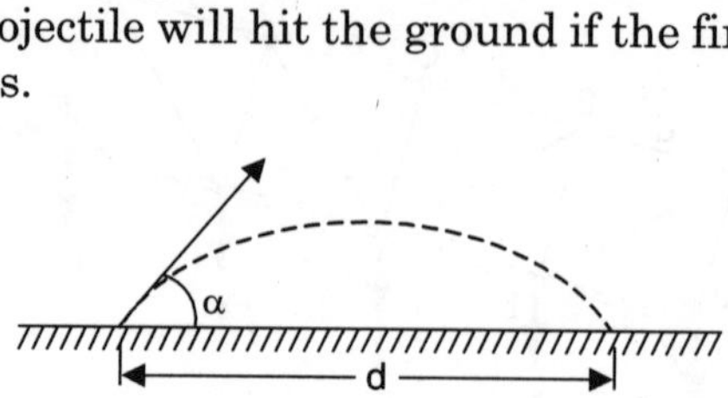

Fig. PS3.16

PS3.17 Figure PS3.17 shows an approximating the general spandrel using a series of n rectangles, each of width ΔA and of the form $bcc'b'$. Write MATLAB program to calculate the coordinates of the centroid of the area. Determine the centroid when (*a*) $m = 2, A = 5$ in, $H = 5$ in, (*b*) m = 5, A = 5 in, $H = 15$ in.

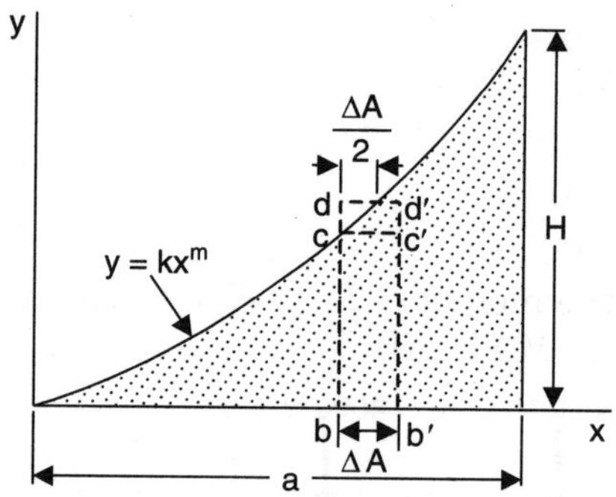

Fig.PS3.17

PS3.18 Figure PS3.18 shows the loading on a beam. At $x = 0$, the load is increasing at the rate of 80 lb/ft. Write a MATLAB program to plot

(*a*) the distributed load $w(x)$ for $w_0 = 0$, 100, 200, 300 and 400 lb/ft

(*b*) the reactions at the two supports as a function of w_0 for $0 \le w_0 \le 400$ lb/ft.

The unit load is $w = w_0$ lb/ft and the load per foot of beam length varies as shown in Fig. PS3.18, namely

$w = k_1x - k_2x^2$.

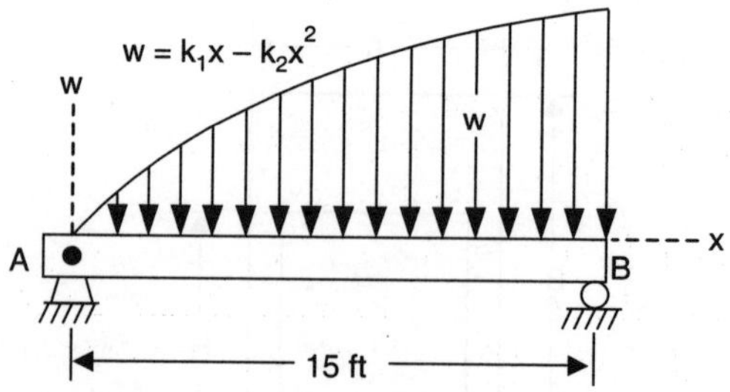

Fig. PS3.18

PS3.19 The following table gives the measurements of force and gauge length from a tension test with a particular specimen. The radius of the round-cross section before the test is 6.5 mm. The initial gauge length is $l_0 = 26$ mm. The schematic of the tensile test set up is shown in Fig. PS3.19. The force P required to pull the specimen and the length l of a gauge section are measured in the test. The engineering stress σ_e and strain ϵ_e are given by

$$\sigma_e = \frac{P}{A_0}$$

and $$\epsilon_e = \frac{l - l_0}{l_0}$$

where, l_0 and A_0 are the initial gauge length and the initial cross-sectional area of the specimen respectively. The true stress σ_t and strain ϵ_t are given by

$$\sigma_t = \frac{Pl}{A_0 l_0}$$

and $$\epsilon_t = \ln\left(\frac{l}{l_0}\right)$$

The initial gauge length l_0 = 27 mm.

P(N)	0	13300	26500	40350	42800	43600	44500	45000
l(mm)	27	25.09	25.09	25.12	25.13	25.131	25.133	25.144
P(N)	45380	46200	48000	49100	50200	53200	56170	58400
l(mm)	25.165	25.210	25.403	25.646	26.075	27.399	29.153	29.785

Write a MATLAB program and plot the engineering and true-stress strain curves.

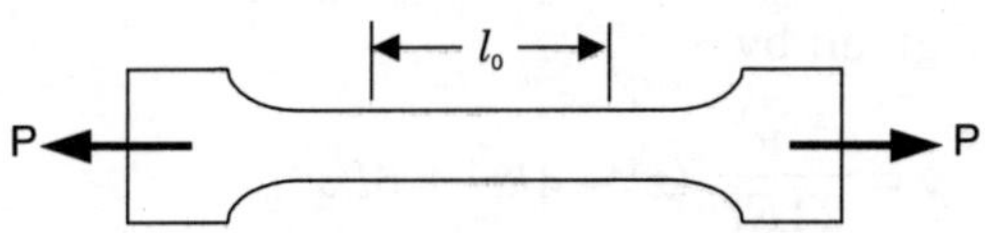

Fig. PS3.19

PS3.20 Write a MATLAB program to determine the volume and the surface area of the solid obtained by rotating the area shown in Fig. PS3.18 about the y-axis when A = 100 mm and (a) n = 1, (b) n = 2, (c) n = 3.

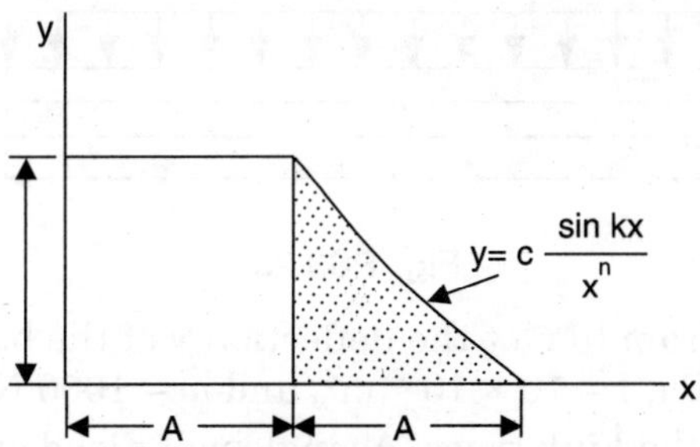

Fig. PS3.20

PS3.21 The following equation can be used to approximately model the orbits of the planets around the sun:

$$r = \frac{eP}{1 - e\cos\theta}$$

Write a MATLAB program to plot the orbits of four planets whose values of the constants *P* and *e* are given in the following table.

Planet	*P* (× 10^6 *m*)	*e*
Earth	8964	0.0167
Mars	2421	0.0934
Mercury	269.2	0.2060
Venus	15913	0.0068

PS3.22 Figure PS3.22 shows a beam subjected to the loading. Write a MATLAB program to plot the magnitude of the vertical reactions at supports *A* and *B* as functions of distance *d* for $0 \le d \le 5$ m.

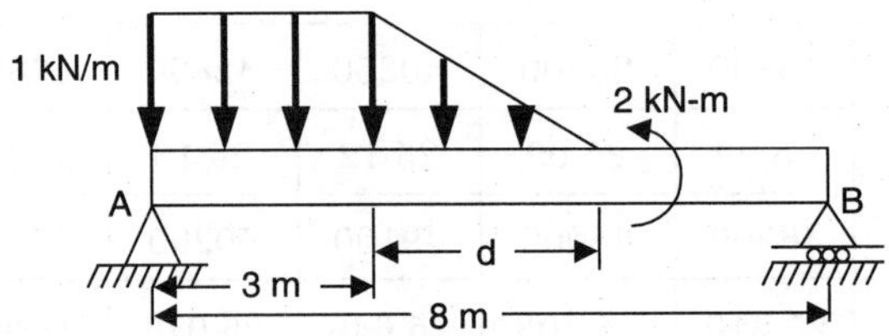

Fig. PS3.22

PS3.23 Figure PS3.23 shows a cantilever beam. The deflection *y* at point *x* with a uniformly distributed load *w* is given by

$$y = \frac{-w}{24EI}(x^4 - 4lx^3 + 6l^2x^2)$$

where *E* is the Young's modulus of elasticity
I is the moment of inertia
l is the length of the beam.

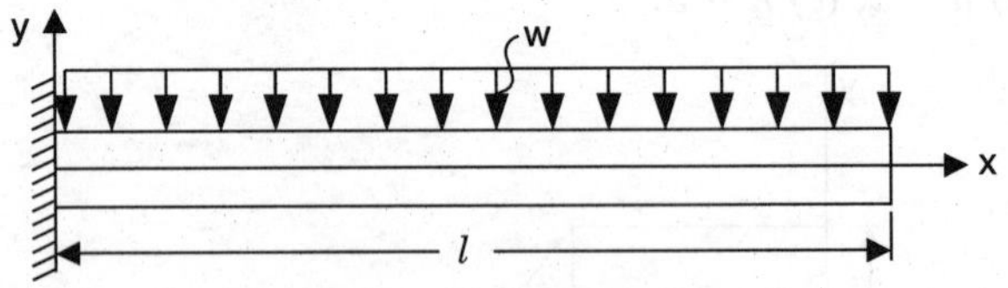

Fig. PS3.23

Write a MATLAB program to plot the deflection *y* of the beam as a function of *x*. Given $l = 10$ m, $E = 70 \times 10^9$ Pa, $I = 10 \times 10^{-6}$ m^4, and $w = 1000$ N/m.

PS3.24 Figure PS3.24 shows the Fink truss. A single point load of 4 kips is to be applied to the top chord of the truss at one of the joints 1, 2, ..., 9. Write a MATLAB program to calculate the force in member *MN* as the load is successively applied at joints 1, 2, ..., 9.

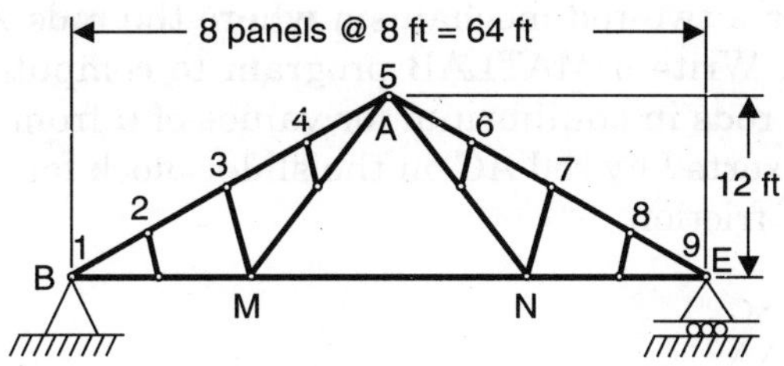

Fig. PS3.24

PS3.25 In Figure PS3.25, the a uniform pole of length l and density ρ' is shown where one end is firmly secured at A to the bottom of a tank. The density of the liquid is ρ and the depth is H. Write a MATLAB program and determine

(*a*) the angle α assumed by the pole for $\rho' < \rho$

(*b*) plot a as α function of the density ratio ρ'/ρ for H/l = 0.2, 0.3, 0.4, 0.5, 0.6 and 0.7.

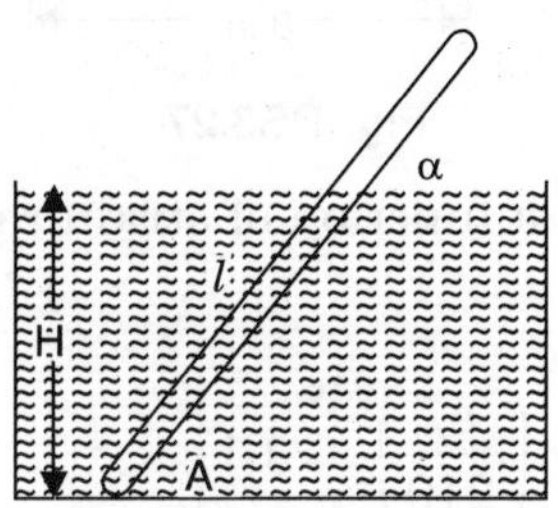

Fig. PS3.25

PS3.26 Figure PS3.26 shows a simply supported beam with a constant distributed load w over half of its length.

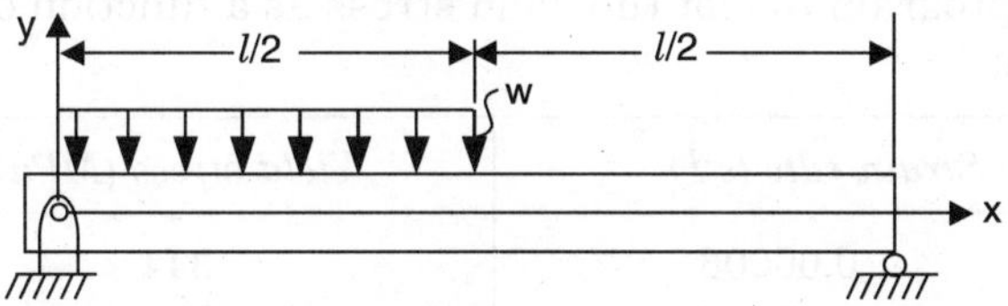

Fig. PS3.26

The deflection y as a function of x is given by

$$y = \frac{-wx}{384EI}(16x^3 - 24lx^2 + 9l^3) \qquad \text{for} \quad 0 \le x \le \frac{l}{2}$$

$$y = \frac{-wx}{384EI}(8x^3 - 24lx^2 + 17l^2x - l_3) \quad \text{for} \quad \frac{l}{2} \le x \le l$$

where

E = Young's modulus of elasticity.

I = moment of inertia.

l = length of the beam.

Write a MATLAB program to plot the deflection of the beam y as a function of x. Given l = 25 m, $E = 200 \times 10^9$ Pa, $I = 350 \times 10^{-6}$ m^4, and $w = 6 \times 10^3$ N/m.

PS3.27 Figure PS3.27 shows a two-rod mechanism where the rods *AC* and *BD* are connected by a slider block *D*. Write a MATLAB program to compute and plot the couple M_A required to hold the rods in equilibrium for values of α from 0 to 120°. Plot the magnitude of the force *F* exerted by rod *AC* on the slider block for values of α from 0 to 120°. Neglect the effect of friction.

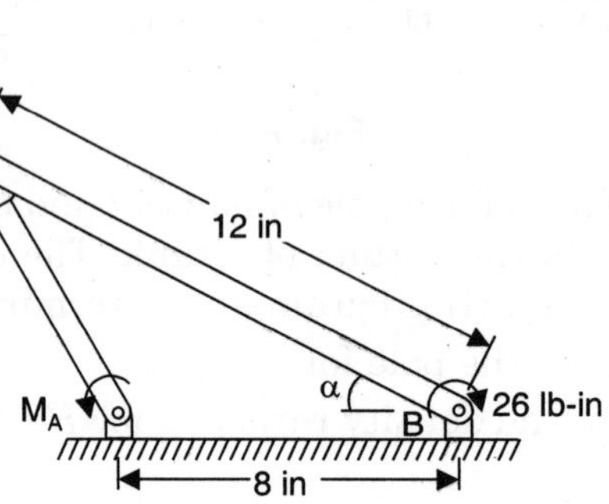

Fig. PS3.27

PS3.28 The yield stress of metals as a function of strain rate can be modeled by the following equation:

$$\sigma_{yld} = 350\left[\left(\frac{\varepsilon_t}{210}\right)^{0.16} + 1\right]$$

where

σ_{yld} = yield stress (MPa)

ε_t = strain rate (s^{-1})

Write a MATLAB program to plot the yield stress as a function of the strain rate for the following test data:

Strain rate (s^{-1})	*Yield stress (MPa)*
0.00006	344
0.00025	365
0.04890	417
0.79950	453
4.15890	483
215.84800	636
3501.04323	833

PS3.29 Derive the equations for the shear and bending moment curves for the beam shown in Fig. PS3.29. Write a MATLAB program to plot the shear and bending moment diagrams for the beam with w_0 = 20 kN/m and l = 4 m.

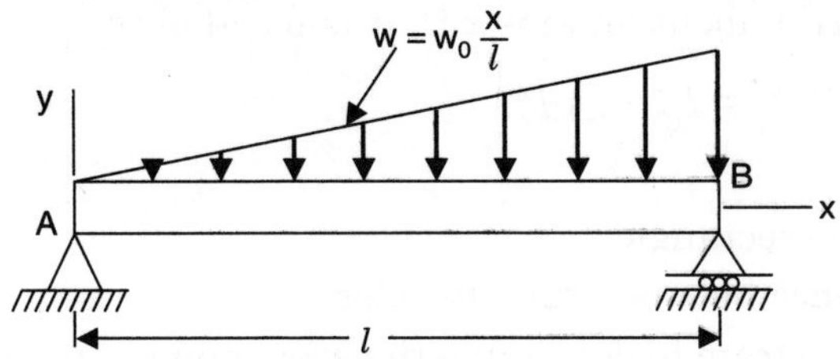

Fig. PS3.29

PS3.30 Figure PS3.30 shows a planar mechanism with two uniform links OQ and OP each of mass m and are connected and constrained as shown. When a horizontal force F is applied, the angle α between the links increases, the light rod connected at M and passing through a pivoted collar at N compresses the spring. The stiffness of the spring is k lb/in. Determine

(*a*) the force F which will produce equilibrium at the angle α if the spring is uncompressed in the position when $\alpha = 0$

(*b*) Write a MATLAB program to determine the equilibrium value of α corresponding to a given force F

(*c*) the equilibrium angle α for $F = 150$ lb, $mg = 60$ lb, $k = 25$ lb/in, and $l = 7$ inches.

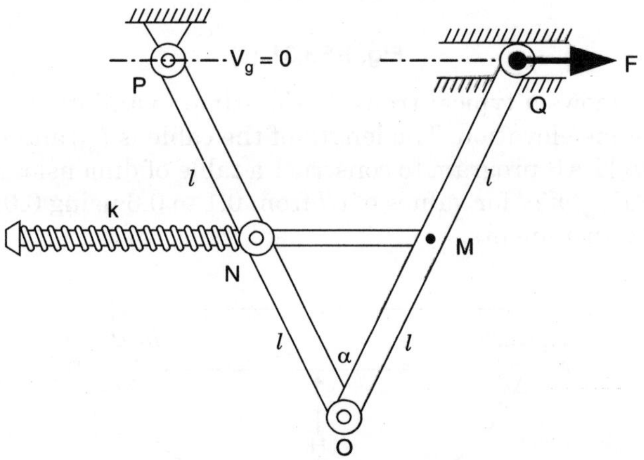

Fig. PS3.30

PS3.31 The area moment of inertia I_{x0} of a rectangle shown in Fig. PS3.31 (*a*) about the axis x_0 that passes through the centroid is given by

$$I_{x0} = \frac{1}{12} BH^3$$

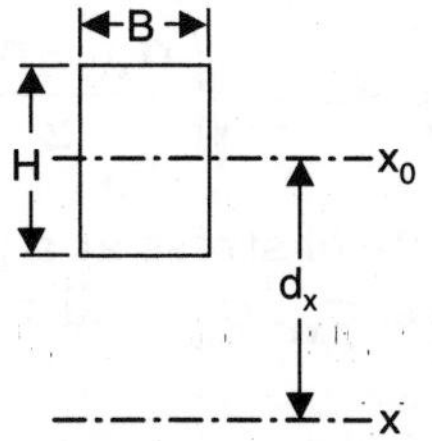

Fig. PS3.31 (*a*)

The moment of inertia about an axis x that is parallel to x_0 is given by

$$I_x = I_{x0} + Ad_x^2$$

where

A = area of the rectangle

d_x = the distance between the two axes.

Write a MATLAB program to determine the area moment of inertia of an "*I*" beam about an axis that passes through its centroid as shown in Fig. PS3.31 (*b*). Given $W = 250$ mm, $H = 350$ mm and $t = 25$ mm.

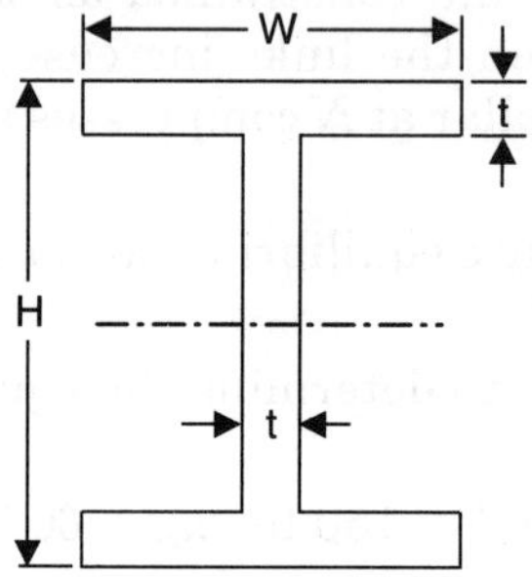

Fig. PS3.31 (*b*)

PS3.32 Figure PS3.32 shows a typical transmission-time installation suspended between two points at the same elevation. The length of the cable is l_{AB} and mass per unit length is m'. Write a MATLAB program to construct a table of dimensionless quantities H/l, l_{AB}/l, $T_0/m'gl$ and $T_{max}/m'gl$ for values of c/l from 0.1 to 0.6 using 0.01 increments and from 1 to 5 using 0.2 increments.

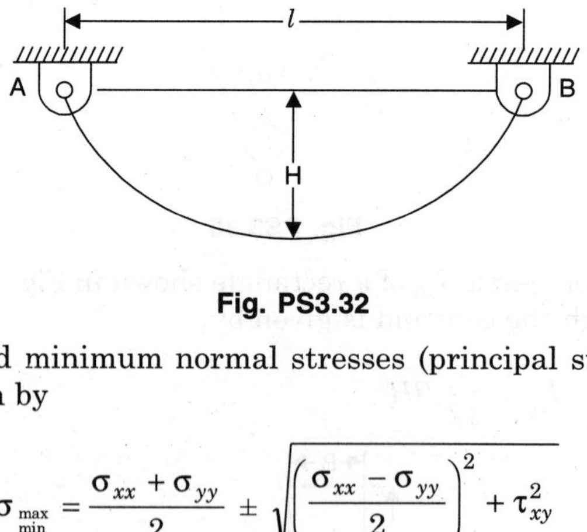

Fig. PS3.32

PS3.33 The maximum and minimum normal stresses (principal stresses) at the point, σ_{max} and σ_{min}, are given by

$$\sigma_{\substack{max \\ min}} = \frac{\sigma_{xx} + \sigma_{yy}}{2} \pm \sqrt{\left(\frac{\sigma_{xx} - \sigma_{yy}}{2}\right)^2 + \tau_{xy}^2}$$

where for a two-dimensional state of stress at a point in a loaded material is defined by the three components of stresses: σ_{xx}, σ_{yy}, and τ_{xy}.

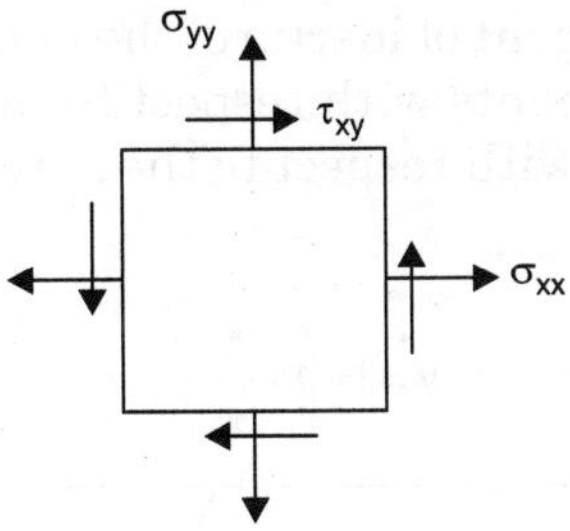

Fig. PS3.33

Write a MATLAB program to determine the principal stresses from the three stress components for the following states of stresses:

(*a*) $\sigma_{xx} = 150$ MPa, $\sigma_{yy} = -45$ MPa, and $\tau_{xy} = 90$ MPa

(*b*) $\sigma_{xx} = -15$ MPa, $\sigma_{yy} = -20$ MPa, and $\tau_{xy} = -10$ MPa

(*c*) $\sigma_{xx} = 200$ MPa, $\sigma_{yy} = -70$ MPa, and $\tau_{xy} = 120$ MPa

PS3.34 Figure PS3.34 shows the position of the rod *AB* controlled by the block which is slowly moved to the left by the force *P*. The weight of the rod *AB* is 30 lbs and that of the block is 5 lbs. The coefficient of kinetic friction between all surfaces contact is 0.3. Write a MATLAB program to plot the magnitude *P* of the force as a function of *x* for values of *x* from 55 in to 5 in. Find the maximum value of *P* and the corresponding values of *x*.

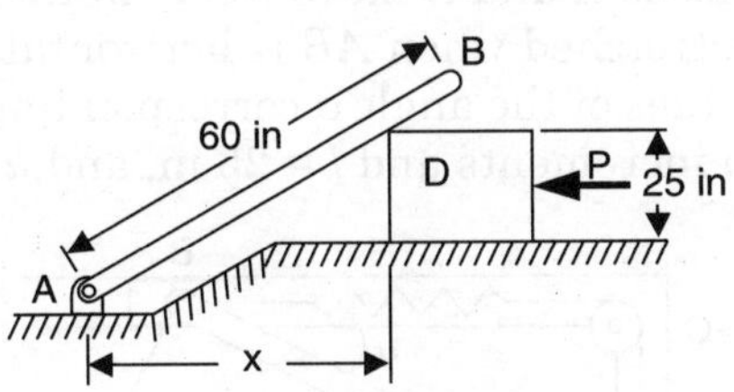

Fig. PS3.34

PS3.35 Figure PS3.35 shows the panel which forms the end of a trough filled with water to line *AA*′.

(*a*) Obtain the depth *d* of the point of application of the resultant of the hydrostatic forces acting on the panel (center of pressure).

(*b*) Write a MATLAB program to plot the depth of the center of pressure *d* as a function of *n*. Given: $H = 500$ mm, $B = 200$ mm and $A = nB$, where *n* varies from 1 to 10.

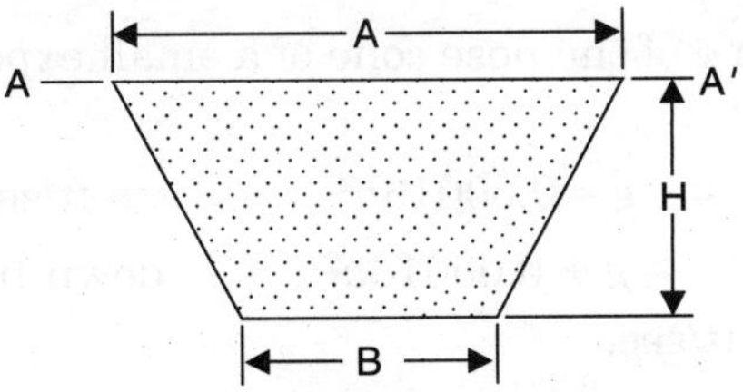

Fig. PS3.35

PS3.36 Approximate the mass moment of inertia of the homogeneous wire shown in Fig. PS3.36 using 12 straight line segments with respect to the x-axis. Write a MATLAB program to determine I_x of the wire with respect to the x-axis when $A = 25$ mm, $l = 250$ mm, and $H = 100$ mm.

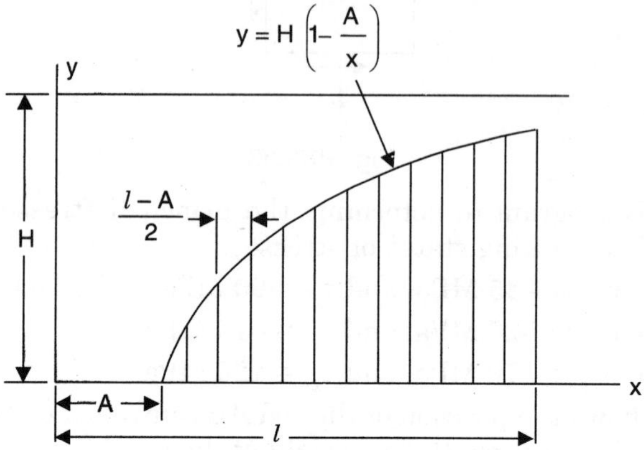

Fig. PS3.36

PS3.37 Figure PS3.37 shows a slender rod AB attached to blocks A and B. The weight of the rod AB is W and the blocks A and B move freely in the guides. The spring constant is k and the spring is unstretched when AB is horizontal. Write a MATLAB program to determine the three values of the angle α corresponding to equilibrium for values of W from 1 to 10 lbs in 1 lb increments and $l = 25$ in, and $k = 1.5$ lb/in.

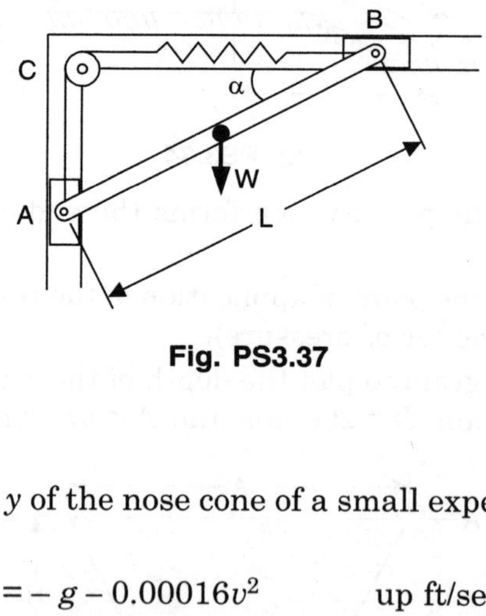

Fig. PS3.37

DYNAMICS

PD3.1 The total acceleration y of the nose cone of a small experimental rocket as it moves up and down is given by

$$\ddot{y} = -g - 0.00016v^2 \qquad \text{up ft/sec}^2$$

$$-g + 0.00016v^2 \qquad \text{down ft/sec}^2$$

where v = velocity in ft/sec.

(*a*) Derive an expression for the speed of the cone as a function of height as the cone moves up to its maximum height and as it returns to the ground. The nose cone is projected vertically from the ground with an initial velocity of 400 ft/s.

(*b*) Write a MATLAB program to plot the speed of the cone as a function of the height for the upward motion and for the downward motion.

PD3.2 The height of the ball at any time t after it is thrown is given by

$$y(t) = y_0 + v_{y0}t + \frac{1}{2}gt^2$$

where y_0 = the initial height of the object above the ground
v_{y0} = the initial vertical velocity of the object
g = acceleration due to gravity

The horizontal distance traveled by the ball after it is thrown is given by

$$x(t) = (x_0 + v_{x0}t)$$

where x_0 = the initial horizontal position of the ball on the ground
v_{x0} = the initial horizontal velocity of the ball

In the above, air friction and earth's curvature are neglected. Assume that the ball is initially thrown from $(x_0, y_0) = (0, 0)$ with 25 m/s at an initial angle of θ degrees.

(*a*) Write a MATLAB program to plot the trajectory of the ball and obtain the horizontal distance traveled before it touches the ground.
Range of θ: $0 \le \theta \le 90°$

(*b*) Determine the angle θ that maximizes the range of the ball and plot the trajectory.

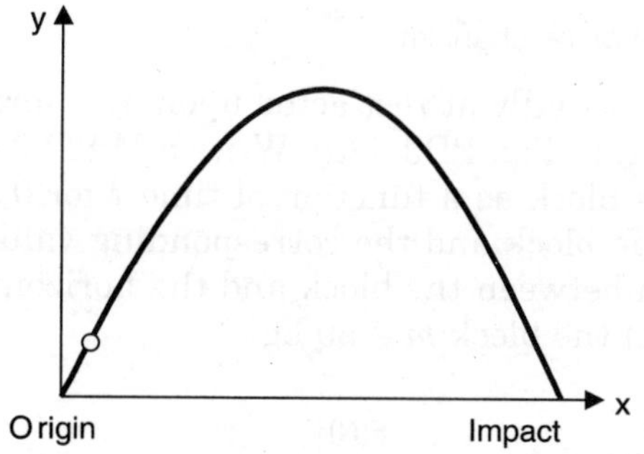

(*a*)

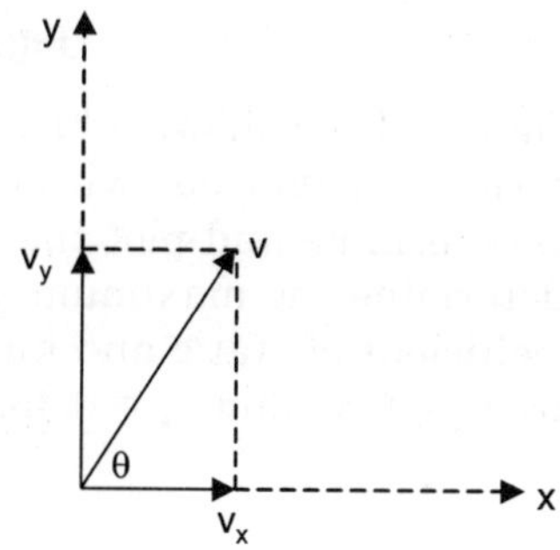

(*b*)

Fig. PD3.2 (*a*) Parabolic trajectory, (*b*) Components of vector v.

PD3.3 Figure PD3.3 shows a block of mass attached to a spring of constant k. The block is released from rest when the spring is in a horizontal and undeformed position. Write a MATLAB program to determine (*a*) the length of the spring, the magnitude and direction of velocity of the block as the block passes directly under the point of suspension of the spring for k/m = 14 s^{-2}, 22s^{-2}, and 25 s^{-2}, (*b*) the value of k/m for which that velocity is horizontal. Assume r_0 = 5 ft.

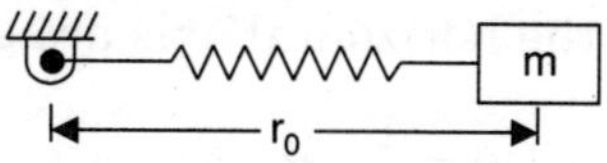

Fig. PD3.3

PD3.4 The velocities of links 3 and 4 of the four-bar mechanism shown in Fig. PD3.4 are given by

$$\begin{bmatrix} -r_3 \sin\theta_3 & r_4 \sin\theta_4 \\ r_3 \cos\theta_3 & -r_4 \cos\theta_4 \end{bmatrix} \begin{bmatrix} w_3 \\ \omega_4 \end{bmatrix} \begin{bmatrix} \omega_2 r_2 \sin\theta_2 \\ -\omega_2 r_2 \sin\theta_2 \end{bmatrix}$$

The link lengths and angles are given below:

Link	*Length, r(cm)*	*Angle, θ (degrees)*
1	13.0	0
2	5.0	45
3	11.0	25
4	8.0	91

The velocity of the input link, ω_2 is given as 110 rad/second. Write a MATLAB program to solve for the velocities of links 3 and 4.

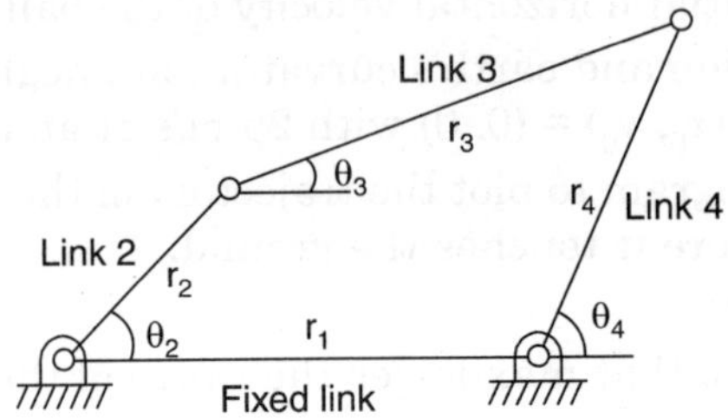

Fig. PD3.4 Four-bar mechanism.

PD3.5 Figure PD3.5 shows a block of mass m initially at rest acted upon by a force F which varies in magnitude with time as shown in Fig. PD3.5 (*a*). Write a MATLAB program to determine and plot the speed of the block as a function of time t for $0 \le t \le 20$ s. Determine the maximum velocity of the block and the corresponding value of t. The coefficient of static and kinetic friction between the block and the horizontal surface are $\mu_s = 0.45$ and $\mu_k = 0.45$. The mass of the block $m = 30$ kg.

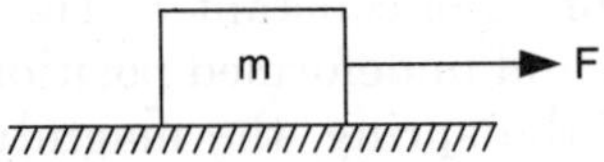

Fig. PD3.5

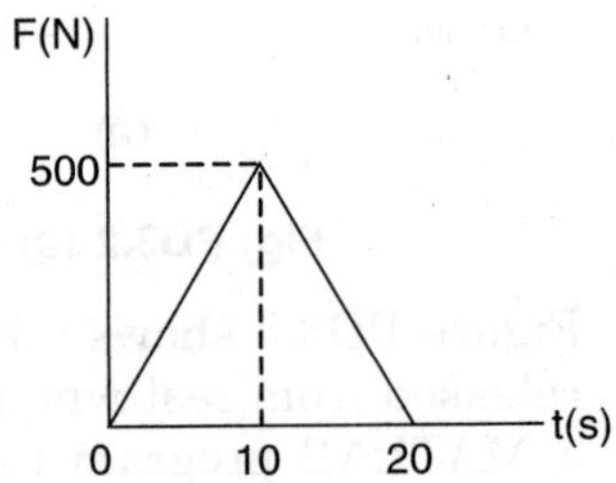

Fig. PD3.5 (*a*)

PD3.6 The height, h, horizontal distance, x, and speed of a projectile, v, launched with a speed, v, at an angle θ to the horizontal axis are given by

$$h(t) = vt \sin\theta - \frac{1}{2} gt^2$$

$$x(t) = vt \cot\theta$$

$$v(t) = \sqrt{v^2 - 2vgt \sin\theta + g^2t^2}$$

The projectile will strike the ground when $h(t) = 0$, and the time of the hit is given by

$$t_{hit} = 2\,\frac{v}{g}\sin\theta$$

Assuming $\theta = 35°$, $v = 50$ m/s, and $g = 9.81$ m/s^2, write a MATLAB program using logical operators to find (*a*) the height is no less than 16 m, (*b*) the height is no less than 16 m and the speed is no greater than 35 m/sec.

PD3.7 Figure PD3.7 shows two hemispheres held together by a cord only which maintain the spring under compression. Write a MATLAB program to calculate and plot the magnitude of the resulting velocity of each hemisphere as a function of θ for $10° \le \theta \le 150°$ knowing that the cord is severed causing the hemispheres to fly apart. The potential energy of the compressed spring is 100 ft-lb and the assembly has an initial velocity of magnitude $v_0 = 25$ ft/s.

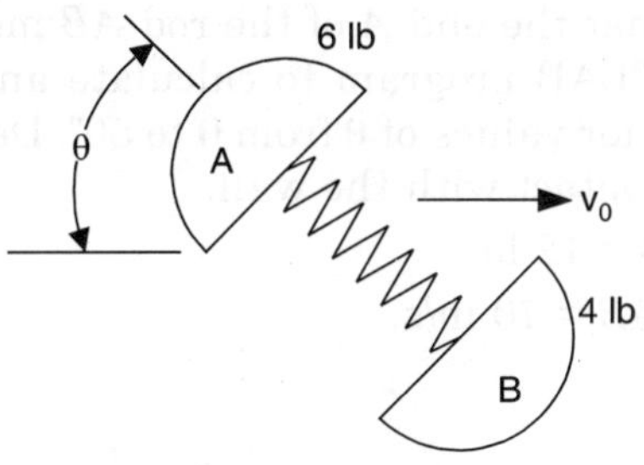

Fig. PD3.7

PD3.8 Figure PD3.8 shows two rotating rods connected by a slider *P*. Write a MATLAB program to determine and plot for values of θ from 0 to 180°. The rod *BP* rotates counterclockwise with a constant angular velocity of 10 rad/s.

(*a*) the angular velocity and angular acceleration of rod AE

(*b*) determine the value of θ for which the angular acceleration α_{AE} of rod *AE* is maximum and the corresponding value of α_{AE}.

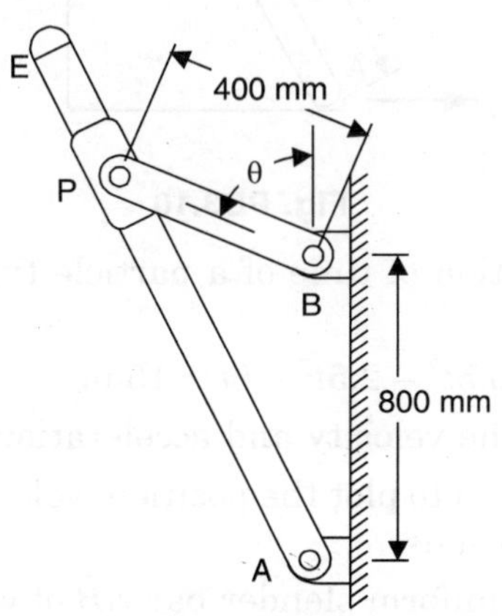

Fig. PD3.8

PD3.9 The variation of acceleration due to gravity g with altitude y is given by

$$g = \frac{R^2}{(R+y)^2} g_0$$

where R = radius of the earth (6371 km)

g_0 = the acceleration due to gravity at sea level (9.81 m/s^2)

The change in the gravitational potential energy, ΔU of an object that is raised up from the earth is given by

$$\Delta U = \int_0^n mgdy$$

Write a MATLAB program to determine the change in the potential energy of a satellite with a mass of 600 kg that is raised from the earth's surface to a height of 900 km.

PD3.10 Figure PD3.10 shows that the end A of the rod AB moved to the left with a constant speed v_A. Write a MATLAB program to calculate and plot the normal reactions at ends A and B of the rod for values of θ from 0 to 50°. Determine the value of θ at which end B of the rod loses contact with the wall.

The mass of the rod AB = 12 lb.

The constant speed of AB = 70 in/s.

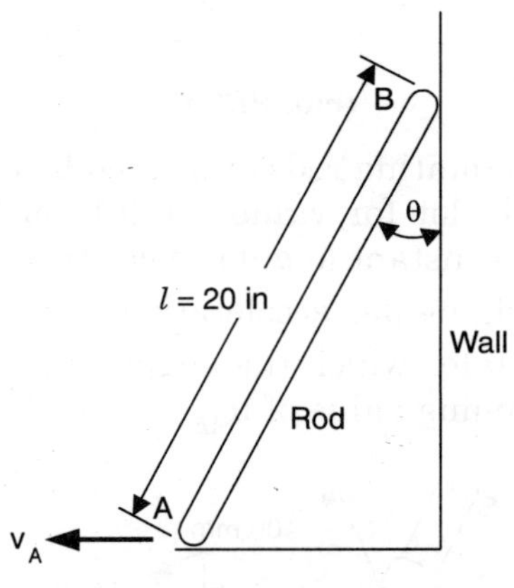

Fig. PD3.10

PD3.11 The position x as a function of time of a particle that moves along a straight line is given by

$$x(t) = 0.5t^2 - 2.5t^2 - 6t + 15 \text{ m}$$

(*a*) Derive expressions for the velocity and acceleration of the particle.

(*b*) Write a MATLAB program to plot the position, velocity and acceleration as a function of time for $0 \le t \le 10$ seconds.

PD3.12 Figure PD3.12 shows a uniform slender bar AB of weight W suspended from springs AC and BD. Write a MATLAB program to calculate and plot the accelerations of ends A and B immediately after spring AC has broken for values of θ from 0 to 90°.

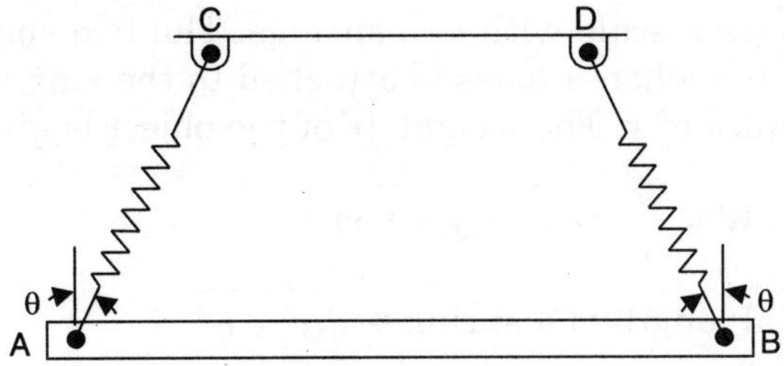

Fig. PD3.12

PD3.13 Refer to Problem PD3.4. Use the link lengths for the four-bar mechanism and write a MATLAB program to determine the angle θ_3 and θ_4 for a full rotation of the input link (link 2), at 1° intervals. Plot the angles θ_3 and θ_4 versus θ_2.

PD3.14 Figure PD3.14 shows a box of mass *m* pulled by a rope. The force required to move the box is given by

$$F = \frac{\mu m g}{\cos\theta + \mu \sin\theta}$$

where μ = coefficient of friction

m = mass of the box

g = acceleration due to gravity (9.81 m/s^2)

θ = angle from horizontal x-axis.

Given: mass of box, m = 25 kg

Coefficient of friction, μ = 0.40

Write a MATLAB program to determine the angle θ when the pulling force *F* is 100 N.

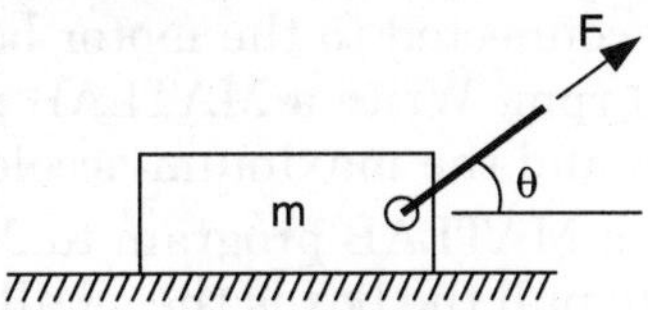

Fig. PD3.14

PD3.15 Figure PD3.15 shows a thin homogeneous disk of weight *W* and radius *r* mounted on the horizontal axle *AB*. The constant angular velocity of the axle is ω. Angle β is the angle, the plane of the disks forms with the plane normal to the axle. Write a MATLAB program (for $0 \le \beta \le 90°$) to (*a*) determine and plot θ formed by the axle and the angular momentum of the disk about *C*, (*b*) the kinetic energy of the disk, (*c*) the magnitude of the rate of change of the angular momentum of the disk about *C*.

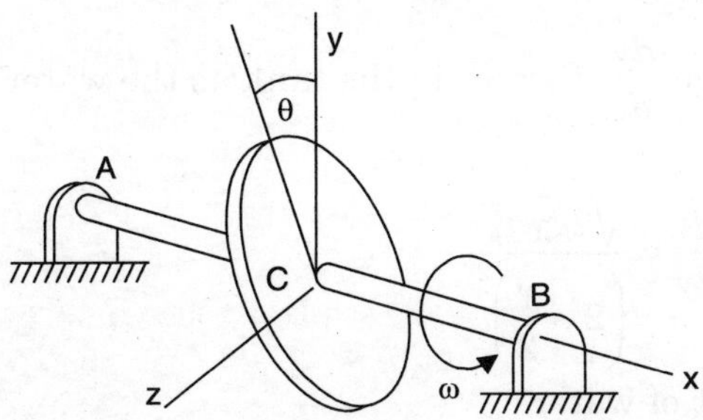

Fig. PD3.15

PD3.16 Figure PD3.16 shows a scale with two springs. The two springs are unstretched initially and will stretch when a mass is attached to the ring and the ring will displace downwards a distance of x. The weight W of the object is given by

$$W = \frac{2k}{l} = (l - l_0)(b + x)$$

where l_0 = initial length of a spring = $\sqrt{a^2 + b^2}$

l = the stretched length of the spring = $\sqrt{a^2 + (b + x)^2}$

k = spring constant

Write a MATLAB program to determine the distance x when a 350 N object is attached to the scale. Given: $a = 0.16$ m, $b = 0.045$ m, and the spring constant $k = 3000$ N.m.

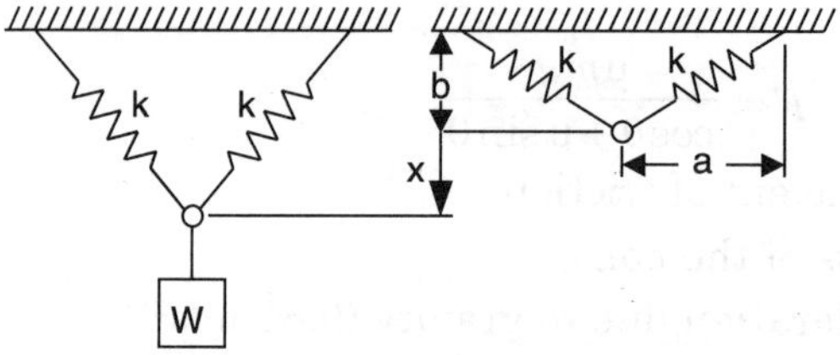

Fig. PD3.16

PD3.17 A 50 lb motor is supported by four springs, each of spring constant 400 lb/in. The unbalance of the motor is equivalent to a weight of 0.05 lb located 6 in. from the axis of rotation. The motor is constrained to move vertically. A dashpot with a coefficient of damping $c = 150$ lb s/ft is connected to the motor base and to the ground. The motor speeds are from 0 to 2500 rpm. Write a MATLAB program to calculate and plot the amplitude of the vibration and the maximum acceleration of the motor.

PD3.18 In Problem PD3.14, write a MATLAB program to determine (*a*) the angle θ at which the force that is required to pull the box is the smallest, (*b*) the magnitude of the force in (*a*).

PD3.19 Figure PD3.19 shows a water tank (shaped as an inverted frustrum cone with a circular hole at the bottom on the side). The velocity of water discharged through the hole is given by

$$v = \sqrt{2gh}$$

where h = height of the water

g = acceleration due to gravity (9.81 m/s^2)

The rate of discharge $\frac{dy}{dt}$ of water in the tank as the water discharges out through the hole is given by

$$\frac{dt}{dy} = \frac{\sqrt{2gy}\, r_h^2}{\left(2 - \frac{y}{2}\right)^2}$$

where y = height of water

r_h = radius of the hole.

Write a MATLAB program to

(*a*) solve the differential equation

(*b*) determine the time when $y = 0.15$ m

(*c*) plot y as a function of time.

The initial height of the water is 2.5 m.

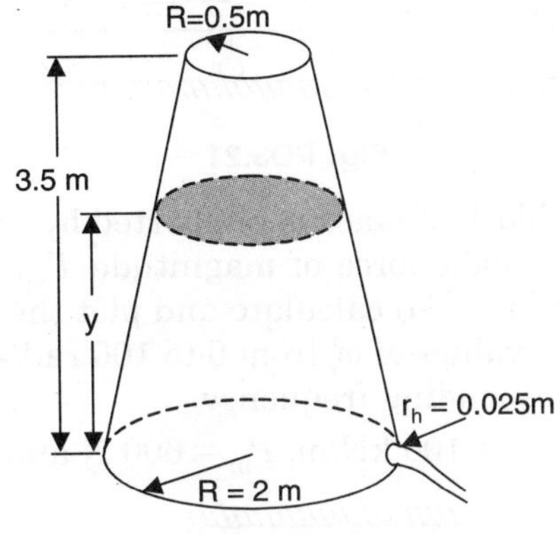

Fig. PD3.19

PD3.20 Figure PD3.20 shows a machine supported by springs and connected to a dashpot and subjected to a periodic force of magnitude P. Write a MATLAB program to calculate and plot the value of T_m for frequency ratios ω_f/ω_n from 0 to 6 and for damping factors c/c_c equal to 0.1, 0.3, 0.5, 0.8 and 1.0. The transmissibility, T_m, of the system is defined as the ratio F_m/P_m of the maximum value F_m of the fluctuating periodic force transmitted to the foundation to the maximum value P_m of the periodic force applied to the machine.

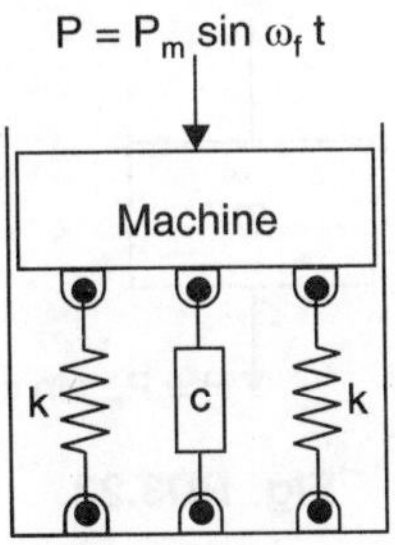

Fig. PD3.20

PD3.21 Figure PD3.21 shows an airplane which uses a parachute as a means of braking on the runway after landing. The acceleration is given by

$$a = -\,0.0045v^2 - 3 \text{ m/s}^2$$

and $$\frac{dv}{dt} = -\,0.0035v^2 - 3 \text{ m/s}^2$$

(*a*) Write a MATLAB program to solve the differential equation and plot the velocity from $t = 0$ seconds until the airplane stops

(*b*) plot the distance x the airplane travels as a function of time.

The velocity of airplane is 350 km/h that opens its parachute and starts decelerating at $t = 0$ seconds.

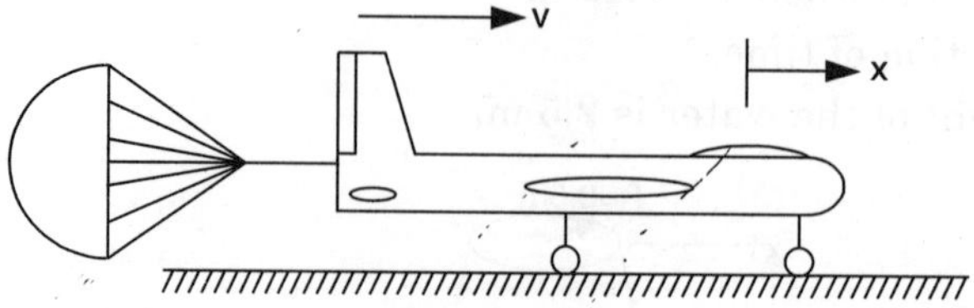

Fig. PD3.21

PD3.22 Figure PD3.22 shows a block of mass m supported by two springs of constants k_1 and k_2 and subjected to a periodic force of magnitude P_m. The damping coefficient is c. Write a MATLAB program to (*a*) calculate and plot the amplitude of the steady state vibration of the block for values of ω_f from 0 to 100 rad/s, (*b*) determine the maximum amplitude and the corresponding frequency.

Given: $m = 100$ kg, $k_1 = k_2 = 100$ kN/m, $P_m = 600$ N and $c = 2600$ N s/m.

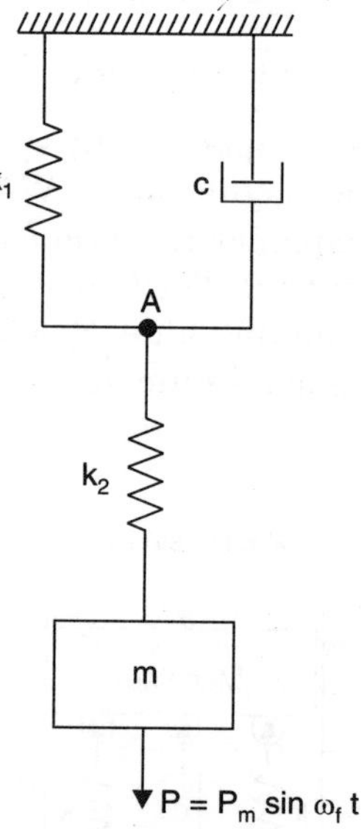

Fig. PD3.22

MECHANICAL VIBRATIONS

PM3.1 A safety bumper placed at the end of a race track to stop out-of-control cars is shown in Fig. PM3.1. The bumper is designed such that the force that the bumper applies to the car is a function of the velocity v and the displacement x of the front end of the bumper given by the equation:

$$F = Kv^3(x + 1)^3$$

where $K = 35$ kg-s/m^5 (a constant).

A car with a mass of 2000 kg hits the bumper at a speed of 100 km/h. Determine and plot the velocity of the car as a function of its position for $0 \le x \le 5$ m.

Fig. PM3.1

PM3.2 The 10 kg body is moved 0.25 m to the right of the equilibrium position and released from rest at $t = 0$ as shown in Fig. PM3.2. Plot the displacement as a function of time for four cases: $c = 10$, 40, 50 and 60 N.s/m. The stiffness of the spring is 40 N/m.

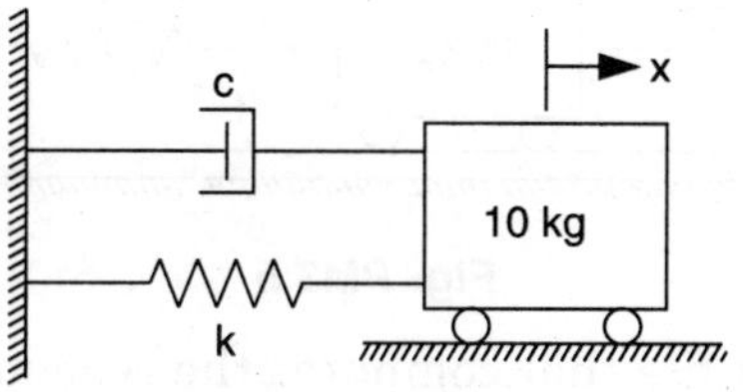

Fig. PM3.2

PM3.3 An airplane uses a parachute and other means of braking as it slows down on the runway after landing. The acceleration of the airplane is given by

$$a = -\,0.005\, v^2 - 4 \text{ m/s}^2$$

Consider an airplane with a velocity of 500 km/h that opens its parachute and starts decelerating at $t = 0$ s.

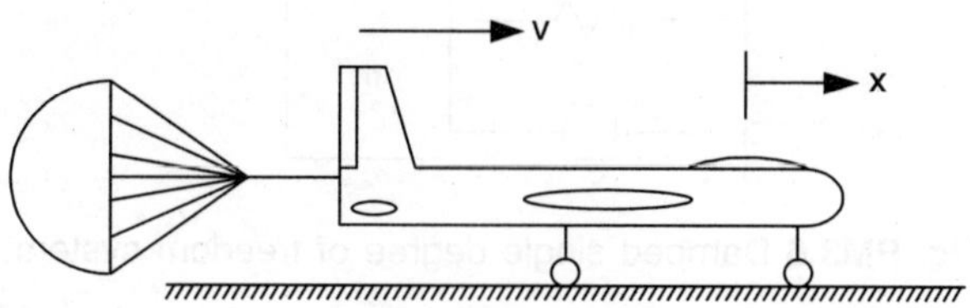

Fig. PM3.3

PM3.4 The piston of 150 lb is supported by a spring of modulus $k = 250$ lb/in. A dashpot of damping coefficient $c = 100$ lb.sec/ft acts in parallel with the spring. A fluctuating pressure $p = 0.75 \sin 30t$ (psi) acts on the piston, whose top surface area is 100 in^2. Plot the response of the system for initial conditions $x_0 = 0.06$ ft and $\dot{x}_0 = 6$, 0, and -6 ft./sec.

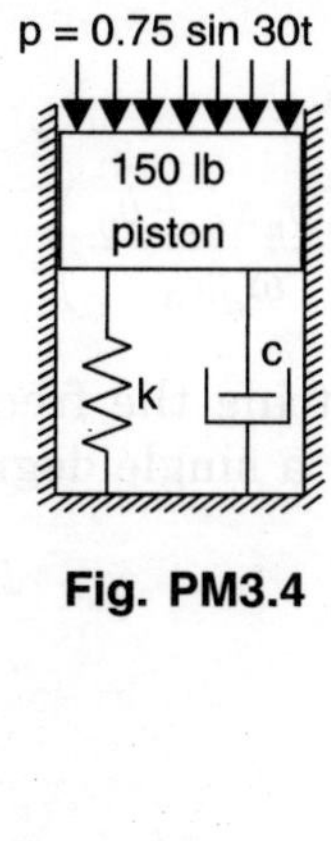

Fig. PM3.4

PM3.5 The 15 kg oscillator contains an unbalanced motor whose speed is N rpm as shown in Fig. PM3.5. The stiffness of the spring $k = 1100$ N/m. The oscillator is also restrained by a viscous damper whose piston is resisted by a force of 50 N when moving at a speed of 0.6 m/s. Determine

(*a*) the viscous damping factor

(*b*) plot the magnification factor for motor speeds from 0 to 350 rpm

(*c*) the maximum value of the magnification factor and the corresponding motor speed.

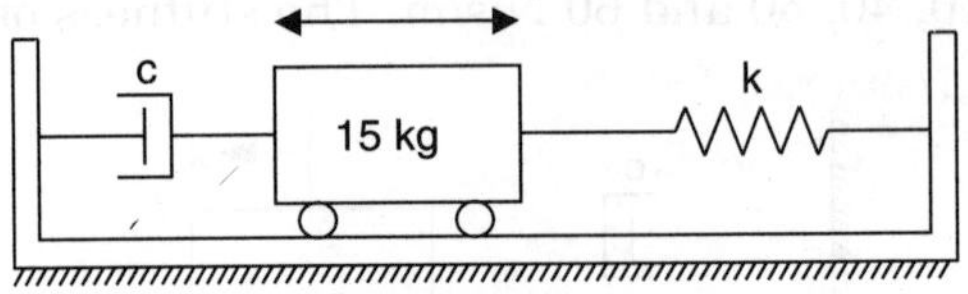

Fig. PM3.5

PM3.6 Write a MATLAB script file that computes the response of a single degree of freedom under damped system shown in Fig. PM3.6 to initial excitations. Use the program to determine and plot the response for the following data:

Initial conditions:

$$x(0) = 0,\ \dot{x}_0 = v_0 = 30 \text{ cm./sec},$$

$$\omega_n = 6 \text{ rad/s, and } \xi = 0.05, 0.1, 0.2, 0.30.$$

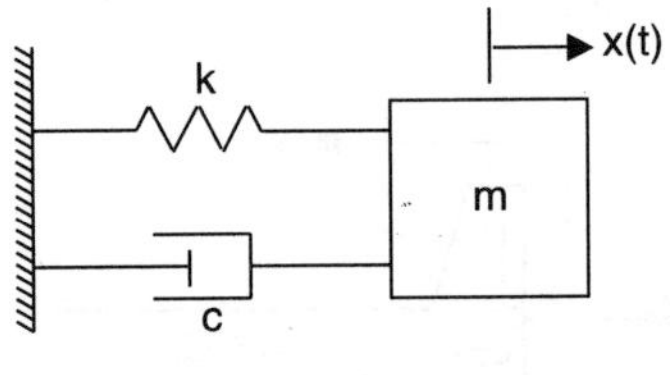

Fig. PM3.6 Damped single degree of freedom system.

The response of the under damped single degree of freedom system is given by

$$x(t) = Ae^{-\xi\omega_n t}\cos(\omega_d t - \phi)$$

where A and ϕ represent the amplitude and phase angle of the response respectively. There are

$$A = \sqrt{x_0^2 + \left(\frac{\xi\omega_n x_0 + v_0}{\omega_d}\right)^2}$$

$$\omega_d = \sqrt{1-\xi^2}\,\omega_n$$

and

$$\phi = \tan^{-1}\left(\frac{\xi\omega_n x_0 + v_0}{\omega_d x_0}\right)$$

PM3.7 Write a MATLAB script for plotting the frequency response magnitude and phase angle using complex notation for a single degree of freedom system given by

$$G(i\omega) = \frac{1-\left(\dfrac{\omega}{\omega_n}\right)^2 - i\,2\xi\left(\dfrac{\omega}{\omega_n}\right)}{\left[1-\left(\dfrac{\omega}{\omega_n}\right)^2\right]+\left(2\xi\dfrac{\omega}{\omega_n}\right)^2}$$

and $$\phi(\omega) = \tan^{-1}\left[\frac{-\operatorname{Im}G(i\omega)}{\operatorname{Re}G(i\omega)}\right] = \tan^{-1}\left[\frac{2\zeta\dfrac{\omega}{\omega_n}}{1-\left(\dfrac{\omega}{\omega_n}\right)^2}\right]$$

PM3.8 Consider the force-free, viscously damped single degree of freedom system shown in Fig. PM3.8.

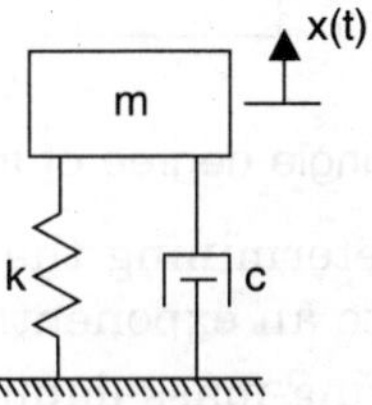

Fig. PM3.8

Plot the response of the system using MATLAB over the interval $0 \le t \le 10$ s to the initial conditions $x(0) = 3$ cm, $\dot{x}(0) = 0$ for the values of the damping factor $\zeta = 0.05, 0.1, 0.5$. The frequency of the undamped oscillation have the values $\omega_n = 15$ rad/s.
The expression for the response of an damped single degree of freedom system in Fig. PM3.8 to initial displacement and velocity is given by

$$x(t) = C\,e^{-\zeta\omega_n t}\cos(\omega_d t - \phi)$$

where C and ϕ represent the amplitude and phase angle of the response, respectively having the values

$$C = \sqrt{x_0^2 + \left(\frac{\zeta\omega_n x_0 + v_0}{\omega_d}\right)^2}$$

$$\phi = \tan^{-1}\left(\frac{\zeta\omega_n x_0 + v_0}{\omega_d x_0}\right)$$

and $$\omega_d = \sqrt{1-\zeta^2}\;\omega_n$$

PM3.9 Write a MATLAB script to obtain the motion of the mass subjected to the initial condition. There is no external forcing function acting on the system. The single degree of freedom system is shown in Fig. PM3.9 and the parameters are given as $m = 3$ kg, $k = 6$ N/m, and $C = 5$ N-s/m. The displacement of the mass is measured from the equilibrium position and at $t = 0$, $x(0) = 0.04$ m and $\dot{x}(0) = 0.10$ m/s.

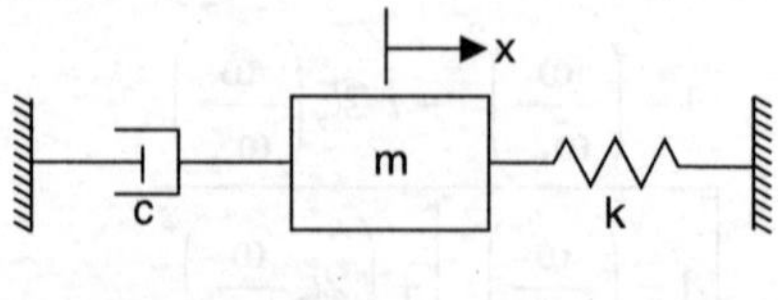

Fig. PM3.9 Single degree of freedom system.

PM3.10 Determine and plot the response of the single degree of freedom system shown in Fig. PM3.10 using MATLAB when 25 N of force (step input) is applied to the mass m. The system is at rest initially and the displacement of the mass m is measured from equilibrium position. The parameters of the system are given as $m = 3$ kg, $c = 25$ N-s/m, and $k = 200$ N/m. The initial conditions are $x(0) = \dot{x}(0) = 0$.

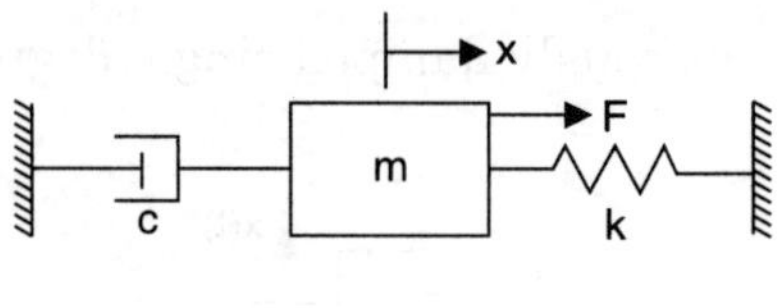

Fig. PM3.10 Single degree of freedom system.

PM3.11 Write a MATLAB script for determining the response of a single degree of freedom system with viscous damping to an exponential excitation $F(t) = e^{-\alpha t}$.

PM3.12 A single degree of freedom spring-mass-damper model has following properties: m = 15 g, c = 25 Ns/m and k = 3500 N/m. If it is subjected to a triangular pulse of amplitude 1000 N for 0.1 seconds, compute the time-domain response and plot the same in MATLAB.

The excitation function is shown in Fig. PM3.12.

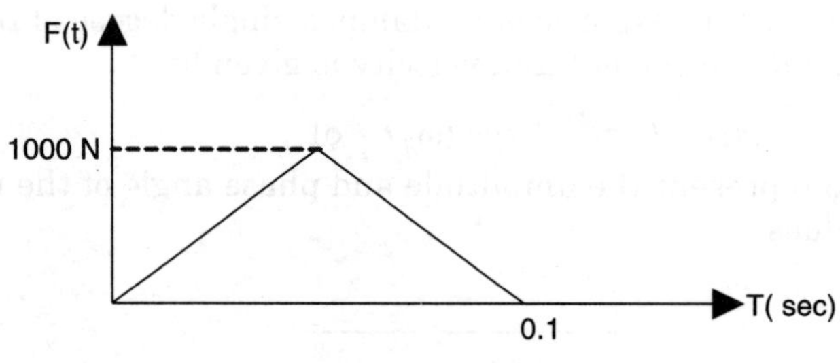

Fig. PM3.12

PM3.13 Determine and plot the response of the system shown in Fig. PM3.13 using MATLAB. The response is $x_0(t)$ when the input $x_i(t)$ is a unit step displacement input. The parameters of the system are k_1 = 15 N/m, k_2 = 25 N/m, c_1 = 7 N-s/m, c_2 = 15 N-s/m.

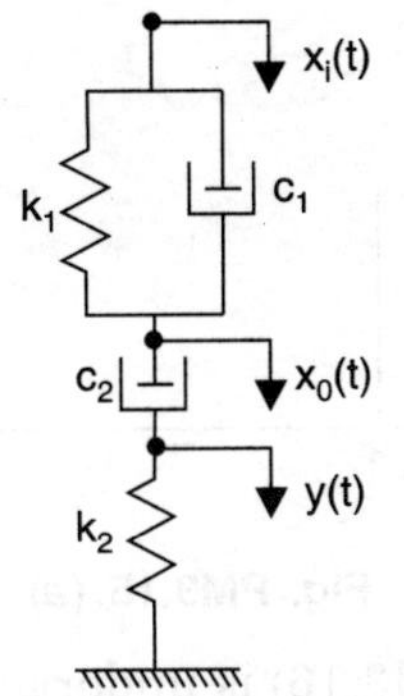

Fig. PM3.13

PM3.14 A two-degree of freedom torsional system shown in Fig. PM3.14 is subjected to initial excitation $\theta_1(0) = 0$, $\theta_2(0) = 2$, $\dot{\theta}_1(0) = 2\sqrt{\frac{GJ}{IL}}$ and $\dot{\theta}_2(0) = 0$. Write MATLAB program and plot the response of the system. Assume $I = 1$ and $GJ = l = 1$.

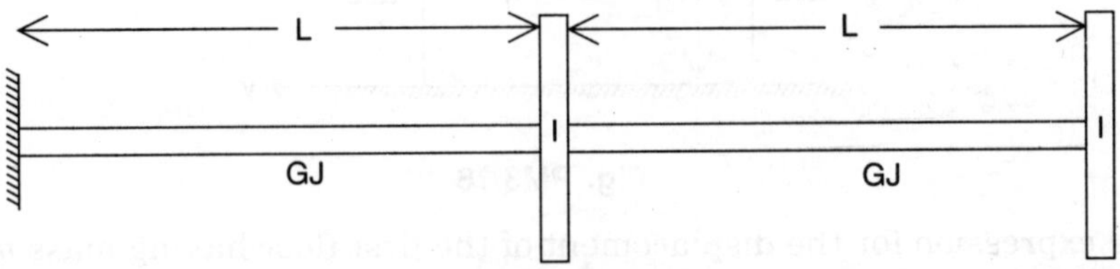

Fig. PM3.14

PM3.15 The mass m_2 in a 2-degree of freedom system shown in Fig. PM3.15 is subjected to a force in the form of saw-tooth pulse of amplitude 1.5 N for duration of 1.5 second. Obtain the response in terms of two coordinates $x_1(t)$ and $x_2(t)$. Assume $k_1 = k_2 = 15$N/m and $m_1 = m_2 = 2$kg.

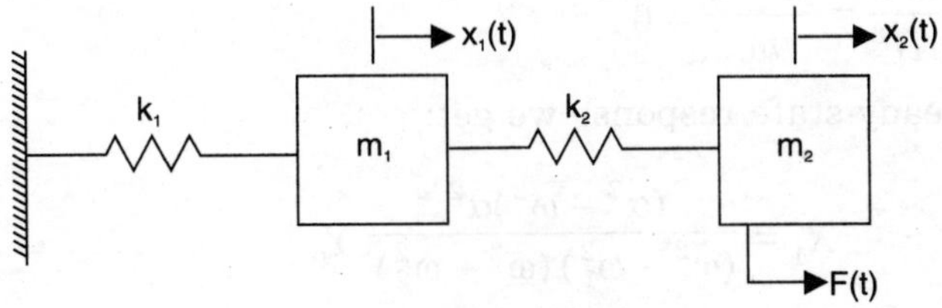

Fig. PM3.15

The mass and stiffness matrices of the system for given system are given as

$$M = \begin{bmatrix} m_1 & 0 \\ 0 & m_2 \end{bmatrix} \quad \text{and} \quad K = \begin{bmatrix} k_1 + k_2 & -k_2 \\ -k_2 & k_2 \end{bmatrix}$$

Force vector is:

$$F = \begin{Bmatrix} F(t) \\ 0 \end{Bmatrix}$$

The saw-tooth pulse takes the form as shown in Fig. PM3.15 (*a*).

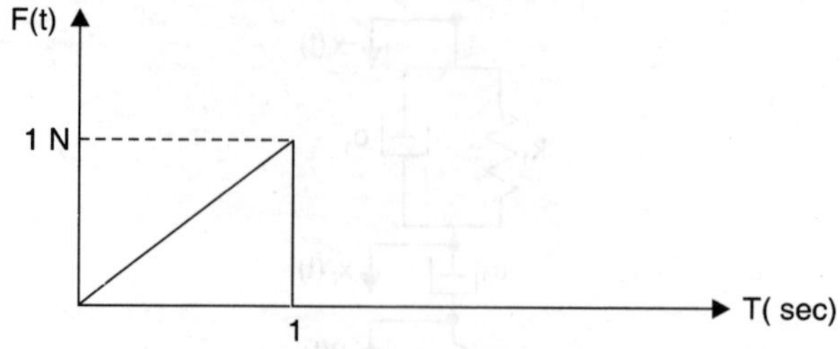

Fig. PM3.15 *(a)*

PM3.16 A two-storey building (Fig. PM3.16) is undergoing a horizontal motion $y(t) = Y_0 \sin \omega t$.

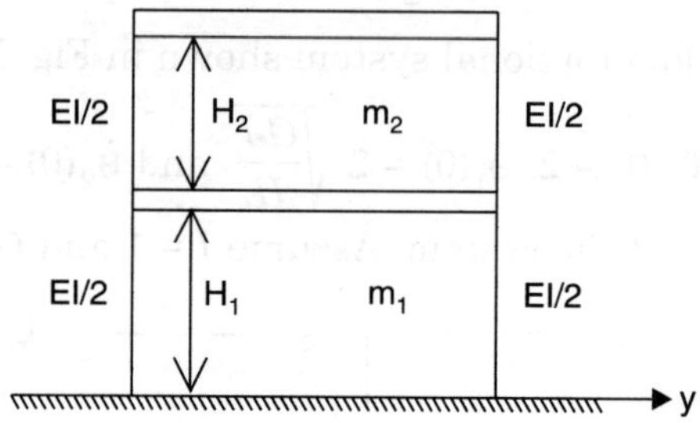

Fig. PM3.16

Derive expression for the displacement of the first floor having mass m_1. Assume $m_1 = m_2 = 4$, $EI = 2$ and $H = 1$m.

The equations of motion for building can be written as:

$$\begin{bmatrix} 4 & 0 \\ 0 & 4 \end{bmatrix} \begin{Bmatrix} \ddot{x}_1 \\ \ddot{x}_2 \end{Bmatrix} + \alpha^2 \begin{bmatrix} 2 & -1 \\ -1 & 1 \end{bmatrix} \begin{Bmatrix} x_1 \\ x_1 \end{Bmatrix} = \alpha^2 \begin{Bmatrix} Y_0 \sin \omega t \\ 0 \end{Bmatrix}$$

where $\alpha^2 = \dfrac{12EI}{mH^3} = \dfrac{12 \times 2}{m} = 6$

Solving for steady-state response we get:

$$X_1 = \frac{(\alpha^2 - \omega^2)\alpha^2}{(\omega^2 - \omega_1^2)(\omega^2 - \omega_2^2)} Y_0$$

$$X_2 = \frac{\alpha^4}{(\omega^2 - \omega_1^2)(\omega^2 - \omega_2^2)} Y_0$$

These values are to be plotted against various values of ω.

PM3.17 Derive the response of the system shown in Fig. PM3.17 in discrete time and plot the response. Given $F(t) = e^{-\alpha t}$

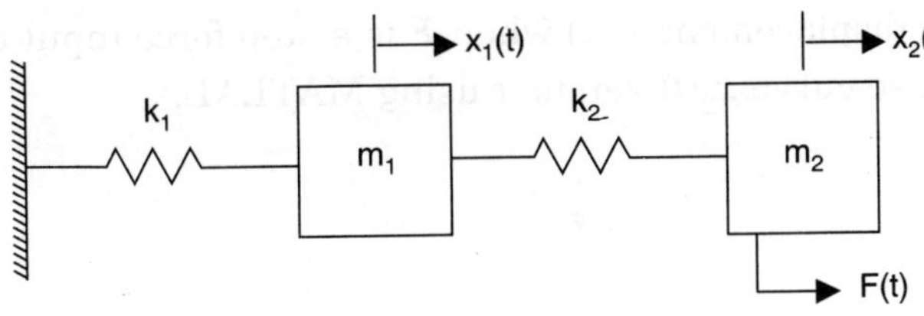

Fig. PM3.17

PM3.18 Consider the system with $M = \begin{bmatrix} 3 & 0 \\ 0 & 2 \end{bmatrix}$, $K = \begin{bmatrix} 6 & -4 \\ -4 & 5 \end{bmatrix}$ with arbitrary viscous damping. Find the eigenvalues and normalized eigenvectors.

PM3.19 For the vibrating system shown in Fig. EM3.19, a mass of 5 kg is placed on mass m at $t = 0$ and the system is at rest initially (at $t = 0$). Given that $m = 20$ kg, $k = 600$ N/m, and $c = 60$ Ns/m. Plot the response curve $x(t)$ versus t using MATLAB.

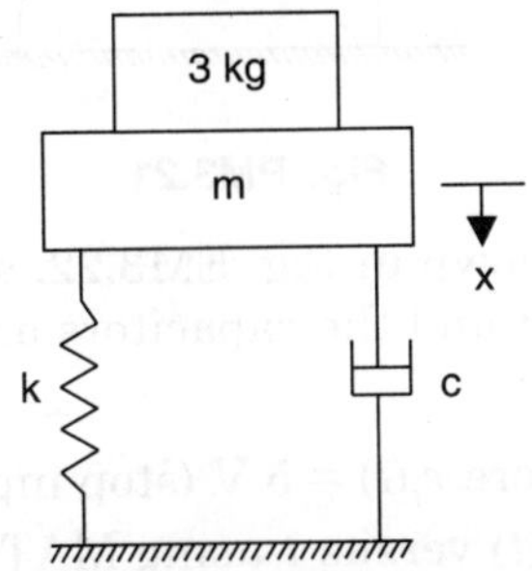

Fig. PM3.19

PM3.20 For the mechanical vibrating system shown in Fig. PM3.20, using MATLAB assume that $m = 3$ kg, $k_1 = 15$ N/m, $k = 25$ N/m, and $c = 10$ N-s/m. Plot the response curve $x(t)$ versus t when the mass m is pulled slightly downward and the initial conditions are $x(0) = 0.05$ m and $\dot{x}(0) = 0.8$ m/s.

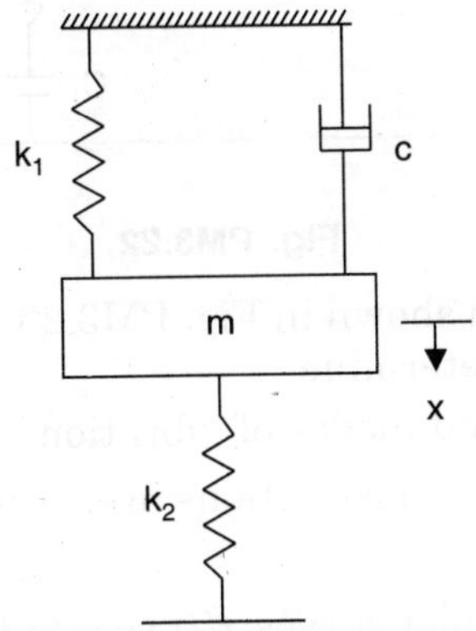

Fig PM3.20

PM3.21 For the mechanical vibrating system shown in Fig. PM3.21, $k_1 = 10$ N/m, $k_2 = 30$ N/m, $c_1 = 3$ N-s/m, and $c_2 = 25$ N-s/m.

(*a*) Determine the displacement $x_2(t)$ when F is a step force input of 4 N.
(*b*) Plot the response curve $x_2(t)$ versus t using MATLAB.

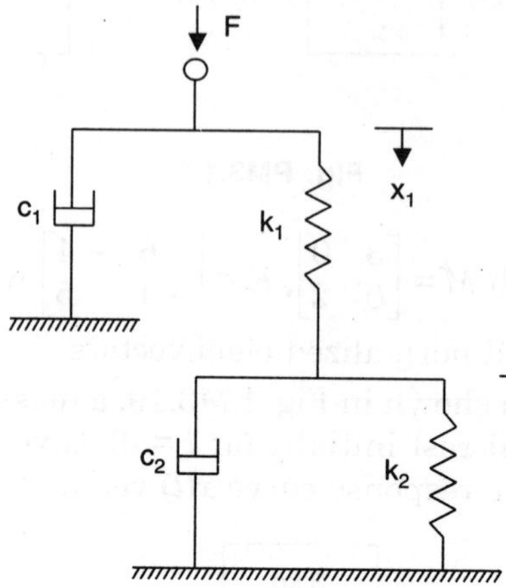

Fig. PM3.21

PM3.22 For the electrical system shown in Fig. EM3.22, assume that R_1 = 2 Ω, R_2 = 1 MΩ, C_1 = 0.75 μF, and C_2 = 0.25 μF and the capacitors are not charged initially and $e_0(0) = 0$ and $\dot{e}_0$ (0) = 0.

(*a*) Find the response $e_0(t)$ where $e_t(t)$ = 5 V (stop input) is applied to the system.
(*b*) Plot the response curve $e_0(t)$ versus t using MATLAB.

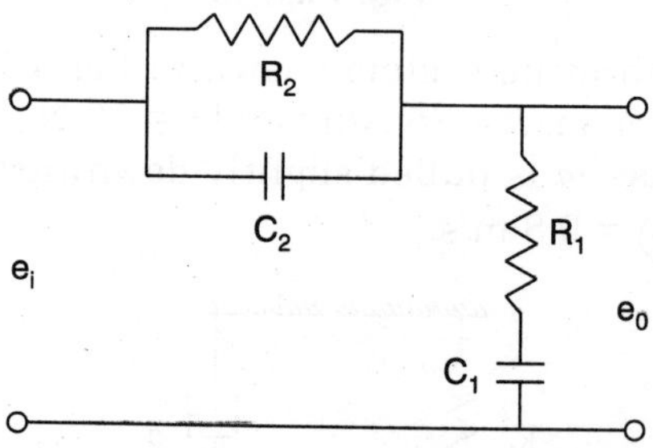

Fig. PM3.22

PM3.23 For the mechanical system shown in Fig. PM3.23, assume m = 3 kg, M= 25 kg, k_1= 25 N/m, and k_2 = 300 N/m. Determine

(*a*) the natural frequencies and modes of vibration
(*b*) the vibration when the initial conditions are: $x(0)$ = 0.05 m, $\dot{x}(0)$ = 0 m/s, $y(0)$ = 0 m, and $\dot{y}(0)$ = 0 m/s.

Use MATLAB program to plot curves $x(t)$ versus t and $y(t)$ versus t.

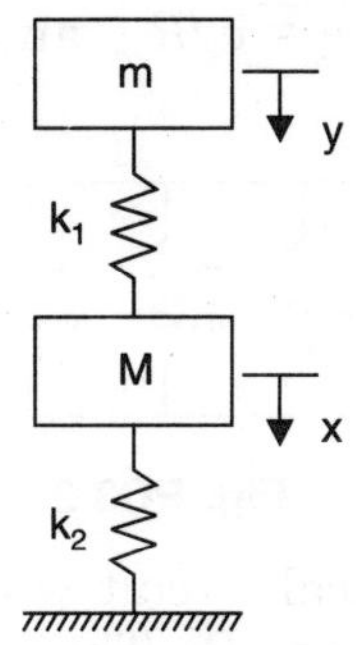

Fig. PM3.23

ELECTRICAL CIRCUITS

PC3.1 Figure PC3.1 shows an electrical circuit with resistors and voltage sources. Write a MATLAB program to determine the current in each resistor, using the mesh current method based on Kirchhoff's second voltage law. Given V_1 = 37V, V_2 = 19V, V_3 = 25V, R_1 = 16Ω, R_2 = 19Ω, R_3 = 11Ω, R_4 = 10Ω, R_5 = 6Ω, R_6 = 15Ω, R_7 = 9Ω, R_8 = 14Ω, R_9 = 6Ω and R_{10} = 3Ω.

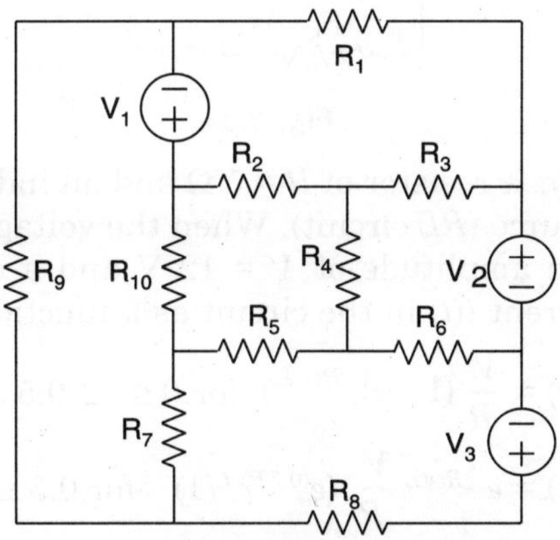

Fig. PC3.1

PC3.2 Write a MATLAB program in a script file that computes the current through each resistor and the power dissipated in each circuit that has resistors connected in parallel as shown in Fig. PC3.2. Use the script file for the circuit shown in Fig. PC3.2.

Note that when several resistors are connected in a circuit in parallel, the current through each of them is given by

$$i_n = \frac{v_s}{R_n}$$

where i_n and R_n are the current through resistors n and its resistance, respectively, and v_s is the source voltage. The equivalent resistance, R_{eq}, is given by

$$\frac{1}{R_{eq}} = \frac{1}{R_1} + \frac{1}{R_2} + \ldots + \frac{1}{R_n}$$

The source current i_s given by: $i_s = v_s/R_{eq}$, and the power, P_n, dissipated in each resistor is given by: $P_n = v_s i_n$.

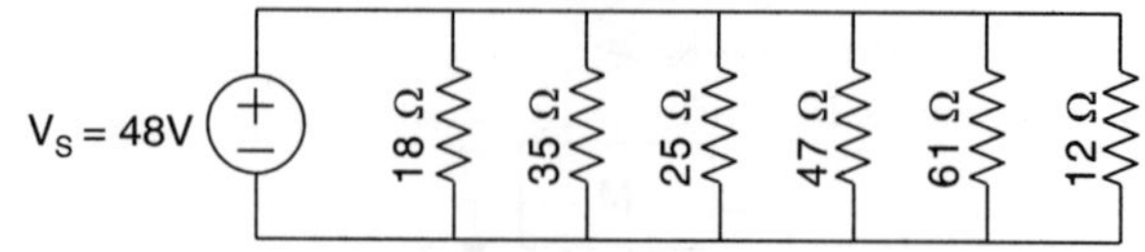

Fig. PC3.2

PC3.3 Figure PC3.3 shows an electrical circuit with a voltage source v_s with an internal resistance r_s, and a load resistance R_L. The power P dissipated in the load is given by

$$P = \frac{v_s^2 R_L}{(R_L + r_s)^2}$$

Write a MATLAB program to plot the power P as a function of R_L for $1 \le R_L \le 12\ \Omega$ given that $v_s = 12$ V, and $r_s = 3\ \Omega$.

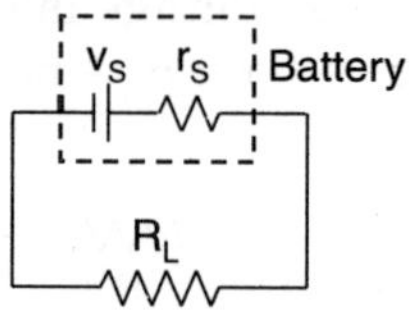

Fig. 3.3

PC3.4 Figure PC3.4 (*a*) shows a resistor of $R = 5\ \Omega$ and an inductor $L = 1.4$ H connected in a circuit to a voltage source (*RL* circuit). When the voltage source applies a rectangular voltage pulse with an amplitude of $V = 12$ V and a duration of 0.5 s as shown in Fig.PC3.4 (*b*), the current $i(t)$ in the circuit as a function of time is given by

$$i(t) = \frac{V}{R}(1 - e^{(-Rt)/L}) \quad \text{for } 0 \le t \le 0.5 \text{ s}$$

$$i(t) = e^{-(Rt)/L}\frac{V}{R}(e^{(0.5R)/L}/1) \quad \text{for } 0.5 \le t \text{ s}$$

Write a MATLAB program to plot the current as a function of time for $0 \le t \le 3$ s.

(*a*) (*b*)

Fig. PC3.4

PC3.5 The ratio of the magnitude of the voltage in a low-pass *RC* filter shown in Fig. PC3.5 is given by

$$RV = \left|\frac{V_o}{V_i}\right| = \frac{1}{\sqrt{1 + (\omega RC)^2}}$$

where ω is the frequency of the input signal.

Write a user-defined MATLAB function that calculates the magnitude ratio. Write a program in script file that uses the lowpass function to generate a plot of RV as a function of ω for $10^{-2} \le \omega \le 10^{6}$ rad/s. Run the script file with $R = 1300\ \Omega$, and $C = 9$ mF.

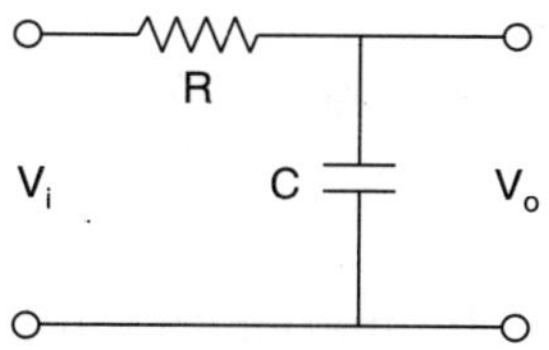

Fig. PC3.5

PC3.6 Figure PC3.6 shows an RLC with an alternating voltage source. The source voltage v_s is given by $v_s = v_m \sin(\omega_d t)$ where $\omega_d = 2\pi f_d$ in which f_d is the driving frequency. The amplitude of the current, I, is given by

$$I = \frac{v_m}{\sqrt{R^2 + (\omega_d L - 1/(\omega_d C))^2}}$$

where R and C are the resistance of the resistor and capacitance of the capacitor, respectively. For the circuit in the figure $C = 16 \times 10^{-6}$ F, $L = 250 \times 10^{-3}$ H, $v_m = 25$ V. Write a MATLAB program to make

(*a*) a 3-D plot of I (z axis) as a function of ω_d (x axis) for $60 \le f \le 110$ Hz, and a function of R (y axis) for $10 \le R \le 40\ \Omega$

(*b*) a plot that is a projection on the x-z plane. Estimate from this plot the natural frequency of the circuit. Compare the estimate with the calculated value of $1/(2\pi\sqrt{LC})$.

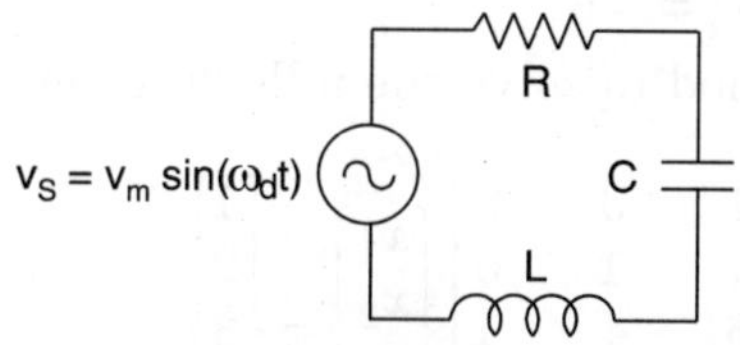

Fig. PC3.6

PC3.7 Figure PC3.7 shows an RC circuit which includes a voltage source v_s, a resistor $R = 50\ \Omega$ and a capacitor $C = 0.001$ F. The differential equation that describes the response of the circuit is given by

$$\frac{dv_c}{dt} + \frac{1}{RC}v_c = \frac{1}{RC}v_s$$

where v_c is the voltage of the capacitor. Initially, $v_s = 0$, and then at $t = 0$ the voltage source is changed. Find the response of the circuit for the following three cases:

(*a*) $v_s = 12$ V for $t \ge 0$.

(*b*) $v_s = 12 \sin(2.60\pi t)$ V for $t \ge 0$.

(*c*) $v_s = 12$ V for $0 \le t \le 0.01$ s, and then $v_s = 0$ for $t \ge 0.01$ s.

Solve each case for $0 \le t \le 0.2$ s. For each case plot v_s and v_c *vs.* time using a MATLAB program.

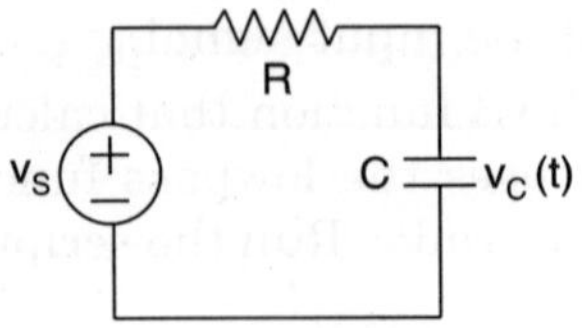

Fig. PC3.7

NUMERICAL METHODS

PN3.1 Use the method of Gaussian elimination to solve the following system of linear equations:

$$x_1 + x_2 + x_3 - x_4 = 2$$
$$4x_1 + 4x_2 + x_3 - x_4 = 11$$
$$x_1 - x_2 - x_3 + 2x_4 = 0$$
$$2x_1 + x_2 + 2x_3 - 2x_4 = 2$$

PN 3.2 Use Gaussian elimination method to solve the system of equations $[A]\{x\} = \{b\}$ where

$$A = \begin{bmatrix} 1 & 1 & 1 & 1 \\ 2 & -1 & 3 & 0 \\ 0 & 2 & 0 & 3 \\ -1 & 0 & 2 & 1 \end{bmatrix} \qquad b = \begin{Bmatrix} 3 \\ 3 \\ 1 \\ 0 \end{Bmatrix}$$

PN3.3 Solve the following set of equations by Gauss-Jordon method.

$$2x_1 + x_2 - 3x_3 = 11$$
$$4x_1 - 2x_2 + 3x_3 = 8$$
$$-2x_1 + 2x_2 - x_3 = -6$$

PN3.4 Use Gauss-Jordon method to solve the following set of equations.

$$\begin{bmatrix} 5 & -4 & 1 & 0 & 0 \\ -4 & 6 & -4 & 1 & 0 \\ 1 & -4 & 6 & -4 & 1 \\ 0 & 1 & -4 & 6 & -4 \\ 0 & 0 & 1 & -4 & -5 \end{bmatrix} \begin{Bmatrix} x_1 \\ x_2 \\ x_3 \\ x_4 \end{Bmatrix} = \begin{Bmatrix} 1 \\ 2 \\ 3 \\ 2 \\ 1 \end{Bmatrix}$$

PN3.5 Solve the following system of equations using Cholesky's factorizations.

$$x_1 + x_2 + x_3 - x_4 = 2$$
$$x_1 - x_2 - x_3 + 2x_4 = 0$$
$$4x_1 + 4x_2 + x_3 + x_4 = 11$$
$$2x_1 + x_2 + 2x_3 - 2x_4 = 2$$

PN3.6 Use Cholesky's method of solution for Problem PN3.2.

PN3.7 Use Jacobi iterative scheme to obtain the solutions of the following system of equations.

$$x + 2y + z = 0$$
$$3x + y - z = 0$$
$$x - y + 4z = 3$$

PN3.8 Use Jacobi iterative scheme to obtain the solution for Problem PN3.1.

PN3.9 Use Gauss-Seidel method to solve the following system of equations in Problem PN3.7.

PN3.10 Solve the system of equations in Problem PN3.2 using the Gauss-Seidel method.

PN3.11 Solve the following set of equations in Problem PN3.5 using Householder's factorization method.

PN3.12 Use the Householder reduction to transfer the following matrix A into tridiagonal form and solve the set of equations $Ax = b$, where

$$A = \begin{bmatrix} 7 & 2 & 3 & -1 \\ 2 & 8 & 5 & 1 \\ 3 & 5 & 12 & 9 \\ -1 & 1 & 9 & 7 \end{bmatrix} \qquad b = \begin{bmatrix} 2 \\ -3 \\ 5 \\ 7 \end{bmatrix}$$

PN3.13 Minimize $f(x_1, x_2) = x_1 - 0.5x_2 + 2x_1^2 + 2x_1x_2 + x_2^2$ using Newton's method. Starting point: (0, 0).

PN3.14 Use Newton's method to find the maximum of $f(x) = 5 \sin x - \frac{8}{19} x^2$ with the initial guess of $x_0 = 1, 8$.

PN3.15 Fit a polynomial by quadratic approximation and find the value of x at which $F(x)$ is minimum.

x	$F(x)$
1	12
2	7
3	20

PN3.16 Fit a polynomial by quadratic approximation and find the value of x at which $F(x)$ is minimum.

x	$F(x)$
1	20
2	−17
3	15

PN3.17 Find the minimum of the following function using Powell's method:

$$F(x) = 4x_1^2 + 3x_2^2 - 5x_1x_2 - 8x_1$$

Starting from (0, 0).

PN3.18 Find the minimum of the following function using Powell's method:

$$F(x) = 7x_1^2 + 5x_2^2 - 8x_1x_1 - 5x_1$$

Starting from (0, 0).

PN3.19 Repeat Problem PN3.17 using Fletcher-Reeves method.

PN3.20 Repeat Problem PN3.18 using Fletcher-Reeves method.

PN3.21 Minimize $f(x_1,x_2) = x_1 - x_2 + 2x_1^2 + 2x_1x_2 + x_2^2$ starting from the point (0, 0) using Hooke and Jeeves method.

PN3.22 Repeat Problem PN3.13 using Hooke and Jeeves method.

PN3.23 Minimize $f(x_1,x_2) = (x_1 - 1)^2 + (x_2 - 1)^2$

Subject to
$$g_1(x_1,x_2) = -x_1 \leq 0$$
$$g_2(x_1,x_2) = -x_2 \leq 0$$
$$g_3(x_1,x_2) = x_1^2 + x_1^2 - 1 \leq 0$$

PN3.24 Minimize $f(x_1,x_2) = (x_1 - 1)^2 + x_2^2$

Subject to
$$g_2(x_1,x_2) = -x_2 \leq 0$$
$$g_3(x_1,x_2) = x_2 - (1 - x_1)^2 \leq 0$$

PN3.25 Determine the eigenvalues and eigenvectors of the following matrix using Jacobi method.

$$[A] = \begin{bmatrix} 11 & 2 & 8 \\ 2 & 2 & -10 \\ 9 & -10 & 5 \end{bmatrix}$$

PN3.26 Use Jacobi method to compute the eigenvalues and the corresponding eigenvectors of the following matrix A:

$$[A] = \begin{bmatrix} 4 & 3 & 2 & 1 \\ 3 & 4 & 1 & 2 \\ 2 & 1 & 4 & 3 \\ 1 & 2 & 3 & 4 \end{bmatrix}$$

PN3.27 Use the MATLAB *fminbnd* function to find the minimum of the function in Problem PN3.21.

PN3.28 Minimize $f(x_1,x_2) = 90(x_2 - x_1)^2 + (1 - x_1)^2$ using the starting point ($x_1 = -1.1$, $x_2 = 1.0$). Use MATLAB built-in function *fminsearch.*

PN3.29 (*a*) Minimize $(x_1 - x_2)^2 + (x_2 - x_3)^4$.

Initial guess: (– 2.5, 2, 2)

(*b*) Minimize $x_1x_4(x_1 + x_2 + x_3) + x_3$

Initial guess: (1, 4, 5, 1).

PN3.30 Minimize $x_1 + 2x_2 + 3x_3 + \exp(x_1x_4)$

Initial guess: (1, 2, 1, 2).

PN3.31 Minimize $f(x) = x_1 + 54x_2 + 3x_3$

Subject to
$$x_1 + 5x_2 - x_3 \geq 4$$
$$-x_1 + x_2 + 2x_3 \geq 1$$
$$-x_1 + 3x_2 + 3x_3 \geq 5$$
$$-3x_1 + 8x_2 - 5x_3 \geq 3$$

Initial guess: $x^0 = (0, 1, 1)$.

PN3.32 Minimize $f(x) = x_1 + 5x_2$

Subject to
$$x_1 + x_2 \leq 3$$
$$-x_1 + x_2 \leq 1.5$$

Initial guess: $x^0 = (-1, 1)$.

PN3.33 Solve Problem PN3.31 with $x_1, x_2, x_3 \geq 0$.

PN3.34 Solve Problem PN3.32 with $x_1, x_2 \geq 0$.

PN3.35 Minimize $f(x) = 100(x_2 - x_1^2)^2 + (1 - x_1)^2$

Subject to

$$x_1 + 1 \geq 0$$
$$1 - x_2 \geq 0$$
$$5x_2 - x_1 - 1 \geq 0$$
$$2 - x_1 - 2x_2 \geq 0$$

Initial guess: [–1, 1].

PN3.36 Minimize $f(x) = x_1^2 + x_2^2 + x_3^2$

Subject to

$$1 - x_2x_3 \geq 0$$
$$x_1 - x_3 \geq 0$$
$$x_1 - x_2^2 + x_2x_3 - 5 = 0$$
$$0 \geq x_1 \geq 5$$
$$0 \geq x_2 \geq 3$$
$$0 \geq x_3 \geq 4$$

Initial guess: [–1, 1].

PN3.37 Minimize $f(x) = 2x_1 - 3(x_2 - 4)^2$

Subject to

$$x_1^2 + x_1^2 - 10 \geq 0$$
$$x_1^2 + (x_2 - 4)^4 \geq 0$$

Initial guess: [–1, 1].

PN3.38 Minimize $f(x) = (x_1 - x_2)^2 + x_2$

Subject to

$$1 - x_1 + x_2 \geq 1$$
$$x_1 + 2x_2 \geq 4$$
$$x_1 \geq 0$$
$$x_2 \geq 0$$

Initial guess: [2, 0].

CHAPTER 4

DIRECT NUMERICAL INTEGRATION METHODS

4.1 INTRODUCTION

The behavior of many dynamic systems undergoing time-dependant changes (transients) can be described by ordinary differential equations. When the solution to the differential equation(*s*) of motion of a dynamic system cannot be obtained in closed form, a numerical procedure is warranted. Many numerical integration methods are available for the approximate solution of such equation(*s*) of motion. All the numerical integration methods have two basic characteristics. First, they do not satisfy the differential equation(*s*) at all time t, but only at discrete time intervals, say Δt apart. Secondly, within each time interval Δt, a specific type of variation of the displacement X, velocity $\dot{X}$, and acceleration $\ddot{X}$ is assumed. Thus several numerical integration schemes are available depending on the type of variation assumed for $X, \dot{X}$ and $\ddot{X}$ within each time interval Δt.

In this chapter we discuss several widely used step-by-step numerical integration schemes for solutions of both single and multi degree of freedom systems. A brief description of these methods is presented for linear dynamic response analysis and their application is illustrated by several examples.

4.2 SINGLE DEGREE OF FREEDOM SYSTEM

The general equation of a viscously damped single degree of freedom dynamical system, which is linear, can be expressed in the following general form:

$$M\ddot{X} + C\dot{X} + KX = F(t) \tag{4.1}$$

where M, C and K are the mass, damping and stiffness of the system; $F(t)$ is the applied force; and X, $\dot{X}$ and $\ddot{X}$ are the displacement, velocity and acceleration of the system.

4.2.1 Finite Difference Method

If the equilibrium relation (4.1) is regarded as an ordinary differential equation with constant coefficients, it follows that any convenient *finite difference* expressions to approximate the velocities and accelerations in terms of displacements can be used. The central idea in the *finite difference* method is to use approximations to derivatives. Hence, the general differential equation such as (4.1) and the associated boundary conditions, if any, are replaced by the corresponding finite difference equations. The continuous variable t is replaced by the discrete variable t_i and the differential equation is solved progressively in time increments $h = \Delta t$ starting from known initial conditions. The solution obtained is approximate but by suitably selecting the time increment the accuracy of the solution can be improved.

In this method, we replace the solution domain with a finite number of points, known as mesh or grid points, and obtain the solution at these points. The mesh or grid points are generally equally spaced along the independent coordinate as shown in Fig. 4.1. Central difference method is based on the Taylor's series expansion of X_{i+1} and X_{i-1} about the grid point i.

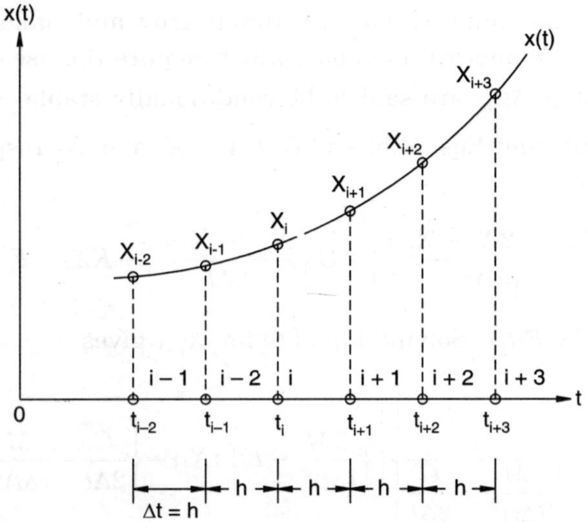

Fig. 4.1

$$X_{i+1} = X_i + h\dot{X}_i + \frac{h_2}{2}\ddot{X}_i + \frac{h^3}{6}\dddot{X}_i + \ldots \tag{4.2}$$

$$X_{i-1} = X_i - h\dot{X}_i + \frac{h_2}{2}\ddot{X}_i - \frac{h^3}{6}\dddot{X}_i + \ldots \tag{4.3}$$

where $X_i = X(t = t_i)$ and the interval $h = t_{i+1} - t_i = \Delta t$. By taking the first two terms only and subtracting Eq. (4.3) from (4.2), we obtain

$$\dot{X}_i = \frac{1}{2h}(X_{i+1} - X_{i-1}) \tag{4.4}$$

Adding Eqs. (4.3) and (4.2), we get

$$\ddot{X}_i = \frac{1}{h^2}(X_{i-1} - 2X_i + X_{i+1}) \tag{4.5}$$

Although there exist a number of finite difference schemes, here we consider only two methods selected for their simplicity. They are the central difference method and the Runge-Kutta method.

4.2.2 Central Difference Method

Let the duration over which the numerical solution of Eq. (4.1) is required be divided into n equal parts of interval $h = \Delta t$ each. The initial conditions are assumed to be given by

$$X(t = 0) = X_o \text{ and } \dot{X}(t = 0) = \dot{X}_0.$$

The accuracy of the solution always depends on the size of the time step. The critical time step is given by $\Delta t_{cri} = \tau_n / \pi$, where τ_n is the natural period of the system. If Δt is selected to be larger than Δt_{cri}, the method becomes unstable, that is the truncation of higher order terms or rounding-off in the computer causes errors to grow and makes the dynamic response calculations meaningless. Numerical methods, which require the use of time, step Δt smaller than the critical time step Δt_{cri} are said to be conditionally stable. A safe rule to use is to choose $h \le \tau_n / 10$. Substituting Eqs. (4.5) and (4.4) for $\ddot{X}_i$ and $\dot{X}_i$ respectively in Eq. (4.1) at mesh or grid point i give

$$M\left\{\frac{X_{i+1} - 2X_i + X_{i-1}}{(\Delta t)^2}\right\} + C\left\{\frac{X_{i+1} - X_{i-1}}{2\Delta t}\right\} + KX_i = F_i \tag{4.6}$$

where $X_i = X(t_i)$ and $F_i = F(t_i)$. Solving Eq. (4.6) for X_{i+1} gives

$$X_{i+1} = \left\{\frac{1}{\dfrac{M}{(\Delta t)^2} + \dfrac{C}{2\Delta t}}\right\}\left[\left\{\frac{2M}{(\Delta t)^2} - K\right\}\{X_i\} + \left\{\frac{C}{2\Delta t} - \frac{M}{(\Delta t)^2}\right\} X_{i-1} + F_i\right] \tag{4.7}$$

which is known as the recurrence formula.

Thus the displacement of the mass, X_{i+1}, can be calculated using Eq. (4.7) if we know the previous displacements, X_i, X_{i-1} and the present external force F_i. Repeated application of Eq. (4.7) gives us the response time history of the behavior of the system. Since the solution of X_{i+1} given in Eq. (4.7) is based on the previous displacement X_i given in Eq. (4.6), the integration procedure is known as an explicit integration method. Note that in order to compute X_1, both X_o and X_{-1} are required but the initial conditions provide only the values of X_o and $\dot{X}_o$. Therefore the central difference method is not self-starting. However, the value of X_{-1} can be obtained from Eqs. (4.4) and (4.5) as follows. First calculate the value of $\ddot{X}_o$ by substituting the given values of X_o and $\dot{X}_o$ in Eq. (4.1) as follows,

$$\ddot{X}_o = \frac{1}{M}[F(t = 0) - C\dot{X}_o - KX_o] \tag{4.8}$$

The value of X_{-1} is then obtained by the application of Eqs. (4.4) and (4.5) at $i = 0$.

$$X_{-1} = X_o - \Delta t\, \dot{X}_o + \frac{(\Delta t)^2}{2} \ddot{X}_o \tag{4.9}$$

4.2.3 Runge-Kutta Method

In the Runge-Kutta method, an approximation to $X_{(t+\Delta t)}$ is obtained from X_t in such a way that the power series expansion of the approximation coincides, up to terms of a certain order $(\Delta t)^N$ in the time interval Δt, with the actual Taylor series expansion of $(t + \Delta t)$ in powers of Δt. The method is based on the assumption that the higher derivatives exist at points required.

The Runge-Kutta method is self-starting and has the advantage that no initial values are needed beyond the prescribed values. A brief discussion of its basis is represented here. In the Runge-Kutta method the second-order differential equation is first reduced to two first-order equations. Consider the differential equation for the single degree of freedom system given in Eq. (4.1). Equation (4.1) can be rewritten as

$$\ddot{X} = \frac{1}{M}[F(t) - C\dot{X} - KX] = f\,(X, \dot{X}, t) \tag{4.10}$$

By letting $X_1 = X$ and $X_2 = \dot{X}$, Eq. (4.10) can be reduced to the following two first-order equations:

$$\begin{aligned} \dot{X}_1 &= X_2 \\ \dot{X}_2 &= f(X_1, X_2, t) \end{aligned} \tag{4.11}$$

By defining

$$\overline{X}(t) = \begin{Bmatrix} X_1\,(t) \\ X_2\,(t) \end{Bmatrix}$$

and

$$\overline{F}(t) = \begin{Bmatrix} X_2 \\ f(X_1, X_2, t) \end{Bmatrix}$$

the following recurrence formula is obtained to find the values of $\overline{X}(t)$ at mesh or grids points t_i according to the fourth order Runge-Kutta method. We omit the details of the derivation of the method.

$$\overline{X}_{i+1} = \overline{X}_i + \frac{1}{6}[\overline{K}_1 + 2\overline{K}_2 + 2\overline{K}_3 + \overline{K}_4]$$

where $\overline{K}_1 = h\overline{F}(\overline{X}_i, t_i)$

$$\begin{aligned} \overline{K}_2 &= h\overline{F}(\overline{X}_i + \frac{1}{2}\overline{K}_1, t_i + \frac{1}{2}h) \\ \overline{K}_3 &= h\overline{F}(\overline{X}_i + \frac{1}{2}\,K_2, t_i + \frac{1}{2}h) \end{aligned} \tag{4.12}$$

and

$$\overline{K}_4 = h\overline{F}(\overline{X}_i + \frac{1}{2}\overline{K}_3, t_{i+1})$$

Although the Runge-Kutta method does not require the computation of derivatives beyond the first, its higher accuracy is obtained by four evaluations of the first derivatives to obtain agreement with the Taylor series solution through terms of order h^4. Since the fourth order Runge-Kutta method is an explicit method, the maximum time step is usually governed by stability considerations. The change in time step can be easily implemented between iterations and hence the method can be considered as an inherently stable method. The main drawback of the method is that each forward step requires several computations of the functions thus increasing the computational cost. The Runge-Kutta method is applicable and extendable to a system of differential equations.

4.3 MULTI DEGREE OF FREEDOM SYSTEM

The general form of the equations of motion for a multi-degree of freedom system are written as

$$[M]\{\ddot{X}\} + [C]\{\dot{X}\} + [K]\{X\} = \{F(t)\} \tag{4.13}$$

where $[M], [C]$, and $[K]$ are the mass, damping and stiffness matrices for the system and $[\ddot{X}], [\dot{X}]$, and $[X]$ refer to the acceleration, velocity, and displacement vectors, respectively. $\{F(t)\}$ is the force vector. Several numerical direct integrating schemes are available to determine the approximate solution of a system of equations of motion. For a linear dynamic system, matrices $[M], [C]$ and $[K]$ are independent of time and therefore remain unchanged during the integration procedure. These matrices vary with time for a nonlinear dynamic system and must be modified during the integration of equations of motion. For the solution of equations of motion for a linear dynamic system, either the normal mode superposition method of dynamic analysis or direct numerical integration methods can be used. However, for the solution of nonlinear equations of motion, direct numerical integration methods are generally recommended.

In a direct integration method, the system of equations of motion is integrated successively by using a step-by-step numerical procedure. No transformation of the equations of motion is needed prior to integration and using difference formulas that involve one or more increments of time usually approximates time derivatives. Basically there are two principal approaches used in the direct integration method: explicit and implicit schemes. In an explicit scheme, the response quantities are expressed in terms of previously determined values of displacement, velocity and acceleration. In an implicit scheme, the difference equations are combined with the equations of motion, and the displacements are calculated directly by solving the equations.

In this section, only selected numerical integration schemes widely used for linear and nonlinear dynamic analyses are considered. Three explicit and four implicit direct integration schemes are examined. A brief description of these schemes is presented and their application is illustrated. The explicit schemes presented are the central difference method, two-cycle iteration with trapezoidal rule, and fourth-order Runge-Kutta. The implicit schemes include the Houbolt, Wilson-Theta, Newark-Beta, and Park Stiffly stable methods. The accuracy, stability, and efficiency of these schemes are examined by comparing the results for sample problems.

4.4 EXPLICIT SCHEMES

As mentioned earlier, in an explicit formulation, the response quantities are expressed in terms of previously determined values of displacement, velocity and acceleration.

4.4.1 Central Difference Method

The procedure indicated for the case of a single degree of freedom system can be directly extended to this case. Consider a displacement time history curve as shown in Fig. 4.2. At the middle of the time interval Δt, the velocity is given by

$$\dot{X}_{i+\frac{1}{2}} = \frac{X_{i+1} - X_i}{\Delta t} \tag{4.14}$$

and

$$\dot{X}_{i-\frac{1}{2}} = \frac{X_i - X_{i-1}}{\Delta t} \tag{4.15}$$

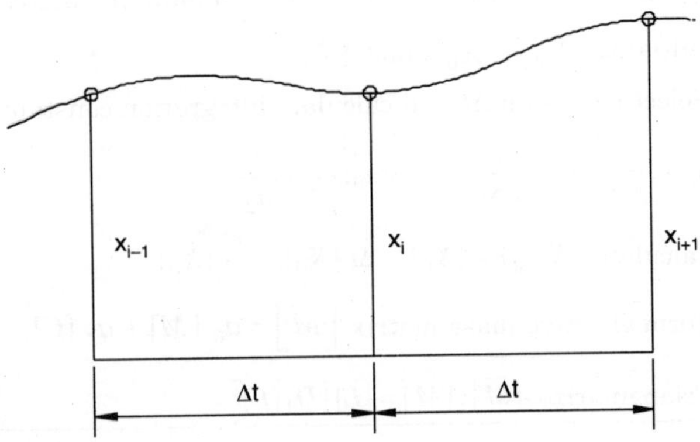

Fig 4.2

The acceleration is

$$\ddot{X}_i = \frac{\dot{X}_{i+\frac{1}{2}} - \dot{X}_{i-\frac{1}{2}}}{\Delta t} \tag{4.16}$$

Substituting $\dot{X}_{i+\frac{1}{2}}$ and $\dot{X}_{i-\frac{1}{2}}$ from the Eqs. (4.14) and (4.15) into Eq. (4.16), we get

$$\ddot{X}_i = \frac{1}{\Delta t^2}(X_{i+1} - 2X_i + X_{i-1}) \tag{4.17}$$

The difference formulas in the central difference method for velocity and acceleration are written in terms of displacement as

$$\{\dot{X}_t\} = \frac{1}{2\Delta t}[\{X_{t+\Delta t}\} - \{X_{t-\Delta t}\}] \tag{4.18}$$

$$\{\ddot{X}_t\} = \frac{t}{\Delta t^2}[\{X_{t+\Delta t}\} - 2\{X_t\} + \{X_{t-\Delta t}\}] \tag{4.19}$$

Substituting $\{\dot{X}_t\}$ and $\{\ddot{X}_t\}$ from Eqs. (4.18) and (4.19), respectively into Eq. (4.13), we get

$$[\overline{M}]\{X_{t+\Delta t}\} = \{\overline{F}_t\} \tag{4.20}$$

where $[\overline{M}]$, the effective mass matrix, and $[\overline{F}_t]$, the effective force vector, is given by

$$[\overline{M}] = \frac{1}{\Delta t^2}[M]\frac{1}{2\Delta t}[C] \tag{4.21}$$

$$[\overline{F}_t] = \{F_t\} - ([K] - \frac{2}{\Delta t^2}[M])\{X_t\} - (\frac{1}{\Delta t^2}[M] - \frac{1}{2\Delta t}[C])\{X_{t-\Delta t}\} \tag{4.22}$$

At time $t + \Delta t$, the displacements $\{X_{t+\Delta t}\}$ can be computed by solving Eq. (4.20), and the velocities and accelerations at time t are determined by substituting these values of $\{X_{t+\Delta t}\}$ into Eqs. (4.18) and (4.19). Note that the calculation of $\{X_{t+\Delta t}\}$ involves $\{X_t\}$ and $\{X_{t-\Delta t}\}$. Hence, in order to obtain the solution at time Δt, a special starting procedure is needed. Table 4.1 summarizes the time integration schedule as suitable for integration in the computer.

Table 4.1

Algorithms based on the central-difference method

(*a*) Initial computations: 1. Form stiffness $[K]$, mass $[M]$, and damping $[C]$ matrices. 2. Initialize $\{X_0\}$, $\{\dot{X}_0\}$, and $\{\ddot{X}_0\}$. 3. Select time step Δt and calculate integration constants a_i: $a_0 = \frac{1}{\Delta t^2}; a_1 = \frac{1}{2\Delta t}; a_2 = 2a_0; a_3 = \frac{1}{a_2}.$ 4. Calculate $\{X_{-\Delta t}\} = \{X_0\} - \Delta t\,\{\dot{X}_0\} + a_3\,\{\ddot{X}_0\}$. 5. Form effective mass matrix $\lfloor \overline{M} \rfloor = a_0\,[M] + a_1\,[C]$. 6. Triangularize $[\overline{M}] : [\overline{M}] = [L][D][L]^T$
(*b*) For each time step: 1. Calculate effective force vector at time t: $\{\overline{F}\} = \{F_t\} - ([K] - a_2\,[M])\,\{X_t\} - (a_0\,[M] - a_1\,[C])\,\{X_{t-\Delta t}\}$ 2. Solve for displacements at time $t + \Delta t$: $\lfloor \overline{M} \rfloor \{X_{t+\Delta t}\} = \lfloor \overline{F}_t \rfloor$ 3. Calculate $\{\dot{X}\}$ and $\{\ddot{X}\}$ at time t: $\{\dot{X}_t\} = a_1(-\{X_{t-\Delta t}\} - \{X_{t+\Delta t}\})$ $\{\ddot{X}_t\} = a_0\,(\{X_{t-\Delta t}\} - 2\{X_t\} + \{X_{t+\Delta t}\})$

The local truncation error of this method is of the order of Δt^2. An important consideration in the use of the central difference method is that the integration method requires that the time step Δt smaller than a critical value, Δt_{cr}, which is limited by the highest frequency of the discrete system $\omega_{\max}$, where

$$\Delta t \le \Delta t_{cr} \le \frac{2}{\omega_{\max}} \qquad \text{...(4.23)}$$

If the time step is longer than Δt_{cr}, the integration is unstable, meaning that any errors resulting from the numerical integration of round off in the computer grow and make the dynamic response calculations questionable.

4.4.2 Two-Cycle Iteration with Trapezoidal Rule

The equations of motion at any time t are expressed in the incremental form as

$$[M]\,\{\Delta\ddot{X}_t\} = \{\Delta F_t\} - [K]\,\{\Delta X_t\} - [C]\,\{\Delta\dot{X}_t\} \qquad (4.24)$$

The increments in the velocities and displacements are estimated by the use of the following relationships in the first iteration cycle:

For the first time step:

$$\{\Delta\dot{X}_t\} = \Delta t\,\{\ddot{X}_{t-\Delta t}\} \tag{4.25}$$

For other time steps:

$$\{\Delta\dot{X}_t\} = 2\Delta t\,\{\ddot{X}_{t-\Delta t}\} - \{\Delta\dot{X}_{t-\Delta t}\} \tag{4.26}$$

$$\{\dot{X}_t\} = \{\dot{X}_{t-\Delta t}\} + \{\Delta\dot{X}_t\} \tag{4.27}$$

$$\{\Delta X_t\} = \frac{1}{2}\Delta t\,[\{\ddot{X}_{t-\Delta t}\} + \{\dot{X}_t\}] \tag{4.28}$$

By substituting the relations $\{\Delta\dot{X}_t\}$ and $\{\Delta X_t\}$ from Eqs. (4.26) and (4.28) into Eq. (4.24) we obtain the increments in the accelerations as

$$\{\Delta\ddot{X}_t\} = [M]^{-1}\,(\{\Delta F_t\} - [K]\,\{\Delta X_t\} - [C]\,\{\Delta\dot{X}_t\}) \tag{4.29}$$

These are then employed to obtain the estimate of the acceleration at time t as

$$\{\ddot{X}_t\} = \{\ddot{X}_{t-\Delta t}\} + \{\Delta\ddot{X}_t\} \tag{4.30}$$

In the second iteration cycle, the increments in the velocities and the accelerations are refined as

$$\{\Delta\dot{X}_t\} = \frac{1}{2}\Delta t(\{\ddot{X}_{t-\Delta t}\} + \{\ddot{X}_t\}) \tag{4.31}$$

$$\{\dot{X}_t\} = \{\dot{X}_{t-\Delta t}\} + \{\Delta\dot{X}_t\} \tag{4.32}$$

$$\{\Delta X_t\} = \frac{1}{2}\Delta t(\{\dot{X}_{t-\Delta t}\} + \{\dot{X}_t\}) \tag{4.33}$$

Finally, the relations for $\{\Delta\dot{X}_t\}$ and $\{\Delta X_t\}$ in Eqs. (4.31) and (4.33) are substituted into Eq. (4.29) to compute the new increments in the accelerations. These are then used in Eq. (4.30) to calculate the accelerations at time t. The computational algorithm based on this method is summarized in Table 4.2.

Table 4.2

Algorithm based on two-cycle iteration with trapezoidal rule

1. Form stiffness [K], mass [M] and damping [C] matrices
2. Initialize $\{X_0\}, \{\dot{X}_0\}$ and $\{\ddot{X}_0\}$
3. Select time step Δt and calculate for the first time step in the first iteration cycle: $\{\Delta\dot{X}_t\} = \Delta t\,\{\ddot{X}_{t-\Delta t}\}$
4. For other time steps: $\{\Delta\dot{X}_t\} = 2\Delta t\,\{\ddot{X}_{t-\Delta t}\} - \{\Delta\dot{X}_{t-\Delta t}\}$; $\{\dot{X}_t\} = \{\dot{X}_{t-\Delta t}\} + \{\Delta\dot{X}_t\}$; $\{\Delta X_t\} = \frac{\Delta t}{2}\,(\{\dot{X}_{t-\Delta t}\} - \{\dot{X}_t\})$

5. Compute $\{\Delta\ddot{X}_t\}$ and $\{\ddot{X}_t\}$ at time t: $$\{\Delta\ddot{X}_t\} = [M]^{-1}(\{\Delta F_t\} - [K]\{\Delta X_t\} - [C]\{\Delta\dot{X}_t\})$$ $$\{\ddot{X}_t\} = \{\ddot{X}_{t-\Delta t}\} + \{\Delta\ddot{X}_t\}$$
6. Second iteration cycle: $$\{\Delta\dot{X}_t\} = \frac{\Delta t}{2}(\{\ddot{X}_{t-\Delta t}\} + \{\ddot{X}_t\})$$ $$\{\dot{X}_t\} = \{\dot{X}_{t-\Delta t}\} + \{\Delta\dot{X}_t\}$$ $$\{\Delta X_t\} = \frac{\Delta t}{2}(\{\dot{X}_{t-\Delta t}\} + \{\dot{X}_t\})$$
7. Compute $\{\Delta\ddot{X}_t\}$ in step 5 using $\{\Delta\dot{X}_t\}, \{\dot{X}_t\}$ and $\{\Delta X_t\}$ from step 6.
8. Finally, compute $\{\ddot{X}_t\}$ from step 5, using $\{\ddot{X}_t\}$ and $\{\Delta\ddot{X}_t\}$ from step 7.

4.4.3 Fourth Order Runge-Kutta Method

In the fourth order Runge-Kutta method, the system of second order differential Eq. (4.13) are converted into state variable form. That is, both the displacements and velocities are treated as unknowns $\{y\}$ defined by

$$\{y\} = \begin{Bmatrix} \{X\} \\ \{\dot{X}\} \end{Bmatrix} \tag{4.34}$$

Using Eq. (4.33), Eq.(4.13) can be rewritten as

$$\{\ddot{X}\} = -[M]^{-1}[K]\{X\} - [M]^{-1}[C]\{\dot{X}\} + [m]^{-1}\{F(t)\} \tag{4.35}$$

Using the identity

$$\{\dot{y}\} = \{\dot{y}\} \tag{4.36}$$

we obtain from Eqs. (4.35) and (4.36)

$$\{\dot{y}\} = \begin{Bmatrix} \{\dot{X}\} \\ \{\ddot{X}\} \end{Bmatrix} = \begin{bmatrix} [0] \\ -[M]^{-1}[K] \end{bmatrix}\begin{bmatrix} [I] \\ -[M]^{-1}[C] \end{bmatrix}\begin{Bmatrix} \{X\} \\ \{\dot{X}\} \end{Bmatrix} + \begin{Bmatrix} 0 \\ [M]^{-1}\{F(t)\} \end{Bmatrix} \tag{4.37}$$

or

$$\{\dot{y}\} = [E]\{y\} + \{F^*(t)\} \tag{4.38}$$

That is

$$\{\dot{y}\} = \{f(t, y)\} \tag{4.39}$$

In the Runge-Kutta method, an approximation to $\{y_{y+\Delta t}\}$ is obtained from $\{y_t\}$ in such a way that the power series expansion of the approximation coincides, upto the terms of a certain order $(\Delta t)^N$ in the time interval Δt, with the actual Taylor series expansion of $(t + \Delta t)$ in powers of Δt. This method has the advantage that no initial values are required beyond the prescribed ones. The general fourth order algorithms are based on formulas of the form

$$\{y_{t+\Delta t}\} = \{y_t\} + \Delta t(aK_1 + bK_2 + cK_3 + dK_4) \tag{4.40}$$

where a, b, c and d are constants and K_1, K_2, K_3, and K_4, are the approximate derivative values computed in the interval $t_K < t < t_{K+\Delta t}$. Several fourth order algorithms have been proposed. The following is due to Runge-Kutta and we omit presenting the details of its derivation.

$$\{y_{t+\Delta t}\} = \{y_t\} + \frac{\Delta t}{6}\,[K_1 + 2K_2 + 2K_3 + K_4] \tag{4.41}$$

in which

$$K_1 = f(t, y_t)$$

$$K_2 = f(t + \frac{\Delta t}{2}, y_t + K_1\frac{\Delta t}{2})$$

$$K_3 = f(t + \frac{\Delta t}{2}, y_t + K_2\frac{\Delta t}{2})$$

$$K_4 = f(t + \Delta t, y_t + K_3\Delta t) \tag{4.42}$$

The Runge-Kutta algorithm does not require the calculation of higher derivatives. This method is completely self-starting and the step size can be changed easily between iterations and hence the method can be considered inherently stable. The truncation error e_t for the fourth order Runge-Kutta scheme is of the form

$$e_t = c(\Delta t)^5 \tag{4.43}$$

where c is constant, which depends on $f(t, y)$ and its higher-order partial derivatives. Runge-Kutta method generates an artificial damping that unduly suppresses the amplitude of the response of a dynamic system to some extent.

4.5 IMPLICIT SCHEMES

In an implicit scheme, the difference equations are combined with the equations of motion, and the displacements are calculated directly by solving the equations.

4.5.1 Houbolt Method

The Houbolt method is based on third-order interpolation of displacements X_t, and the multistep implicit formulas for $\dot{X}_t$ and $\ddot{X}_t$ are obtained in terms of X_t by using backward differences.

The difference formulas are summarized in the following with reference to Fig. 4.3. By considering a cubic curve that passes through the four successive ordinates, the following equations can be obtained:

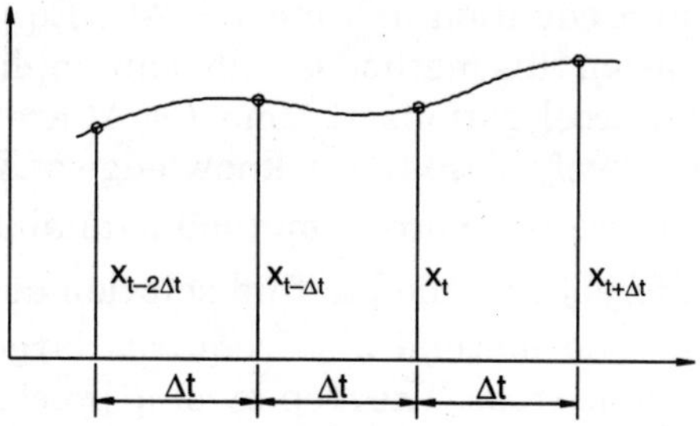

Fig. 4.3

$$X_t = X_{t-\Delta t} - \Delta t\dot{X}_{t+\Delta t} + \frac{\Delta t^2}{2}\ddot{X}_{t+\Delta t} - \frac{\Delta t^3}{6}\dddot{X}_{t+\Delta t} \tag{4.44}$$

$$X_{t-\Delta t} = X_{t+\Delta t} - (2\Delta t)\dot{X}_{t+\Delta t} + \frac{(2\Delta t)^2}{2}\ddot{X}_{t+\Delta t} - \frac{(2\Delta t)^3}{6}\dddot{X}_{t+\Delta t} \tag{4.45}$$

$$X_{t-2\Delta t} = X_{t+\Delta t} - (3\Delta t)\ \dot{X}_{t+\Delta t} + \frac{(3\Delta t)^2}{2}\ddot{X}_{t+\Delta t} - \frac{(3\Delta t)^3}{6}\dddot{X}_{t+\Delta t} \tag{4.46}$$

Solving Eqs. (4.44) to (4.46) for $\ddot{X}_{t+\Delta t}$ and $\dot{X}_{t+\Delta t}$ we get

$$\ddot{X}_{t+\Delta t} = \frac{1}{\Delta t^2}(2X_{t+\Delta t} - 5X_t + 4X_{t-\Delta t} - X_{t-2\Delta t}) \tag{4.47}$$

$$\dot{X}_{t+\Delta t} = \frac{1}{6\Delta t}(11X_{t+\Delta t} - 18X_t + 9X_{t-\Delta t} - 2X_{t-2\Delta t}) \tag{4.48}$$

The difference formulas in the Houbolt algorithm are therefore given by

$$\{\ddot{X}_{t+\Delta t}\} = \frac{1}{\Delta t^2}[2\{X_{t+\Delta t}\} - 5\{X_t\} + 4\{X_{t-\Delta t}\} - \{X_{t-2\Delta t}\}] \tag{4.49}$$

$$\{\dot{X}_{t+\Delta t}\} = \frac{1}{6\Delta t}[11\{X_{t+\Delta t}\} - 18\{X_t\} + 9\{X_{t-\Delta t}\} - 2\{X_{t-2\Delta t}\}] \tag{4.50}$$

By substituting the expressions for $\{\ddot{X}_{t+\Delta t}\}$ and $\{\dot{X}_{t+\Delta t}\}$ from (4.49) and (4.50), respectively, into (4.13), we get

$$\lfloor \overline{M} \rfloor \{X_{t+\Delta t}\} = \{\overline{F}_{t+\Delta t}\} \tag{4.51}$$

where $\lfloor \overline{M} \rfloor$ is the effective mass matrix and $\{\overline{F}_{t+\Delta t}\}$ is the effective force vector.

$$[\overline{M}] = \frac{2}{\Delta t^2}[M] + \frac{11}{6\Delta t}[C] + [K] \tag{4.52}$$

$$\begin{aligned}\{\overline{F}_{t+\Delta t}\} &= \{F_{t+\Delta t}\} + (\frac{5}{\Delta t^2}[M] + \frac{3}{\Delta t}[C])\{X_t\} \\ &- (\frac{4}{\Delta t^2}[M] + \frac{3}{2\Delta t}[C])\{X_{t-\Delta t}\} + (\frac{1}{\Delta t^2}[M] + \frac{1}{3\Delta t}[C])\{X_{t-2\Delta t}\}\end{aligned} \tag{4.53}$$

Note that the equilibrium equation at time $t + \Delta t$, Eq. (4.51) is used in finding the solution for $\{X_{t+\Delta t}\}$. For this reason, this method is called an implicit integration method. It can be seen that the velocities and accelerations at time $t + \Delta t$ are obtained by substituting for $\{X_{t+\Delta t}\}$ in (4.50) and (4.49) respectively. Also that a knowledge of X_t, $X_{t-\Delta t}$, and $X_{t-2\Delta t}$ is needed to find solution for $\{X_{t+\Delta t}\}$. Since there is no direct method available to find $\{X_{t-\Delta t}\}$ and $\{X_{t-2\Delta t}\}$, initially we use the central difference method to find solution at time Δt and $2\Delta t$. This makes the method non-self-starting. The method also requires large computer storage to store displacements for the previous time steps. The step-by-step procedure to be used in the Houbolt method is summarized in Table 4.3. A basic difference between the Houbolt method in Table 4.3 and the central difference method in Table 4.2 is the appearance of the stiffness matrix $\boldsymbol{K}$ as a factor to the required displacements $X_{t+\Delta t}$. The term $KX_{t+\Delta t}$ appears because the equilibrium is considered at time $t + \Delta t$ and not at time t as in the central-difference method. There is no critical time-step limit, and Δt can in general be selected much larger than that given for the central-difference method.

Table 4.3

Algorithm based on Houbolt Method

(*a*) Initial Computations:

1. Form stiffness [*K*], mass [*M*] and damping [*C*] matrices
2. Initialize $\{X_0\}, \{\dot{X}_0\}$ and $\{\ddot{X}_0\}$.
3. Select time step Δt and calculate integration constants:

$$a_0 = \frac{2}{\Delta t^2};\ a_1 = \frac{11}{6\Delta t};\ a_2 = \frac{5}{\Delta t^2};\ a_3 = \frac{3}{\Delta t};\ a_4 = -2a_0;\ a_5 = \frac{-a_3}{2};$$

$$a_6 = \frac{a_0}{2};\ a_7 = \frac{a_3}{9}.$$

4. Use special starting procedure, such as central-difference method to calculate $\{X_{\Delta t}\}$ and $\{X_{2\Delta t}\}$.
5. Calculate effective stiffness matrix:

$$\left\lfloor \overline{K} \right\rfloor = [K] + a_0\,[M] + a_1\,[C]$$

6. Triangularize $[\overline{K}]: [\overline{K}] = [L][D][L]^T$

(*b*) For each time step:

1. Calculate effective force vector at time $t + \Delta t$:

$$\{\overline{F}_{t+\Delta t}\} = \{F_{t+\Delta t}\} + [M]\,(a_2\,\{X_t\} + a_4\,\{X_{t-\Delta t}\} + a_6\,\{X_{t-2\Delta t}\})$$
$$+[C]\,(a_3\{X_t\} + a_5\,\{X_{t-\Delta t}\} + a_7\,\{X_{t-2\Delta t}\})$$

2. Solve for displacements at time $t + \Delta t$

$$\left\lfloor \overline{K} \right\rfloor \{X_{t+\Delta t}\} = \{\overline{F_{t+\Delta t}}\}$$

3. Calculate $\{\dot{X}\}$ and $\{\ddot{X}\}$ at time $t + \Delta t$:

$$\{\dot{X}_{t+\Delta t}\} = a_1\,\{X_{t+\Delta t}\} - a_3\,\{X_t\} - a_5\,\{X_{t-\Delta t}\} - a_7\,\{X_{t-2\Delta t}\}$$

$$\{\ddot{X}_{t+\Delta t}\} = a_0\,\{X_{t+\Delta t}\} - a_2\,\{X_t\} - a_4\,\{X_{t-\Delta t}\} - a_6\,\{X_{t-2\Delta t}\}$$

4.5.2 Wilson Theta Method

The Wilson Theta method assumes that the acceleration of the system varies linearly between two instants of time. Referring to Fig. 4.4, the acceleration is assumed to be linear from time *t* between $t_i = i\Delta t$ to time $t_{i+\theta} = t_i + \theta\Delta t$, where $\theta \geq 1.0$. Because of this reason, the method is known as the Wilson Theta method.

If τ is the increase in time *t* between *t* and $t + \theta\Delta t$ $(0 \leq \tau \leq \theta\Delta t)$, then for time interval *t* to $t + \theta\Delta t$, it can be assumed that

$$\ddot{X}_{t+\tau} = \ddot{X}_t + \frac{\tau}{\theta\Delta t}\,(\ddot{X}_{t+\theta\Delta t} - \ddot{X}_t) \tag{4.54}$$

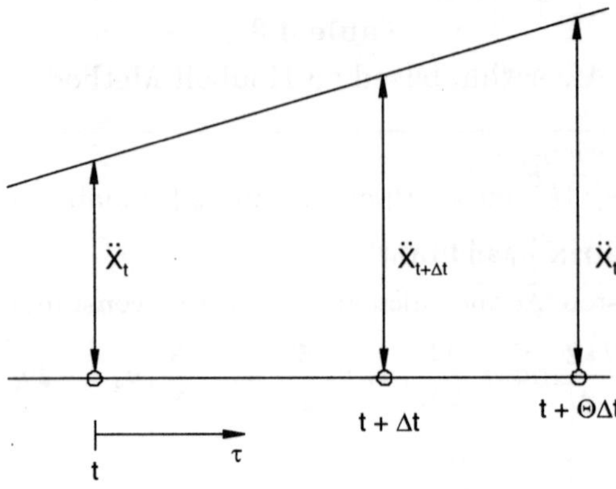

Fig. 4.4

Successive integration for Eq. (4.54) gives the following expressions for $\ddot{X}_{t+\tau}$ and $X_{t+\tau}$:

$$\dot{X}_{t+\tau} = \dot{X}_t + \ddot{X}_t\,\tau + \frac{\tau^2}{2\theta\Delta t}(\ddot{X}_{t+\theta\Delta t} - \ddot{X}_t) \tag{4.55}$$

$$X_{t+\tau} = X_t + \dot{X}_t\,\tau + \frac{1}{2}\ddot{X}_t\,\tau^2 + \frac{\tau^3}{6\theta\Delta t}(\ddot{X}_{t+\theta\Delta t} - \ddot{X}_t) \tag{4.56}$$

Substituting $\tau = \theta\Delta t$ into the above Eqs. (4.55) and (4.56), we obtain the following expressions for $\dot{X}$ and X at time $t + \theta\Delta t$:

$$\dot{X}_{t+\theta\Delta t} = \dot{X}_t + \frac{\theta\Delta t}{2}(\ddot{X}_t + \ddot{X}_{t+\theta\Delta t}) \tag{4.57}$$

$$X_{t+\theta\Delta t} = X_t + \theta\Delta t\,\dot{X}_t + \frac{\theta^2\Delta t^2}{6}(\ddot{X}_{t+\theta\Delta t} + 2\ddot{X}_t) \tag{4.58}$$

Solving Eqs. (4.57) and (4.58) for $\ddot{X}_{t+\theta\Delta t}$ and $\dot{X}_{t+\theta\Delta t}$ in terms of $X_{t+\theta\Delta t}$, we get

$$\ddot{X}_{t+\theta\Delta t} = \frac{6}{\theta^2\Delta t^2}(X_{t+\theta\Delta t} - X_t) - \frac{6}{\theta\Delta t}(\dot{X}_t) - 2\,(\ddot{X}_t)$$

$$\dot{X}_{t+\theta\Delta t} = \frac{3}{\theta\Delta t}(X_{t+\theta\Delta t} - X_t) - 2\dot{X}_t - \frac{\theta\Delta t}{2}\ddot{X}_t \tag{4.59}$$

The difference formulas in the Wilson Theta algorithm are then given by

$$\{\ddot{X}_{t+\theta\Delta t}\} = \frac{6}{\theta^2\Delta t^2}(\{X_{t+\theta\Delta t}\} - \{X_t\}) - \frac{6}{\theta\Delta t}\{\dot{X}_t\} - 2\,\{\ddot{X}_t\} \tag{4.60}$$

$$\{\ddot{X}_{t+\theta\Delta t}\} = \frac{3}{\theta\Delta t}(\{X_{t+\theta\Delta t}\} - \{X_t\}) - 2\,\{\dot{X}_t\} - \frac{\theta\Delta t}{2}\{\ddot{X}_t\} \tag{4.61}$$

We employ Eq. (4.13) at time $\tau + \Delta t$ to obtain a solution for displacement, velocity, and acceleration at time $t + \Delta t$. Since accelerations vary linearly, a linear projected force vector is used such that

$$[M]\,\{\ddot{X}_{t+\theta\Delta t}\} + [C]\,\{\dot{X}_{t+\theta\Delta t}\} + [K]\,\{X_{t+\theta\Delta t}\} = \{F_{t+\theta\Delta t}\} \tag{4.62}$$

where $\{F_{t+\theta\Delta t}\} = \{F_t\} + \theta\Delta t\,(\{F_{t+\theta\Delta t}\} - \{F_t\})$.

By substituting the expressions for $\{\ddot{X}_{t+\theta\Delta t}\}$ and $\{\dot{X}_{t+\theta\Delta t}\}$ from Eq. (4.60) and (4.61), respectively, into Eq.(4.62), we get

$$\lfloor \overline{M} \rfloor \{X_{t+\theta\Delta t}\} = \{\overline{F}_{t+\Delta t}\} \tag{4.63}$$

where the effective mass matrix $\lfloor \overline{M} \rfloor$ and the effective force vector $\{\overline{F}_{t+\Delta t}\}$ are given by

$$[\overline{M}] = \frac{6}{\theta^2 \Delta t^2}[M] + \frac{3}{\theta\Delta t}[C] + [K] \tag{4.64}$$

$$\{\overline{F}_{t+\theta\Delta t}\} = \{F_{t+\theta\Delta t}\} + (\frac{6}{\theta^2\Delta t^2}[M] + \frac{3}{\theta\Delta t}[C])\{X_t\}$$

$$+(\frac{6}{\theta\Delta t}[M] + 2[C])\{\dot{X}_t\} + (2[M] + \frac{\theta\Delta t}{2}[C])\{\ddot{X}_t\} \tag{4.65}$$

The solution of Eq. (4.63) gives $\{X_{t+\theta\Delta t}\}$ which is then substituted into the following relationships to obtain the displacements, velocities, and accelerations at time $t + \Delta t$.

$$\{\ddot{X}_{t+\Delta t}\} = \frac{6}{\theta^2\Delta t^2}(\{X_{t+\theta\Delta t}\} - \{X_t\}) - \frac{6}{\theta^2\Delta t}\{\dot{X}_t\} + (1 - \frac{3}{\theta})\{\ddot{X}_t\} \tag{4.66}$$

$$\{\dot{X}_{t+\Delta t}\} = \{\dot{X}_t\} + \frac{\Delta t}{2}(\{\ddot{X}_{t+\Delta t}\} - \{\ddot{X}_t\}) \tag{4.67}$$

$$\{X_{t+\Delta t}\} = \{X_t\} + \Delta t\,(\{\dot{X}_t\} + \frac{\Delta t^2}{6}\{\ddot{X}_{t+\Delta t}\} - 2\{\ddot{Y}_t\}) \tag{4.68}$$

When $\theta = 1.0$, the method reduces to the linear acceleration scheme. The method is unconditionally stable for linear dynamic systems when $\theta \geq 1.37$, and a value of $\theta = 1.4$ is often used for nonlinear dynamic systems. It may also be noted that no special starting procedures are needed, since $X, \dot{X}$ and $\ddot{X}$ are expressed at time $\Delta + \Delta t$ in terms of the same quantities at time t only. The complete algorithm used in the Wilson Theta method is given in Table 4.4.

Table 4.4

Algorithm based on Wilson Theta method

(*a*) Initial Computations:

1. Form stiffness $[K]$, mass $[M]$ and damping $[C]$ matrices
2. Initialize $\{X_0\}\{\dot{X}_0\}$ and $\{\ddot{X}_0\}$.
3. Select time step Δt and calculate integration constants, $\theta = 1.4$ (say):

$$a_0 = \frac{6}{(\theta\Delta t)^2};\ a_1 = \frac{2}{\theta\Delta t};\ a_2 = 2a_1;\ a_3 = \frac{\theta\Delta t}{2};\ a_4 = \frac{a_0}{\theta};\ a_5 = \frac{-a_2}{\theta};$$

$$a_6 = 1 - \frac{3}{\theta};\ a_7 = \frac{\Delta t}{2};\ a_8 = \frac{\Delta t^2}{6}$$

4. Form effective stiffness matrix: $\lfloor \overline{K} \rfloor = [K] + a_0[M] + a_1[C]$
5. Triangularize $[\overline{K}]$: $[\overline{K}] = [L][D][L]^T$

(*b*) For each time step:

1. Calculate effective force vector at time $t + \Delta t$:

$$\{\overline{F_{t+\theta\Delta t}}\} = \{F_t\} + \theta\,(\{F_{t+\Delta t}\} - \{F_t\}) + [M]\,(a_0\,\{X_t\} + a_2\,\{\dot{X}_t\} + 2\,\{\ddot{X}_t\})$$
$$+ [C]\,(a_1\,\{X_1\} + 2\,\{\dot{X}_t\} + a_3\,\{\ddot{X}_t\})$$

2. Solve for displacements at time $t + \theta\,\Delta t$:

$$\lfloor \overline{K} \rfloor \{X_{t+\theta\Delta t}\} = \{\overline{F_{t+\theta\Delta t}}\}$$

3. Calculated $\{X\}, \{\dot{X}\}$ and $\{\ddot{X}\}$ at time $t + \Delta t$:

$$\{\ddot{X}_{t+\Delta t}\} = a_4\,(\{X_{t+\theta\Delta t}\} - \{X_t\}) + a_5\,\{\dot{X}_t\} + a_6\,\{\ddot{X}_t\}$$

$$\{\dot{X}_{t+\Delta t}\} = \{\dot{X}_t\} + a_7\,(\{\ddot{X}_{t+\Delta t}\} + \{\ddot{X}_t\})$$

$$\{X_{t+\Delta t}\} = \{X_t\} + \Delta t\,\{\dot{X}_t\} + a_8\,(\{\ddot{X}_{t+\Delta t}\} + 2\,\{\ddot{X}_t\})$$

4.5.3 Newmark Beta Method

The Newmark Beta integration method is also based on the assumption that the acceleration varies linearly between two instants of time. Two parameters α and β are used in this method, which can be changed to suit the requirements of a particular problem. The expressions for velocity and displacements are given by

$$\dot{X}_{t+\Delta t} = \dot{X}_t + [(1-\alpha)\ddot{X}_t + a\ddot{X}_{t+\Delta t}]\,\Delta t \tag{4.69}$$

$$X_{t+\Delta t} = X_t + \dot{X}_t\,\Delta t + \left[(\frac{1}{2} - \beta)\ddot{X}_t + \beta\ddot{X}_{t+\Delta t}\right]\Delta t^2 \tag{4.70}$$

The parameters α and β indicate how much the acceleration enters into the velocity and displacement equations at the end of the interval Δt. Therefore, α and β are chosen to obtain the desired integration accuracy and stability. When α =1/6 and β =1/2, Eqs. (4.69) and (4.70) correspond to the linear acceleration method (which can also be obtained using $\theta = 1$ in Wilson method). When $\alpha = 1/2$ and $\beta = 0$, the acceleration is constant and equal to $\ddot{X}_t$ during each time interval Δt. If α =1/2 and β = 1/8, the acceleration is constant from the beginning as $\ddot{X}_t$ and then changes to $\ddot{X}_{t+\Delta t}$ in the middle of the time interval Δt. When α =1/2 and β = 1/4, this corresponds to the assumption that the acceleration remains constant at an average value of $(\ddot{X}_t + \ddot{X}_{t+\Delta t})/2$. The finite difference formulas for the Newmark Beta scheme are

$$\{\ddot{X}_{t+\Delta t}\} = \frac{1}{\beta\Delta t^2}\,(\{X_{t+\Delta t}\} - \{X_t\}) - \frac{1}{\beta\Delta t}\,\{\dot{X}_t\} - (\frac{1}{2\beta} - 1)\,\{\ddot{X}_t\} \tag{4.71}$$

$$\{\dot{X}_{t+\Delta t}\} = \frac{1}{\beta\Delta t}\,(\{X_{t+\Delta t}\} - \{X_t\}) - \left(\frac{\alpha}{\beta} - 1\right)\{\dot{X}_t\} - \Delta t\left(\frac{\alpha}{2\beta} - 1\right)\{\ddot{X}_t\} \tag{4.72}$$

Equation (4.13) can be employed to obtain a solution for displacements, velocity and accelerations at time $t + \Delta t$. Therefore, by substituting the expressions for $\{\ddot{X}_{t+\Delta t}\}$ and $\{\dot{X}_{t+\Delta t}\}$ from Eqs. (4.71) and (4.72), respectively, into Eq. (4.13), we get

$$\lfloor \overline{M} \rfloor \{X_{t+\Delta t}\} = \{\overline{F}_{t+\Delta t}\} \tag{4.73}$$

where the effective mass matrix $\lfloor \overline{M} \rfloor$ and the effective force vector $\{\overline{F}_{t+\Delta t}\}$ are given by

$$[\overline{M}] = \frac{1}{\beta \Delta t^2} [M] + \frac{\alpha}{\beta \Delta t} [C] + [K] \tag{4.74}$$

$$\{\overline{F}_{t+\Delta t}\} = \{F_{t+\Delta t}\} + \left[(\frac{1}{2\beta} - 1) [M] + \Delta t \, (\frac{\alpha}{2\beta} - 1) [C] \right] \{\ddot{X}_t\}$$
$$+ \left[\frac{1}{\beta \Delta t} [M] + (\frac{\alpha}{\beta} - 1) [C] \right] \{\dot{X}_t\}$$
$$+ \left[\frac{1}{\beta \Delta t^2} [M] + \frac{\alpha}{\beta \Delta t} [C] \right] \{X_t\} \tag{4.75}$$

Solution of Eq. (4.73) gives $\{X_{t+\Delta t}\}$, which is then substituted into (4.71) and (4.72) in order to obtain the accelerations and velocities at time $t + \Delta t$. One of the features of Newmark Beta method is that for linear systems the amplitude is conserved ar d the response is unconditionally stable, provided that $\alpha \geq \frac{1}{2}$ and $\beta \geq \frac{1}{4} (\alpha + \frac{1}{2})^2$. For values of $\alpha = \frac{1}{2}$ and $\beta = \frac{1}{4}$ the largest truncation errors occur in the frequency of the response as opposed to other β values. It is also important to note that unless $\beta = \frac{1}{2}$, there is a spurious damping introduced, proportional to $(\beta - \frac{1}{2})$. If $\beta = 0$, a negative damping results; this involves a self-excited vibration arising solely from the numerical procedure. In a likewise manner, if β is greater than $\frac{1}{2}$, a positive damping is introduced which reduces the magnitude of the response even without real damping in the problem. For a multidegree of freedom system in which a number of modes constitute the total response, the peak amplitude may not be correct. The complete algorithm using the Newmark Beta integration method is given in Table 4.5.

Table 4.5

Algorithm based on Newmark Beta method

(*a*) Initial Computations:

1. Form stiffness $[K]$, mass $[M]$ and damping $[C]$ matrices
2. Initialize $\{X_0\}, \{\dot{X}_0\}$ and $\{\ddot{X}_0\}$.
3. Select time step Δt, parameters α and β, and calculate integration constants,

   $\beta \geq 0.5; \alpha \geq 0.25(0.5 + \beta)^2$

$$a_0 = \frac{1}{\beta(\Delta t)^2};\ a_1 = \frac{\alpha}{\beta\Delta t};\ a_2 = \frac{1}{\beta\Delta t};\ a_3 = \frac{1}{2\beta};\ a_3 = \frac{1}{2\beta} - 1;\ a_4 = \frac{\alpha}{\beta} - 1;$$

$$a_5 = \frac{\Delta t}{2}\left(\frac{\alpha}{\beta} - 2\right);\ a_6 = \Delta t\,(1 - \beta);\ a_7 = \beta\Delta t$$

4. Form effective stiffness matrix:

$$\lfloor \bar{K} \rfloor = [K] + a_0\,[M] + a_1\,[C]$$

5. Triangularize $[\bar{K}]$: $[\bar{K}] = [L]\,[D]\,[L]^T$

(*b*) For each time step:

1. Calculate effective force vector at time $t + \Delta t$:

$$\{\bar{F}_{t+\Delta t}\} = \{F_{t+\Delta t}\} + [M]\,(a_0\,\{X_t\} + a_2\,\{\dot{X}_t\} + a_2\,\{\dot{X}_t\} + a_3\,\{\ddot{X}_t\})$$
$$+ [C]\,(a_1\,\{X_t\} + a_4\,\{\dot{X}_t\} + a_5\,\{\ddot{X}_t\})$$

2. Solve for displacements at time $t + \Delta t$

$$\lfloor \bar{K} \rfloor \{X_{t+\Delta t}\} = \{\bar{F}_{t+\Delta t}\}$$

3. Calculate $\{\dot{X}\}$ and $\{\ddot{X}\}$ at time $t + \Delta t$:

$$\{\ddot{X}_{t+\Delta t}\} = a_0\,(\{X_{t+\Delta t}\} - \{X_t\}) - a_2\,\{\dot{X}_t\} - a_3\,\{\ddot{X}_t\}$$

$$\{\dot{X}_{t+\Delta t}\} = a_1\,(\{X_{t+\Delta t}\} - \{X_t\} - a_4\,\{\dot{X}_t\} - a_5\,\{\ddot{X}_t\}$$

4.5.4 Park Stiffly Stable Method

The Park Stiffly Stable method is an accurate method for low frequency ranges and stable for all higher-frequency components. Using a linear combination of the following two difference formulas derives the velocity in the Park Stiffly Stable method:

$$\dot{X}_{t+\Delta t} = \frac{1}{6\Delta t}\,[11X_{t+\Delta t} - 18X_t + 9X_{t-\Delta t} - 2X_{t-2\Delta t}] \tag{4.76}$$

$$\dot{X}_{t+\Delta t} = \frac{1}{2\Delta t}\,[2X_{t+\Delta t} - 4X_t + X_{t-\Delta t}] \tag{4.77}$$

The linear combination of (4.76) and (4.77) gives

$$\dot{X}_{t+\Delta t} = \frac{1}{4\Delta t}\,[3X_{t+\Delta t} - 4X_t + X_{t-\Delta t})$$
$$+ \frac{1}{12\Delta t}\,(11X_{t+\Delta t} - 18X_t + 9X_{t-\Delta t} - 2X_{t-2\Delta t}) \tag{4.78}$$

or

$$\dot{X}_{t+\Delta t} = \frac{1}{6\Delta t}\,(10X_{t+\Delta t} - 15X_t + 6X_{t-\Delta t} - X_{t-2\Delta t}) \tag{4.79}$$

Similarly, for the acceleration we obtain

$$\ddot{X}_{t+\Delta t} = \frac{1}{6\Delta t}\,(10\dot{X}_{t+\Delta t} - 15\dot{X}_t + 6\dot{X}_{t-\Delta t} - \dot{X}_{t-2\Delta t}) \tag{4.80}$$

The difference formulas in the Park Stiffly method are given by

$$\{\ddot{X}_{t+\Delta t}\} = \frac{1}{6\Delta t}[10\{\dot{X}_{t+\Delta t}\} - 15\{\dot{X}_t\} + 6\{\dot{X}_{t-\Delta t}\} - \{\dot{X}_{t-2\Delta t}\}] \tag{4.81}$$

$$\{\dot{X}_{t+\Delta t}\} = \frac{1}{6\Delta t}[10\{X_{t+\Delta t}\} - 15\{X_t\} + 6\{X_{t-\Delta t}\} - \{X_{t-2\Delta t}\}] \tag{4.82}$$

We consider Eq. (4.13) to obtain solution for the displacements, velocities and accelerations at time $t + \Delta t$. By substituting the expressions for $\{\ddot{X}_{t+\Delta t}\}$ and $\{\dot{X}_{t+\Delta t}\}$ from (4.81) and (4.82), respectively, into (4.13), we get

$$\lfloor \overline{M} \rfloor \{X_{t+\Delta t}\} = \{\overline{F}_{t+\Delta t}\} \tag{4.83}$$

where the effective mass matrix $\lfloor \overline{M} \rfloor$ and the effective force vector $\{\overline{F}_{t+\Delta t}\}$ are given by

$$[\overline{M}] = \frac{100}{36\Delta t^2}[M] - \frac{10}{6\Delta t}[C] + [K] \tag{4.84}$$

$$\begin{aligned}\{\overline{F}_{t+\Delta t}\} &= \frac{15}{6\Delta t}[M]\{\dot{X}_t\} - \frac{1}{\Delta t}[M]\{\dot{X}_{t-\Delta t}\} + \frac{1}{6\Delta t}[M]\{\dot{X}_{t-2\Delta t}\} \\ &+ (\frac{150}{36\Delta t^2}[M] + \frac{15}{6\Delta t}[C])\{X_t\} - (\frac{10}{6\Delta t^2}[M] + \frac{1}{\Delta t}[C])\{X_{t-\Delta t}\} \\ &+ (\frac{1}{36\Delta t^2}[M] + \frac{1}{6\Delta t}[C])\{X_{t-2\Delta t}\}\end{aligned} \tag{4.85}$$

The solution of Eq.(4.83) gives $\{X_{t+\Delta t}\}$, which is then substituted in Eq.(4.82) to obtain velocities. The values of $\{\ddot{X}_{t+\Delta t}\}$ are then obtained by the use of Eq.(4.81). Note that in the Park Stiffly stable method, the calculation of $\{X_{t+\Delta t}\}$ requires the displacements and velocities at t, $t - \Delta t$ and $t - 2\Delta t$. Therefore, in order to obtain the solution at time Δt and $2\Delta t$, a special starting procedure is needed, which makes the method non-self-starting. The complete algorithm based on Park Stiffly stable method used in the integration is given in Table 4.6. The method requires a large computer memory in order to store the displacements and velocities for the two previous time steps.

Table 4.6

Algorithm based on Park Stiffly stable method

(*a*) Initial Computations:

1. Form stiffness $[K]$, mass $[M]$, and damping $[C]$ matrices
2. Initialize $\{X_0\}, \{\dot{X}_0\}$, and $\{\ddot{X}_0\}$.
3. Select time step Δt and calculate integration constants:

$$a_0 = \frac{10}{6\Delta t}; a_1 = \frac{-15}{6\Delta t}; a_2 = \frac{1}{\Delta t}; a_3 = \frac{-1}{6\Delta t}$$

4. Form effective stiffness matrix:

$$\lfloor \overline{K} \rfloor = a_0^2\,[M] - a_0\,[C] + [K]$$

5. Triangularize $[\overline{K}]$: $[\overline{K}] = [L]\,[D]\,[L]^T$

(*b*) For each time step:

1. Calculate effective force vector at time $t + \Delta t$:

$$\{\overline{F}_{t+\Delta t}\} = (-a_1\{\dot{X}_t\} - a_2\{\dot{X}_{t-\Delta t}\} - a_3\{\dot{X}_{t-2\Delta t}\} - a_0a_1\{X_t\} - a_0a_2\{X_{t-\Delta t}\} + a_3^2\{X_{t-2\Delta t}\})[M] - (a_1\{X_t\} + a_2\{X_{t-\Delta t}\} + a_3\{X_{t-2\Delta t}\})[C]$$

2. Solve for displacements at time $t + \Delta t$

$$\lfloor \overline{M} \rfloor \{X_{t+\Delta t}\} = \{\overline{F}_{t+\Delta t}\}$$

3. Calculate $\{\dot{X}\}$ and $\{\ddot{X}\}$ at time $t + \Delta t$:

$$\{\dot{X}_{t+\Delta t}\} = a_0\{X_{t+\Delta t}\} + a_1\{X_t\} + a_2\{X_{t-\Delta t}\} + a_3\{X_{t-2\Delta t}\}$$

$$\{\ddot{X}_{t+\Delta t}\} = a_0\{\dot{X}_{t+\Delta t}\} + a_1\{\dot{X}_t\} + a_2\{\dot{X}_{t-\Delta t}\} + a_3\{\dot{X}_{t-2\Delta t}\}$$

4.6 SUMMARY

In this chapter we have briefly reviewed the direct numerical integration methods for the solution of a single or system of differential equations. Many numerical methods are available for the solutions of the response of dynamic systems. We have discussed several widely used step-by-step numerical integration methods for linear dynamic response analysis. A brief description of these integration methods is presented and their application is illustrated. The integration schemes considered were three explicit and four implicit methods. They are the explicit schemes (the central difference method, two-cycle interaction with trapezoidal rule and fourth order Runge-Kutta method) and the implicit schemes(Houbolt method, Wilson Theta method, Newmark Beta method and the Park Stiffly stable method). Application of these direct numerical integration methods is illustrated with numerical examples for a linear dynamic system.

The use of a particular integration method is mainly dependent on the nature of the problem and is often dictated by the desired solution accuracy.

4.7 EXAMPLE PROBLEMS AND SOLUTIONS USING MATLAB

Example 4.1. *Find the response of a viscously damped single degree of freedom system subjected to a force*

$$F(t) = F_0\left(1 - \sin\frac{\pi t}{2t_0}\right)$$

with the following data: $F_0 = 2N$, $t_0 = \pi$ *seconds,* $m = 2$ *kg,* $c = 0.3$ *Ns/m and* $k = 1$ *N/m. The values of the displacement and velocity of the mass at* $t = 0$ *are zero. Use the central difference method. Choose* $\Delta t = 1$, 0.1 *and* 0.5 *seconds and compare the results.*

Solution:

This is a single-degree of freedom system problem with all initial conditions zero. The following MATLAB program is executed to obtain the results.

```
% INITIAL VALUES
```

```
m=2;k=1;c=0.3;dt=0.1;
x0=0;x0d=0;
F0=2;
T=5;
x0dd=inv(m)*(F0-c*x0d-k*x0);
xprev=x0-(dt*x0d)+((dt^2)*x0dd/2);
a0=1/dt^2;a1=1/(2*dt);a2=2*a0;
mbar=a0*m+a1*c;
t=0;
v(1)=x0d;a(1)=x0dd;
i=1;
for t=0:dt:T+dt
X(i)=x0;
f=F0*(1-sin(0.5*t));
fbar=f+(a2*m-k)*x0+(a1*c-a0*m)*xprev;
x=inv(mbar)*fbar;
xprev=x0;
x0=x;
i=i+1;
p=i;
end
for i=2:p-1
  if i<p-1
  v(i)=(X(i+1)-X(i-1))/(2*dt);
  a(i)=(X(i+1)-2*X(i)+X(i-1))/dt^2;
  end
end
  fprintf('\ntime\t\tdisplacement\tvelocity\tacceleration\n');
  i=1;
  for t=0:dt:T
    fprintf('%f\t%f\t%f\t%f\n',t,X(i),v(i),a(i));
    i=i+1;
  end
  t=[0:dt:T+dt];
  plot(t,X,'-p');
  xlabel('time(s)');
```

The output of the program is shown below for various values of Δt.

When $\Delta t = 1$ sec, the following results are obtained

time	displacement	velocity	acceleration
0.000000	0.000000	0.000000	1.000000
1.000000	0.500000	0.590965	0.181930
2.000000	1.181930	0.433220	– 0.497419
3.000000	1.366441	– 0.144974	– 0.658969
4.000000	0.891982	– 0.606607	– 0.264297
5.000000	0.153226	– 0.536092	0.405328

When $\Delta t = 0.5$ sec, the following results are obtained

Time	displacement	velocity	acceleration
0.000000	0.000000	0.000000	1.000000
0.500000	0.125000	0.407252	0.629008
1.000000	0.407252	0.620473	0.223877
1.500000	0.745473	0.638891	– 0.150209
2.000000	1.046143	0.491762	– 0.438307
2.500000	1.237235	0.231599	– 0.602342
3.000000	1.277742	– 0.075255	– 0.625078
3.500000	1.161980	– 0.359295	– 0.511081
4.000000	0.918447	– 0.558261	– 0.284782
4.500000	0.603719	– 0.625966	0.013962
5.000000	0.292481	– 0.538461	0.336057

When $\Delta t = 0.1$ sec, the following results are obtained

time	displacement	velocity	acceleration
0.000000	0.000000	0.000000	1.000000
0.100000	0.005000	0.096651	0.933023
0.200000	0.019330	0.186429	0.862537
0.300000	0.042286	0.269009	0.789068
0.400000	0.073132	0.344120	0.713147
0.500000	0.111110	0.411543	0.635310
0.600000	0.155441	0.471113	0.556092
0.700000	0.205332	0.522719	0.476028
0.800000	0.259985	0.566303	0.395644
0.900000	0.318593	0.601858	0.315459
1.000000	0.380356	0.629430	0.235982
1.100000	0.444479	0.649114	0.157706
1.200000	0.510179	0.661055	0.081110
1.300000	0.576690	0.665443	0.006652
1.400000	0.643268	0.662514	– 0.065229
1.500000	0.709193	0.652547	– 0.134117
1.600000	0.773777	0.635860	– 0.199624
1.700000	0.836365	0.612810	– 0.261384
1.800000	0.896339	0.583787	– 0.319064
1.900000	0.953122	0.549216	– 0.372359

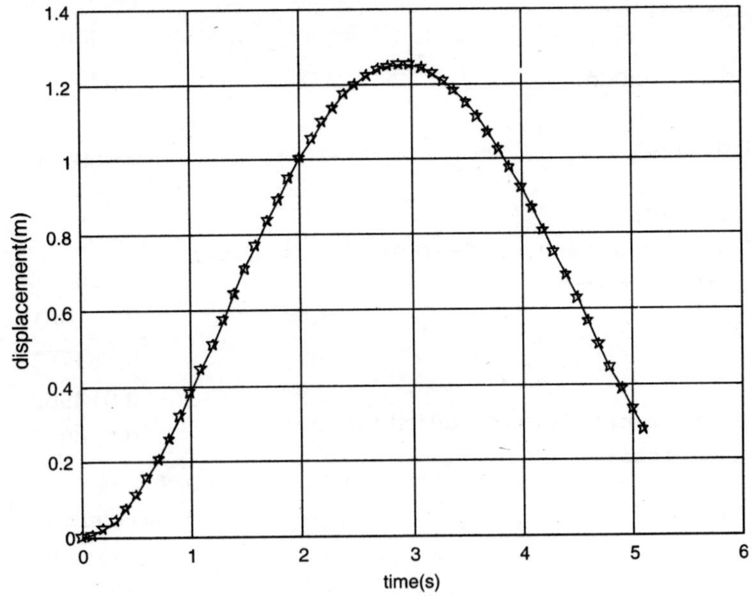

Fig. E4.1 Displacement response plot for $\Delta t = 0.1$ sec.

Example 4.2. *Find the solution of the equation* $5\ddot{X} + 2.5\dot{X} + 4000X = F(t)$, *where F(t) is as shown in Fig. E4.2 for the duration* $0 \le t \le 1$. *Assume that* $X_0 = \dot{X}_0 = 0$ *and* $\Delta t = 0.05$. *Use the central difference method.*

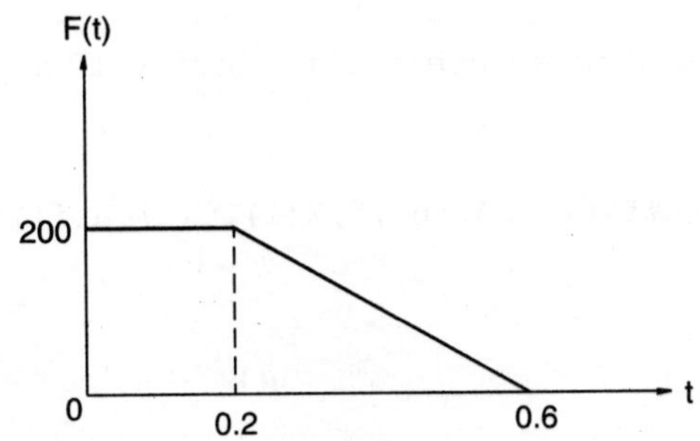

Fig. E4.2

Solution:

MATLAB program for this is given below:

```
% INITIAL VALUES
    m=5;k=4000;c=2.5;dt=0.05;
    x0=0;x0d=0;
    F0=200;
    T=1;
    x0dd=inv(m)*(F0-c*x0d-k*x0);
    xprev=x0-(dt*x0d)+((dt^2)*x0dd/2);
```

```
  a0=1/dt^2;a1=1/(2*dt);a2=2*a0;
  mbar=a0*m+a1*c;
t=0;
v(1)=x0d;a(1)=x0dd;
i=1;
for t=0:dt:T+dt
X(i)=x0;
if t<=0.2 f=F0;
else if (t>0.2 & t<=0.6) f=-(F0/0.4)*(t-0.6);
   else if t>0.6 f=0;
      end
   end
end
  fbar=f+(a2*m-k)*x0+(a1*c-a0*m)*xprev;
x=inv(mbar)*fbar;
xprev=x0;
x0=x;
i=i+1;
p=i;
end
for i=2:p-1
   if i<p-1
    v(i)=(X(i+1)-X(i-1))/(2*dt);
    a(i)=(X(i+1)-2*X(i)+X(i-1))/dt^2;
   end
end
 fprintf('\ntime\t\tdisplacement\tvelocity\tacceleration\n');
i=1;
for t=0:dt:T
    fprintf('%f\t%f\t%f\t%f\n',t,X(i),v(i),a(i));
   i=i+1;
end
t=[0:dt:T+dt];
plot(t,X,'-p');
  xlabel('time(s)');
  ylabel('displacement(m)');
  grid on;
```

The output is as follows:

time	displacement	velocity	acceleration
0.000000	0.000000	0.000000	40.000000
0.050000	0.050000	0.987654	– 0.493827
0.100000	0.098765	– 0.000000	– 39.012346
0.150000	0.050000	– 0.963268	0.481634
0.200000	0.002439	0.000000	38.049078
0.250000	0.050000	0.816027	– 5.408013

0.300000	0.084041	– 0.246914	– 37.109594
0.350000	0.025309	– 1.042791	5.274482
0.400000	– 0.020238	– 0.006097	36.193308
0.450000	0.024699	0.770130	– 5.144248
0.500000	0.056775	– 0.240967	– 35.299646
0.550000	0.000602	– 0.998028	5.017230
0.600000	– 0.043028	– 0.011896	34.428050
0.650000	– 0.000587	0.849928	0.044924
0.700000	0.041965	0.011602	– 33.577975
0.750000	0.000573	– 0.828943	– 0.043815
0.800000	– 0.040929	– 0.011316	32.748889
0.850000	– 0.000559	0.808475	0.042733
0.900000	0.039918	0.011036	– 31.940274
0.950000	0.000545	– 0.788513	– 0.041678
1.000000	– 0.038933	– 0.010764	31.151626

Fig. E4.2 (*a*) shows the response history.

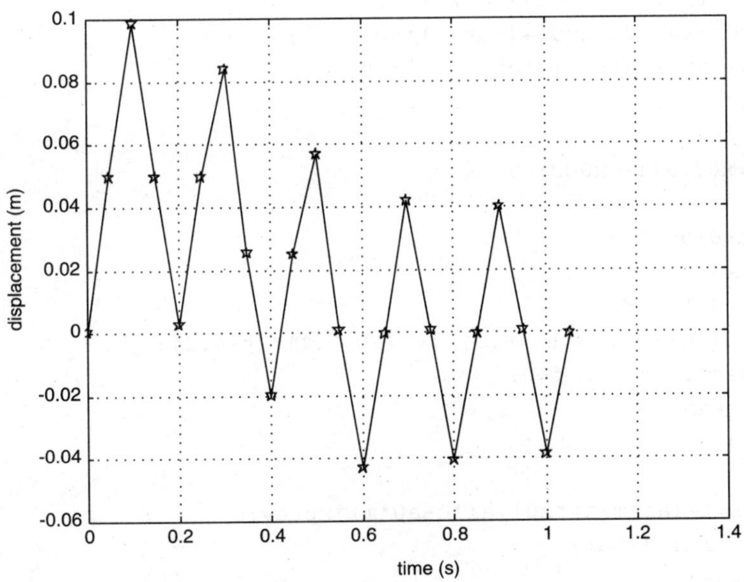

Fig. E4.2 (*a*) MATLAB output for Δt = 0.05 s.

Example 4.3. *Solve numerically the differential equation* $4\ddot{X} + 2000\,X = F(t)$ *with the initial conditions* $X_0 = \dot{X}_0 = 0$ *and forcing function F(t) as shown in Fig. E4.3. Use central difference method with* $\Delta t = 0.02$ *sec.*

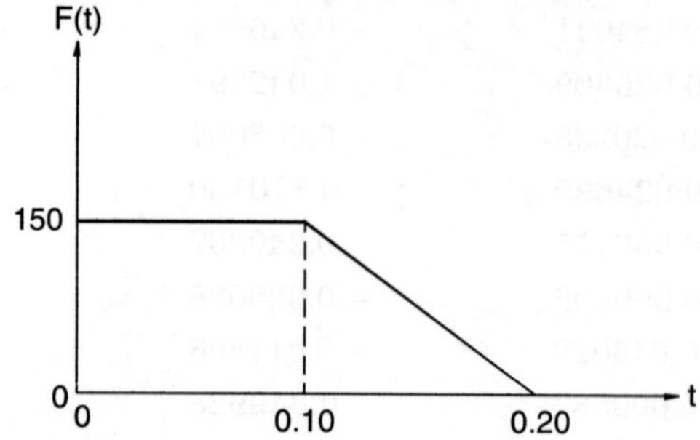

Fig. E4.3

Solution:

Here the complete MATLAB program is shown below:

```
% INITIAL VALUES
    m=4;k=2000;c=0;dt=0.02;
    x0=0;x0d=0;
    F0=150;
    T=1;
    x0dd=inv(m)*(F0-c*x0d-k*x0);
    xprev=x0-(dt*x0d)+((dt^2)*x0dd/2);
    a0=1/dt^2;a1=1/(2*dt);a2=2*a0;
    mbar=a0*m+a1*c;
  t=0;
  v(1)=x0d;a(1)=x0dd;
  i=1;
  for t=0:dt:T+dt
  X(i)=x0;
  if t<=0.1 f=F0;
  else if (t>0.1 & t<=0.2) f=-(F0/0.1)*(t-0.2);
     else if t>0.2 f=0;
        end
     end
  end
    fbar=f+(a2*m-k)*x0+(a1*c-a0*m)*xprev;
  x=inv(mbar)*fbar;
  xprev=x0;
  x0=x;
  i=i+1;
  p=i;
  end
  for i=2:p-1
     if i<p-1
     v(i)=(X(i+1)-X(i-1))/(2*dt);
     a(i)=(X(i+1)-2*X(i)+X(i-1))/dt^2;
     end
```

```
end
 fprintf('\ntime\t\tdisplacement\tvelocity\tacceleration\n');
i=1;
for t=0:dt:T
    fprintf('%f\t%f\t%f\t%f\n',t,X(i),v(i),a(i));
   i=i+1;
end
t=[0:dt:T+dt];
plot(t,X,'-p');
  xlabel('time(s)');
  ylabel('displacement(m)');
  grid on;
```

The output is given below:

time	displacement	velocity	acceleration
0.000000	0.000000	0.000000	37.500000
0.020000	0.007500	0.712500	33.750000
0.040000	0.028500	1.282500	23.250000
0.060000	0.058800	1.596000	8.100000
0.080000	0.092340	1.590300	– 8.670000
0.100000	0.122412	1.266540	– 23.706000
0.120000	0.143002	0.614472	– 41.500800
0.140000	0.146991	– 0.310490	– 50.995440
0.160000	0.130582	– 1.323355	– 50.290992
0.180000	0.094057	– 2.221548	– 39.528346
0.200000	0.041720	– 2.825432	– 20.860030
0.220000	– 0.018961	– 2.939229	9.480291
0.240000	– 0.075849	– 2.465181	37.924555
0.260000	– 0.117568	– 1.498096	58.783907
0.280000	– 0.135773	– 0.231392	67.886478
0.300000	– 0.126824	1.081590	63.411753
0.320000	– 0.092509	2.178254	46.254678
0.340000	– 0.039693	2.839268	19.846667
0.360000	0.021061	2.932428	– 10.530677
0.380000	0.077604	2.439102	– 38.801886
0.400000	0.118625	1.457956	– 59.312718
0.420000	0.135922	0.185219	– 67.961006
0.440000	0.126034	– 1.124562	– 63.017093
0.460000	0.090940	– 2.209431	– 45.469761
0.480000	0.037657	– 2.852413	– 18.828477
0.500000	– 0.023157	– 2.924913	11.578502
0.520000	– 0.079340	– 2.412430	39.669781
0.540000	– 0.119654	– 1.417461	59.827104
0.560000	– 0.136038	– 0.139000	68.019006
0.580000	– 0.125214	1.167261	62.607106
0.600000	– 0.089348	2.240070	44.673786
0.620000	– 0.035611	2.864865	17.805708

0.640000	0.025247	2.916687	– 12.623511
0.660000	0.081056	2.385171	– 40.528028
0.680000	0.120654	1.376622	– 60.326940
0.700000	0.136121	0.092748	– 68.060463
0.720000	0.124364	– 1.209676	– 62.181894
0.740000	0.087734	– 2.270164	– 43.866946
0.760000	0.033557	– 2.876620	– 16.778609
0.780000	– 0.027331	– 2.907751	13.665450
0.800000	– 0.082753	– 2.357333	41.376419
0.820000	– 0.121624	– 1.335447	60.812104
0.840000	– 0.136171	– 0.046473	68.085368
0.860000	– 0.123483	1.251797	61.741559
0.880000	– 0.086099	2.299706	43.049438
0.900000	– 0.031495	2.887675	15.747429
0.920000	0.029408	2.898109	– 14.704065
0.940000	0.084429	2.328921	– 42.214747
0.960000	0.122565	1.293948	– 61.282479
0.980000	0.136187	0.000186	– 68.093715
1.000000	0.122572	– 1.293613	– 61.286209

Fig. E4.3 (*a*) shows the response history.

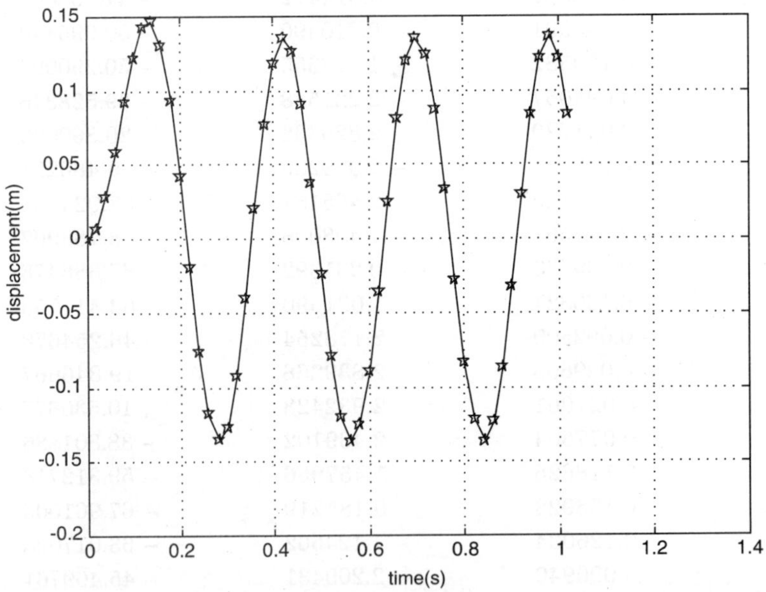

Fig. E4.3(*a*) Response history

Example 4.4. *Solve numerically the solution to the problem of a spring mass system excited by a triangular impulse. The differential equation of motion and the initial conditions are given as*

$$0.5\ddot{X} + 8\pi^2 X = F(t)$$

with $$X_1 = \dot{X}_1 = 0$$

The triangular force is defined in Fig. E4.4. Use central difference method with $\Delta t = 0.05$ *sec.*

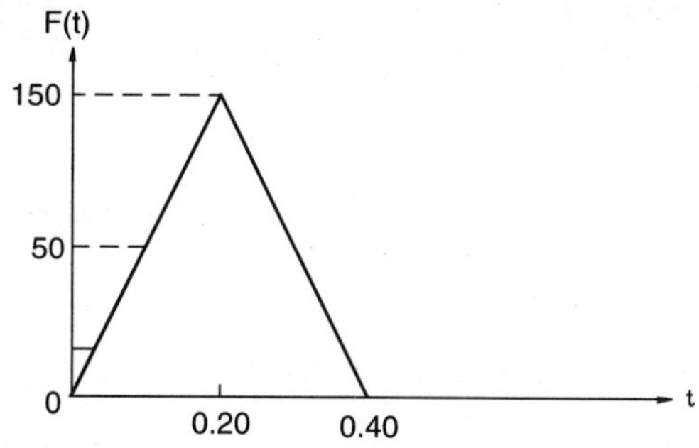

Fig. E4.4

Solution:

The following MATLAB program is developed:

```
% INITIAL VALUES
    m=0.5;k=8*pi^2;c=0;dt=0.05;
    x0=0;x0d=0;
    F0=0;F=150;
    T=1;
    x0dd=inv(m)*(F0-c*x0d-k*x0);
    xprev=x0-(dt*x0d)+((dt^2)*x0dd/2);
    a0=1/dt^2;a1=1/(2*dt);a2=2*a0;
    mbar=a0*m+a1*c;
  t=0;
  v(1)=x0d;a(1)=x0dd;
  i=1;
  for t=0:dt:T+dt
  X(i)=x0;
  if t<=0.2 f=(F*t/0.2);
  else if (t>0.2 & t<=0.4) f=-(F/0.2).*(t-0.4);
          else if t>0.4 f=0;
          end
     end
  end
    fbar=f+(a2*m-k)*x0+(a1*c-a0*m)*xprev;
  x=inv(mbar)*fbar;
  xprev=x0;
  x0=x;
  i=i+1;
  p=i;
  end
```

```
for i=2:p-1
   if i<p-1
    v(i)=(X(i+1)-X(i-1))/(2*dt);
    a(i)=(X(i+1)-2*X(i)+X(i-1))/dt^2;
   end
end
 fprintf('\ntime\t\tdisplacement\tvelocity\tacceleration\n');
i=1;
for t=0:dt:T
    fprintf('%f\t%f\t%f\t%f\n',t,X(i),v(i),a(i));
   i=i+1;
end
t=[0:dt:T+dt];
plot(t,X,'-p');
  xlabel('time(s)');
  ylabel('displacement(m)');
  grid on;
```

The output is given below:

time	displacement	velocity	acceleration
0.000000	0.000000	0.000000	0.000000
0.050000	0.000000	1.875000	75.000000
0.100000	0.187500	6.759780	120.391187
0.150000	0.675978	12.725905	118.253838
0.200000	1.460091	17.418045	69.431745
0.250000	2.417782	15.233816	– 156.800902
0.300000	2.983472	3.285518	– 321.131034
0.350000	2.746334	– 13.709851	– 358.683716
0.400000	1.612487	– 29.042788	– 254.633742
0.450000	– 0.157945	– 34.785091	24.941603
0.500000	– 1.866022	– 26.794791	294.670399
0.550000	– 2.837424	– 8.226331	448.067983
0.600000	– 2.688655	13.589754	424.575418
0.650000	– 1.478448	30.040819	233.467197
0.700000	0.315427	34.632244	– 49.810180
0.750000	1.984776	25.551408	– 313.423286
0.800000	2.870567	6.383280	– 453.301838
0.850000	2.623104	– 15.304866	– 414.223998
0.900000	1.340081	– 30.950893	– 211.617078
0.950000	– 0.471985	– 34.377997	74.532916
1.000000	– 2.097719	– 24.233212	331.258494

Figure E4.5 shows the plot of the response versus time.

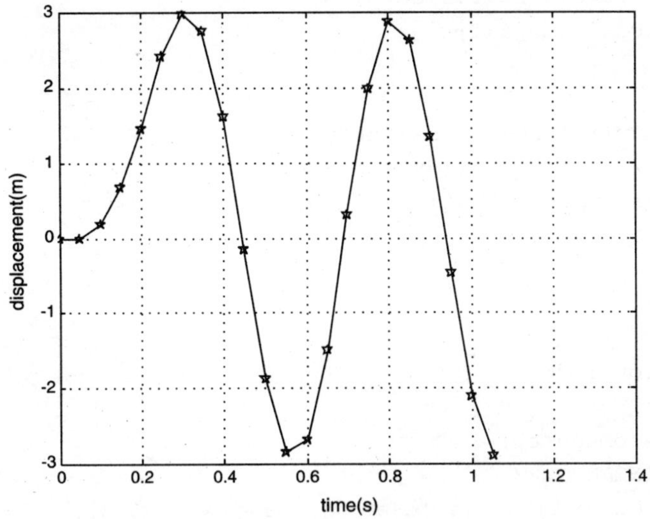

Fig. P 4.5 (*a*) MATLAB output for $\Delta t = 0.05$ s.

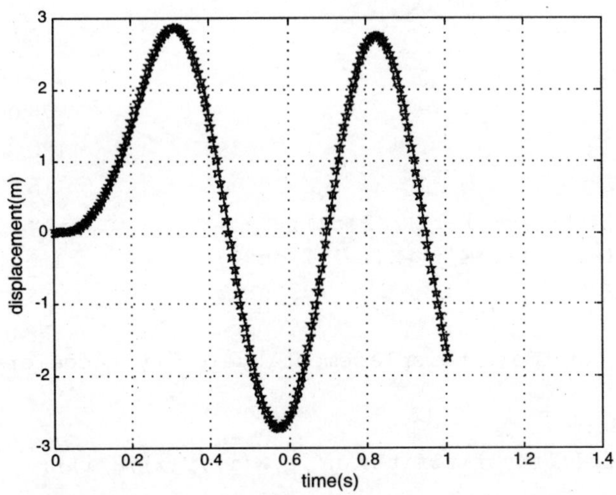

Fig. P 4.5 (*b*) MATLAB output for $\Delta t = 0.005$ s.

Example 4.5 *Solve the following nonlinear vibration problem, using the central difference method.*

$$M\ddot{X} + C\dot{X} + KX + K^*X^3 = F \cos \omega t$$

with $M = 1.0$, $C = 0.5$, $K = 1.0$, $K^* = 0.5$, $\Delta t = 0.05$, $t_{max} = 5.0$, *and the initial conditions* $X_0 = \dot{X}_0 = 0$. *Plot the variation of X with t. Take* $\omega = 1$ *and* $F = 10$.

Solution:

Here in X_{i+1} an additional term with $-K^*X_i^3$ will come and other things will remain same. Assuming $F = 10$ N, the following MATLAB program is developed.

```
% INITIAL VALUES
    m=1;k=1;c=0.5;ks=0.5;dt=0.05;
    x0=0;x0d=0;omega=1;
    F0=10;
    T=5;
    x0dd=inv(m)*(F0-c*x0d-k*x0);
    xprev=x0-(dt*x0d)+((dt^2)*x0dd/2);
    a0=1/dt^2;a1=1/(2*dt);a2=2*a0;
    mbar=a0*m+a1*c;
  t=0;
  v(1)=x0d;a(1)=x0dd;
  i=1;
  for t=0:dt:T+dt
  X(i)=x0;
  f=F0*cos(omega*t);
  % NON-LINEAR TERM
   fbar=f+(a2*m-k)*x0+(a1*c-a0*m)*xprev-ks*(x0^3);
  x=inv(mbar)*fbar;
  xprev=x0;
  x0=x;
  i=i+1;
  p=i;
  end
  for i=2:p-1
      if i<p-1
      v(i)=(X(i+1)-X(i-1))/(2*dt);
      a(i)=(X(i+1)-2*X(i)+X(i-1))/dt^2;
      end
  end
   fprintf('\ntime\t\tdisplacement\tvelocity\tacceleration\n');
  i=1;
  for t=0:dt:T
      fprintf('%f\t%f\t%f\t%f\n',t,X(i),v(i),a(i));
      i=i+1;
  end
  t=[0:dt:T+dt];
  plot(t,X,'-p');
    xlabel('time(s)');
```

The output of the program is given below:

Time	displacement	velocity	acceleration
0.000000	0.000000	0.000000	10.000000
0.050000	0.012500	0.493210	9.728397
0.100000	0.049321	0.971789	9.414766
0.150000	0.109679	1.433672	9.060536

0.200000	0.192688	1.876835	8.665983
0.250000	0.297362	2.299209	8.229010
0.300000	0.422609	2.698528	7.743753
0.350000	0.567215	3.072103	7.199215
0.400000	0.729819	3.416537	6.578158
0.450000	0.908869	3.727404	5.856518
0.500000	1.102560	3.998908	5.003655
0.550000	1.308760	4.223595	3.983832
0.600000	1.524919	4.392175	2.759342
0.650000	1.747977	4.493550	1.295680
0.700000	1.974274	4.515166	– 0.431058
0.750000	2.199494	4.443769	– 2.424815
0.800000	2.418651	4.266666	– 4.659319
0.850000	2.626160	3.973458	– 7.069001
0.900000	2.815997	3.558129	– 9.544163
0.950000	2.981973	3.021179	– 11.933829
1.000000	3.118115	2.371360	– 14.058926
1.050000	3.219109	1.626489	– 15.735916
1.100000	3.280764	0.812907	– 16.807359
1.150000	3.300400	– 0.036584	– 17.172266
1.200000	3.277105	– 0.886080	– 16.807592
1.250000	3.211792	– 1.700621	– 15.774049
1.300000	3.107043	– 2.450079	– 14.204274
1.350000	2.966784	– 3.112116	– 12.277189
1.400000	2.795832	– 3.673705	– 10.186361
1.450000	2.599413	– 4.131135	– 8.110872
1.500000	2.382718	– 4.488775	– 6.194716
1.550000	2.150536	– 4.757067	– 4.536959
1.600000	1.907011	– 4.950278	– 3.191475
1.650000	1.655508	– 5.084393	– 2.173152
1.700000	1.398572	– 5.175400	– 1.467123
1.750000	1.137968	– 5.238034	– 1.038230
1.800000	0.874769	– 5.284965	– 0.839002
1.850000	0.609472	– 5.326325	– 0.815407
1.900000	0.342136	– 5.369468	– 0.910322
1.950000	0.072525	– 5.418854	– 1.065097
2.000000	– 0.199749	– 5.475975	– 1.219747
etc.			

Figure 4.5 (*c*) shows the MATLAB response.

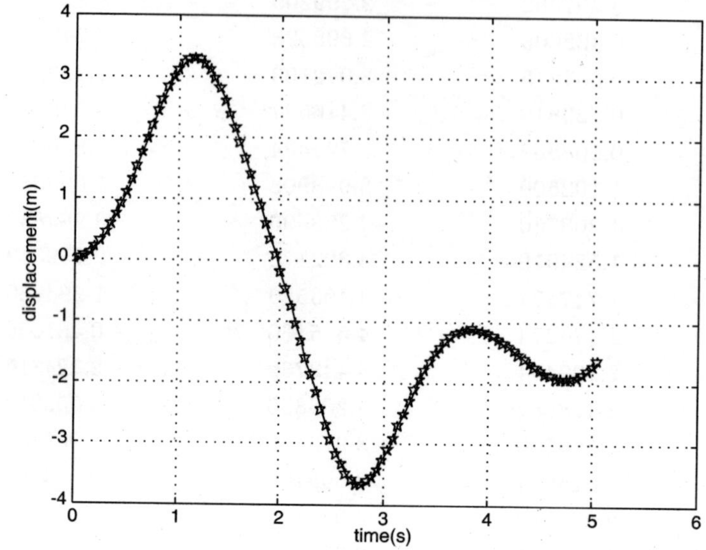

Fig. E4.5 (*c*) MATLAB output for $\Delta t = 0.05$ s.

Example 4.6 *Solve Example 4.1 using the fourth order Runge-Kutta method.*
Solution:

Here $\dot{Y} = f(x_1, x_2, t)$ is a vector of functions
For single degree of freedom system, it contains

$$\dot{Y} = \begin{Bmatrix} \dot{x} \\ \ddot{x} \end{Bmatrix} = f(x_1, x_2, t) = \begin{Bmatrix} x_2 \\ \dfrac{1}{m}(F(t) - kx_1 - cx_2) \end{Bmatrix}$$

Final solution takes the form

$$Y_{i+1} = Y_i + \frac{\Delta t}{6}\,[K_1 + 2K_2 + 2K_3 + K_4], \text{ where}$$

$$K_1 = f(x_1, x_2, t) = \begin{Bmatrix} p \\ q \end{Bmatrix}$$

$$K_2 = f(x_1 + p/2,\, x_2 + q/2,\, t + \Delta t/2) = \begin{Bmatrix} r \\ s \end{Bmatrix}$$

$$K_3 = f(x_1 + r/2,\, x_2 + q/2,\, t + \Delta t/2) = \begin{Bmatrix} u \\ v \end{Bmatrix}$$

$$K_4 = f(x_1 + u,\, x_2 + v,\, t + \Delta t) = \begin{Bmatrix} m \\ n \end{Bmatrix}$$

Complete MATLAB program for computing the response and its derivative in every time step is given below:

```
dt=0.5;T=10;
h=dt;
x1=0;
x2=0;
i=1;
for t=0:h:T
f1=h*f(t,x1,x2);  g1=h*g(t,x1,x2);
f2=h*f((t+h/2),(x1+f1/2),(x2+g1/2));g2=h*g((t+h/2),(x1+f1/2),(x2+g1/2));
f3=h*f((t+h/2),(x1+f2/2),(x2+g2/2));g3=h*g((t+h/2),(x1+f2/2),(x2+g2/2));
f4=h*f((t+h),(x1+f3),(x2+g3));  g4=h*g((t+h),(x1+f3),(x2+g3));
x1=x1+((f1+f4)+2*(f2+f3))/6.0;
x2=x2+((g1+g4)+2*(g2+g3))/6.0;
X(i)=x1;
Y(i)=x2;
i=i+1;
end
t=[0:h:T];
plot(t,X,'-p',t,Y,'-*');
grid on;
xlabel('time(s)');
legend('displacement(m)','velocity(m/s)',2)
%%%%%%%%%%%%%%%%%%%%%%%%%%%%%%%%%
```

This program is executed with two other separate programs f.m and g.m given below:

```
%%%% file f.m  %%%%%%%
          function  v1=f(t,x1,x2)
          v1=x2;
          %%%%%%%%%%%%%%%
          %%%% file g.m %%%%%%%
          function  v2=g(t,x1,x2)
          k=1; m=1; c=0;
F=100*(1-cos(t));
v2=(F-k*x1-c*x2)/m;
%%%%%%%%%%%%%%%
```

The output values of the data are presented below:

time	displacement	velocity
0.000000	0.000000	0.000000
0.500000	0.110437	0.411687
1.000000	0.379029	0.629789
1.500000	0.707386	0.653171
2.000000	1.004189	0.510442
2.500000	1.198540	0.253323
3.000000	1.249282	– 0.052604
3.500000	1.149336	– 0.338363
4.000000	0.924789	– 0.541854

4.500000	0.629191	– 0.616591
5.000000	0.334076	– 0.537864

The output of the program is shown in Fig. E4.6.

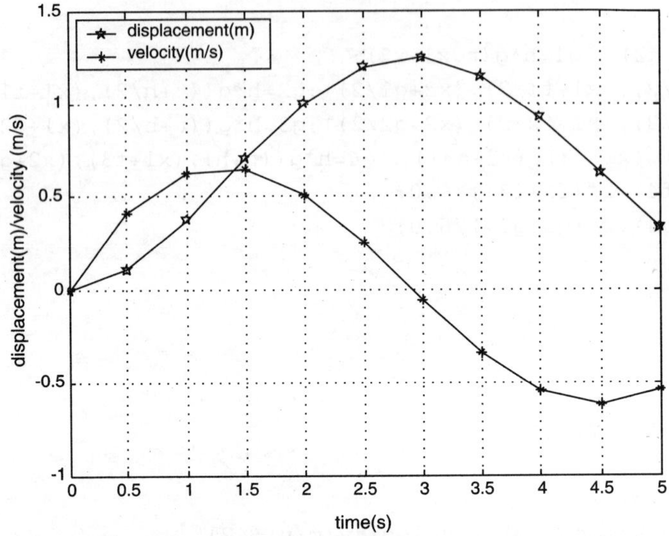

Fig. E4.6 MATLAB output

Example 4.7. *Solve Example 4.2 using the fourth order Runge-Kutta method.*

Solution:

MATLAB program for this problem is given below. Only change occurs in defining the function g. Here $dt = 0.05$ s and $T = 1$ s.

```
dt=0.05;T=1;
h=dt;
x1=0;              % displacement
x2=0;              % velocity
i=1;
fprintf('time\t\tdisplacement\tvelocity\n');
for t=0:h:T
fprintf('%f\t%f\t%f\n',t,x1,x2);
X(i)=x1;
Y(i)=x2;
f1=h*f(t,x1,x2);  g1=h*g(t,x1,x2);
f2=h*f((t+h/2),(x1+f1/2),(x2+g1/2));g2=h*g((t+h/2),(x1+f1/2),(x2+g1/2));
f3=h*f((t+h/2),(x1+f2/2),(x2+g2/2));g3=h*g((t+h/2),(x1+f2/2),(x2+g2/2));
f4=h*f((t+h),(x1+f3),(x2+g3));  g4=h*g((t+h),(x1+f3),(x2+g3));
x1=x1+((f1+f4)+2*(f2+f3))/6.0;
x2=x2+((g1+g4)+2*(g2+g3))/6.0;
i=i+1;
end
```

```
time = [0:h:T];
plot(time, X, '- p');
grid on;
xlabel('time(s)');
ylabel('displacement(m)')

function g.m is given below:
%%%%%%%%%%%%%%%%%%%%%%%%%%%%%%%%%%
function v2=g(t,x1,x2)
        k=4000; m=5; c=2.5;
    if t<=0.2 F=200;
    else if (t>0.2 & t<=0.6) F=-(200/0.4)*(t-0.6);
    else if t>0.6 F=0;
    end
    end
    end
   v2=(F-k*x1-c*x2)/m;
%%%%%%%%%%%%%%%%%%%%%%%%%%%%%%%%%%
```

The output is as follows:

time	displacement	velocity
0.000000	0.000000	0.000000
0.050000	0.041253	1.316874
0.100000	0.091824	0.439092
0.150000	0.071773	– 1.031937
0.200000	0.019836	– 0.736985
0.250000	0.018390	0.574514
0.300000	0.056157	0.655850
0.350000	0.060285	– 0.490566
0.400000	0.018109	– 0.945601
0.450000	– 0.009407	– 0.071506
0.500000	0.009400	0.627115
0.550000	0.030533	0.077915
0.600000	0.010993	– 0.730337
0.650000	– 0.022121	– 0.405278
0.700000	– 0.017213	0.518375
0.750000	0.014055	0.535490
0.800000	0.020088	– 0.285294
0.850000	– 0.005878	– 0.574286
0.900000	– 0.019935	0.063796
0.950000	– 0.001387	0.535146
1.000000	0.017375	0.121352

The displacement response is shown in Fig. E4.7

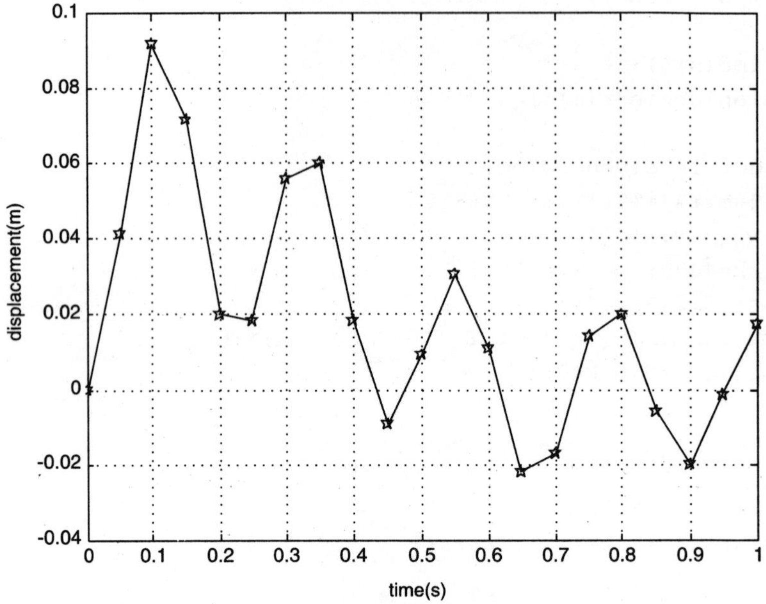

Fig. E4.7 MATLAB output.

Example 4.8. *Solve Example 4.3 by the Runge-Kutta method.*

Solution:

MATLAB program for this problem is given below. Only change occurs in defining the function g. Here $dt = 0.05$ s and $T = 1$ s.

```
dt=0.02;T=1;
h=dt;
x1=0;              % displacement
x2=0;              % velocity
i=1;
fprintf('time\t\tdisplacement\tvelocity\n');
for t=0:h:T
fprintf('%f\t%f\t%f\n',t,x1,x2);
X(i)=x1;
Y(i)=x2;
f1=h*f(t,x1,x2);  g1=h*g(t,x1,x2);
f2=h*f((t+h/2),(x1+f1/2),(x2+g1/2));g2=h*g((t+h/2),(x1+f1/2),(x2+g1/2));
f3=h*f((t+h/2),(x1+f2/2),(x2+g2/2));g3=h*g((t+h/2),(x1+f2/2),(x2+g2/2));
f4=h*f((t+h),(x1+f3),(x2+g3));  g4=h*g((t+h),(x1+f3),(x2+g3));
x1=x1+((f1+f4)+2*(f2+f3))/6.0;
x2=x2+((g1+g4)+2*(g2+g3))/6.0;
i=i+1;
end
time=[0:h:T];
```

```
plot(time,X,'-p');
grid on;
xlabel('time(s)');
ylabel('displacement(m)')
```

The function g.m defining the force signal is given below:

```
function v2=g(t,x1,x2)
    k=2000; m=4; c=0;
     if t<=0.1 F=150;
     else if (t>0.1 & t<=0.2) F=-(150/0.1)*(t-0.2);
     else if t>0.2 F=0;
     end
     end
     end
v2=(F-k*x1-c*x2)/m;
```

The output of the program is as follows:

time	displacement	velocity
0.000000	0.000000	0.000000
0.020000	0.007375	0.725000
0.040000	0.028041	1.307417
0.060000	0.057936	1.632787
0.080000	0.091181	1.637183
0.100000	0.121242	1.319778
0.120000	0.141711	0.669243
0.140000	0.146114	– 0.260185
0.160000	0.130641	– 1.285790
0.180000	0.095386	– 2.205969
0.200000	0.044333	– 2.839867
0.220000	– 0.014931	– 2.989164
0.240000	– 0.071253	– 2.550901
0.260000	– 0.113564	– 1.611284
0.280000	– 0.133548	– 0.355057
0.300000	– 0.127280	0.970822
0.320000	– 0.095995	2.105735
0.340000	– 0.045845	2.826625
0.360000	0.013311	2.991841
0.380000	0.069845	2.568967
0.400000	0.112643	1.641187
0.420000	0.133296	0.390919
0.440000	0.127747	– 0.936053
0.460000	0.097088	– 2.078891
0.480000	0.047349	– 2.812982
0.500000	– 0.011691	– 2.994079
0.520000	– 0.068427	– 2.586645
0.540000	– 0.111707	– 1.670828

0.560000	– 0.133025	– 0.426696
0.580000	– 0.128194	0.901173
0.600000	– 0.098165	2.051764
0.620000	– 0.048845	2.798940
0.640000	0.010071	2.995880
0.660000	0.067001	2.603933
0.680000	0.110755	1.700204
0.700000	0.132735	0.462383
0.720000	0.128622	– 0.866188
0.740000	0.099228	– 2.024359
0.760000	0.050333	– 2.784500
0.780000	– 0.008450	– 2.997242
0.800000	– 0.065566	– 2.620828
0.820000	– 0.109788	– 1.729309
0.840000	– 0.132425	– 0.497977
0.860000	– 0.129031	0.831104
0.880000	– 0.100275	1.996680
0.900000	– 0.051812	2.769666
0.920000	0.006829	2.998167
0.940000	0.064122	2.637329
0.960000	0.108805	1.758141
0.980000	0.132097	0.533471
1.000000	0.129421	– 0.795925

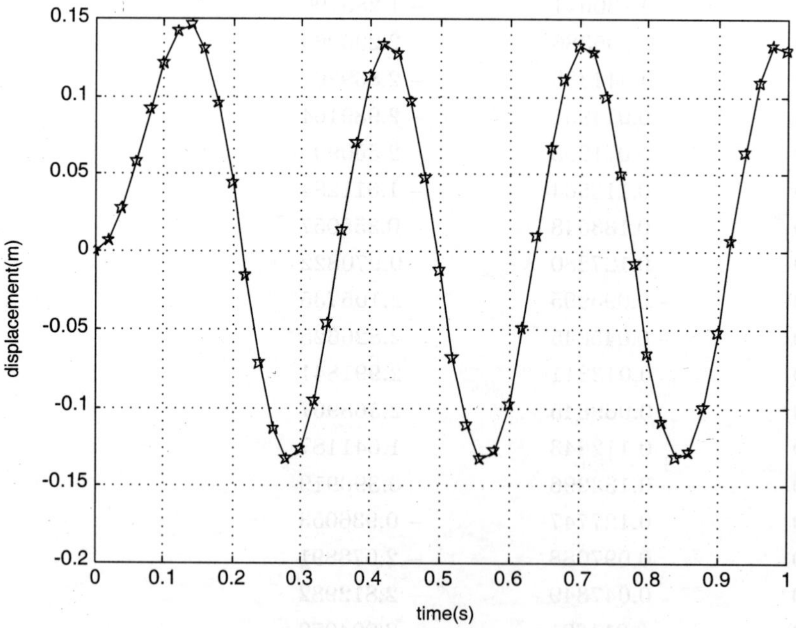

Fig. E4.8 MATLAB output.

Example 4.9. *Solve Example 4.4 by the Runge-Kutta method.*

Solution:

Triangular pulse is defined with the following MATLAB m function

```
%%%%%%%%%%% g.m  %%%%%%%%%%%
function v2=g(t,x1,x2)
k=8*pi^2;m=0.5;  c=0;
if t<=0.2 F=(150*t/0.2);
else if (t>0.2 & t<=0.4) F=-(150/0.2)*(t-0.4);
else if t>0.4 F=0;
       end
       end
       end
       v2=(F-k*x1-c*x2)/m;
```

Here dt=0.05s and T=1 s. The output is shown below:

time	displacement	velocity
0.000000	0.000000	0.000000
0.050000	0.031250	1.813315
0.100000	0.231900	6.553226
0.150000	0.706315	12.411526
0.200000	1.454472	17.155388
0.250000	2.309559	15.351753
0.300000	2.826498	4.081899
0.350000	2.627673	– 12.352850
0.400000	1.608467	– 27.686924
0.450000	0.008154	– 34.265866
0.500000	– 1.593966	– 27.784720
0.550000	– 2.587509	– 10.723309
0.600000	– 2.594445	10.409650
0.650000	– 1.612934	27.559532
0.700000	– 0.017719	34.195743
0.750000	1.582951	27.798535
0.800000	2.579242	10.815730
0.850000	2.592074	– 10.273897
0.900000	1.617357	– 27.432202
0.950000	0.027244	– 34.125340
1.000000	– 1.571956	– 27.811835

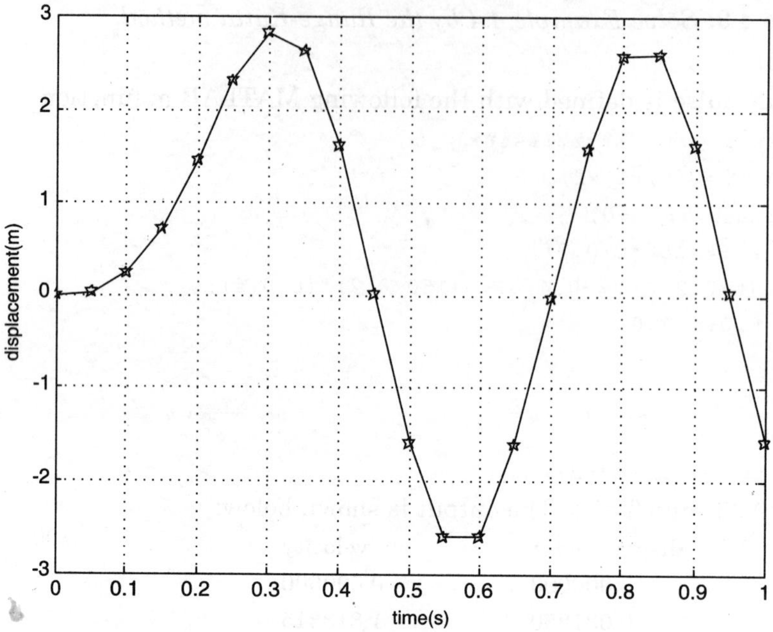

Fig. E4.9 MATLAB output.

Example 4.10. *Solve Example 4.5 by the Runge-Kutta method.*

Solution:

Here the function defining the system g.m is given below:

```
%%%%%%%%%% g.m %%%%%%%%%%
function v2=g(t,x1,x2)
k=1;m=1; c=0.5;omega=1;
ks=0.5 % CUBIC STIFFNESS
F=10*cos(omega*t);
v2=(F-k*x1-c*x2-ks*x1^3)/m;
%%%%%%%%%%%%%%%%%%%%%%
```

The output is given below for $dt = 0.05$ s and $T = 5$ s

time	displacement	velocity
0.000000	0.000000	0.000000
0.050000	0.012391	0.493389
0.100000	0.049095	0.972141
0.150000	0.109327	1.434195
0.200000	0.192202	1.877534
0.250000	0.296734	2.300105
0.300000	0.421830	2.699660
0.350000	0.566273	3.073539
0.400000	0.728700	3.418380
0.450000	0.907555	3.729795
0.500000	1.101028	4.002029
0.550000	1.306982	4.227665
0.600000	1.522864	4.397443

0.650000	1.745611	4.500284
0.700000	1.971567	4.523624
0.750000	2.196426	4.454170
0.800000	2.415222	4.279150
0.850000	2.622401	3.988046
0.900000	2.811982	3.574682
0.950000	2.977828	3.039377
1.000000	3.114021	2.390700
1.050000	3.215294	1.646314
1.100000	3.277487	0.832466
1.150000	3.297926	-0.018051
1.200000	3.275671	-0.869257
1.250000	3.211578	-1.686035
1.300000	3.108154	-2.438054
1.350000	2.969237	-3.102755
1.400000	2.799563	-3.666911
1.450000	2.604294	-4.126654
1.500000	2.388573	-4.486248
1.550000	2.157169	-4.756089
1.600000	1.914225	-4.950445
1.650000	1.663116	-5.085340
1.700000	1.406414	-5.176825
1.750000	1.145909	-5.239713
1.800000	0.882705	-5.286762
1.850000	0.617325	-5.328190
1.900000	0.349853	-5.371441

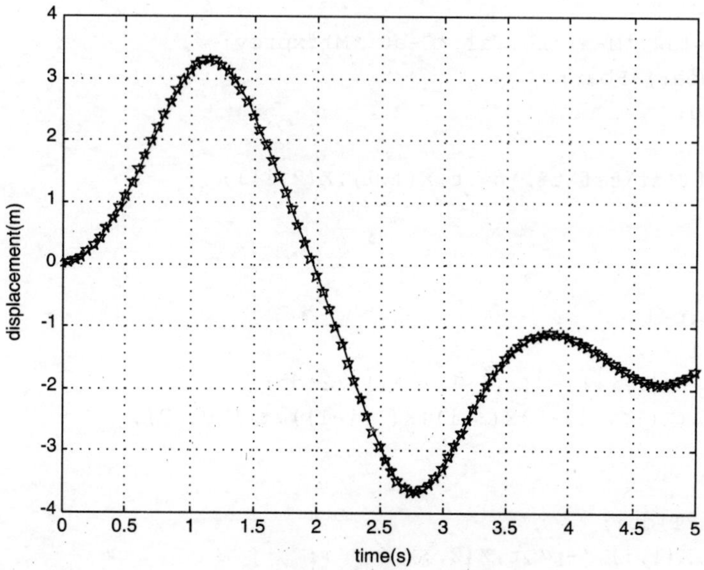

Fig. E4.10 MATLAB output.

Example 4.11. *Find the response of the two degree of freedom system when $F_1(t)=0$ and $F_2(t)=10$, using the central difference method. The mass, stiffness and damping matrices for this system are given as*

$$[M]=\begin{bmatrix}1 & 0\\0 & 10\end{bmatrix},\ [K]=\begin{bmatrix}11 & -1\\-1 & 1\end{bmatrix},\ [C]=\begin{bmatrix}0.5 & -0.1\\-0.1 & 0.1\end{bmatrix}$$

All the initial conditions are given as zero. Use $\Delta t = 0.05$.

Solution:

The procedure is modified for matrices instead of scalars. Complete program is given below:

```
% INITIAL VALUES
M=[1 0;0 10];
K=[21 -1;-1 1];
C=[0.5 -0.1;-0.1 0.1];
dt=0.05;
x0=[0;0];x0d=[0;0];
F0=[0;10];
T=2;
x0dd=inv(M)*(F0-C*x0d-K*x0);
xprev=x0-(dt.*x0d)+((dt^2).*(x0dd/2));
a0=1/dt^2;a1=1/(2*dt);a2=2*a0;
mbar=(a0.*M)+(a1.*C);
t=0;
v(:,1)=x0d;a(:,1)=x0dd;
i=1;
fprintf('time\t\tX(1)\t\tX(2)\n');
for t=0:dt:T+dt
X(:,i)=x0;
F=F0;
Fbar=F+(a2.*M-K)*x0+(a1.*C-a0.*M)*xprev;
x=inv(mbar)*Fbar;
xprev=x0;
x0=x;
fprintf('%f\t%f\t%f\n',t,X(1,i),X(2,i));
i=i+1;
p=i;
end
for i=2:p-1
if i<p-1
v(:,i)=(X(:,i+1)-X(:,i-1)).*(1/(2*dt));
a(:,i)=(X(:,i+1)-2*X(:,i)+X(:,i-1)).*(1/dt^2);
end
end
t=[0:dt:T+dt];
plot(t,X(1,:),'-p',t,X(2,:),'-*');
```

```
xlabel('time(s)');
ylabel('displacement(m)');
legend('DOF-1','DOF-2',2);
grid on;
```

The output of the program is as follows

time	X(1)	X(2)
0.000000	0.000000	0.000000
0.050000	– 0.000000	0.001250
0.100000	0.000015	0.004998
0.150000	0.000067	0.011243
0.200000	0.000178	0.019981
0.250000	0.000376	0.031210
0.300000	0.000688	0.044924
0.350000	0.001142	0.061120
0.400000	0.001762	0.079793
0.450000	0.002571	0.100937
0.500000	0.003586	0.124545
0.550000	0.004821	0.150611
0.600000	0.006281	0.179128
0.650000	0.007969	0.210088
0.700000	0.009880	0.243481
0.750000	0.012004	0.279301
0.800000	0.014325	0.317535
0.850000	0.016824	0.358176
0.900000	0.019481	0.401212
0.950000	0.022271	0.446631
1.000000	0.025170	0.494422

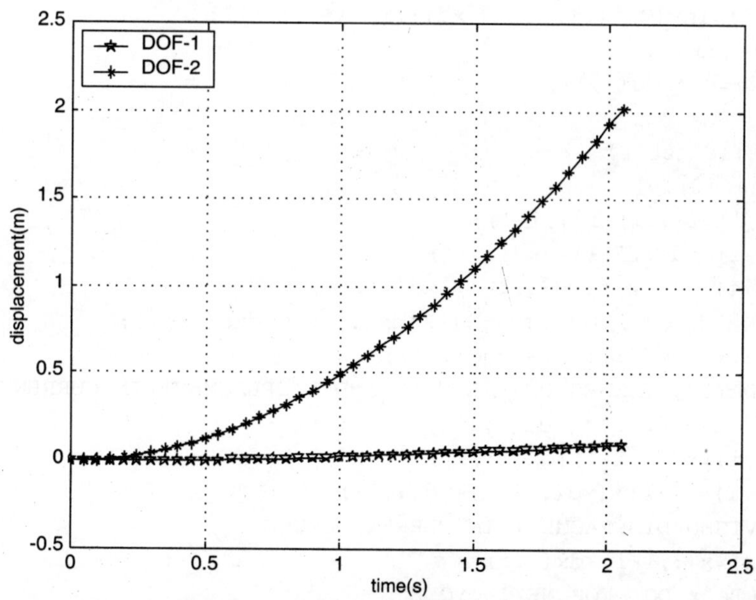

Fig. E4.11 MATLAB Output response.

Example 4.12. *Solve Example 4.11 using the two-cycle iteration with trapezoidal rule.*

Solution:

For an undamped system the following equations are applicable.

We now have:

$$\left(\frac{4}{\Delta t^2}[M]+[K]\right)U_{n+1}=R_{n+1}[M]\left(\frac{4}{\Delta t^2}U_n+\frac{4}{\Delta t}\dot{U}_n+\ddot{U}_n\right)$$

The initial conditions are : $U_0 = 0$, $\dot{U}_0 = 0$, $\ddot{U}_0 = [M]^{-1}(R_0 - [K]\ U_0)$.

$$\ddot{U}_{n+1}=\frac{4}{\Delta t^2}[U_{n+1}-U_n]-\frac{4}{\Delta t}\dot{U}_n-\ddot{U}_n,\ \dot{U}_{n+1}=\frac{\Delta t}{2}[\ddot{U}_{n+1}+\ddot{U}_n]+\dot{U}_n$$

$$=\frac{2}{\Delta t}[U_{n+1}-U_n]-\dot{U}_n$$

Complete program for damped vibrating system is given below:

```
%%%%%%%%%%%%%%%%%%%%%%
% INITIAL VALUES
 M=[1 0;0 10];
 K=[21 -1;-1 1];
 C=[0.5 -0.1;-0.1 0.1];
 dt=0.05;
 T=2;dt=0.05;
 t=[0:dt:T];
 i=1;
 x(:,i)=[0;0];xd(:,i)=[0;0];
 f(:,i)=[0;10];
 xdd(:,i)=inv(M)*(f(:,i)-C*xd(:,i)-K*x(:,i));
 % FIRST time step
 dxd(:,2)=dt*xdd(:,1);

 for i=2:length(t)
   f(:,i)=[0;10];
   df(:,i)=f(:,i)-f(:,i-1);
   xd(:,i)=xd(:,i-1)+dxd(:,i);
   dx(:,i)=(dt/2)*(xd(:,i-1)+xd(:,i));
   dxdd(:,i)=inv(M)*(df(:,i)-K*dx(:,i)-C*dxd(:,i));
   xdd(:,i)=xdd(:,i-1)+dxdd(:,i);
   % UPDATING VALUES OF VELOCITY AND DISPLACMENT IN CURRENT CYCLE
   dxd(:,i)=(dt/2)*(xdd(:,i-1)+xdd(:,i));
   xd(:,i)=xd(:,i-1)+dxd(:,i);
   dx(:,i)=(1/2)*(xd(:,i-1)+xd(:,i));
   % REVISED DISPLACMENT IN CURRENT CYCLE
   x(:,i)=x(:,i-1)+dx(:,i);
   % DELTA x DOT FOR NEXT CYCLE
   dxd(:,i+1)=2*dt*xdd(:,i)-dxd(:,i);
 end
```

```
fprintf('time\t\tX(1)\t\tX(2)\n');
p=1
for time=0:dt:T
   fprintf('%f\t%f\t%f\n',time,x(1,p),x(2,p));
  p=p+1;
end
 plot(t,x(1,:),'-p',t,x(2,:),'-*');
  xlabel('time(s)');
 ylabel('displacement(m)');
 legend('DOF-1','DOF-2',2);
  grid on;
```

The output is as follows:

Time	X(1)	X(2)
0.000000	0.000000	0.000000
0.050000	0.000078	0.024992
0.100000	0.000492	0.099950
0.150000	0.001642	0.224830
0.200000	0.004000	0.399578
0.250000	0.008087	0.624125
0.300000	0.014442	0.898392
0.350000	0.023594	1.222290
0.400000	0.036037	1.595716
0.450000	0.052204	2.018558
0.500000	0.072446	2.490690
0.550000	0.097016	3.011977
0.600000	0.126057	3.582274
0.650000	0.159599	4.201422
0.700000	0.197561	4.869252

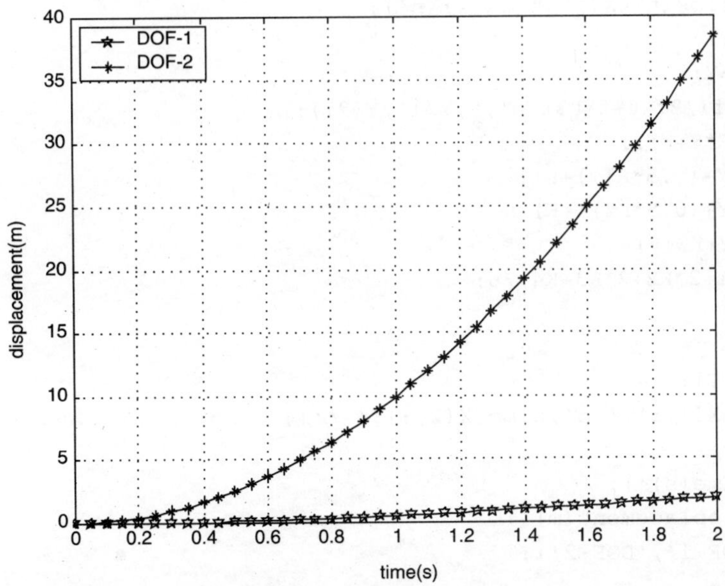

Fig. E4.12 MATLAB output.

Example 4.13. *Solve Example 4.11 using the fourth-order Runge-Kutta method.*

Solution:

Here $\dot{Y} = f(x_1, x_2, x_3, x_4, t)$, where $Y = \begin{Bmatrix} x_1 \\ x_2 \\ x_3 \\ x_4 \end{Bmatrix}$ and $f(x_1, x_2, x_3, x_4, t) = \begin{Bmatrix} x_3 \\ x_4 \\ \dot{x}_3 \\ \dot{x}_4 \end{Bmatrix}$

Where $\begin{Bmatrix} \dot{x}_3 \\ \dot{x}_4 \end{Bmatrix} = -[M]^{-1}[K]\begin{Bmatrix} x_1 \\ x_2 \end{Bmatrix} - [M]^{-1}[C]\begin{Bmatrix} x_3 \\ x_4 \end{Bmatrix} + [M]^{-1}F(t)$

In total it can be written as $\begin{Bmatrix} \dot{x}_1 \\ \dot{x}_2 \\ \dot{x}_3 \\ \dot{x}_4 \end{Bmatrix} = \begin{bmatrix} [0] & [I] \\ -[M]^{-1}[K] & -[M]^{-1}[C] \end{bmatrix}\begin{Bmatrix} x_1 \\ x_2 \\ x_3 \\ x_4 \end{Bmatrix} + \begin{bmatrix} [0] \\ [M]^{-1}F(t) \end{bmatrix}$

Or
$$\dot{Y} = [E]\,Y + F$$

Thus instead of defining functions as in SDOF system, here entire thing will come in one program. The following program is developed based on the above equations.

```
dt=0.05;T=1;
h=dt;
x1=0;       x2=0;% displacements
x3=0;       x4=0;% velocity
M=[1 0;0 10];K=[21 -1;-1 1];C=[0.5 -0.1;-0.1 0.1];f=[0;10];
E=[zeros(size(M)) eye(size(M));-inv(M)*K -inv(M)*C];
F=[0;0;inv(M)*f];
i=1;
Y=[x1;x2;x3;x4];
fprintf('time\t\tX(1)\t\tX(2)\n');
for t=0:h:T
   X(:,i)=Y;
    fprintf('%f\t%f\t%f\n',t,Y(1),Y(2));
   K1=h*(E*Y+F);
   K2=h*(Y+(0.5*K1))+F);
   K3=h*(Y+(0.5*K2))+F)
   K4=h*(Y+K3+F)
   Y=Y+(K1+2*K2+2*K3+K4)/6;
   i=i+1;
end
time=[0:h:T];
plot(time,X(1,:),'-p',time,X(2,:),'-*');
grid on;
xlabel('time(s)');
ylabel('displacement(m)')
legend('DOF-1','DOF-2');
```

The output is shown below:

time	X(1)	X(2)
0.000000	0.000000	0.000000
0.050000	0.000002	0.001250
0.100000	0.000020	0.004998
0.150000	0.000074	0.011242
0.200000	0.000188	0.019980
0.250000	0.000388	0.031208
0.300000	0.000702	0.044922
0.350000	0.001156	0.061118
0.400000	0.001776	0.079790
0.450000	0.002584	0.100933
0.500000	0.003597	0.124541
0.550000	0.004828	0.150606
0.600000	0.006285	0.179122
0.650000	0.007969	0.210081
0.700000	0.009874	0.243474
0.750000	0.011993	0.279292
0.800000	0.014309	0.317526
0.850000	0.016805	0.358166
0.900000	0.019458	0.401200
0.950000	0.022246	0.446618
1.000000	0.025144	0.494409

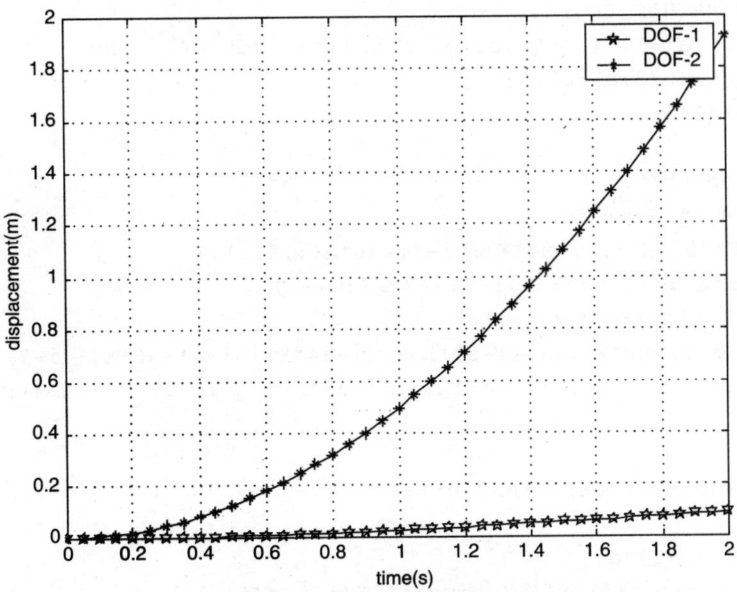

Fig. E4.13 MATLAB output.

Example 4.14. *Solve Example 4.11 using the Houbolt method.*

Solution:

In Houbolt's method for obtaining first two displacements $X\Delta t$ and $X2t$, central difference method is employed. Then three step Houbolt's algorithm is used. Velocities and accelerations are likewise defined. Complete MATLAB program is given below:

```
%%%%%%%%%%%%%%%%%%
K=[21 -1;-1 1];
M=[1 0;0 10];
C=[0.5 -0.1;-0.1 0.1];
dt=0.05;T=2;
X0=[0;0];X0d=[0;0];F=[0;10];
t=[0:dt:T];
X(:,2)=X0;
X0dd=inv(M)*(F-C*X0d-K*X0);
% USING CENTRAL DIFFERENCE METHOD TO OBTAIN PREVIOUS 3 VALUES
Xprev=X0-(dt*X0d)+((dt^2)*(X0dd/2));
a0=1/dt^2;a1=1/(2*dt);a2=2*a0;
mbar=(a0*M)+(a1*C);
kbar=(K-a2*M);
cbar=(a0*M-a1*C);
X(:,1)=X0;
Fbar=F-kbar*X0-cbar*Xprev;
X(:,2)=inv(mbar)*Fbar;
Fbar=F-kbar*X(:,2)-cbar*X0;
X(:,3)=inv(mbar)*Fbar;
% HOUBOLT METHOD BEGINS
a0=2/(dt^2);a1=11/(6*dt);a2=5/(dt^2);a3=3/dt;a4=-2*a0;
a5=-a3/2;a6=a0/2;a7=a3/9;
Kb=K+a0*M+a1*C;
p=3;
for i=3:length(t)
    F=[0;10];% F(t+2dt)
     Fb=F+M*(a2*X(:,i)+a4*X(:,i-1)+a6*X(:,i-2))+
       C*(a3*X(:,i)+a5*X(:,i-1)+a7*X(:,i-2));
    X(:,i+1)=inv(Kb)*Fb;
     Xdd(:,i+1)=a0*X(:,i+1)-a2*X(:,i)-a4*X(:,i-1)-a6*X(:,i-2);
     Xd(:,i+1)=a1*X(:,i+1)-a3*X(:,i)-a5*X(:,i-1)-a7*X(:,i-2);
    p=p+1;
end
fprintf('\ntime\t\tX1\t\tX2\n');
for i=1:p
    time(i)=(i-1)*dt;
     fprintf('%f\t%f\t%f\n',time(i),X(1,i),X(2,i))
end
plot(time,X(1,:),'-p',time,X(2,:),'-*');
```

```
grid on;
xlabel('time(s)');
ylabel('displacement (m)');
legend('DOF-1','DOF-2');
%%%%%%%%%%%%%%%%%%%%%%%%%%%%%%%%%%%%%%%
```

The output of the program is given below:

time	$X1$	$X2$
0.000000	0.000000	0.000000
0.050000	– 0.000000	0.001250
0.100000	0.000015	0.004998
0.150000	0.000069	0.011243
0.200000	0.000184	0.019980
0.250000	0.000387	0.031207
0.300000	0.000704	0.044920
0.350000	0.001161	0.061114
0.400000	0.001782	0.079784
0.450000	0.002588	0.100924
0.500000	0.003595	0.124528
0.550000	0.004817	0.150590
0.600000	0.006260	0.179102
0.650000	0.007927	0.210055
0.700000	0.009814	0.243443
0.750000	0.011912	0.279255
0.800000	0.014208	0.317483
0.850000	0.016686	0.358115
0.900000	0.019325	0.401142
0.950000	0.022104	0.446553
1.000000	0.025000	0.494334

Fig. E4.14 shows the plot of histories of two degrees of freedom.

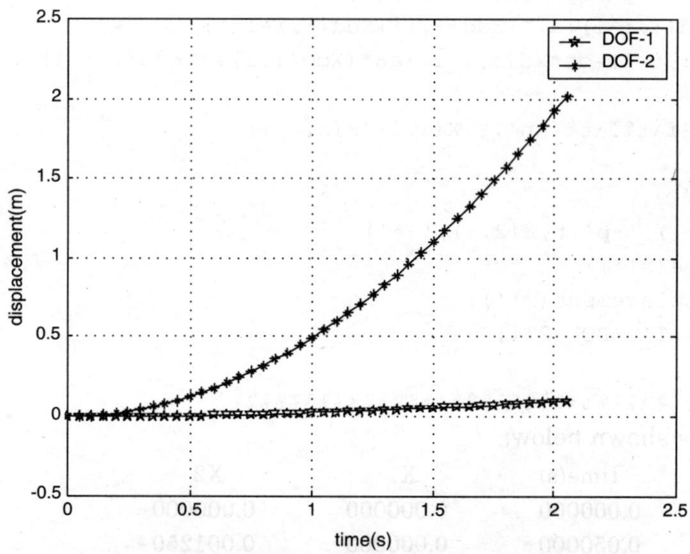

Fig. E4.14 MATLAB output (Houbolt's method).

Example 4.15. *Solve Example 4.11 by the Wilson Theta method.*

Solution:

Here theta is chosen as 1.4 and followed the algorithm. The complete MATLAB program is given below:

```
%%%%%%%%%%%%%%%%%%%%%%%%%%%%%%
K=[21 -1;-1 1];
M=[1 0;0 10];
C=[0.5 -0.1;-0.1 0.1];
dt=0.05;T=2;
X0=[0;0];X0d=[0;0];F0=[0;10];
X0dd=inv(M)*(F0-C*X0d-K*X0);
theta=1.4;
a0=6/(theta*dt)^2;a1=3/(theta*dt);a2=2*a1;
a3=2;a4=(1/2*theta*dt);a5=-a2/theta;
a6=1-3/theta;
a7=dt/2;a8=dt^2/6;
Kb=K+a0*M+a1*C;
i=1;
X(:,1)=X0;Xd(:,1)=X0d;Xdd(:,1)=X0dd;t=0;
fprintf('time(s)\t\tX1\t\tX2\n');
fprintf('%f\t%f\t%f\n',t,X(1,1),X(2,1));
for t=dt:dt:T
 i=i+1;
 F=[0;10];
 Ftb=F0+M*(a0*X(:,i-1)+a2*Xd(:,i-1)+a3*Xdd(:,i-1))+C*(a1*X(:,i-1)+a3*Xd(:,i-
1)+a4*Xdd(:,i-1))+theta*(F-F0);
 Xt(:,i)=inv(Kb)*Ftb;
 Xdd(:,i)=(a0/theta)*(Xt(:,i)-X(:,i-1))+a5*Xd(:,i-1)+a6*Xdd(:,i-1);
 Xd(:,i)=Xd(:,i-1)+a7*(Xdd(:,i)+Xdd(:,i-1));
 X(:,i)=X(:,i-1)+dt*Xd(:,i-1)+a8*(Xdd(:,i)+2*Xdd(:,i-1));
 F0=F;
 fprintf('%f\t%f\t%f\n',t,X(1,i),X(2,i));
end
t=[0:dt:T];
plot(t,X(1,:),'-p',t,X(2,:),'-*')
xlabel('time(s)');
ylabel('displacement(m)');
legend('DOF-1','DOF-2');
grid on;
%%%%%%%%%%%%%%%%%%%%%%%%%%%%%%%%%%%%%%%%%%%%%%%%%%
```

The output is shown below:

time(s)	X1	X2
0.000000	0.000000	0.000000
0.050000	0.000003	0.001250
0.100000	0.000022	0.004998
0.150000	0.000079	0.011242

0.200000	0.000197	0.019979
0.250000	0.000400	0.031206
0.300000	0.000717	0.044919
0.350000	0.001173	0.061113
0.400000	0.001793	0.079784
0.450000	0.002598	0.100926
0.500000	0.003607	0.124531
0.550000	0.004831	0.150595
0.600000	0.006279	0.179109
0.650000	0.007952	0.210065
0.700000	0.009847	0.243455
0.750000	0.011954	0.279271
0.800000	0.014259	0.317502
0.850000	0.016745	0.358138
0.900000	0.019392	0.401169
0.950000	0.022176	0.446584
1.000000	0.025074	0.494370
1.050000	0.028063	0.544516
1.100000	0.031122	0.597008
1.150000	0.034233	0.651833
1.200000	0.037379	0.708978
1.250000	0.040549	0.768426
1.300000	0.043738	0.830164
1.350000	0.046942	0.894174
1.400000	0.050165	0.960442
1.450000	0.053413	1.028950
1.500000	0.056697	1.099681

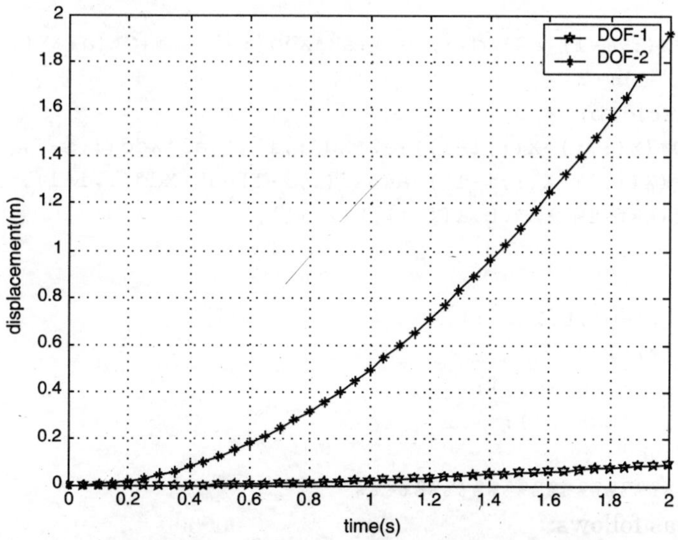

Fig. E4.15 MATLAB response (Wilson-theta).

Example E4.16 Solve Example 4.11 by the Newmark Beta method.

Solution:

Table 1. The Newmark scheme

- $u_{n+1} = u_n + hv_n + h^2(1/2 - \beta)a_n + h^2 \beta a_{n+1}$
- $v_{n+1} = v_n + h(1 - \gamma)a_n + h\gamma a_{n+1}$
- $Ma_{n+1} + Cv_{n+1} + Ku_{n+1} = F_{n+1}$

Here for $\beta = 0.5$ and $\gamma = 1/6$, the following values are obtained. The MATLAB program is shown below:

```
%%%%%%%%%%%%%%%%%%%%%%%%%%%%%%%%%%%
K=[21 -1;-1 1];
M=[1 0; 0 10];
C=[0.5 -0.1;-0.1 0.1];
dt=0.05;T=2;
X0=[0;0];X0d=[0;0];F=[0;10];
X0dd=inv(M)*(F-C*X0d-K*X0);
beta=0.5;gama=1/6;%0.25*(0.5+beta);
a0=1/(beta*dt^2);a1=gama/(beta*dt);a2=1/(beta*dt);
a3=(1/2*beta)-1;a4=(gama/beta-1);a5=0.5*(gama/beta-2)*dt;
a6=dt*(1-beta);a7=beta*dt;
Kb=K+a0*M+a1*C;
i=1;
X(:,1)=X0;Xd(:,1)=X0d;Xdd(:,1)=X0dd;t=0;
fprintf('time(s)\t\tX1\t\tX2\n');
fprintf('%f\t%f\t%f\n',t,X(1,1),X(2,1));
for t=dt:dt:T
 i=i+1;
 F=[0;10];
 Fb=F+M*(a0*X(:,i-1)+a2*Xd(:,i-1)+a3*Xdd(:,i-1))+C*(a1*X(:,i-1)+a4*Xd(:,i-
1)+a5*Xdd(:,i-1));
 X(:,i)=inv(Kb)*Fb;
 Xdd(:,i)=a0*(X(:,i)-X(:,i-1))-a2*Xd(:,i-1)-a3*Xdd(:,i-1);
 Xd(:,i)=a1*(X(:,i)-X(:,i-1))-a4*Xd(:,i-1)-a5*Xdd(:,i-1);
 fprintf('%f\t%f\t%f\n',t,X(1,i),X(2,i));
end
t=[0:dt:T];
plot(t,X(1,:),'-p',t,X(2,:),'-*')
xlabel('time(s)');
ylabel('displacement(m)');
legend('DOF-1','DOF-2');
grid on;
%%%%%%%%%%%%%%%%%%%%%%%%%%%%%%%%%%%
```

The output is as follows:

time(s)	X1	X2
0.000000	0.000000	0.000000
0.050000	0.000005	0.000312

0.100000	0.000017	0.002654
0.150000	0.000045	0.007027
0.200000	0.000102	0.013427
0.250000	0.000205	0.021853
0.300000	0.000372	0.032303
0.350000	0.000622	0.044774
0.400000	0.000976	0.059264
0.450000	0.001454	0.075768
0.500000	0.002075	0.094284
0.550000	0.002854	0.114806
0.600000	0.003808	0.137330
0.650000	0.004947	0.161851
0.700000	0.006278	0.188365
0.750000	0.007804	0.216864
0.800000	0.009526	0.247345
0.850000	0.011437	0.279799
0.900000	0.013529	0.314220
0.950000	0.015790	0.350601
1.000000	0.018203	0.388935

Fig. E4.16 shows the plot of histories of two degree of freedom.

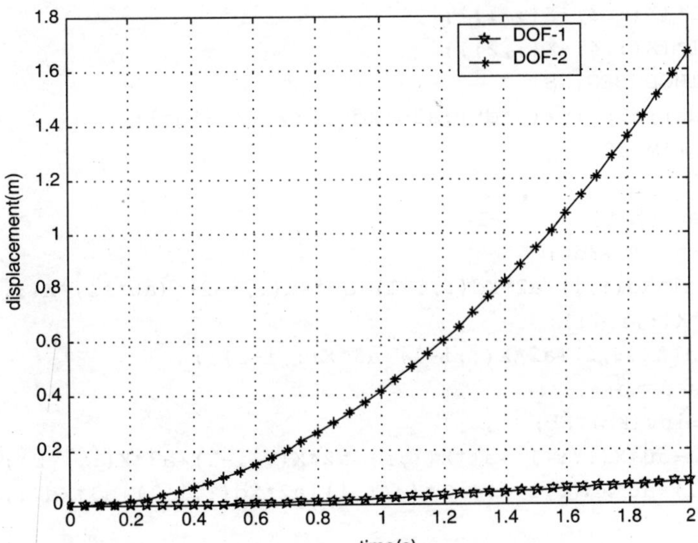

Fig. E4.16 MATLAB output-Newmark Method.

Example 4.17 *Solve Example 4.11 by the Park Stiffly stable method.*

Solution:

It is another implicit method like Houbolt's method. Based on the algorithm, the following MATLAB program is developed. It also is not a self starting method.

```
%%%%%%%%%%%%%%%%%%%%%%%%%%%%%%%%%%
K=[21 -1;-1 1];
M=[1 0;0 10];
C=[0.5 -0.1;-0.1 0.1];
dt=0.05;T=2;
X0=[0;0];X0d=[0;0];F=[0;10];
t=[0:dt:T];
X(:,2)=X0;
X0dd=inv(M)*(F-C*X0d-K*X0);
% USING CENTRAL DIFFERENCE METHOD TO OBTAIN PREVIOUS 3 VALUES of X AND Xd
Xprev=X0-(dt*X0d)+((dt^2)*(X0dd/2));
a0=1/dt^2;a1=1/(2*dt);a2=2*a0;
mbar=(a0*M)+(a1*C);
kbar=(K-a2*M);
cbar=(a0*M-a1*C);
X(:,1)=X0;Xd(:,1)=X0d;
Fbar=F-kbar*X0-cbar*Xprev;
X(:,2)=inv(mbar)*Fbar;
Fbar=F-kbar*X(:,2)-cbar*X0;
X(:,3)=inv(mbar)*Fbar;
Fbar=F-kbar*X(:,3)-cbar*X0;
X(:,4)=inv(mbar)*Fbar;
Xd(:,2)=a1*(X(:,3)-X(:,1));
Xd(:,3)=a1*(X(:,4)-X(:,2));
% PARK METHOD BEGINS
a0=10/(6*dt);a1=-15/(6*dt);a2=1/dt;a3=-1/(6*dt);
Kb=K+(a0^2)*M-a0*C;
p=3;
for i=3:length(t)
 F=[0;10];% F(t+3dt)
 mass=M*(-a1*Xd(:,i)-a2*Xd(:,i-1)-a3*Xd(:,i-2)-(a0*a1)*X(:,i)-(a0*a2)*X(:,i-
1)+(a3^2)*X(:,i-2));
  damp=C*(a1*X(:,i)+a2*X(:,i-1)+a3*X(:,i-2));
  Fb=F+mass-damp;
  X(:,i+1)=inv(Kb)*Fb;
  Xd(:,i+1)=a0*X(:,i+1)+a1*X(:,i)+a2*X(:,i-1)+a3*X(:,i-2);
  Xdd(:,i+1)=a0*Xd(:,i+1)+a1*Xd(:,i)+a2*Xd(:,i-1)+a3*Xd(:,i-2);
 p=p+1;
end
fprintf('\ntime\t\tX1\t\tX2\n');
for i=1:p
 time(i)=(i-1)*dt;
 fprintf('%f\t%f\t%f\n',time(i),X(1,i),X(2,i))
end
plot(time,X(1,:),'-p',time,X(2,:),'-*');
```

```
grid on;
xlabel('time(s)');
ylabel('displacement (m)');
legend('DOF-1','DOF-2');
%%%%%%%%%%%%%%%%%%%%%%%%%%%%%%%
```

The output is as follows:

time(s)	X1	X2
0.000000	0.000000	0.000000
0.050000	– 0.000000	0.001250
0.100000	0.000015	0.004998
0.150000	0.000010	0.010911
0.200000	– 0.000103	0.017760
0.250000	– 0.000428	0.024395
0.300000	– 0.001067	0.029618

Fig. E4.17 shows the MATLAB plot.

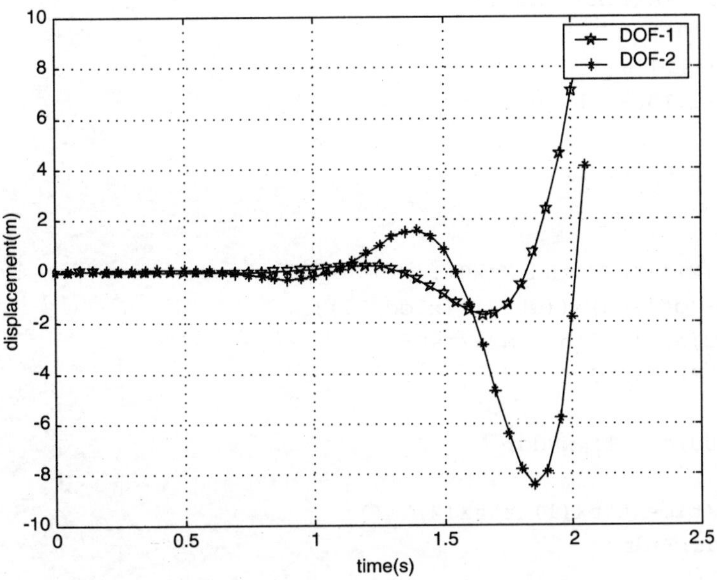

Fig. E4.17 MATLAB output.

Example E4.18 *The numerical values of the mass, damping and stiffness are chosen as* $M_1=1$, $M_2 = 10$, $C_1 = 0$, $C_2 = 0.15$, $K_1 = 19$ *and* $K_2 = 1$. *The initial conditions are selected as zero and the forcing vector is*

$$\begin{Bmatrix} F_1(t) \\ F_2(t) \end{Bmatrix} = \begin{Bmatrix} 0 \\ 5 \end{Bmatrix} \quad \text{for } t > 0$$

and

$$\begin{Bmatrix} F_1(t) \\ F_2(t) \end{Bmatrix} = \begin{Bmatrix} 0 \\ 0 \end{Bmatrix} \quad \text{for } t < 0$$

In matrix motion, these equations may be written as

$$\begin{bmatrix} M_1 & 0 \\ 0 & M_2 \end{bmatrix}\begin{bmatrix} \ddot{X}_1 \\ \ddot{X}_2 \end{bmatrix} + \begin{bmatrix} C_1 + C_2 & -C_2 \\ -C_2 & C_2 \end{bmatrix}\begin{bmatrix} X_1 \\ X_2 \end{bmatrix}$$

$$+\begin{bmatrix} K_1 + K_2 & -K_2 \\ -K_2 & K_2 \end{bmatrix}\begin{Bmatrix} X_1 \\ X_2 \end{Bmatrix} + \begin{Bmatrix} -0.5K_2(X_2 - X_1)^3 \\ 0.5K_2(X_2 - X_1)^3 \end{Bmatrix} = \begin{Bmatrix} F_1 \\ F_2 \end{Bmatrix}$$

Take $\Delta t = 0.05$ sec. Use the central difference method and compute the response of the system.

Solution:

Substituting the given values the following matrices is obtained:

$$M = \begin{bmatrix} 1 & 0 \\ 0 & 10 \end{bmatrix}, C = \begin{bmatrix} 0.15 & -0.15 \\ -0.15 & 0.15 \end{bmatrix}, K = \begin{bmatrix} 20 & -1 \\ -1 & 1 \end{bmatrix}, F = \begin{Bmatrix} 0 \\ 5 \end{Bmatrix}$$

So the nonlinear spring force is added to the external force equation without confusion.

Complete MATLAB program is given below:

```
%%%%%%%%%%% INITIAL VALUES
M=[1 0;0 10];
K=[20 -1;-1 1];
C=[0.15 -0.15;-0.15 0.15];
dt=0.05;
x0=[0;0];x0d=[0;0];
F0=[0;5];
T=2;
x0dd=inv(M)*(F0-C*x0d-K*x0);
xprev=x0-(dt*x0d)+((dt^2)*(x0dd/2));
a0=1/dt^2;a1=1/(2*dt);a2=2*a0;
mbar=(a0*M)+(a1*C);
t=0;
v(:,1)=x0d;a(:,1)=x0dd;
i=1;
fprintf('time\t\tX(1)\t\tX(2)\n');
for t=0:dt:T+dt
X(:,i)=x0;
% NONLINEAR SPRING
Fr=[-0.5*1*(X(2,i)-X(1,i))^3;0.5*1*(X(2,i)-X(1,i))^3];
F=F0+Fr;
Fbar=F+(a2*M-K)*x0+(a1*C-a0*M)*xprev;
x=inv(mbar)*Fbar;
xprev=x0;
x0=x;
fprintf('%f\t%f\t%f\n',t,X(1,i),X(2,i));
i=i+1;
p=i;
end
for i=2:p-1
```

```
 if i<p-1
  v(:,i)=(X(:,i+1)-X(:,i-1))*(1/(2*dt));
  a(:,i)=(X(:,i+1)-2*X(:,i)+X(:,i-1))*(1/dt^2);
 end
 end
  t=[0:dt:T+dt];
  plot(t,X(1,:),'-p',t,X(2,:),'-*');
xlabel('time(s)');
 ylabel('displacement(m)');
 legend('DOF-1','DOF-2',2);
grid on;
%%%%%%%%%%%%%%%%%%%%%%%%%%%
```

The output is as follows:

time	X(1)	X(2)
0.000000	0.000000	0.000000
0.050000	-0.000000	0.000625
0.100000	0.000011	0.002499
0.150000	0.000046	0.005620
0.200000	0.000121	0.009988
0.250000	0.000251	0.015599
0.300000	0.000453	0.022452
0.350000	0.000743	0.030543
0.400000	0.001134	0.039871
0.450000	0.001640	0.050433
0.500000	0.002270	0.062224

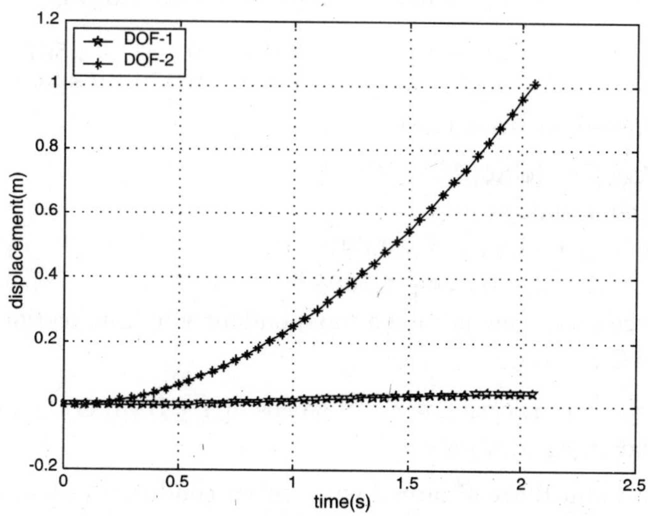

Fig. E4.18 MATLAB output.

Example 4.19 *Consider a simple system for which the governing equilibrium equations are*

$$[M] = \begin{bmatrix} 2 & 0 \\ 0 & 1 \end{bmatrix}$$

$$[K] = \begin{bmatrix} 4 & -1 \\ -1 & 3 \end{bmatrix}$$

$$[C] = \begin{bmatrix} 0 & 0 \\ 0 & 0 \end{bmatrix}$$

and $$F_1(t) = 0,\ F_2(t) = 10$$

(a) calculate the transformation matrix for this system and thus establish the decoupled equations of equilibrium on the basis of mode shape vectors.

(b) compute the exact response by integrating each of the two decoupled equilibrium equations.

Solution:

(*a*) Transformation matrix of the system is obtained from eigenvalue problem.

Solving the undamped eigenvalue equation, we obtain eigenvector as

$$X = \begin{bmatrix} -0.8069 & 0.3437 \\ -0.5907 & -0.9391 \end{bmatrix} \text{ corresponding to eigenvalues } \lambda = \begin{bmatrix} 1.634 \\ 3.366 \end{bmatrix}$$

Find modal masses $m_1 = \text{sqrt}(X_1'MX_1) = 1.2850$

$m_2 = \text{sqrt}(X_2'MX_2) = 1.0574$

Now the transformation matrix is

$$\phi = \begin{bmatrix} -0.8069/1.2850 & 0.3437/1.0574 \\ -0.5907/1.2850 & -0.9391/1.0574 \end{bmatrix} = \begin{bmatrix} -0.628 & 0.325 \\ -0.4597 & -0.8881 \end{bmatrix}$$

$$\phi'F = \begin{bmatrix} -0.628 & 0.325 \\ -0.4597 & -0.8881 \end{bmatrix}' \begin{bmatrix} 0 \\ 10 \end{bmatrix} = \begin{bmatrix} -4.597 \\ -8.8808 \end{bmatrix}$$

Thus the decoupled equations are

$$[\phi' M\phi]\ddot{U} + [\phi' K\phi]\, U = \phi' F$$

That is modal equations are:

$$\ddot{u}_1 + 1.634\, u_1 = -4.597$$

$$\ddot{u}_2 + 3.366\, u_2 = -8.8808$$

(*b*) Each of these equations is solved independently by integration as follows:

Solution is u = C.F + P.I.,

where C.F.= complementary function = $e^{-\xi\omega_n t}$ $(A \cos \omega_d t + B \sin \omega_d t)$ and

P.I. = Particular integral = $f/\omega_n^{\,2}$

The constants A and B are obtained with initial condition $u(0)$ and $\dot{u}(0)$

$$u(t) = e^{-\xi\omega_n t}\,(A \cos \omega_d t + B \sin \omega_d t) + f/\omega_n^{\,2}$$

$$\dot{u} = e^{-\xi\omega_n t}\,((B\omega_d - \xi\omega_n A) \cos \omega_d t - (A\omega_d + \xi\omega_n B)\sin \omega_d t)$$

Hence $u(0) = A + f/\omega_n^2$ or $A = u(0) - f/\omega_n^2$

and $\dot{u}(0) = (B\omega_d - \xi\omega_n A)$ or $B = [\dot{u}(0) + \xi\omega_n A]/\omega_d$

If there is no damping $\xi = 0$, thus $B = \dot{u}(0)/\omega_n$

Also with zero initial conditions we can write the solutions as follows:

$$u_1 = f_1/\omega_{n1}^2 (1 - \cos \omega_{n1}t) \text{ and } u_2 = f_2/\omega_{n2}^2 (1 - \cos \omega_{n2}t)$$

or $u_1 = [-4.597/1.634]\,(1 - \cos 1.2783t)$

and $u_2 = [-8.8808/3.366]\,(1 - \cos 1.8347t)$

$$= -2.8133\,(1 - \cos 1.2783t) \quad = -2.6384\,(1 - \cos 1.8347t)$$

Finally the response in original coordinates is obtained as follows:

$$X = \phi U = \begin{bmatrix} -0.628 & 0.325 \\ -0.4597 & -0.8881 \end{bmatrix} \begin{bmatrix} -2.8133\,(1 - \cos 1.2783t) \\ -2.6384\,(1 - \cos 1.8347t) \end{bmatrix}$$

$$= \begin{bmatrix} 0.9091\,(1 - \cos 1.2783t) \\ 3.6364\,(1 - \cos 1.8347t) \end{bmatrix}$$

The following program is used to obtain the transformation matrix:

```
M=[2 0;0 1];
K=[4 -1;-1 3];
F=[0;10];
[u,W]=eig(K,M);
for i=1:2
     wn(i)=sqrt(W(i,i));
end
[w,I]=sort(wn);
for j=1:2
U(:,j)=u(:,I(j)); % ARRANGE MODAL VECTORS IN ASCENDING ORDER
end
% MODAL MASSES
m1=sqrt(U(:,1)'*M*U(:,1));
m2=sqrt(U(:,2)'*M*U(:,2));
phi=[U(:,1)/m1;U(:,2)/m2]; % transformation matrix
% PLOT RESPONSES
t=[0:0.1:5];
u1=-2.8133*(1-cos(1.2783*t)); u2=-2.6384* (1-cos (1.8347*t));
x1=0.9091*(1- cos(1.2783*t)); x2=3.6364* (1-cos (1.8347*t));
subplot(2,1,1);                    %SUB-PLOT-1
plot(t,u1,'-p',t,u2,'-*');
title('\bfPlot of Modal amplitudes');
xlabel('\bfTime(s)');
legend('At frequency-1=1.2783 rad/s','At frequency-2=1.8347 rad/
grid on;
subplot(2,1,2);                    % SUB-PLOT-2
```

```
plot(t,x1,'-p',t,x2,'-*');
title('\bfPlot of Original amplitudes');
xlabel('\bfTime(s)');
legend('At DOF-1', 'At DOF-2');
grid on;
%%%%%%%%%%%%%%%%%%%%%%%%%%
```

The output of the program is shown in Fig. E4.19.

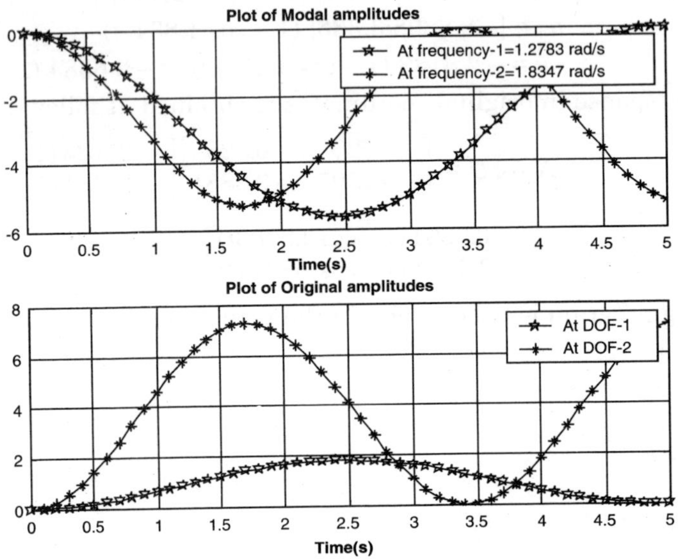

Fig. E4.19 Output of the MATLAB program.

PROBLEMS

4.1 Determine the vibratory response of an undamped single degree of freedom system with $M = 2$, and $K = 2$. Assume the initial conditions for the velocity and displacement at $t = 0$ to be equal to 1. Use the Runge-Kutta method with a time step of 0.5 seconds.

4.2 Determine the free vibratory response of a viscously damped single degree of freedom system with $M = C = K = 2$. Assume the initial conditions for the velocity and displacement at $t = 0$ to be equal to 1. Use the fourth order Runge-Kutta method with time step of 0.5 seconds.

4.3 Solve Problem 4.2 with $C = 3$ using the fourth order Runge-Kutta method.

4.4 Find the solution of the differential equation for an undamped single degree of freedom system with $M = 5$, $C = 2$, and $K = 300$ and the forcing function is defined as

$$F(t) = 200 \text{ for } 0 < t < 0.2$$

and

$$F(t) = (1000 - 2000t)/3 \text{ for } 0.2 < t < 0.5$$

with the initial conditions of displacement and velocity as zero. Use the fourth order Runge-Kutta method.

4.5 Find the vibratory response of a single degree of freedom system with $M = 1$, $C = 0.2$, $K = 1$, and $F(t) = 5$ with the initial conditions for velocity and displacement as zero. Take time step of 0.5 seconds. Use the fourth order Runge-Kutta method.

4.6 Find the response of the two degree of freedom system when $F_1(t) = 0$ and $F_2(t) = 10$, using the central difference method. The mass, stiffness and damping matrices for this system as shown in Fig. 4.5 are given as

$$[M] = \begin{bmatrix} 1 & 0 \\ 0 & 10 \end{bmatrix},\ [K] = \begin{bmatrix} 21 & -1 \\ -1 & 1 \end{bmatrix},\ [C] = \begin{bmatrix} 0.5 & -0.1 \\ -0.1 & 0.1 \end{bmatrix}$$

All the initial conditions are given as zero and use the central difference method with $\Delta t = 0.001$ seconds.

4.7 Solve Problem 4.6 using the fourth-order Runge-Kutta method.

4.8 Solve Problem 4.6 using the Houbolt method.

4.9 Solve Problem 4.6 by the Park Stiffly Stable method.

4.10 Solve Problem 4.6 by the Wilson Theta method with theta = 1.5.

4.11 Solve Problem 4.6 by the Newmark Beta method with alpha = 0.5 and beta = 1/6.

4.12 Solve Problem 4.6 using the two-cycle iteration with trapezoidal rule.

4.13 Find the response of the two degree of freedom system when $F_1(t) = 0$ and $F_2(t) = 10$, using the central difference method. The mass, stiffness and damping matrices for this system are given as

$$[M] = \begin{bmatrix} 1 & 0 \\ 0 & 2 \end{bmatrix},\ [K] = \begin{bmatrix} 6 & -2 \\ -2 & 8 \end{bmatrix},\ [C] = \begin{bmatrix} 0 & 0 \\ 0 & 0 \end{bmatrix}$$

All the initial conditions are given as zero and $\Delta t = 0.241$ second.

4.14 Find the response of the two degree of freedom system when $F_1(t) = 0$ and $F_2(t) = 0$, using the Wilson-theta method. The mass, stiffness and damping matrices for this system are given as

$$[M] = \begin{bmatrix} 1 & 0 \\ 0 & 1 \end{bmatrix},\ [K] = \begin{bmatrix} 2 & 2 \\ 2 & 5 \end{bmatrix},\ [C] = \begin{bmatrix} 0 & 0 \\ 0 & 0 \end{bmatrix}$$

Assuming initial conditions to be

$$x_0 = \begin{Bmatrix} 1 \\ 0 \end{Bmatrix},\ \dot{x}_0 = \begin{Bmatrix} 0 \\ 0 \end{Bmatrix} \text{ and } \Delta t = 0.25 \text{ seconds.}$$

4.15 Solve Problem 4.14 using Newmark's beta method; alpha = 0.25 and beta = 0.5.

4.16 Find the response of the two degree of freedom system when $F_1(t) = \sin 2t$ and $F_2(t) = \sin 2t$, using the fourth order Runge-Kutta method. The differential equations of this system are given as

$$\begin{bmatrix} 2 & 0 \\ 0 & 1 \end{bmatrix} \begin{Bmatrix} \ddot{x}_1 \\ \ddot{x}_2 \end{Bmatrix} + \begin{bmatrix} 3 & -0.5 \\ -0.5 & 0.5 \end{bmatrix} \begin{Bmatrix} \dot{x}_1 \\ \dot{x}_2 \end{Bmatrix} + \begin{bmatrix} 3 & -1 \\ -1 & 1 \end{bmatrix} \begin{Bmatrix} x_1 \\ x_2 \end{Bmatrix} = \begin{bmatrix} 1 \\ 1 \end{bmatrix} \sin 2t$$

with the initial conditions $x_0 = \begin{Bmatrix} 0 \\ 0.1 \end{Bmatrix},\ \dot{x}_0 = \begin{Bmatrix} 1 \\ 0 \end{Bmatrix}$ m/s.

Bibliography

There are several outstanding text and reference books on control systems, engineering mechanics: statics and dynamics, mechanics of materials, mechanical vibrations, electrical circuits, and MATLAB that merit consultation for those readers who wish to pursue these topics further. The following list is but a representative sample of the many excellent books.

Control Systems

Beards, C.F., *Vibrations and Control System,* Ellis Horwood, 1988.

Chesmond, C.J., *Basic Control System Technology,* Edward Arnold, 1990.

Dorf, R.C., and Bishop, R.H., *Modern Control Systems,* 9th ed., Prentice Hall, Upper Saddle River, NJ, 2001.

Dorsey, John., *Continuous and Discrete Control Systems,* McGraw Hill, New York, NY, 2002.

Dukkipati, R.V., *Control Systems,* Narosa Publishing House, New Delhi, India, 2005.

Dukkipati, R.V., *Engineering System Dynamics,* Narosa Publishing House, New Delhi, India, 2004.

Franklin, G.F., David Powell, J., and Abbas Emami-Naeini, *Feedback Control of Dynamic Systems,* 3rd ed., Addison Wesley, Reading, MA, 1994.

Johnson, C., and Malki, H., *Control Systems Technology,* Prentice Hall, Upper Saddle River, NJ, 2002.

Kuo, B.C., *Automatic Control Systems,* 6th ed., Prentice Hall, Englewood Cliffs, NJ, 1991.

Nise, Norman, S., *Control Systems Engineering,* 3rd ed., Wiley, New York, NY, 2000.

Ogata, K., *Modern Control Engineering,* 3rd ed., , Prentice Hall, Englewood Cliffs, NJ, 1997.

Palm III, W.J., *Control Systems Engineering,* Wiley, New York, NY, 1986.

Phillips, C.L., and Harbour, R.D., *Feedback Control Systems,* 4th ed., Prentice Hall, Upper Saddle River, NJ, 2000.

Shinners, S. M., *Modern Control System Theory and Design,* 2nd ed., Wiley Inter Science, New York, NY, 1998.

Engineering Mechanics

Bedford, A., and Fowler, W., *Engineering Mechanics: Statics & Dynamics,* 4th ed., Prentice Hall, New Jersey, 2005.

Beer, F.B., Russell Johnston, E., and Eisenberg, E.R., *Vector Mechanics for Engineers-Statics and Dynamics,* 8th ed., McGraw Hill, New York, 2007.

Boresi, A., and Schmidt, R.J., *Engineering Mechanics: Statics,* 2nd ed., Brooks/Cole Publishing Company, New York, 2001.

Hibbler, R.C., *Engineering Mechanics: Statics & Dynamics,* Prentice Hall, Englewood Cliffs, NJ, 1995.

Meriam, J.L., and Kraige, L.G., *Engineering Mechanics: Dynamics,* 5th ed., Wiley, New York, 2002.

Nelson, E.W., Best, C.L., and McLean, W.G., *Schaum's Outlines: Engineering Mechanics-Statics and Dynamics,* 5th ed., McGraw Hill, New York, 1998.

Pytel, A., and Kiusalaas, J., *Engineering Mechanics: Statics,* 2nd ed., Brooks/Cole Publishing Company, New York, 1999.

Shelley, J.F., *Vector Mechanics for Engineers Volume I: Statics,* Schaum's Solved Problems Series, McGraw Hill, New York, 1990.

Mechanics of Materials

Beer, F.B., Russell Johnston, E., and DeWolf, J.T., *Mechanics of Materials,* 3rd ed., McGraw Hill, New York, 2001.

Cernica, J.N., *Strength of Materials,* Holt, Rinehart and Winston, Inc., New York, 1966.

Nash, W., *Schaum's Outlines: Statics and Mechanics of Materials,* McGraw Hill, New York, 1992.

Nash, W., *Schaum's Outlines: Strength of Materials,* 4th ed., McGraw Hill, New York, 1998.

Riley, W., Sturges, L., and Morris, D., *Mechanics of Materials,* Wiley, New York, 2007.

Mechanical Vibrations

Beards, C.F., *Structural Vibration Analysis,* Ellis Harwood, U.K, 1983.

Bishop, R.E.D., *Vibration,* Cambridge University Press, Cambridge, England, 1979.

Dimarogonas, A.D., *Vibration for Engineers,* 2nd ed., Prentice Hall, Englewood Cliffs, NJ, 1996.

Dukkipati, R.V., *Advanced Mechanical Vibrations,* Narosa Publishing House, New Delhi, India, 2006.

Dukkipati, R.V., and Srinivas, J., *A Text Book of Mechanical Vibrations,* Prentice Hall of India, New Delhi, India, 2005.

Dukkipati, R.V., and Srinivas, J., *Vibrations: Problem Solving Companion,* Narosa Publishing House, New Delhi, India, 2006.

Dukkipati, R.V., *Engineering System Dynamics,* Narosa Publishing House, New Delhi, India, 2005.

Dukkipati, R.V., *Vibration Analysis,* Narosa Publishing House, New Delhi, India, 2005.

Kelly, S.G., *Fundamentals of Mechanical Vibration,* McGraw Hill, New York, NY, 1993.

Kelly, S.G., *Theory and Problems of Mechanical Vibrations,* Schaum's Outline Series, McGraw Hill, New York, NY, 1996.

Meirovitch, L., *Elements of Vibration Analysis,* 2nd ed., McGraw Hill, New York, NY, 1986.

Seto, W.W., *Theory and Problems of Mechanical Vibrations,* Schaum series, McGraw Hill, New York, NY, 1964.

Steidel, R.F., *An Introduction to Mechanical Vibrations,* 3rd ed., Wiley, New York, NY, 1981.

Thomson, W.T., and Dahleh, M.D., *Theory of Vibrations with Applications,* 5th ed., Prentice Hall, Englewood Cliffs, NJ, 199.

Tse, F.S., Morse, I.E., and Hinkle, R.T., *Mechanical Vibrations,* Allyn and Bacon, Boston, MA, 1963.

Vierck, R.K., *Vibration Analysis,* 2nd ed., Harper & Row, New York, NY, 1979.

Electrical Circuits

Cogdell, J.R., *Foundations of Electrical Circuits,* Prentice Hall, Englewood Cliffs, NJ, 1999.

Dorf, R.C, and Svoboda, J.A., Introduction *to Electric Circuits,* Wiley, New York, NY, 2006.

Fogiel, M, and Ogden, J.R., *The Electric Circuits Problem Solver: A Complete Solution Guide to Any Text Book,* Research & Education Association, 1998.

Hayt, W.H., and Kemmerly, J.E., *Engineering Circuit Analysis,* 5th ed., McGraw Hill, New York, NY, 1993.

Johnson, D.E, Johnson, J.R, Hilburn, and J.L, Scott, P.D., *Electric Circuit Analysis,* 3rd ed., Wiley, New York, 1997.

Johnson, D.E., *Basic Electric Circuit Analysis,* 5th ed., Wiley, New York, 2006.

Johnson, D.E., Hilburn, J.L., Johnson, J.R., and Scott, P.D., *Electric Circuit Analysis,* 3rd ed., Prentice Hall, Englewood Cliffs, NJ, 1990.

Nahvi, M., *Schaums Outline of Electrical Circuits,* 4th ed., McGraw Hill, New York, NY, 2002.

Nilsson, J.W., and Riedel, S.A., *Electric Circuits,* 6th ed, Prentice Hall, Englewood Cliffs, NJ, 1999.

Nilsson, J.W., *Electrical Circuits,* 7th ed, Prentice Hall, Englewood Cliffs, NJ, 2004.

Paul, C.R., *Fundamentals of Electric Circuit Analysis,* Wiley, New York, 2000.

Singh Guru, B, and Warrier, R., *Electric Circuits-Analysis and Design,* Oxford University Press, Oxford, 2005.

Smith, R.J., and Dorf, R.C., Circuits, Systems, and Devices, 5th ed., Wiley, New York, 1992.

Starr, A.T., *Electrical Circuits and Wave Filters,* Pitman Publishing, New York, 1938.

Numerical Methods

Abramowitz, M., and Stegun, I., *Handbook of Mathematical Functions,* Dover Press, New York, 1964.

Akai, T.J., *Applied Numerical Methods for Engineers,* Wiley, New York, NY, 1993.

Al-Khafaji, A.W., and Tooley, J.R., *Numerical Methods in Engineering Practice,* Holt, Rinehart and Winston, New York, 1986.

Allen, M., and Issacson, E., *Numerical Analysis for Applied Science,* Wiley, New York, 1998.

Ames, W.F., *Numerical Methods for Partial Differential Equations,* 3rd ed., Academic Press, New York, 1992.

Ascher,U., Mattheij, R., and Russell, R., *Numerical Solution of Boundary Value Problems for Ordinary Differential Equations,* Prentice Hall, Englewood Cliffs, NJ, 1988.

Atkinson, K., and Han, W., *Elementary Numerical Analysis,* 3rd ed., Wiley, New York, 2004.

Atkinson, K.E., *An Introduction to Numerical Analysis*, 2nd ed., Wiley, New York, NY, 1993.

Atkinson, L.V., and Harley, P.J., *Introduction to Numerical Methods with PASCAL*, Addison Wesley, Reading, MA, 1984.

Atkinson, L.V., Harley, P.J., and Hudson, J.D., *Numerical Methods with FORTRAN 77*, Addison Wesley, Reading, MA, 1989.

Axelsson, K., *Iterative Solution Methods,* Cambridge University Press, Cambridge, UK, 1994.

Ayyub, B.M., and McCuen, R.H., *Numerical Methods for Engineers*, Prentice Hall, Upper Saddle River, New Jersey, NJ, 1996.

Baker, A.J., *Finite Element Computational Fluid Mechanics*, McGraw Hill, New York, 1983.

Balagurusamy, E., *Numerical Methods*, Tata McGraw Hill, New Delhi, India, 2002.

Bathe, K.J., and Wilson, E.L., *Numerical Methods in Finite Element Analysis*, Prentice Hall, Englewood Cliffs, NJ, 1976.

Bhat, R.B., and Chakraverty, S., *Numerical Analysis in Engineering*, Narosa Publishing House, New Delhi, India, 2004.

Bhat, R.B., and Gouw, G.J., *Numerical Methods in Engineering*, Simon and Schuster Custom Publishing, Needham Heights, MA, 1996.

Bjorck, A., *Numerical Methods for Least Squares Problems*, Society for Industrial and Applied Mathematics (SIAM), Philadelphia, PA, 1996.

Booth, A.D., *Numerical Methods*, Academic Press, New York, 1958.

Brice, C., Luther, H.A and Wilkes, J. O., *Applied Numerical Methods*, New York, NY, 1969.

Buchanan, J.L., and Turner, P.R., *Numerical Methods and Analysis,* McGraw Hill, New York, 1992.

Burden, R.L., and Faires, J.D., *Numerical Analysis,* 6th ed., Brooks/Cole, Pacific Grove, 1997.

Carnahan, B., Luther, A., and Wilkes, J.O., *Applied Numerical Methods,* Wiley, New York, 1969.

Chapra, S.C, and Canale, R.P., *Introduction to Computing for Engineers*, 2nd ed., McGraw Hill, New York, 1994.

Chapra, S.C, and Canale, R.P., *Numerical Methods for Engineers with Personal Computers*, McGraw Hill, New York, 1985.

Chapra, S.C., *Applied Numerical Methods with MATLAB for Engineers and Scientists*, McGraw Hill, New York, 2005.

Chapra, S.C., *Numerical Methods for Engineers with Software and Programming Applications*, 4th ed., McGraw Hill, New York, NY, 2002.

Cheney, W., and Kincaid, D., *Numerical Mathematics and Computing*, 2nd ed., Brooks/ Cole, Monterey, CA, 1994.

Chui, C., *An Introduction to Wavelets*, Academic Press, Burlington, MA, 1992.

Consatantinides, A., *Applied Numerical Methods with Personal Computers*, McGraw Hill, New York, 1987.

Conte, S.D., and DeBoor, C.W., *Elementary Numerical Analysis: An Algorithm Approach,* 2nd ed., McGraw Hill, New York, NY, 1972.

Dahlquist, G., and Bjorck, A., *Numerical Methods,* Prentice Hall, Englewood Cliffs, NJ, 1974.

Davis, P., and Rabinowitz, P., *Methods of Numerical Integration,* Academic Press, 2nd ed., New York, 1998.

Demmel, J.W., *Applied Numerical Linear Algebra,* Society for Industrial and Applied Mathematics (SIAM), Philadelphia, PA, 1997.

Dennis, J.E., and Schnabel, R.B., *Numerical Methods for Unconstrained Optimization and Nonlinear Equations,* Society for Industrial and Applied Mathematics (SIAM), Philadelphia, PA, 1996.

Epperson, J.F., *An Introduction to Numerical Methods and Analysis,* Wiley, New York, NY, 2001.

Fadeev, D.K., and Fadeeva, V.N., *Computational Methods of Linear Algebra,* Freeman, San Francisco, 1963.

Fadeeva, V.N., (Trans. Curtis D. Benster), *Computational Methods of Linear Algebra,* Dover, New York, 1959.

Fatunla, S.O., *Numerical Methods for Initial Value Problems in Ordinary Differential Equations,* Academic Press, San Diego, 1988.

Ferziger, J.H., *Numerical Methods for Engineering Applications,* 2nd ed., Wiley, New York, NY, 1998.

Forbear, C. E., *Introduction to Numerical Analysis,* Addison Wesley, Reading, MA, 1969.

Forsythe, G.E., and Wasow, W.R., *Finite-Difference Methods for Partial Differential Equations,* Wiley, New York, 1960.

Forsythe, G.E., Malcolm, M.A., and Moler, C.B., *Computer Methods for Mathematical Computation,* Prentice Hall, Englewood Cliffs, NJ, 1977.

Froberg, C.E., *Introduction to Numerical Analysis,* Addison-Wesley, Reading, MA, 1965.

Gautschi, W., *Numerical Analysis: An Introduction,* Birkhauser, Boston, MA, 1997.

Gear, C.W., *Numerical Initial Value Problems in Ordinary Differential Equations,* Prentice Hall, Englewood Cliffs, NJ, 1971.

Gerald, C.F., and Wheatley, P.O., *Applied Numerical Analysis,* 5th ed., Addison-Wesley, Reading, MA, 1994.

Gladwell, J., and Wait, R., *A Survey of Numerical Methods of Partial Differential Equations,* Oxford University Press, New York, 1979.

Goldberg, D.E., *Genetic Algorithms in Search, Optimization and Machine Learning,* Addison-Wesley, Reading, MA, 1989.

Golub, G.H., and Van Loan, C.F., *Matrix Computations,* 3rd ed., Johns Hopkins University Press, Baltimore, MD, 1996.

Greenbaum, A., *Iterative Methods for Solving Linear Systems,* Society for Industrial and Applied Mathematics (SIAM), Philadelphia, PA, 1997.

Griffiths, D.V., and Smith, I.M., *Numerical Methods for Engineers,* Oxford University Press, 1991.

Guest, P.G., *Numerical Methods of Curve Fitting*, Cambridge University Press, New York, 1961.

Hager, W.W., *Applied Numerical Algebra*, Prentice Hall, Upper Saddle River, NJ, 1998.

Hamming, R.W., *Numerical Methods for Scientists and Engineers*, 2nd ed., McGraw Hill, New York, 1973.

Henrici, P.H., *Elements of Numerical Analysis,* Wiley, New York, 1964.

Higham, N.J., *Accuracy and Stability of Numerical Algorithms*, Society for Industrial and Applied Mathematics (SIAM), Philadelphia, PA, 1996.

Hildebrand, F.B., *Introduction to Numerical Analysis*, 2nd ed., McGraw Hill, New York, NY, 1974.

Hoffman, J., *Numerical Methods for Engineers and Scientists*, McGraw Hill, New York, 1992.

Hornbeck, R.W., *Numerical Methods*, Quantum, New York, 1975.

Householder, A.S., *Principles of Numerical Analysis*, McGraw Hill, New York, 1953.

Householder, A.S., *The Theory of Matrices in Numerical Analysis*, Blaisdell, New York, 1964.

Iserles, A., *A First Course in the Numerical Analysis of Differential Equations,* Cambridge University Press, New York, 1996.

Issaccson, E., and Keller, H.B., and Bishop, H., *Analysis of Numerical Methods*, Wiley, New York, 1966.

Jacobs, D. (ed.), *The State of the Art in Numerical Analysis*, Academic Press, London, 1977.

Jacques, I., and Colin, J., *Numerical Analysis*, Chapman and Hall, New York, 1987.

Jain, M.K., *Numerical Analysis for Scientists and Engineers*, S.B.W. Publishers, New Delhi, India, 1971.

James, M.L., Smith, G.M., and Wolford, J.C., *Applied Numerical Methods for Digital Computations with FORTRAN and CSMP*, 3rd ed., Harper & Row, New York, 1985.

Johnson, L.W., Riess, R.D., *Numerical Analysis*, 2nd ed., Addison-Wesley, Reading, MA, 1982.

Johnston, R.L., *Numerical Methods: A Software Approach*, Wiley, New York, 1982.

Kahaneer, D., Moher, C., and Nash, S., *Numerical Methods and Software*, Prentice Hall, Englewood Cliffs, NJ, 1989.

Keller, H.B., *Numerical Methods for Two-Point Boundary Value Problems*, Wiley, New York, 1968.

Kelley, C.T., *Iterative Methods of Optimization*, Society for Industrial and Applied Mathematics (SIAM), Philadelphia, PA, 1999.

Kharab, A., and Guenther, R.B., *An Introduction to Numerical Methods-A MATLAB Approach,* CRC Press, Boca Raton, FL, 2001.

Kincaid, D., and Cheney, W., *Numerical Analysis: Mathematics of Scientific Computing,* Brooks/Cole, Pacific Grove, CA, 1996.

Kress, R., *Numerical Analysis*, Springer-Verlag, New York, 1998.

Krishnamurthy, E.V., and Sen, S.K., *Numerical Algorithms*, East West Publishers, New Delhi, India, 1986.

Krommer, A. R., and Ueberhuber, C.W., *Computational Integration*, Society for Industrial and Applied Mathematics (SIAM), Philadelphia, PA, 1998.

Lambert, J.D., *Numerical Methods for Ordinary Differential Equations–The Initial Value Problems,* Wiley, New York, NY, 1991.

Lapidus, L., and Pinder, G.F., *Numerical Solution of Ordinary Differential Equations in Science and Engineering*, Wiley, New York, 1981.

Lapidus, L., and Seinfield, J.H., *Numerical Solution of Partial Differential Equations,* Academic Press, New York, 1971.

Lastman, G.J., and Sinha, N.K., *Microcomputer Based Numerical Methods for Science and Engineering,* Saunders College Publishing, New York, NY, 1989.

Levy, H., and Baggott, E.A., *Numerical Solutions of Differential Equations,* Dover, New York, 1950.

Maron, M.J., *Numerical Analysis, A Practical Approach*, Macmillan, New York, 1982.

Mathews, J.H., *Numerical Methods for Mathematics, Science and Engineering,* 2nd ed., Prentice Hall of India, New Delhi, India, 1994.

Milne, W.E., *Numerical Solution of Differential Equations*, Wiley, New York, 1953.

Moin, P., *Fundamentals of Engineering Numerical Analysis*, Cambridge University Press, New York, 2001.

Morton, K.W., and Mayers, D.F., *Numerical Solution of Partial Differential Equations: An Introduction,* Cambridge University Press, Cambridge, UK, 1994.

Myron, A., and Issacson, E.L., *Numerical Analysis for Applied Science*, Wiley, Hoboken, NJ, 1998.

Na, T.Y., *Computational Methods in Engineering Boundary Value Problems*, Academic Press, New York, 1979.

Nakamura, S., *Computational Methods in Engineering and Science,* Wiley, New York, NY, 1977.

Nielson, K.L., *Methods in Numerical Analysis,* Macmillan Company, New York, 1964.

Noble, B., *Numerical Methods, Vol. 2*, Oliver and Boyd, Edinburgh, 1964.

Nocedal, J., and Wright, S.J., *Numerical Optimization*, Springer-Verlag, New York, 1999.

Ortega, J.M., *Numerical Analysis-A Second Course*, Academic Press, New York, NY, 1972.

Powell, M., *Approximation Theory and Methods*, Cambridge University Press, Cambridge, UK, 1981.

Press, W.H., Teukolsky, S.A., Vetterling, W.T., and Flannery, B.P., *Numerical Recipes: The Art of Scientific Computing*, 2nd ed., Cambridge University Press, New York, 1992.

Quarteroni, A., Sacco, R., and Saleri, F., *Numerical Mathematics,* Springer Verlag, New York, 2000.

Ralston, A., and Rabinowitz, P., *A First Course in Numerical Analysis*, 2nd ed., McGraw Hill, New York, 1978.

Ralston, A., and Wilf, H.S., eds., *Mathematical Methods for Digital Computers,* Vol. 1 and 2, Wiley, New York, 1967.

Rao, K.S., *Numerical Methods for Scientists and Engineers,* Prentice Hall, New Delhi, India, 2001.

Rao, S.S., *Applied Numerical Methods for Engineers and Scientists*, Prentice Hall, Upper Saddle River, New Jersey, NJ, 2002.

Ratschek, H., and Rokne, J., *Computer Methods for the Range of Functions*, Ellis Horwood, Chichester, 1984.

Rice, J.R., *Numerical Methods, Software and Analysis*, McGraw Hill, New York, 1983.

Sastry, S.S., *Introductory Methods of Numerical Analysis*, Prentice Hall of India, New Delhi, India, 2001.

Scarborough, J.B., *Numerical Mathematical Analysis*, 6th ed., Johns Hopkins Press, Baltimore, MD, 1966.

Scheid, F., *Schaum's Outline of Theory and Problems in Numerical Analysis*, 2nd ed., Schaum's Outline Series, McGraw Hill, New York, 1988.

Schiesser, W.E., *Computational Mathematics in Engineering and Applied Science*, CRC Press, Boca Raton, FL, 1994.

Shampine, L.F., *Numerical Solution of Ordinary Differential Equations*, Chapman and Hall, New York, 1994.

Sharma, J.N., *Numerical Methods for Engineers and Scientists,* Narosa Publishing House, New Delhi, India, 2004.

Smith, G.D., *Numerical Solution of Partial Differential Equations: Finite Difference Methods*, 3rd ed., Oxford University Press, Oxford, 1985.

Smith, W.A., *Elementary Numerical Analysis*, Prentice Hall, Englewood Cliffs, NJ, 1986.

Snyder, M.A., *Chebyshev Methods in Numerical Approximation*, Prentice Hall, Englewood Cliffs, NJ, 1966.

Stanton, R.G., *Numerical Methods for Science and Engineering*, Prentice Hall of India, New Delhi, India, 1967.

Stark, P.A., *Introduction to Numerical Methods*, Macmillan, New York, 1970.

Stewart, G.W., *Matrix Algorithms, Vol. 1, Basic Decompositions,* Society for Industrial and Applied Mathematics (SIAM), Philadelphia, PA, 1998.

Stoer, J., and Bulirsch, R., *Introduction to Numerical Analysis*, Springer-Verlag, New York, 1980.

Stroud, A., and Secrets, D., *Gaussian Quadrature Formulas*, Prentice Hall, Englewood Cliffs, 1966.

Stroud, A.H., *Numerical Quadrature and Solution of Ordinary Differential Equations*, Springer-Verlag, New York, 1974.

Taylor, J.R., *An Introduction to Error Analysis*, University Science Books, Mill Valley, CA, 1982.

Traub, J.F., *Iterative Methods for the Solution of Equations*, Prentice Hall, Englewood Cliffs, NJ, 1964.

Trefethen, L.N., and Bau, D., *Numerical Linear Algebra*, Society for Industrial and Applied Mathematics (SIAM), Philadelphia, PA, 1997.

Tyrtyshnikov, E.E., *A Brief Introduction to Numerical Analysis*, Birkhauser, Boston, 1997.

Ueberhuber, C.W., *Numerical Computation 1: Methods, Software, and Analysis*, Springer-Verlag, New York, 1997.

Ueberhuber, C.W., *Numerical Computation 2: Methods, Software, and Analysis*, Springer-Verlag, New York, 1997.

Vemuri, V., and Karplus, W.J., *Digital Computer Treatment of Partial Differential Equations,* Prentice Hall, Englewood Cliffs, NJ, 1981.

Vichnevetsky, R., *Computer Methods for Partial Differential Equations, Vol. 1: Elliptic Equations and the Finite Element Method*, Prentice Hall, Englewood Cliffs, NJ, 1981.

Vichnevetsky, R., *Computer Methods for Partial Differential Equations, Vol. 2: Initial Value Problems,* Prentice Hall, Englewood Cliffs, NJ, 1982.

Wendroff, B., *Theoretical Numerical Analysis*, Academic Press, New York, 1966.

Wilkinson, J.H., *Rounding Errors in Algebraic Processes*, Dover, New York, 1994.

Yokowitz, S., and Szidarovsky, F., *An Introduction to Numerical Computation*, Macmillan, New York, 1986.

Yong, D.M., and Gregory, R.T., *A Survey of Numerical Mathematics, Vol. 1 and 2*, Addison-Wesley, Reading, MA, 1972.

Young, D., *Iterative Solution for Large Linear Systems,* Academic Press, New York, 1971.

MATLAB

Chapman, S.J., *MATLAB Programming for Engineers,* 2nd ed., Brooks/Cole, Thomson Learning, Pacific Grove, CA, 2002.

Chapra, S.C., *Applied Numerical Methods with MATLAB*, 2nd ed., McGraw Hill, New York, 2008.

Dabney, J.B., and Harman, T.L., *Mastering SIMULINK 4,* Prentice Hall, Upper Saddle River, NJ, 2001.

Daku, B.L.F., *MATLAB Tutor CD-Learning MATLAB Superfast*, Wiley, New York, 2006.

Djaferis, T.E., *Automatic Control–The Power of Feedback using MATLAB*, Brooks/Cole, Thomson Learning, Pacific Grove, CA, 2000.

Dukkipati, R.V., *Analysis and Design of Control Systems Using MATLAB*, New Age International Publishers, New Delhi, India, 2006.

Dukkipati, R.V., *MATLAB for Electrical Engineers*, New Age International Publishers, New Delhi, India, 2007.

Dukkipati, R.V., and Srinivas, J., *Solving Engineering Mechanics Problems with MATLAB*, New Age International Publishers, New Delhi, India, 2007.

Dukkipati, R.V., *MATLAB for Engineers*, New Age International Publishers, New Delhi, India, 2006.

Dukkipati, R.V., *MATLAB: An Introduction and Applications,* New Age Publishers, New Delhi, India.

Dukkipati, R.V., *Solving Engineering System Dynamics Problems with MATLAB*, New Age International Publishers, New Delhi, India, 2006.

Dukkipati, R.V., *Solving Vibration Analysis Problems with MATLAB*, New Age International Publishers, New Delhi, India, 2006.

Etter, D.M., *Engineering Problem Solving with MATLAB,* Prentice Hall, Englewood Cliffs, NJ, 1993.

Gardner, J.F., *Simulation of Machines using MATLAB and SIMULINK*, Brooks/Cole, Thomson Learning, Pacific Grove, CA, 2001.

Gilat, A., and Subramaniam, V., *Numerical Methods for Engineers and Scientists: An Introduction with Applications using MATLAB,*Wiley, New York, 2007.

Hanselman, D., and Littlefield, B., *Mastering MATLAB 7*, Prentice Hall, Upper Saddle River, NJ, 2005.

Harper, B. D., *Solving Dynamics Problems in MATLAB*, 5th ed., Wiley, New York, 2002.

Herniter, M.E., *Programming in MATLAB,* Brooks/Cole, Pacific Grove, CA, 2001.

Karris, S.T., *Signals and Systems with MATLAB Applications*, Orchard Publications, Fremont, CA, 2001.

Kiusalaas,J., *Numerical Methods in Engineering with MATLAB,* Cambridge University Press, Cambridge, UK, 2005.

Leonard, N.E., and Levine, W.S., *Using MATLAB to Analyze and Design Control Systems*, Addison-Wesley, Redwood City, CA, 1995.

Lyshevski, S.E., *Engineering and Scientific Computations Using MATLAB*, Wiley, New York, 2003.

Moore, H., *MATLAB for Engineers*, Prentice Hall, Upper Saddle River, NJ, 2007.

Ogata, K., *Designing Linear Control Systems with MATLAB***,** Prentice Hall, Upper Saddle River, NJ, 1994.

Ogata, K., *Solving Control Engineering Problems with MATLAB***,** Prentice Hall, Upper Saddle River, NJ, 1994.

Palm, W.J. III., *Introduction to MATLAB 7 for Engineers,* McGraw Hill, New York, 2005.

Pratap, Rudra., *Getting Started with MATLAB- A Quick Introduction for Scientists and Engineers,* Oxford University Press, New York, NY, 2002.

Recktenwald, G., *Numerical Methods with MATLAB*, Prentice Hall, Upper Saddle River, NJ, 2000.

Saadat, Hadi., *Computational Aids in Control Systems Using MATLAB*, McGraw Hill, New York, NY, 1993.

Sigman, K., and Davis, T.A., *MATLAB Primer*, 6th ed., Chapman & Hall/CRC Press, Boca Raton, FL, 2002.

The MathWorks, Inc., *SIMULINK, Version 3,* The MathWorks, Inc., Natick, MA, 1999.

The MathWorks, Inc., *MATLAB*: *Application Program Interface Reference Version 6*, The MathWorks, Inc., Natick, 2000.

The MathWorks, Inc., *MATLAB*: *Control System Toolbox User's Guide, Version 4,* The MathWorks, Inc., Natick, 1992-1998.

The MathWorks, Inc., *MATLAB*: *Function Reference,* The MathWorks, Inc., Natick, 2000.

The MathWorks, Inc., *MATLAB*: *Symbolic Math Toolbox User's Guide, Version 2,* The MathWorks, Inc., Natick, 1993-1997.

The MathWorks, Inc., MATLAB: *Using MATLAB Graphics, Version 6*, The MathWorks, Inc., Natick, 2000.

The MathWorks, Inc., *MATLAB*: *Using MATLAB, Version 6*, The MathWorks, Inc., Natick, 2000.